Turbulence and Instabilities in Magnetised Plasmas, Volume 2

Gyrokinetic theory and gyrofluid turbulence

IOP Series in Plasma Physics

About the series

The IOP Plasma Physics ebook series aims at comprehensive coverage of the physics and applications of natural and laboratory plasmas, across all temperature regimes. Books in the series range from graduate and upper-level undergraduate textbooks, research monographs and reviews.

The conceptual areas of plasma physics addressed in the series include:

- Equilibrium, stability and control
- Waves: fundamental properties, emission, and absorption
- Nonlinear phenomena and turbulence
- Transport theory and phenomenology
- Laser-plasma interactions
- Non-thermal and suprathermal particle populations
- Beams and non-neutral plasmas
- High energy density physics
- Plasma-solid interactions, dusty, complex and non-ideal plasmas
- Diagnostic measurements and techniques for data analysis

The fields of application include:

- Nuclear fusion through magnetic and inertial confinement
- Solar-terrestrial and astrophysical plasma environments and phenomena
- Advanced radiation sources
- Materials processing and functionalisation
- Propulsion, combustion and bulk materials management
- Interaction of plasma with living matter and liquids
- Biological, medical and environmental systems
- Low temperature plasmas, glow discharges and vacuum arcs
- Plasma chemistry and reaction mechanisms
- Plasma production by novel means

Turbulence and Instabilities in Magnetised Plasmas, Volume 2

Gyrokinetic theory and gyrofluid turbulence

Bruce Scott
Max-Planck Institute for Plasma Physics, Garching, Germany

IOP Publishing, Bristol, UK

ISBN 978-0-7503-3855-4 (ebook)
ISBN 978-0-7503-3853-0 (print)
ISBN 978-0-7503-3856-1 (myPrint)
ISBN 978-0-7503-3854-7 (mobi)

DOI 10.1088/978-0-7503-3855-4

Version: 20211101

IOP ebooks

British Library Cataloguing-in-Publication Data: A catalogue record for this book is available from the British Library.

Published by IOP Publishing, wholly owned by The Institute of Physics, London

IOP Publishing, Temple Circus, Temple Way, Bristol, BS1 6HG, UK

US Office: IOP Publishing, Inc., 190 North Independence Mall West, Suite 601, Philadelphia, PA 19106, USA

This work is dedicated to the memory of Professor Dieter Pfirsch, one of the founding directors of the Max-Planck-Institut für Plasmaphysik, and a colleague and friend, 1927–2004.

Contents

Preface

Preface to both volumes

I wrote this because of the large number of concepts more than 30 years of work have developed on this subject, which too large a fraction of the community hasn't learned about yet.

The 1970s picture of a plasma is that at the macroscopic level you have a conducting fluid interacting with magnetic fields. There is one density, one velocity, and one pressure, and there is the current which is the source for an embedded magnetic field. The electric field itself is submerged in ways that will be explained in detail herein. This is the picture of magnetohydrodynamics, or MHD for short. This serves the conceptual structure in an analogous way to fluid dynamics (the technical term is hydrodynamics), wherein an ideal or perfect fluid does the dynamics and some kinetic mechanism is added on as a series of diffusion or friction coefficients to do the dissipation. The plasma is expected to follow MHD on large scales and then one turns to kinetic mechanisms for the dissipation, with the addition that these mechanisms are often the cause of instabilities. This is the picture as it developed into the late 1970s and it is still the picture most people have today.

Even in dissipation free environments, a much more useful picture allows the electrons to be a separate fluid to the ions, even in the case of a 'single component' plasma (one positive charge carrier). The electrons are no longer a massless species behaving like electricity, they are another fluid with finite mass density and pressure, just with a disparate mass ratio (1845 for protons or normal hydrogen, 3670 for deuterium). Now, we have two densities, two velocities, and two pressures, to go with the electric and magnetic fields. This can also be treated in an ideal (dissipation free) way, together with Maxwell's equations.

The complication now is charge neutrality. Sort of automatic in MHD, it has to be treated carefully when there are at least in principle two densities. So at low frequencies, the system of a 'quasineutral, two-fluid plasma' is what has to be developed. There are rigorous ways to incorporate this concept, which means you have zero charge density but you can still have an electric field. This is a result of finite polarisability, so it is formulated as one density (electrons) and then instead of another density you have the ion polarisation as the other variable. This also takes care of the electric field. At low frequency, which means slower than electrodynamics but also slower than compressional MHD processes, we know how to treat this and it forms the theoretical substance of this work. There are two approaches to it, one more fluidlike, and one with more of a kinetic basis, and we will do both.

This is the case with magnetised plasma turbulence, not plasma turbulence and not turbulence. So we have to say first what a magnetised plasma is (strong background magnetic field, but what are the conditions of that statement?). Then we have to say what turbulence is. Then we have to build up a look and feel for the turbulence and understand the significance of a self-consistent, statistical process. Then we do the first version of the underlying theory (more problems, but easier to understand, and with a more obvious capture of MHD as a subset). After that, and only then, do we start addressing waves and instabilities. When the concepts, especially polarisation become familiar, then we can do the second version of the theory, gyrokinetic and gyrofluid theory, which is more rigorous and can treat a wider range of situations. At the end, we go back and make some still remaining connections to the standard picture of MHD instabilities.

It is important, very important, to understand this subject as turbulence first, and instabilities second. The reason for that is that linear theory for small fluctuations has a range of applicability which is much narrower than is often realised, and can give an incorrect interpretive physical picture (cartoon) to what the energetics and dynamics actually is.

It takes a lot of effort to do this, and I would not have done so had I not felt a need for it. If everybody were already familiar with this material and could teach it to their students, then this work would not have been necessary.

So I did my best. One can always do better. You are never finished. You are never ready. You are never certain. If you think you are, you are in the wrong field.

Author biography

Bruce Scott

B Scott is a research plasma physicist having graduated with a Doctorate from the University of Maryland in 1985 and with the German Habilitation from the Heinrich-Heine-Universität Düsseldorf in 2001. He is a Fellow of the American Physical Society with membership since 1979. He has several tens of first author papers in peer reviewed journals in the field of theoretical plasma physics.

IOP Publishing

Turbulence and Instabilities in Magnetised Plasmas, Volume 2
Gyrokinetic theory and gyrofluid turbulence
Bruce Scott

Chapter 1

Prelude to volume two

This work has been split into two volumes due to its size, which was forced by the need to include extra introductory material in Volume One. We will briefly recap what was covered therein before proceeding. The point of division is the treatment of temperature gradients and temperature fluctuation dynamics *per se* as part of the turbulence. One also finds new classes of instabilities with ∇T present. Volume One considered the pressure gradient, ∇p, generically, with the density equal in background, gradient, and fluctuation dynamics for both electrons and ions. Due to the nature of the dynamics, one of the densities gets replaced by the ExB vorticity, which is due to the curl of the ExB velocity, but which also accounts for the polarisation current whose role is simultaneously to provide for charge conservation and for turbulent vorticity nonlinearity which maintains all the energy transfer channels and thereby the routes to dissipation. This property is called quasineutrality and it is a central element to low frequency dynamics in magnetised plasmas and specifically to this type of turbulence.

1.1 Plasma, magnetised, parameters

We consider the plasma as a set of fluids made up of charged particles. Each has a specific charge Ze and mass m, with e being the fundamental unit of charge and Z the charge state. Each species has a thermal velocity determined by its temperature (always in units of energy),

$$V_e^2 = \frac{T_e}{m_e}, \qquad V_i^2 = \frac{T_i}{m_i} \tag{1.1}$$

with subscripts e and i for electrons and ions, respectively, assuming a single component plasma which is also fully ionised. Each species also has a density, n_e and n_i in this case. We usually assume a neutral background, with $n_e = Zn_i$ in this case. Dissipation processes enter through the collisional frequencies given by

$$\nu_e = \frac{4\sqrt{2\pi}\, n_e \lambda e^4}{3m_e^{1/2} T_e^{3/2}}, \qquad \nu_i = \frac{4\sqrt{\pi}\, n_i \lambda e^4}{3m_i^{1/2} T_i^{3/2}} \tag{1.2}$$

where the factor of λ is the Coulomb logarithm, the logarithm of the ratio between the Debye length and the distance of closest approach, usually the de Broglie wavelength of the two particles in the centre of momentum frame. The ion to electron temperature ratio is denoted $\tau_i = T_i/T_e$, allowing for the possibility that the ions are cold, in which case τ_i is small. It is usually taken to be either unity or zero unless a particular experiment or situation is being represented.

In the presence of a bath of charged particles, the electric field due to one particular particle is screened by the reaction of its neighbours to that field. Considering a homogeneous plasma with electron charge density $-n_e e$, assume a small disturbance in the electric field, through the electrostatic potential $\tilde{\phi}$, is balanced by a small disturbance in the electron pressure, $T_e \tilde{n}_e$, with the tilde symbols denoting a small departure from the equilibrium state. The charge density disturbance is given by the force balance

$$n_e e\tilde{\phi} = T_e \tilde{n}_e \qquad \text{hence} \quad e\tilde{n}_e = \frac{n_e e^2}{T_e}\tilde{\phi} \tag{1.3}$$

This is placed into the Maxwell equation for charge

$$\varepsilon_0 \nabla \cdot \mathbf{E} = \rho_{ch} \tag{1.4}$$

with the assumption of an electrostatic response, so that

$$\left(\frac{n_e e^2}{T_e} - \varepsilon_0 \nabla_\perp^2\right)\tilde{\phi} = e\delta(\mathbf{r} - \mathbf{r}_0) \tag{1.5}$$

with the charge density due to the test particle at $\mathbf{r} = \mathbf{r}_0$ on the right hand side, where $\delta()$ is the Dirac delta function. The solution of this away from the test particle is a Yukawa potential

$$\phi(r) = \frac{1}{r}\exp\left(-\frac{r}{\lambda_D}\right) \tag{1.6}$$

where r is the magnitude of $\mathbf{r} - \mathbf{r}_0$. The quantity λ_D is the Debye length, and together with the electron plasma frequency ω_{pe}, it is given by

$$\lambda_D^2 = \varepsilon_0 \frac{T_e}{n_e e^2}, \qquad \omega_{pe}^2 = \frac{n_e e^2}{\varepsilon_0 m_e} \tag{1.7}$$

so that $\omega_{pe} = V_e/\lambda_D$. This frequency is the one at which an initialised charge density disturbance will oscillate in the absence of driving. In a neutral plasma at low frequency, these so-called space charge effects are not present.

Dynamics forced at a much lower frequency, $\omega \ll \omega_{pe}$, does not excite space charge effects; rather, the electrons evolve in quasistatic force balance in which charge neutrality is maintained. One has a fluid in which the electrons are electrically coupled to the ions, and in the absence of a magnetic field, the plasma only supports acoustic waves. If the ion temperature is small, $T_i \ll T_e$ one is in the cold-ion regime, and the electrons provide the mobility and the ions the inertia. The resulting thermal motion involves the sound speed, c_s, which is given by

$$c_s^2 = \frac{T_e}{m_i} \tag{1.8}$$

in which we note the similarity to the thermal velocities. The nature of this type of response is that the electrons as carriers of the current relax the charge density on the fast time scale represented by ω_{pe}, so that on slower time scales represented by a wavelength divided by c_s, disturbances are only in the mass density given by

$$\rho = n_i m_i + n_e m_e \tag{1.9}$$

and the flow velocity. This is the situation in a neutral plasma in the absence of magnetic fields, either background or self generated.

In a magnetised plasma the charged particles follow gyromotion in the plane perpendicular to the field. We usually consider a magnetised plasma as the same set of charged fluids as above, but threaded by a strong background magnetic field which is considered to be a guide field. The magnetic field is **B**, with strength B and direction unit vector **b**. The gyromotion follows from the Lorentz force, yielding a gyrofrequency and a gyroradius for each species, given by

$$\Omega_e = \frac{eB}{m_e}, \qquad \rho_e = \frac{V_e}{\Omega_e} \tag{1.10}$$

$$\Omega_i = \frac{ZeB}{m_i}, \qquad \rho_i = \frac{V_i}{\Omega_i} \tag{1.11}$$

for the electrons and ions, respectively. As parameters, these are not signed; otherwise, the electron gyrofrequency would appear with a minus sign. These frequencies are still extremely fast, even though Ω_i is much slower than ω_{pe}. Dynamics driven by pressure gradients typically proceeds at a range of frequencies up to

$$\omega \lesssim \frac{c_s}{L_\perp} \qquad \text{where} \quad L_\perp = |\nabla \log n_e|^{-1} \tag{1.12}$$

where this version of perpendicular scale length $L_\perp$ holds in an isothermal situation. The plasma is considered magnetised if the flow eddy vorticity of the dynamics or turbulence is much smaller than Ω_i and the perpendicular scale length of the background is much larger than the ion gyroradius,

$$\omega \ll \Omega_i, \qquad \rho_i \ll L_\perp \tag{1.13}$$

where the ions are used because they have the slowest gyrofrequency and the largest gyroradius. Under the spatial scale separation, vorticity dynamics up to $\omega \sim c_s/L_\perp$ also satisfies the frequency separation. This is what is meant by a magnetised plasma.

We will have magnetohydrodynamics (MHD) phenomena in the general dynamics together with ExB fluid and thermal effects. Compressional and shear Alfvén waves proceed according to

$$\left(\frac{\partial^2}{\partial t^2} - [v_A^2 + v_s^2]\nabla_\perp^2\right)\tilde{B} = 0, \qquad \left(\frac{\partial^2}{\partial t^2} - v_A^2\nabla_\parallel^2\right)\tilde{\mathbf{B}}_\perp = 0 \tag{1.14}$$

where $\tilde{B}$ denotes a disturbance in the magnetic field strength, $\tilde{\mathbf{B}}_\perp$ denotes a magnetic disturbance perpendicular to $\mathbf{B}$, and v_s is the thermal sound speed of an adiabatic compression. The Alfvén velocity and plasma beta are given by

$$v_A^2 = \frac{B^2}{\mu_0 \rho}, \qquad \beta = \frac{2\mu_0 p}{B^2} \tag{1.15}$$

where in this expression p is the sum of all the partial pressures. One of the conditions of the plasma being magnetised is that $\beta \ll 1$, which signals $v_s \ll v_A$ and defines the Alfvénic MHD regime. The shear Alfvén waves are transverse waves, with velocity perpendicular to and the gradient parallel to the guiding background magnetic field. As such they do not exist in the unmagnetised case.

1.2 Low frequency, flute mode ordering

The time and space scales of the turbulence satisfy

$$\omega \sim k_\parallel v_A \ll k_\perp v_A, \qquad k_\parallel \ll k_\perp \tag{1.16}$$

where ω is a frequency standing in for the dynamical strength of the turbulence, and $k_\parallel$ and $k_\perp$ are parallel and perpendicular wavenumbers (relative to the guide field). This is the statement of what is called flute mode ordering. The parallel and perpendicular dynamics separate, with the background guide field ordering the dynamics. Perpendicular and parallel gradients are formally defined by

$$\nabla_\parallel = \mathbf{b} \cdot \nabla, \qquad \nabla_\perp = -\mathbf{b} \times \mathbf{b} \times \nabla \tag{1.17}$$

with the unit vector $\mathbf{b}$ that of the background guide field. In the dynamics, $\nabla_\parallel$ is redefined to include the magnetic disturbances, as given below.

Low frequency refers to

$$\omega \ll \Omega_i, \qquad \omega \ll k_\perp v_A \tag{1.18}$$

so that the frequency range encountered is always much slower than either the slowest of the gyrofrequencies (i.e., the ion one) or the compressional Alfvén frequency at the scales of motion encountered. As a result, particles move according to their gyrocentre drifts rather than their instantaneous velocity, and the turbulence is never strong enough to dynamically compress the magnetic field. Quasistatic compression in a curved background magnetic field still occurs, but not with wavelike responses.

With perpendicular Alfvénic compression not excited, the electric field is static in the perpendicular direction and inductive only along the magnetic field. So it is expressed as

$$\mathbf{E} = -\frac{\partial A_\parallel}{\partial t}\mathbf{b} - \nabla\phi \tag{1.19}$$

that is, in terms of ϕ, the electrostatic potential, and $A_\parallel$, the parallel magnetic potential. The magnetic field is given by the background guide field plus disturbances represented by $A_\parallel$ according to

$$\mathbf{B}_T = \mathbf{B} + \nabla \times \left(A_\parallel \mathbf{b}\right) \tag{1.20}$$

It is always formulated, or re-formulated, such that its divergence exactly vanishes. Under low frequency conditions, we face shear Alfvén responses but not compressional ones. The flow across the field is dynamically incompressible, and the responses along the field are compressible but wavelike, with a spectral spread essentially caused by the turbulence such that the ordering in Eq. (1.16) is satisfied.

The parallel gradient under these conditions becomes redefined in terms of $\mathbf{B}_T \cdot \nabla$, always keeping the new form as a pure divergence. The operator $\nabla_\parallel$ is then given by dividing by a factor of B, so that $B\nabla_\parallel$ becomes the pure divergence. The nonlinear component of this, due to $A_\parallel$, are referred to variously as the parallel gradient, or simply magnetic, nonlinearity, and, especially when involving gradient drive or turbulent fluxes, the 'magnetic flutter' effect. Specific forms of this depend on the model geometry in any setting.

1.3 Drifts, ExB flow, currents

The ExB velocity and vorticity are defined as

$$\mathbf{v}_E \equiv \frac{\mathbf{E}\times\mathbf{B}}{B^2} = \frac{1}{B}\mathbf{b}\times\nabla\phi, \qquad \Omega_E \equiv \frac{1}{B}\nabla_\perp^2\phi \tag{1.21}$$

where $\mathbf{E}$ and $\mathbf{B}$ are the electric and magnetic field and E and B their magnitudes, and ϕ is the electrostatic potential. The vorticity is the baseline strength of $\nabla \times \mathbf{v}_E$ (with the variation in $\mathbf{B}$ much weaker).

The ExB velocity has this form because of the role of the Lorentz force in the equation of motion for one of the fluids comprising the plasma,

$$nm\frac{d\mathbf{u}}{dt} = -\nabla p + nZe(\mathbf{E} + \mathbf{u}\times\mathbf{B}) \tag{1.22}$$

where m and Ze are the specific mass and charge of the species, n and p are its density and pressure, and $\mathbf{u}$ is its velocity. The partial mass density is nm and the partial charge density is nZe, and d/dt is the advective derivative with the species velocity,

$$\frac{d}{dt} = \frac{\partial}{\partial t} + \mathbf{u}\cdot\nabla \tag{1.23}$$

Since the dynamics is slow compared to the gyrofrequency, the velocity on the right hand side is the one solved for, and the inertia term is retained as a correction,

$$\mathbf{u} = u_\parallel \mathbf{b} + \frac{\mathbf{E}\times\mathbf{B}}{B^2} + \frac{1}{nZe}\frac{\mathbf{B}\times\nabla p}{B^2} + \frac{m}{Ze}\frac{\mathbf{B}}{B^2}\times\frac{d\mathbf{u}}{dt} \tag{1.24}$$

This gives the flow vector in terms of the parallel velocity, the ExB velocity, the diamagnetic (pressure drift) velocity, and the polarisation as an inertial correction on the gyromotion for the perpendicular drifts. The dependent variables are the $u_\parallel$, the p (hence n for an isothermal model), and the ϕ present within $\mathbf{E}$. The velocity vector $\mathbf{u}$ itself is not a dependent variable anymore, but is given in terms of the drifts. This is the defining statement of drift ordering. Due to the low frequency regime, with respect to $k_\perp v_A$ as well as to Ω_i, the electric field is given in terms of the potentials ϕ and $A_\parallel$ as described in Eq. (1.19). Hence the ExB velocity involves ϕ only.

1.4 Polarisation and quasineutrality

The polarisation drift as just calculated is usually thought of as small because of the factor Ω_i^{-1} in front of d/dt in the expression. With $\mathbf{u}$ on the left hand side, the appearance of $\Omega_i^{-1}(d\mathbf{u}/dt)$ indicates a small correction. This is actually needed to justify solving for the drifts in this fashion. But the necessity of keeping the polarisation drift is understood by the appearance of $\nabla\cdot\mathbf{u}$ in all the continuity equations (density and temperature of each species). Though the correction is small, its divergence is not. The divergence of the other contributions carries the spatial scale of the magnetic field, usually much larger than even $L_\perp$. So when the divergence is taken the polarisation drift is promoted by one order in ω/Ω_i, so that all the resulting terms are formally comparable.

The polarisation drift has a very important role in the dynamics, which would not function without it. It maintains the balance among the currents. We can show this by writing the drift velocity for both the electrons and ions, multiplying each by the species charge (in the usual case, $\pm e$ for ions and electrons), and summing them. This yields the expression for the plasma current,

$$\mathbf{J} = J_\parallel \mathbf{b} + \rho_{ch}\frac{\mathbf{E}\times\mathbf{B}}{B^2} + \frac{\mathbf{B}\times\nabla p_T}{B^2} + \frac{\rho}{B^2}\mathbf{B}\times\frac{d\mathbf{u}}{dt} \tag{1.25}$$

where $J_\parallel$ is the parallel current, p_T is the total (species sum of) pressure, and we keep the charge density $\rho_{ch} = n_i Ze - n_e e$ for the following demonstration. The continuity equation for charge is given by the species sum of the specific charge times the continuity equation for density,

$$\frac{\partial\rho_{ch}}{\partial t} + \nabla\cdot\mathbf{J} = 0 \tag{1.26}$$

The expression for $\mathbf{J}$ in Eq. (1.25) is inserted here to get the result. For scaling purposes we assume that $\mathbf{u}$ is comparable to $\mathbf{v}_E$ and use

$$\mathbf{B}\times\mathbf{v}_E = \mathbf{E}_\perp \tag{1.27}$$

in the inertia term, to arrive at

$$\frac{\partial\rho_{ch}}{\partial t} + \nabla\cdot(\rho_{ch}\mathbf{v}_E) + \nabla\cdot\left(J_{\|}\mathbf{b} + \frac{\mathbf{B}\times\nabla p_T}{B^2}\right) + \nabla\cdot\frac{\rho}{B^2}\frac{d\mathbf{E}_\perp}{dt} = 0 \tag{1.28}$$

In this form we are assuming that the scale of **B** is long enough that it can be commuted past the perpendicular derivatives in the last term. Using Eq. (1.4), we may compare the first and last terms, also assuming that the first two terms are comparable (they form the ExB advective derivative of ρ_{ch} plus corrections). The comparison is

$$\nabla\cdot\left(\varepsilon_0\frac{d\mathbf{E}}{dt}\right) \leftrightarrow \nabla\cdot\left(\frac{\rho}{B^2}\frac{d\mathbf{E}_\perp}{dt}\right) \tag{1.29}$$

with or without the divergence. With the flute mode ordering on **E**, each expression is essentially of the same form, except for the coefficients. So if

$$\varepsilon_0 \ll \frac{\rho}{B^2} \tag{1.30}$$

then the charge density terms are neglected. This means that

$$\nabla\cdot\mathbf{J} = 0 \qquad \text{in the presence of} \qquad \nabla\cdot\mathbf{E} \neq 0 \tag{1.31}$$

Indeed, a finite ExB vorticity implies this through Eq. (1.21).

This situation is referred to as quasineutrality and one can see now how it underlies all the dynamics due to ExB vorticity, which is the expression of turbulence involving the ExB plasma drift flow. ExB eddies imply ExB vorticity, and turbulence implies statistical nonlinearity in the process. Quasineutrality holds in low frequency dynamics anytime Eq. (1.30) is satisfied. If this holds, then $v_A \ll c$ and $\lambda_D \ll \rho_i$ by the same measure (assuming $\tau_i \sim 1$). Ultimately, this is why we can neglect space charge effects including the Debye length and the Maxwell displacement current in this type of dynamics.

1.5 Turbulence

Turbulence as such was the subject of Chapter 2 of Volume One. Briefly, one refers to nonlinearity, but statistical nonlinearity, in small scale turbulence on a background with a gradient. The scale of motion is down to and including ρ_i, but smaller than $L_\perp$, so that a mixing ordering on the nonlinearities is possible. One assumes that

$$\frac{\tilde{v}_E}{c_s} \sim \frac{\tilde{n}_e}{n_e} \sim \frac{\Delta}{L_\perp} \sim \frac{\omega}{\Omega_i} \ll 1 \tag{1.32}$$

where Δ is the scale of motion. With the resulting scalings given by

$$\mathbf{v}_E\cdot\nabla \sim \frac{\tilde{v}_E}{\Delta} \sim \frac{c_s}{L_\perp} \sim \omega, \qquad \nabla\tilde{n}_e \sim \frac{\tilde{n}_e}{\Delta} \sim \frac{n_e}{L_\perp} \sim \nabla n_e \tag{1.33}$$

then nonlinear advection and any forcing by gradients are comparable. This also means

$$\mathbf{v}_E \cdot \nabla \sim \Omega_E \sim \frac{c_s}{L_\perp} \ll \Omega_i \tag{1.34}$$

which is the most fundamental way to express the ordering. The strength of the dynamics is represented by the RMS vorticity, defined as the square root of the spatial-domain average of the square of the vorticity. The RMS of the ExB vorticity gives the overall strength of the ExB advection.

1.5.1 Energy and enstrophy

Which way free energy and other quantities are passed between scales is constrained by the conservation laws. If only energy is conserved, then the transfer proceeds from large scale to small, since there are more available states per scale range at smaller scale. The degrees of freedom are free energy at a given wavenumber, $\mathbf{k}$, and the condition according to which they interact is the three wave condition $\mathbf{k} + \mathbf{k}' + \mathbf{k}'' = 0$, a consequence of the quadratic nonlinearity represented by $\mathbf{b} \times \nabla\phi \cdot \nabla f$, after the ExB advection of a scalar quantity represented by f. A range of scales is represented by $\log k$ falling within a constant interval, so that in two dimensions (perpendicular to the guide field) the density of states is proportional to k, i.e., there are more of them at higher k. The set of wavenumbers is effectively quantised by the system size, L, which corresponds to $k_0 = 2\pi/L$, a fundamental wavenumber. The set of available $\mathbf{k}$ is a set of integer multiples of k_0 in each direction.

Considering the most basic model of turbulence in an incompressible flow in two dimensions, we have

$$\mathbf{u} = \mathbf{b} \times \nabla\phi, \qquad \mathbf{W} = \nabla \times \mathbf{u} = \varpi\mathbf{b} \tag{1.35}$$

so that the curl of the equation of motion yields

$$\frac{\partial\varpi}{\partial t} + [\phi, \varpi] = 0, \qquad \varpi = \nabla_\perp^2\phi \tag{1.36}$$

where the perpendicular symbol denotes the plane perpendicular to $\mathbf{b}$. In this case the units are arbitrary. The only important consideration is the structure. The symbol $[\phi, \varpi]$ is a spatial Poisson bracket given by

$$[\phi, \varpi] = \mathbf{b} \cdot \nabla\phi \times \nabla\varpi = \frac{\partial(\phi, \varpi)}{\partial(x, y)} = \frac{\partial\phi}{\partial x}\frac{\partial\varpi}{\partial y} - \frac{\partial\varpi}{\partial x}\frac{\partial\phi}{\partial y} \tag{1.37}$$

where xy denote a Cartesian coordinate set on the perpendicular plane. This is the basic form of a quadratic nonlinearity. The properties of the Poisson bracket are antisymmetry, the Leibnitz rule and the pure divergence property,

$$[f, g] = -[g, f], \qquad [f, hg] = h[f, g] + g[f, h], \qquad \int d\mathcal{V}\,[f, g] = 0 \tag{1.38}$$

where $d\mathcal{V}$ is the volume element of the spatial domain. It also satisfies the Jacobi identity, though we delay discussion of that until Chapter 17.

The main point of this is that there are two conserved quantities. Since $[\phi, \phi]$ vanishes, multiplication of Eq. (1.36) by $-\phi$ yields a continuity equation,

$$\frac{\partial U}{\partial t} + \nabla \cdot \left(-\phi \frac{\partial}{\partial t} \nabla_\perp \phi\right) + [\phi, (-\phi \varpi)] = 0 \tag{1.39}$$

in which the Leibnitz property is used on the Poisson bracket. The volume integration over this yields conservation of the flow energy,

$$\frac{\partial}{\partial t} \int d\mathcal{V}\, U = 0, \qquad U = \frac{1}{2} |\nabla_\perp \phi|^2 \tag{1.40}$$

here using the divergence property of the Poisson bracket. Multiplication of Eq. (1.36) by ϖ also yields a continuity equation,

$$\frac{\partial W}{\partial t} + [\phi, W] = 0 \tag{1.41}$$

over which the volume integration yields conservation of the flow enstrophy,

$$\frac{\partial}{\partial t} \int d\mathcal{V}\, W = 0, \qquad W = \frac{1}{2} \varpi^2 \tag{1.42}$$

A simple calculation shows that if energy is transferred from larger to smaller scale, and it is conserved, then the enstrophy must increase. Since it cannot do so, one concludes that if there is a definite direction of scale transfer, then enstrophy passes to smaller scale while energy passes to larger. This is the dual cascade of two-dimensional turbulence. It works because the mechanism of vortex tube stretching, the process by which a three-dimensional flow can generate enstrophy, is not present in two dimensions because the vorticity vector is orthogonal to the plane of motion, in which both the flow and any gradients lie. This yields the specialised properties of three-wave coupling transfer, statistical equipartition, and disparate rates of decay of energy and enstrophy when there is dissipation. All of this was covered in detail in the first few Chapters in Volume One.

1.5.2 Magnetic field responses

In a plasma, one also faces the dynamics of self generated magnetic fields and currents. The simplest model is two-dimensional MHD, given by

$$\frac{d\varpi}{dt} = \nabla_\| J_\|, \qquad \varpi = \frac{\rho}{B^2} \nabla_\perp^2 \phi \tag{1.43}$$

$$\frac{\partial A_\|}{\partial t} = -\nabla_\| \phi, \qquad J_\| = -\mu_0^{-1} \nabla_\perp^2 A_\| \tag{1.44}$$

The dependent variables are ϖ and $J_\parallel$, and the fields ϕ and $A_\parallel$ are given by the equations on the right in each line. There are now two types of nonlinearity, advection with ϕ and the nonlinear parallel gradient with $A_\parallel$,

$$\frac{d}{dt} = \frac{\partial}{\partial t} + [\phi,], \qquad \nabla_\parallel = [-A_\parallel,] \tag{1.45}$$

and the bracket is defined with units as

$$[f, g] = \frac{1}{B}\left(\frac{\partial f}{\partial x}\frac{\partial g}{\partial y} - \frac{\partial g}{\partial x}\frac{\partial f}{\partial y}\right) \tag{1.46}$$

The equation for $A_\parallel$ is the ideal MHD version of the Ohm's law, equivalent to the statement that $E_\parallel$ vanishes. One can see the native velocity v_A given by the coefficients, according to the appearances of ρ/B^2 and μ_0 in the combination given by Eq. (1.15). Here, B is the strength of the background guide field, and it is assumed that there is no variation along the guide field, just in the perpendicular plane.

This system has the coupling process given by field line bending, $\nabla_\parallel J_\parallel$, in the vorticity equation, and the parallel static electric field, $-\nabla_\parallel\phi$, in the Ohm's law. Multiplying by $-\phi$ and $J_\parallel$, respectively, yields

$$\frac{\partial U_E}{\partial t} + \nabla\cdot\left(-\phi\frac{\rho}{B^2}\frac{\partial}{\partial t}\nabla_\perp\phi\right) + [\phi, (-\phi\varpi)] + \nabla_\parallel(\phi J_\parallel) = J_\parallel\nabla_\parallel\phi \tag{1.47}$$

$$\frac{\partial U_B}{\partial t} + \nabla\cdot\left(-\frac{\partial A_\parallel}{\partial t}\frac{1}{\mu_0}\nabla_\perp A_\parallel\right) = -\, J_\parallel\nabla_\parallel\phi \tag{1.48}$$

where

$$U_E = \frac{1}{2}\frac{\rho}{B^2}\,|\nabla_\perp\phi|^2, \qquad U_B = \frac{1}{2}\frac{1}{\mu_0}\,\left|\nabla_\perp A_\parallel\right|^2 \tag{1.49}$$

give the ExB and magnetic energy densities. The terms in Eq. (1.47) represent some of the currents as given in Eq. (1.28). The Alfvénic energy transfer effect is given by $J_\parallel\nabla_\parallel\phi$, appearing on the right hand side of both equations with opposite sign. The quantities $\nabla_\parallel(\phi J_\parallel)$ and $[\phi, \phi\varpi]$ are transport terms, signified by the pure divergence. The other divergence terms are field terms, representing pieces of the Poynting energy flux in this model. The total energy (integral over $U_E + U_B$) is conserved, but the Alfvénic transfer, representing field line bending, allows energy to pass back and forth. If the system starts with magnetic energy only, then the fluid energy acquires half of the initial energy and then they decay together (only partially so if dissipation is strong enough to enter earlier). This is equipartition. On the other hand, the magnetic flux ($A_\parallel$) is conserved, so the quadratic quantity $A_\parallel^2/2$ also follows a continuity equation,

$$\frac{\partial}{\partial t}\frac{A_\parallel^2}{2} + \frac{1}{2}[\phi, A_\parallel^2] = 0 \tag{1.50}$$

This property limits how much magnetic energy can grow from the fluid energy, since it can only grow by shifting the flux to smaller scale. So the equipartition is essentially one-way. This was covered in detail in Chapters 3 and 5 of Volume One.

MHD expressed in terms of vorticity and parallel current like this is referred to as Reduced MHD (RMHD, covered in Volume One). Its basis is the current balance represented by the vorticity equation, and the parallel force balance (in this case just $E_{\|} = 0$) represented by the equation for $A_{\|}$, called the Ohm's law. Generalised to three dimensions, to turbulence, and to two-fluid dynamics as introduced below, these two equations still form the basis of the responses in low frequency plasma dynamics. They have their counterparts in the gyrokinetic and gyrofluid descriptions, as covered in detail in these Chapters.

1.6 Turbulence in magnetised plasmas

In three dimensions in a magnetised plasma, we have the above processes active, and then also wavelike (shear Alfvénic) coupling in the direction of the guide field, which represents the third direction. The speed of v_A relative to c_s, in turn larger than the v_E of the turbulence, results by Eq. (1.16) in a narrow $k_{\|}$ spectrum relative to $k_{\perp}$, so that 3D really means 2D + 1D in the dynamics. So we do not find the well known results of three-dimensional fluid or even MHD turbulence. Instead, we have an essentially 2D MHD turbulence in an exchange with the parallel Alfvénic responses. There are now four dependent variables, with n and $u_{\|}$, thermal and parallel velocity fluctuations, adding to ϖ and $J_{\|}$. We recall that pressure is density times a background temperature in an isothermal model. But before we add complexity, there was another major process to refamiliarise with, the presence of parallel pressure gradient forcing in the Ohm's law.

1.6.1 Adiabatic response

We briefly touched on the idea of force balance in the description of the Debye shielding process which defines the plasma. This is also active in general in a magnetised plasma. The MHD model neglects the pressure force effect in the Ohm's law, considering only the Lorentz force in the absence of dissipation in its construction. In general it is always present, in addition to electromagnetic induction, nonlinear electron inertia, and electrical resistivity as a friction between electron and ion parallel motion. We have at minimum an Ohm's law given by

$$\frac{\partial A_{\|}}{\partial t} + \frac{m_e}{n_e e^2}\frac{dJ_{\|}}{dt} + \eta_{\|} J_{\|} = \frac{1}{n_e e}\nabla_{\|} p_e - \nabla_{\|}\phi \tag{1.51}$$

where $\eta_{\|}$ is the parallel resistivity and d/dt is the ExB advective derivative, and $\nabla_{\|}$ is the nonlinear parallel gradient. These latter three are given by

$$\eta_{\|} = 0.51\frac{m_e \nu_e}{n_e e^2}, \qquad \frac{d}{dt} = \frac{\partial}{\partial t} + [\phi, \,], \qquad B\nabla_{\|} = B^s\frac{\partial}{\partial s} + [-A_{\|}B, \,] \tag{1.52}$$

Here, ν_e is the collision frequency of the electrons, and B and B^s are the strength and parallel coordinate component of the guide field. We are careful to write $B\nabla_\parallel$, not $\nabla_\parallel$ itself, to satisfy the Leibnitz rule and the pure divergence property exactly at any level of approximation, with the distinction important along the parallel coordinate.

The statement on energetic consistency with any of these effects in the parallel dynamics is

$$\phi\left(B\nabla_\parallel \frac{J_\parallel}{B}\right) + J_\parallel \nabla_\parallel \phi = B\nabla_\parallel \frac{\phi J_\parallel}{B} \tag{1.53}$$

that is, the parallel divergence in a continuity equation and the parallel gradient in a forcing equation combine in such a way as to conserve the free energy represented by the dependent variables. The strength of the guide field can have spatial dependence, and the degree to which such effects are kept in the model always has to be taken into account in the formulation of all the operators.

There are also magnetic divergences in the equations, defined by representing the divergence of the ExB velocity,

$$\nabla \cdot \mathbf{v}_E = \nabla \cdot \frac{\mathbf{B} \times \nabla\phi}{B^2} \equiv -\mathcal{K}(\phi) \tag{1.54}$$

in terms of a curvature operator $\mathcal{K}$ which also satisfies the Leibnitz rule and the pure divergence property, so that

$$\tilde{\phi}\mathcal{K}(\tilde{p}_e) + \tilde{p}_e\mathcal{K}(\tilde{\phi}) = \mathcal{K}\left(\tilde{\phi}\tilde{p}_e\right) \tag{1.55}$$

also represents an energetically conservative process, in this case interchange forcing in the vorticity equation and ExB compression in the pressure equation. With warm ions and the temperature dynamics, there are more terms in each of these cases but they always satisfy energetic consistency.

1.6.2 Two-fluid dynamics

The main point of the Ohm's law in Eq. (1.51) is that it is a proper two-fluid Ohm's law, in which the current is being treated like a velocity. The velocity of the ions along the magnetic field is $u_\parallel$ and the corresponding velocity of the electrons is $v_\parallel$, with the current given by $J_\parallel = n_i Zeu_\parallel - n_e ev_\parallel$. Two-fluid sometimes means two temperatures, but it always means two velocities. The ion and electron continuity equations are equivalent, but only because one contains the polarisation current and the other contains the parallel current, respectively. Their difference is the vorticity equation. All of this ensures a self-consistent response, not least with the compressibility in the polarisation flow in the ions and the parallel current flow in the electrons. The forces on the right hand side of Eq. (1.51) are the parallel pressure and static electric forces on the electrons. The response to any imbalance in these is to drive parallel currents, and it is called the adiabatic response. A quasistatic evolution with the forces always in balance would mean what is called adiabatic electrons. But in contrast to the situation in Eq. (1.3), due to the large plasma frequency $\omega_{pe} \gg \omega$, we nearly always have a finite range in $k_{\|}$ over which the responses are concurrent, hence an adiabatic response on nonadiabatic electrons. This means we have to watch out for what the energetic exchanges involving these forces are in general. These subtleties were treated in detail in Chapters 11 through 14 of Volume One and are summarised again in the next Chapter.

1.6.3 Energetics

We always have to ensure exact energetic consistency when treating driven and dissipative turbulence in saturation. With the equations satisfying this property, and with an adequate numerical scheme, one can hold the turbulence in statistical saturation indefinitely. Saturation is defined in terms of the effective nonlinear growth rate, given by

$$\Gamma_T = \frac{1}{2E_T}\frac{\partial E_T}{\partial t} \tag{1.56}$$

where E_T is the total energy, always expressed as a volume average over the spatial domain. In saturation Γ_T should exhibit a mean close to zero with a standard deviation comparable to the rate given by drive, transfer, and dissipation all in balance. The mean over a finite time interval (at least several hundred $L_\perp/c_s$) should be at least an order of magnitude smaller than the standard deviation, for saturation to be judged as sufficient for measurement purposes. Even if running out of a nonlinear start to $t = 1000L_\perp/c_s$ is sufficient to demonstrate saturation, one should always run an exemplary case for much longer just to confirm the issue. If the equations were not energetically consistent and/or the numerical scheme were even marginally unstable, it would not be possible to do these demonstrations.

The energy theorem of the equations always has positive definite expressions for the content, for E_T as well as for a set of contributions representing each dependent variable. In the absence of drive and dissipation E_T is conserved even if there is vigorous exchange between the pieces. In the 2D MHD example given above, neither E_E nor E_B are conserved (the volume averages of U_E and U_B, respectively), but their sum $E_E + E_B$ is so. Then, the set of drive terms (background gradients, or currents, or in some cases flows) and another set of dissipation terms (i.e., resistivity and also subgrid dissipation) appear on the right hand side of the energy theorem and are directly measured to assess the capture by the numerics of the energetics represented by the equations. The dissipation terms also form a set of terms representing positive definite damping, meaning a positive definite term preceded by a minus sign. The drive and transfer terms are of indeterminate sign, set self-consistently by the dynamics.

Subgrid dissipation refers to the necessity of an energetic sink at the high-k end of the spectrum to contain the cascade dynamics. Thermal free energy and enstrophy always transfer preferentially to high-k, even if ExB flow energy does not. This was covered in detail in Chapters 4 and 5 of Volume One, before the three-dimensional models were addressed after Chapter 10. We do not detail this here since we will go through it in the next Chapter. Since the electron temperature by itself does not change the mode structure and energetics qualitatively, this will also serve as a recap of what was covered using the isothermal models in Volume One.

1.6.4 Total-f versus delta-f models

We encountered the concepts of 'mixing level ordering' or 'drift ordering' or 'delta-f' and "total-f' models. In a magnetised plasma the gyroradius scale is small compared to the scale length. So with turbulence, especially the vorticity, mostly at smaller scales, the gradient of a pressure disturbance is of similar size to the background gradient already for small amplitude. This is mixing level ordering and it is the basis of drift ordering, which assigns a fluctuation level in any of $e\tilde{\phi}/T_e$ or $\tilde{n}_e/n_e$, or similar, to the level of $\rho_s/L_\perp$, where $\rho_s = c_s/\Omega_i$ is the drift scale. If the entire model is fully nonlinear without approximations it retains such effects as $u_\parallel\nabla_\parallel$ in d/dt, or the variable p_e in $p_e\nabla \cdot \mathbf{v}_E$, or the variable n_e in $n_e e\nabla_\parallel\phi$, then it is a total-f model. No amplitude separation approximation has been made between background and fluctuating quantities. Under drift ordering, one can drop $u_\parallel\nabla_\parallel$ as small by $\rho_s/L_\perp$ with respect to the rest of d/dt as given in equation (1.52) and replace the other two example expressions by $-n_e T_e\mathcal{K}(\tilde{\phi})$ and $n_e e\nabla_\parallel\tilde{\phi}$, with n_e and T_e now given by background parameters. The equations are linearised everywhere except in $[\tilde{\phi},\ ]$ and $[-\tilde{A}_\parallel,\ ]$ in d/dt and $\nabla_\parallel$, respectively. This is the description of a delta-f model.

The main difference in the physical framework is that a total-f model conserves energy and a delta-f model conserves free energy. The free energy is a positive definite quantity quadratic in each of the dependent variables. For example,

$$\frac{3}{2}p_e \qquad \text{is replaced by} \qquad \frac{1}{2}n_e T_e\left(\frac{\tilde{n}_e^{\,2}}{n_e} + \frac{3}{2}\frac{\tilde{T}_e^{\,2}}{T_e}\right) \tag{1.57}$$

if temperature dynamics is present in the model. On the other hand, it was shown in Chapter 7 of Volume One that the energy transfer dynamics in the two models proceeds through the same channels given by the same quadratic expression in both cases (e.g., $\tilde{J}_\parallel \nabla_\parallel \tilde{p}_e$). For this reason, one can study the same physics with a delta-f model if the background parameters do not have to be treated as stratified. The basic physics of the turbulence is the same. We will return to this concept herein, specifically by building the gyrokinetic theory (see below) in both total-f and delta-f forms and once again showing their correspondence. The free energy of a delta-f model will be shown to be the second variation of the thermodynamic free energy of a total-f model and to be positive definite.

1.6.5 Mode structure

The mechanisms behind the energetics of the turbulence, drive and transfer especially, are all in competition with each other in the general mix. They can be separated out by means of control cases, or by direct diagnosis within a turbulence computation. They can also be isolated in the control cases and then if definite signatures in what we call the mode structure appear, these can be at least in principle be measured in experiments. Control cases are those in which one physical mechanism is isolated. Examples are the 2D neutral fluid and MHD cases, or the 3D pure drift wave or interchange cases, in which competing mechanisms are neglected in the model. Direct measurement of an energy transfer term is usually not possible experimentally. But measurement of the spectral regions in which some of the free energy components tend to collect, or most especially cross correlation and cross phase information between two quantities, are possible to acquire and then to compare across the control cases. The control cases are in turn not always possible to realise experimentally. But in their computations one can find signatures which then can be at least in principle measured in the general setting, even in an experiment. Some of the signatures do not distinguish very well, such as the transport flux spectrum which tends to look the same across many of these models. On the other hand, phase correlations between state variables such as density and potential, or with a flux variable like the current, are quite different in both form and result between the cases. The variables can be in or out of phase, or be randomly phased, with shifts whose distributions are wide in some cases (i.e., uncorrelated turbulence) or narrow in others (i.e., adiabatic electrons). These also distinguish well between turbulence and any underlying linear instabilities. Chapter 10 of Volume One gave an overview of these diagnostics, which were then examined in detail in all the following studies of the three-dimensional cases. The contrast between linear and nonlinear mode structure is very important in this context, since it is often the case that the mechanisms active in the turbulence are either absent or subdominant in the instabilities, and vice versa.

1.7 Kinetic theory, turbulence, and MHD instabilities

Once again, this will be reviewed extensively herein as each new physical mechanism is added. We start with the electron temperature, then the ion temperature which gives additional pathways which remain active if the electrons are nearly adiabatic. Then we move on to gyrokinetic theory. This is basically kinetic theory for low frequency dynamics. It is not merely an averaging over gyromotion time scales. The main concept is a change of representation. Instead of particles we treat gyrocentres. Polarisation is separated away from the gyrocentre motion by treating it as a density, not a current or a flow. The equations are not only more tractable, they have a more solid anchor in basic theory than the fluid drift ones do. Overlying this is gyrofluid theory, essentially a matter of changing the representation at the fluid level, or equivalently a matter of making the low frequency approximation first and then the kinetic to fluid one second. The way these two representations end up solving the same problem with different methods is detailed by proving correspondence in the large scale limit. Afterward, we recover some textbook kinetic theory results by taking the small gyroradius limit and exploring the wavelike responses. Ultimately, all this is a way of preserving tractable computations for turbulence which also preserve exact energetic consistency while allowing exploration of thermal anisotropy and the scale range below the ion gyroradius.

After this necessary background is developed we explore the effects of the temperature anisotropy on turbulence and flows. We continue with a bounce averaged trapped electron model which captures the effects mirror trapping in an axisymmetric magnetic field. After that we consider some MHD processes which are driven by currents rather than the thermal free energy gradients. Then, some additional theoretical issues are collected at the end, mostly to motivate future work which is still needed. It is very much not true that all the relevant physics of this kind of turbulence has already been uncovered and the only thing still needed should be bigger (and maybe more wasteful) computations. What is actually needed is more insight, and in some cases better theory.

Further Reading

Fundamental processes and parameters in plasma physics are treated in basic texts such as F F Chen, *Introduction to Plasma Physics and Controlled Fusion*, 3rd edn (Springer, 2016); and R Fitzpatrick, *Plasma Physics: An Introduction* (CRC Press, 2014); and K Nishikawa and M Wakatani, Plasma Physics: Basic Theory with Fusion Applications, 3rd ed, (Springer, 2013); R J Goldston with P H Rutherford, Introduction to Plasma Physics (CRC Press, 1995); and N Krall and A Trivelpiece, *Principles of Plasma Physics* (McGraw-Hill, 1973). The standard reference for a long time in the USA was L Spitzer, *Physics of Fully Ionized Gases* (Interscience, New York, 1956). Those with internet access can find the lectures by R Fitzpatrick, Plasma Physics, at http://farside.ph.utexas.edu/teaching/plasma/lectures/Plasmahtml.html

For magnetohydrodynamics (MHD), see *Ideal Magnetohydrodynamics*, by J P Freidberg (Plenum Press, New York, 1987); and *Cosmical Magnetic Fields*, by E N Parker (Clarendon Press, Oxford, 1979); and *Theory of Tokamak Plasmas*, by R B White (North Holland, Amsterdam, 1989). Of more relevance to our situation is *Nonlinear Magnetohydrodynamics*, by D Biskamp (Cambridge, 1993). Reduced MHD was introduced by H R Strauss, in 'Nonlinear, three-dimensional magnetohydrodynamics of noncircular tokamaks,' *Phys. Fluids* **19** (1976) 134.

For further background regarding both drifts and turbulence one can consult the Further Reading sections in Chapters 1, 2, 6, and 7 of Volume One.

IOP Publishing

Bruce Scott

Chapter 2

Effects of the electron temperature

In the exposition of turbulence in magnetised plasmas in Volume One, we kept to the simplest model of the basic physics which is still capable of capturing the principal effects of fluid drift Alfvén turbulence driven by the pressure gradient. In an isothermal model, pressure and density are synonymous, with the temperature as proportionality constant between them. With each state variable, there is a corresponding flux variable as part of a self consistent dynamical system. We have the ExB vorticity and parallel electric current as one pair, and the density and parallel plasma velocity as the other. If the ions are cold, that is, the temperature ratio $\tau_i = T_i/T_e$ is arbitrarily small, the model is almost isomorphic to the reduced MHD case. The main difference is very important, namely, the adiabatic response just covered in the Prelude. When the parallel gradient in the electron pressure is comparable to the parallel electric field, one has two-fluid dynamics and then also the corresponding compression processes in the continuity equations. The polarisation current flow is in the ion continuity equations, the parallel current part of the electron flow is in the electron continuity equations, and their compression combines with the electron forcing to form the adiabatic response.

When the temperature goes in with comparable dynamics to the density, the heat flux should go in with comparable dynamics to the parallel velocity. This means in terms of specific forms, $J_\parallel/n_e e$ and $q_{e\parallel}/p_e$ in units of velocity, we should have them at comparable magnitude with no smallness ordering relative to each other. With this step we move beyond the scope of conventional fluid dynamics or the collisional plasma model, in which heat flux is always ordered small compared with the flow. If the flow dynamics is reactive, so must be the heat flux dynamics as well. Dissipation is then also treated at the same level of ordering for both quantities.

2.1 Introduction—electron temperature

The basic concepts having been covered, we begin to introduce more complexity, one step at a time. Herein we bring in the electron temperature gradient as a

doi:10.1088/978-0-7503-3855-4ch2

potential source of free energy, and the fluctuations in both the electron temperature and parallel heat flux as additional dynamical variables. In almost all regimes of interest in magnetised plasmas, the time scale spectrum of the turbulence overlaps the collisional time scale, so that the heat flux response to the temperature is time dependent. The parallel heat flux therefore enters as a dependent variable with its own dynamics as well.

We consider the electron and ion temperatures separately since they have very different effects. The electron temperature participates in the adiabatic response, parallel Alfvén responses tying the current to the electron pressure in general. In combination with the parallel heat flux it introduces thermal conduction dissipation. Its gradient can be a strong and sometimes the principal drive for the turbulence. In an electromagnetic setting the magnetic flutter effects, parallel heat flux down perturbed field lines, can become strong and sometimes stabilising. The electron temperature fluctuations are most often smaller than the density fluctuations, but by virtue of being purely nonadiabatic, they are more efficiently lined up with the fluctuating ExB velocity for free energy access. On the other hand, the mechanisms of free energy transfer between electron pressure and ExB flow disturbances do not exhibit strong qualitative changes with the addition of the electron temperature. So we can use this Chapter as a bridge between Volumes One and Two.

For the present purpose we will stick to the collisional regime in which the heat flux, even if time dependent, effects dissipation through collisions. Extension to kinetic effects such as Landau damping will be addressed after the necessary background is developed in later Chapters. The ion temperature is left to the next Chapter since both temperatures bring in enough new physics to address them separately.

2.2 Conservative effects

The electron temperature has a number of dissipationless effects, which we can think of as those which alter the thermodynamics without introducing actual dissipation. These are in the parallel heat flux dynamics, which enters the kinetic shear Alfvén dynamics, and the diamagnetic heat flux, which in the fluid model is purely reactive. The diamagnetic fluxes, which arise from inertial corrections to the gyromotion at the particle level, enter through the perpendicular dynamics so we discuss them separately after the parallel responses.

2.2.1 Parallel dynamics and energetics

We studied the kinetic shear Alfvén wave properties in terms of the adiabatic response in Chapter 8 of Volume One. The force potential is $(\tilde{p}_e - n_e e\tilde{\phi})$, and the response is in $\tilde{J}_{\|}$. When the temperature is present, the normalised pressure disturbance consists of both density and temperature contributions. In a delta-f setting, the pressure is linearised under any time or space derivatives, so that we represent it as

$$\tilde{p}_e = T_e\tilde{n}_e + n_e\tilde{T}_e \tag{2.1}$$

with n_e and $\tilde{n}_e$ now always explicitly indicated. In general $\tilde{T}_e$ can react to the dynamics differently than does $\tilde{n}_e$, so we must consider what happens to it separately. To have this discussion we need the field equations

$$\varpi = \frac{\rho}{B^2}\nabla_\perp^2\phi, \qquad J_\parallel = -\frac{1}{\mu_0}\nabla_\perp^2 A_\parallel \tag{2.2}$$

which relate ϕ and $A_\parallel$ to the first two of the plasma variables, ϖ and $J_\parallel$. In the cold-ion case ϖ is ρ/B times the actual ExB vorticity.

In the parallel electron dynamics describing the adiabatic response, we have

$$\frac{\partial A_\parallel}{\partial t} + \frac{m_e}{n_e e^2}\frac{dJ_\parallel}{dt} + \eta_\parallel J_\parallel = \frac{1}{n_e e}\nabla_\parallel p_e - \nabla_\parallel\phi \tag{2.3}$$

as discussed in the Prelude. Expressed in terms of the disturbances in scaled units (pull out a factor of T_e/e on the right hand side), this is

$$\frac{\partial \tilde{A}_\parallel}{\partial t} + \frac{m_e}{n_e e^2}\frac{d\tilde{J}_\parallel}{dt} + \eta_\parallel \tilde{J}_\parallel = \frac{T_e}{e}\nabla_\parallel\left(\frac{\tilde{n}_e}{n_e} + \frac{\tilde{T}_e}{T_e} - \frac{e\tilde{\phi}}{T_e}\right) \tag{2.4}$$

giving the relevant parallel forcing term for the parallel current and magnetic induction. The quantity given in parentheses,

$$\tilde{h}_e = \frac{\tilde{n}_e}{n_e} + \frac{\tilde{T}_e}{T_e} - \frac{e\tilde{\phi}}{T_e} \tag{2.5}$$

can be regarded as the nonadiabatic force potential.

The counterpart to this is the presence of the parallel current divergence not only in the ExB vorticity equation,

$$\frac{d\varpi}{dt} = \cdots + B\nabla_\parallel\frac{\tilde{J}_\parallel}{B} \tag{2.6}$$

but also in the continuity equations for the electrons,

$$\frac{d\tilde{n}_e}{dt} = \cdots + \frac{1}{e}\,B\nabla_\parallel\frac{\tilde{J}_\parallel}{B} \tag{2.7}$$

$$\frac{3}{2}\frac{d\tilde{T}_e}{dt} = \cdots + \frac{T_e}{n_e e}\,B\nabla_\parallel\frac{\tilde{J}_\parallel}{B} \tag{2.8}$$

not only for density but also for temperature. If we divide these by $n_e e$, n_e, and T_e, respectively, we have

$$\begin{aligned}
\frac{d}{dt}\frac{\varpi}{n_e e} &= \cdots + \frac{1}{n_e e}\,B\nabla_\parallel\frac{\tilde{J}_\parallel}{B}\\
\frac{d}{dt}\frac{\tilde{n}_e}{n_e} &= \cdots + \frac{1}{n_e e}\,B\nabla_\parallel\frac{\tilde{J}_\parallel}{B}\\
\frac{3}{2}\frac{d}{dt}\frac{\tilde{T}_e}{T_e} &= \cdots + \frac{1}{n_e e}\,B\nabla_\parallel\frac{\tilde{J}_\parallel}{B}
\end{aligned} \tag{2.9}$$

One should already be able to see the symmetry in these three equations and Eq. (2.4). Together, they comprise the adiabatic response of the plasma to any parallel force imbalances in the electrons.

Two main points are to be made. The first is that the same term is sitting in the right hand side of each of these equations. So given the form of the vorticity variable in Eq. (2.2), we use the T_e/e scaling in the adiabatic response to recast the scaled version for disturbances as

$$\frac{\varpi}{n_e e} = \rho_s^2 \nabla_\perp^2 \frac{e\tilde{\phi}}{T_e} \qquad \text{where} \quad \rho_s^2 = \frac{m_i T_e}{e^2 B^2} \tag{2.10}$$

where $m_i = \rho/n_e$ defines the effective ion mass in any situation. The scale ρ_s is not a true gyroradius but is often called the ion sound gyroradius due to its form and the dependence of the sound speed on T_e/m_i. It is also variously called the ion inertial drift scale or simply the drift scale. It makes sense only after the parallel dynamics has chosen T_e/e as the relevant scaling for ϕ, and then it represents the polarisability of the plasma (recall that $\rho_s \gg \lambda_D$ is one of the statements of quasineutrality). The main point is that if $k_\perp \rho_s < 1$ then $\tilde{\phi}$ reacts the fastest to restore parallel force imbalances. So we do not have electrons thermalising within a given potential well. We rather have the parallel kinetic Alfvén responses, whose relaxed state after dissipation is adiabatic electrons with $\tilde{h}_e$ and $\tilde{J}_\parallel$ vanishing.

The second point to be made is how these forcing and continuity processes combine to conserve free energy. Multiplying Eqs (2.9) by $p_e(e\tilde{\phi}/T_e)$, $p_e(\tilde{n}_e/n_e)$, and $p_e(\tilde{T}_e/T_e)$, respectively, and Eq. (2.4) by $\tilde{J}_\parallel$, and using the field equations (Eqs 2.2), yields

$$\begin{aligned} &\frac{1}{2}\frac{\partial}{\partial t}\left[\rho\,|\mathbf{v}_E|^2 + \frac{1}{\mu_0}|\mathbf{B}_\perp|^2 + n_e m_e \left|\frac{\tilde{J}_\parallel}{n_e e}\right|^2 + p_e\left(\left|\frac{\tilde{n}_e}{n_e}\right|^2 + \frac{3}{2}\left|\frac{\tilde{T}_e}{T_e}\right|^2\right)\right] \\ &- p_e B\nabla_\parallel\left(\frac{1}{B}\tilde{h}_e\frac{\tilde{J}_\parallel}{n_e e}\right) = \cdots - \eta_\parallel \tilde{J}_\parallel^2 \end{aligned} \tag{2.11}$$

where

$$\mathbf{v}_E = \frac{1}{B}\mathbf{b}\times\nabla\tilde{\phi}, \qquad \mathbf{B}_\perp = -\frac{1}{B}\mathbf{b}\times\nabla\tilde{A}_\parallel \tag{2.12}$$

appear in the ExB and magnetic energy of the disturbances. There follows the parallel electron kinetic energy and then the thermal free energy. The term with $B\nabla_\parallel$ represents the parallel divergence combination and hence an overall transport term. Note that $-\tilde{J}_\parallel/n_e e$ is the parallel electron velocity in the absence of a parallel mass flow. On the right hand side we have the resistivity in a term representing positive definite damping (a positive definite quantity preceded by a minus sign). The process is therefore energetically consistent and this is our definition of what that means. This part of the energy theorem suffices as demonstration of the necessity to keep the polarisation drift and its divergence in the continuity equations. With the electron dynamics we see the parallel current divergence, but this implies that the ion density equation, which should be formally identical to the electron one, has the polarisation divergence in the same place. One finds the ion equation by subtracting the vorticity equation from the electron equation,

$$\frac{d\tilde{n}_i}{dt} = \cdots + \frac{1}{e}\frac{d\varpi}{dt} \tag{2.13}$$

(noting $\tilde{n}_i = \tilde{n}_e$), which in the currents cancels the parallel divergence but replaces it with the polarisation divergence. The same will hold for the ion temperature equation when we introduce it in the next Chapter.

One additional observation is the scale implied by the presence of the electron inertia. Due to Ampére's law (second relation in 2.2), we have

$$A_\| + \frac{m_e}{n_e e^2}J_\| = (1 - d_e^2\nabla_\perp^2)A_\| \qquad \text{with} \quad d_e^2 = \frac{m_e}{\mu_0 n_e e^2} \tag{2.14}$$

which defines d_e as the electron inertial skin depth. This scale often falls within or near the spectrum, so it is always retained in the electron dynamics. Then, the results decide if it is relevant or not.

With the temperature dynamics present we also have the parallel heat flux, representing another dissipation channel. In the electron temperature equation we have

$$\frac{3}{2}n_e\frac{d\tilde{T}_e}{dt} = \cdots - B\nabla_\|\frac{\tilde{q}_{e\|}}{B} \tag{2.15}$$

which adds to the velocity compression processes just discussed. In the equation for the heat flux itself we have the parallel gradient of just the temperature,

$$\cdots = -\frac{5}{2}\frac{p_e}{m_e}\nabla_\|\tilde{T}_e \tag{2.16}$$

(noting that $p_e/m_e = n_e V_e^2$) where we will give the details and put in the left hand side later on. Multiplying them by $\tilde{T}_e/T_e$ and $(2/5)m_e\tilde{q}_{e\|}/p_e T_e$, respectively, these combine into a free energy term, whose heat flux contribution we will put in later, and a total divergence,

$$\frac{3}{2}p_e\frac{d}{dt}\left|\frac{\tilde{T}_e}{T_e}\right|^2 + p_e B\nabla_\|\left[\frac{1}{B}\left(\frac{\tilde{T}_e}{T_e}\frac{\tilde{q}_{e\|}}{p_e}\right)\right] = \cdots - (\text{dissipation}) \tag{2.17}$$

This part of the dynamics conserves free energy in the same way as the flows do, with the $\tilde{T}_e$ and $\tilde{q}_{e\|}$ combining in the same way as $\tilde{h}_e$ and $\tilde{J}_\|$ in Eq. (2.11). When the dissipation in the heat flux is included we will find another positive definite damping term on the right hand side, representing thermal conduction dissipation. In a collisional model $\tilde{q}_{e\|}$ is expressed as a dissipative flux leading to a diffusion term, which is how thermal conduction is usually treated. In our case, however, we will have to discuss the heat flux as a time dependent variable on the same footing as the parallel flow and current. More on this later, but for now we note that these are conservative terms which will modify the kinetic shear Alfvén responses. Dissipation will appear as a damping rate on the parallel heat flux in its equation, not as a temperature diffusion term, since it is a weak correction.

2.2.2 Diamagnetic fluxes and energetics

The other dissipation free effect are the diamagnetic flow and heat flux, which were derived in Chapters 6 and 7 of Volume One. Briefly, one takes a moment hierarchy over the Lorentz force on the right hand side and the gradients on the left hand side of the kinetic equation,

$$\mathbf{v}\cdot\nabla f+\frac{Ze}{m}(\mathbf{E}+\mathbf{v}\times\mathbf{B})\cdot\frac{\partial f}{\partial\mathbf{v}}=0 \tag{2.18}$$

in which to lowest order in the drift approximation the partial time derivative is dropped. Moments are integrals over $d^3\mathbf{v}$ times factors of $m\mathbf{v}$ and $v^2/2$, with the background Maxwellian

$$f_0=F^M(n,T)=n(2\pi T/m)^{-3/2}\exp\left[-mv^2/2T\right],\qquad V^2=T/m \tag{2.19}$$

inserted into the first term. The moments over 1, $\mathbf{v}$, $mv^2/3$, $mv^2\mathbf{v}/2$, and $mv^4/2$ yield the density n, the flow $\mathbf{u}$, the pressure $p=nT$, the total heat flux $\mathbf{q}+(5/2)p\mathbf{u}$, and, to close, the heat capacity $(5/2)pT/m$, respectively, with even ones taken over the Maxwellian.

In the equation of motion, found by taking the moment over $m\mathbf{v}$, we obtain

$$\nabla p=nZe(\mathbf{E}+\mathbf{u}\times\mathbf{B}) \tag{2.20}$$

This is solved for $\mathbf{u}$ to obtain the drift formulae

$$\mathbf{u}=\mathbf{v}_E+\mathbf{u}_*,\qquad \mathbf{v}_E=\frac{1}{B}\mathbf{b}\times\nabla\phi,\qquad \mathbf{u}_*=\frac{1}{nZe}\frac{1}{B}\mathbf{b}\times\nabla p \tag{2.21}$$

in which the ExB velocity $\mathbf{v}_E$ arises from $\mathbf{E}$ and the diamagnetic velocity $\mathbf{u}_*$ arises from ∇p. The latter is ordered small in a conventional MHD model, but in a two-fluid setting we always have $nZe\mathbf{E}$ and ∇p comparable, consistent with $\mathbf{u}$ and $\mathbf{J}/nZe$ comparable. The diamagnetic current is found by taking the species sum of partial charge density nZe times velocity $\mathbf{u}$ to obtain

$$\mathbf{J}_*=\frac{1}{B}\mathbf{b}\times\nabla p_T \tag{2.22}$$

where p_T is the total pressure (in a cold ion case $p_T=p_e$). Under the divergence, the three currents, polarisation, parallel, and diamagnetic, yield the vorticity equation, in which $\mathbf{J}_*$ contributes in an inhomogeneous guide field,

$$\nabla\cdot\mathbf{J}_*=\nabla\cdot\frac{1}{B}\mathbf{b}\times\nabla p_T\equiv-\mathcal{K}(p_T) \tag{2.23}$$

in the same way as the ExB and diamagnetic flow divergences do in the continuity equations. In the latter for the electrons, the polarisation is ordered small, due to the mass ratio.

The next odd velocity moment, $(mv^2/2)m\mathbf{v}$, yields the heat flux equation,

$$\nabla\left(\frac{5}{2}pT\right) = \frac{5}{2}Ze\, p\mathbf{E} + Ze\left(\frac{5}{2}p\mathbf{u} + \mathbf{q}\right) \times \mathbf{B} \tag{2.24}$$

where $\mathbf{q}$ is the conductive part of the heat flux. After subtracting $(5/2)T$ times Eq. (2.20) and then solving for $\mathbf{q}$, we find

$$\mathbf{q}_\wedge = \frac{5}{2}\frac{p}{Ze}\frac{1}{B}\mathbf{b} \times \nabla T \tag{2.25}$$

which gives the diamagnetic heat flux. The corresponding magnetic divergence term adds to the temperature equation. We have

$$\frac{3}{2}n\mathbf{u}_* \cdot \nabla T + nT\nabla \cdot \mathbf{u}_* + \nabla \cdot \mathbf{q}_\wedge + \frac{T}{Ze}\,\mathcal{K}(p) + \frac{5}{2}\frac{p}{Ze}\,\mathcal{K}(T) = 0 \tag{2.26}$$

representing the diamagnetic cancellation in the temperature equation, so that the terms on its right hand side are

$$\frac{3}{2}n\frac{dT}{dt} = \cdots + p\,\mathcal{K}(\phi) + \frac{T}{Ze}\,\mathcal{K}(p) + \frac{5}{2}\frac{p}{Ze}\,\mathcal{K}(T) = 0 \tag{2.27}$$

with the first one representing ExB compression in this equation. The diamagnetic cancellation removes the diamagnetic velocity from direct advection, leaving d/dt as (solely) the ExB advective derivative.

In a two-dimensional interchange model (starting with the Eulerian ExB vorticity equation, and passive scalar equations for density and temperature), these curvature terms become conservative coupling terms in the equations for the vorticity and electron density and temperature disturbances (with $Z = -1$). The three equations are

$$\frac{d\varpi}{dt} = -\mathcal{K}(\tilde{p}_e) \tag{2.28}$$

$$\frac{d\tilde{n}_e}{dt} = n_e\mathcal{K}(\tilde{\phi}) - \frac{1}{e}\mathcal{K}(\tilde{p}_e) \tag{2.29}$$

$$\frac{3}{2}n_e\frac{d\tilde{T}_e}{dt} = p_e\mathcal{K}(\tilde{\phi}) - \frac{T_e}{e}\mathcal{K}(\tilde{p}_e) - \frac{5}{2}\frac{p_e}{e}\mathcal{K}(\tilde{T}_e) \tag{2.30}$$

The energetics is found in the same way as for the parallel dynamics. Multiply the three equations by $-\tilde{\phi}$, $T_e\tilde{n}_e$ and $\tilde{T}_e/T_e$, respectively, to obtain

$$\begin{aligned} &\frac{1}{2}\frac{\partial}{\partial t}\left[\rho\,|\mathbf{v}_E|^2 + p_e\left(\left|\frac{\tilde{n}_e}{n_e}\right|^2 + \frac{3}{2}\left|\frac{\tilde{T}_e}{T_e}\right|^2\right)\right] \\ &+ \frac{1}{2}\left[p_e\frac{T_e}{e}\mathcal{K}\left(|\tilde{n}_e|^2\right) + \frac{5}{2}\frac{n_e}{e}\mathcal{K}\left(|\tilde{T}_e|^2\right)\right] + \cdots = 0 \end{aligned} \tag{2.31}$$

always treating the background n_e and T_e as constant parameters, and with $\tilde{h}_e$ the nonadiabatic force potential given in Eq. (2.5). This serves to show how the curvature terms conserve energy collectively. Since $\tilde{T}_e/T_e$ and $\tilde{h}_e$ are comparable, it follows that $\mathbf{q}_\wedge$ and $p\mathbf{u}_*$ must in general be treated as comparable.

We note the scaling with T_e/e, which combined with R, the radius of magnetic curvature, the B in the drifts, and the perpendicular wavenumber $k_\perp$ arising from ∇ on the disturbances, gives the native frequency of these effects as $2(T_e/eBR)k_\perp$. Using the definition of ρ_s in Eq. (2.10), this is expressed as $2c_s/R$ times $k_\perp\rho_s$. The factor of two reflects combination of both $\nabla \log B$ and $\nabla \times \mathbf{b}$, which usually add in a magnetically confined plasma. In a purely poloidal axisymmetric magnetic field, we also have these equal up to $O(\beta)$ corrections. With the statement of MHD equilibrium

$$\mathbf{J} \times \mathbf{B} = \nabla p \tag{2.32}$$

and with ∇p ordered small, we have

$$\mathbf{B} \cdot \nabla\mathbf{B} = \nabla\frac{B^2}{2} \quad \Longrightarrow \quad \mathbf{b} \cdot \nabla\mathbf{b} = \nabla_\perp \log B \tag{2.33}$$

using Ampère's law $\mu_0\mathbf{J} = \nabla \times \mathbf{B}$ and a vector identity. The vector $\mathbf{b} \cdot \nabla\mathbf{b}$ has magnitude $1/R$ and direction towards the centre of curvature. The second vector then implies $B \propto R^{-1}$. This is why the magnetic divergence terms are usually called curvature terms It is only in a tokamak-like magnetic field (mostly toroidal, with finite poloidal corrections) that this R is about the same as the toroidal major radius.

2.2.3 Gradient forcing

There are two gradients available for free energy drive in this case. These are parameterised as

$$\omega_n = |L_\perp \nabla \log n_e|, \qquad \omega_t = |L_\perp \nabla \log T_e| \tag{2.34}$$

Now $L_\perp$ is an arbitrary perpendicular scale length. It is usually set to the shorter of the two logarithmic gradient scales, indicating the steeper of the two gradients in relative terms. In a tokamak this is almost always $\nabla \log T_e$, though in many settings such as in astrophysics and small laboratory experiments the density gradient is principal. With these present the drift wave dynamics, with for example

$$\frac{d\tilde{n}_e}{dt} \rightarrow \frac{\partial\tilde{n}_e}{\partial t} + [\tilde{\phi}, \tilde{n}_e + n_{e0}] \tag{2.35}$$

for an ExB advective derivative and

$$\nabla_\parallel\tilde{n}_e \rightarrow \mathbf{b} \cdot \nabla\tilde{n}_e + [-\tilde{A}_\parallel, \tilde{n}_e + n_{e0}] \tag{2.36}$$

explicitly containing the gradient term (in which n_{e0} containing the gradient is distinct from n_e the constant parameter). In normalised equations these are expressed with logarithmic gradients. The part of the parallel gradient due to $\tilde{A}_\parallel$ is most often referred to as the magnetic flutter effect, especially when it acts on the background gradient.

If the scaling from the adiabatic response is put into the density equation, we have

$$\frac{\partial \tilde{n}_e}{\partial t} \sim n_e \frac{e}{T_e} \frac{\partial \tilde{\phi}}{\partial t} \sim \frac{1}{B} \frac{\partial n_{e0}}{\partial x} \frac{\partial \tilde{\phi}}{\partial y} \sim -\frac{1}{B} \frac{n_e}{L_\perp} \frac{\partial \tilde{\phi}}{\partial y} \tag{2.37}$$

for the linear advection term and the time derivative. This implies

$$\omega \sim \omega_* = \frac{T_e k_y}{eBL_\perp} = \frac{c_s}{L_\perp} (k_y \rho_s) \tag{2.38}$$

This is the origin of the $c_s/L_\perp$ scaling for magnetised plasma turbulence. The corresponding linear wave dynamics is drift waves, which were covered in Chapter 9 of Volume One. The similarity in the magnitudes of linear and nonlinear advection leads to RMS vorticity values with this scaling. Once again, the RMS is measured as the square root of the volume average of the square.

2.3 Dissipative effects

Dissipative effects in this context arise from the electron collision frequency which for an isothermal model manifests as electrical resistivity, a friction between electron and ion flows which dissipates the electric current. In a magnetised plasma at reasonable temperature, the parallel resistivity in the Ohm's law constrains the nonadiabatic component of the dynamics for a given level of parallel current which has been excited by the ExB vorticity nonlinearity. The electron temperature brings in thermal conductivity. In the conventional model the thermal conductivity is assumed to be faster than the dynamics so it enters as a diffusive dissipation along field lines directly on the electron temperature. We briefly discuss this first. Then, we note that thermal conduction, even by electrons along magnetic field lines, is not so fast in many regimes. The treatment has to be modified so that it looks much more like what we have already been doing for resistivity: even if the dissipation due to collisions is faster, the inertial response is always present, so that the dynamical processes in the equations decide whether one is actually deep enough into the diffusion regime for diffusion to be a reasonable model. They often decide otherwise, and the time dependence turns out to be relevant. The thermal conduction effect is rather slower, relative to this type of turbulence, than is most often thought.

2.3.1 Conventional treatment

In the Ohm's law for $J_\parallel$ the resistivity enters as a damping term, $\eta_\parallel J_\parallel$ in an isothermal setting. The parallel heat flux, by contrast is brought in as a diffusive flux given by a conductivity rather than a damping rate. These are given as

$$\frac{\partial \tilde{A}_\parallel}{\partial t} + \nabla_\parallel \tilde{\phi} = -\eta_\parallel \tilde{J}_\parallel = -0.51 \frac{m_e \nu_e}{n_e e^2} \tilde{J}_\parallel \tag{2.39}$$

for a reduced MHD model not treating inertia or the adiabatic response, and

$$\tilde{q}_{e\parallel} = -n_e\kappa_{\parallel e}\nabla_\parallel\tilde{T}_e = -\kappa_e\frac{V_e^2}{\nu_e}\,n_e\nabla_\parallel\tilde{T}_e \tag{2.40}$$

where the coefficients 0.51 and $\kappa_e = 3.2$ give the numerical values of the Spitzer (1953) resistivity and thermal conduction, respectively. Recall that in $\nabla_\parallel\tilde{p}_e$ and $\nabla_\parallel\tilde{T}_e$, it is important to remember the derivative of the background gradient as well as the fluctuation, in case these are treated separately, since $\nabla_\parallel$ is nonlinear and includes the piece in the radial direction due to the magnetic flutter effect.

Since the collisions are Coulomb collisions, which depend on the kinetic velocity of interaction, thermal effects enter with a thermal force in Ohm's law and a thermoelectric effect in the heat flux, given by

$$\frac{\partial\tilde{A}_\parallel}{\partial t} + \nabla_\parallel\tilde{\phi} = \frac{1}{n_e e}\nabla_\parallel\tilde{p}_e + \frac{\alpha_e}{e}\nabla_\parallel\tilde{T}_e - \eta_\parallel\tilde{J}_\parallel \tag{2.41}$$

$$\tilde{q}_{e\parallel} = -\kappa_e\frac{V_e^2}{\nu_e}\,n_e\nabla_\parallel\tilde{T}_e - \alpha_e\frac{T_e}{e}\tilde{J}_\parallel \tag{2.42}$$

respectively. To be reasonable we specify a two-fluid Ohm's law retaining adiabatic response, since the pressure, temperature, and electrostatic potential in the scaled units of Eq. (2.5) should always be assumed comparable in any ordering. The coefficient $\alpha_e = 0.71$ gives the Braginskii (1965) thermal force coefficient. This is how these dissipative effects are put into fluid plasma equations in what can be called the standard Braginskii model. It assumes that V_e is arbitrarily fast compared with the dynamics, so that even when ν_e is faster than the time changes, thermal conduction is faster still. But the experience as collected in Volume One has shown that V_e is not so fast. In fact we have to treat the regime where not only v_A is comparable to V_e but also where the drift frequency $c_s/L_\perp$, giving the native level for RMS vorticity in the turbulence, is faster than $K_\parallel V_e$, the native transit frequency of electron communication over the connection length $2\pi/K_\parallel$ of the field lines. This means that we can have both $k_\parallel V_e > \omega$ and $k_\parallel V_e < \omega$ within the overall spectrum in the same system. Under these conditions we have to do better with the parallel electron heat flux and electron thermal conduction.

2.3.2 Time dependent heat flux

In the usual form of the dissipation due to resistivity and thermal conductivity the first one is put into an Ohm's law and the second one becomes a diffusive term in the temperature equation. As pointed out by Hassam (1980) this is not entirely consistent. When one includes the electron inertia in the Ohm's law, one should include the time dependence of the heat flux in its own equation as well, since these appear together in practically any ordering. The native frequencies of the parallel electron dynamics are between $k_\parallel v_A$ and $k_\parallel V_e$, representing both the kinetic shear Alfvén and parallel electron heat flux responses. The native frequency of the turbulence is with the RMS vorticity similar to $c_s/L_\perp$. The parameter situation of greatest interest for drift Alfvén turbulence is with all three of these of similar magnitude. It makes sense that if we treat one of these effects, induction or inertia in the Ohm's law, we have to treat all of them, including the electron thermal inertia in the equation for the heat flux. The same applies to the collisional dissipation. Then, if this maximal drift Alfvén regime is departed, the equations should relax into the various limits naturally.

The Ohm's law with electron inertia can be put into scaled units according to

$$\frac{\partial \tilde{A}_\parallel}{\partial t} + \frac{m_e}{e}\frac{d}{dt}\frac{\tilde{J}_\parallel}{n_e e} = \frac{T_e}{e}\nabla_\parallel\left(\frac{\tilde{p}_e}{p_e} + \alpha_e\frac{\tilde{T}_e}{T_e} - \frac{e\tilde{\phi}}{T_e}\right) - 0.51\nu_e\frac{m_e}{e}\frac{J_\parallel}{n_e e} \tag{2.43}$$

where d/dt is the ExB advective derivative as we have seen before, with the α_e thermal force term the only addition aside from $\tilde{T}_e$ in the pressure. We start by putting the dissipative formula for the heat flux, Eq. (2.42), into the same scaling, namely, T_e/e times $\nabla_\parallel$ on a normalised forcing term and $m_e\nu_e/e$ on the parallel flux in units of velocity, so that

$$0 = -\frac{T_e}{e}\nabla_\parallel\left(\frac{\tilde{T}_e}{T_e}\right) - \frac{1}{\kappa_e}\nu_e\frac{m_e}{e}\left(\frac{\tilde{q}_{e\parallel}}{p_e} + \alpha_e\frac{J_\parallel}{n_e e}\right) \tag{2.44}$$

using $V_e^2 = T_e/m_e$ to convert the formula, and setting the thermoelectric term $\alpha_e\tilde{J}_\parallel$ next to the conductive heat flux.

Now, we can solve this equation for $\nabla_\parallel(\tilde{T}_e/T_e)$ and put the result into the Ohm's law where we see the thermal force. This yields

$$\frac{\partial}{\partial t}\left(\tilde{A}_\parallel + \frac{m_e}{e}\frac{\tilde{J}_\parallel}{n_e e}\right) = \frac{T_e}{e}\nabla_\parallel\left(\frac{\tilde{p}_e}{p_e} - \frac{e\tilde{\phi}}{T_e}\right) - \frac{m_e}{e}\nu_e\left(0.51\frac{J_\parallel}{n_e e} + \frac{\alpha_e}{\kappa_e}\left(\frac{\tilde{q}_{e\parallel}}{p_e} + \alpha_e\frac{J_\parallel}{n_e e}\right)\right) \tag{2.45}$$

This result shows us that the usual formula for the heat flux is just the limit in which inertia can be neglected, and the thermal force coefficient is now the indicator of how the dissipation matrix is mixed. Instead of conductivities we explicitly speak of damping rates, and a dissipation matrix whose off-diagonal coefficients are finite. This is called 'mixing of the moments', a feature of the dependence of the collision process on kinetic velocity, one of the main points of the 1965 review by Braginskii, also discussed in Chapter 6 of Volume One. If the factor $(V_e/v)^3$, the cube of ratio of thermal velocity to kinetic velocity, in the collision integral is set to unity, the thermal force coefficient α_e is found to be zero. Stated another way, the collision process would be diagonal with respect to the fluid moments. So we have $\alpha_e = 0.71$ to cover both the thermal force and thermoelectric effect through the same origin, the dependence of the Coulomb collision cross section upon relative particle velocity.

The time dependence is found by considering the third moment level in the list given in section 2.2.2, with the same basic assumptions except that the parallel components $u_\parallel$ and $q_\parallel$ are the ones considered for the vectors. The moment over

$mv^2v_\parallel/2$ is taken over the gradient term as before, but now instead of the Lorentz force we have the collision term. The details are in the Braginskii (1965) reference, with the simpler illustrative version from Hassam (1980) having been given in Volume One. The result under the drift approximation is

$$\frac{m_e}{e}\frac{d}{dt}\frac{\tilde{q}_{e\parallel}}{p_e} + \frac{5}{2}\frac{T_e}{e}\nabla_\parallel\left(\frac{\tilde{T}_e}{T_e}\right) = -\frac{m_e}{e}\frac{5/2}{\kappa_e}\nu_e\left(\frac{\tilde{q}_{e\parallel}}{p_e} + \alpha_e\frac{J_\parallel}{n_e e}\right) \tag{2.46}$$

as the heat flux equation in which nonlinear advection can act on $\tilde{q}_{e\parallel}$ as it does on the other dependent variables. Indeed, if $K_\parallel V_e$ and ν_e are both less than $c_s/L_\perp$ then the nonlinear mixing by turbulence is actually the fastest process in the equation. The second term in this equation is the one written in Eq. (2.16). If the collision frequency is the fastest process in the equation, then the ExB advective derivative is small and the collisional limit given by Braginskii is recovered.

We will see in section 2.6 that this remains a set of equations with exact energetic consistency in the conservative coupling terms (parallel dynamics, curvature) and positive definite damping in the dissipation. As such, although it like any other model is an approximation, it is a legitimate way to treat thermal conduction with finite response times.

The general problem of a moment hierarchy can be seen in this case in that the equation for moment level n involves contributions for moments $n + 1$ through the $\nabla_\parallel$ terms, where the level n refers to the power of kinetic velocity by which the distribution function is multiplied in the integration over velocity space. The specific thermal capacity $r_\parallel$ would subtract the $(5/2)pT/m$ and other lower-moment effects off, leaving the fourth-moment piece by itself. The effective closure in the dissipation model is to effectively recast the higher moments not kept in terms of moments which are kept, in this case replacing $\nabla_\parallel r_\parallel$ in the equation for $q_\parallel$ in terms of the dissipation term. We will derive a more systematic method for this moment chain and introduce a relevant closure model in Chapter 9 on gyrofluid theory after the necessary predecessors in gyrokinetic theory just beforehand. We will find that the gyrofluid model in the collisional dissipation regime recovers Eq. (2.46) as a subset.

2.4 Equations for magnetised plasma turbulence

With the combination of the parallel dynamics and the curvature, and dissipation, listed above, the toroidal drift Alfvén equations start with the dynamics of the vorticity, density, parallel current, and parallel flow. They expand to acquire the effects of the electron temperature fluctuations in the pressure forcing and diamagnetic flow, and then the electron temperature and parallel equations themselves. The derivation of these equations was given in Volume One.

We have two-dimensional (2D) turbulent dynamics perpendicular to the background magnetic field, coupled to kinetic shear Alfvén and acoustic wavelike dynamics in the direction along the field, so the three-dimensional (3D) situation is more a 2D plus 1D case than a general 3D case. The range of parallel wavenumbers excited is given by the constraining against the turbulence driven by the gradients. The energetics covers frequencies up to $c_s/L_\perp$, which becomes the baseline parameter for the RMS vorticity of the turbulence. Then, the parallel wavenumber space satisfies

$$\omega \sim \frac{c_s}{L_\perp} \sim k_\parallel v_A \sim k_\parallel V_e \tag{2.47}$$

for the electron energetics, where we are assuming that the electron dynamical beta and the electron to ion mass ratio are in a maximal ordering, which means we allow general values and assume they are comparable,

$$\beta_e \equiv \frac{\mu_0 p_e}{B^2}, \qquad \mu_e = \frac{m_e}{m_i}, \qquad \beta_e \sim \mu_e \ll 1 \tag{2.48}$$

This also means that

$$k_\parallel L_\perp \ll 1, \qquad k_\parallel \ll k_\perp \tag{2.49}$$

which is the statement of flute mode ordering. The scale of the magnetic field is much larger than $L_\perp$ but comparable to $k_\parallel^{-1}$, so that while B is assumed constant across field lines except for the existence of the curvature terms, it does not commute past the unperturbed parallel gradient. This leads to the treatment of field line geometry, given below. For now, we assume coordinates $\{xy\}$ to cover the perpendicular plane and s to follow the field lines, with only s involved in the unperturbed parallel gradient.

The vorticity equation consists of the polarisation, parallel, and diamagnetic current divergences, given, respectively, by

$$\frac{d\varpi}{dt} = B\nabla_\parallel \frac{\tilde{J}_\parallel}{B} - \mathcal{K}(\tilde{p}_e) \tag{2.50}$$

noting the positions of the factors of B in the parallel divergence. All three of the nonlinear part of the ExB advective derivative d/dt, the parallel divergence $B\nabla_\parallel$, and the magnetic divergence $\mathcal{K}$, are assumed to satisfy the Leibnitz property and the divergence property exactly.

The electron density equation is a continuity equation, with the advective derivative, the divergence of the parallel electron flow, and the magnetic divergences of the ExB and diamagnetic flow, given, respectively, by

$$\frac{1}{n_e}\frac{d\tilde{n}_e}{dt} + B\nabla_\parallel \frac{\tilde{u}_\parallel}{B} = \frac{1}{n_e e} B\nabla_\parallel \frac{\tilde{J}_\parallel}{B} + \mathcal{K}(\tilde{\phi}) - \frac{1}{n_e e}\mathcal{K}(\tilde{p}_e) \tag{2.51}$$

with polarisation kept for the ions in the current but not for electrons. The nonlinear time derivative also includes the advection of the background gradient.

The electron temperature equation is done similarly, with the parallel and diamagnetic heat fluxes also appearing,

$$\begin{aligned} \frac{3/2}{T_e}\frac{d\tilde{T}_e}{dt} + B\nabla_\parallel \frac{\tilde{u}_\parallel}{B} + \frac{1}{p_e} B\nabla_\parallel \frac{\tilde{q}_{e\parallel}}{B} &= \frac{1}{n_e e} B\nabla_\parallel \frac{\tilde{J}_\parallel}{B} \\ &+ \mathcal{K}(\tilde{\phi}) - \frac{1}{n_e e}\mathcal{K}(\tilde{p}_e) - \frac{5/2}{e}\mathcal{K}(\tilde{T}_e) \end{aligned} \tag{2.52}$$

In each case the specific flows denoted by $\tilde{u}_\parallel$, $\tilde{J}_\parallel/n_e e$, and $\tilde{q}_{e\parallel}/p_e$ are all assumed to be comparable, along with the $e\tilde{\phi}/T_e$, $\tilde{n}_e/n_e$, and $\tilde{T}_e/T_e$ appearing in the curvature terms.

The three state variables ϖ, $\tilde{n}_e$, and $\tilde{T}_e$ have the three flux variables $\tilde{J}_\parallel$, $\tilde{u}_\parallel$, and $\tilde{q}_{e\parallel}$ appearing through their parallel divergences. The equations for these flux variables close the system. The Ohm's law consists of magnetic induction and electron inertia, the force potential, and the resistive and thermal force dissipation, given, respectively, by

$$\frac{\partial \tilde{A}_\parallel}{\partial t} + \frac{m_e}{n_e e^2}\frac{d\tilde{J}_\parallel}{dt} = \frac{T_e}{e}\nabla_\parallel\left(\frac{\tilde{n}_e}{n_e} + \frac{\tilde{T}_e}{T_e} - \frac{e\tilde{\phi}}{T_e}\right) - \eta_\parallel\left[\tilde{J}_\parallel + \frac{\alpha_e}{0.51\kappa_e}\left(\frac{e}{T_e}\tilde{q}_{e\parallel} + \alpha_e\tilde{J}_\parallel\right)\right] \quad (2.53)$$

The heat flux equation consists of inertia, the parallel temperature gradient, and the conductive and thermoelectric effect dissipation, given, respectively, by

$$\frac{2}{5}\frac{m_e}{p_e e}\frac{d\tilde{q}_{e\parallel}}{dt} = -\frac{T_e}{e}\nabla_\parallel\left(\frac{\tilde{T}_e}{T_e}\right) - \frac{1}{0.51\kappa_e}\eta_\parallel\left(\frac{e}{T_e}\tilde{q}_{e\parallel} + \alpha_e\tilde{J}_\parallel\right) \quad (2.54)$$

The parallel momentum equation consists of ion inertia, the parallel pressure gradient, and dissipation by parallel viscosity, given, respectively, by

$$n_i m_i\frac{d\tilde{u}_\parallel}{dt} = -\nabla_\parallel\tilde{p}_e + B\nabla_\parallel\frac{\mu_\parallel}{B}\nabla_\parallel\tilde{u}_\parallel \quad (2.55)$$

in which the electron contribution to the inertia is neglected.

The main dissipation coefficients are

$$\eta_\parallel = 0.51\frac{m_e\nu_e}{n_e e^2}, \qquad \mu_\parallel = 0.96 n_i m_i\frac{T_i}{m_i\nu_i} \quad (2.56)$$

with $\nu_{e,i}$ being the Braginskii collision frequencies for electrons and ions. These numerical coefficients together with the

$$\alpha_e = 0.71, \qquad \kappa_e = 3.2, \qquad \kappa_i = 3.9 \quad (2.57)$$

where for use in the next Chapter we also give the ion value, are also the Braginskii values.

The advective and parallel derivatives acting upon $\tilde{n}_e$ or $\tilde{T}_e$ or their combination $\tilde{p}_e$ carry contributions due to the background gradients, which comprise the set of gradient drive terms for the energetics. The dissipation terms give the physical sinks.

2.5 Parameters, normalised equations, geometry

We often do the actual work with normalised equations containing dimensionless dependent variables and parameters. This is done not only for computational purposes but also to make the symmetries in the equations most visible, especially when considering the energetics. In the end, one should become familiar enough with the scaling of the equations to be able to slide back and forth between the physical and normalised equations. Even before they are normalised, the physical equations are often cast in scaled form as we did above especially to elucidate the adiabatic response and motivate the relevant space and time scales. The effectiveness of the responses is controlled by several normalised parameters, which give either scale ratios or frequency ratios. The parameters are presented first, then the normalisation scheme, then the equations, and finally the representation of the magnetic geometry. In the meantime, we note only that the split between the fluidlike dynamics in the perpendicular plane and the wavelike responses along the field sets the 2D plus 1D representation with the advection and curvature terms on the one hand and the parallel gradients and divergences on the other.

2.5.1 Parameters

The normalised parameters controlling the scale ratios in the turbulence are

$$\delta_0 = \frac{\rho_s}{L_\perp}, \qquad \hat{\epsilon} = \left(K_\parallel L_\perp\right)^{-2} \tag{2.58}$$

the drift parameter and the square of the parallel-to-perpendicular scale ratio. These are assumed to be small and large, respectively, but not so much so that they would not enter. The normalised parameters controlling the electron dynamics are

$$\hat{\beta} = \beta_e\left(K_\parallel L_\perp\right)^{-2}, \qquad \hat{\mu} = \mu_e\left(K_\parallel L_\perp\right)^{-2}, \qquad C = 0.51\hat{\mu}\frac{\nu_e}{c_s/L_\perp} \tag{2.59}$$

controlling the inductive, inertial, and resistive reactions in the adiabatic response, with $\hat{\mu}$ and C also controlling the inertial and collisional response of the heat flux to the thermal fluctuations. These have a maximal ordering, which means that the equations are set up as if they were all of order unity.

2.5.2 Normalisation

The equations are for six variables: the vorticity ϖ and current J, the density n and parallel velocity u, and the electron temperature T parallel heat flux V_q, as three pairs of state and associated flux variables. The normalisation scheme is

$$t \leftarrow c_s t/L_\perp, \qquad \{xy\} \leftarrow \{xy\}/L_\perp, \qquad s \leftarrow K_\parallel s \tag{2.60}$$

for the time and space scales, where $L_\perp$ is the perpendicular gradient scale and $K_\parallel$ is also the fundamental wavenumber of the field line connection. For the dependent variables, the normalised quantities are

$$\varpi \leftarrow \frac{\varpi}{n_e e}, \qquad n \leftarrow \frac{\tilde{n}_e}{n_e}, \qquad T \leftarrow \frac{\tilde{T}_e}{T_e} \tag{2.61}$$

for the state variables,

$$J \leftarrow \frac{\tilde{J}_\parallel}{n_e e c_s}(K_\parallel L_\perp), \qquad u \leftarrow \frac{\tilde{u}_\parallel}{c_s}(K_\parallel L_\perp), \qquad V_q \leftarrow \frac{\tilde{q}_{e\parallel}}{p_e c_s}(K_\parallel L_\perp) \tag{2.62}$$

for the flux variables, and

$$\phi \leftarrow \frac{e\tilde{\phi}}{T_e}, \qquad A \leftarrow \frac{\tilde{A}_\parallel}{B\rho_s}(K_\parallel L_\perp) \tag{2.63}$$

for the field variables. As with u and J, V_q is normalised to c_s, not V_e, and therefore has a factor of $\hat{\mu}$ in its inertia. Gradient drive terms are given explicitly in terms of the ω_n and ω_t parameters given above. The geometric operators, including the Laplacian, acquire factors of δ_0 as given below.

2.5.3 Equations

The normalised equations are

$$\frac{d\varpi}{dt} = B\nabla_\parallel \frac{J}{B} - \mathcal{K}(n + T) \tag{2.64}$$

$$\frac{\partial A}{\partial t} + \hat{\mu}\frac{dJ}{dt} = \nabla_\parallel(n_0 + T_0 + n + T - \phi) - C\left(J + \frac{\alpha_e}{\kappa_e}\left(V_q + \alpha_e J\right)\right) \tag{2.65}$$

$$\frac{d}{dt}(n_0 + n) = B\nabla_\parallel \frac{J - u}{B} + \mathcal{K}(\phi - n - T) \tag{2.66}$$

$$\hat{\varepsilon}\frac{du}{dt} = -\nabla_\parallel(n_0 + T_0 + n + T) - B\nabla_\parallel \frac{P}{B} \tag{2.67}$$

$$\frac{3}{2}\frac{d}{dt}(T_0 + T) = B\nabla_\parallel \frac{J - u - V_q}{B} + \mathcal{K}(\phi - n - T) - \frac{5}{2}\mathcal{K}(T) \tag{2.68}$$

$$\frac{2}{5}\hat{\mu}\frac{dV_q}{dt} = -\nabla_\parallel(T_0 + T) - \frac{1}{\kappa_e}C\left(V_q + \alpha_e J\right) \tag{2.69}$$

for the dependent variable evolution equations, where the nonlinearities are

$$\frac{d}{dt} = \frac{\partial}{\partial t} + [\phi,], \qquad \nabla_\parallel = b^s\frac{\partial}{\partial s} + [-A,] \tag{2.70}$$

The gradient drive terms are given by

$$\frac{\partial n_0}{\partial x} = -\omega_n, \qquad \frac{\partial T_0}{\partial x} = -\omega_t \tag{2.71}$$

Note that $\partial/\partial t$ on n_0 or T_0 vanishes. The equations for the fields and the parallel viscosity are

$$\varpi = \nabla_\perp^2\phi, \qquad J = -\hat{\beta}^{-1}\nabla_\perp^2 A, \qquad P = -X\nabla_\parallel u_\parallel \tag{2.72}$$

in which P is the viscous correction to the pressure and X is its parameter. The collisional coefficients are all normalised in terms of C in terms of $\eta_\parallel$ with the 0.51 coefficient. So we have $\kappa_e = 1.6$ instead of 3.2 while $\alpha_e = 0.71$ is maintained.

2.5.4 Magnetic geometry

The magnetic geometry is simplified, essentially a minimal model for a sheared magnetic field with toroidal topology. The x-direction is in the direction of the minor radius, the s-direction is along $\mathbf{b}$, aligned to the magnetic field lines of the equilibrium, and the y-direction is the electron drift direction parallel to $-\mathbf{b} \times \nabla \log p$ of the equilibrium. In other words, $\nabla x \times \nabla y \cdot \nabla s > 0$ form a right handed coordinate system. The curvature of the magnetic field lines is represented only through the finite divergence terms represented by the magnetic divergences. The shear of the magnetic field lines is represented by an x-dependent shift in the y-direction, in the finite differences taken in the s-direction.

The metric coefficients for the grid are

$$g^{xx} = 1, \qquad g^{xy} = S(s - s_k), \qquad g^{yy} = 1 + S^2(s - s_k)^2, \qquad \sqrt{g} = 1 \tag{2.73}$$

where S is the magnetic shear, s_k is a constant giving a reference location along the field lines, and $\sqrt{g}$ is the volume element,

$$\mathcal{V} = \int d\mathcal{V}, \qquad d\mathcal{V} = \sqrt{g}\; dx\; dy\; ds \tag{2.74}$$

The magnetic field strength and nonvanishing component are

$$B = 1, \qquad B^s = 1, \qquad b^s = 1 \tag{2.75}$$

in normalised units. The perpendicular Laplacian at $s = s_k$ is therefore the 2D Cartesian one in the xy plane.

The spatial brackets and curvature operator are given by

$$[a, b] = \frac{1}{B_0}\left(\frac{\partial a}{\partial x}\frac{\partial b}{\partial y} - \frac{\partial b}{\partial x}\frac{\partial a}{\partial y}\right) \tag{2.76}$$

$$\mathcal{K} = \frac{2}{I_0}\left(\sin s\frac{\partial}{\partial x} + (\cos s + S(s - s_k)\sin x)\frac{\partial}{\partial y}\right) \tag{2.77}$$

including the units as shown, with $I_0 = B_0 R_0$ in a tokamak model, in which R_0 is the major radius and B_0 is the reference magnetic field strength. In normalised units these become

$$[a, b] = \delta_0\left(\frac{\partial a}{\partial x}\frac{\partial b}{\partial y} - \frac{\partial b}{\partial x}\frac{\partial a}{\partial y}\right) \tag{2.78}$$

$$\mathcal{K} = \delta_0\omega_B\left(\sin s\frac{\partial}{\partial x} + (\cos s + S(s - s_k)\sin x)\frac{\partial}{\partial y}\right) \tag{2.79}$$

with

$$\nabla_\perp^2 = \delta_0^2\left(\frac{\partial^2}{\partial x^2} + 2S(s - s_k)\frac{\partial^2}{\partial x \partial y} + (1 + S(s - s_k)^2)\frac{\partial^2}{\partial y^2}\right) \tag{2.80}$$

for the perpendicular Laplacian. Note that $1/B_0$ is replaced by δ_0, ρ_s^2 is replaced by δ_0^2, and $2/I_0$ is replaced by $\delta_0\omega_B$, with $\omega_B = 2L_\perp/R_0$ the curvature forcing parameter set to its nominal value. With these geometric elements, the geometry is concentric cylindrical except for the existence of the curvature operator.

The perpendicular terms (everything except $\partial/\partial s$) are evaluated at $s = s_k$ on each drift (xy) plane. The parallel operations follow the property of field line connection. The derivative $\partial/\partial s$ is taken across drift planes each with its own coordinate system,

$$2h_s\frac{\partial f}{\partial s}(x, y_k, s_k) = f(x, y_{k+1} - \Delta^+, s_{k+1}) - f(x, y_{k-1} - \Delta^-, s_{k-1}) \tag{2.81}$$

with the shifts given by

$$\Delta^\pm = \pm\ Sxh_s \tag{2.82}$$

In discrete numerics, this is equivalent to taking the shift implied in the periodicity conditions,

$$f(x, y_k + L_y, s) = f(x, y_k, s), \qquad f(x, y_k + 2\pi Sx, s + 2\pi) = f(x, y_k, s) \tag{2.83}$$

over an entire periodicity length (2π in normalised units) and distributing it among all the s-nodes. The parallel boundary condition becomes

$$f(x, y_{k+N}, s_{k+N}) = f(x, y_k, s_k) \tag{2.84}$$

where N is the number of s-nodes, and therefore $s_{k+N} - s_k$ is 2π. Each location along a field line is now treated numerically equivalently, which is in order with the property of the slab geometry that all locations along the field line are equivalent. The combination of periodicity and shear constrains the allowable range of $k_\parallel$ to values of order $K_\parallel$ and larger, which is what is meant by field line connection. In a tokamak-like magnetic field, $K_\parallel = 1/qR_0$, where q is the toroidal to poloidal pitch of the field line connectivity. In a different model, $K_\parallel = 2\pi/L_\parallel$, where $L_\parallel$ is the effective periodicity length of the field lines.

The Fourier transform of Eqs (2.83) in the y-domain produces a spectrum of wavenumbers given by integer multiples $k_y = lk_0$ of the fundamental wavenumber $k_0 = 2\pi/L_y$ of the domain. The periodicity condition in s (the second expression in 2.83) then yields a phase shift of $2\pi Sk_y x$ of each of these wavenumbers over a parallel distance $L_\parallel$. This yields a set of 'rational surfaces' which are locations where $Sk_y x$ takes integer values. The name rational surfaces arises from a tokamak-like magnetic field, in which q takes rational values at these locations.

We will also have to do with what are called 'flux surface averages,' which are defined over an infinitesimal volume centred upon a particular value of x, which is the same a the surface of constant equilibrium magnetic flux in a magnetised plasma with a pressure gradient. These are also called 'zonal' averages or components. In our case with $\sqrt{g} = 1$ this is a simple average over the ys-plane, and with equidistant grids in the coordinates it is a discrete average over all the points in y and s at a given x. Note that once y is summed up, the shifts s_k do not affect the average. In the same sense, any quantity $f = f(x)$ which is independent of y and s is called a 'flux function'.

2.5.5 Numerics

The equations are advanced on a uniform grid in all three coordinates with the time step given by the Karniadakis (1991) scheme. Briefly, given

$$\frac{\partial u}{\partial t} = S(u), \qquad u_n = u(t - n\,\tau), \qquad S_n = S(u_n) \tag{2.85}$$

for dependent variables u, and timestep τ, then the scheme is given by

$$\frac{11}{6}\,u_0 = 3u_1 - \frac{3}{2}u_2 + \frac{1}{3}u_3 + \tau(3S_1 - 3S_2 + S_3) \tag{2.86}$$

After the variables ϖ and $A + \hat{\mu}J$ have been advanced, ϕ and A, and J, are solved for using the field equations (Eq. 2.2). The combination $A + \hat{\mu}J$ is solved according to

$$\left(1 - \hat{\beta}^{-1}\nabla_\perp^2\right)A = A + \hat{\mu}J \equiv A^*, \qquad J = \hat{\mu}^{-1}(A^* - A) \tag{2.87}$$

We will see in later Chapters that A^* generalises to the effective canonical momentum in the model, here for electrons and later for each species.

The spatial discretisation for the Poisson bracket (e.g., for $[\phi,\ \varpi]$) is given by the fourth-order Arakawa (1966) scheme. Given

$$\begin{aligned} J^{++} = &\left(\phi_{i+1,j} - \phi_{i-1,j}\right)(\varpi_{i,j+1} - \varpi_{i,j-1}) \\ &- (\varpi_{i+1,j} - \varpi_{i-1,j})\left(\phi_{i,j+1} - \phi_{i,j-1}\right) \end{aligned} \tag{2.88}$$

$$\begin{aligned} J^{+\times} = &\ \phi_{i+1,j}(\varpi_{i+1,j+1} - \varpi_{i+1,j-1}) - \phi_{i-1,j}(\varpi_{i-1,j+1} - \varpi_{i-1,j-1}) \\ &- \phi_{i,j+1}(\varpi_{i+1,j+1} - \varpi_{i-1,j+1}) + \phi_{i,j-1}(\varpi_{i+1,j-1} - \varpi_{i-1,j-1}) \end{aligned} \tag{2.89}$$

$$\begin{aligned} J^{\times+} = &\ \phi_{i+1,j+1}(\varpi_{i,j+1} - \varpi_{i+1,j}) - \phi_{i-1,j-1}(\varpi_{i-1,j} - \varpi_{i,j-1}) \\ &- \phi_{i-1,j+1}(\varpi_{i,j+1} - \varpi_{i-1,j}) + \phi_{i+1,j-1}(\varpi_{i+1,j} - \varpi_{i,j-1}) \end{aligned} \tag{2.90}$$

$$\begin{aligned} J^{\times\times} = &\left(\phi_{i+1,j+1} - \phi_{i-1,j-1}\right)(\varpi_{i-1,j+1} - \varpi_{i+1,j-1}) \\ &- \left(\phi_{i-1,j+1} - \phi_{i+1,j-1}\right)(\varpi_{i+1,j+1} - \varpi_{i-1,j-1}) \end{aligned} \tag{2.91}$$

$$\begin{aligned} J_2^{+\times} = &\ \phi_{i+2,j}(\varpi_{i+1,j+1} - \varpi_{i+1,j-1}) - \phi_{i-2,j}(\varpi_{i-1,j+1} - \varpi_{i-1,j-1}) \\ &- \phi_{i,j+2}(\varpi_{i+1,j+1} - \varpi_{i-1,j+1}) + \phi_{i,j-2}(\varpi_{i+1,j-1} - \varpi_{i-1,j-1}) \end{aligned} \tag{2.92}$$

$$\begin{aligned} J_2^{\times+} = &\ \phi_{i+1,j+1}(\varpi_{i,j+2} - \varpi_{i+2,j}) - \phi_{i-1,j-1}(\varpi_{i-2,j} - \varpi_{i,j-2}) \\ &- \phi_{i-1,j+1}(\varpi_{i,j+2} - \varpi_{i-2,j}) + \phi_{i+1,j-1}(\varpi_{i+2,j} - \varpi_{i,j-2}) \end{aligned} \tag{2.93}$$

the combination

$$[\phi,\varpi]=\frac{\delta_0}{12h_xh_y}\left[2\left(J^{++}+J^{+\times}+J^{\times+}\right)-\frac{1}{2}\left(J^{\times\times}+J_2^{+\times}+J_2^{\times+}\right)\right] \tag{2.94}$$

(given in normalised units) satisfies the antisymmetry, Leibnitz and divergence properties to machine accuracy and the Jacobi identity to high accuracy.

Subgrid dissipation is applied wherever nonlinear ExB advection acts,

$$D()=\nu_0\left(\frac{h_x^2}{4}\frac{\partial^2}{\partial x^2}+\frac{h_y^2}{4}\frac{\partial^2}{\partial y^2}\right)^2+\nu_0\left(\frac{h_s^2}{4}\frac{\partial^2}{\partial s^2}\right)^2+s_fp_f \tag{2.95}$$

including the boundary dissipation given by the last term, in which

$$p_f(x)=\exp\left[-100\left(\frac{x}{L_x}+\frac{1}{2}\right)^2\right]+\exp\left[-100\left(\frac{x}{L_x}-\frac{1}{2}\right)^2\right] \tag{2.96}$$

where L_x is the length of the domain in the x-direction. Nominal values are $\nu_0=1.0$ and $s_f=0.3$ for the coefficients.

The boundary conditions in the perpendicular plane are Dirichlet in the x-direction and periodic in the y-direction,

$$f(x=\pm L_x/2,y,s)=0,\qquad f(x,y+L_y,s)=f(x,y,s) \tag{2.97}$$

where L_x and L_y are the respective domain sizes given by

$$L_x=L_\perp=64h_x,\qquad L_y=4L_\perp=256h_y,\qquad h_y/h_x=1 \tag{2.98}$$

where $h_{x,y}$ are the step sizes. This follows the 4×1 rule for the domain aspect ratio, covered in Chapter 11 of Volume One. All of the models used in this book have the same numerical treatments as outlined above, so we will not have to repeat the description with the introduction of each new model.

2.6 Energetics

This also follows the usage developed in Volume One. Multiply Eqs (2.64–2.69) by $-\phi$ and J and n and u and T and V_q, respectively, to obtain

$$\begin{aligned}&\frac{1}{2}\frac{\partial}{\partial t}|\nabla_\perp\phi|^2-\nabla\cdot\left(\phi\frac{\partial}{\partial t}\nabla_\perp\phi\right)+[\phi,-\phi\varpi]+B\nabla_\parallel\frac{\phi J}{B}-\mathcal{K}(\phi(n+T))\\&\quad=J\nabla_\parallel\phi-(n+T)\mathcal{K}(\phi)\end{aligned} \tag{2.99}$$

$$\begin{aligned}&\frac{1}{2}\frac{\partial}{\partial t}\left(\hat\beta^{-1}|\nabla_\perp A|^2+\hat\mu J^2\right)-\nabla\cdot\left(\hat\beta^{-1}\frac{\partial A}{\partial t}\nabla_\perp A\right)+\frac{1}{2}\hat\mu[\phi,J^2]\\&\quad=J\nabla_\parallel(n_0+T_0+n+T-\phi)-C\,J^2-\frac{1}{\kappa_e}\alpha_eJ\,C(V_q+\alpha_eJ)\end{aligned} \tag{2.100}$$

$$\frac{1}{2}\frac{\partial}{\partial t}n^2 + \frac{1}{2}[\phi, n^2] + B\nabla_\| \frac{n(u-J)}{B} + n\mathcal{K}(n+T) \\ = (u-J)\nabla_\| n + n\mathcal{K}(\phi) \tag{2.101}$$

$$\frac{1}{2}\frac{\partial}{\partial t}\hat{\varepsilon}u^2 + \frac{1}{2}\hat{\varepsilon}[\phi, u^2] + B\nabla_\| \frac{uP}{B} \\ = -u\nabla_\|(n_0 + T_0 + n + T) - X(\nabla_\| u)^2 \tag{2.102}$$

$$\frac{1}{2}\frac{\partial}{\partial t}\frac{3}{2}T^2 + \frac{1}{2}\frac{3}{2}[\phi, T^2] + B\nabla_\| \frac{T(u + V_q - J)}{B} + T\mathcal{K}(n+T) + \frac{1}{2}\frac{5}{2}\mathcal{K}(T^2) \\ = (u + V_q - J)\nabla_\| T + T\mathcal{K}(\phi) \tag{2.103}$$

$$\frac{1}{2}\frac{2}{5}\frac{\partial}{\partial t}\left(\hat{\mu}V_q^2\right) + \frac{1}{2}\frac{2}{5}\hat{\mu}\left[\phi, V_q^2\right] \\ = -V_q\nabla_\|(T_0 + T) - \frac{1}{\kappa_e}V_q\, C(V_q + \alpha_e J) \tag{2.104}$$

in which each appearance of $\nabla_\perp$ carries a factor of δ_0 as in Eq. (2.80).

In the individual energetic pieces, terms appearing in two equations with opposite sign represent the transfer channels, while isolated terms reflect the sources and sinks. The thermal conduction sink is only positive definite when added up, understandable due to the off-diagonal damping terms between J and V_q due to the thermal force and thermoelectric effect. Some of the transfer effects also appear in such combination; an example is the kinetic Alfvén transfer dynamics, in which the pieces of $(n + T)$, representing $\tilde{p}_e$ under the parallel gradient, multiplying J in the equations for n and T combine to balance $J\nabla_\|(n + T)$ in the Ohm's law. There is also a curvature mediated transfer between n and T through the diamagnetic current, but this keeps free energy in the overall thermal content, with no contact to the ExB energy.

The volume integration yields the total energy equation,

$$\frac{\partial}{\partial t}\mathcal{V}^{-1}\int d\mathcal{V}\,\frac{1}{2}\left(|\nabla_\perp\phi|^2 + \hat{\beta}^{-1}|\nabla_\perp A|^2 + n^2 + \frac{3}{2}T^2 + \hat{\mu}J^2 + \hat{\varepsilon}u^2 + \frac{2}{5}V_q^2\right) \\ = \mathcal{V}^{-1}\int d\mathcal{V}\left(n[n_0, \phi] + \frac{3}{2}T[T_0, \phi] + (J-u)[n_0, A] + (J - u - V_q)[T_0, A] \right. \\ \left. - CJ^2 - \frac{1}{\kappa_e}C(V_q + \alpha_e J)^2 - X(\nabla_\| u)^2\right) \tag{2.105}$$

expressed as a volume average. All the terms on the right hand side represent gradient drive and dissipative sink effects, showing that the total free energy E_T is conserved in their absence.

The total free energy is the quantity on the left hand side, with the seven pieces labelled E_e for the ExB energy, E_B for the magnetic energy, E_n and E_t for the density and temperature contributions to the thermal free energy, E_j and E_u for the Ohmic and acoustic contributions to the parallel kinetic energy, and E_q for the heat flux contribution to the free energy. It is common to combine $E_B + E_j = E_b$ into a group, since in the Ohm's law A and $\hat{\mu}J$ appear together under the time derivative.

The quantities on the right hand side are in two groups, giving the gradient drive and collisional dissipation, respectively. One sees that curvature forcing and parallel dynamics are in and of themselves conservative, as they integrate out of the volume average. The collisional sinks—resistivity, thermal conduction, and viscosity—all represent positive definite damping. Subgrid dissipation adds to this in the total energetics.

The total energy is

$$E_T = E_e + E_b + E_n + E_u + E_t + E_q \tag{2.106}$$

and the nonlinear growth rate is

$$\Gamma_T = \frac{1}{2E_T}\frac{\partial E_T}{\partial t} = \Gamma_n + \Gamma_t + \Gamma_v - \Gamma_c - \Gamma_k - \Gamma_s \tag{2.107}$$

with the terms on the right hand side representing the drive and dissipation channels. The computational error in this equation is denoted Γ_E and called the energetic error.

The gradient drive is

$$\Gamma_n = \frac{1}{2E_T}\mathcal{V}^{-1}\int d\mathcal{V}\,\omega_n n v_E^x, \qquad \Gamma_t = \frac{1}{2E_T}\mathcal{V}^{-1}\int d\mathcal{V}\,\frac{3}{2}\omega_t T v_E^x \tag{2.108}$$

for the ExB flux pieces, and

$$\Gamma_v = \frac{1}{2E_T}\mathcal{V}^{-1}\int d\mathcal{V}\left(\omega_t V_q b^x + (\omega_n + \omega_t)(u - J)b^x\right) \tag{2.109}$$

for the magnetic flutter pieces. The radial ExB velocity and magnetic fluctuation components are

$$v_E^x = [\phi, x], \qquad b^x = [-A, x] \tag{2.110}$$

in normalised units in the Poisson bracket notation with x the radial coordinate. It is important to note that the contribution by J to the magnetic flutter Γ_v is zero in volume average neglecting surface terms, because

$$\hat{\beta}Jb^x = -(\nabla_\perp^2 A)b^x = \nabla\cdot(-(\nabla_\perp A)b^x) - \frac{1}{2}[|\nabla_\perp A|^2, x] \tag{2.111}$$

is a pure divergence. So Γ_v is mainly due to V_q in this model, with a small piece due to u. The same occurs for any appearance of ϖu^x whenever the stream function for $\mathbf{u}$ is proportional to the potential in the definition of ϖ. This applies in any model of this type.

The resistive and thermal conductive sinks for the electrons are

$$\Gamma_c = \frac{1}{2E_T}\mathcal{V}^{-1}\int d\mathcal{V}\,C\,J^2, \qquad \Gamma_k = \frac{1}{2E_T}\mathcal{V}^{-1}\int d\mathcal{V}\,\frac{1}{\kappa_e}C\,(V_q + \alpha_e J)^2 \tag{2.112}$$

and viscous sink for the parallel flow is

$$\Gamma_s = \frac{1}{2E_T}\mathcal{V}^{-1}\int d\mathcal{V}\, X(\nabla_\| u)^2 \tag{2.113}$$

The subgrid dissipation rate is given by

$$\Gamma_w = \frac{1}{2E_T}\mathcal{V}^{-1}\int d\mathcal{V}\bigg(-\phi\; D(\varpi) + n\; D(n) + \frac{3}{2}T\; D(T) \\ + \hat{\mu}J\; D(J) + \frac{2}{5}\hat{\mu}V_q\; D(V_q) + \hat{\varepsilon}u\; D(u)\bigg) \tag{2.114}$$

where D is defined in Eq. (2.95). These are all normalised as growth or damping rates.

We also measure the amplitude functionals of ϕ and ϖ, given by

$$A_p = \mathcal{V}^{-1}\int d\mathcal{V}\frac{1}{2}\,\phi^2, \qquad A_w = \mathcal{V}^{-1}\int d\mathcal{V}\frac{1}{2}\,\varpi^2 \tag{2.115}$$

for comparison purposes. The multiplier 1/2 facilitates amplitude comparison between ϕ and n through A_p versus E_n.

2.7 Heat flux and kinetic shear Alfvén waves

The linearised equations for the electrons as a dissipative fluid and the ions as an inertial/acoustic anchor are what we find by neglecting the nonlinear terms in d/dt and $\nabla_\|$ and setting the gradients and curvature to zero.

In the ideal (dissipation-free) fluid limit, with C and X set to zero and also neglecting V_q, we can add the density and temperature to form the pressure,

$$\frac{\partial\varpi}{\partial t} = \frac{\partial J}{\partial s}, \qquad \frac{\partial}{\partial t}(A + \hat{\mu}J) = \frac{\partial}{\partial s}(n + T) - \frac{\partial\phi}{\partial s} \tag{2.116}$$

$$\frac{\partial}{\partial t}(n + T) = \Gamma\frac{\partial}{\partial s}(J - u), \qquad \hat{\varepsilon}\frac{\partial u}{\partial t} = -\frac{\partial}{\partial s}(n + T) \tag{2.117}$$

with $\Gamma = 5/3$ coming from the factor $c_f = 3/2$ and $\Gamma = (1 + c_f)/c_f$, with $2c_f$ giving the number of degrees of freedom for the isotropic temperature. The perpendicular Laplacian including the δ_0 factors is parameterised as $-K^2$, so that $\varpi = -K^2\phi$ with $K = k_\perp\rho_s$ giving the perpendicular scale of the disturbances.

These equations yield the four basic modes, namely, two kinetic shear Alfvén waves and two sound waves. Assuming $\hat{\varepsilon} \gg \hat{\beta}$, we find the low frequency side by setting the electrons adiabatic, $h = n - \phi = 0$ and $T = 0$, forming a set for n and u. These give the sound (acoustic) waves,

$$\omega = \pm k_\| V_s, \qquad V_s^2 = \hat{\varepsilon}^{-1}\frac{\Gamma}{1 + \Gamma K^2} \tag{2.118}$$

where $\Gamma = 1$ since $T = 0$. We find the high frequency side by setting $u = 0$, allowing J to respond to the nonzero h. This yields the kinetic shear Alfvén waves,

$$\omega = \pm k_\| V_a, \qquad V_a^2 = \hat{\beta}^{-1}\frac{1 + \Gamma K^2}{\beta_e + \mu_e K^2} \tag{2.119}$$

where $\Gamma = 5/3$ if T is allowed to evolve naturally. These are in units of $c_s/L_\perp$, which reverting to physical units yields

$$V_s^2 = \frac{\Gamma}{1+\Gamma K^2}\, c_s{}^2, \qquad V_a^2 = \frac{1+\Gamma K^2}{1+(\mu_e/\beta_e)K^2}\, v_A^2 \tag{2.120}$$

for the velocities, noting that β_e is the square of c_s/v_A, and μ_e/β_e is the square of d_e/ρ_s, with d_e the skin depth given in Eq. (2.14).

Since there are five equations there is a fifth mode, which we can find by subtracting the temperature from the density equation to find that $\partial/\partial t$ on $n - (3/2)T$ vanishes. This mode is damped by thermal conduction, but in the ideal model it remains with its initial value. So to this point all $\tilde{T}_e$ does is set Γ to 5/3, following the factor of 3/2 in the temperature equation. $\partial\tilde{T}_e/\partial t$. If this factor is changed to an arbitrary constant c_f, then Γ becomes $(1 + c_f)/c_f$ accordingly.

If we put the thermal conduction and the dissipation coefficients back in, then we have a well behaved thermodynamic system. Setting the derivatives $\partial/\partial t \to -i\omega$ and $\nabla_\parallel \to ik_\parallel$, neglecting the nonlinearities, gradient drive, and curvature, we have a linear system with six eigenmodes. Rather than going through this, we display the numerical analysis of the initial value problem, with

$$n_k(s) = a_0\, e^{-s^2} \qquad \text{at} \quad t = 0 \tag{2.121}$$

with all other variables zero, and with a choice of one $K = k_\perp \rho_s$ in each case. The equations are advanced on a uniform grid of 128 in s with fourth-order centred differences, and with the time step scheme given above. In doing this we ascertain the relative roles of T and the nonadiabatic density $h = n - \phi$, and of the resistive versus thermal conductive dissipation.

As in Volume One for this problem, we take two large-scale cases, nominally within the MHD regime. Two types of cases are showcased. A warm plasma can be regarded as fairly inductive and collisional, meaning $\beta_e > \mu_e$ and $\nu_e > K_\parallel v_A$, but not extremely so. This regime makes the adiabatic response very visible without the electrons being too kinetic, i.e., resistivity is still quite active. A hot plasma can be regarded as mostly adiabatic for any electrostatic dynamics but still MHD-like if β_e/μ_e is high enough, in either case with negligible collisionality. The reason that $K^2 \ll 1$ is MHD-like is that the reaction of the pressure to the parallel current is much smaller than that of the vorticity. So the waves are as they would be in MHD, except that they are initially launched by the parallel pressure gradient.

The warm plasma case is done with $K_\parallel L_\perp = 0.01$ and has

$$\hat{\beta} = 5, \quad \hat{\mu} = 2.72, \quad C = 13.9, \quad \hat{\varepsilon} = 10^4, \quad X = 1 \tag{2.122}$$

with timestep $\tau = 10^{-2}$, and the hot plasma case, done with $K_\parallel L_\perp = 0.1$, has

$$\hat{\beta} = 1, \quad \hat{\mu} = 0.0272, \quad C = 0.00139, \quad \hat{\varepsilon} = 100, \quad X = 1 \tag{2.123}$$

with timestep $\tau = 10^{-3}$, with the factor of $K_\parallel$ folded into the parameters, *e.g.*, with $\hat{\beta}$ instead of β_e, with the time scale reflecting its normalisation in terms of $L_\perp$ rather

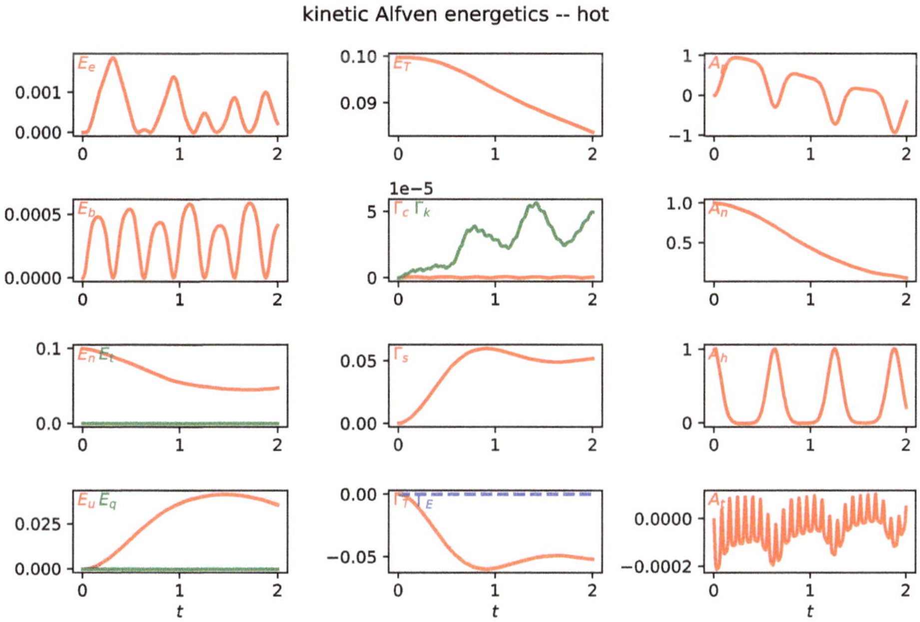

Figure 2.1. Time evolution of the energetic quantities in a kinetic shear Alfvén wave oscillation, for the hot plasma case with $K = 0.1$, on the Alfvén time scale. The quantities E_e, E_b, E_n, E_u, E_t, and E_q give the contributions of ϕ, A, n, u, T, and V_q to the total energy (E_T), respectively. The electron temperature adds E_t for the temperature contribution to the thermal free energy and E_q for the heat flux contribution to the parallel and magnetic energy. Both are very small for this case, which does not excite an appreciable thermal content.

than $K_\|^{-1}$. The timestep follows $\hat{\mu}$, taking different values according to $K_\| L_\perp$ despite the mass ratio always being 2.72×10^{-4}, assuming deuterium ions. The scale $K = 0.1$ is taken in both cases.

For this purpose we restrict to the initial responses from $t = 0$ to $t = 200$. In addition to the energetics, comparisons among the variables are done by measuring the time traces of the RMS amplitudes of all the variables, averaged over 21 time points spanning this range.

The initial response in the hot plasma case is shown in figure 2.1. With this small value of $\hat{\mu}$ and C we find that T oscillates about small values but V_q sets up on the sound wave scale and is as large as u. As in the isothermal case, the Alfvén responses do not damp effectively. In these cases both collisionality channels are very small, while the oscillatory activity in the thermal channels is very fast ($K_\| V_e$) compared with the actual Alfvén responses ($K_\| v_A$), following the square root of $\hat{\beta}/\hat{\mu}$ which is about 6 in this case. This indicates the need of additional physics to treat kinetic electron damping mechanisms, which we will do by introducing the gyrokinetic and gyrofluid models later on.

On the other hand, the initial response in the warm plasma case, in figure 2.2, shows fairly significant damping of the Alfvén responses, as in the isothermal case. The temperature has roughly 10 percent the amplitude of the density on the slower acoustic time scale but only about one percent of the oscillation amplitude in the

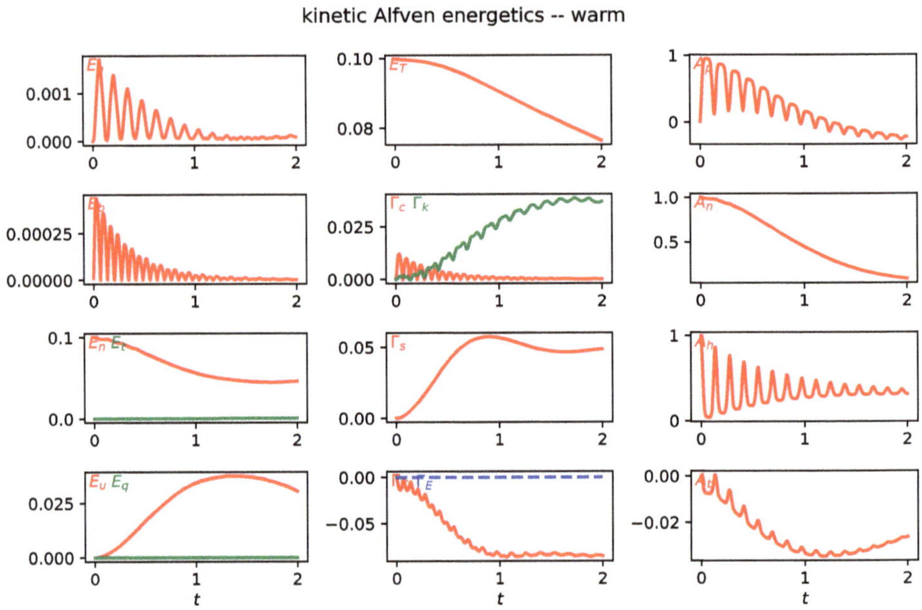

Figure 2.2. Time evolution of the energetic quantities in a kinetic shear Alfvén wave oscillation, for the warm plasma case with $K = 0.1$, on the Alfvén time scale. The quantities E_e, E_b, E_n, E_u, E_t, and E_q give the contributions of ϕ, A, n, u, T, and V_q to the total energy (E_T), respectively. The electron temperature adds E_t for the temperature contribution to the thermal free energy and E_q for the heat flux contribution to the parallel and magnetic energy. Also in this case, the thermal component to the Alfvén oscillation remains subdominant, but the thermal conduction dissipation is larger than the resistive.

Alfvén response. The thermal conduction adds about 20 percent to the damping rate. The damping rate persists because the state after the Alfvén transient has damped still has the small temperature response, but as a result heat flux is larger than the current and comparable to the mass flow, with the RMS about 0.2 in velocity units normalised to the sound speed.

Comparison with the Braginskii collisional regime is done by measuring the time traces of the RMS amplitudes of all the variables, averaged over 11 time points from $t = 1$ to 2. Besides V_q we measure the Braginskii defect, which is

$$B_q = V_q + 0.71J + 1.6C^{-1}\nabla_{\|}T \tag{2.124}$$

in these units. If $B_q \ll V_q$ then we can say the Braginskii form for the parallel heat flux is satisfied. The results are given in figure 2.3 for the warm plasma case with C set to various values across the transition given by $\nu_e/K_{\|}V_e = 1$, or with $\nu_e L_{\perp}/c_s = 0.61$ in these units. Below this value, the collisional regime as measured by $B_q \ll V_q$ breaks down.

The weak temperature responses found for the response to a density disturbance may be controlled against by initialising the activity in a different way. In this case, we start with all variables set to zero and then excite the temperature with a short-lived heat source, with time and space profiles given by

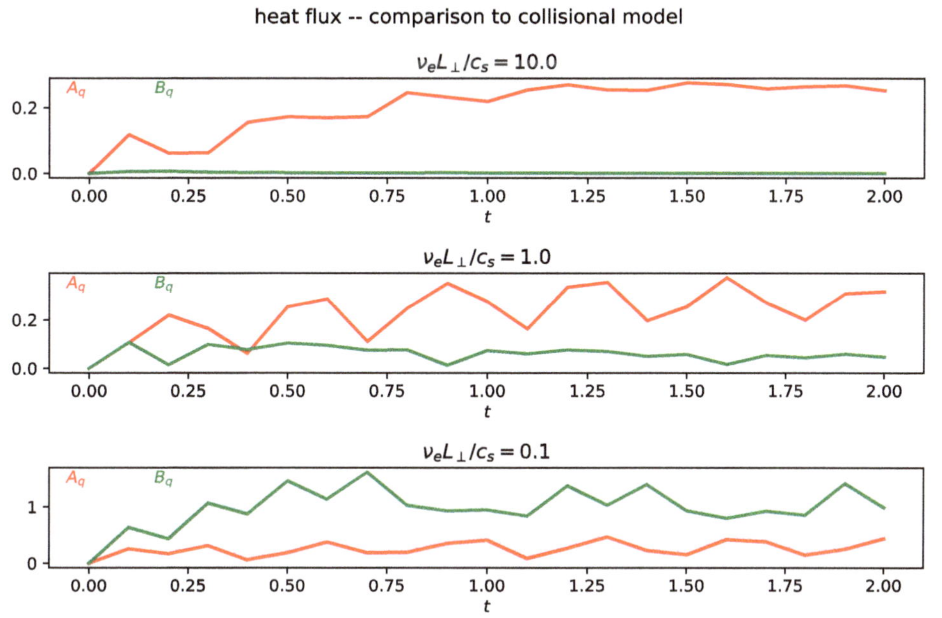

Figure 2.3. Comparison of the heat flux dependent variable V_q with the Braginskii defect B_q, given by the difference between the actual V_q to the form given by the Braginskii collisional model (Eq. 2.124), for collisionality values across the transition to the kinetic regime. In the first case, the RMS levels averaged over $1 < t < 2$ are $V_q = 0.257 \pm 0.014$ and $B_q = (6.69 \pm 4.03) \times 10^{-4}$.

$$Q = f(t)\, p(s), \qquad f(t) = 1 + \cos \pi \frac{t - 0.5}{0.3}, \qquad p(s) = \exp[-s^2] \tag{2.125}$$

where $f(t)$ is set to zero outside of the interval $0.2 < t < 0.8$. This does not create much of an Alfvén oscillation but serves to illustrate the transition away from the collisional regime. Figure 2.4 shows the results for the same C values as before. For yet lower values V_q becomes oscillatory, indicating that more physical content is needed for the model.

In cases with ν_e comparable to $K_\| V_e$ the Braginskii relation is therefore at least not far wrong and remains usable. On the other hand, the use of a time dependent heat flux equation is successful, not only energetically but also in retaining the collisional result in the collisional regime. Since it is easier to handle numerically, due to the absence of second derivatives, a time dependent equation for the heat flux with collisional damping it offers a useful way to treat thermal conduction without facing the attendant difficulties. Indeed, use of the collisional diffusion model becomes problematic even for the transitional cases, since it takes over the Courant limit. This limit for a diffusion operator $\nabla_\| D \nabla_\|$ is $\tau < D^{-1}(h_s/b^s)^2$ up to a numerical factor dependent on the scheme. For high resolution cases an implicit method—evaluating this term at the forward time step—becomes necessary and is particularly problematic given that $\nabla_\|$ is nonlinear. Any splitting method would damage the alignment of the diffusion process to the magnetic field. In our case, the time dependent and damped formulation avoids these problems and continues to treat the linear and

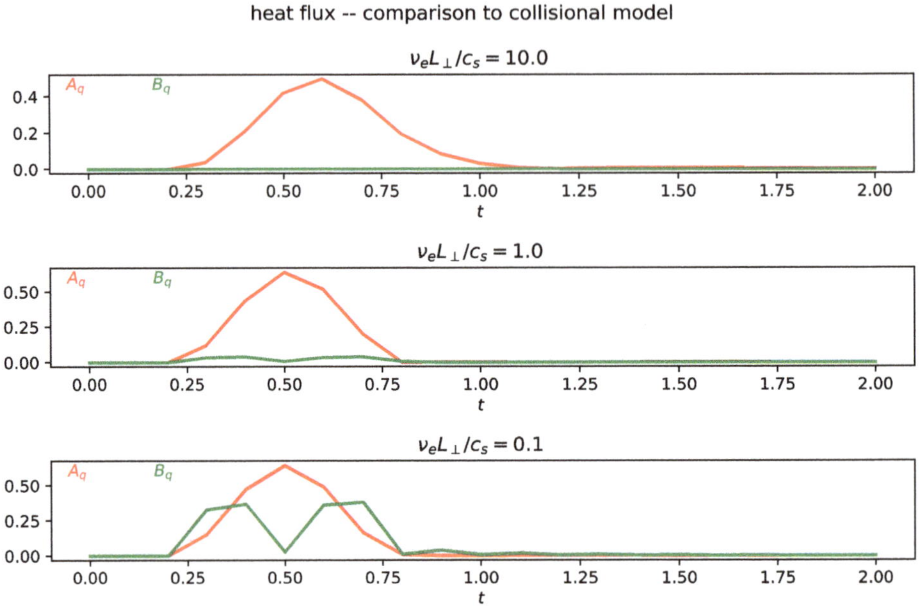

Figure 2.4. Comparison of the heat flux dependent variable V_q with the Braginskii defect B_q, given by the difference between the actual V_q to the form given by the Braginskii collisional model (Eq. 2.124), for a kinetic shear Alfvén oscillation resulting from the heat source described by Eq. (2.125). The collisionality values reflect the transition towards the kinetic regime. In the first case, the RMS levels averaged over $0.25 < t < 0.75$ are $V_q = (9.04 \pm 7.46) \times 10^{-3}$ and $B_q = (0.758 \pm 1.23) \times 10^{-4}$, and in the second case they are $V_q = (7.79 \pm 3.63) \times 10^{-4}$ and $B_q = (6.71 \pm 4.00) \times 10^{-5}$.

nonlinear contributions together. This is one of the reasons such high energetic accuracy, measured by the small Γ_E, is possible in general cases.

The time dependent heat flux model with this collisional damping closure remains useful in the transition regime where ν_e is smaller than $K_\| V_e$ but not much smaller. On the other hand, we find large departures in deeply weak collisionality regimes. For $\nu_e / K_\| V_e$ of order 10^{-2} the heat flux response becomes oscillatory. What is happening is that the kinetic details of the same dissipation process are more active with decreasing C. The thermal conduction process continuously merges with and rolls over into a related process in the kinetic theory which is called Landau damping. We will explore this, also touching base with the textbook cases, in Chapter 8 after the gyrokinetic theory has been developed in the meantime (this reduces to the kinetic theory if both the finite gyroradius and the polarisation are neglected for electrons). The gyrofluid model to be introduced in Chapter 9 will do a much better job than even this extension to the fluid closure for $C \ll \hat{\mu}$, so we defer the hot plasma cases with thermal electron dynamics until then.

2.8 Drift Alfvén turbulence

The electron temperature physics generally adds to the basic drift wave turbulence in a sheared magnetic field. Whether the turbulence is stronger or weaker depends on the competition between different processes which are involved. This is due to the fact that while the temperature gradient is added to the sources, thermal conduction is added to the dissipation. The temperature gradient also adds to the pressure gradient which enters the parallel dynamics due to the magnetic fluctuations. In most experimental situations, except for colder discharges which are not fully ionised, the temperature gradient is at least as steep as the density one, and the resulting thermal physics is felt. When this is the case, we are required to keep the electron temperature physics in order to treat drift waves and their associated turbulence, by extension drift Alfvén turbulence.

The extra gradient drive in a temperature gradient tends by itself to strengthen the turbulence. One often finds that for equal gradient scale lengths the temperature gradient has more of the total drive despite the smaller temperature fluctuations in relative units compared with the density fluctuations. This results from the phase shifts between each of these quantities and the electrostatic potential. The adiabatic response forces the density and potential towards each other but merely forces the temperature towards zero without imposing any relationship to the potential. These three, $\tilde{T}_e$ and $\tilde{n}_e$ and $\tilde{\phi}$, form the set of state variables whose relative gradient in the form of $\tilde{p}_e - n_e e\tilde{\phi}$ drives the current and the pathways to dissipation. The heat flux $\tilde{q}_{e\|}$ enters due to the part of this parallel gradient dynamics on $\tilde{T}_e$ alone. Although collisions are generally damping, they act directly on $\tilde{q}_{e\|}$ so that if this latter becomes small there is less dissipation on the temperature itself. This works the same way as resistivity does, so that the collisionality dependence is much the same as in the isothermal case.

Although the temperature has different structure from the density, it is similar to the nonadiabatic part of the density. In most cases these two show the same morphology and spectra. Therefore, the temperature physics doesn't change the mode structure appreciably and to the principal effects are quantitative, even though it is always interesting to see a smaller $\tilde{T}_e$ doing a larger relative share of the gradient drive with $\nabla \log n_e$ and $\nabla \log T_e$ similar, and especially if the latter is larger (a factor of two is common in experiments).

For low-beta cases at stronger collisionality the density fluctuations in the turbulence are slightly weaker than in the isothermal case but the overall pressure fluctuations, given by $T_e\tilde{n}_e + n_e\tilde{T}_e$ when temperature dynamics is present, are slightly stronger for equal gradient scale length. As beta is increased the magnetic flutter effect enters more strongly, since both gradients add to the effect. Magnetic induction and flutter refer to the $\partial\tilde{A}_\|/\partial t$ and $\nabla_\| \log p_e$, both linear terms, respectively. In the isothermal model, the magnetic flutter effect was found to largely cancel the inductive effect, and could be stabilising if artificially increased. With the electron temperature dynamics now present, the temperature gradient essentially provides this enhancement. For equal gradient strengths the combined $\nabla \log p_e$ is a factor of two larger, so that the magnetic flutter effect is definitively stabilising to the basic drift waves.

In toroidal geometry, however, the onset of interchange drive brings in a new mode, covered in Chapter 13 of Volume One. This is the ideal ballooning mode, so named as the interchange drive is positive if the curvature vector lines up with the pressure gradient. As this drive is localised to the outer midplane region of the torus, it is regarded as 'ballooning' in the mode structure. The ideal ballooning mode threshold occurs when $K_\parallel v_A$, representing the field line bending restoring force $(\nabla_\parallel \tilde{J}_\parallel)$, is overcome by γ_I, the pure interchange growth rate ($\gamma_I^2 = 2c_s^2/L_\perp R$ if $L_\perp$ is the pressure scale length and R is the radius of curvature). This threshold is nominally $\alpha_M = \hat{\beta}\omega_B(\omega_n + \omega_t) > 0.66\ S$ in these units. In the turbulence, this drive process enters progressively with increasing $\hat{\beta}$, and the turbulence strength follows.

At very low collisionality, in the regime of a hot plasma of order $T_e = 1$ keV or higher, not much of this applies since the adiabatic coupling is too strong. We will see that change when adding the ion temperature and its gradient in the next Chapter, and especially the thermal anisotropy later on. Hence, this situation of purely electron drift Alfvén turbulence is most relevant to conditions with $\hat{\mu} > 1$, especially $C > 1$, namely thermal or nominally hydrodynamic electrons, and more so if also collisional. The adiabatic response always enters, though, even at high collisionality, since the $k_\parallel$-spectrum widens to become limited when $k_\parallel V_e$ reaches the dynamical frequency of the turbulence given by the RMS vorticity.

2.8.1 Computational examples

We repeat some of the warm plasma toroidal drift Alfvén cases of Chapter 14 of Volume One to underly the above points and show the mode structure one finds in general and specifically in the temperature and heat flux dynamics. We should like to compare the resulting heat flux spectra and level with what would result by taking the results given by the form of the other quantities and using the Braginskii form for V_q as briefly done for the shear Alfvén cases.

The above nonlinear equations are used and the cases are run in the way we did for the kinetic shear Alfvén wave cases, setting all variables except $n(x, y, s)$ to zero in the initial state. This is the usage from Volume One. We start with a 2D random-phase density field determined by

$$n_{\mathbf{k}} = A_0\, k\, (1 + (k/k_1)^6)^{-1} \exp(i\Theta), \qquad n_{2D} = \sum_{\mathbf{k}} n_{\mathbf{k}}\, e^{i\mathbf{k}\cdot\mathbf{x}} \tag{2.126}$$

with the expression involving k giving the isotropic spectrum and Θ a random number on $[0, 2\pi]$ giving the phases. The constant A_0 chosen such that the RMS amplitude is set to δ_0. The $n_{\mathbf{k}}$ are Fourier transformed back to xy-space and labelled n_{2D}. It is then multiplied by an envelope function in the parallel direction so that

$$n(x, y, s) = n_{2D}(x, y) \exp[-(s/2.5)^2] \tag{2.127}$$

gives the 3D initialisation. It is aligned to the field (hence s), globally on the coordinate system with $s_k = 0$, and then on each xy-plane it is shifted (Eq. 2.81) onto the system with s_k equal to the grid location of that plane. Then, the run is started and run to $t = 1000$ for these cases. In all of this, the mode structure and role of the temperature and heat flux in the energetics is the main point, not anything about the scaling, which is similar to what we have seen in the isothermal model.

2.9 Mode structure

The concept of mode structure was developed across Volume One, with Chapter 10 dedicated to it, and the relevant sections of Chapters 12, 13, and 14, showing the details for the drift (Alfvén) wave and resistive (interchange) ballooning control cases, and the toroidal drift Alfvén cases, respectively. This together with the energetic analysis is what carries the most information about the physical mechanisms behind the turbulence in a driven/dissipative statistical equilibration. In all types of spectra, the quantities are Fourier transformed accordingly (one can always do this in a diagnostic, even if the boundary conditions disallow it in a computation). Then, the relevant wavenumber is assigned to a bin depending on its nearest integer multiple of a fundamental wavelength, k_0, which is the minimum k_y in a wavenumber spectrum, the minimum $k_\perp$ in an isotropic perpendicular spectrum, or $K_\parallel$ in a parallel wavenumber spectrum. Then, the spectrum is averaged over all other coordinates, and time within a given interval, and normalised such that the sum over modes of a given quantity gives the volume average of that quantity. We recall the mode structure overview herein. The case with $\hat{\beta} = 5$ and $C = 13.9$ (arises from $\hat{\mu} = 2.72$ and $\nu_e L_\perp / c_s = 10$) is referred to as nominal and is showcased.

The portion of the time traces of the energy channels and the drive and dissipation channels (energetics) up to $t = 1000$ is shown in figure 2.5. With the nonlinear start, saturation and development of the turbulence is very rapid. Due to the parallel localisation, E_n starts smaller at about 4×10^{-5}, but this level is already nonlinear; once the parallel domain fills out the level settles well higher than 10^{-3}, with

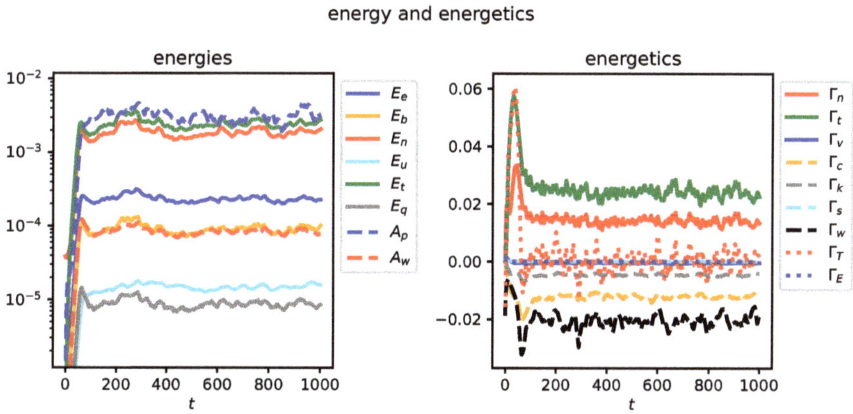

Figure 2.5. Time traces of (left) energy pieces and amplitudes of ϕ and ϖ (dashed lines), and (right) energetic drive and dissipation, nonlinear growth rate, and error. The amplitudes of n and ϕ, measured through E_n and A_p, respectively, are comparable, but ϕ is larger because of the contribution of T to the pressure (recall the 3/2 factor in E_t). The enstrophy (A_w) is smaller than the ExB energy(E_e), in turn smaller than A_p or E_n or thermal free energy E_n, reflecting the various spectral ranges where these peak. The gradient drive is larger for ∇T than for ∇n but both are temporally similar. The total drive rate (Γ_T, dotted) follows the drive in the early phase but saturates well before $t = 500$. The sinks, with resistive larger than conductive, are temporally smoother as they are active at smaller scale. The subgrid dissipation (Γ_w) makes up the balance. The magnetic flutter drive (Γ_v) is very small, as is the error (Γ_E).

$\delta_0^2 = 2.25 \times 10^{-4}$ the comparison baseline. We actually have E_t slightly higher than E_n, by a bit less than the 3/2 multiplier. The amplitude A_p is commensurate with the adiabatic response following $n + T - \phi$. The energy content of the flux variables is much less, with $E_b > E_u > E_q$. The ExB energy E_e is much less than A_p due to the scale being larger than ρ_s, and for the same reason A_w is still lower. We have $E_b < E_e$ for this case, for which $\hat{\beta}$ is about a factor of two lower than the nominal MHD ballooning boundary. This threshold is at $\alpha_M = (\omega_n + \omega_t)\hat{\beta}\omega_B > 0.66\ S$, translating to $\hat{\beta} = 11$ for these values $\omega_B = 0.03$ and $S = 1$.

The energetics shows how saturation works up to about $t = 100$, with the drive rates Γ_n and Γ_t developing first, with the visible sink rates Γ_w, Γ_c, and Γ_k (stronger to weaker) following. For this display the sinks representing positive definite damping are signed negative. The total growth rate Γ_T follows the drive before falling to zero into saturation, wandering visibly about the zero line thereafter. For this case both Γ_v and Γ_s are too small for visibility. The energetic error Γ_E is very small. The statistics measured over $500 < t < 1000$ versus those measured in the same run over $2000 < t < 4000$ are

$$\begin{aligned}
\Gamma_n &= 1.40 \pm 0.115 && \text{versus} && 1.33 \pm 0.132 \times 10^{-2} \\
\Gamma_t &= 2.40 \pm 0.176 && \text{versus} && 2.30 \pm 0.204 \times 10^{-2} \\
\Gamma_v &= -5.38 \pm 0.609 && \text{versus} && -5.33 \pm 0.679 \times 10^{-4} \\
\Gamma_c &= 1.22 \pm 0.074 && \text{versus} && 1.17 \pm 0.099 \times 10^{-2} \\
\Gamma_k &= 0.458 \pm 0.028 && \text{versus} && 0.445 \pm 0.037 \times 10^{-2} \\
\Gamma_s &= 1.61 \pm 0.138 && \text{versus} && 1.56 \pm 0.162 \times 10^{-6} \\
\Gamma_w &= 2.04 \pm 0.198 && \text{versus} && 1.96 \pm 0.202 \times 10^{-2} \\
\Gamma_T &= 0.241 \pm 3.07 && \text{versus} && 0.032 \pm 3.24 \times 10^{-3} \\
\Gamma_E &= -0.214 \pm 2.76 && \text{versus} && 0.215 \pm 3.16 \times 10^{-6}
\end{aligned} \tag{2.128}$$

The energetics is in statistical balance, with all channels in agreement over the two phases. The only number with significant difference is the mean of Γ_T, the total growth rate. The standard deviation is also in agreement, but the mean is one order of magnitude smaller in the shorter phase and two orders smaller in the longer phase. We then have confirmation that the physics is well saturated even over the shorter phases. The error is very small, with a standard deviation larger than the mean and more than three orders of magnitude less than the standard deviations for the main dissipation channels. For the main drive and dissipation channels, the mean is well larger than the standard deviation (don't forget this is in volume average).

The isotropic perpendicular ($k_\perp$) spectra are shown in figure 2.6 for amplitudes (top left), energetics (bottom left), and the four ExB nonlinear transfer channels (right side, combining $[\phi, n + T]$ and $[\phi, J + 0.4V_q]$). These and other spatial mode structure diagnostics are all averaged over $2000 < t < 4000$ in the late saturation phase. The signatures are those which led to the arrangement of the time traces. The energy content of the state variables follows these amplitude curves, and they are

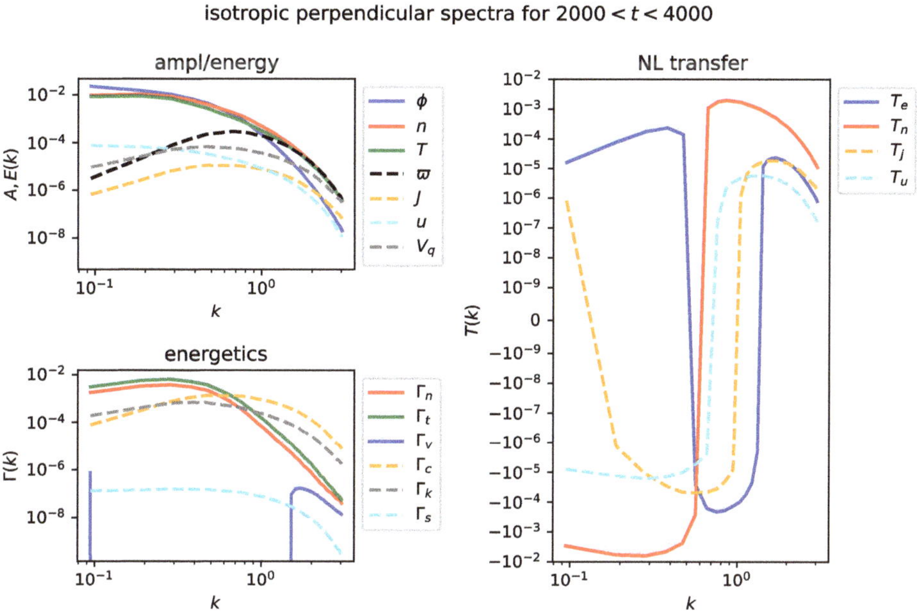

Figure 2.6. Isotropic perpendicular spectra for the nominal case during $2000 < t < 4000$, in the saturated phase. The binning of the quantities into nearest-to intervals is explained in the text. The state variable spectra are weighted towards larger scale, with the vorticity (ϖ) near $k_\perp\rho_s \sim 1$ and very robust. The current and heat flux are also more active at small scale. The energetics has the ExB gradient drive (solid lines) at larger scale than the dissipation channels (dashed lines). The magnetic flutter drive (Γ_v) is small and mostly negative. The nonlinear transfer is shown on the right hand side. The symmetric-log format allows both positive and negative values to be shown together. The transfer in the ExB vorticity nonlinearity (T_e) is from smaller scales (about $k_\perp > 0.5$) to larger for the while all the others are opposite, with T_n combining both n and T representing the state variables by far the largest. These transfer effects together with the adiabatic response ties the spectrum together as a unit, so that there is no single dominant mode. This is basic drift wave or drift Alfvén mode structure.

weighted towards larger scale, especially in ϕ. The three which are collected near $k_\perp\rho_s = 1$ are ϖ, J, and V_q, all of which represent the adiabatic response to the turbulence. The energy at large and vorticity at small scale is the usual situation. The same is reflected in the energetics, in which the drive channels are at larger scale while the dissipation is arranged with the vorticity and parallel electron flux variable amplitudes. The sound wave viscous dissipation is very small, and the magnetic flutter drive is also negative. The energy transfer spectra show depletion in the middle range and enhancement in both larger and smaller scale, for the ExB channel, and a robust, direct, transfer from large scale to small for the thermal free energy channels. Together with the drive and dissipation spectra, this shows that the turbulence is driven at larger scale, mostly tansferred to smaller, and then dissipated through the parallel electron dynamics.

The drift direction wavenumber (k_y) spectra are shown in figure 2.7 for the same amplitudes and energetics as in figure 2.6. Arranged in k_y, the vorticity and related spectra are flat up to $k_y\rho_s = 1$, while the state variables, especially ϕ, are collected

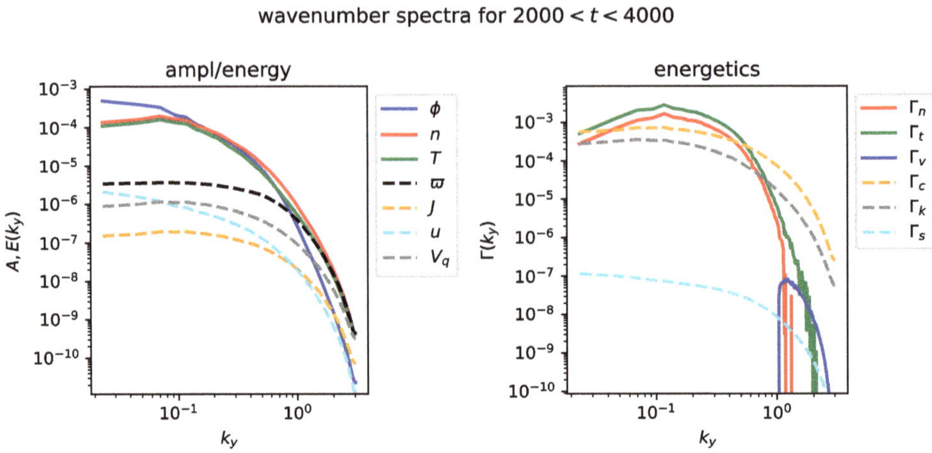

Figure 2.7. (left) Wavenumber spectra for the nominal case during $2000 < t < 4000$, in the saturated phase. The labelling and format are as in figure 2.6, except for k_y instead of $k_\perp$ on the horizontal axis. The state variables are collected more toward long wavelength, with the gradient drive slightly shorter. The flux variables and the dissipation very flat to both sides, following the vorticity. The amplitude of ϕ is the most strongly shifted toward long wavelength, following the energetic transfer dynamics shown in the other diagnostics.

towards long wavelength. The energetics shows the gradient drive peaking in the $0.1 < k_y\rho_s < 0.2$ vicinity, with the dissipation flatter to both sides.

The parallel wavenumber spectra are shown in figure 2.8, for amplitudes, energetics, and the four ExB nonlinear transfer channels, done as for isotropic perpendicular spectra in figure 2.6. The standout feature is the wide vorticity spectrum compared with the state variables. Here, too, ϖ follows the flux variables for the electrons while ϕ follows the other state variables. The energetics shows the drive channels, with Γ_t slightly higher, in the same form as the state variables, while the electron dynamical sinks ($\Gamma_c > \Gamma_k$ slightly) follow the flux variables. The energy transfer spectra all show depletion in the low-$k_\parallel$ range and enhancement, mostly through the pressure, in the range $k_\parallel > 3$ or so in magnitude. Together with the drive and dissipation spectra, this shows that the turbulence is driven at low-$k_\parallel$, transferred to higher, and then dissipated through the parallel electron dynamics, consistent with the isotropic perpendicular spectra.

The parallel envelopes (in the s-direction) are shown in figure 2.9 for the squared amplitudes of the three state variables with $h = n - \phi$ and the transport fluxes. The dynamics is moderately ballooned in the outer part of the torus ($-\pi/2 < s < \pi/2$). With this moderately large $\hat{\beta}\omega_B$ combination, ϕ is higher than either of n or T, while h is lower than either and also much flatter in envelope. The degree of ballooning in the fluxes is only about two to one, which is much less than the corresponding linear mode structure (Chapter 14 of Volume One). The magnetic flutter drive M_e is small and also slightly negative, as the inductivity is not large enough for this drive mechanism to set up its own phase shift (between V_q and A); it is overshadowed by

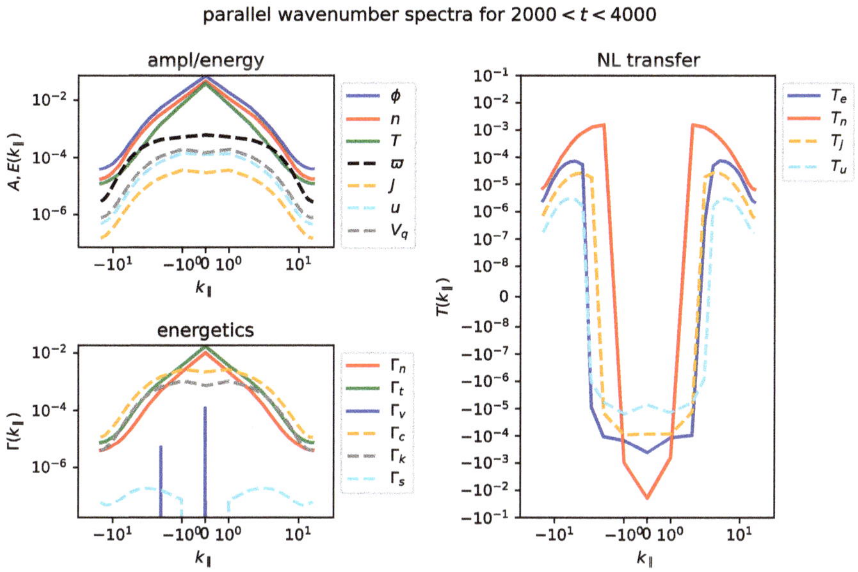

Figure 2.8. Parallel wavenumber spectra for the nominal case, averaged over $2000 < t < 4000$, in saturation. The labelling and format are as in figure 2.6, except the horizontal scale is also symmetric-log. The peak of the amplitudes and gradient drive in the flute mode region ($k_{\|} \sim 0$) reflects the much weaker adiabatic response there. The vorticity (ϖ) has the flattest spectrum, while the state variables (ϕ, n, T) follow each other. Following ϖ, the parallel dissipation channels are also flatter. The transfer spectra are all from the flute mode region outward. The ExB transfer (T_e) shows an inverse transfer into the flute mode region in the isothermal case, but not here. This is also basic drift wave mode structure.

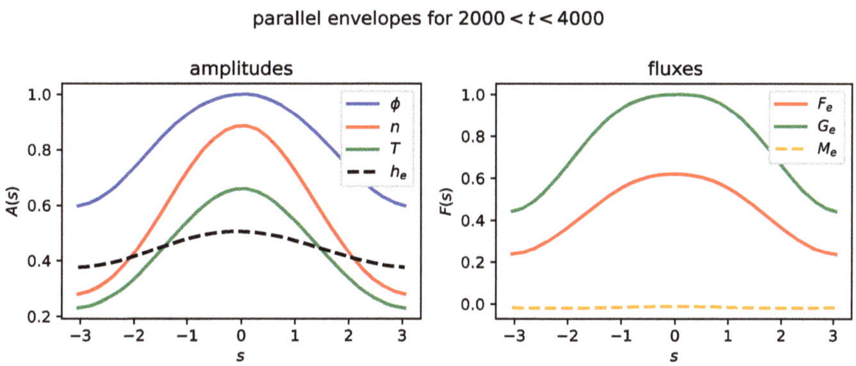

Figure 2.9. Parallel envelopes of the squared amplitudes (left) and turbulent ExB and magnetic flutter transport flux (right), which are averages over x and y and time $2000 < t < 4000$ of each quantity. The curves are more moderately ballooned, and the amplitudes (left) of the state variables follow the relationships shown in the spectra, with ϕ the strongest. For electrostatic cases ϕ is the most ballooned (localised towards $s = 0$), but for this inductivity ($\hat{\beta} = 5$) ϕ is flatter. However, $\hat{\beta}$ is not high enough to make the flutter drive M_e go positive. See also figure 2.20, below. The nonadiabatic response $h = n - \phi$ is robust but smaller than either of ϕ or n. This is a result of the Alfvén dynamics constraining h rather than ϕ alone. That the transport flux is ballooned is not by itself a revealing indicator.

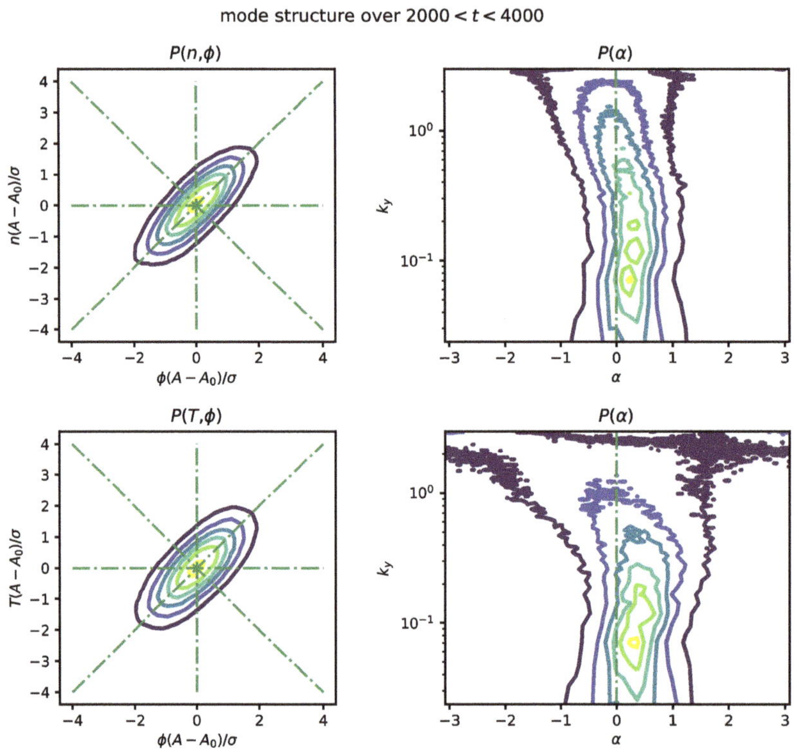

Figure 2.10. Cross coherence (left), and phase shift distributions (right) between n and ϕ (top) and between T and ϕ (bottom), for the nominal case over $2000 < t < 4000$, in saturation. Both diagnostics show drift wave mode structure, with significant correlation between n or T and ϕ and narrower phase shift distributions centred not on $\pi/2$ but closer to zero, despite the robustness of the dynamics and $h = n - \phi$, the nonadiabatic response.

the ExB advection drive and the resulting setting of the phase structure by the adiabatic response.

These spatial signatures all indicate that the vorticity is more involved in stirring up the current and heat flux rather than participating in the drive mechanism.

The cross coherence and phase shift distributions are shown in figure 2.10. The top frames are for the statistics between n and ϕ, and the bottom frames give these for T and ϕ. The phase shifts are positive, as they must be to reflect a positive gradient drive, but the width of the distributions is narrow, the peak values (mean phase shifts) are close to lining up across the k_y-spectrum, and most of the activity is safely within the $\alpha < \pi/2$ range, where $\alpha = \text{Im}[\log n^*\phi]$ for n and ϕ (switch $n \rightarrow T$ for the T to ϕ case). This indicates a strong adiabatic response. In addition the cross coherence indicates that transported quantity, n or T, is visibly correlated to the transporting flow potential ϕ. By contrast, a passive scalar or hydrodynamic character in the responses would exhibit wider $P(\alpha)$ distributions centred upon $\pi/2$, and an uncorrelated signature with nearly circular contours in the cross coherences. Note that the phase shift of T to ϕ is slightly larger than the one for n to ϕ.

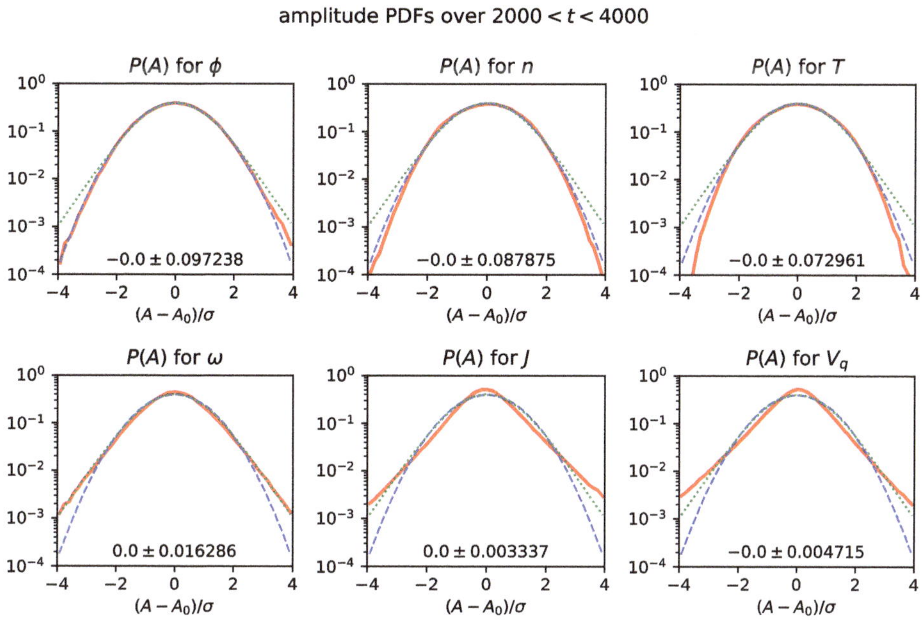

Figure 2.11. Amplitude probability distributions of the state variables and ϖ, J, and V_q. The blue and green dashed curves show Gaussian and passive scalar distributions for comparison. The state variables are closest to Gaussian. The vorticity is close to a passive scalar distribution, with stronger large-event tails than the Gaussian. The extreme spatial intermittency seen in pure fluid turbulence is absent, due to the coupling in the adiabatic response.

The amplitude probability distributions are shown in figure 2.11. The main observation is that the vorticity has the most pronounced tails in comparison to the Gaussian curves, nearly following a passive scalar character. This shows the strength of its nonlinearity (polarisation current, ExB vorticity self advection). The distributions in the flux variables follow, since they are excited by the vorticity. By contrast, the state variables, especially ϕ show more of a Gaussian character reflective as straightforward randomised turbulence without strong intermittency in the spatial structure.

The spatial morphology in the central drift plane ($s = 0$) is shown in figure 2.12, at $t = 1000$ for this case. Intermittency is most visibly in the current, somewhat less in the vorticity. This is a strong inductivity signature ($\hat{\beta} > 1$). The state variables show a more regular structure albeit still with visible scale distribution. Noteworthy is the similarity between ϕ, n, and T, especially at large scales, and the still-visible sheet structures in the vorticity. This is the ExB vorticity, $\Omega_E = \delta_0^{-1}\varpi$ in normalised units, with an RMS amplitude of order unity. This structure is the result of the moderate scale separation between different pieces of the dynamics, namely, more drive at large scale and more dissipation at small, with the adiabatic response active at all scales.

The three-dimensional morphology of the density fluctuations is shown in a snapshot at $t = 1000$ in figure 2.13 for this case. There is enough of a large-scale

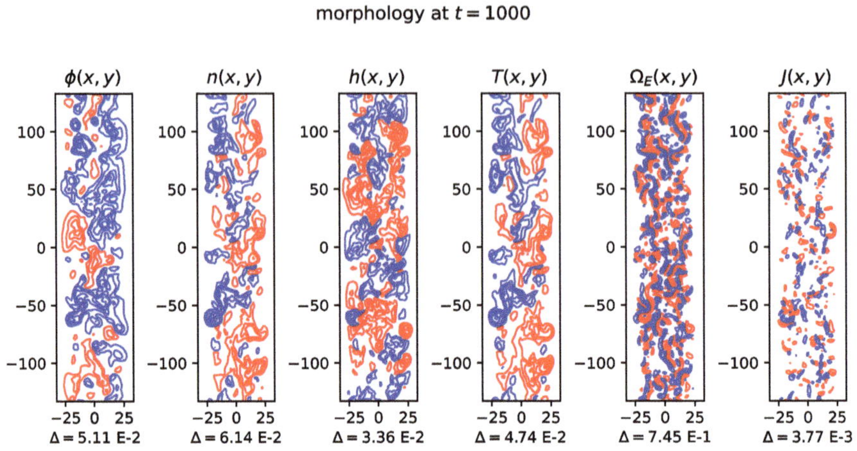

Figure 2.12. Spatial morphology at $t = 1000$ of the state variables and the parallel current, for the case with $\hat{\beta} = 5$ and $C = 13.9$. The ExB vorticity in units of $c_s/L_\perp$ is given by Ω_E, which for this model is $\delta_0^{-1}\varpi$ in normalised units. Positive contours are in red; negative, blue. The nonadiabatic density, $h = n - \phi$, is comparable to the temperature T, and both are a reasonable fraction of ϕ and n. The similar size of the latter two is a drift wave signature, while the robustness of the former two reflects the nonadiabatic energetics active in the turbulence. Note how well correlated n and T are, even this far below the nominal MHD boundary which is at $\hat{\beta} = 11$ for these parameters. The vorticity and current show the same differences to the other variables as in the isothermal model, emphasising that different parts of the dynamics are at different scales.

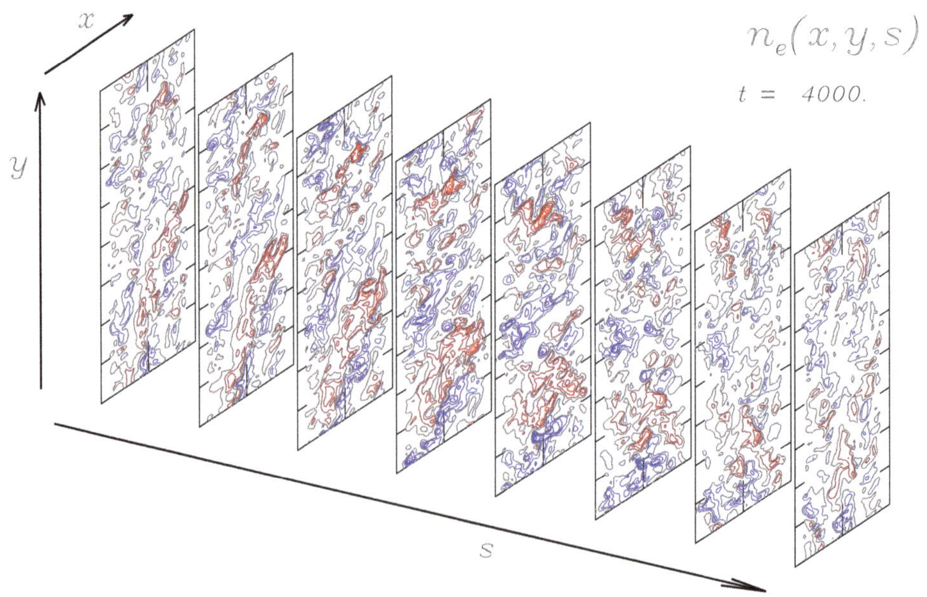

Figure 2.13. Three-dimensional structure of n for the nominal case. Of the 32 planes in s in the computation, 8 are shown. Some twist with the shear of the magnetic field is visible in the largest structures. If the entire structure is ever really visible on this type of plot, then the computation was likely too quasilinear.

component to make individual structures which twist with the shear of the magnetic field lines identifiable. These last two figures illustrate the basic look and feel of the manifestation of drift wave mode structure in the morphology, but by themselves they do not discern the physical processes, for which we need the other diagnostics, and especially the energetics.

2.9.1 Details of the energetics

This set of diagnostics is representative of essentially the same systematics found for the isothermal case in Volume One. We find more or less the same when looking at the details of the energetics through the isotropic spectra of not only the drive and dissipation but also the transfer channels. These are displayed next.

The isotropic spectra of the state variable energetics are shown in figure 2.14. These are for the ExB energy E_e on the left and thermal free energy $E_p = E_n + E_t$ on the right. The definitions of the channels are

$$T_e = -\phi(-[\phi, \varpi]), \quad T_j = -\phi\left(\nabla_{\|} J\right), \quad T_P = \phi \mathcal{K}(n + T) \tag{2.129}$$

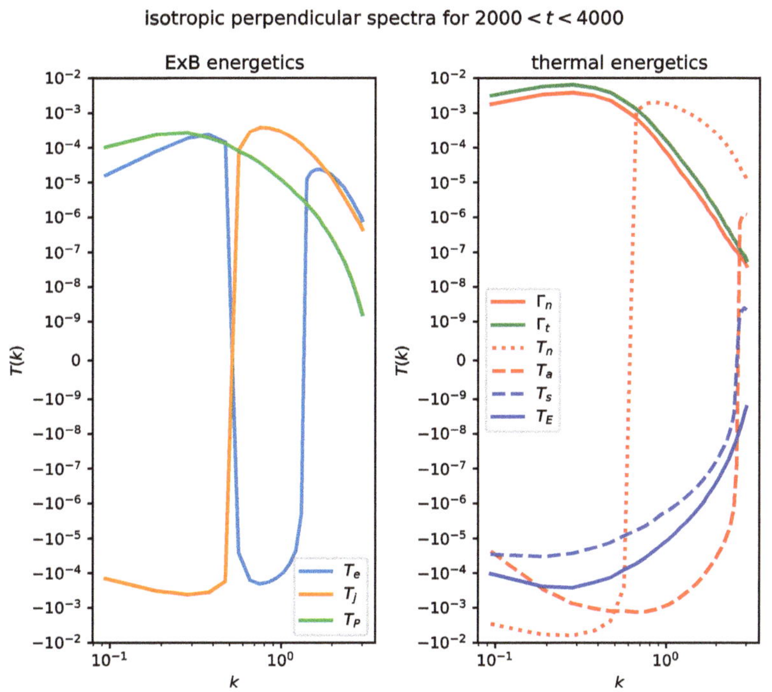

Figure 2.14. State variable energetics, for the ExB (left) and thermal free energy (right). Labels and definitions are as given in the text. The ExB energy has no sources, but is supported by incoming transfer from the current via Alfvén transfer (T_j) at small scale, and the ExB vorticity nonlinearity (T_e) and the diamagnetic current (T_P) at large scale. The thermal free energy ($E_n + E_t$) is supported by the gradient drive ($\Gamma_n + \Gamma_t$) at all scales, balanced mostly by nonlinear advective transfer (T_n) at larger scale and adiabatic transfer (T_a) to the current at smaller scale. The large scale signature due to T_e and the small scale one due to T_j and T_n would not occur if the turbulence were to exhibit the same mode structure as for a linear instability.

for E_e and

$$\begin{aligned} T_n &= n(-[\phi, n]) + \frac{3}{2}T(-[\phi, T]) \\ T_a &= (n + T)(\nabla_\| J), \qquad T_s = (n + T)(-\nabla_\| u) \\ T_E &= (n + T)\mathcal{K}(\phi) \end{aligned} \tag{2.130}$$

along with Γ_n and Γ_t for E_p. All of these are defined in volume average as in Eq. (2.105). The nonlinear transfer channels T_e and T_n are the same ones as in figure (2.6), and the spectra are computed in the same way. The ExB nonlinear transfer (T_e) is large enough to overcome the MHD interchange forcing (T_P), to balance the Alfvénic coupling (T_j) except at the largest scales. In the thermal free energy the gradient drive (Γ_n, Γ_t) acts as a source of all the dynamics. In this equation it is balanced by the nonlinear ExB advection (T_n) at large scale, but at small scales T_n is the main source, balanced principally by the adiabatic compression (T_a). Hence the adiabatic coupling is the larger transfer mechanism sending energy towards the ExB side. This appears as T_j in the ExB flow energetics, and it is what compensates the nonlinear transfer in the vorticity nonlinearity. This is why the ExB vorticity is so much higher than the total gradient drive rate for this type of turbulence, by contrast to the usual model for strong hydrodynamic turbulence which assumes $\gamma \sim k_\perp v_E$ from the linear physics given the growth rate γ and the inverse scale $k_\perp$of the corresponding instability. The adiabatic and Alfvénic coupling transfer is mediated through the magnetic and parallel electron kinetic energy as seen below. The transfer in from E_e through the ExB (curvature) compression (T_E) and out into the sound waves by acoustic compression (T_s) is small in comparison. There are no physical sinks in either of these state variable equations.

The isotropic spectra of the flux variable energetics are shown in figure 2.15. These are for the Ohmic (magnetic and parallel electron kinetic energy, E_b), acoustic (E_u), thermal (E_q), channels, from left to right, respectively. The definitions of the channels are

$$T_j = J(-\hat{\mu}[\phi, J]), \quad T_u = u(-[\phi, u]), \quad T_q = \frac{2}{5}V_q\left(-\hat{\mu}[\phi, V_q]\right) \tag{2.131}$$

for the nonlinear transfer, and

$$T_A = J\left(-\nabla_\| \phi\right), \quad T_a = J\left(\nabla_\|(n + T)\right), \quad T_s = u\left(-\nabla_\|(n + T)\right), \quad T_k = -V_q \nabla_\| T \tag{2.132}$$

for the Alfvénic, adiabatic, acoustic, and conductive transfer channels, respectively. The solid red curves give the sinks (Γ_c, Γ_s, Γ_k), and the solid green curves give the pieces of magnetic flutter (Γ_v, due to $-J(n\omega_n + T\omega_t)$, $u(n\omega_n + T\omega_t)$, and $V_q(T\omega_t)$, respectively). The magnetic flutter gradient drive is labelled Γ_v. The three effects appear on separate graphs and so there is no confusion. The resistive dissipation curve is the lowest one in the left frame, and the thermal conduction dissipation curve is the lowest one on the right, respectively.

The magnetic flutter gradient drive in the Ohmic energy E_b averages to zero (due to Jb^x) and is small everywhere. In the acoustic energy E_u it is simply small. In the

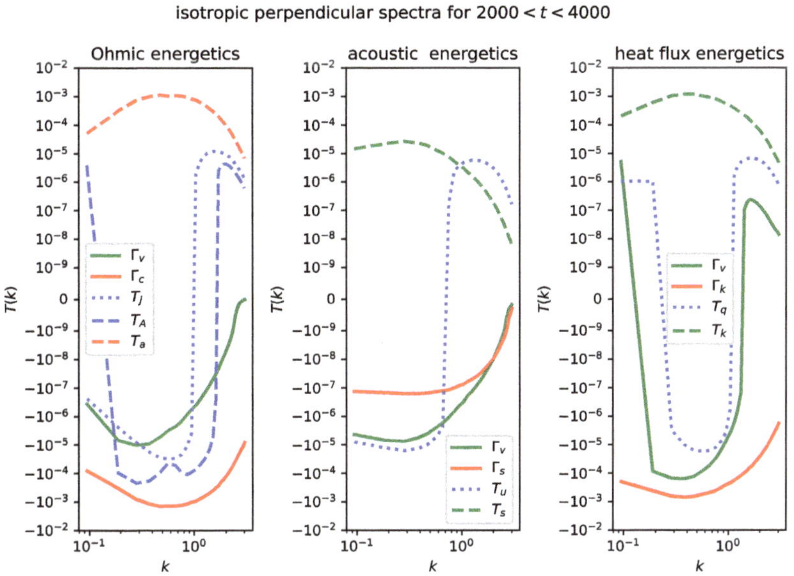

Figure 2.15. Flux variable energetics, for the electron (Ohmic, left; thermal conductive, right) and ion channels (acoustic, centre). Labels and definitions are as given in the text. Sinks are signed and therefore negative. The Ohmic (including magnetic) energy (E_b) is supported at all scales by adiabatic transfer (T_a) incoming from the thermal free energy, balancing both resistive dissipation (Γ_c) and Alfvénic transfer (T_A) to the ExB energy (E_e). The acoustic energy (E_u) is supported mostly by acoustic transfer (T_s) from the combined pressure, balancing nonlinear transfer at large scale (T_u) and mostly viscous dissipation (Γ_s) at small. The conductive energy (E_q) is supported by incoming thermal transfer (T_k) which balances conductive dissipation (Γ_k).

thermal energy E_q it is still small but somewhat larger in comparison and becoming positive at larger scale.

The sinks Γ_c and Γ_s represent positive definite damping. The main sources for all three pieces are the incoming transfer through parallel dynamics, adiabatic (T_a) transfer for the Ohmic, sound wave transfer (T_s) for the acoustic, and conductive (T_k) for the thermal channels, respectively. These are positive everywhere and constitute the main source effect for the flux variable free energy channels. The main sinks are the dissipation channels for all three quantities. Nonlinear transfer is significant only in the acoustic energy where it transfers from large to small scale, balancing the incoming acoustic transfer at large scale, and the viscous dissipation at small scale. This is also the main effect limiting the size of the parallel gradient in the adiabatic part of the system.

These details give a picture of a unified, self-consistent spectrum of nonlinear dynamics, depicted in figure 2.16 (the same as figure 10.13 from Chapter 10 of Volume One). The energy transfer is mostly circular: Start with thermal free energy at low-$k_\perp$ and follow how it is thrown to small scale, transfers into the Ohmic energy, then into the ExB energy, through both adiabatic and Alfvénic coupling, then back to large ExB eddy scales, then through Alfvénic coupling to the current and then adiabatic coupling out of it, and thereby back to the thermal free energy. The net drive due mostly to ExB advection of the background gradient is needed to support

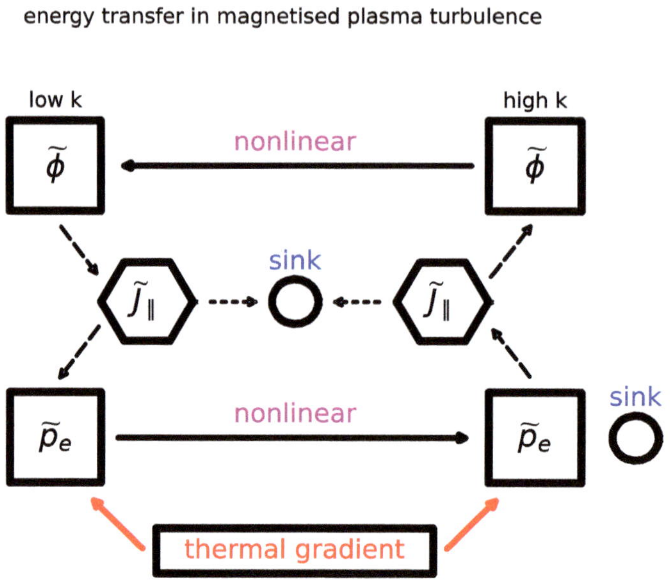

Figure 2.16. (Figure 10.13 from Volume One) diagram of the principal dynamics of energy transfer in magnetised plasma turbulence. The thermal gradient drives electron pressure disturbances ($\tilde{p}_e$) and the parallel current ($\tilde{J}_\parallel$) is dissipative at all scales. Nonlinear transfer via unsteady ExB advection is between scales, shown with solid lines. The scale transfer is mostly through $k_\perp$ even in three dimensions, since the guide field makes the parallel dynamics wavelike. Linear transfer is mostly through the adiabatic response, directed as $\tilde{p}_e \to \tilde{J}_\parallel \to \tilde{\phi}$ into the ExB energy at small scales and vice versa at large scales. The parallel current acts as a mediator between the two state variables ($\tilde{p}_e$, $\tilde{\phi}$). Due to the bi-directional nonlinear transfer, the sink spectrum is wider than the drive spectrum, with most of the energetics in an intermediate spectral range. Where this range falls depends on the relative strengths of all the interactions, self-consistently, eventually determined by the parameters. The entire spectrum is energetically tied together, and reacts to changes as a unit.

all of this, but the result of it is that the RMS ExB vorticity is much larger than the net drive by at least one order of magnitude. The typical level of RMS ϖ is order few times δ_0 and the drive is order few times δ_0^2, translating into an ExB vorticity Ω_E of order few times $c_s/L_\perp$ and a drive rate or order few times $v_D/L_\perp$ in physical units. This falsifies any model which assumes that a nonlinear decorrelation rate balances a linear drive. Interchange transfer is concurrent with this, especially for higher $\hat{\beta}$ values, but it only changes the picture when the MHD boundary is crossed.

2.10 Dependence on parameters

2.10.1 Electromagnetic induction

The dependence upon inductivity at two values of collisionality is considered. The parameters controlling the electrons are

$$1 < \hat{\beta} < 50, \qquad \hat{\mu} = 2.72, \qquad 1.39 < C < 13.9 \tag{2.133}$$

and the scale ratios are

$$\hat{\epsilon} = 10\,000, \qquad \delta_0 = 0.015, \qquad L_y/L_x = 4 \tag{2.134}$$

For this study we set equal gradient strengths

$$\omega_n = 1, \qquad \omega_t = 1 \tag{2.135}$$

giving the density and temperature equal chances to drive the turbulence.

The envelopes of the squared amplitudes of the three state variables and the transport fluxes in the s-direction is shown in figure 2.20, for three values of inductivity, giving the transition into the MHD regime (nominal boundary at $\hat{\beta} = 11$). For $\hat{\beta} = 1$ the indication is of drift wave mode structure, with h and T the flattest of the curves, and ϕ the largest and most ballooned. At this collisionality h is not so small, but both h and T are smaller than ϕ. With equal gradient forcing, the pressure fluctuations balancing ϕ are split between n and T, which are of roughly equal magnitude. At $\hat{\beta} = 5$ the degree of ballooning increases, while at $\hat{\beta} = 20$ into the MHD regime we start to pick up the MHD mode structure given by (for $\omega_n = \omega_t$) equal magnitude and structure for n and T, and ϕ larger so that h is larger than either n or T. The MHD regime signifies the statistical independence of $\tilde{\phi}$ versus the other state variables, which is felt in all the mode structure measurements.

The cross coherence and phase shift distributions between n and ϕ are shown for the same three cases in figure 2.21, with the same format as in figure 2.10. In these the temperature has close to the same relationship to the potential as the density, so it is not shown. The transition away from drift wave mode structure as $\hat{\beta}$ crosses the critical value of $\hat{\beta} = 11$ is visible. The long-wavelength component separates and moves to a value closer to $\pi/2$ in the phase shifts, and the variables become less correlated as the nonlinearities competing with the adiabatic response are assisted by the slowing down of the latter with finite inductivity. This is the same result found for isothermal cases (Chapter 14) in Volume One, with the drift wave (Chapter 12) and ballooning mode structure (Chapter 13) serving as the control cases.

2.10.2 The heat flux versus its collisional form

The Braginskii model replaces the parallel heat flux with a thermal diffusive damping on the temperature, keeping the thermoelectric effect. As in the Alfvén response cases, we compare them by evaluating the Braginskii heat flux defect in Eq. (2.124), which with the temperature gradient present is

$$B_q = V_q + \alpha_e J + C^{-1}\kappa_e \nabla_\parallel (T_0 + T) \tag{2.136}$$

noting that the parallel gradient is nonlinear and acts upon the profile T_0 as well as the disturbance. That is, we have to keep the parallel gradient of the entire temperature in nonlinear cases.

To do this comparison, the RMS values of V_q and B_q are measured. For $\hat{\beta} = 5$ and $C = 13.9$ the result is $(3.99 \pm 0.16) \times 10^{-3}$ for V_q and $(3.81 \pm 0.29) \times 10^{-4}$ for B_q. For the high collisionality cases the Braginskii model is fairly closely represented. This corresponding to $\nu_e L_\perp / c_s = 10$, it is to be expected. If we neglect to include the magnetic flutter effect arising from $\nabla_\parallel T_0$ in the measurement, then the defect is of order unity, also indicting a strong balance between linear and nonlinear components of $\nabla_{\|}(T + T_0)$. For $\hat{\beta} = 5$ and $C = 1.39$ the result is $(2.59 \pm 0.68) \times 10^{-3}$ for V_q and $(1.32 \pm 0.57) \times 10^{-3}$ for B_q. Although with $\nu_e L_\perp / c_s = 1$ the collisions are faster than most of the dynamics, the trans-collisional nature of the fastest part of the dynamics is far enough over into the transition regime that the Braginskii heat flux is no longer accurate—its implied neglect of the (nonlinear advective) time derivative is no longer a good assumption. For this case a value of $(4.47 \pm 1.04) \times 10^{-3}$ is found for ϖ, which implies 0.298 ± 0.069 for Ω_E, the ExB vorticity. Although the collision frequency is more than three times as large as Ω_E, the departures from the Braginskii model are significant. The k_y and isotropic $k_\perp$ spectra both show similar shapes in V_q and B_q except where subgrid dissipation affects V_q so that it drops somewhat more with inverse scale below ρ_s. In all, not only is the computational maintenance of the time dependent heat flux model somewhat easier, due to the lack of double nonlinear parallel derivatives, it is also more accurate.

2.10.3 Thermal effect on electromagnetic responses

The run of amplitudes among the state variables with inductivity ($\hat{\beta}$) are shown in figure 2.17, for two values of collisionality (C). These are the RMS values for $e\tilde{\phi}/T_e$ and $\tilde{n}_e/n_e$ and $\tilde{T}_e/T_e$ and $\varpi/n_e e$, and the magnetic field $\left| \nabla_\perp \tilde{A}_{\|} \right| / B_0$, respectively. The volume average is measured at each time point, then the time average with 51 samples over $500 < t < 1000$ and then the square root are taken. At these parameters the results are qualitatively the same as in the isothermal model. As the inductivity increases, there is at first a light drop in most of the channels from $\hat{\beta} = 1$ to 2, but thereafter one finds a monotonic rise. At both collisionality values the state

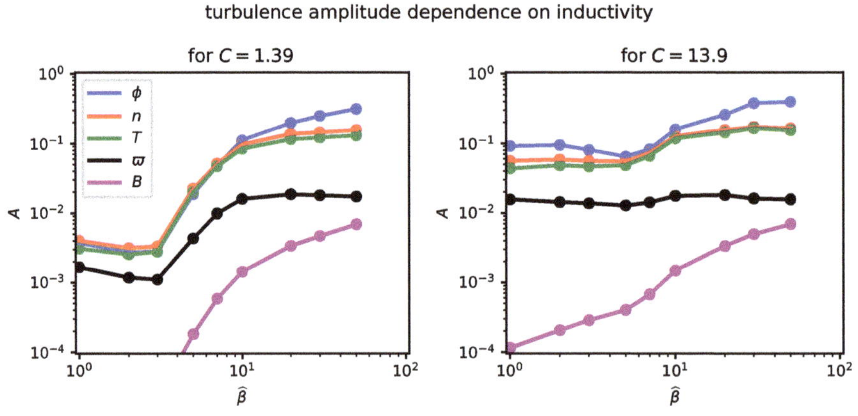

Figure 2.17. Dependence of the RMS fluctuation amplitudes, for lower and higher collisionality, C, upon finite beta, or inductivity, given by $\hat{\beta}$ as the square of the ratio between the nominal drift wave vorticity and the Alfvén transit frequency. As $\hat{\beta}$ increases the nominal MHD ballooning threshold is reached at $\alpha_M = (\omega_n + \omega_t)\hat{\beta}\omega_B = 0.66\ S$, which means $\hat{\beta} = 11$ for these cases. The amplitude of ϕ relative to n and T in these units increases the fastest among state variables. The magnetic fluctuation amplitude relative to the background is also given. The increase of this with $\hat{\beta}$ is to be expected. All are signatures of the increasing MHD character in the mode structure.

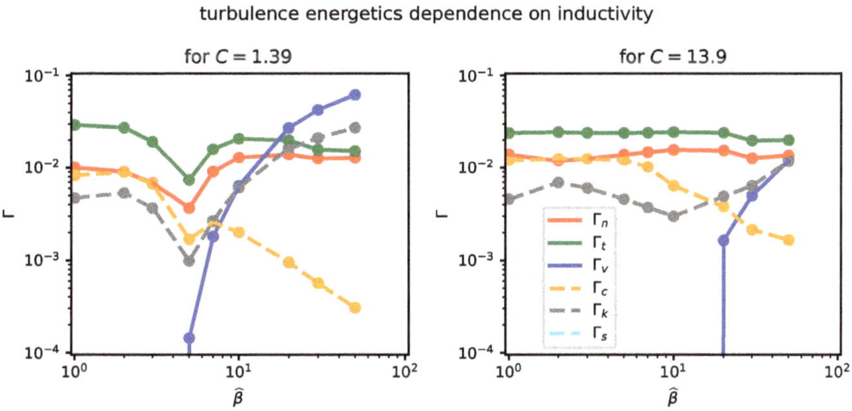

Figure 2.18. Dependence of the energetics in the turbulence upon inductivity, $\hat{\beta}$, for lower and higher collisionality, C. The gradient drive is due to ExB advection in the density (Γ_n) and temperature (Γ_t) gradients, and magnetic flutter due to the fluctuating heat flux in the presence of the temperature gradient (Γ_v). The parallel dissipation channels are resistivity (Γ_c), thermal conduction (Γ_k), and parallel ion viscosity (Γ_s). The latter is always small, in the range 10^{-5} to 10^{-6} in these units (as rates in terms of $c_s/L_\perp$). Thermal conduction increases in significance in the region where the magnetic fluctuations are significant, a result of electromagnetic thermal forcing. The temperature gradient is always the strongest drive even for equal relative gradients ($\omega_n = \omega_t$).

variables all rise together, until about $\hat{\beta} = 10$ the amplitude of $e\tilde{\phi}/T_e$ separates and rises higher. This is the onset of the MHD regime, whose benchmark is $\alpha_M = (\omega_n + \omega_t)\hat{\beta}\omega_B$ the critical value of $0.66\,S$ at $\hat{\beta} = 11$ for these values $\omega_B = 0.03$ and $S = 1$. It is interesting to note that the transition towards MHD is gradual and even at the highest $\hat{\beta}$ at the higher C is not complete.

The energetic processes in the turbulence are measured through the Γ contributions given in Eqs (2.108–2.113) and shown in figure 2.18. The ExB temperature gradient forcing Γ_t is the strongest everywhere except at the lower C where into the MHD regime the magnetic flutter drive due to $\omega_t V_q b^x$ becomes larger. The magnetic flutter drive is small and actually negative at low $\hat{\beta}$ and doesn't enter until the inductivity is large enough to break up the drift wave phase shift relations linking J and V_q to n, T, and ϕ. The parallel dissipation channels mediate the physics but the subgrid dissipation due to the strong forward cascade dynamics in both n and T actually takes up more of the free energy. The thermal conduction sink becomes the stronger among the parallel dynamics when the magnetic fluctuations and their thermal gradient drive are significant.

For these relative gradients with $\omega_n = \omega_t$ the fluctuation amplitudes in n and T are comparable (we return to normalised units for clarity). This is not the case in the transport channels, however. The normalised ExB particle and conductive heat transport fluxes and the and the magnetic flutter conductive heat transport flux are given by

$$F_e = \langle n v_E^x \rangle, \quad G_e = \frac{3}{2}\langle T v_E^x \rangle, \quad M_e = \left\langle V_q b^x \right\rangle, \quad Q_e = \frac{3}{2}F_e + G_e \tag{2.137}$$

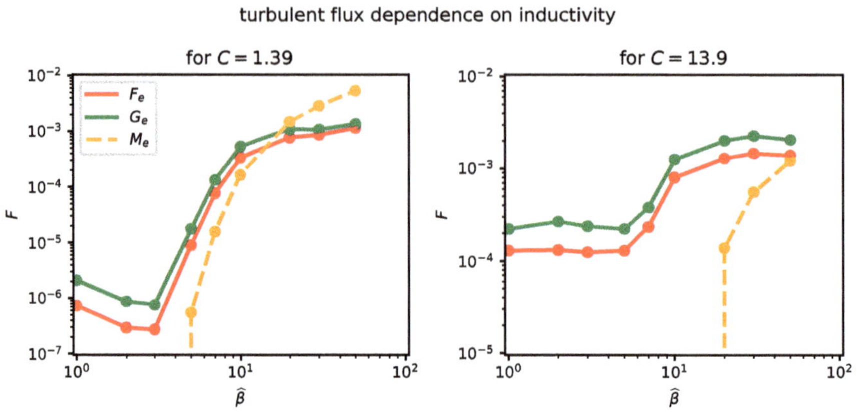

Figure 2.19. Dependence of the turbulent transport fluxes upon inductivity, $\hat{\beta}$, for lower and higher collisionality, C. The ExB fluxes F_e and G_e, tracking the density and temperature fluctuations, respectively, fall or rise together. The magnetic flutter transport flux M_e is negative at lower $\hat{\beta}$ and rises to significance during the transition into the MHD regime, nominally at $\hat{\beta} = 11$. For lower C, the dominant flux is actually M_e at the highest $\hat{\beta}$.

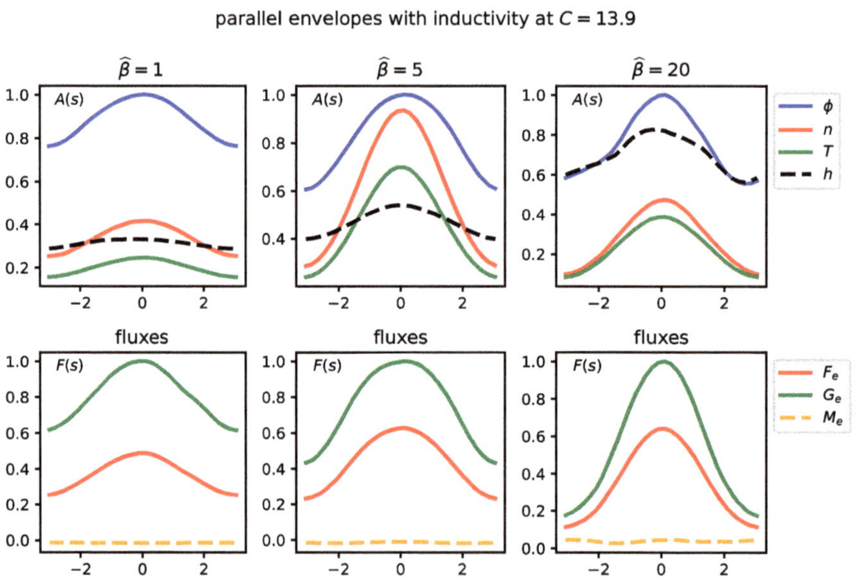

Figure 2.20. Parallel envelopes of the squared amplitudes (top row) and turbulent ExB transport flux (bottom), which are averages of each quantity over x, y and time, for $500 < t < 1000$. Amplitudes for ϕ, n, T, and $h = n - \phi$, and fluxes for F_e, G_e, and M_e (Eq. 2.137) are labelled accordingly. The parameters are $C = 13.9$, and $\hat{\beta} = 1$, 5, and 20. The critical $\alpha_M = 0.66\ S$ occurs at $\hat{\beta} = 11$. At low-$\hat{\beta}$ the curves for T and h are flatter, while ϕ is the most ballooned of all. The transition into the MHD regime is seen in the rise of h indicating departure of ϕ from n and T. Note that T plays the same role in the physics as h, part of the overall nonadiabatic dynamics, with $p_e - \phi = h + T$ the adiabatic response force potential.

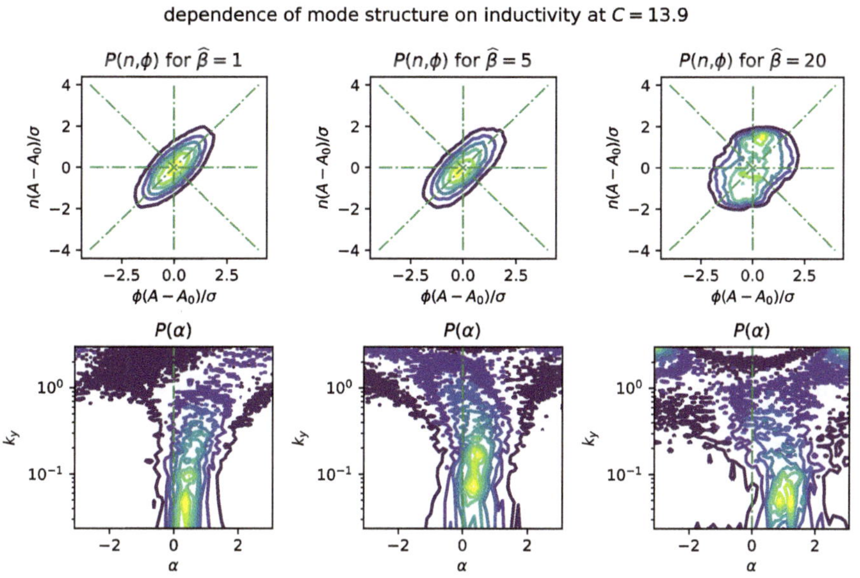

Figure 2.21. Cross coherence (top row) and phase shift distributions (bottom) between n and ϕ, at $C = 13.9$ for $\hat{\beta} = 1$, 5, and 20. The nominal MHD ballooning boundary $\alpha_M = 0.66\ S$ occurs at $\hat{\beta} = 11$. At low-$\hat{\beta}$ the phase shifts are closer to zero, and the cross correlation is higher as indicated by narrower ellipses. This is the signature of drift wave mode structure. Into the MHD regime the long wavelength component separates away and migrates towards larger phase shifts (note $\pi/2$ is the optimal value for free energy release), producing a higher degree of randomness in the cross correlation (circular contours would indicate a lack of correlation). These changes are towards the idealised hydrodynamic or MHD mode structure This is a very similar result to that found for the isothermal cases in Volume One. For equal gradient drive ($\omega_n = \omega_t$) these diagnostics are similar for n and T versus ϕ.

with Q_e the total ExB heat flux including the convective piece, in normalised units. The dominant contribution to the magnetic flutter transport is due to V_q, as noted above (Eq. 2.109). In physical units M_e is simply the parallel heat flux $\tilde{q}_{e\|}$ times the radial component of the fluctuating magnetic field, whose correlation is nonzero and therefore contributes to a total heat flux. The fluxes are shown for various $\hat{\beta}$ at the two C values in figure 2.19. The particle flux in this model at lower $\hat{\beta}$ is smaller than it was in the isothermal case. The conductive heat flux, while larger than F_e here, is still smaller than the result in the isothermal case. This is due to the combination of magnetic flutter stabilisation in the phase shifts to which is now added the thermal conduction dissipation. As the transition into the MHD regime is entered, the two ExB flux channels approach each other more, with the eventual relationship $G_e = (3/2)(\omega_t/\omega_n)F_e$ the expected result for passive ExB advection solely in the presence of the gradients. In these cases however we always have slightly $T < n$ but accounting for relatively more transport, even at the highest $\hat{\beta}$ considered.

The magnetic flutter transport yields an interesting result. For α_M less than critical, we find M_e is actually negative. This happens because the magnetic flutter transport is not a simple advective process in a gradient. The magnetic fluctuations do not have control of the dynamics, but are a passive result of the current fluctuations as part of the drift wave physics. The phase shifts among the state variables ϕ, n, and T are determined as a result of this physics, with J determined by the resulting phase shifts. The relationship between J and V_q is then a result of the chain of parallel dynamics ending in the resistive and thermal conductive dissipation sinks. Whatever happens between V_q and T is a passive result with no back reaction to the rest of the system. This produces the negative M_e due to the resulting phase shift between V_q and b^x. When the transition toward MHD occurs, the energetic effect of a finite M_e begins to compete with the drift wave physics, and positive values then self select when their effect overcomes the parallel dynamics chain in setting up the phase shifts. At higher $\hat{\beta}$ the magnetic flutter heat transport rises high enough to compete with and, for extreme values, overcome the ExB advection. Before this happens, the RMS density and temperature n and T rise to higher values than 10 percent of the background parameters, and the physical regime of stationary, saturated turbulence in the presence of at most weakly perturbed gradients is departed.

2.11 Summary

The addition of the electron temperature dynamics to the turbulence at first is as an enhancement of the parallel adiabatic response, as for the parallel dynamics the temperature fluctuation $\tilde{T}_e$ has a similar role to the nonadiabatic density fluctuation $\tilde{h}_e = \tilde{n}_e/n_e - e\tilde{\phi}/T_e$ in the kinetic shear Alfvén responses. Already at the inductivity represented by $\hat{\beta} = 1$, the stabilising influence of the magnetic fluctuation in the presence of the background gradient is felt. This is the fluctuating $\nabla_\parallel$ acting upon $\nabla \log p_e$ in the Ohm's law. Recall that in the isothermal case this stabilisation was marginal, essentially countering the destabilisation represented by the induction effect in the parallel electric field, in breaking up the static force balance tending to tie $\tilde{h}_e$ down to smaller values. This is the time derivative $\partial A/\partial t$ in the Ohm's law. With the temperature gradient present, the action of $\nabla_\parallel$ on $\nabla \log p_e$ is stronger and more stabilising. The addition of the heat flux fluctuation $\tilde{q}_{e\parallel}$ to the dynamics brings in thermal conduction dissipation, acting as an effective friction on $\tilde{q}_{e\parallel}$, not a diffusion on $\tilde{T}_e$, in a time dependent setting.

When the inductivity and therefore the electromagnetic responses are stronger, the additional energetics of the fluctuating heat flux, $\tilde{q}_{e\parallel}$ and the fluctuating radial magnetic disturbance, b^x, combine in the presence of the temperature gradient $\nabla \log T_e$ to form another significant drive. For electrostatic and weakly electromagnetic cases this effect is small and negative. Small because b^x is small. Negative because the phase correlation between $\tilde{q}_{e\parallel}$ and b^x is set by the dissipation chain tying $\tilde{J}_\parallel$ and $\tilde{q}_{e\parallel}$ to $\tilde{h}_e$ and $\tilde{T}_e$, and with the latter two following each other, the negative sign in Ampère's law makes the magnetic flutter flux $M_e = \tilde{q}_{e\parallel} b^x$ and hence its gradient drive net negative. Only when the magnetic fluctuations are larger, enough to break up this phase relation chain, does this turn around and the sign of M_e become positive and the energetic effect eventually become significant.

The nominal transition into the MHD regime through the ballooning stability parameter $\alpha_M = q^2 R_0 \, |\nabla \log p|$ occurs at the critical value $\alpha_M = 0.66\ S$. With actual collisionality, the turnaround into higher electromagnetic amplitude actually occurs at lower values than this, due to all the non-MHD additional physics. At higher collisionality the rise of the turbulence is less pronounced, since it is already much stronger on the electrostatic side, but the change in the energetics is still obvious, with M_e becoming positive and then at least concurrent to the ExB gradient drive due to the combined turbulent heat flux $Q_e = (3/2)\tilde{p}_e v_E^x$. By this time, however, the RMS amplitude of the turbulence is rather large for a scenario of saturated turbulence and transport in the presence of quasi-stationary gradients.

The time dependent model for the parallel heat flux is actually part of the physics, not an artificial construct to assist the numerics in a computation. The neglect of the ExB advective time derivative in favour of the collision frequency is what produces the standard Spitzer formula for thermal conduction, which is also used in the Braginskii two-temperature model (i.e., one each for electrons and ions). Even for fairly moderate collisionality in confined plasma cases, this is at best a marginal assumption. Breaking it and leaving the time derivative in the model, carrying $\tilde{q}_{e\|}$ along as one among the set of dependent variables, actually produces a better model. Of course, at lower collisionality still, the responses become problematic and one needs kinetic corrections. We will get these with the gyrofluid model later on, but before that we will consider the processes brought in by the ion temperature, in the next Chapter.

Further Reading

The standard model for collisional dissipation effects in a magnetised plasma is the review given by S I Braginskii, 'Transport processes in a plasma,' *Rev. Plasma Phys.* **1** (1965) 205. It builds on earlier work with thermal conductivity by Robert S Cohen and Lyman Spitzer Jr and Paul McR Routly, in 'The Electrical Conductivity of an Ionized Gas,' *Phys. Rev.* **80** (1950) 230; and Lyman Spitzer Jr and Richard Härm, in 'Transport phenomena in a completely ionized gas,' *Phys. Rev. B* **89** (1953) 977.

Long mean free path treatments for the electron heat flux were given by J F Luciani and P Mora and J Virmont, in 'Nonlocal heat transport due to steep temperature gradients,' *Phys. Rev. Lett.* **51** (1983) 1664, and J R Albritton and E A Williams and I B Bernstein and K P Swartz, in 'Nonlocal electron heat transport by not quite Maxwell–Boltzmann distributions,' *Phys. Rev. Lett.* **57** (1986) 1887, and Holstein (1987), E M Epperlein, in 'Kinetic theory of laser filamentation in plasmas,' *Phys. Rev. Lett.* **65** (1990) 2145. and in *Phys. Fluids B* **3** (1991) 3092.

The time dependent heat flux arises when the time scale of changes to the temperature fluctuations due to other means (e.g., turbulence) are fast enough that communication down the field line has a finite time delay. The fundamentals were given by A B Hassam, in 'Higher-order Chapman–Enskog theory for electrons,' *Phys. Fluids* **23** (1980) 38.

The fact that the curvature terms with diamagnetic fluxes collectively conserve free energy was shown by H Nordman and J Weiland, in 'Transport due to toroidal η_i mode turbulence in tokamaks,' *Nucl. Fusion* **29** (1989) 251.

Temperature effects in drift wave turbulence were studied by B Scott, in 'Self-sustained collisional drift-wave turbulence in a sheared magnetic field,' *Phys. Rev. Lett.* **65** (1990) 3289. The role of the temperature and the time-dependent heat flux response in drift Alfvén turbulence by B Scott, in 'Three-dimensional computation of drift Alfvén turbulence,' *Plasma Phys. Contr. Fusion* **39** (1997) 1635.

The result that the transport directly due to the disturbances of the magnetic field lines is negligible compared with that due to the ExB eddies was found by R E Waltz, in 'Numerical simulation of electromagnetic turbulence in tokamaks,' *Phys. Fluids* **28** (1985) 577. The main point is that drift wave phase shifts set the responses if the gradients are below the MHD threshold.

IOP Publishing

Turbulence and Instabilities in Magnetised Plasmas, Volume 2

Gyrokinetic theory and gyrofluid turbulence

Bruce Scott

Chapter 3

Effects of the ion temperature

3.1 Introduction—ion temperature as independent

In the work with the electron temperature from the previous Chapter, we neglected the effects of the ion temperature. We now turn to the ions. In contrast to the electron temperature disturbances, whose properties are the same as those of the nonadiabatic part of the density disturbances, the ion temperature is independent of the other state variables in the parallel dynamics. Alone among state variables at this level, it does not participate in Alfvén dynamics, and therefore not in the adiabatic response. There is little or no coupling to the other state variables through the parallel dynamics, just the density through the acoustic dynamics on that time scale, rather slower than the turbulence.

The ion temperature couples to the other state variables through interchange dynamics, but just as in the case of the electron temperature all the coupling constants are comparable and so the result for pure interchange dynamics is not qualitatively different from a two-variable model. The interesting effects occur when both magnetic curvature and parallel dynamics are present. Unlike in an MHD situation, the parallel dynamics couple the other variables to each other efficiently, leaving the ion temperature to do different things. As a dependent variable, it introduces new eigenmodes. Rather than the general pressure gradient, these have specifically the ion temperature gradient as their free energy source, and a coupling between the ion temperature disturbances and the ExB eddies as their defining transfer mechanism setting up their mode structure. These modes are called ITG modes in the literature, with ITG denoting 'ion temperature gradient driven' modes. It is possible to get these modes in a straight magnetic field in a slab layer geometry, but the real interest is when this occurs with magnetic curvature.

The classic realm of these ITG modes is in the case where the electrons are adiabatic, leaving the ion temperature gradient as the only remaining energy source, and the ion temperature disturbance as the only state variable not in a prescribed relationship to the electrostatic potential (the density is adiabatic, and the electron temperature is zero). As such, they are the dominant process in hot plasma cases maintained by thermal sources, so that both temperature gradients are equally steep, but with the electrons nearly adiabatic the ion temperature gradient is free to drive ITG turbulence. In terms of mode structure the linear and turbulent regimes are similar, so this has often been the type of low frequency drift turbulence that is the easiest to understand. However, even in the simple case it is helpful to recall how far from an MHD situation this ITG turbulence is in. With the plasma beta low enough to avoid the MHD stability boundary, we still find robust turbulence in a regime and of a type that does not exist in an MHD picture. The adiabatic response, in shutting down the electrons and with them the Alfvén response, opens up the ions such that the resulting dynamics is greatly stronger than anything else except for MHD ballooning, to which it is comparable.

In the case of turbulence driven by steep enough gradients to keep the electrons nonadiabatic, the mode structure contains both drift wave and ITG mode aspects. But before that we have to discuss the ion temperature as such more fundamentally, for this is the place where we can no longer avoid the issue of finite gyroradius effects, and the need to have these treated by the fluid model in a reasonable way. It has been abundantly clear that the drift scale ρ_s has to be treated in any of these turbulence models because, irrespective of any result in linear theory, that is where most of the vorticity is. The vorticity dynamics and its role in enabling parallel currents in the turbulence is the reason these scales enter, not because there is a new mode there. And when the ion and electron temperatures are comparable, because $\rho_i^2 = (T_i/T_e)\rho_s^2$, it follows that we not only have to face the existence of ρ_i we also have to resolve it in any reasonable computation which also contains electron dynamics. The fluid drift model is an expansion in finite but low $k_\perp\rho_i$, and the nonlinearities involved ultimately break down if, as we have seen, it is necessary to take $k_\perp\rho_i$ somewhat into the sub-gyroradius range in the spectrum.

To do this well, we will have to wait until the gyrofluid model is developed in Chapter 9. For now, we will present an energetic analysis which shows why the offending diamagnetic polarisation nonlinearity has to be kept, and then defer the result until its time. In the hot-plasma ITG section we will lay down a result which will then have to be modified, but the physics discussion will ultimately have to include the gyrofluid concepts.

3.2 The Larmor radius and gyro averaging

If dynamics at the drift scale, ρ_s, is active in drift wave turbulence, and the electron and ion temperatures are comparable, this means we also have dynamics at the scale of the ion Larmor radius, or gyroradius, ρ_i. These scales are given by

$$\rho_s^2 = \frac{M_i T_e}{e^2 B^2}, \qquad \rho_i^2 = \frac{M_i T_i}{e^2 B^2} \tag{3.1}$$

and are related by the parameter controlling the ion/electron temperature ratio,

$$\tau_i \equiv \frac{T_i}{T_e}, \qquad \rho_i = \tau_i^{1/2}\rho_s \tag{3.2}$$

We therefore have to face the physical effects of a finite gyroradius in the dynamics we are considering, and this actually puts pressure on the validity of the fluid model. This model formally breaks down if for perpendicular wavenumbers $k_\perp$ we have $k_\perp\rho_i \sim 1$.

The fluid model treats finite Larmor radius (FLR) effects by retaining the off diagonal components of the pressure tensor, whose elements which are independent of the collision frequency we called the gyroviscosity, noting that it is actually a diamagnetic momentum flux. The relationship between this effect and actual gyroaveraging of the vorticity dynamics is something we will address after developing the gyrofluid model, which emphasises the latter rather than the former. For now we look heuristically to a picture of the ExB velocity for a particle with a finite gyroradius, in a potential fluctuation ϕ of finite size in the perpendicular plane. A picture of this is given in figure 3.1. If ϕ has no significant structure at the level of ρ_i, the picture we draw of the dynamics leading to the ExB velocity contains smooth equipotential lines, and over the gyroradius the electric field is nearly constant. As the size of the gyroradius approaches the scale of the fluctuation, the variations of the perpendicular electric field over the trajectory of the particle become non-negligible. If we can do this perturbatively we introduce the effects scaling as $k_\perp^2\rho_i^2$ as corrections, so we have to require that the scale of motion $\Delta_\perp \gg \rho_i$. When the motion of the centre of this orbit is considered rather than the orbit of the particle itself, we expect to have a drift of the centre which can be modelled by using the average of the potential over the circle of the gyro-orbit in the formula for the ExB velocity. However, in doing this, we change the representation, and more than this orbit have to be taken into account. We will mean different things when we say 'density' or 'vorticity' in that case. Chapters 5 on the gyrokinetic theory through 9 on the gyrofluid model contain what we have to do in that case.

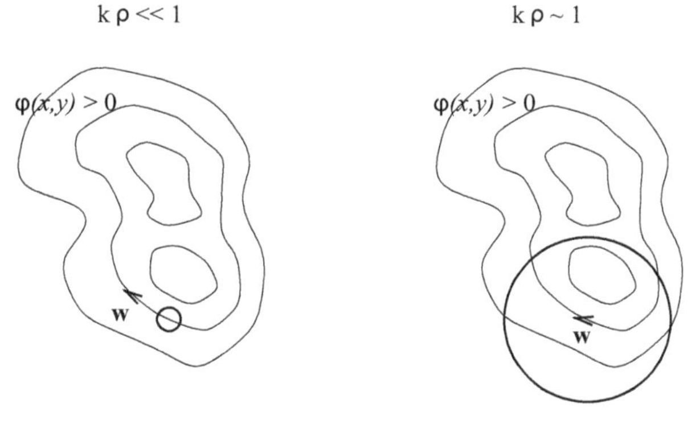

Figure 3.1. Gyroaveraging due to finite Larmor radius of an ion in a potential field. The circles give the size of the Larmor radius. The sense of the ExB drift for an ion is clockwise. If the Larmor radius is small (left), then the drift velocity of the gyrocentre will follow the ExB drift, instantaneously along a contour of $\tilde{\phi}$. If the Larmor radius is comparable to the size of the disturbance, then the particle will feel an average electric force corresponding to the various places along its gyro orbit. Its velocity will typically be slower than the nominal ExB velocity, and may be in a different direction.

3.3 Gyroaveraging versus gyroviscosity

In the fluid model, the effects of a finite Larmor radius are introduced by the diamagnetic fluxes. In the equation of motion whose perpendicular components define the drifts, the relevant diamagnetic flux is the momentum flux, commonly known as gyroviscosity. The differences between the ExB motion of the electrons, whose gyroradius is neglected, and the ions are now being treated in the non-linearities of the polarisation drift. The purely perpendicular part of this gyroviscosity contains three pieces: one which cancels the diamagnetic velocity $\mathbf{u}_*$ out of the advection, one which properly cancels divergences of the diamagnetic flow and heat flux, leaving only negligible magnetic divergence terms which are small by order ρ_i/L_B, and one which is a true finite gyroradius reduction of all the flows themselves. In a homogeneous magnetic field we can write these as

$$
\begin{aligned}
-(\nabla_\perp \cdot \mathbf{\Pi}^*)_\perp &= \mathbf{b} \times \nabla_\perp \frac{p_i}{\Omega_i} \cdot \nabla \mathbf{u}_\perp \\
&\quad - \mathbf{b} \times \nabla_\perp \left(\frac{p_i}{2\Omega_i} \nabla_\perp \cdot \mathbf{u}_\perp \right) + \nabla_\perp \left(\frac{p_i}{2\Omega_i} \nabla \cdot \mathbf{u}_\perp \times \mathbf{b} \right)
\end{aligned}
\tag{3.3}
$$

respectively, where $p_i/\Omega_i = n_i M_i (T_i/eB)$ signifies the action of p_i and gyroviscosity as an FLR effect, once the first term has cancelled $\mathbf{u}_* \cdot \nabla \mathbf{u}_\perp$. In a case where the gradient $\nabla \log T_i$ or the fluctuation $\tilde{T}_i$ is significant (relative to $\tilde{\phi}$ or $\tilde{n}_e$), then the gyroviscosity acquires extra terms arising from the diamagnetic heat flux, which is the same size as the diamagnetic velocity if the temperature is large fraction (or all of) the gradient or the fluctuation of the pressure (see Further reading; this is detailed in Chapter 9). The second term drops down to higher order in $k_\perp L_B$ as a result, where L_B is the scale of the magnetic field. The third term appears as a finite Larmor radius correction to the Lorentz force,

$$
\nabla_\perp \left(\frac{p_i}{2\Omega_i} \nabla \cdot \mathbf{u}_\perp \times \mathbf{b} \right) = \nabla_\perp \left[n_i e \frac{\rho_i^2}{2} \nabla_\perp \cdot \left(\frac{\mathbf{u} \times \mathbf{B}}{c} \right) \right]
\tag{3.4}
$$

However, taking the drift operation $-\nabla \cdot \mathbf{b}\times$ on Eq. (3.3) as we do to construct the fluid drift equations, we find that it is given by the perpendicular gradient of a scalar. This gyroreduction piece is annihilated by the drift divergence up to inhomogeneous magnetic field effects, leaving

$$
\begin{aligned}
&\nabla \cdot \mathbf{b} \times [(\nabla_\perp \cdot \mathbf{\Pi}^*)_\perp + n_i M_i \mathbf{u}_* \cdot \nabla \mathbf{u}_\perp] \\
&= \nabla_\perp^2 \left(\frac{p_i}{2\Omega_i} \nabla_\perp \cdot \mathbf{u}_\perp \right) + \nabla \cdot \mathbf{b} \times \nabla \left(\frac{p_i}{2\Omega_i} \nabla \cdot \mathbf{u}_\perp \times \mathbf{b} \right)
\end{aligned}
\tag{3.5}
$$

Both terms on the right side, representing contributions to the polarisation drift equation, are small by measure of ρ_i/L_B, and so as a finite-T_i gyroviscosity effect we only have the cancellation of $\mathbf{u}_* \cdot \nabla \mathbf{u}_\perp$, with no additional corrections.

This serves to highlight the fluid equations as an expansion in $k_\perp \rho_i$, which the gyroviscosity ultimately is: if we use the Chapman–Enskog procedure to derive the diamagnetic fluxes we find they come from the expansion of the kinetic equation to first order in $k_\perp^2\rho_i^2$. The true gyroaveraging corrections only come in at higher order. Attempts to continue the moment expansion approaches normally used, end in failure for the finite gyroradius effects. The expansion to first order usually breaks down for $k_\perp\rho_i \approx 1/2$. The expansion to second order will do better but still fail before $k_\perp\rho_i = 1$. Further effort must come from a different approach. Instead of deriving nonlinear fluid equations and then taking the low-frequency drift approximations, the kinetic theory itself is done under low-frequency drift approximations. The kinetic particle Lagrangian is transformed into a different one for what are now called gyrocentres. This makes the treatment more rigorous but it also changes the representation in fundamental ways. In short, some of the detail in the nonlinear polarisation is shifted from the particles to the fields, so that the gyrocentres which are left do not describe this motion, but the reaction of the fields to the particles does so. In the end, numerical treatments of the gyrofluid model are better behaved in the region $k_\perp\rho_i \sim 1$ than the fluid model. We defer this to Chapter 9 and later.

For now we will retain the fluid drift equations as a model which we simplify at high-$k_\perp$ for computability. We will only argue validity for the later model, along the way showing exact correspondence back to the fluid model up through $O(k_\perp^2\rho_i^2)$ corrections. It is necessary to introduce what the ion temperature dynamics will bring to the physics of the turbulence first, especially in hot plasmas where all of the parameters controlling the electron responses, namely the dynamical beta, the inertia, and the resistive dissipation, are small.

3.4 Effects on cold ion dynamics

As with the electron temperature, taking into account the ion temperature changes the ratio of specific heats for the ion fluid, and therefore the sound velocity changes. In contrast to a neutral fluid, the sound waves in a plasma can exist when the ions are cold, because the electrons to which they are coupled through the electric field still have a finite temperature, and this leads to the sound speed c_s. The ion temperature enters the parallel dynamics of the ions exactly as T_e does for the electrons as discussed in the previous Chapter, with the result that idealised sound waves with adiabatic electrons have a propagation velocity corresponding to the remaining available degrees of freedom. The effects of the two fluids add, because their pressures add in the total force on fluid elements, and so the propagation velocity is given by

$$\left(V_s^2\right)_{\text{adiabatic}} = \frac{T_e + (5/3)T_i}{M_i} \tag{3.6}$$

where the factor of 5/3 reflects the action of $\tilde{T}_i$ in the dynamics. The electrons add a factor of unity because they are isothermal due to the fact that they are adiabatic. Recall that adiabatic in the context of parallel dynamics means force balance, with $\tilde{p}_e \to n_e e\tilde{\phi}$ and $\tilde{T}_e \to 0$, which is different from the concept of adiabaticity in thermodynamics. Thermodynamic adiabaticity in this context is for an ideal fluid, in which dissipation mechanisms are small. In magnetised plasma dynamics, adiabatic electrons refers to quasistatic parallel force balance.

The temperature effects on ideal interchange dynamics are the same for the ions as for the electrons. If the scale of motion is larger than ρ_i, then the picture of the ideal interchange remains. As in the case of the electrons, we can simply sum up all the pieces of the total pressure disturbance,

$$\tilde{p} = \tilde{p}_e + \tilde{p}_i = (T_e + T_i)\tilde{n}_e + n_e(\tilde{T}_e + \tilde{T}_i) \tag{3.7}$$

assuming $\tilde{n}_i = \tilde{n}_e$ as well as $n_i = n_e$, and then work with a model using only the two variables $\tilde{p}$ and $\tilde{\phi}$. In the same way as for the disturbances, the pieces of the pressure gradients add,

$$\nabla p = \nabla p_e + \nabla p_i = (T_e + T_i)\nabla n_e + n_e(\nabla T_e + \nabla T_i) \tag{3.8}$$

Using a profile scale $L_\perp$, the normalised total pressure gradient is then

$$\frac{L_\perp}{L_p} = \omega_n + \omega_t + \tau_i(\omega_n + \omega_i) \tag{3.9}$$

with the three normalised gradients defined as

$$\omega_n = -L_\perp\nabla \log n_e, \qquad \omega_t = -L_\perp\nabla \log T_e, \qquad \omega_i = -L_\perp\nabla \log T_i \tag{3.10}$$

and with the ion /electron temperature ratio

$$\tau_i = \frac{T_i}{T_e} \tag{3.11}$$

All of these become parameters in local drift equations. With the ideal interchange model neglecting diamagnetic velocities and fluxes in the pressure equation, we have the same model as in a simple MHD case, as the density and temperature equations are isomorphic and can be combined in to a single pressure equation. The 2D interchange dynamics does not change appreciably, since we still have $\gamma_I^2 = c_s^2/L_B L_p$, just with the new pressure scale length, and the coupling coefficients among the curvature terms are all similar. This is due to the neglect of fluid compressibility in the simplest MHD model.

The ion temperature also has an indirect role in kinetic shear Alfvén dynamics, but only near $k_\perp\rho_i \sim 1$. This is due to the the divergence of the ion polarisation velocity appearing in the ion density and temperature equations. We recall that the interesting effects in parallel Alfvén responses enter when $k_\perp^2\rho_s^2$ becomes appreciable, since the reaction of the density becomes finite. The finite ion polarisation divergence, entering at first through ρ_s and not ρ_i, is a result of the adiabatic response of the electrons. Due to quasineutrality the ion and electron densities are the same; the effects on the Alfvénic activity follow equivalently from the divergence of the parallel current (electrons) and the divergence of the polarisation current (ions), which ultimately also couples the ion temperature to the parallel electron dynamics. Since $\tilde{T}_i$ enters the equation for $\tilde{\phi}$ scaling with ρ_i (factors of τ_i in normalised units), it adds true FLR corrections to the kinetic shear Alfvén dynamics. As is the case with the electron temperature, however, the qualitative nature of those effects is not different from those of the $\tilde{n}_e \leftrightarrow \tilde{\phi}$ coupling already present in the four-field model, and so the physics of the waves does not change (we will address the analysis within the kinetic theory).

The ion and electron pressures also combine in the standard ballooning mode theory. Adding all the pressure effects together, the ballooning parameter becomes

$$\alpha_M = q^2 R_0 \, |\nabla \beta| = \hat{\beta}\omega_B[\omega_n + \omega_t + \tau_i(\omega_n + \omega_i)] \tag{3.12}$$

in terms of the parameters we have been using. The nominal threshold for MHD ballooning instability is the same,

$$\alpha_M > 0.66\, S, \qquad \text{for instability} \tag{3.13}$$

but with all the gradients in the model, it is given with the total pressure gradient including both temperature gradients and the density gradient. For equal scale lengths and $\tau_i = 1$, this makes α_M four times larger for the same $\hat{\beta}$ as in the isothermal model. So if $\omega_B = 0.03$ and $S = 1$ the threshold for $\hat{\beta}$, which was 22 in the cold-ion isothermal model, drops to 11 in either the cold-ion model with $\nabla \log T_e$ or the $\tau_i = 1$ isothermal model, and with $\nabla \log T_i$ now down to 5.5 in the case of equal temperature and gradients. Hence, the MHD regime transition is a stronger issue in this case with both temperatures, and we will see its effect especially in the warm ion cases.

3.5 Ion temperature gradient (ITG) modes

The situation gets interesting when we consider not yet dissipation, but the fact that we have a new regime possible with the combination of the interchange dynamics and the parallel electron responses. Suppose that the plasma is hot enough to keep the electrons nearly adiabatic. In a cold ion model the interchange turbulence becomes drift wave turbulence in a toroidal magnetic field at still very high values of collisionality and then is reduced to low amplitude as the collision frequency goes to zero. The interchange effects on the turbulence are mostly absent since they are effectively shorted out by the adiabatic response which tends to reduce all of the effective potentials, upon which the curvature terms operate), to zero.

When the electrons become adiabatic, the ion density is also removed from the energetics because of quasineutrality, $Zn_i = n_e$ and $Z\tilde{n}_i = \tilde{n}_e$ for charge state Z. But now we have the ion temperature, and that means we now have a state variable with a background gradient acting as a free energy source, but which does not feel the adiabatic response if the electrons are adiabatic. The ion temperature can still participate in interchange dynamics even if the electrons are adiabatic, and the ion parallel dynamics might still be too slow to alter the results. The instability we find in this regime is called the ITG-interchange mode, or in a toroidal magnetic field the toroidal ITG mode, where ITG stands for ion temperature gradient since ∇T_i is now the only available free energy source.

The simplest form of the ITG model combines the two-dimensional adiabatic case of the drift wave model and the two-dimensional interchange model. The electrons are assumed to be kept adiabatic by an arbitrarily fast Alfvén dynamics, while the ions are assumed to remain hydrodynamic due to an arbitrarily slow sound wave dynamics, as in the classic drift wave ordering, $k_\parallel V_a \gg \omega \gg k_\parallel V_i$, with V_a given in Eq. (2.120) of the previous Chapter. The equations are those for the ion density and temperature, with the ExB flow disturbances determined by the adiabatic constraint. We may simplify to a singly charged fluid and use n_e for n_i, and hence $\tilde{\phi}$ for $\tilde{n}_i$. Keeping the polarisation drift in both equations, which we scale in units of equations for charge densities, we have

$$\frac{\partial}{\partial t}\left(n_e e\frac{e}{T_e}\tilde{\phi}-\varpi\right) + \frac{n_e e}{B}[\tilde{\phi}, \log n_e] = \mathcal{K}\left(n_e e\tilde{\phi} + \tilde{p}_i\right) \tag{3.14}$$

$$\frac{\partial}{\partial t}\left(\frac{3}{2}n_e e\frac{\tilde{T}_i}{T_i} - \varpi\right) + \frac{3}{2}\frac{n_e e}{B}[\tilde{\phi}, \log T_i] = \mathcal{K}\left(n_e e\tilde{\phi} + \tilde{p}_i + \frac{5}{2}n_e \tilde{T}_i\right) \tag{3.15}$$

in a linearised model, with the vorticity given by

$$\varpi = \frac{\rho}{B^2}\nabla_\perp^2\left(\tilde{\phi}+\frac{1}{n_e e}\tilde{p}_i\right) \tag{3.16}$$

with pressures

$$\tilde{p}_e = n_e e\tilde{\phi}, \qquad \tilde{p}_i = \frac{T_i}{T_e}n_e e\tilde{\phi} + n_e\tilde{T}_i \tag{3.17}$$

and explicit drive terms given by logarithmic gradients specified by a constant parameter times $-x/L_\perp$. The gradient terms from ϖ, which contains $\tilde{T}_i$, are neglected since they enter through third derivatives (small by two factors of $\rho_i/L_\perp$). All quantities except $\tilde{\phi}$ and $\tilde{T}_i$ and the gradients are constants.

These equations linearise into

$$\frac{\partial F_n}{\partial t} = -i\omega_* \phi + i\omega_B\omega_*\, W \tag{3.18}$$

$$\frac{\partial F_T}{\partial t} = -i\frac{3}{2}\eta_i\omega_*\phi + i\omega_B\omega_*\,(W + 2.5\tau_i\, T) \tag{3.19}$$

where the quantities advanced under the time derivative are

$$F_n = \phi + K^2 W, \qquad F_T = \frac{3}{2}T + K^2 W \tag{3.20}$$

the total ion stream function is

$$W = (\phi + \tau_i(\phi + T)) \tag{3.21}$$

and the wavenumber and gradient parameters are

$$K = k_y\rho_s, \qquad \omega_* = \omega_n K, \qquad \eta_i = \omega_i/\omega_n, \qquad \omega_B = 2L_n/R \tag{3.22}$$

with $\phi = e\tilde{\phi}/T_e$ and $T = \tilde{T}_i/T_i$, and scales normalised as before to $c_s/L_\perp$ and ρ_s. Factors of τ_i arise from ρ_i in terms of ρ_s, and ϕ and T reflecting T_e and T_i, respectively. The gradient drive parameters ω_n and ω_i are defined in Eq. (3.10). After F_n and F_T are advanced, they are solved simultaneously for W and then ϕ and T.

Interchange instabilities usually involve only the largest scales possible in a given domain, due to the fact that the purely hydrodynamic ideal interchange mode has the same growth rate for all k_y as $k_y \to 0$. But we recall the inhibition of large scale activity in $\tilde{\phi}$ in the adiabatic limit of the drift wave turbulence. The same result takes place here. The presence of a factor of order unity in addition to the perpendicular Laplacian in both of Eqs (3.18, 3.19) has this effect since all forcing terms scale with k_y, with the result of a growth rate proportional to k_y as $k_y \to 0$.

We investigate this briefly with two values of K. We set $\omega_n = \tau_i = 1$ and $\omega_B = 0.1$, and vary ω_i to find the instability. Alternatively we set $\omega_i = 1$ and vary ω_n, as either way the control parameter is $\eta_i = \omega_i/\omega_n$, the ratio, but ω_B provides differences. The resulting run of mode frequencies and growth rates with ω_i at $\omega_n = 1$ is shown figure 3.2. The first case is with $K = 0.3$, in the range at which we have been finding

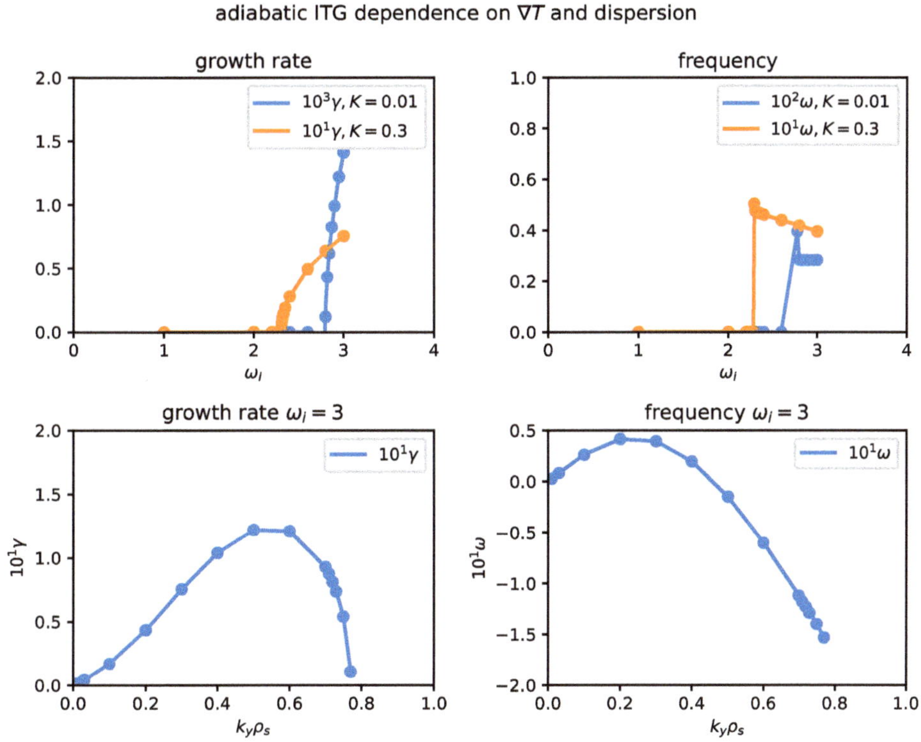

Figure 3.2. (top row) Adiabatic ITG-interchange growth rate and mode frequency as a function of $\eta_i = \omega_i/\omega_n$, here given by ω_i with $\omega_n = 1$, for normalised wavenumber $K = k_y\rho_s$ of 0.01 and 0.3, the latter of which is much more robust. The threshold is similar but sharper for higher K. (bottom) For $\omega_i = 3$, the run of growth rate and mode frequency with K, showing the robust region $K \sim 0.5$, and the threshold before $K = 1$ is reached. Results are shown only for $\gamma > 0$ as otherwise no eigenmode is captured.

the strongest free energy source rates in drift wave turbulence. The second, $K = 0.01$, represents the very largest scale available in a typical system. We find no instability whatsoever for η_i below about 2. For $\omega_n = 1$ the threshold for instability is $\omega_i = 2.29$ for $K = 0.3$ and $\omega_i = 2.8$ for $K = 0.01$. For $\omega_i = 1$ the threshold for stability is $\omega_n = 0.58$ for $K = 0.3$ and $\omega_n = 0.56$ for $K = 0.01$, with lower values of ω_n yielding growth. For $K = 1$ even $\omega_n = 0.1$ and higher are stable. The reason for the apparent differences between holding ω_i or ω_n fixed is the role of the curvature, which we leave fixed. The ExB compression terms in the density and temperature equations have effect and are retained for consistency.

This is a situation akin to buoyant convection in a neutral fluid; due to subtle details in the energetics there is no free energy gain for exchanging fluid elements until a threshold is reached. The threshold is not dissipative, with γ smoothly going through zero, but it is a case where ω^2 changes from real to complex values. Unlike a pure oscillation or growth problem where ω^2 is always real, the gradient drive terms yield complex ω in general. Eqs (3.18, 3.19) may be Fourier transformed in time and solved for the eigenmodes, and this yields a quadratic equation for ω. The threshold is reached when the discriminant of that equation becomes negative. This always has some manifestation in the energetics, and in this case the threshold results from the adiabatic constraint on the electrons, which is what makes this situation different from the ideal interchange instability. The adiabatic constraint results in an initial disturbance being shifted away in the y-direction due to the diamagnetic drift, before it can liberate sufficient energy from ∇T_i. Here, we note that for the adiabatic ITG-interchange mode ∇T_i is the sole free energy source, so there is a competition between a radial flow tapping the free energy, and its tendency to drift away from the $\tilde{T}_i$ disturbances it causes before they can grow. For $\eta_i = 1$ the adiabatic ITG-interchange mode is nearly always stable.

We can see the effect of the adiabatic constraint on the threshold by relaxing it, and taking the case of a general situation containing both drift wave and interchange dynamics, plus the parallel ion dynamics. Doing the same thing as above, except dropping the adiabaticity constraint, we go back to the 2D dissipative coupling model for the electrons. Instead of the ion density equation we use the electron one. Recall (cf. Section 7.8, Volume One) that we can use either one since they are equal for one ion and the +1 charge state. Now we do the same for the ion temperature, by adding the vorticity equation to not only the ion density but also the ion temperature equations. This cancels the ion compression terms and replaces them with the electron ones, except for the extra factor $(5/2)\tilde{T}_i$ in the ion diamagnetic heat flux.

In physical units this yields

$$\frac{\partial \varpi}{\partial t} = \frac{1}{T_e} D_{\|}(n_e e\tilde{\phi} - T_e\tilde{n}_e) - \mathcal{K}(\tilde{p}_e + \tilde{p}_i) \tag{3.23}$$

$$e\frac{\partial \tilde{n}_e}{\partial t} + \frac{n_e e}{B}[\tilde{\phi}, \log n_e] = \frac{1}{T_e} D_{\|}(n_e e\tilde{\phi} - T_e\tilde{n}_e) + \mathcal{K}\left(n_e e\tilde{\phi} - \tilde{p}_e\right) \tag{3.24}$$

$$\frac{3}{2}\frac{n_e e}{T_i}\frac{\partial \tilde{T}_i}{\partial t} + \frac{3}{2}\frac{n_e e}{B}[\tilde{\phi}, \log T_i] = \frac{1}{T_e}D_{\|}(n_e e\tilde{\phi} - T_e\tilde{n}_e) + \mathcal{K}\left(n_e e\tilde{\phi} - \tilde{p}_e + \frac{5}{2}n_e\tilde{T}_i\right) \tag{3.25}$$

with vorticity and linearised pressures given by

$$\varpi = \frac{\rho}{B^2}\nabla_\perp^2\left(\tilde{\phi} + \frac{1}{n_e e}\tilde{p}_i\right), \quad \tilde{p}_e = T_e\tilde{n}_e, \quad \tilde{p}_i = T_i\tilde{n}_e + n_e\tilde{T}_i \tag{3.26}$$

All of this is with the same usage as above. These equations linearise into

$$-K^2\frac{\partial W}{\partial t} = D(\phi - n) - i\omega_B\omega_*(n + \tau_i(n + T)) \tag{3.27}$$

$$\frac{\partial n}{\partial t} = D(\phi - n) - i\omega_*\phi + i\omega_B\omega_*(\phi - n) \tag{3.28}$$

$$\frac{3}{2}\frac{\partial T}{\partial t} = D(\phi - n) - i\frac{3}{2}\eta_i\omega_*\phi + i\omega_B\omega_*(\phi - n + 2.5\tau_i\, T) \tag{3.29}$$

The energetics is $\Gamma_T = \Gamma_n + \Gamma_i - \Gamma_c$, with Γ_i arising from $\tau_i T[\phi, (3/2)T_0]$ in normalised units with $T_0 = -\omega_i x$. Since this is a linear problem we can choose the timestep τ as low as needed for a given K and D, requiring $\tau\ D/K^2 < 0.2$ for the Courant limit for this scheme.

Now let us examine the same cases we did above, and sweep the dissipative coupling parameter D for $K = 0.1$ and $\omega_n = \omega_i = 1$, which is stable in the adiabatic case. A timestep $\tau = 10^{-4}$ covers us well up to $D = 10$. The case with $D = 10$ shows oscillatory behaviour in both squared amplitudes and the real part of the frequency. The case with $D = 1$ is still stable. But $D = 0.1$ shows robust growth with both drive rates settling to final values after abut $t = 150$. Averaged over the final 1/3 of the interval up to $t = 250$, they are $\Gamma_n = 0.018 \pm 3.4 \times 10^{-5}$ for the density gradient and $\Gamma_t = 0.011 \pm 2.1 \times 10^{-4}$ for the temperature. Already at this point the density is the stronger drive, while in the adiabatic case Γ_n vanishes by construction. The energetic error is $\Gamma_E = 5 \times 10^{-7} \pm 10^{-8}$, demonstrating energetic consistency of the equations and the computation.

One further point of note is that there is also a version of this ITG instability in slab geometry, in terms of which the ITG mode was first discovered. It is a combination of an adiabatic sound wave and the ion temperature gradient and disturbance effects. Assuming an ideal fluid at large scales, we neglect the polarisation drift entirely, so that

$$\frac{\partial\tilde{\phi}}{\partial t} = -i\omega_n k_y\tilde{\phi} - ik_{\|}\tilde{u}_{\|} \tag{3.30}$$

$$\hat{\varepsilon}\frac{\partial\tilde{u}_{\|}}{\partial t} = -ik_{\|}[(1 + \tau_i)\tilde{\phi} + \tilde{T}_i] \tag{3.31}$$

$$\frac{\partial \tilde{T}_i}{\partial t} = -i[(1 + \tau_i)\omega_n + \tau_i\omega_i]k_y\tilde{\phi} - \frac{2}{3}ik_\parallel\tilde{u}_\parallel \tag{3.32}$$

Taking the Fourier transform in time as well, we can manipulate this into the following cubic equation for ω,

$$\left[\omega - \omega_n k_y\right]\left[\hat{\epsilon}\omega^2 - \left(1 + \frac{5}{3}\tau_i\right)k_\parallel^2\right] = [1 + \tau_i(1 + \eta_i)]\omega_n k_y k_\parallel^2 \tag{3.33}$$

The two quantities in brackets on the left side represent the adiabatic drift wave and sound waves in the low-k_y limit, with a coupling term on the right side. Due to this coupling term there is a threshold in η_i, before this term can compete with the ω-independent term on the left side, enough to force two of the roots into the upper and lower complex plane, off the real axis. This gives the slab branch of the ITG mode, unstable for $\eta_i = 2/3$ for three degrees of freedom (isotropic temperature). For one degree of freedom (a kinetic situation with a parallel temperature disturbance only), the threshold is at $\eta_i = 2$.

3.6 Warm-ion toroidal drift Alfvén model

Before writing down the model equations, we briefly note that besides the finite Larmor radius effects, we also have the same problem with the parallel heat flux as for the electrons. We treat this with the same sort of time dependent heat flux as for the electrons, and compare with the Braginskii form in the results. This is done the same way as for the electrons, except that there is no thermoelectric effect. In general, there are thermal forces among the ions when more than one species is present ($\nabla_\parallel T$ due to other species in any one's parallel velocity equation), but not in a single component plasma. We therefore have

$$\frac{m_i}{e}\frac{d}{dt}\frac{\tilde{q}_{i\parallel}}{p_i} + \frac{5}{2}\frac{T_i}{e}\nabla_\parallel\left(\frac{\tilde{T}_i}{T_i}\right) = -\frac{m_i}{e}\frac{5/2}{\kappa_i}\nu_i\left(\frac{\tilde{q}_{i\parallel}}{p_i}\right) \tag{3.34}$$

in physical units, with coefficient $\kappa_i = 3.9$ for ions, written with the drive term due to $\nabla_\parallel \log T_i$ understood as part of $\nabla_\parallel\tilde{T}_i$, analogously to Eq. (2.46) from Chapter 2. In this case, since the parallel ion dynamics has frequencies on the acoustic scale ($V_i = \tau_i^{1/2}c_s$), the nonlinear advective ExB derivative is the largest term in the equation by at least one order of magnitude and sometimes two, depending on the disparity of the $K_\parallel L_\perp$ scale ratio. The ions are never in the collisional regime by time in this type of turbulence.

3.6.1 Equations

Putting the ion temperature/heat flux combination into the 3D toroidal model from the previous Chapter, there are now eight dependent variables. Due to the notation problem with T_e and T_i being background parameters and what we would use as well for normalised dependent variables, we leave these in physical units. Curvature and parallel dynamics terms go in according to the same procedure we just did for the cold ion model in the previous Chapter (the systematic derivation is given in Chaper 7 of Volume One).

The equations are

$$\nabla \cdot \frac{\rho}{B^2}\frac{d}{dt}\nabla_\perp \tilde{W}_i = B\nabla_\parallel \frac{\tilde{J}_\parallel}{B} - \mathcal{K}(\tilde{p}_e + \tilde{p}_i) \tag{3.35}$$

$$n_e e\frac{\partial \tilde{A}_\parallel}{\partial t} + \frac{m_e}{e}\frac{d\tilde{J}_\parallel}{dt} = \nabla_\parallel\left(\tilde{p}_e - n_e e\tilde{\phi}\right) - \frac{m_e \nu_e}{e}\left(0.51\tilde{J}_\parallel + \frac{\alpha_e}{\kappa_e T_e}\left(e\tilde{q}_{e\parallel} + \alpha_e T_e \tilde{J}_\parallel\right)\right) \tag{3.36}$$

$$e\frac{d\tilde{n}}{dt} = B\nabla_\parallel \frac{\tilde{J}_\parallel - n_e e\tilde{u}_\parallel}{B} + \mathcal{K}\left(n_e e\tilde{\phi} - \tilde{p}_e\right) \tag{3.37}$$

$$n_e m_i \frac{du}{dt} = -\nabla_\parallel(\tilde{p}_e + \tilde{p}_i) - B\nabla_\parallel \frac{P}{B} \tag{3.38}$$

$$\frac{3}{2}n_e e\frac{d\tilde{T}_e}{dt} = B\nabla_\parallel \frac{T_e\tilde{J}_\parallel - p_e e\tilde{u}_\parallel - e\tilde{q}_{e\parallel}}{B} + \mathcal{K}\left(n_e e\tilde{\phi} - \tilde{p}_e - 2.5n\tilde{T}_e\right) \tag{3.39}$$

$$m_e \frac{d\tilde{q}_{e\parallel}}{dt} = -\frac{5}{2}p_e \nabla_\parallel \tilde{T}_e - \frac{5/2}{\kappa_e e} m_e \nu_e \left(e\tilde{q}_{e\parallel} + \alpha_e T_e \tilde{J}_\parallel\right) \tag{3.40}$$

$$n_e e\frac{d\tilde{T}_i}{dt} = B\nabla_\parallel \frac{T_i\tilde{J}_\parallel - p_i e\tilde{u}_\parallel - e\tilde{q}_{i\parallel}}{B} + \mathcal{K}\left(n_e e\tilde{\phi} - \tilde{p}_e + 2.5n\tilde{T}_i\right) \tag{3.41}$$

$$m_i \frac{d\tilde{q}_{i\parallel}}{dt} = -\frac{5}{2}p_i \nabla_\parallel \tilde{T}_i - \frac{5/2}{\kappa_i} m_i \nu_i \tilde{q}_{i\parallel} \tag{3.42}$$

with auxiliaries

$$\frac{\rho}{B^2}\nabla_\perp^2 \tilde{W}_i = \varpi, \qquad \tilde{W}_i = \tilde{\phi} + \frac{1}{n_e e}\tilde{p}_i, \qquad -\nabla_\perp^2 \tilde{A}_\parallel = \mu_0 \tilde{J}_\parallel \tag{3.43}$$

$$\frac{d}{dt} = \frac{\partial}{\partial t} + [\tilde{\phi}, \,], \qquad \nabla_\parallel = b^s \frac{\partial}{\partial s} + [-\tilde{A}_\parallel, \,] \tag{3.44}$$

$$\tilde{p}_e = T_e \tilde{n}_e + n_e \tilde{T}_e \qquad \tilde{p}_i = T_i \tilde{n}_e + n_e \tilde{T}_i, \qquad P = -\mu_\parallel \nabla_\parallel \tilde{u}_\parallel \tag{3.45}$$

The factors ρ, $n_e = n_i$, T_e, and T_i along with the masses and charge $e = e_0$ are constant parameters. Gradients of $\tilde{n}_e$, $\tilde{T}_e$, and $\tilde{T}_i$ under d/dt or $\nabla_\parallel$ include drive terms according to

$$\frac{\partial \tilde{n}_e}{\partial x} \to \frac{\partial \tilde{n}_e}{\partial x} - n_e \omega_n, \qquad \frac{\partial \tilde{T}_e}{\partial x} \to \frac{\partial \tilde{T}_e}{\partial x} - T_e \omega_t, \qquad \frac{\partial \tilde{T}_i}{\partial x} \to \frac{\partial \tilde{T}_i}{\partial x} - T_i \omega_i \tag{3.46}$$

with the drive parameters ω_n, ω_t, ω_i defined in Eq. (3.10). The coordinates and flux surface geometry ($\{xy\}$ and s, spatial Poisson bracket, and $\nabla_\perp^2$, b^s, B^2, $\mathcal{K}$, and the metric) and computational scheme and domain remain as before. The factors of ν_e and ν_i in the equations are the Braginskii collision frequencies as before.

One thing deserves further comment: the treatment of the gyroviscosity. The diamagnetic cancellation results in the total ion drift velocity being advected by the ExB velocity. Among other things, this breaks vorticity conservation. Energy conservation is preserved with the total inertial drift energy proportional to $|\nabla_\perp \tilde{W}_i|^2$. In the numerical treatment we can no longer simply advect the vorticity as another state variable using an ExB characteristic equation. While this is part of the treatment we use, it has to be corrected to give the entire term on the left side of Eq. (3.35). This term is evaluated as

$$\nabla \cdot \frac{\rho}{B^2}\frac{d}{dt}\nabla_\perp \tilde{W}_i = \frac{d\varpi}{dt} + \frac{\rho}{n_e e B^2}[\nabla_\perp \tilde{\phi}; \nabla_\perp \tilde{p}_i] \tag{3.47}$$

The last term on the right side is called the gyroviscous correction. The semicolon indicates contraction between the $\nabla_\perp$ operators within the spatial Poisson bracket. For the cold ion models this correction vanishes because for $T_i = 0$ in the background we have $\tilde{W}_i = \tilde{\phi}$, and then the combination of derivatives vanishes by antisymmetry. So in the correction we replace $\tilde{W}_i$ with $(1/n_e e)\tilde{p}_i$, applying the gyroviscous correction as a forcing term. We note that each factor of $\tilde{\phi}$ and $\tilde{p}_i$ in this equation appears under two derivatives, and so there are no gradient drive terms

3.6.2 Energetics

The energetics of the model is now expanded to include the ion temperature. All of the electron based source, transfer, and sink mechanisms are still present, and we wrote them in the previous Chapter. The electron collisional sinks generally scale with $m_e\nu_e$, and the ion sinks scale with $m_i\nu_i$. The part of the energetics added due to the ion temperature is

$$\frac{\rho}{B^2}\frac{1}{2}\frac{\partial}{\partial t}|\nabla_\perp \tilde{W}_i|^2 - \nabla \cdot (\cdots) = \cdots + \frac{1}{n_e e}\tilde{J}_\parallel \nabla_\parallel \tilde{p}_i - \tilde{p}_i \mathcal{K}(\tilde{\phi}) + \frac{1}{n_e e}\tilde{p}_i \mathcal{K}(\tilde{p}_e) \tag{3.48}$$

$$\begin{aligned}\frac{T_e + T_i}{n_e}\frac{1}{2}\frac{\partial}{\partial t}(\tilde{n}_e)^2 = \cdots &- \frac{T_i}{n_e e}\tilde{J}_\parallel \nabla_\parallel \tilde{n}_e + T_i \tilde{u}_\parallel \nabla_\parallel \tilde{n}_e \\ &+ T_i \tilde{n}_e \mathcal{K}(\tilde{\phi}) - \frac{T_i}{n_e}\tilde{n}_e \mathcal{K}(\tilde{p}_e) + T_i \omega_n \tilde{n}_e \tilde{v}_E^x\end{aligned} \tag{3.49}$$

$$n_i m_i \frac{1}{2}\frac{\partial}{\partial t}(\tilde{u}_\parallel)^2 = \cdots - \tilde{u}_\parallel \nabla_\parallel \tilde{p}_i + (\omega_n + \omega_i)\, p_i \tilde{u}_\parallel \tilde{B}^x \tag{3.50}$$

$$\frac{3}{2}\frac{n_i}{T_i}\frac{1}{2}\frac{\partial}{\partial t}(\tilde{T}_i)^2 = -\frac{1}{e}\tilde{J}_{\|}\nabla_{\|}\tilde{T}_i + \frac{1}{T_i}\tilde{q}_{i\|}\nabla_{\|}\tilde{T}_i + n_i\tilde{u}_{\|}\nabla_{\|}\tilde{T}_i$$
$$+ n_e\tilde{T}_i\mathcal{K}(\tilde{\phi}) - \tilde{T}_i\mathcal{K}(\tilde{p}_e) + \frac{3}{2}n_i\omega_n\,\tilde{T}_i\tilde{v}_E^x \tag{3.51}$$

$$\frac{2}{5}\frac{n_i m_i}{p_i^2}\frac{1}{2}\frac{\partial}{\partial t}(\tilde{q}_{i\|})^2 = -\frac{1}{T_i}\tilde{q}_{i\|}\nabla_{\|}\tilde{T}_i + \omega_i\,\tilde{q}_{i\|}\tilde{B}^x - \frac{n_i m_i}{\kappa_i p_i^2}\nu_i(\tilde{q}_{i\|})^2 \tag{3.52}$$

Factors of $T_i\tilde{n}_e$ and $n_e\tilde{T}_i$ combine into the $\tilde{p}_i$ terms that transfer energy between $\tilde{u}_{\|}$ and $\tilde{p}_i$ through sound wave channels, and between $\tilde{W}_i$ and $\tilde{p}_i$ through the ExB and polarisation compression. These effects are relatively negligible for drift wave turbulence. But the ion sources represented by ω_n and ω_i operating on $\tilde{n}_e\tilde{v}_E^x$ and $\tilde{T}_i\tilde{v}_E^x$ are significant, with $p_i\tilde{u}_{\|}\tilde{B}^x$ and $\tilde{q}_{i\|}\tilde{B}^x$ less so. The transfer channel over the curvature on $n_i\tilde{T}_i\mathcal{K}(\tilde{\phi})$ is important as the mechanism for the ITG-interchange drive in transferring the ion temperature free energy into ExB turbulence. The transfer terms involving $\tilde{J}_{\|}$ and $\nabla_{\|}\tilde{p}_i$ represent transfer between $\tilde{\phi}$ and $\tilde{p}_i$ through polarisation compression. In slab geometry this compression can become significant, but in an inhomogeneous magnetic field it is overshadowed by the interchange coupling. The ion dissipation channel involving $\tilde{q}_{i\|}$ and $\tilde{T}_i$, namely thermal conduction, scales similarly to the sound waves.

Overall, the important source effects are the ExB gradient drive given by

$$\Gamma_n = (T_e + T_i)\,\omega_n\mathcal{V}^{-1}\int d\mathcal{V}\,\tilde{n}_e\tilde{v}_E^x \tag{3.53}$$

for the electron and ion densities, and

$$\Gamma_t = \frac{3}{2}n_e\,\omega_t\mathcal{V}^{-1}\int d\mathcal{V}\,\tilde{T}_e\tilde{v}_E^x, \qquad \Gamma_i = \frac{3}{2}n_e\,\omega_i\mathcal{V}^{-1}\int d\mathcal{V}\,\tilde{T}_i\tilde{v}_E^x \tag{3.54}$$

for the electron and ion temperatures. We also have the magnetic flutter gradient drive, whose significant contributions are

$$\Gamma_v = \mathcal{V}^{-1}\int d\mathcal{V}\,(\omega_t\tilde{q}_{e\|} + \omega_i\tilde{q}_{i\|})\tilde{B}^x \tag{3.55}$$

mostly the electrons, coming from heat flow down the gradient. The contributions due to $\tilde{J}_{\|}$ and $\tilde{u}_{\|}$ are negligible (as in Eq. 2.109) although both are kept to evaluate the energetic error. In these expressions,

$$\tilde{v}_E^x = [\tilde{\phi}, x], \qquad \tilde{B}^x = [-\tilde{A}_{\|}, x] \tag{3.56}$$

are the ExB flow and the magnetic flutter component in the radial (x) direction, down the gradient.

The important sinks are all in the electron parallel dynamics,

$$\Gamma_c = \mathcal{V}^{-1}\int d\mathcal{V}\,\eta_{\|}\big(\tilde{J}_{\|}\big)^2, \qquad \Gamma_k = \mathcal{V}^{-1}\int d\mathcal{V}\,\frac{m_e\nu_e}{\kappa_e}\left(\tilde{q}_{e\|} + \alpha_e\frac{T_e}{e}\tilde{J}_{\|}\right)^2 \tag{3.57}$$

giving resistive and electron thermal conductive dissipation as before, plus the ion channels, collectively

$$\Gamma_s = \mathcal{V}^{-1}\int d\mathcal{V}\,\frac{n_i m_i}{\kappa_i p_i^2}\nu_i(\tilde{q}_{i\|})^2 + \mathcal{V}^{-1}\int d\mathcal{V}\,\mu_\|\left(\nabla_\|\tilde{u}_\|\right)^2 \tag{3.58}$$

and also in the ExB direct cascades in $\tilde{n}_e$, $\tilde{T}_e$, and $\tilde{T}_i$, absorbed by subgrid dissipation, which has the same form as used before, with the operator acting on whichever quantity is ExB advected in each equation.

The normalised ExB particle and electron and ion conductive heat transport fluxes are given by

$$F_e = \frac{1}{n_e c_s}\,\tilde{n}_e\tilde{v}_E^x, \qquad G_e = \frac{3/2}{T_e c_s}\,\tilde{T}_e\tilde{v}_E^x, \qquad G_i = \frac{3/2}{T_i c_s}\,\tilde{T}_i\tilde{v}_E^x \tag{3.59}$$

and the magnetic flutter conductive heat fluxes are given by

$$M_e = \frac{1}{p_e c_s}\,\tilde{q}_{e\|}\tilde{B}^x, \qquad M_i = \frac{1}{p_i c_s}\,\tilde{q}_{i\|}\tilde{B}^x \tag{3.60}$$

all in terms of the volume average. The total ion and electron heat transport channels are then

$$Q_e + M_e = p_e c_s\left(\frac{3}{2}F_e + G_e + M_e\right), \qquad Q_i + M_i = p_i c_s\left(\frac{3}{2}F_e + G_i + M_i\right) \tag{3.61}$$

in physical units, noting the factor of τ_i to get the latter. The ExB advective heat fluxes are the $Q_{e,i}$ as combinations of the density and temperature pieces. The temperature dependence of the normalising factor is $T_{e,\,i}^{3/2}$ for the given species. This should be remembered when interpreting the scaling in any physical situation.

3.7 Electromagnetic ITG turbulence in a hot plasma

In a hot plasma the collisionality (C) is always very low. We will consider $\nu_e L_\perp/c_s = 0.1$, low enough that the parallel dissipation is very weak while avoiding pathologies incurred by setting it to zero without any kinetic dissipation mechanisms considered (we leave these to the gyrofluid studies in later Chapters). The definition of the hot plasma ITG regime is that all three of the parameters controlling the electrons,

$$\hat{\mu} \ll 1, \qquad \hat{\beta} \ll 1, \qquad \nu_e L_\perp/c_s \ll 1 \tag{3.62}$$

are small, so that the parallel responses are strong enough to keep the non-adiabaticity in the electrons at minimal levels. Note that the first of these is a scale ratio limitation, meaning essentially that

$$K_\| L_\perp \gg (m_e/m_i)^{1/2} \tag{3.63}$$

where $K_{\|}$ is the fundamental wavenumber of the field line connection. In a tokamak $K_{\|} = 1/qR_0$ is a function of the flux surface, and the perpendicular scale length is the same as the temperature scale length. With deuterium ions this means that

$$qR_0/L_T \ll 60 \tag{3.64}$$

We usually have values in the vicinity of $1 < q < 2$ and $R_0/L_T \sim 10$, so this requirement is satisfied. Taking $\hat{\mu} = 0.1$, we can expect $q \sim 2$ and $R_0/L_T \sim 10$, which are reasonable values about two thirds of the way out radially from the magnetic axis to the edge (LCFS: last closed flux surface). If we take $\hat{\beta} = 0.1$ as well, we are assuming that $\beta_e = m_e/m_i$, which is rather low, so there is a lot of leeway. We will sweep $\hat{\beta}$ between 0.03 and 1.0, and we will see that this range suffices to illustrate the conventional ITG mode structure at the low end and the transition into MHD ballooning mode structure at the high end. Furthermore, the ITG regime requires $\omega_i > \omega_n$, or $L_T < L_n$. In a tokamak in this region, a value of $\omega_n = 0.3$ is reasonable with $\omega_t = \omega_i = 1$, and $\tau_i = 1$ is also reasonable. The only parameters to vary with this choice of scale ratios are $\hat{\beta}$ and $L_T/L_\perp$. The first does the ITG/MHD transition and the second will find the ITG threshold. Reasonable choices for shear range between $S = 0.3$ which is fairly low to 1 which is fairly high. We will take $S = 0.5$. This choice of scale ratios also sets $\omega_B = 2L_\perp/R_0 = 0.2$ and $\hat{\varepsilon} = 367$. The nominal MHD boundary $\alpha_M = 0.66\,S$, with α_M given in Eq. (3.12), is then reached at $\hat{\beta} = 0.63$ and we will see the mode structure make the transition in this vicinity. This parameter set roughly corresponds to

$$R_0 = 1.65\ \text{m}, \quad L_\perp = 16.5\ \text{cm}, \quad q = 1.9, \quad S = 0.5 \tag{3.65}$$

$$n_e = 4 \times 10^{19}\ \text{m}^{-3}, \quad T_e = T_i = 800\ \text{eV}, \quad B_0 = 2.5\ \text{T} \tag{3.66}$$

as a physical example scenario. The domain size $L_x = L_\perp$ is about $100\rho_s$ so the parameter $\delta_0 = 0.01$ is set.

The run of amplitudes among the state variables with inductivity ($\hat{\beta}$) are measured as in the previous Chapter (now including $\tilde{T}_i$) and are shown in the left frame of figure 3.3. These are the RMS values for $e\tilde{\phi}/T_e$ and $\tilde{n}_e/n_e$ and $\tilde{T}_e/T_e$ and $\tilde{T}_e/T_i$

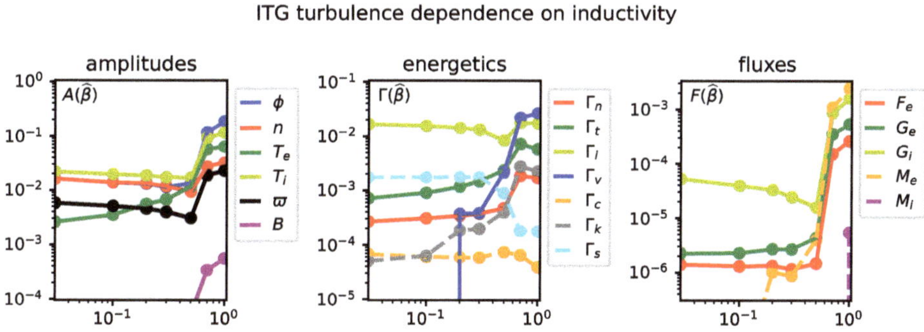

Figure 3.3. Dependence of the turbulence energetic and flux channels upon inductivity, $\hat{\beta}$, for the hot plasma ITG cases. As $\hat{\beta}$ increases the nominal MHD ballooning threshold is reached at the critical α_M, at about $\hat{\beta} = 0.63$ for these cases.

and $\varpi / n_e e$, and the magnetic field $B_\perp = B_0^{-1} \left| \nabla_\perp \tilde{A}_\| \right|$, respectively. The volume average is measured at each time point, then the time average with 51 samples over $500 < t < 1000$ and then the square root are taken. Before the regime transition, up to $\hat{\beta} = 0.5$, there is little dependence upon $\hat{\beta}$ except for slight rise in $\tilde{T}_e$, slight fall in $\tilde{n}_e$ and ϖ, and a linear rise in $B_\perp$ as a passive quantity. The MHD regime transition starts at $\hat{\beta} = 0.63$ whereupon all quantities rise sharply into a different structure whose main feature in the amplitudes is the relative size of $\tilde{\phi}$ compared with all the others. The magnetic fluctuations become very strong by this point as well.

The energetic processes in the turbulence that are measured through the Γ contributions given in Eqs (3.53–3.58) are shown in the centre frame of figure 3.3. The drive terms are done with solid lines and the sink terms are done with dashed lines. The ITG drive is the strongest for all cases until the MHD regime is reached, when the magnetic flutter drive is the largest. Almost all of this latter is due to the electrons, with the ion contribution several orders of magnitude smaller (see the fluxes, next). Of interest is that the ITG drive falls lightly with $\hat{\beta}$, while the electron channels are all rising. Magnetic induction is clearly an accelerator for the electron dynamics in this type of turbulence. The magnetic flutter drive rises the fastest, followed by the thermal conduction sink, and then the other electron channels. A halfway regime is noted between the deep ITG cases at low $\hat{\beta}$ and the actual MHD transition for $\hat{\beta} > 0.5$, indicated by the positive magnetic flutter drive most clearly for the $\hat{\beta} = 0.2$, 0.3, and 0.5 cases. We will see more of this in the mode structure.

The run of the various turbulent transport fluxes, given in Eqs (3.59, 3.60), are shown in the right frame of figure 3.3. The ExB advective fluxes, F_e and G_e and G_i are done with solid lines, and the magnetic flutter heat fluxes, M_e and M_i, are done with dashed lines. Note that M_i is negligible for all cases except the last, and even there it is small compared with the others. The fall in the ITG transport flux and the weak rise in the electron thermal flux follow the energetics. Of special interest is the way that the electron magnetic flutter heat flux begins rising and finds a plateau well before the onset of the MHD regime, nominally at $\hat{\beta} = 0.63$. We will see its effect on the mode structure in the $\hat{\beta} = 0.5$ case, which is where M_e overcomes G_e as the dominant electron channel. In the MHD regime M_e is in fact the largest of the flux channels.

3.7.1 Mode structure

As we look at the various diagnostics we will start with the impression that the onset of the transition to the MHD regime induced by $\hat{\beta}$ is gradual. The way the parallel envelopes and the cross coherences change will seem to confirm this. But the sudden change at $\hat{\beta} = 0.5$, before the nominal MHD regime, becomes visible in the phase shifts, and then we will have to investigate other elements in the structure not previously encountered. It is interesting how these effects are all induced by the electrons, and by electromagnetic responses, in what is usually seen as a fundamentally electrostatic, ion-based, process of physics.

The case with $\hat{\beta} = 0.03$ exemplifies the conventional, nearly electrostatic mode structure of the conventional picture of ITG turbulence and is showcased as a control case. Its parallel envelope structure is shown in figure 3.4, with the drift-plane averaged squared relative amplitudes and turbulent transport fluxes as functions of s, and averaged over $500 < t < 1000$ with the same 51 samples as before. The nonadiabatic electron quantities $\tilde{T}_e$ and $\tilde{h}_e$ are very low and flat, $\tilde{T}_i$ is the largest and most ballooned fluctuation in relative terms, with $\tilde{\phi}$ and $\tilde{n}_e$ nearly equal and still ballooned but less so. Only the ExB ion advection among the transport fluxes is significant.

The cross coherence and phase shift distributions between $\tilde{n}_e$ and $\tilde{\phi}$ are shown in figure 3.5. The fluctuations are sampled as pairs, in xy-space for the first and in x- and k_y-space for the second, at the central drift plane ($s = 0$), and at the same 51 time points as before. As far as these two variables are concerned, this shows deep drift wave mode structure, with nearly adiabatic electrons. This means nearly complete correlation and sharp phase shift distributions near zero phase shift for all wavelengths.

We now look at the effect of the run of inductivity on this mode structure, illustrating the transition into the MHD regime (nominal boundary at $\hat{\beta} = 0.63$). Each diagnostic is done in the same way as just done for $\hat{\beta} = 0.03$, now with the other values. The parallel envelopes of the squared amplitudes and transport fluxes are shown in figures 3.6 and 3.7, first for $\hat{\beta} = 0.1$ to 0.3, and then for 0.5 to 1.0. The rise of the electron channels with inductivity is at first gradual (first three), and then sudden. The MHD regime is exemplified by $\tilde{\phi}$ as the largest fluctuation with similar parallel structure to $\tilde{T}_i$, which in turn remains larger than either of $\tilde{T}_e$ or $\tilde{n}_e$. The curve for $\tilde{h}_e$ loses meaning in the MHD regime but it is visually helpful to see $\tilde{\phi}$ separate

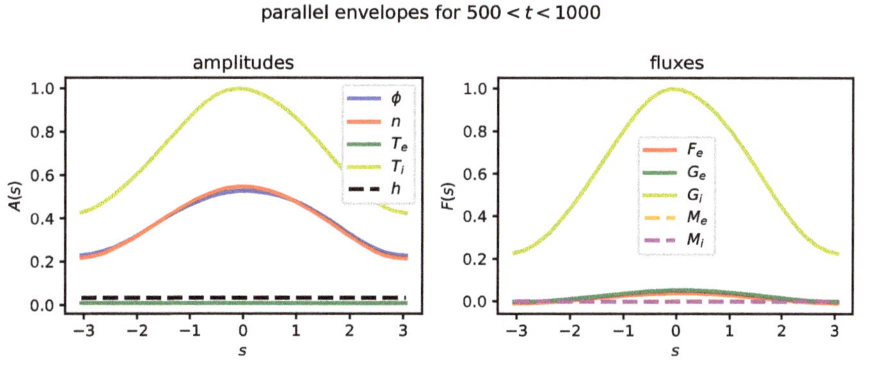

Figure 3.4. Parallel envelopes of the squared amplitudes (left) and turbulent ExB transport flux (right), which are averages over x and y and time over $500 < t < 1000$. The normalised squared amplitudes of $\tilde{\phi}$, $\tilde{n}_e$, $\tilde{T}_e$, $\tilde{T}_i$, and $\tilde{h}_e$ are labelled accordingly. The conventional, nearly electrostatic mode structure is shown by this lowest-beta $\hat{\beta} = 0.03$ case. The nonadiabatic electron quantities $\tilde{T}_e$ and $\tilde{h}_e$ are very low and flat, $\tilde{T}_i$ is the largest and most ballooned fluctuation in relative terms, with the others in between. The normalised ExB fluxes F_e, G_e, and G_i and the magnetic flutter fluxes M_e, and M_i are labelled accordingly. Among the fluxes only the ion thermal transport is significant. This case serves as a control case against finite beta effects.

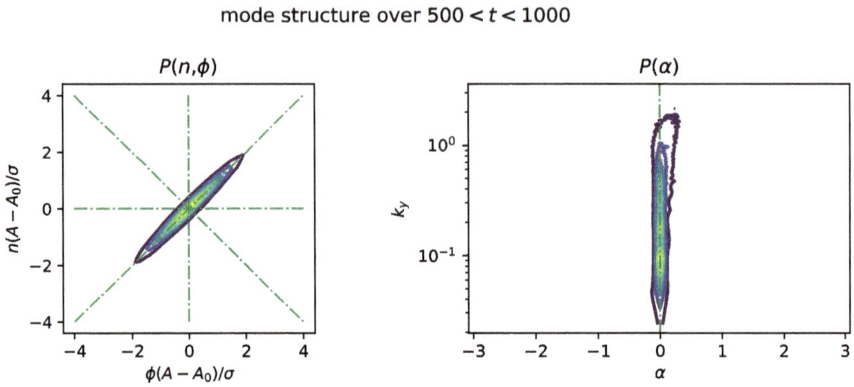

Figure 3.5. Cross coherence (left) and phase shift distributions (right) between n and ϕ, for the lowest-beta $\hat{\beta} = 0.03$ case. Both diagnostics show fairly strict drift wave mode structure, with near-unity correlation between n and ϕ and narrow phase shift distributions centred on zero, following from the nearly adiabatic electrons. Along with $C = 0.139$ and $\hat{\mu} = 0.1$, the inductivity is similarly weak so no process to oppose the adiabatic response is effective.

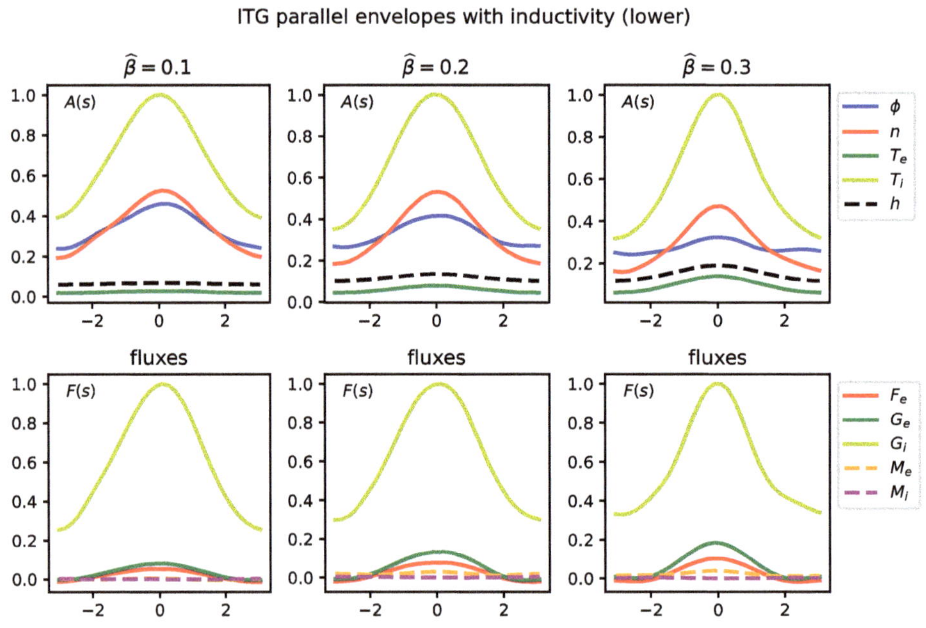

Figure 3.6. Parallel envelopes of amplitudes (top row) and turbulent fluxes (bottom) in hot plasma ITG ($\nabla \log T_i$) turbulence, for inductivity $\hat{\beta} = 0.1$, 0.2 and 0.3 (left to right), labelled as in figure 3.4. As the inductivity comes in, the nonadiabaticity increases without appreciable effect on the ions. The electron fluxes become more appreciable compared with the main ITG one. The magnetic flutter fluxes remain small.

from $\tilde{n}_e$ and $\tilde{h}_e$ to go with it, from the initial form with $\tilde{\phi}$ and $\tilde{n}_e$ coincident and $\tilde{h}_e$ small. When the MHD regime is reached, $\tilde{\phi}$ and $\tilde{T}_i$ then switch places. In the fluxes, the ExB channels stay together in the same $G_i > G_e > F_e$ hierarchy, but the main effect is the emergence of and then takeover by the electron magnetic flutter heat

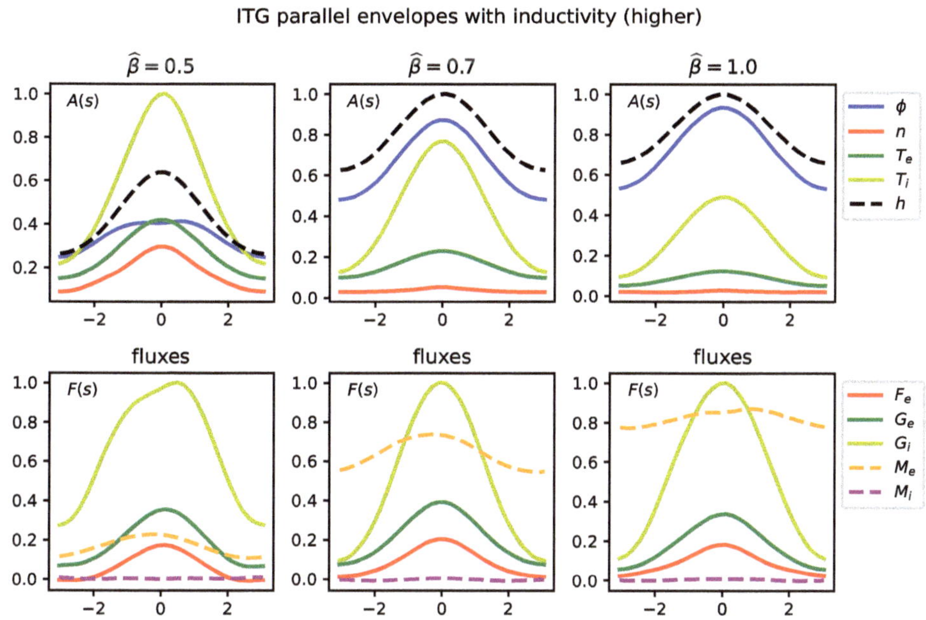

Figure 3.7. Continuing from the previous figure, these parallel envelopes of normalised squared amplitudes (top row) and fluxes (bottom), for inductivity $\hat{\beta} = 0.5$, 0.7, and 1.0 (left to right) show the transition into the MHD regime, nominally at $\hat{\beta} = 0.63$. The signature of this is that $\tilde{\phi}$ becomes the largest fluctuation in relative terms The exchange of position among the curves reflects the changes in the energetics. The ExB transport channels are similarly ballooned with the ion thermal one still the largest, and the electron thermal magnetic flutter heat transport becomes the largest overall and remains the least ballooned.

flux, which also remains very flat. The magnetic flutter fluxes for particles and ion heat remain small.

The cross coherence and phase shift distributions between $\tilde{n}_e$ and $\tilde{\phi}$ are shown in figure 3.8, for $\hat{\beta} = 0.2$, 0.5, and 1.0, measured the same way as before. At first the only change is for both distributions to widen, as the delays in the adiabatic response caused by inductivity (not collisions) reduce the tightness of the coupling between these two state variables. At $\hat{\beta} = 0.5$ the long and short wavelengths begin to separate and then into the MHD regime the long wavelengths separate and their phase shifts migrate to $\pi/2$. The manifestation of the ion temperature is that the short wavelengths migrate not to zero but to $\pm\pi$, that is, towards an antiphased relationship. The cross coherence, now dominated by the long wavelengths, shows an uncorrelated relationship as in a hydrodynamic setting. It is interesting to note that cross correlation is lost already at $\hat{\beta} = 0.5$, that is, before the transition into MHD. At this point the parallel envelopes show strong nonadiabaticity but still ITG mode structure. So this is a hybrid case.

3.7.2 Microtearing and the electron energetics

In the cases involving substantial magnetic flutter energetics and transport, we have a new drive mechanism in play. This is the parallel heat flux and the radial magnetic

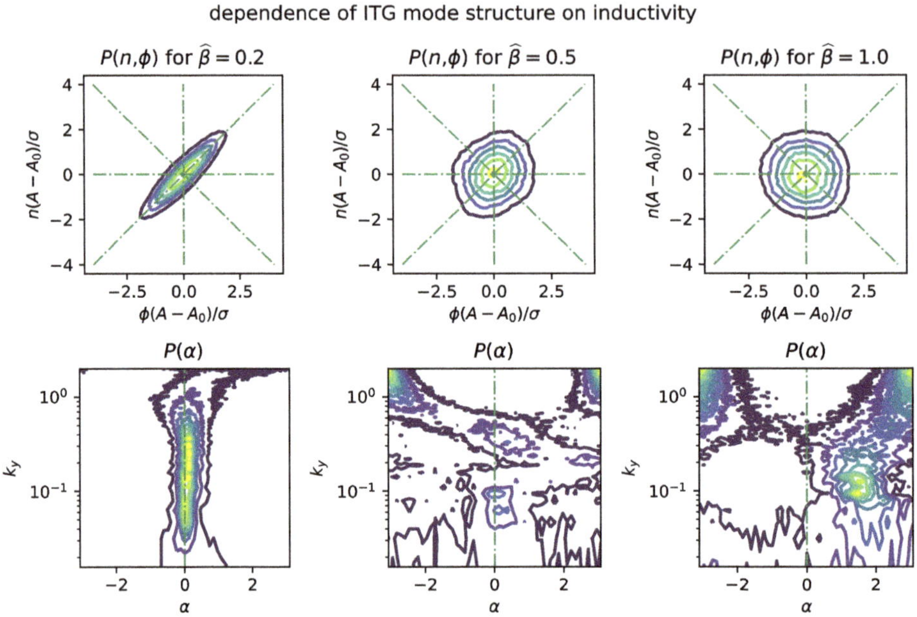

Figure 3.8. Cross coherence (top row) and phase shift distributions (bottom) between n and ϕ, for $\hat{\beta} = 0.2, 0.5$, and 1.0 (left to right), showing the transition into the MHD regime. Both distributions widen as the adiabatic coupling is more effectively opposed by inductivity, and then the long wavelengths separate and migrate to $\pi/2$ as the cross correlation is lost, once the MHD regime is entered.

flutter component down the temperature gradient as a drive, the parallel heat flux divergence and the parallel temperature gradient as a coupling, and then the parallel current divergence into the Ohm's law where the parallel electron temperature gradient completes the coupling back to the drive of magnetic fluctuations. This mechanism, energetically is given by

$$\frac{2}{5}\tilde{q}_{e\|}\frac{\partial \tilde{q}_{e\|}}{\partial t} + \tilde{q}_{e\|}\nabla_{\|}\tilde{T}_e = -\tilde{q}_{e\|}\tilde{B}^x\nabla \log T_e - \frac{C}{1.6}\tilde{q}_{e\|}\left(\tilde{q}_{e\|} + 0.71\tilde{J}_{\|}\right) \tag{3.67}$$

$$\tilde{T}_e\frac{3}{2}\frac{\partial \tilde{T}_e}{\partial t} + \tilde{T}_e\nabla_{\|}\tilde{q}_{e\|} - \tilde{T}_e\nabla_{\|}\tilde{J}_{\|} = 0 \tag{3.68}$$

$$\tilde{J}_{\|}\frac{\partial}{\partial t}\left(\tilde{A}_{\|} + \hat{\mu}\tilde{J}_{\|}\right) - \tilde{J}_{\|}\nabla_{\|}\tilde{T}_e = -C\tilde{J}_{\|}\left[\tilde{J}_{\|} - \frac{C}{1.6}0.71\tilde{J}_{\|}\left(\tilde{q}_{e\|} + 0.71\tilde{J}_{\|}\right)\right] \tag{3.69}$$

in normalised units, with the drive and dissipation on the right hand side and the couplings on the left. The magnetic flutter term due to $\tilde{J}_{\|}\tilde{B}^x$ in the Ohm's law (third equation) is not written since it vanishes under the volume integral. The transfer terms involving the parallel gradient are written in pairs which combine into total divergences, and the dissipation sums into the negative of a positive definite quantity. This leaves the drive, which is not in the temperature equation (where the ExB drive is) but in the heat flux equation. This is the energetic process by which the electron thermal dynamics frees itself of the adiabatic constraint at sufficient inductivity in electromagnetic ITG turbulence. This mechanism is effective in a hot plasma where the collisional dissipation acting upon it is small. By itself, it produces anaemic growth rates, but as part of the ITG energetics it becomes a significant secondary player.

To show how nonlinear this microtearing process is, we can re-do some of the cases with an initial amplitude of $10^{-6}\delta_0$ instead of the usual unity times δ_0 we have been using. A snapshot is taken during the linear phase in which the mode structure has settled without further change, and then again after saturation has occurred and the turbulence has become fully developed. These are shown for $\hat{\beta} = 0.2$ in figure 3.9, and $\hat{\beta} = 0.5$ in figure 3.10. They can be compared with figures 3.6 and 3.7. We see that in each case the electron temperature is nowhere near as present in the linear phase as it is once the turbulence has saturated. The ExB processes in the ITG turbulence also excite $\tilde{T}_e$, which then reinforces $\tilde{q}_{e\parallel}$ such that the magnetic flutter energetics hence drive is increased. This identifies the microtearing we see in the turbulence as a nonlinear process.

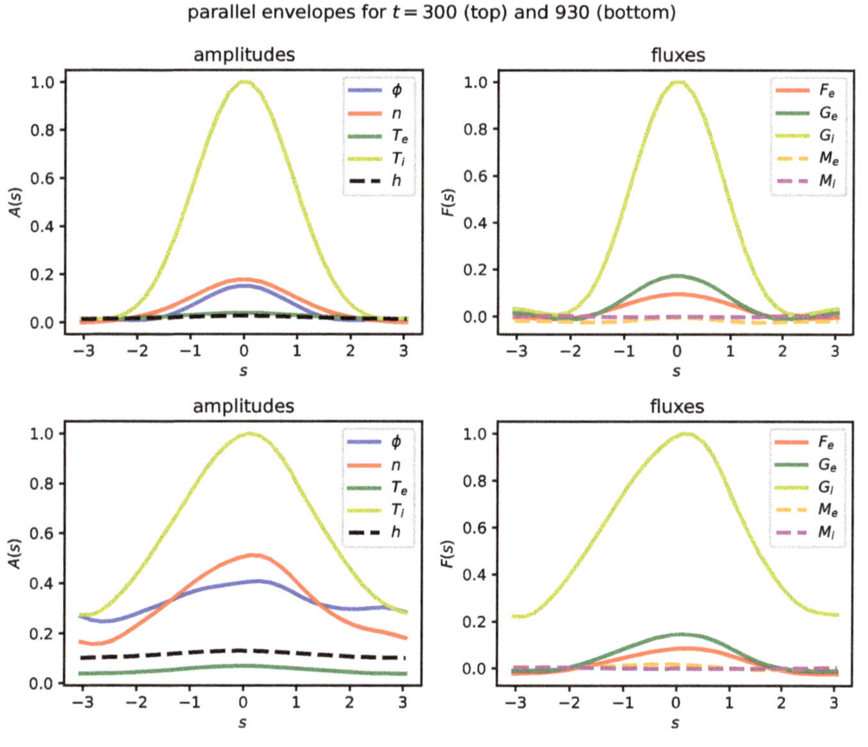

Figure 3.9. Comparison of parallel envelopes between the linear (top) and saturated turbulence (bottom) stages of the same ITG case with lower inductivity $\hat{\beta} = 0.2$. In the linear stage the electron dynamics is weaker, and the ion temperature and thermal ExB flux envelopes are more ballooned.

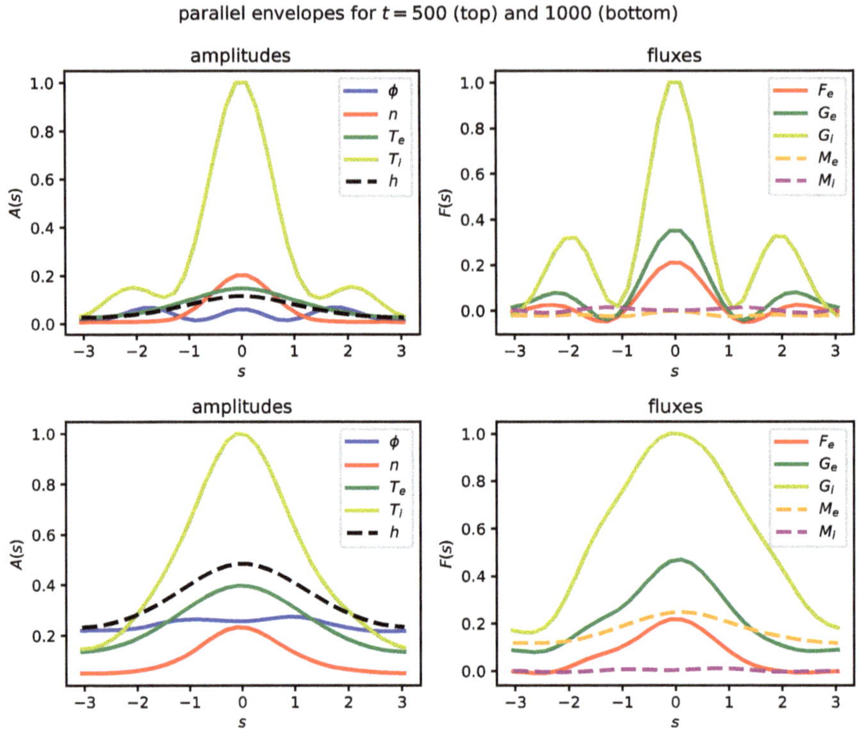

Figure 3.10. Comparison of parallel envelopes between the linear (top) and saturated turbulence (bottom) stages of the same ITG case with higher inductivity $\hat{\beta} = 0.5$. In the linear stage the ballooning structure has more than one lobe, while it is more generic and has different amplitude ratios after the turbulence has become fully developed.

3.7.3 Gradient threshold

This section will be very contingent on the use of this model. We are missing the thermal effects arising from kinetic theory we will need in this parameter regime. The results of the gyrofluid study in Chapter 11 will be more definitive (with thermal anisotropy and kinetic damping mechanisms). But we briefly report the situation with the gradient threshold here. For more collisional cases the threshold is washed out (see the next section), but these hot plasma cases will exhibit one.

The first sweep is on the temperature gradient at fixed density gradient, with $\omega_t = \omega_i$ varying holding $\omega_n = 0.3$ fixed. The inductivity is set at $\hat{\beta} = 0.2$ for all these gradient sweep cases. The time traces of the ion ExB advective flux Q_i are shown in figure 3.11 for these cases. On the logarithmic scale one has a relatively ambiguous threshold, but on a linear scale it would effectively be at about $\omega_t = 0.7$, which translates to $\eta_i \approx 2.3$, in line with the simplified linear cases done above. The jumps from 0.55 to 0.6 and then again to 0.7 are about a factor of three each. However, only the 0.5 case definitively decays over time. The ExB fluxes for density, and electron and ion heat, together with the magnetic flutter electron heat transport, are shown in the frame on the left. At this value of $\hat{\beta}$ the magnetic piece M_e comes up into reasonable level for the highest $\omega_t = 1.8$ case. It is still an order of magnitude below Q_i but it is competitive with the electron ExB heat flux Q_e. For the nominal case $\omega_t = 1.0$, however, M_e is rather smaller than Q_e.

We then vary the density gradient at fixed temperature gradient, so $\omega_t = \omega_i = 1$ with $\omega_n = 0.1$ up to 1. The time traces of the ion ExB advective flux Q_i are shown in figure 3.12

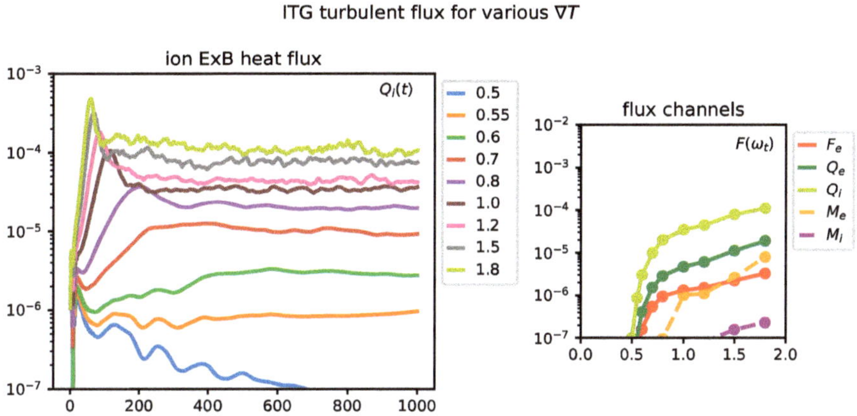

Figure 3.11. (left) Time traces of the ExB advective ion heat flux at various values of temperature gradient (both species, $\omega_t = \omega_i$), at fixed density gradient $\omega_n = 0.3$ and other parameters. One finds straightforward evolution, with an effective threshold near 0.7, translating to $\eta_i = \omega_i/\omega_n = 2.3$. (right) Time averaged ($500 < t < 1000$) fluxes at the various ∇T values, with the ExB advective fluxes for density (F_n) and electron (Q_e) and ion heat (Q_i) channels, and the electron magnetic flutter heat flux (M_e). With $\hat{\beta} = 0.2$, the latter rises to reasonable amplitude in the steepest ∇T cases.

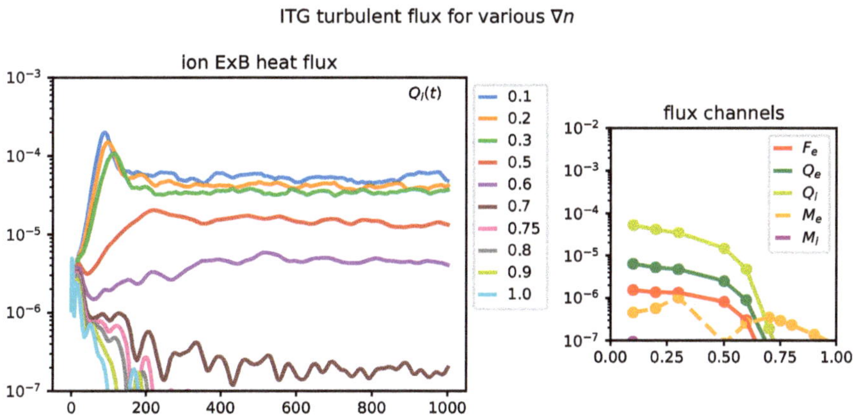

Figure 3.12. (left) Time traces of the ExB advective ion heat flux at various values of density gradient ω_n at fixed temperature gradient $\omega_t = \omega_i = 1$ and other parameters. One finds straightforward evolution, independent of ∇n until η_i reduces to two, and then dropping away for smaller $\eta_i = \omega_i/\omega_n$ values. (right) Time averaged ($500 < t < 1000$) fluxes at the various ∇n values, with the ExB advective fluxes for density (F_n) and electron (Q_e) and ion heat (Q_i) channels, and the electron magnetic flutter heat flux (M_e). As in the scan with various ∇T, the ExB fluxes change together, but for these cases M_e remains small.

for these cases. The first three cases show little dependence on ω_n, but for $\omega_n = 0.5$, with $\eta_i = 2$, the turbulence drops by about a factor of three. The case with 0.6 can be considered as at or below threshold, and 0.7 and higher decay with time. As for the ω_t sweep, the main four transport flux channels are shown together in the frame on the right. For these cases the magnetic flutter electron heat flux remains small.

What one does not see is any temporal shifts in the result, namely any mode structure changes away from the forms shown above. Moreover, the cases that grow robustly in the linear stage remain at finite amplitude in the nonlinear stage. We will find this changes when the gyrofluid model, to which thermal anisotropy (parallel and perpendicular) is native, is considered in Chapter 11.

3.8 Warm ion drift Alfvén turbulence

The effects of the ITG physics and of the ion temperature in general on the warm plasma cases is also of interest, since in a plasma confinement experiment one tries to optimise temperature among other things. Reasonable cases always have a temperature gradient which is at least concurrent with the density, as we considered for the electron temperature in the previous Chapter. And with reasonable densities the collisional exchanges between the temperatures ensures an ion temperature at least in the same order of magnitude as the electron one. So continuing with the minimal assumptions for the warm ion cases we set not only $\tau_i = 1$ but also $\omega_n = \omega_t = \omega_i$, namely, equal temperatures and gradients across species and state variables.

We repeat the essential cases from the previous Chapter, adding only the ion parameters. Once more we set

$$\hat{\mu} = 2.72, \qquad \hat{\epsilon} = 10\,000, \qquad \omega_B = 0.03, \qquad S = 1 \tag{3.70}$$

with the domain aspect ratio and size

$$L_x = L_\perp, \qquad L_y = 4L_x, \qquad \delta_0 = 0.015 \tag{3.71}$$

and the temperature ratio and normalised gradients

$$\tau_i = 1, \qquad \omega_n = \omega_t = \omega_i = 1 \tag{3.72}$$

together with various values of $\hat{\beta}$ and C. Nominally we start with $\hat{\beta} = 1$ and $C = 1.39$ as before—this results from $\nu_e L_\perp / c_s = 1$ and the other parameters. The ion collision frequency is slower by the factor $\sqrt{m_e/2m_i} = 1/85$, as prescribed by the Braginskii model. This is also commensurate with the slower parallel dynamics due to V_i and not V_e. The range of $\hat{\beta}/\hat{\mu}$ values involves both $v_A > V_e$ and $v_A < V_e$ sides of the kinetic shear Alfvén regimes. As with the electron dynamics we compare the structure of the ion heat flux relative to the temperature with what using the Braginskii form for $\tilde{q}_{i\parallel}$ would yield, given the same results.

3.8.1 Inductivity

The first set of cases is at $C = 1.39$ with $\hat{\beta}$ ranging from 1 to 10. The nominal MHD ballooning transition is given by $\alpha_M = 0.66\ S$ (Eq. 3.12), which for these parameters is

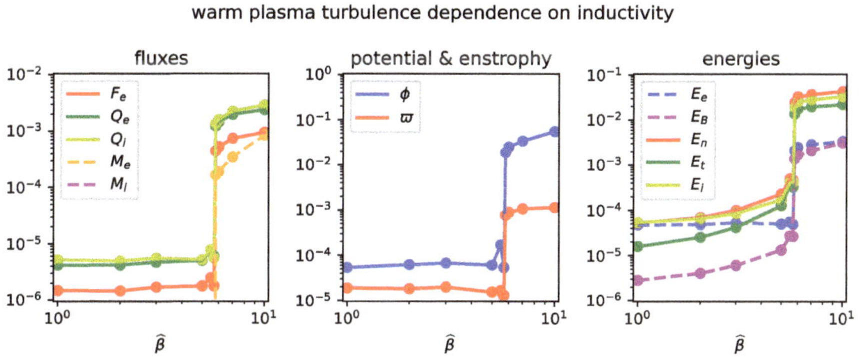

Figure 3.13. Dependence of the turbulent transport fluxes (left), the half-squared potential and vorticity (center), the ExB, magnetic, and thermal free energy components (right), all in volume average, against magnetic inductivity $\hat{\beta}$, at $C = 1.39$. There is nearly no $\hat{\beta}$-dependence in the fluxes until the MHD regime is sharply reached. The nominal transition is at $\hat{\beta} = 5.5$, the actual one is between the two cases $\hat{\beta} = 5.7$, still on the drift wave side, and $\hat{\beta} = 5.8$, the first case in the MHD regime. In contrast to the fluxes and any quantity following $\tilde{\phi}$, the thermal free and magnetic energies rise with inductivity in the drift wave regime. The ExB and magnetic energies are approximately equipartitioned on the MHD side.

at $\hat{\beta} = 5.5$, so several additional cases are done around 5.5 to find the effective transition. The run of three sets of quantities with the inductivity is given in figure 3.13. The first frame gives the turbulent fluxes, including the electron and ion magnetic flutter heat fluxes. The ion one turns out to be small and negative for all these cases so does not appear. The second frame gives A_p and A_w, the half-squared amplitudes of $e\tilde{\phi}/T_e$ and $\varpi/n_e e$. The third frame gives the thermal free energy components due to the density and the electron and ion temperatures, and the ExB and magnetic energies. The last frame gives the ExB and magnetic energies. All are in volume average and time averaged over $500 < t < 1000$ as done previously. All quantities following the potential (ExB energy and fluxes, and enstrophy) are very flat with increasing $\hat{\beta}$ until the MHD regime is sharply reached between $5.7 < \hat{\beta} < 5.8$, compared with the nominal threshold at 5.5. The jump in the fluxes is more than two orders of magnitude at the transition. The thermal quantities and the magnetic energy all rise with $\hat{\beta}$ on the drift wave side. The electron magnetic flutter heat flux is small and negative on the drift wave side and positive but not dominant on the MHD side. The ion counterpart is small and negative for all these cases.

With the ion temperature dynamic present, the turbulence acquires a more prominent longer wavelength component ($0.02 < k_y\rho_s < 0.3$) which with finite inductivity makes itself felt in the energetics, both the free energy drive due to the gradients and the energy content in the state variables. As the inductivity $\hat{\beta}$ increases, the scale of the state variables increases while the vorticity remains near ρ_s. This is shown in figure 3.14, giving the isotropic perpendicular spectra for $\hat{\beta} = 1$ through 10. The solid/dashed lines are for the state/flux variables, respectively, except the black dashed line for the vorticity. For $\hat{\beta} = 7$ and 10 the MHD ballooning mode structure is seen by contrast. The plot scale changes for the larger inductivity values, following the amplitude increase on the MHD side.

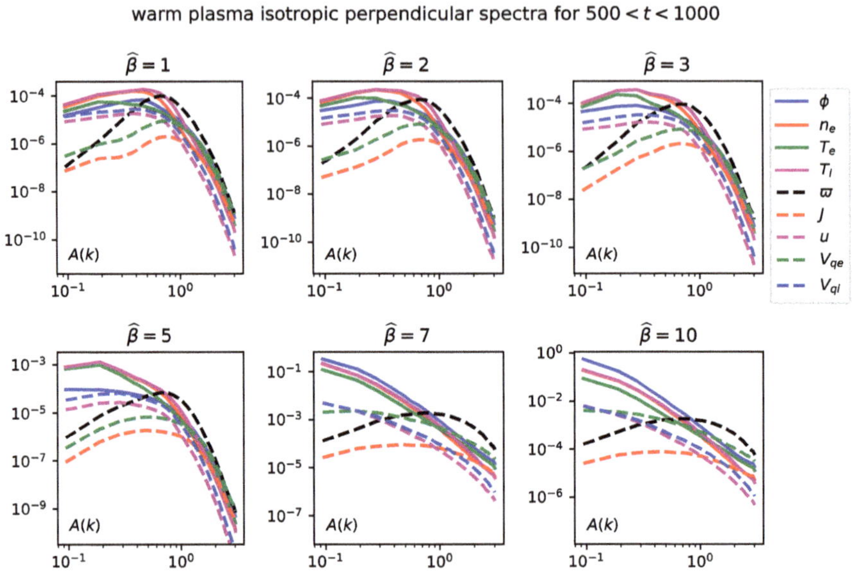

Figure 3.14. Isotropic perpendicular ($k_\perp$) spectra of the fluctuations, for $\hat{\beta}$ = 1 to 10, at C = 1.39, showing the transition from drift wave mode structure to MHD ballooning, near $\hat{\beta}$ = 5.8 as indicated in figure 3.13. The solid lines are for the state variables, and the dashed lines are for the flux variables. The main feature is that while the state variables jump over to large scale at the MHD transition, the vorticity remains at the ρ_s scale. Note the change in the scale of the vertical axis, highlighting the jump in turbulence strength into the MHD side.

The changes in the parallel wavenumber spectrum for increasing inductivity are shown in figure 3.15, for the same cases. Note how the vorticity (strong dashed line) becomes much flatter on the MHD side. The same holds for the current (red dashed). The potential (blue solid) becomes dominant without any broadening on the MHD side. The most flute mode like of any state variable is $\tilde{T}_e$, which broadens only slightly on the MHD side. On the drift wave side the vorticity is about as wide as the ion temperature. The green dashed line is for $\tilde{h}_e$, which is similar to $\tilde{T}_e$ on the drift wave side and to $\tilde{\phi}$ (since it is greater than $\tilde{n}_e$) on the MHD side.

The mode structure, cross correlations and phase shift distributions, is shown in figure 3.16, for the cases with $\hat{\beta}$ = 1, 3, and 7, from this set of runs.

3.8.2 Collisionality

The second set of cases is at $\hat{\beta}$ = 1 with C ranging from 1.39 to 13.9, following $\nu_e L_\perp / c_s$ = 1 to 10. The run of three sets of quantities with the collisionality is given in figure 3.17. The three frames give the same quantities with the same format as in figure 3.13. The main point here is that while the turbulence greatly increases in strength with collisionality, all the dependence is monotonic and gradual, with all curves in the same form and little change of any local dependence along the curves. This is indicative of the same basic set of processes collectively active, with a single overall dependence on collisionality.

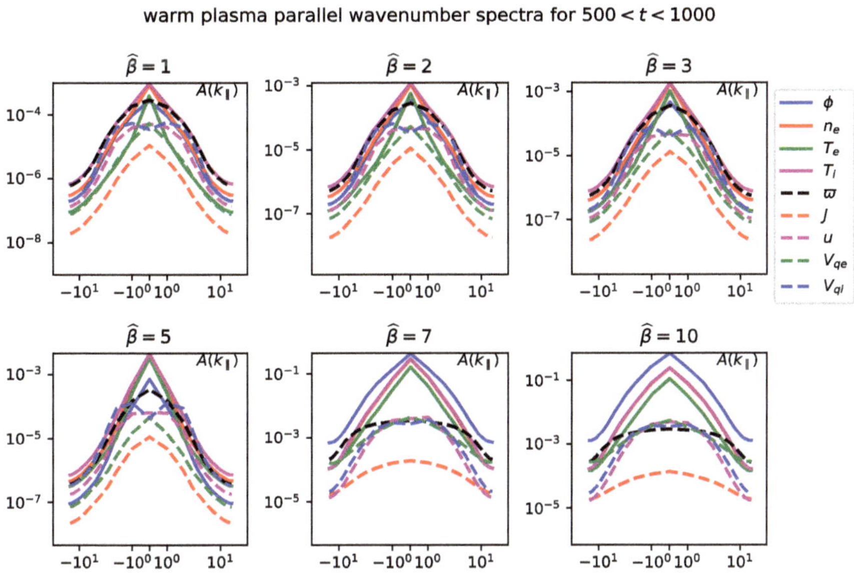

Figure 3.15. Parallel wavenumber ($k_\parallel$) spectra of the fluctuations, for for $\hat{\beta} = 1$ to 10, at $C = 1.39$, showing the transition from drift wave mode structure to MHD ballooning, near $\hat{\beta} = 5.8$ as before. The solid lines are for the state variables, and the dashed lines are for the flux variables. On the drift wave side the vorticity has the same shape as $\tilde{T}_i$, while on the MHD side it is flatter, along with the current. The electron temperature is the narrowest. Note the change in the scale of the vertical axis, highlighting the jump in turbulence strength into the MHD side.

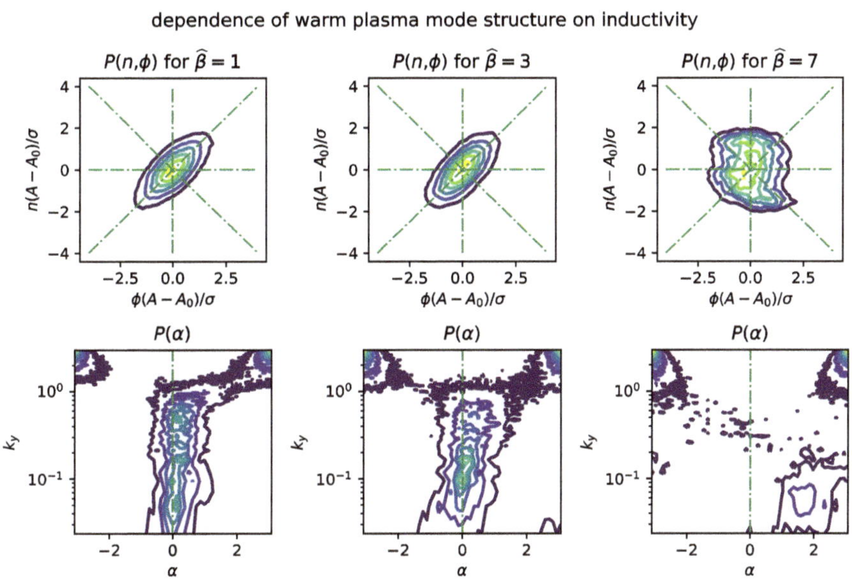

Figure 3.16. Cross coherence (top row) and phase shift distributions (bottom) between n and ϕ, for $C = 1.39$ and $\hat{\beta} = 1$, 3, and 7 (left to right), which one can compare with the cold ion cases in figure 2.10 from Chapter 2. The finite ion temperature yields the $\alpha = \pm\pi$ signature for $k_y\rho_i > 1$. Its gradient and fluctuation dynamics together increase the clarity of the transition to hydrodynamic mode structure in the MHD ballooning regime, here for $\hat{\beta} > 5.7$.

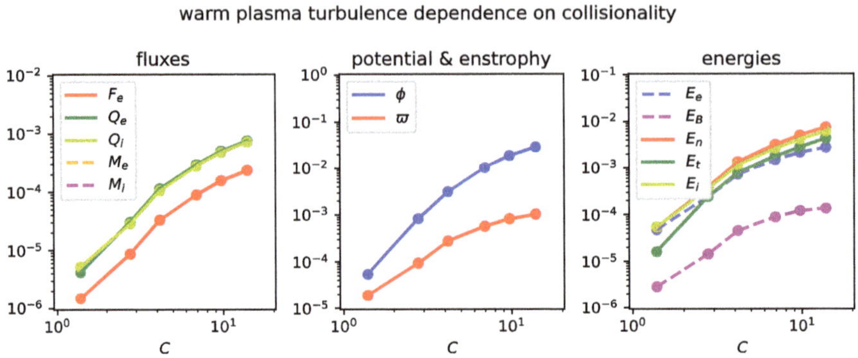

Figure 3.17. Dependence of the turbulent transport fluxes (left), the half-squared potential and vorticity (center), the ExB, magnetic, and thermal free energy components (right), all in volume average, against collisionality C, for $\hat{\beta} = 1$. The general increase in the strength of the turbulence is gradual with no clear indication of a change of regime. All quantities rise together with the same form of the curve.

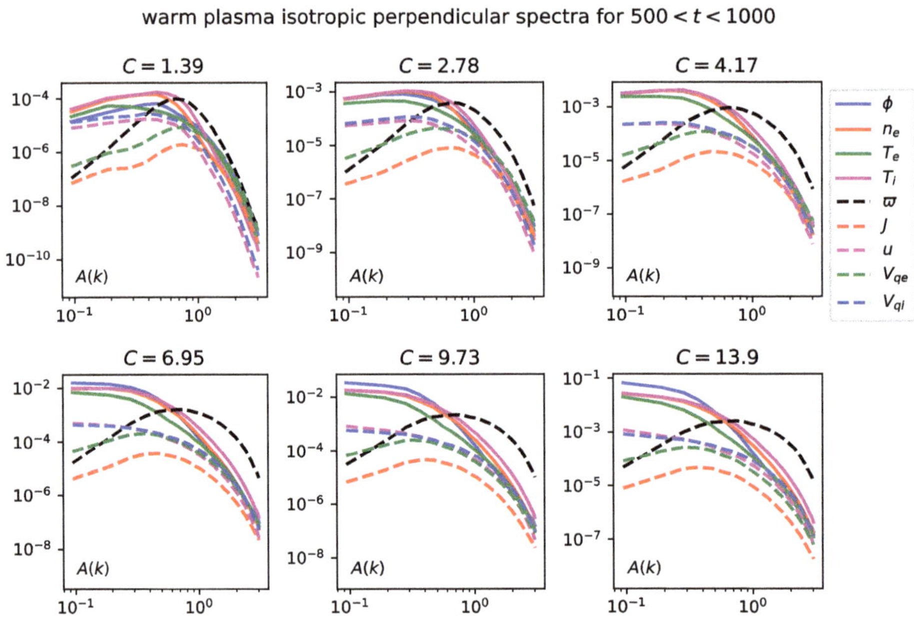

Figure 3.18. Isotropic perpendicular ($k_\perp$) spectra of the fluctuations, for $C = 1.39$ through 13.9 at $\hat{\beta} = 1$, showing the general increase of the turbulence strength with collisionality. The solid lines are for the state variables, and the dashed lines are for the flux variables. The changes from low to high collisionality are less pronounced as for the inductivity variation in figure 3.14, but the increasing dominance of the larger scales is unmistakeable. The vorticity remains near the ρ_s scale.

The isotropic perpendicular spectra are shown in figure 3.18, for the run of $C = 1.39$ through 13.9. The format is the same as in shown in figure 3.14, above. As the collisionality C increases, the scale of the state variables increases gradually while the vorticity remains near ρ_s. The shapes of the state variable spectra, including

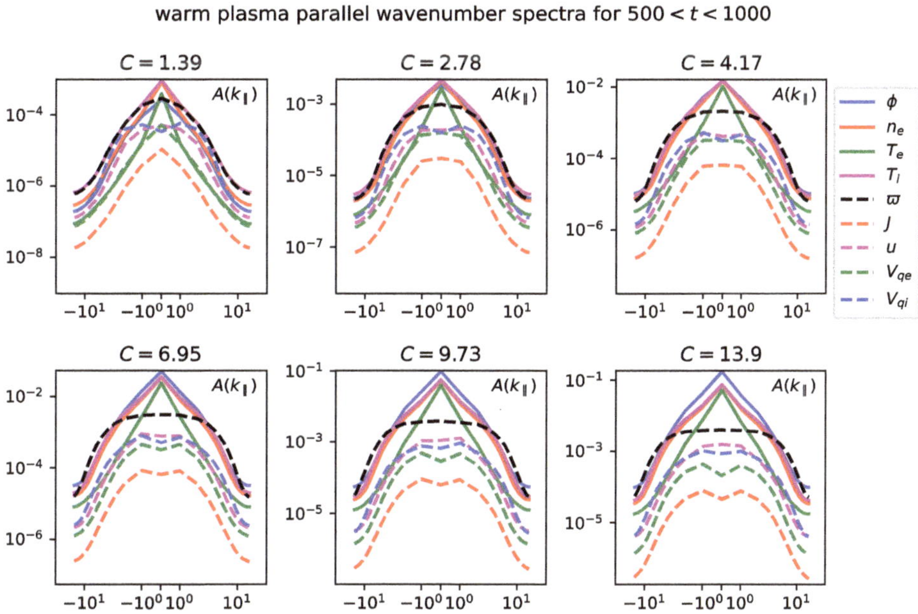

Figure 3.19. Parallel wavenumber ($k_{\|}$) spectra of the fluctuations, for $C = 1.39$ through 13.9 at $\hat{\beta} = 1$, showing the general increase of the turbulence strength with collisionality. The solid lines are for the state variables, and the dashed lines are for the flux variables. The changes from low to high collisionality are less pronounced as for the inductivity variation in figure 3.15, but the vorticity spectrum still widens towards the collisional side. The electron temperature is the narrowest throughout. The scale of the plot is held fixed, highlighting the rise in turbulence strength towards the collisional side.

that of $\tilde{\phi}$, are all similar. This is how the drift wave mode structure looked in toroidal geometry, where the increased resistive ballooning forcing added to the turbulence without changing the relationships among the energetics.

The changes in the parallel wavenumber spectrum for increasing inductivity are shown in figure 3.19, for the same cases. The nonadiabatic channels are broadened with higher collisionality, but the spectra of $\tilde{\phi}$ and $\tilde{T}_e$ remain similar in both amplitude and form. The ion temperature and vorticity also broaden with the increased strength of the turbulence.

The mode structure, cross correlations and phase shift distributions, is shown in figure 3.20, for the cases with $C = 1.39$, 4.17, and 13.9, from this set of runs. In contrast to the dependence upon inductivity given in figure 3.16, there is no change in the basic form, just the broadening one expects in the distributions given the increase in turbulence strength seen in the previous two figures. The correlation between $\tilde{n}_e$ and $\tilde{\phi}$ remains effective, and the phase shifts remain near zero. So these cases with increasing collisionality all show drift wave mode structure.

3.8.3 Gradient synergy

The run of the turbulent transport fluxes and the energetics with the temperature gradient for the case with $\hat{\beta} = 1$ and $C = 1.39$ at fixed pressure gradient is given in

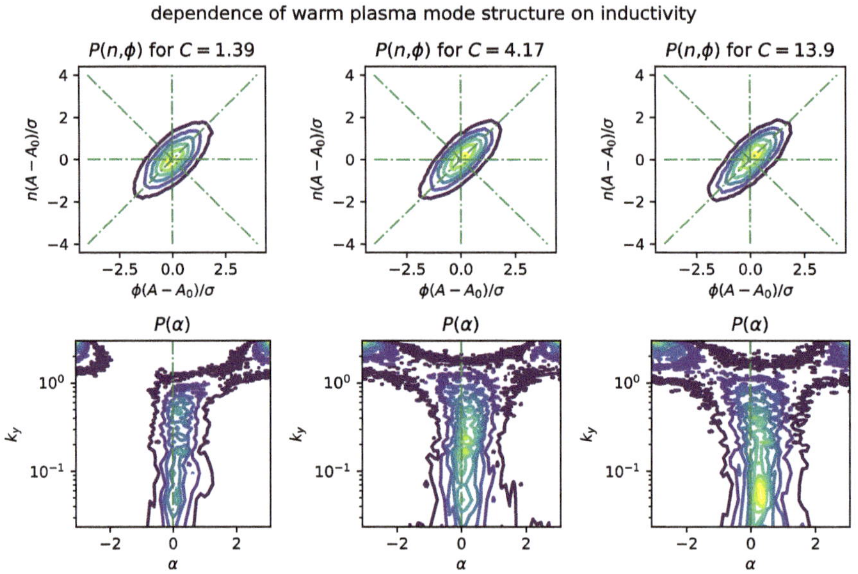

Figure 3.20. Cross coherence (top row) and phase shift distributions (bottom) between n and ϕ, for $\hat{\beta} = 1$ and $C = 1.39$, 4.17, and 13.9 (left to right), which one can compare with the run with inductivity in figure 3.16. Here, although the turbulence strength greatly increases with C, the mode structure is seen to remain in the drift wave relationships, with effective correlation and phase shift distributions, while broadening with C, remain near zero.

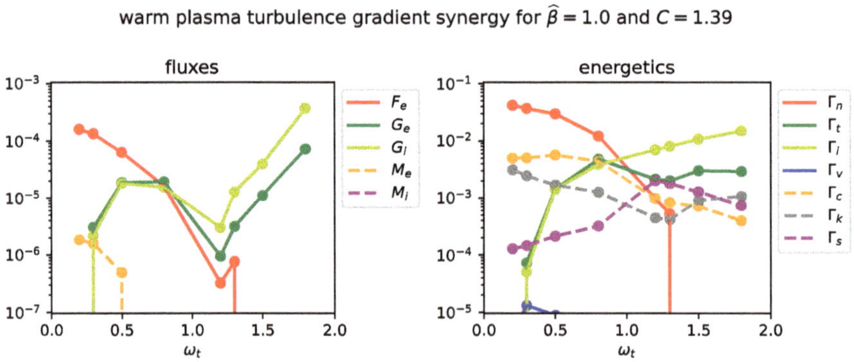

Figure 3.21. Gradient synergy in the turbulent fluxes (left) and energetics (right) with $\omega_n + \omega_t = 2.0$, for the $\hat{\beta} = 1$ and $C = 1.39$ case. The ∇n drive drops out for $\omega_n < 0.8$ and the ∇T (both species together) for $\omega_t < 0.2$. Note how the temperature drive persists over a wider range. The weakest turbulence is near equal gradients, at which the dissipation and cascade dynamics is maximal, which overcomes the positive synergy in the gradients.

figure 3.21. Equal temperature values and gradients for ions and electrons are assumed for the background. In contrast to the hot plasma cases there is no value at which the turbulence disappears. This is due to the action of the drift wave dynamics for strong enough nonadiabatic electron dynamics for $C > 1$. The sharpest decrease is not on near $\omega_t = 2$ at which $\omega_n = 0$, but slightly to the temperature gradient side of

$\omega_n = \omega_t = 1$. Here, the energetics shows that all dissipation channels remain robust and the sound wave dissipation channel actually peaks in this range. The turbulent cascade dynamics, remaining robust whenever there is finite free energy, maintains the dissipation.

This feature, which can be considered a 'hole' in the turbulence, depends on how well the density and temperature energetic channels, electron and ion, exchange among one another. Its occurrence can be understood if we check its dependence upon the increase of nonadiabaticity in the electrons. This can be either through the plasma beta (scaling: $n_e T_e$), the collision frequency (normalised scaling: n_e/T_e^2), or the scale ratio $K_\| L_\perp$, which affects all of the electron response parameters: the inductivity $\hat{\beta}$, the collisionality C, and the effective electron inertia $\hat{\mu}$, which affects the Ohm's law to some extent but in this case especially the electron heat flux response, making it more time dependent hence in turbulence more statistical.

The effect of the collisionality is found by increasing C by ten. The run of the turbulent transport fluxes and the energetics with the temperature gradient for this case with $\hat{\beta} = 1$ and $C = 13.9$ with the temperature gradient at fixed pressure gradient is given in figure 3.22. The main effect is the general strengthening of the drift wave side as found above. The dependence of the turbulence upon the details of the forcing is much less direct, as any drive more easily affects all the channels with $\nabla_\| \tilde{J}_\|$ more open than before. The dynamics with the higher C is at larger scale, so more of the spectrum can participate. The dependence on the fraction of ∇T within ∇p is now gradual, and the turbulence is comparably robust with similar energetics for the entire range of $0.2 < d \log T / d \log p < 1.8$ investigated. Note how flat the dashed curves are (for the sinks), compared with the run of the drives, which have the expected form, dropping out at the end of the range one would expect in each case.

The effect of the inductivity is found by increasing $\hat{\beta}$ by three, still below the MHD transition at the 5.8 value found above. The run of the turbulent transport fluxes and

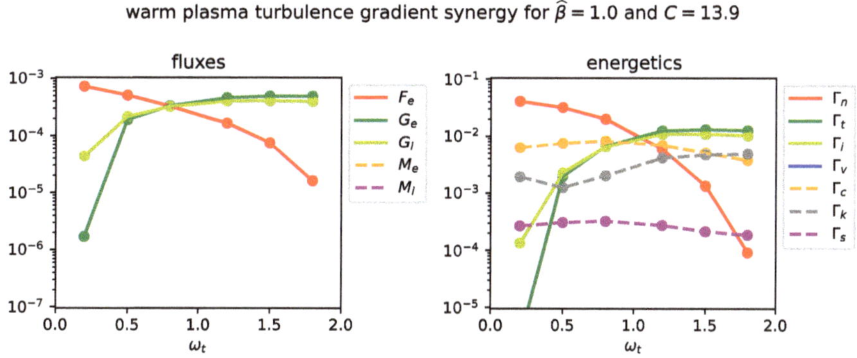

Figure 3.22. Gradient synergy in the turbulent fluxes (left) and energetics (right) with $\omega_n + \omega_t = 2.0$, for the $\hat{\beta} = 1$ and $C = 13.9$ case, higher collisionality than the previous figure. The ∇n and ∇T drives are still present or absent in the same way as before. But the greatly increased robustness due to the larger $\nabla_\| \tilde{J}_\|$ allows the turbulence to rearrange so as to keep roughly the same overall strength (transport, energetic throughput) across this range of variation. Note the flatness of the dissipation curves (dashed lines).

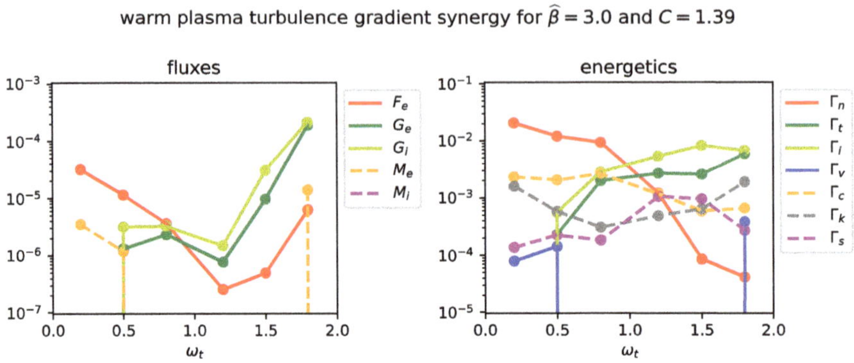

Figure 3.23. Gradient synergy in the turbulent fluxes (left) and energetics (right) with $\omega_n + \omega_t = 2.0$, for the $\hat{\beta} = 3$ and $C = 1.39$ case, higher beta than the previous figures, but still below the MHD ballooning threshold, at about $\hat{\beta} = 5.8$ for these cases. The ∇n and ∇T drives are still present or absent in the same way as before. The density gradient side ($\omega_t \to 0$) is more stabilised by the magnetic flutter effect, as any additional drive from the ions is mostly absent. On the temperature side ($\omega_t \to 1$) the ions provide drive but now so does the magnetic flutter due to the stronger inductivity. The hole in the middle of the range is still present and still as deep.

the energetics with the temperature gradient for the case with $\hat{\beta} = 3$ and $C = 1.39$ at fixed pressure gradient is given in figure 3.23. The difference to the cases with $\hat{\beta} = 1$ is mainly that the density gradient driven side ($\omega_t \to 0$) is diminished with $\hat{\beta}$. Here, by contrast to the temperature side ($\omega_t \to 1$), there is no ion drive channel to maintain the turbulence against the magnetic flutter stabilisation effect. But the hole in the equal gradient region is still present and just as deep. So these investigations find that this feature depends more on collisionality.

3.8.4 The heat flux versus its collisional form

As in the previous Chapter, we briefly compare the ion and electron parallel heat flux dependent variables against what their forms would be under the Braginskii collisional model, given the same temperature fluctuation field. The Braginskii heat flux defect is given by

$$B_{qe} = \tilde{q}_{e\|} + \alpha_e \frac{T_e}{e} \tilde{J}_\| + \frac{\kappa_e p_e}{m_e \nu_e} \nabla_\| (\tilde{T}_e + T_e), \qquad B_{qi} = \tilde{q}_{i\|} + \frac{\kappa_i p_i}{m_i \nu_i} \nabla_\| (\tilde{T}_i + T_i) \tag{3.73}$$

respectively, for ions and electrons. In these expressions the factor of $T_{e,i}$ acted upon by $\nabla_\|$ denotes the magnetic flutter term due to the background gradient. That is, we have to keep the parallel gradient of the entire temperature in nonlinear cases.

We measure this on the $\hat{\beta} = 1$ and $C = 1.39$ case with equal background gradients and temperatures as noted above. The RMS average of any of these quantities is measured as the square root of the average square, where the average is over the volume and over time between $500 < t < 1000$, as with other measurements. For this case the result is $(1.28 \pm 0.049) \times 10^{-3}$ for $\tilde{q}_{e\|}/p_e c_s$ and $(2.29 \pm 0.19) \times 10^{-4}$ for B_{qe}. For the case of $\hat{\beta} = 5$ the result is very similar, 1.19 versus 0.142. This is a closer match than the cold ion case of the previous Chapter. For the ions it is very different, with $(3.58 \pm 0.011) \times 10^{-5}$ for B_{qi}, and 4.99 versus 17.9 respectively (times 10^{-5}) in the case of $\hat{\beta} = 5$. For the ions, due to the low value of $\nu_i L_\perp / c_s$, the Braginskii model would overestimate the result by nearly an order of magnitude in some of these cases. The collisionality in conventional terms is similar for both species, since it is reckoned with the ratio of parallel transit, $VK_\parallel$ for thermal velocity V, versus the collision frequency. These are similar since the relevant temperature ratio is the same and the mass ratio is common to both ν and V in each case. However, in the case of turbulence, by contrast, the comparison is not the parallel transit but the overall dynamical frequency for each species. And since the ExB nonlinear advection is the dominant piece for the ions, the evaluation is very different for them than it is for the electrons, since for the latter the collision, ExB advection, and parallel transit frequencies are all comparable. So while the Braginskii model is marginal for electrons, it is not valid at all for ions in turbulence driven by thermal gradients, for which ExB advective mixing is the dominant nonlinear process. This is one of the things which drives us to go back to the kinetic theory and eventually to the gyrofluid equations. Another process doing this is gyroviscosity and the role of dynamics near ρ_s hence ρ_i, the ion gyroradius. We turn to this next.

3.9 On gyroviscosity

One can treat warm ion dynamics with or without gyroviscosity and still have an energetically consistent system. One can also go back to the form dependent upon MHD ordering that was standard in the 1970s (references given below), but then all the physics of the adiabatic coupling and response has to be neglected to recover exact energetic consistency. Then, in the resulting turbulence, the effects of gyroviscosity can be found by comparing to the model in which it is not treated. Both conserve the same energy, but the dynamics of the vorticity is different, especially in the diamagnetic part of it, and if gyroviscosity is kept then the ExB nonlinearity no longer conserves enstrophy. Finally, gyroviscosity is found to have a large effect on the strength of the turbulence, especially on the ITG side. We treat each of these concepts in turn.

3.9.1 Gyroviscous cancellation in the brackets

The formulation of the vorticity equation in normalised units without gyroviscosity is given by

$$\frac{\partial \varpi}{\partial t} + [\tilde{W}_i, \varpi] = (\text{RHS}) \tag{3.74}$$

where for these purposes we won't explicitly write the right hand side (RHS). In the cold ion or MHD models the contribution due to $\tilde{p}_i$ in $\tilde{W}_i$ is assumed to be small, so that $\tilde{W}_i \to \tilde{\phi}$ and we are back to the pure ExB vorticity advection which represents the 2D Euler equation in this language. Either way, if we multiply by $-\tilde{W}_i$ we find

$$\frac{1}{2}\frac{\partial}{\partial t}\frac{\rho}{B^2}\,|\nabla_\perp \tilde{W}_i|^2 - \nabla\cdot\left(\tilde{W}_i\frac{\partial}{\partial t}\frac{\rho}{B^2}\nabla_\perp \tilde{W}_i\right) + [\tilde{W}_i, (-\tilde{W}_i\varpi)] = -\tilde{W}_i\,(\text{RHS}) \quad (3.75)$$

or if we multiply by ϖ we find

$$\frac{\partial}{\partial t}\frac{1}{2}\varpi^2 + \frac{1}{2}[\tilde{W}_i, \varpi^2] = \varpi\,(\text{RHS}) \quad (3.76)$$

In both cases we have the density of a conserved quantity under $\partial/\partial t$, a pure divergence, and then the effects of the RHS which are assumed to be consistent. So far, this is the same as the operations in the 2D fluid models to show energy and enstrophy conservation, just that now we have $\tilde{W}_i$ instead of $\tilde{\phi}$. So far, so good.

Now we put in gyroviscosity. In this formulation it is given by

$$\nabla\cdot\mathbf{\Pi}^* = -\nabla\cdot\frac{\rho}{n_e eB^2}\left[\tilde{p}_i, \nabla_\perp\tilde{W}_i\right] \quad (3.77)$$

so that putting it in we have the form in Eq. (3.35) since the role of the gyroviscosity as a diamagnetic momentum flux is to cancel the factor of $\tilde{p}_i/n_e e$ in $\tilde{W}_i$ to produce $\tilde{\phi}$ in the nonlinear part of d/dt. The energy conservation still works by multiplying by $-\tilde{W}_i$ so that

$$\frac{1}{2}\frac{\partial}{\partial t}\frac{\rho}{B^2}\,|\nabla_\perp \tilde{W}_i|^2 - \nabla\cdot\left(\tilde{W}_i\frac{d}{dt}\frac{\rho}{B^2}\nabla_\perp \tilde{W}_i\right) + \frac{\rho}{2B^2}[\tilde{W}_i, |\nabla_\perp\tilde{W}_i|^2] = -\tilde{W}_i\,(\text{RHS}) \quad (3.78)$$

in which the only differences are in the transport flux terms, which are generally small compared with those in other equations (density, temperatures).

If we do this in normalised units for increased clarity, the cancellation is

$$[W, \nabla_\perp W] - \left[p_i, \nabla_\perp W\right] = [\phi, \nabla_\perp W] \quad (3.79)$$

so that the vorticity equation is the one we have in Eq. (3.35), which in normalised units is

$$\frac{\partial\varpi}{\partial t} + \nabla\cdot[\phi, \nabla_\perp W] = \cdots, \qquad \varpi = \nabla_\perp^2 W, \qquad W = \phi + p_i \quad (3.80)$$

Now, it is possible to show that

$$\frac{\partial}{\partial t}\nabla_\perp^2 p_i + \nabla\cdot\left[\phi, \nabla_\perp p_i\right] = \nabla\cdot\left[p_i, \nabla_\perp\phi\right] + \nabla_\perp^2\left(\frac{dp_i}{dt}\right) \quad (3.81)$$

using some bracket algebra such as

$$\left[\phi, \nabla_\perp p_i\right] = \nabla_\perp\left[\phi, p_i\right] - \left[\nabla_\perp\phi, p_i\right] = \nabla_\perp\left[\phi, p_i\right] + \left[p_i, \nabla_\perp\phi\right] \quad (3.82)$$

What the MHD formulation (Hinton and Horton 1971) does is to order the last piece in Eq. (3.81) small, assuming incompressible flow or small FLR due to the τ_i factor in p_i multiplied by the implied ρ_s^2 in $\nabla_\perp^2$. Then we have

$$\frac{\partial \varpi}{\partial t} + [W, \varpi] + \nabla \cdot \mathbf{\Pi}^* = \frac{\partial \varpi'}{\partial t} + \nabla \cdot [W, \nabla_\perp \phi] + \text{(neglected terms)}$$
$$\varpi' = \nabla_\perp^2 \phi \tag{3.83}$$

with the prime symbol on ϖ' to denote the pure ExB form. Note the difference between the right hand side of Eq. (3.83) and the left hand side of Eq. (3.80), whose consequence under the demand of energetic consistency is discussed next.

3.9.2 Effect on energetics

If the MHD formulation is used, then the ion flow energy is assumed to be the ExB one not the total one effectively including diamagnetic flow. In other words, while we have the latter in Eq. (3.75), in the MHD form we have to use

$$\frac{1}{2}\frac{\partial}{\partial t}\frac{\rho}{B^2}|\nabla_\perp \tilde{\phi}|^2 - \nabla \cdot \left(\tilde{\phi}\frac{\partial}{\partial t}\frac{\rho}{B^2}\nabla_\perp \tilde{\phi}\right) + [\tilde{\phi}, (-\tilde{\phi}\varpi')] = -\tilde{\phi}\,(\text{RHS}) \tag{3.84}$$

so that the flow energy is the ExB one, and the same for the flow vorticity. This difference propagates all the way through the system since it is now $\tilde{\phi}$ and not $\tilde{W}_i$ which multiplies the RHS to produce transfer terms which have to cancel when all the energetic pieces including those in other equations are summed up.

In the electron density equation we have the form given in Eq. (3.49), with T_i multiplying $\tilde{J}_\| \tilde{n}_e$. This term now has nowhere to go. If we neglect the T_i contribution multiplying $\partial \tilde{n}_e / \partial t$, then the sound wave transfer channels fail to balance, because the $T_i \tilde{u}_\| \nabla_\| \tilde{n}_e$ term needed to balance $\tilde{u}_\| \nabla_\| \tilde{p}_i$ in Eq. (3.50) is now missing. Only if the $\tilde{J}_\| \nabla_\| \tilde{n}_e$ term can be neglected in this equation can we save the situation while leaving T_i in. In normalised units, an MHD isothermal model would look like

$$\frac{\partial \varpi'}{\partial t} + \nabla \cdot [W, \nabla_\perp \phi] = B\nabla_\| \frac{J}{B} - \mathcal{K}(p_e + p_i) \tag{3.85}$$

$$\frac{\partial A}{\partial t} + \hat{\mu}\frac{dJ}{dt} = \nabla_\|(-\phi) - C\,J \tag{3.86}$$

$$\frac{dn}{dt} = -B\nabla_\| \frac{u}{B} + \mathcal{K}(\phi) \tag{3.87}$$

$$\hat{\epsilon}\frac{du}{dt} = -\nabla_\|(p_e + p_i) - B\nabla_\| \frac{P}{B} \tag{3.88}$$

where $p_i = \tau_i n$ and $p_e = n$ and $P = -X\nabla_\| u$ is parallel viscosity. Note the absence of $B\nabla_\|(J/B)$ in the density equation. If we multiply these by $-\phi$, J, $(1 + \tau_i)n$, and u, respectively, we get a closed energy theorem (try it), with the two sinks due to C and P we had in previous models. Note that $(1 + \tau_i)n = p_e + p_i$ in this case, which helps with the transfer terms.

In a warm ion isothermal drift Alfvén model these equations appear as

$$\frac{\partial \varpi}{\partial t} + \nabla \cdot [\phi, \nabla_\perp W] = B\nabla_\parallel \frac{J}{B} - \mathcal{K}(p_e + p_i) \tag{3.89}$$

$$\frac{\partial A}{\partial t} + \hat{\mu}\frac{dJ}{dt} = \nabla_\parallel(p_e - \phi) - C\,J \tag{3.90}$$

$$\frac{dn}{dt} = -B\nabla_\parallel \frac{u - J}{B} + \mathcal{K}(\phi - n) \tag{3.91}$$

$$\hat{\varepsilon}\frac{du}{dt} = -\nabla_\parallel(p_e + p_i + P) \tag{3.92}$$

The only difference to the MHD model outside the vorticity equation is the addition of $\nabla_\parallel p_e$ in the Ohm's law and $B\nabla_\parallel(J/B)$ and $\mathcal{K}(n)$ in the density equation, adding those coupling terms Now we multiply these by $-W$ (not $-\phi$), J, $(1 + \tau_i)n$, and u, respectively, we get the same closed energy theorem (try it), with the two sinks due to C and P we had in previous models, just that now we have the adiabatic transfer $J\nabla_\parallel n$ between the Ohm's law and the density equation, and the diamagnetic compression in the density equation. In the vorticity equation we have the differences discussed above. Each of these models is perfectly energetically consistent, but the price of using the MHD one is using the MHD forms for the other equations and the neglect of the adiabatic response. In Chapter 14 of Volume One, we have already seen that procedure cause damage to the results. Neglecting n against ϕ by assumption, multiplied by τ_i or not, yields a turbulence result which always violates the assumption. Moreover, the mode structure and energetics are very different. So the *a priori* to *a posteriori* test always fails, and the model with MHD ordering has to be regarded as obsolete.

3.9.3 Effect on the turbulence

We run illustrative cases with $C = 1.39$ and $C = 13.9$ while varying $\hat{\beta}$ as we did previously. Here the gradients are chosen in the ITG regime while keeping the other parameters of the warm plasma case. We have $\hat{\beta} = 1$ to 7 with $\omega_t = \omega_i = \tau_i = 1$ and $\omega_n = 0.5$, which emphasises the ∇T drive, nominally mostly for the ions but at these parameters also for the electrons. One set of runs keeps gyroviscosity, as in Eq. (3.35). The second neglects is entirely, so that Eq. (3.35) is replaced by

$$\frac{\partial \varpi}{\partial t} + [\tilde{W}_i, \varpi] = B\nabla_\parallel \frac{\tilde{J}_\parallel}{B} - \mathcal{K}(\tilde{p}_e + \tilde{p}_i) \tag{3.93}$$

with ϖ and $\tilde{W}_i$ defined in Eq. (3.43). This is a control test to isolate the effect of gyroviscosity on the turbulence.

The isotropic perpendicular spectra for the case with $\hat{\beta} = 1$ and $C = 13.9$ with and without gyroviscosity are shown in figure 3.24. The relevant vorticity nonlinearity is the one in Eq. (3.35), for the nominal case, and for the case not treating gyroviscosity it is the one in Eq. (3.93). The main difference is in the vorticity spectrum. The effect

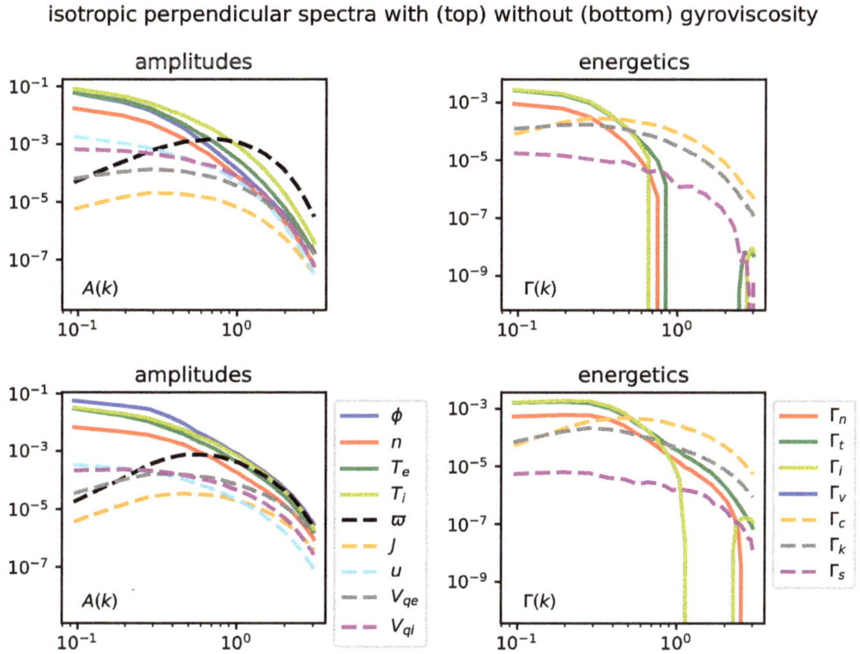

Figure 3.24. Isotropic perpendicular spectra for the case with $\hat{\beta} = 1$ and $C = 13.9$, with and without gyroviscosity, during $500 < t < 1000$, in the saturated phase. The case with gyroviscosity is nominal for warm ions. The relevant vorticity nonlinearity is the one in Eq. (3.35). The neglect of gyroviscosity leads to an underestimate of the vorticity dynamics in the $k_\perp \rho_i \sim 1$ region. The difference results from whether or not enstrophy is conserved by the vorticity nonlinearity (see text).

of gyroviscosity is to provide the lowest order (in $k_\perp \rho_i$) finite gyroradius correction to the vorticity dynamics. When this is done, there is more enstrophy in this small scale range (the ϖ amplitude). Its main effect is to reduce the energy content in the state variables, especially in this range. An additional effect is to reduce the amplitude of $\tilde{\phi}$ in the entire spectral range. In the absence of gyroviscosity, the first two terms in Eq. (3.93) are isomorphic to the 2D Euler equation, so that the dual energy/enstrophy cascade tendency is pristine at all scales. By contrast, when gyroviscosity is restored, the overall nonlinearity still conserves energy but it no longer conserves enstrophy, which can now grow. Less energy is shifted to larger scale in the ExB channel, while the increased vorticity shifts more small scale thermal energy towards high-$k_\perp$ dissipation. The net result is a strong reduction in the overall turbulence amplitude, especially when the nonadiabatic nonlinear dynamics is robust, i.e., at higher collisionality.

This is a large qualitative difference in the dynamics and is the reason gyroviscosity is fundamental to warm-ion dynamics. It does not help to maintain that ρ_s scales are small and hence gyroviscosity can be neglected. We find the same nonlinear transfer tendency everywhere in the spectrum in either case. So the fact that enstrophy conservation is broken has its effect at large as well as small scales.

3.10 Summary

The principal effect of the ion temperature on fluid drift dynamics is to add a channel of energetics which is independent of the kinetic shear Alfvén waves and the adiabatic force balance and response. The ion temperature couples directly only to the acoustic dynamics and the ion parallel heat flux as both a dynamical quantity and a dissipation mechanism. Its relationship to the rest of the system is primarily independent of the electrons, although its energetic process is still the nonlinear ExB advection process which is indirectly coupled back to the electrons and the parallel current.

A new type of instability occurs, the so-called eta-ion, or ion temperature gradient (ITG) mode, which is sensitive to $\eta_i = d \log T_i / d \log n_e$ as a parameter. The classical version is the one with adiabatic electrons and an accessibility mechanism for the ion temperature gradient. There is a slab branch, functioning as a thermal sound wave, and an interchange branch as a radial flow mode existing in an inhomogeneous magnetic field. In both cases, a threshold exists for the relative gradient strength ratio, η_i. The threshold results from the tendency for an ExB eddy to drift out of the way of an emerging ion temperature disturbance, for smaller values of η_i. For strongly nonadiabatic regimes, the threshold disappears, but the transition to an ITG mode structure characterised by $\tilde{T}_i$ being the largest normalised disturbance still persists. This can emerge in turbulence as well, but investigation of that regime has just started.

For fluid drift turbulence in toroidal geometry, the mode structure is either drift wave or ballooning like, depending mostly on proximity to the ideal ballooning threshold given in Eq. (3.13). The resistive side is a more gradual transition to longer wavelengths and more robust amplitudes which, eventually at arbitrarily high collisionality would produce a more MHD type of dynamics. This would occur at a collision frequency greater than nominal RMS vorticity by several orders of magnitude due to arbitrarily large scale of background. In laboratory plasmas with a toroidal magnetic field and a gradient scale length comparable to the scale of the device, however, such a large separation does not occur, and the dynamics remains characterised by a limited scale range, two orders of magnitude at most, between large and gyroradius scales of motion. In this case the drift and gyroradius scales and the gradient scale length represent two regions of the spectrum that remain in constant energetic contact, such that the result cannot ignore either one of them. The result is a self-consistent product of all processes in this scale range.

Two things have emerged about the fluid model. The collisionality regime that would allow replacement of the parallel heat flux by a collisional formula given by the parallel gradient and the collision parameters is not satisfied. The electrons can be marginal, but the ions are never this collisional. The parallel heat fluxes must therefore be carried as dependent variables, which eventually opens the question of whether that is enough detail, and even if so, which kinetic processes enter to overcome the, now weak, collisional dissipation. The drift scale ρ_s, is not a proper gyroradius but an effective ion inertia scale, c_s/Ω_i, as the low frequency, perpendicular force balance counterpart to the Alfvén inertial scale, v_A/Ω_i, of Hall MHD. But due to the drift wave turbulence dynamics, this is where the vorticity tends to collect, regardless of whether other forcing mechanisms produce larger scale motion or not. So it always has to be faced. Now, since $T_i \sim T_e$ in usual situations, ρ_s is equivalent to ρ_s, and ρ_i is a proper gyroradius. We can treat the corresponding corrections to the fluid model, gyroviscosity, that go with this. But the treatment eventually breaks down since the powers of $k_\perp$ involved guarantee that the nonlinear responses are not bounded. We will need proper finite gyroradius effects at arbitrary order, not as a correction. It is to treat these processes properly that we will need the gyrofluid model and the kinetic theory that grounds it. One other process will emerge that has large effect, especially for ions: that of thermal anisotropy. The perpendicular and parallel temperature, essentially weighted averages of the kinetic energy in particle gyromotion on the one hand and streaming motion on the other, have different roles in the dynamics. Each appears together with its own parallel heat flux. So there are perp-parallel and parallel-parallel heat fluxes, each as dependent variables. These will also bring important changes to the results, especially in the hot plasma ITG dynamics. We now turn to all this in the next set of Chapters. We also delay the issues of flows in systems dominated by temperature dynamics until then.

Further Reading

Early work laying the groundwork for this type of ion temperature gradient instability is reviewed by B B Kadomtsev and O P Pogutse, in 'Turbulence in toroidal systems,' *Rev. Plasma Phys.* **5** (1970) 249. Salient refences are L I Rudakov and R Z Sagdeev, 'On the instability of a nonuniform rarefied plasma in a strong magnetic field,' *Dokl. Akad. Nauk. SSSR* **138** (1961) 581 [English translation: Sov. Phys. Dokl. 6 (1961) 415]; and B Coppi, 'Fluidlike electron and ion modes in inhomogeneous plasmas,' *Phys. Rev. Lett.* **21** (1968) 1055.

The toroidal version was given by P N Guzdar, Liu Chen, W M Tang, and P H Rutherford, in 'Ion-temperature-gradient instability in toroidal plasmas,' *Phys. Fluids* **26** (1983) 673. Toroidal curvature destabilisation on collisionless fluid like modes was also considered by B Coppi and F Pegoraro, in 'Theory of the ubiquitous mode,' *Nucl. Fusion* **17** (1977) 969.

The complications of the finite ion Larmor radius for drift instabilities and turbulence have been known for a long time, and are also discussed in the two reviews mentioned earlier, by W M Tang, in *Nucl. Fusion* **18** (1978) 1089, and by P C Liewer, in *Nucl. Fusion* **25** (1985) 543. Several models for finite Larmor radius fluid equations were in use by that time, and one of the most important is the one by F L Hinton and C W Horton Jr, in *Phys. Fluids* **14** (1971) 116. It is important to note that in those days MHD ordering was still common, as noted in section 3.9.

The Hinton and Horton version of the normalised polarisation equation is Eq. (3.83), their Eqs (11–13), which was noted by A Hasegawa and K Mima, *Phys. Rev. Lett.* **39** (1977) 205.

Three dimensional ITG computations were done by R E Waltz, 'Three-dimensional global numerical simulation of ion temperature gradient mode turbulence,' *Phys. Fluids* **31** (1988) 1962, and X Garbet *et al*, 'Global simulations of ion turbulence with magnetic shear reversal,' *Phys. Plasmas* **8** (2001) 2793.

The version of polarisation with gyroviscosity at finite ion temperature in Eq. (3.80) appeared in Eq. (48) of B Scott, 'Three-dimensional computation of collisional drift wave turbulence and transport in tokamak geometry,' *Plasma Phys. Contr. Fusion* **39** (1997) 471, in Eq. (2) of B N Rogers and J F Drake, 'Enhancement of turbulence in Tokamaks by magnetic fluctuations,' *Phys. Rev. Lett.* **79** (1997) 229, and in Eq. (31) of A Zeiler, J F Drake, and B N Rogers, 'Nonlinear reduced Braginskii equations with ion thermal dynamics in toroidal plasma,' *Phys. Plasmas* **4** (1997) 2134.

The gyrofluid model mentioned in the text will be covered in detail in Chapter 9, where the main references are cited. Similar for the gyrokinetic model on which the gyrofluid one is based, covered in Chapters 5 and 7. To do this properly, one has to start with the general formulation of the theory, and then re-derive the equations in turn. We begin this with the next Chapter.

IOP Publishing

Turbulence and Instabilities in Magnetised Plasmas, Volume 2
Gyrokinetic theory and gyrofluid turbulence
Bruce Scott

Chapter 4

Lagrangian field theory and drifts

The idea behind this Chapter as an interlude is to introduce the basics of Maxwell/plasma electrodynamics as a form of classical field theory, to assemble the necessary concepts before moving on to gyrokinetic theory. We review the basics of drift motion of charged particles in an electric and a magnetic field, in which the motion can be solved exactly by conventional means. Allowing a straight electric field to be slowly time dependent introduces the frequency ordering and the meaning of polarisation. Due to the difficulties of more involved calculations, we would like to recast the representation so that polarisation enters as a density rather than a flow. So then we review the basics of the Lagrangian theory of motion in a field with a conservative interaction, but in the language in which we will need it later. Canonical form is introduced here. Then we introduce the field theory version so that the self consistent responses are treated properly. This follows textbook classical field theory but with special emphasis on our problem. Maxwell electrodynamics is recovered but at the same time recast in the language or representation in which we will need it to do gyrokinetic theory. The conservation laws are indicated by Noether's theorem, which identifies a conserved quantity to go with the symmetry of the Lagrangian with respect to any coordinate. Lastly, the idea of the background magnetic field as an anchor within the Lagrangian is introduced, and the initial difficulties are identified. This points the way to the procedure we will then need in the next Chapter to do the gyrokinetic theory formally. The present Chapter serves largely as motivation. It is placed here not for mathematical rigour but to gain the necessary physical understanding beforehand.

One could call this Chapter a matter of doing the basics of Maxwell/plasma electrodynamics with phase space Lagrangians before we go on and do gyrokinetic theory.

4.1 Low frequency drifts

We start with the textbook equation of motion for charged particles. Each particle has a mass m and a charge e and moves according to the Lorentz force. The theory is nonrelativistic, and the units are in SI. We have the equation of motion,

$$m\frac{d\mathbf{v}}{dt} = e(\mathbf{E} + \mathbf{v} \times \mathbf{B}) \tag{4.1}$$

where $\mathbf{v}$ is the particle velocity, and $\mathbf{E}$ and $\mathbf{B}$ are the electric and magnetic fields. The scale of the time derivative is the dynamical frequency ω, while that of the Lorentz force yields the gyrofrequency Ω. If this is the sole dynamical content ($\mathbf{E}$ and $\mathbf{B}$ are prescribed) then an equivalent form is

$$\left(\frac{d}{dt} + \Omega\mathbf{b}\times\right)\mathbf{v} = \frac{e}{m}\mathbf{E}, \qquad \Omega \equiv \frac{eB}{m} \tag{4.2}$$

which defines the gyrofrequency in terms of the magnetic field strength B, with directionality given by the cross product with the unit vector $\mathbf{b}$. In the absence of an electric field, the acceleration is always at right angles to the velocity, so the motion is circular, and the coefficient gives the frequency of motion as $\omega = \Omega$. If the magnetic field is constant, then this solution is exact. If a constant electric field is added as a small correction, we can still solve for the motion exactly, and the result is a small drift superposed upon the gyromotion. But we get a better picture by solving in the beginning for that drift.

If we assume that $\mathbf{E}$ and $\mathbf{B}$ are forced upon by other means with their own dynamical frequencies ω, and further assume that $\omega \ll \Omega$, then we solve for the $\mathbf{v}$ in the Lorentz force rather than advancing it through the time derivative. This is the same procedure as done for fluid drifts in Volume One, but now we concentrate on the motion at the particle level. Ultimately this will motivate the concept of a gyrocentre. We operate from the left with the drift combination $B^{-1}\mathbf{b}\times$, use the vector identity and auxiliaries

$$\mathbf{b} \times (\mathbf{v} \times \mathbf{b}) = \mathbf{v}_\perp, \qquad \mathbf{v}_\perp = \mathbf{v} - v_\parallel\mathbf{b}, \qquad v_\parallel = \mathbf{b} \cdot \mathbf{v} \tag{4.3}$$

which defines the perpendicular velocity $\mathbf{v}_\perp$ and the parallel velocity $v_\parallel$, and find

$$\mathbf{v}_\perp = \frac{1}{B^2}\mathbf{E} \times \mathbf{B} + \frac{m}{eB^2}\mathbf{B} \times \frac{d\mathbf{v}}{dt} \tag{4.4}$$

The first term is now the main ExB drift velocity, and the second is the correction, found by successive substitution, which yields

$$\mathbf{v}_\perp = \frac{1}{B^2}\mathbf{E} \times \mathbf{B} + \frac{m}{eB^2}\frac{d\mathbf{E}_\perp}{dt} \tag{4.5}$$

assuming we can commute the B-cross operation past the time derivative. The correction is called the polarisation drift, because it separates charge. This is because *e*/*m* is species dependent, so it is no longer true that the perpendicular motion is the same for every charged particle.

This analysis assumes that scale of variation of **B** is large and its time variation is very small, etc, compared to the scales for **E**. These are necessary assumptions to describe these drifts. To summarise, the dynamical frequency range must be slow compared to Ω, and the time and space variation for **E** must be stronger than for **B**, which has to be able to serve as an anchor. The background magnetic field as an anchor for the motion is the fundamental prescription for the motion as drift motion.

4.1.1 Meaning/significance of ExB drift motion

The main point of this result is that the ExB drift is the same for every charged particle. This is the reason fluid drift models work. Even in the absence of thermalising dissipation, the entire fluid moves together at least in perpendicular motion.

The correction due to polarisation is small, but it is significant in providing charge separation which affects the variation of the electric field. This charge separation is ultimately how we can get electric fields in a neutral plasma.

In general, we start to think even of individual particle motion as being replaced by drift or gyrocentre motion. It is also necessary to remember that the trajectories implicitly described by this motion exist only in an infinitesimal sense. Even the gyrocentres are not tracing out paths, or orbits, in a given set of fields. The fields are themselves in evolution in response to what the gyrocentres do. We allow small disturbances in the magnetic field to accompany the electric field, which in its entirety is considered among the dynamical variables even if it can sometimes have a quasi-steady component describing an ExB flow.

Because we have a set of arbitrarily many gyrocentres and the electric and magnetic disturbances in a self consistent interaction, we treat them as a distribution and dispense with individual orbits in favour of characteristics of the motion. The gyrocentres are described by a distribution function, or equivalently a phase space density, with position and momentum described by coordinates. The collective responses of these then serve as sources for the fields. The response of each piece changes that of all the others, and vice versa and so on, so that they evolve together. Although we do consider the reaction of one to the others as given, we only do so in an infinitesimal sense, so that second order differentials can be neglected. This is done to build the theory, and even to advance the equations computationally, but not to describe the physics.

Instead of setting up drifts in a heuristic manner as we have been doing, we would like a better approach. The one of choice is to set up a system Lagrangian, which describes not only the motion of one gyrocentre, but each of them. It is a field theory, so that we eventually have a system Lagrangian which describes the motion of each gyrocentre and each field degree of freedom simultaneously. It is better to think of this in terms of all of the gyrocentres, rather than in terms of each gyrocentre, in order to form a clearer picture of what a self-consistent system is and how it evolves.

So instead of doing the drift approximations on a set of equations already known, we rather set the approximations in the system Lagrangian. This way, the exact consistency of the system is not changed even though we restrict to the low frequency guide field regime. Then, the derivation process in obtaining the equations of motion from the Lagrangian remains exact despite the use of approximations in the Lagrangian itself. This is why one does field theory in the first place.

The contrast to the conventional approach in terms of fluid systematics is the order of the approximations. It is conventional to derive fluid electrodynamic equations with Maxwell's equations and then take the low frequency approximations. Now, we take the low frequency approximations in the Lagrangian first, and then derive the kinetic theory and at least in principle a form of the fluid theory. Since gyrocentres are being set up first, the fluid version of the theory is now called 'gyrofluid' theory, following the kinetic theory having been recast in terms of a 'gyrokinetic' theory. This is the origin of this nomenclature.

The validity of the gyrokinetic and gyrofluid approach is to be decided *a posteriori* in the sense of showing how the gyrofluid equations and the fluid drift equations become isomorphic in the appropriate limit. In other words, we will have to show that the two representations actually solve the same problem. We will do so after having developed the gyrofluid equations. After that, subsequent Chapters will address its handling of turbulence in a magnetised plasma.

4.1.2 Outline of the rest of this lecture

We start with the particle picture to develop the field theory structure we will need later. For now it is particles and fields, and in the last section on drifts and quasineutrality we start making the moves towards the drifts picture, eventually replacing particles with gyrocentres, and charge density with polarisation.

We will first recover the textbook version of classical electrodynamic equations using the Lagrangian theory in phase space, using an $\mathbf{x}$ and a $\mathbf{p}$ for position and momentum rather than an $\mathbf{x}$ and an $\dot{\mathbf{x}}$ for position and velocity. Doing the derivations both ways helps to understand the meaning of canonical form, by which variation of dynamical fields is kept entirely within the time component of the system Lagrangian. Then we consider a strong background magnetic field, at first in an unspecified measure, through a static but spatially dependent field potential, which is varied only through the effect on $\mathbf{x}$ and $\mathbf{p}$. This will show how that particular term sets up drifts. There arise various difficulties in consistency in keeping the electric field energy to control the evolution of the electric field through its potential variable. We encounter the comparison between static electric field energy and the kinetic energy of ExB motion. This then motivates the idea of quasineutrality, and now we will get that by taking a definite step in the Lagrangian.

Afterward, once all these concepts are in place, we move on to consider the gyrocentre dynamics, at first individually, and then gyrokinetic field theory, to get the entire dynamical system.

4.2 Lagrangian field theory

We follow the derivation path taken by Landau/Lifshitz, and consider an axial vector potential and a charge interaction. We set up a free particle Lagrangian, an interaction Lagrangian, and a free field Lagrangian, and then join them as three pieces of a whole. In a nonrelativistic setting this is

$$L = L_{fp} + L_{int} + L_{free} \tag{4.6}$$

The free particle piece is

$$L_{fp} = \frac{1}{2}m\,|\dot{\mathbf{x}}|^2 \tag{4.7}$$

where the dot denotes d/dt. The action is the integration of this over a time interval, following the particle therefore on the trajectory, with the endpoints fixed. The minimum of the action is found by varying the trajectory in between and setting the variation to zero. The strategy is to isolate the variation $\delta\mathbf{x}$ so that the extremum condition leads to setting its coefficient to zero, with all necessary integrations by parts done along the way. The conditions formed by setting the variations to zero are the 'Euler-Lagrange equations' for the model described by the Lagrangian.

For a total differential, we have

$$L = \frac{dQ}{dt}, \qquad \delta\int L\,dt = \delta\int dQ = \delta Q\Big|_{\text{ends}} = 0 \tag{4.8}$$

so it contributes neither to the action nor to the resulting dynamics. Adding a total differential to the differential action can be thought of as a sort of gauge transformation, to recast some of its pieces. We will make extensive use of this idea later. Some pieces of the differential action are collected into a total differential and then subtracted off, which is basically the same operation.

For the free particle Lagrangian, the variation with respect to $\mathbf{x}$ gives

$$\delta\int L_{fp}\,dt = \int dt\,\frac{d}{dt}(\delta\mathbf{x}\,m\dot{\mathbf{x}}) - \int dt\,\delta\mathbf{x}\cdot\left(m\frac{d\dot{\mathbf{x}}}{dt}\right) = 0 \tag{4.9}$$

The first term vanishes as a pure differential under the integral. The second term yields $m\,d^2\mathbf{x}/dt^2 = 0$, the equation for a free particle.

The interaction Lagrangian is the four-vector dot product between a position and an axial four-vector potential, which in a nonrelativistic setting is

$$L_{int} = e\mathbf{A}\cdot\dot{\mathbf{x}} - e\phi \tag{4.10}$$

The particle Lagrangian is the sum of these two pieces,

$$L_p = L_{fp} + L_{int} \tag{4.11}$$

as the part of the whole Lagrangian dependent on the particles. This is the form for one particle. For the sum of the particles, the part of L dependent on the particles is the sum over all particles of the L_p for each particle.

For the particle motion we vary only $\mathbf{x}$ (still) to obtain

$$\delta\int L_p\,dt = -\int dt\,\delta\mathbf{x}\cdot\left(m\frac{d\dot{\mathbf{x}}}{dt} + e\frac{\partial\mathbf{A}}{\partial t} + e\nabla\phi - e\dot{\mathbf{x}}\times\nabla\times\mathbf{A}\right) = 0 \tag{4.12}$$

no longer writing the total differential. This yields

$$m\frac{d\dot{\mathbf{x}}}{dt} = -e\left(\frac{\partial \mathbf{A}}{\partial t} + \nabla\phi\right) + e\dot{\mathbf{x}} \times \nabla \times \mathbf{A} \tag{4.13}$$

which when defining $\mathbf{v} = \dot{\mathbf{x}}$ we recognise as the Lorentz force as given above, with

$$\mathbf{E} = -\left(\frac{\partial \mathbf{A}}{\partial t} + \nabla\phi\right), \qquad \mathbf{B} = \nabla \times \mathbf{A} \tag{4.14}$$

To get these we have used another vector identity,

$$(\nabla\mathbf{A}) \cdot \dot{\mathbf{x}} - \dot{\mathbf{x}} \cdot (\nabla\mathbf{A}) = \dot{\mathbf{x}} \times \nabla \times \mathbf{A} \tag{4.15}$$

We have the particle motion, and now we get the responses of the field variables.

The field Lagrangian is given in terms of a volume integral over a density,

$$L_{free} = \int d\mathcal{V}\, \frac{1}{2}\left(\varepsilon_0 E^2 - \mu_0^{-1} B^2\right) \tag{4.16}$$

where ε_0 and μ_0 are physical constants. The sign choice is peculiar, and Landau/Lifshitz motivate it using the four-vector language. For now we simply lift this from the texts, and substitute the forms using ϕ and $\mathbf{A}$ into the expression involving $\mathbf{E}$ and $\mathbf{B}$, so that

$$L_{free} = \int d\mathcal{V}\, \frac{1}{2}\left[\varepsilon_0\left(\frac{\partial \mathbf{A}}{\partial t} + \nabla\phi\right)^2 - \mu_0^{-1}(\nabla \times \mathbf{A})^2\right] \tag{4.17}$$

Noting the gauge invariance with respect to

$$\mathbf{A} \to \mathbf{A} - \nabla\chi, \qquad \phi \to \phi + \frac{\partial\chi}{\partial t} \tag{4.18}$$

we can always find a χ which makes the potentials satisfy

$$\frac{\partial\phi}{\partial t} + \nabla \cdot \mathbf{A} = 0 \qquad \text{using} \qquad \frac{\partial^2\chi}{\partial t^2} - \nabla^2\chi = \text{source} \tag{4.19}$$

with the source given by the degree to which this gauge condition is not satisfied. This condition is called the Lorentz gauge and is standard for time-dependent electrodynamics.

The interaction Lagrangian is already specified, but now the particles must be brought into the Lagrangian density. The way to do this is through the Dirac delta function, which satisfies

$$\int d\mathcal{V}\, \delta(\mathbf{x} - \mathbf{x}_p) = 1, \qquad \delta(\mathbf{x} - \mathbf{x}_p) = \begin{cases} \text{indeterminate if } \mathbf{x} = \mathbf{x}_p \\ 0 \text{ otherwise} \end{cases} \tag{4.20}$$

We can then define a charge density and a current

$$\rho_{ch}(\mathbf{x}) = \sum_p e_p\, \delta(\mathbf{x} - \mathbf{x}_p), \qquad \mathbf{J}(\mathbf{x}) = \sum_p e_p \dot{\mathbf{x}}\, \delta(\mathbf{x} - \mathbf{x}_p) \tag{4.21}$$

where the sum is over particles, each of which with its own charge e_p and mass m_p, the latter not appearing in L_{int}.

The field Lagrangian written with both the free field and the interaction pieces is now

$$L_{int} + L_{free} = \int d\mathcal{V}\left[\mathbf{A}\cdot\mathbf{J} - \rho_{ch}\phi + \frac{\varepsilon_0}{2}\left(\frac{\partial\mathbf{A}}{\partial t} + \nabla\phi\right)^2 - \frac{1}{2\mu_0}(\nabla\times\mathbf{A})^2\right] \tag{4.22}$$

Varying the action with respect to $\mathbf{A}$ and ϕ, leaving ρ_{ch} and $\mathbf{J}$ fixed, yields

$$\delta\int\left(L_{int} + L_{free}\right)dt = \int dt\int d\mathcal{V}$$
$$\left[\delta\phi\left(\varepsilon_0\nabla\cdot\mathbf{E} - \rho_{ch}\right) - \frac{1}{\mu_0}\delta\mathbf{A}\cdot\left(\nabla\times\mathbf{B} - \mu_0\mathbf{J} - \mu_0\varepsilon_0\frac{\partial\mathbf{E}}{\partial t}\right)\right] = 0 \tag{4.23}$$

discarding pure divergence terms as well as pure time derivatives. This yields

$$\nabla\times\mathbf{B} = \mu_0\mathbf{J} + \mu_0\varepsilon_0\frac{\partial\mathbf{E}}{\partial t}, \qquad \nabla\cdot\mathbf{E} = \varepsilon_0^{-1}\rho_{ch} \tag{4.24}$$

which are the two Maxwell equations with sources. The other two come from the properties of the Maxwell stress four-tensor, defined by

$$F_{ab} = \nabla_a A_b - \nabla_b A_a \tag{4.25}$$

and the symmetry principle which results from it,

$$\nabla_a F_{bc} + \nabla_c F_{ab} + \nabla_b F_{ca} = 0 \tag{4.26}$$

which yield

$$\nabla\cdot\mathbf{B} = 0, \qquad \nabla\times\mathbf{E} = -\frac{\partial\mathbf{B}}{\partial t} \tag{4.27}$$

by construction. This is how to get both the charged particle equation of motion, with the Lorentz force, and also the Maxwell equations, from the same system Lagrangian.

In general, a Lagrangian field theory has

$$L = \int d\mathcal{V}\,\mathcal{L}, \qquad \mathcal{L} = \mathcal{L}_{fp} + \mathcal{L}_{int} + \mathcal{L}_{free} \tag{4.28}$$

with the same three pieces as before, just with all the particles put onto delta functions. The particle motion is found by varying $\mathbf{x}_p$, and the derivatives of the field potentials are obtained by varying the occurrence of $\mathbf{x}_p$ within the delta function and integrating by parts. The field equations are found by varying each field variable in turn, which we did here using the potentials.

Some notes are in order. The main point is the structure of the theory, more than any individual result. One ends with the gradients of the potentials in the equations describing the dynamics, with the time derivative of the one coupled to velocity implicitly or explicitly stated. The next step is to extend the coordinates to a kinetic phase space, with position and momentum, and a Hamiltonian describing the time component. Then we re-do the same problem as before to become familiar with the new language.

4.3 Canonical representation

On the way to defining canonical form, we change things around to form what becomes known as the canonical representation. We start with a Lagrangian $L(\mathbf{x}, \dot{\mathbf{x}})$ for one particle with position $\mathbf{x}$ and velocity $\dot{\mathbf{x}}$. We define the canonical momentum $\mathbf{p}$ and then the Hamiltonian H as

$$\mathbf{p} \equiv \frac{\partial L}{\partial \dot{\mathbf{x}}}, \qquad H = \mathbf{p} \cdot \dot{\mathbf{x}} - L \tag{4.29}$$

The action can now be written as

$$S = \int dt\ L, \qquad L = \mathbf{p} \cdot \dot{\mathbf{x}} - H \tag{4.30}$$

This looks like a simple re-arrangement, and it is, actually. But the space of coordinates is now enlarged from just $\mathbf{x}$ to $(\mathbf{x}, \mathbf{p})$. The steps obtaining first $\mathbf{p}$ and H and then re-expressing $L(\mathbf{x}, \dot{\mathbf{x}})$ into $L(\mathbf{x}, \mathbf{p})$ are collectively referred to as the Legendre transformation. It is actually a specific example of the more general Legendre transformation in mathematics, but when one hears of it in physics this is what is usually meant.

Any Lagrangian in this form is a canonical Lagrangian, if the canonical momentum is actually being used as the coordinate. The next step is to express the action as a differential. With $\dot{\mathbf{x}}\,dt$ re-expressed as $\mathbf{dx}$, we obtain

$$dS = \mathbf{p} \cdot \mathbf{dx} - H\ dt \equiv \gamma_\mu\ dz^\mu \qquad \text{with} \quad \mu \in \{1, 2, 3, 4, 5, 6, 7\} \tag{4.31}$$

where the index μ covers the dimensions of $\mathbf{dx}$ and $\mathbf{dp}$, and time as the seventh coordinate. One can get more mathematical and have an energy as the eighth coordinate, but we will not do that here. In this and other similar expressions to follow, the Einstein convention is used, with repeated indices, one up (contravariant) and one down (covariant), are implied summed without having to write the sum symbol. We use the symbol explicitly for the sum over species or particles, but not for components of vectors, tensors, or forms. Technically speaking, dS is the differential action and not the Lagrangian, but when using this language we can be cavalier and refer to $\gamma_\mu\ dz^\mu$ as simply the Lagrangian.

In a nonrelativistic setting we can separate time from phase space and rewrite the differential action as

$$dS = \gamma_a\ dz^a - H\ dt \qquad \text{with} \quad a \in \{1, 2, 3, 4, 5, 6\} \tag{4.32}$$

explicitly specifying that any transformation will not apply to time. Either of these forms of the differential action dS (with time separated off or not) are referred to as the Poincaré–Cartan fundamental one-form, and we will use it as a central element of the transform used in gyrokinetic theory later on.

Now we vary the action according to

$$\delta(dS) = \delta\gamma_a\, dz^a + \gamma_a\, \delta(dz^a) - \delta H\, dt \tag{4.33}$$

Use the chain rule to get

$$\delta\gamma_a = \gamma_{a,b}\, \delta z^b, \qquad \delta H = H_{,a}\, \delta z^a, \qquad H_{,a} \equiv \frac{\partial H}{\partial z^a} \tag{4.34}$$

where the subscript after a comma indicates differentiation as noted. These result in

$$\delta(dS) = \gamma_{a,b}(dz^a\, \delta z^b - dz^b\, \delta z^a) - \left(\frac{\partial\gamma_a}{\partial t} + H_{,a}\right)dt\, \delta z^a + d(\gamma_a\, \delta z^a) \tag{4.35}$$

The last term as a total differential vanishes under the integration over the trajectory and can be dropped.

Then to find the equations of motion we set the coefficient of each δz^a to zero. To do this we first switch the indices ab under the sum on the first term so that δz^a becomes common. This yields

$$\delta(dS) = \left[(\gamma_{b,a} - \gamma_{a,b})dz^b - \left(\frac{\partial\gamma_a}{\partial t} + H_{,a}\right)dt\right]\delta z^a + d() \tag{4.36}$$

and then we have

$$(\gamma_{b,a} - \gamma_{a,b})dz^b - \left(\frac{\partial\gamma_a}{\partial t} + H_{,a}\right)dt = 0 \tag{4.37}$$

as our set of six expressions, or

$$(\gamma_{b,a} - \gamma_{a,b})\frac{dz^b}{dt} = \frac{\partial\gamma_a}{\partial t} + H_{,a} \tag{4.38}$$

as our version of Hamilton's equations. The quantity in parentheses on the left hand side is called the symplectic structure, which arranges the dependences of the time derivatives of the coordinates against the coordinate derivatives of H and also any time dependence in the $\gamma_a\, dz^a$ pieces, collectively referred to as the symplectic part of the Lagrangian.

Hamilton's equations are explicitly this result for the canonical Lagrangian. If

$$\gamma_\nu dz^\nu = p_i\, dx^i - H\, dt \tag{4.39}$$

then applying the above we have

$$dx^i = \frac{\partial H}{\partial p_i}\, dt, \qquad dp_i = -\frac{\partial H}{\partial x_i}\, dt \tag{4.40}$$

as the coefficients of dp_i and dx^i, respectively, are set to vanish. These are Hamilton's equations as differentials.

Looking at total differentials again, we note that if we add dQ to our one-form we have

$$\gamma_\nu \, dz^\nu + dQ = \left(\gamma_\nu + Q_{,\nu}\right) dz^\nu \tag{4.41}$$

The exterior derivative of this is

$$(\gamma_{\nu,\mu} - \gamma_{\mu,\nu}) dz^\mu dz^\nu + \left(Q_{,\nu\mu} - Q_{,\mu\nu}\right) dz^\mu dz^\nu \tag{4.42}$$

Since the second term vanishes, we can say that the addition of any total differential to the fundamental one-form leaves the dynamical equations unchanged. This is the simplest form of a gauge transformation, and it can be used to re-cast the Lagrangian into a more convenient form to make the resulting equations more readily computable.

4.3.1 Canonical form

When the time is separated out, we have the equations of motion with the symplectic part on the right hand side and the derivatives of H on the left, accompanied by $\partial\gamma_a/\partial t$, the time derivative of the components of the one-form. Note that we do not get a term due to $\partial H/\partial t$ in this. That is because H is itself the time component, and contributions due to $\gamma_{\mu,\mu}$ cancel out of the sympletic structure due to its antisymmetry.

The time derivative of γ_a comes from the integration by parts of $\gamma_a\, \delta(dz^a)$. In a symplectic representation this piece can include time dependent field variables and so their time derivatives appear here. What we refer to as canonical form is simply that all the dependence on the field variables is absent in the symplectic part. Dynamical field variables therefore appear solely in the time component, which for the particle motion is just H. So we would have γ_a independent of the fields and hence subject to geometrical symmetries, and H dependent on time through the time dependence of the field variables. This is a Lagrangian in canonical form. If $\gamma_a\, dz^a$ is just $\mathbf{p}\cdot\mathbf{dx}$, then the one-form is actually a canonical Lagrangian. In canonical form, the symplectic part is not necessarily $\mathbf{p}\cdot\mathbf{dx}$, it is just independent of the field variables so that it inherits any symmetry in the geometry of the problem.

4.3.2 Canonical representation of electrodynamics

It is useful to learn this by applying it to something for which we already know the answer—particle motion under the Lorentz force. Starting with the particle Lagrangian, we find from Eq. (4.29)

$$\mathbf{p} = \frac{\partial L_p}{\partial \dot{\mathbf{x}}} = e\mathbf{A} + m\dot{\mathbf{x}}, \qquad H = \frac{1}{2m}\, |\mathbf{p} - e\mathbf{A}|^2 + e\phi \tag{4.43}$$

and then from Eq. (4.30) we have

$$dS = \mathbf{p}\cdot\mathbf{dx} - H\, dt \qquad \text{where} \quad \gamma_{1,2,3} = \mathbf{p} \quad \text{and} \quad \gamma_{4,5,6} = 0 \tag{4.44}$$

as above, with this γ_a as our specific example. Now we work through the components of the symplectic structure due to this one-form. Varying **p** we find for the coefficients of δp^i

$$\frac{dx_i}{dt} = \frac{\partial H}{\partial p^i} = \frac{1}{m}(p_i - eA_i) = U_i \tag{4.45}$$

where $\mathbf{U}(\mathbf{x}, \mathbf{p})$ is the velocity which is a function of the phase-space coordinates. Varying **x** we find for the coefficients of δx^i

$$\frac{dp_i}{dt} = -\frac{\partial H}{\partial x^i} = eU^j A_{j,i} - e\phi_{,i} \tag{4.46}$$

This is the canonical representation of the Lorentz force motion where we have p_i under the time derivative instead of U_i. To recover the familiar form we explicitly substitute the expression for $p_i = eA_i + mU_i$ into the time derivative, and use

$$\frac{dA_i}{dt} = \frac{\partial A_i}{\partial t} + A_{i,j}\frac{dx^j}{dt} = \frac{\partial A_i}{\partial t} + A_{i,j}U^j \tag{4.47}$$

This now yields

$$m\frac{dU_i}{dt} = eU^j\left(A_{j,i} - A_{i,j}\right) - e\left(\frac{\partial A_i}{\partial t} + \phi_{,i}\right) \tag{4.48}$$

This is the symplectic representation of the equations of motion, which is the more familiar one. Note the antisymmetry in the expression involving derivatives of **A**. The first term is actually velocity cross magnetic field, if we recognise

$$F_{ij} = A_{j,i} - A_{i,j}, \qquad F_{ij} = \varepsilon_{ijk}B^k, \qquad F_{ij}U^j = (\mathbf{U} \times \mathbf{B})_i \tag{4.49}$$

where the F_{ij} are the space-space components of the Maxwell field tensor, the ε_{ijk} are the components of the rank-3 Levi-Civita pseudotensor, and this is tensor notation for the cross product[1]. This recovers Eq. (4.13).

4.4 Lagrangian field theory in canonical form

We are interested in the Lagrangian of the particle motion but even more in the field theory, which was introduced above with the Lagrangian density dependent upon the set of dynamical field variables. This is married to the particle motion in the way introduced above. We follow the usage of Landau/Lifshitz in specifying a particle

[1] The Levi-Civita tensor contains the permutation $[ijk]$ which is ± 1 or 0 for positive or negative permutation of ijk or repeated indices, respectively. Hence, for example, $[123] = 1$ while $[213] = -1$ and $[133] = 0$. In other words, $[ijk]$ is completely antisymmetric in all its indices. The purely covariant form ε_{ijk} has units $\sqrt{g}$, or $\varepsilon_{ijk} = \sqrt{g}[ijk]$, and is the volume element of the coordinate space. The purely contravariant form is $\varepsilon^{ijk} = g^{-1/2}[ijk]$, and is the quantity also referred to as the Jacobian of the coordinate system, especially in three dimensions. Here, $g = \det g_{ij}$ is the determinant of the covariant metric coefficients. In this form the generalisation to more dimensions is straightforward. See the references in Further Reading for more detail.

Lagrangian, labelled L_p, and a Lagrangian density for the fields, labelled $\mathcal{L}_f$. We already encountered the way the interaction Lagrangian is used both ways, one to get the particle motion and the other to get the responses of the fields. These are to be combined into a single Lagrangian L with density $\mathcal{L}$. We did the version of the fields with $\mathbf{x}$ and $\dot{\mathbf{x}}$, and now we will do it for the phase space Lagrangian, having already done the particles. We will need this general formulation to set up the transformations leading from particle kinetics to gyrocentre gyrokinetics, which will separate polarisation out in a different way from the procedure in the drift approximation as given above. With a single Lagrangian for the whole system, even under approximations, we are guaranteed exact conservation laws for the gyrokinetic representation with the same functionality as the ones for the kinetic representation.

The first part of the procedure is to set up the Lagrangian. We start with the three pieces as given above, and set not only the interaction Lagrangian but the entire particle part $L_{fp} + L_{int}$ for one particle, free particle plus interaction, with differential action $L_p\, dt$. This gives Hamilton's equations in canonical form, with the symplectic structure, the coordinate time derivatives, and the coordinate derivatives of H, but no time derivatives of the symplectic part. Then we apply Liouville's theorem to convert the particle motion into a phase space evolution equation for a continuous variable taking the place of the Lagrangian coordinates as a dependent variable. Denote $f(z^a, t)$ as the time-dependent distribution function in phase space, which follows what the particles do but whose change represents the motion into and out of each infinitesimal region around a given point z^a in phase space. Then since each piece follows a set of particles, the density is advected in phase space by the particle motion. Therefore, it evolves according to

$$\frac{df}{dt} = \frac{\partial f}{\partial t} + \dot{z}^a \frac{\partial f}{\partial z^a} = 0 \tag{4.50}$$

This is why the infinitesimal trajectories dz^a are referred to as the characteristics of the motion. They are the characteristics of the evolution of f. This is the advective form of the evolution equation.

The overall Lagrangian for the particles is, as before,

$$L_{fp} + L_{int} = \sum_p L_p \tag{4.51}$$

It is converted to a density, formally using the delta functions giving the coordinates $\mathbf{z}_p$ of each particle, according to

$$\sum_p L_p = \sum_p \int d\Lambda\, L_p\, \delta(\mathbf{z} - \mathbf{z}_p) = \sum_{sp} \int d\Lambda f\, L_p \tag{4.52}$$

The phase space element $d\Lambda$ is a superset of configuration space $d\mathcal{V}$ and the momentum space $d\mathcal{W}$, which we also refer to as velocity space. Overall, we have

$$d\Lambda = d\mathcal{V} \otimes d\mathcal{W}, \qquad \int d\Lambda = \int d\mathcal{V} \int d\mathcal{W} \tag{4.53}$$

where in general the phase space element is the set of coordinate differentials multiplied by a metric coefficient $\mathcal{J}$,

$$d\Lambda = \mathcal{J}\, dz^1 \ldots dz^6 \tag{4.54}$$

The phase space conservation law is given by

$$\frac{\partial f}{\partial t} + \mathcal{J}^{-1} \frac{\partial}{\partial z^a} \mathcal{J}(f\dot{z}^a) = 0 \tag{4.55}$$

The flow of the phase space motion conserves volumes, so that the advective and conservative, or divergence, form of the evolution equation are equivalent.

The system Lagrangian is now just this piece with the free field piece added. The free field Lagrangian is given in general by

$$L_{free} = \int d\mathcal{V}\, \mathcal{L}_f(\phi_a) \tag{4.56}$$

where the ϕ_a are a set of dynamical field variables which can depend on time. Any sort of differential operators can be involved in this piece. The form of L_{free} used above is a special case of this, for Maxwell's equations. The Lagrangian for the entire system may be written as

$$L = \int d\mathcal{V}\, \mathcal{L}, \qquad \mathcal{L} = \mathcal{L}_p + \mathcal{L}_f \tag{4.57}$$

but in practice we use only the L_p and $\mathcal{L}_f$ pieces in derivations. This is written out as

$$L = \sum_{sp} \int d\Lambda f\, L_p + \int d\mathcal{V}\, \mathcal{L}_f \tag{4.58}$$

with the particle Lagrangian given by

$$L_p = \gamma_a \dot{z}^a - H \tag{4.59}$$

for one particle of each species. We normally work in terms of various species, so that the charge and mass that appear are species constants, with the charge carrying a sign.

The second part of the procedure is obtaining the equations of motion, so that $\dot{z}^a$ is found in terms of phase space geometry and field variables, whose equations are found in terms of spatial geometry and source densities dependent on all the f's, found from velocity space integrations. The particle motion is found from L_p in the way we have done. Given L_p find $\dot{z}^a$. Now, however, the $z^a(t)$ do not form orbits with structural integrity over time and space, but characteristics for the evolution of f. One now has a flow in phase space whose components are the $\dot{z}^a$,

$$\mathbf{Z} = \dot{\mathbf{z}}, \qquad Z^a = \dot{z}^a \tag{4.60}$$

in which the subscript p on $\mathbf{z}_p$ can be dropped. This $\mathbf{Z}$ is now the phase space flow, and appears in the evolution equation for f, either of

$$\frac{\partial f}{\partial t} + Z^a f_{,a} = 0 \qquad \text{or} \qquad \frac{\partial f}{\partial t} + \mathcal{J}^{-1}\frac{\partial}{\partial z^a}(\mathcal{J}Z^a f) = 0 \tag{4.61}$$

with one flow and one distribution function for each species.

4.4.1 Functional derivatives for the fields

The field variables are found by doing their variations separately, as we did above. Varying L with respect to one of the field variables, ϕ_a, is done through the definition of a functional derivative

$$\delta L(\delta\phi_a) = \int d\mathcal{V}\, \delta\phi_a \frac{\delta L}{\delta\phi_a} \tag{4.62}$$

where we recall that pieces due to both L_p and $\mathcal{L}_f$ are involved. The quantity L is called a functional because it is a function of a function or functions (several dependent variables including ϕ_a) which returns a real number, i.e., $\mathcal{L}$ is a (scalar) field and L is a number. The derivative of a functional with respect to a field is formed in the same way as the derivative of a function with respect to a coordinate. We set it up as

$$L(\phi_a + \delta\phi_a) - L(\phi_a) = \delta L(\delta\phi_a) \tag{4.63}$$

The field need not be a scalar (e.g., in Maxwell's equations we have $\mathbf{A}$). But we are varying a field with spatial structure, which means the functional derivative is not a simple derivative of $\mathcal{L}$ with respect to ϕ_a. If there are no spatial derivatives of ϕ_a involved, then it is. But in general, to find δL we have to do some integrations by parts to find $\delta L/\delta\phi_a$. The procedure is the same as that to find the equations of motion for the particles, $\dot{z}_p$, as we did above. Form the variations, and then do the necessary integrations by parts to isolate $\delta\phi_a$, so that one is left with an integral of the form given in the definition of δL. If we have a scalar dependence, then for example

$$L = \int d\mathcal{V}\, \alpha\phi^2, \qquad \delta L = \int d\mathcal{V}\, 2\alpha\phi(\delta\phi), \qquad \frac{\delta L}{\delta\phi} = 2\alpha\phi \tag{4.64}$$

The functional is a number but the functional derivative is a field. In this case the functional derivative of a Lagrangian is a density, since the integration is over a volume. We always assume surface terms do not appear when doing the integrations by parts. If we have a gradient dependence, then for example

$$L = \int d\mathcal{V}\, \alpha\, |\nabla\phi|^2, \qquad \delta L = \int d\mathcal{V}\, 2\alpha\nabla\phi \cdot (\nabla\delta\phi), \qquad \frac{\delta L}{\delta\phi} = -\nabla \cdot 2\alpha\nabla\phi \tag{4.65}$$

Similarly, for a Laplacian,

$$L = \int d\mathcal{V}\,\alpha\left(\nabla^2\phi\right)^2, \quad \delta L = \int d\mathcal{V}\, 2\alpha\nabla^2\phi(\nabla^2\delta\phi), \quad \frac{\delta L}{\delta\phi} = \nabla^2(2\alpha\nabla^2\phi) \tag{4.66}$$

With these dependences we always have

$$L = \int d\mathcal{V}\,\mathcal{L}(\phi, \nabla\phi, \nabla^2\phi), \quad \frac{\delta L}{\delta\phi} = \frac{\partial\mathcal{L}}{\partial\phi} - \nabla\cdot\frac{\partial\mathcal{L}}{\partial\nabla\phi} + \nabla^2\frac{\partial\mathcal{L}}{\partial\nabla^2\phi} \tag{4.67}$$

Note the sign on the second term on the right hand side. It results from haven taken an odd number (one) of integrations by parts. With an Hermitian operator J separating two quantities, one of which is the field, we have

$$L = \int d\mathcal{V}\,\rho_{ch}\,J(\phi), \qquad \frac{\delta L}{\delta\phi} = J(\rho_{ch}) \tag{4.68}$$

And so on. See the text by Gelfand and Fomin referenced below for more detail.

With all of this in hand, the equation satisfied by each field variable is given by setting the functional derivative to zero,

$$\mathcal{L} = \mathcal{L}(\phi_a, \phi_b, \phi_c, \ldots) \qquad \text{then} \qquad \frac{\delta L}{\delta\phi_a} = 0, \quad \frac{\delta L}{\delta\phi_b} = 0, \quad \frac{\delta L}{\delta\phi_c} = 0 \quad \text{etc.} \tag{4.69}$$

This is just the minimisation principle applied to the fields, which is what we were doing above to get the equations satisfied by ϕ and $\mathbf{A}$ for electrodynamics.

4.4.2 For electrodynamics

Applying this procedure to the Maxwell Lagrangian is simply to repeat what we did above using these steps. The canonical Lagrangian is used for the particles so that only H contains dependence on the field variables ϕ and $\mathbf{A}$. To this is added the free field piece from $\mathcal{L}_f$. We have already done the particles by doing the operations on the canonical version of L_p for them. The equation for each distribution function is

$$\frac{\partial f}{\partial t} + \mathbf{U}\cdot\nabla f + [e(\nabla\mathbf{A})\cdot\mathbf{U} - e\nabla\phi]\cdot\nabla_{\mathbf{p}} f = 0 \tag{4.70}$$

where $\nabla_{\mathbf{p}}$ is the gradient in momentum space, whose contravariant components are $\partial/\partial p_i$. This equation looks a bit unusual in the last term, but note that the coordinate is $\mathbf{p}$, not $\mathbf{U}$.

Now we find the functional derivative of L with respect to each field, in this case one scalar and one vector. The variation of ϕ is given by the two pieces

$$\delta(E^2) = -2\mathbf{E}\cdot\nabla(\delta\phi), \qquad \delta(fH) = fe(\delta\phi) \tag{4.71}$$

and this yields

$$\epsilon_0\nabla\cdot\mathbf{E} = \rho_{ch} \qquad \text{where} \qquad \rho_{ch} = \sum_{sp}\int d\mathcal{W} fe \tag{4.72}$$

The variation of $\mathbf{A}$ is given by the three pieces

$$\delta(E^2) = -2\mathbf{E} \cdot \frac{\partial}{\partial t}(\delta\mathbf{A}), \qquad \delta(B^2) = 2\mathbf{B} \cdot \nabla \times (\delta\mathbf{A}), \qquad \delta(fH) = -fe\mathbf{U} \cdot (\delta\mathbf{A}) \tag{4.73}$$

and this yields

$$\nabla \times \mathbf{B} = \mu_0\epsilon_0\frac{\partial \mathbf{E}}{\partial t} + \mu_0\mathbf{J} \qquad \text{where} \qquad \mathbf{J} = \sum_{sp}\int d\mathcal{W} fe\mathbf{U} \tag{4.74}$$

This is useful practice in gaining familiarity with the new language as it is applied to something we already know. Note how only the time component of the Lagrangian is involved in the field equations. This is because we did it in canonical form, without any dependence of the γ_a upon the fields. The gain is minimal now, but will become decisive later on.

4.5 Towards drifts

We just got the equations of motion for the charged particle distribution functions and Maxwell's equations for the fields using the canonical phase space Lagrangian. The result has the fast gyromotion in the equation for each f, through the velocity space derivative. We would like to get rid of this in the equation for the part of f which interests us, the one for gyrocentres rather than particles. The drift approximation is how to do this, but we would like to do it in the Lagrangian rather than directly on the equations. The drift approximation depends on being able to use the background magnetic field as an anchor. The main field is not varied, but its fluctuations are, and perhaps we can find a formulation which will reflect this automatically. We will do this with a formal gauge transform later on, but for now we do a heuristic experiment. We formulate the anchor using a potential $\mathbf{A}_0$ such that

$$L_p = (e\mathbf{A}_0 + \mathbf{p}) \cdot \dot{\mathbf{x}} - H \tag{4.75}$$

We assume $\mathbf{A}_0$ has spatial dependence to form $\mathbf{B}_0 = \nabla \times \mathbf{A}_0$, but no time dependence. For simplicity let us assume electrostatic motion so that

$$H = \frac{p^2}{2m} + e\phi \tag{4.76}$$

We assume that the motion in our case is slow compared to (fast) compressional Alfvén waves ($\omega \ll k_\perp v_A$, nearly always true), and in this case even to (slow) shear Alfvén waves ($\omega \ll k_\| v_A$, only sometimes true). The equations of motion are found in the same way as before, except that the spatial derivatives of $\mathbf{A}_0$ are carried along. The Lagrangian is no longer canonical, but because $\mathbf{A}_0$ does not depend on time or the field variables it still has canonical form.

Varying $\delta\mathbf{x}$ we obtain

$$dS(\delta\mathbf{x}) = \delta\mathbf{x} \cdot \left[e\nabla\mathbf{A}_0 \cdot \dot{\mathbf{x}} - \frac{d}{dt}(e\mathbf{A}_0 + \mathbf{p}) - \nabla H\right]dt + d() \tag{4.77}$$

The Euler–Lagrange equation is found by setting the quantity contracted with $\delta\mathbf{x}$ to zero, neglecting the total differential piece. We construct the tensor $\mathbf{F}$ according to

$$F_{ij} = \nabla_i(A_0)_j - \nabla_j(A_0)_i, \qquad \mathbf{F} = \boldsymbol{\varepsilon} \cdot \mathbf{B} \tag{4.78}$$

We can contract with $\mathbf{F}$ using

$$\mathbf{F} \cdot \mathbf{F} = \mathbf{BB} - B^2\mathbf{g} \qquad \text{hence} \qquad e\mathbf{F} \cdot \mathbf{F} \cdot \dot{\mathbf{x}} = \mathbf{F} \cdot \nabla H + \mathbf{F} \cdot \dot{\mathbf{p}} \tag{4.79}$$

These operations are used to solve the Euler–Lagrange equation for $\dot{\mathbf{x}}$ in terms of $\dot{\mathbf{p}}$ to find

$$m\dot{\mathbf{x}} = -\frac{1}{\Omega B}\mathbf{F} \cdot \nabla H - \frac{1}{\Omega B}\mathbf{F} \cdot \dot{\mathbf{p}} \tag{4.80}$$

noting how $\dot{\mathbf{p}}$ and ∇H simply add. The operator doing the contraction is simply the drift operator again,

$$-\frac{1}{\Omega B}\mathbf{F} \cdot \nabla H = m\,\frac{1}{B^2}\mathbf{B} \times \nabla\phi \tag{4.81}$$

so that using the other Euler–Lagrange equation, $m\dot{\mathbf{x}} = \mathbf{p}$, we have

$$\mathbf{p} = -\frac{1}{\Omega B}\mathbf{F} \cdot \nabla H - \frac{1}{\Omega B}\mathbf{F} \cdot \dot{\mathbf{p}} \tag{4.82}$$

If we express $\mathbf{p}$ to lowest order using the first term, and then substitute that into the second term where $\dot{\mathbf{p}}$ appears, noting that $\nabla H = e\nabla\phi$, we have the lowest order ExB drift and the polarisation correction. All in all, this is the drifts picture again. So we know where we want to go.

There are some problems however (of course there are!). The main one is with the field theory version. We have the particle motion all right, but given the time component of the Lagrangian,

$$L = \cdots - \sum_{sp} \int d\Lambda f\left(\frac{p^2}{2m} + e\phi\right) + \int d\mathcal{V}\,\frac{1}{2}\varepsilon_0\,|\nabla\phi|^2 \tag{4.83}$$

if we substitute the ExB drift into the kinetic energy, we find that

$$\frac{1}{2}\frac{\rho}{B^2}\,|\nabla\phi|^2 \gg \frac{1}{2}\varepsilon_0\,|\nabla\phi|^2 \tag{4.84}$$

by the condition on polarisability, $\varepsilon_0 \ll \rho/B^2$, so it is tempting to drop the field piece entirely. If we do this while keeping the formulation we have, then there is no way to control ϕ, whose equation is now simply that ρ_{ch} should vanish. Previous treatments (up to the years around 1980) did this by adding $\rho_{ch} \to 0$ as a constraint. This can be sufficient for linear analysis, but never works for computational models whose degree of nonlinearity can be as turbulent as we have seen. So how should we control ϕ, in a way similar to how the fluid drift dynamics does it? How do we get a vorticity?

4.6 Quasineutrality

Quasineutrality is the main concept behind any formulation of self-consistent dynamics in this regime. We have magnetised plasma turbulence, at frequencies slow compared to the gyrofrequency of any charged constituent, and we have a low beta plasma so that the perpendicular electric field is electrostatic (dynamics slow compared to compressional waves). The problem is, the dynamics should be quasineutral, which means that the charge density is formally zero but $\nabla \cdot \mathbf{E}$ is finite so we can have an ExB vorticity, with

$$\mathbf{b} \cdot \nabla \times \mathbf{v}_E \approx \Omega_E \equiv \frac{1}{B}\nabla_\perp^2 \phi \tag{4.85}$$

This means that ε_0 is formally taken to zero, so that we do have to neglect $\varepsilon_0 E^2$ in the field Lagrangian. The ExB drift arises from $e\phi$ in the Hamiltonian, so that part is in order. In order to get a field equation out of the Lagrangian, we need a second order piece in the Lagrangian density somewhere, and we just discarded the usual one.

The simplest way to get an energy term from ϕ is the ExB energy. If we just declare it and set the fields part of the Lagrangian to

$$L = \cdots - \sum_{sp} \int d\Lambda f\, e\phi + \int d\mathcal{V} \frac{1}{2}\frac{\rho}{B^2} |\nabla_\perp \phi|^2 \tag{4.86}$$

or even better express ρ in terms of the species mass m and f, so that

$$L = \cdots - \sum_{sp} \int d\Lambda f \left(e\phi - \frac{1}{2}\frac{m}{B^2} |\nabla_\perp \phi|^2 \right) \tag{4.87}$$

then the neglect of $\varepsilon_0 E^2$ poses no problems.

We are almost there. In this representation, polarisation density is not a drift, but a density. Variation of ϕ in this expression leads to

$$\sum_{sp} \int d\mathcal{W} \left[f\, e + \mathcal{J}^{-1}\nabla \cdot \mathcal{J}\frac{\rho}{B^2}\nabla_\perp \phi \right] = 0 \tag{4.88}$$

which is a viable way to get ϕ from f. The polarisation charge density represents the ExB vorticity, in this case through

$$\nabla \cdot \frac{\rho}{B^2}\nabla_\perp \phi \approx \frac{\rho}{B}\Omega_E \tag{4.89}$$

We have already seen this within the fluid drift approximation. Now we almost have our goal within the treatment of kinetic field theory in canonical form. The total charge density is declared zero, with the charge density due to each species re-defined as the quantity in square brackets above. We have re-defined fe as the phase space charge density for gyrocentres, not particles, with the second piece, the polarisation, defined as the part of the charge density not captured by the drift motion. This will put quasineutrality on a firm footing within the kinetic field theory, but we still have work to do. We still have $p^2/2m$ in the Hamiltonian, and so we are double-counting the ExB drift energy. We want something better than a declaration to get the ExB energy into the Lagrangian and the gyromotion out of the coordinates. The motion of the gyrocentre coordinates should yield the drifts rigorously rather than by successive approximation.

The way to do all this is called gyrokinetic theory, and the version with the fields is gyrokinetic field theory. This lecture has served to collect all the basics we need before we proceed with the transform methods used to get gyrokinetic theory. So now we proceed with that in the next one.

4.7 Interlude—Noether's theorem

One of the main reasons for setting up the problem as a classical field theory is the relationship between symmetry of the system and its conservation laws. The theorem covering this is called Noether's Theorem in physics, after the 1918 paper which produced it. The basic statement is that if the system is invariant under a transformation, there is a conserved quantity conjugate to that transformation. For time symmetry, energy conservation follows. For spatial symmetry, momentum conservation follows. For symmetry under rotation through an angle, conservation of angular momentum in that direction follows. The concept extends indefinitely, with the widest scope usually covered in texts on quantum field theory.

4.7.1 Symmetry and total variations

In the operations used to get the Euler–Lagrange equations for the dependent variables (fields and particle Lagrange coordinates), integration by parts under an integration was prominent. The result in each case is a functional derivative. But the divergences which under the integration are dropped are interesting in their own right. The general action is an integral over both time and space of a Lagrangian density dependent on particle positions z_k and field quantities ϕ_a. When varying the particle quantities only, it is an integral over time over their collective Lagrangian. In total, the variation of the Lagrangian is

$$\begin{aligned}\delta L = &\sum_k \left[(\delta z_k)\frac{\partial L}{\partial z_k} + (\delta \dot{z}_k)\frac{\partial L}{\partial \dot{z}_k} \right] \\ &+ \sum_a \int d\mathcal{V} \left[(\delta\phi_a)\frac{\partial \mathcal{L}}{\partial \phi_a} + (\delta\nabla\phi_a)\frac{\partial \mathcal{L}}{\partial \nabla\phi_a} + (\delta\partial_t\phi_a)\frac{\partial \mathcal{L}}{\partial \partial_t\phi_a} \right]\end{aligned} \tag{4.90}$$

here assuming dependence on the ϕ_a through their amplitudes and first derivatives, with ∂_t denoting the partial time derivative. The z_k can be assumed to cover coordinates, particles, and species collectively. Forming the Euler–Lagrangian equations as we did above, we keep the divergences in an effective integration by parts writing $u\ dv$ as $d(uv) - v\ du$ in each case. The Euler–Lagrangian equations are assumed to be satisfied, so the operation leaves only the total derivative pieces

$$\delta L = \sum_k \frac{d}{dt}\left[(\delta z_k)\frac{\partial L}{\partial \dot{z}_k}\right] + \sum_a \int d\mathcal{V}\left[\frac{\partial}{\partial t}\left((\delta\phi_a)\frac{\partial \mathcal{L}}{\partial \partial_t \phi_a}\right) + \nabla\cdot\left((\delta\phi_a)\frac{\partial \mathcal{L}}{\partial \nabla\phi_a}\right)\right] \tag{4.91}$$

When considering δL only, not $\delta\mathcal{L}$, the spatial divergence integrates away, leaving

$$\delta L = \sum_k \frac{d}{dt}\left[(\delta z_k)\frac{\partial L}{\partial \dot{z}_k}\right] + \sum_a \int d\mathcal{V}\,\frac{\partial}{\partial t}\left((\delta\phi_a)\frac{\partial \mathcal{L}}{\partial \partial_t \phi_a}\right) \tag{4.92}$$

This statement gives globally conserved quantities as a result of symmetries. If a shift in the coordinates (not direct variation of the dependent variables) leaves L unchanged, then δL is zero and then the resulting vanishing total time derivative gives the associated conservation law.

For a shift in time δL does not vanish because L is a function of time. But so are all dependent variables, so all are varied the same way. Directly taking the derivative of L and following the same operations as before, we have

$$\frac{dL}{dt} = \frac{\partial L}{\partial t} + \sum_k \frac{d}{dt}\left(\dot{z}_k\frac{\partial L}{\partial \dot{z}_k}\right) + \sum_a \int d\mathcal{V}\,\frac{\partial}{\partial t}\left(\frac{\partial \mathcal{L}}{\partial(\partial_t\phi_a)}(\partial_t\phi_a)\right) \tag{4.93}$$

where the partial time derivative accounts for explicit time dependence, perhaps in terms of a drive through an externally imposed potential with time dependence. The Lagrangian is symmetric in time if $\partial L/\partial t$ vanishes, which means that L depends on time solely through the dependent variables. The partial time derivative is taken out of the spatial integration over the fields, so that

$$\frac{dE_T}{dt} = 0, \qquad E_T = -L + \sum_k (\dot{z}_k)\frac{\partial L}{\partial \dot{z}_k} + \sum_a \int d\mathcal{V}\,(\partial_t\phi_a)\frac{\partial \mathcal{L}}{\partial \partial_t\phi_a} \tag{4.94}$$

If there are no terms in L due to time derivatives of fields, then we obtain the textbook result for the case of particles in a self-consistent potential,

$$E_T = -L + \sum_k (\dot{z}_k)\frac{\partial L}{\partial \dot{z}_k} \tag{4.95}$$

The energy E_T reduces to H for a free particle Lagrangian with $L\,dt = p\,dq - H\,dt$ with $H = p^2/2m$. This determines the sign. The statement is that time symmetry leads to energy conservation, effectively that time and energy are conjugate in the general statement that symmetry in a coordinate leads to conservation of its conjugate quantity.

For a spatial shift the operations are different. The z_k are functions of time only, but the ones describing spatial coordinates are passively shifted. For a shift of the coordinate axis by ε in the ξ-direction, we have

$$\delta\xi_k = -\varepsilon, \qquad \delta\phi_a = \varepsilon\frac{\partial\phi_a}{\partial\xi} \tag{4.96}$$

where ξ_k is the ξ-component of the z_k. The main point is that while the origin is shifted by ε into the gradient (for $\varepsilon > 0$), the positions of Lagrangian coordinates are shifted to more negative values, so that the signs are opposite. The variation of L is given by

$$\delta L = \varepsilon \int d\mathcal{V} \frac{\partial \mathcal{L}}{\partial \xi} + \varepsilon \sum_a \int d\mathcal{V} \frac{\partial}{\partial t}\left(\frac{\partial \phi_a}{\partial \xi} \frac{\partial \mathcal{L}}{\partial \partial_t \phi_a} \right) - \varepsilon \sum_k \frac{d}{dt} p_\xi^* \tag{4.97}$$

where $p_\xi^* = \partial L / \partial \dot{\xi}_k$ is the covariant ξ-component of the particle canonical momentum. The Lagrangian itself is not changed, just the axis, so that the left hand side is zero. Explicit spatial dependence in the ξ-direction is covered by the first term. If this vanishes, then we have

$$\frac{dP_\xi^*}{dt} = 0 \qquad P_\xi^* = \sum_k p_\xi^* - \sum_a \int d\mathcal{V} \left(\frac{\partial \phi_a}{\partial \xi} \frac{\partial \mathcal{L}}{\partial \partial_t \phi_a} \right) \tag{4.98}$$

This gives the conservation of the covariant ξ-component of the total momentum, from particles and fields. In this case, the Lagrangian is symmetric in ξ, and ξ is called a symmetry coordinate. If there is no dependence of L on time derivatives of fields, then the sum of particle canonical momentum in the ξ-direction is conserved. If spatial symmetry is present in all directions, then momentum is conserved as a vector, and

$$\frac{d\mathbf{P}^*}{dt} = 0 \qquad \mathbf{P}^* = \sum_k \mathbf{p}^* - \sum_a \int d\mathcal{V} \left(\frac{\partial \mathcal{L}}{\partial(\partial_t \phi_a)} (\nabla \phi_a) \right) \tag{4.99}$$

and similarly just the $\mathbf{p}_*$ in the simpler case. The statement is that spatial symmetry in a particular coordinate (here, ξ) leads to momentum conservation in direction of that coordinate, with momentum serving as the quantity conjugate to the spatial coordinate.

The version of this result most often seen in classical field theory texts is the one due solely to fields. In that case we have

$$\delta \mathcal{L} = \frac{\partial}{\partial t}\left(\sum_a (\delta \phi_a) \frac{\partial \mathcal{L}}{\partial \partial_t \phi_a} \right) + \nabla \cdot \left(\sum_a (\delta \phi_a) \frac{\partial \mathcal{L}}{\partial \nabla \phi_a} \right) \tag{4.100}$$

which is a four-dimensional divergence inside the overall integration over both time and space. Considering a shift in any of the four coordinates, we can do the variation in terms of derivatives, so that we have

$$\frac{\partial}{\partial t}\left(-\mathcal{L} + \sum_a (\partial_t \phi_a) \frac{\partial \mathcal{L}}{\partial \partial_t \phi_a} \right) + \nabla \cdot \left(\sum_a (\partial_t \phi_a) \frac{\partial \mathcal{L}}{\partial \nabla \phi_a} \right) = 0 \tag{4.101}$$

for energy, and

$$\frac{\partial}{\partial t}\left(\sum_a \frac{\partial \mathcal{L}}{\partial \partial_t \phi_a}(\nabla \phi_a)\right) + \nabla \cdot \left(-\mathcal{L}\mathbf{g} + \sum_a \frac{\partial \mathcal{L}}{\partial \nabla \phi_a}(\nabla \phi_a)\right) = 0 \tag{4.102}$$

for momentum, where **g** is the spatial metric tensor. This is the divergence of a tensor, which is why the field variation as $(\nabla\phi_a)$ is written last. Considering relativistic situations these are combined into a four-dimensional space–time divergence of an energy momentum stress tensor, which for completion has to be symmetrised. For more in that vein see the texts cited below, especially those which treat relativity directly.

4.7.2 Example of a single particle in a field

The simplest case is a Lagrangian of two degrees of freedom, a position q and a momentum p, belonging to a particle moving in a potential field,

$$L\,dt = p\,dq - H\,dt, \qquad H = \frac{p^2}{2m} - e\Phi(q) \tag{4.103}$$

where the potential $\Phi(x, t)$ is a function of both time and position, interacting with the particle through the latter's position, and m and e are constants. The action is $S = \int L\,dt$. Varying p and q and minimising the action produces Hamilton's equations

$$\frac{dq}{dt} = H_{,p} = p, \qquad \frac{dp}{dt} = -H_{,q} = -e\Phi_{,q} \tag{4.104}$$

where $\Phi_{,q}$ is $\Phi_{,x}$ evaluated at $x = q$. Note that no term due to $\Phi_{,t}$ occurs, because it is in the time component of L.

Doing the temporal shift as above, we have the broken symmetry due to time dependence in the external field. The variation yields

$$\frac{d}{dt}\left(\frac{p^2}{2m} + e\Phi\right) = e\frac{\partial \Phi}{\partial t} \tag{4.105}$$

with the term on the right due to $\partial L/\partial t$ in Eq. (4.93). If we use the equations of motion to build this equation we have

$$\frac{p}{m}\left(\frac{dp}{dt} + e\Phi_{,q}\right) = 0 \tag{4.106}$$

Since p/m is $\dot{q}$ we get d/dt on Φ by adding the time derivative to both sides, which recovers Eq. (4.105).

Doing the spatial shift (in x) as above, we first confirm that δL is indeed zero. Noting that the variations of Φ due to δx and δq cancel, due to the rules of the shift, we confirm the part due to Φ vanishes. This merely states that the position of Φ at the particle does not change when we do the coordinate shift. Then, since $\dot{q}$ and p do not change, δL is zero. Doing the shift according to the variations and applying the Euler–Lagrange equations, we have

$$\delta L = \varepsilon\left(\frac{\partial}{\partial x}(-e\Phi) - \frac{dp}{dt}\right) = 0 \tag{4.107}$$

which yields a copy of the equation of motion. So although p is not conserved, its change in value merely follows the equation of motion.

This exercise provides redundant confirmation of the theorem using the equations of motion. It will be helpful for the next Chapter, as there we will have Noether's theorem to tell us what should be conserved, but we will have to re-confirm it using the equations of motion since there is an intermediate step between the equations for individual Lagrangian coordinates and the evolution equation for the distribution.

4.7.3 Self-consistent field

Now, if the free field has its own Lagrangian this is different. Since the field becomes a dependent variable, the Lagrangian is now symmetric. The simplest example is one with an electrostatic field energy,

$$L_{free} = \int dx\ \frac{1}{2}\varepsilon_0(\Phi_{,x})^2 \tag{4.108}$$

where ε_0 is a parameter belonging to the field. Later, we consider a set of fields ϕ_a as written above, in the Maxwell-plasma system.

This doesn't make good sense if we continue to think of one particle and the field, since any effects from that, once it is self-consistent (i.e., not externally specified), are equivalent to self forces. Self forces are only negligible when the number of particles is large, as in a plasma. So we now consider a set of arbitrarily many particles, and express the particle part of the Lagrangian as a sum over single L_k summed over all particles k, each with its own p_k and q_k, and $\Phi(q_k)$ as $\Phi(x)$ evaluated at $x = q_k$ for that particle. We can have (p, q) in one dimension, or $(\mathbf{p}, \mathbf{q})$ in two or three dimensions for each k.

The free particle Lagrangian is the sum over the pieces not dependent upon the field potential $\Phi(x)$, or

$$L_{fp} = \sum_k\left(p_k\ \dot{q}_k - \frac{p_k^2}{2m_k}\right) \tag{4.109}$$

The interaction Lagrangian is equally expressed as a particle coupling or as a field density coupling,

$$L_{int} = \sum_k(-e\Phi)_k = \int dx\ (-\rho_{ch}\Phi) \quad \text{where} \quad \rho_{ch} = \sum_k e_k\delta(x - q_k) \tag{4.110}$$

While varying the p and q we use the first form, and while varying Φ we use the second form. This piece is not double counted because the variations are taken separately.

For varying the particle coordinates (p_k, q_k) we use L_{fp} and L_{int} in the particle language,

$$L_k = p_k \dot{q}_k - \frac{p_k^2}{2m_k} - e_k \Phi(q_k) \tag{4.111}$$

with $\Phi(q_k)$ as above, $\Phi(x)$ with $x = q_k$, for the k'th particle. Note that the m_k and e_k are the mass and charge for each particle, so that the sum covers the sum over species as well. The resulting equations of motion are

$$\frac{dq_k}{dt} = p_k, \qquad \frac{dp_k}{dt} = -e_k \Phi_{,q} \tag{4.112}$$

following Hamilton's equations (Eq. 4.104). For varying the field we obtain the simple Poisson equation

$$-\varepsilon_0 \frac{\partial^2 \Phi}{\partial x^2} = \rho_{ch} \tag{4.113}$$

There are no time derivatives since there is no $\Phi_{,t}$ term in L.

For variations, we can use the fact that L is symmetric in both t and x, so that the same expression can be used for both. Although L_{int} is the same term in both expressions, the variations due to the $(p, q)_k$ and due to Φ add, and can be done separately. This yields, after application of the Euler–Lagrange equations,

$$\delta L = \sum_k \frac{d}{dt}((\delta q_k) p_k) \tag{4.114}$$

namely just the particle part. The important thing to note about the field piece of the overall variation is that only the part due directly to the time variation of Φ enters, and there is none here.

In the particle moving in an external field, the variation of Φ due to coordinate spatio-temporal shifts entered here. But with a self-consistent field the Φ is varied twice, one due to the variation of q, producing $(\delta q)\Phi_{,q}$, and one due to the variation of the field amplitude, $(\delta\Phi)$. Both pieces now drop out of the total variation of L, due to the Euler–Lagrange equation. The only one left which survives integration is the one with the $\Phi_{,t}$ dependence in the Lagrangian. In our example, this piece is absent and we do not get a variation due directly to Φ in the coordinate shifts. This matters a lot when considering Noether's theorem.

Doing the shift in time, we take the time derivative as done before, yielding

$$\frac{dL}{dt} = \frac{d}{dt} \sum_k p_k \dot{q}_k \tag{4.115}$$

again with no direct Φ term. The resulting conserved energy is given by

$$E_T = \sum_k \frac{p_k^2}{2m_k} + \int dx \left(\rho_{ch}\Phi - \frac{1}{2}\varepsilon_0(\Phi_{,x})^2\right) \tag{4.116}$$

where we use the field version of L_{int} to combine the field pieces. This looks deceptive, due to the appearance of potential energy and field energy pieces, with the latter negative, so we eliminate the $\rho_{ch}\Phi$ piece using the equation for the field. Multiplying Eq. (4.113) by Φ and integrating over x, doing one derivative by parts, we find

$$\int dx\ \rho_{ch}\Phi = \int dx\ \varepsilon_0(\Phi_{,x})^2 \tag{4.117}$$

which is just twice the field energy. Using this to eliminate the $\rho_{ch}\Phi$ term in the energy, we find

$$\frac{dE_T}{dt} = 0 \qquad \text{where} \quad E_T = \sum_k \frac{p_k^2}{2m_k} + \int dx\ \frac{1}{2}\varepsilon_0(\Phi_{,x})^2 = 0 \tag{4.118}$$

The total energy is conserved between the particles and the field. There is no potential energy term because the field is self-consistent. Moreover, the total energy is positive definite, since the field energy and particle kinetic energy by themselves are.

For momentum we do the same thing as before with the spatial coordinate shift. Since Φ has its own self-consistent response, the explicit $\delta\Phi$ term only appears if Φ has a time derivative in the Lagrangian, and there is none here. We have already seen that this immediately yields conservation of the total canonical momentum,

$$\frac{d}{dt}\sum_k p_k = 0 \tag{4.119}$$

This is a different result from before. If the field is externally imposed, then it can act as an anchor for momentum if it has a derivative in the same direction as the shift. If not, then x is a symmetry coordinate, and then the momentum is again conserved. In this case, the field is self-consistent and serves only to mediate exchange of momentum between the particles. In the electrostatic case there is no field momentum and we simply have conservation of total particle momentum. Again, it only works out this way if the field is self-consistent.

4.7.4 Maxwell's equations

For Maxwell's equations, several new elements enter. We have two field potentials, with the second having its interaction via the particle momentum. We also have time dependence in the field Lagrangian. Given the canonical Lagrangian, the properties we have taken advantage of (no field dependence in $p\ dq$ or its generalisation, and linear field coupling) still remain. The part of L involving the Lagrangian coordinates is

$$L_{fp} + L_{int} = \mathbf{p}_k \dot{\mathbf{q}}_k - H_k, \qquad H_k = \frac{1}{2m_k} \left|\mathbf{p}_k - e_k \mathbf{A}\right|^2 + e\phi \tag{4.120}$$

for the k-th particle. The part involving the fields is

$$L_{free} + L_{int} = \int d\mathcal{V}\, \mathcal{L}, \qquad \mathcal{L} = \mathbf{A} \cdot \mathbf{J} - \rho_{ch}\phi + \frac{\varepsilon_0}{2}E^2 - \frac{1}{2\mu_0}B^2 \tag{4.121}$$

with

$$\mathbf{E} = -\left(\frac{\partial \mathbf{A}}{\partial t} + \nabla\phi\right), \qquad \mathbf{B} = \nabla \times \mathbf{A} \tag{4.122}$$

with ρ_{ch} and $\mathbf{J}$ as sources. Variation of the particle coordinates produces either form of the Lorentz force motion (Eqs 4.13, 4.46), and variation of the fields ϕ and $\mathbf{A}$ produces the two Maxwell equations with sources (Eq. 4.24).

For Noether's theorem we now have the terms due to $\partial_t A_i$ for the three components of $\mathbf{A}$, so that

$$\frac{\partial \mathcal{L}}{\partial \partial_t \mathbf{A}} = -\varepsilon_0 \mathbf{E} \tag{4.123}$$

Putting this into the time derivative (Eq. 4.93) and using the time symmetry of L since each piece is self-consistent, we find

$$E_T = -L + \sum_k (\dot{\mathbf{q}}_k) \cdot \frac{\partial L}{\partial \dot{\mathbf{q}}_k} + \int d\mathcal{V} \frac{\partial \mathbf{A}}{\partial t} \cdot (-\varepsilon_0 \mathbf{E}) \tag{4.124}$$

Using $\varepsilon_0 \nabla \cdot \mathbf{E} = \rho_{ch}$ to recast the $\rho_{ch}\phi$ potential energy term in L,

$$\rho_{ch}\phi = \phi\nabla \cdot (\varepsilon_0 \mathbf{E}) = \nabla \cdot (\varepsilon_0 \phi \mathbf{E}) - \varepsilon_0 \mathbf{E} \cdot \nabla\phi \tag{4.125}$$

and then combining the last terms in each of these equations, we find

$$E_T = \sum_k m_k \frac{U_k^2}{2} + \frac{1}{2} \int d\mathcal{V} \left(\varepsilon_0 E^2 + \mu_0^{-1} B^2\right), \qquad m_k \mathbf{U}_k = \mathbf{p}_k - e\mathbf{A} \tag{4.126}$$

defining the velocity in terms of the canonical momentum for each particle. This recovers the standard form of kinetic plus field energy in classical electrodynamics with a set of charged particles.

For the spatial shift, we use the symmetry of the Lagrangian, and the time dependent field variation (Eq. 4.123), in the vector shift form (Eq. 4.99) of the conserved momentum, to find

$$\mathbf{P}^* = \sum_k \mathbf{p}_k - \int d\mathcal{V} (\nabla \mathbf{A}) \cdot (-\varepsilon_0 \mathbf{E}) \tag{4.127}$$

as the momentum Noether's theorem says is conserved. As with the canonical representation of the Lorentz force law (Eq. 4.46), we re-familiarise this by substituting the velocity for the canonical momentum. Converting to the momentum density representation, we have

$$\mathbf{P}^{*}=\int d\mathcal{V}\left(\rho_{ch}\mathbf{A}+\rho\mathbf{U}\right)+\int d\mathcal{V}\,(\nabla\mathbf{A})\cdot(\varepsilon_0\mathbf{E}) \tag{4.128}$$

where the $\mathbf{U}$ without subscript is the centre of mass velocity and ρ is the mass density. As above we eliminate ρ_{ch} in favour of $\nabla\cdot\mathbf{E}$,

$$\rho_{ch}\mathbf{A}=(\mathbf{A})\nabla\cdot(\varepsilon_0\mathbf{E})=\nabla\cdot(\varepsilon_0\mathbf{E}\mathbf{A})-\varepsilon_0\mathbf{E}\cdot(\nabla\mathbf{A}) \tag{4.129}$$

and therefore

$$\mathbf{P}^{*}=\int d\mathcal{V}\,\rho\mathbf{U}+\int d\mathcal{V}\,\varepsilon_0((\nabla\mathbf{A})\cdot\mathbf{E}-\mathbf{E}\cdot(\nabla\mathbf{A})) \tag{4.130}$$

We recognise the quantity in parentheses as $\mathbf{E}\times(\nabla\times\mathbf{A})$, according to the vector identity

$$\mathbf{E}\times(\nabla\times\mathbf{A})=(\nabla\mathbf{A})\cdot\mathbf{E}-\mathbf{E}\cdot\nabla\mathbf{A} \tag{4.131}$$

Using this, the field terms combine such that

$$\mathbf{P}^{*}=\int d\mathcal{V}\,(\rho\mathbf{U}+\varepsilon_0\mathbf{E}\times\mathbf{B}) \tag{4.132}$$

This is the conserved momentum. The sum of particle momentum and the field's Poynting momentum is conserved, namely, the momentum of the plasma together with the Poynting momentum from the electromagnetic field. As with the energy, the expected form is re-confirmed once some vector operations are done.

The main point of doing this, besides re-confirming familiar results, is that the relative signs between the Lagrangian coordinates of the particles and the field quantities are the same in the temporal shift and opposite in the spatial shift used in the application of Noether's theorem on a mixed Lagrangian. This assures the correct signs in the particle-field interactions both for externally imposed and self-consistent fields.

4.7.5 Spatial dependence in the background

Anticipating the situation with gyrokinetics, we consider the case in which the fields are still self-consistent but there is spatially dependent geometric information in both the Hamiltonian and the canonical momentum for the particles. Consider that there is a term in H due to the magnetic field strength, so that

$$\nabla H=\nabla H\left(\nabla\phi_a\right)+M\nabla B \tag{4.133}$$

where the ϕ_a are the self-consistent fields, the first term gives the part of the gradient due solely to the spatial dependence of the ϕ_a, M is a constant, and B is the strength of the magnetic field. Consider also that there is a background term in the canonical momentum, which now obtains its general form,

$$\mathbf{p}^{*}=\frac{\partial L}{\partial\dot{\mathbf{q}}} \tag{4.134}$$

The Lagrangian is still assumed to have canonical form, but now the $\mathbf{p}^*$ can depend on coordinates. In our case, it becomes

$$\mathbf{p}^* = \mathbf{P}_0(\mathbf{x}) + \mathbf{p} \rightarrow \mathbf{P}_0(\mathbf{q}) + \mathbf{p} \tag{4.135}$$

where $\mathbf{p}$ denotes the set of particle momentum coordinates. Evaluated at the particle position, $\mathbf{q}$, this is understood with $\mathbf{P}_0(\mathbf{q})$ in the same way as $\Phi(q)$ in the first example. The particle Lagrangian is

$$L_{fp} + L_{int} = \sum_k L_k \tag{4.136}$$

with

$$L_k = (\mathbf{p}^* \cdot \dot{\mathbf{q}} - H)_k, \qquad H_k = H(\mathbf{p}_k, \mathbf{q}_k, \phi_a, B) \tag{4.137}$$

The field Lagrangian is unchanged.

Temporal variations are the same as before, assuming that $\mathbf{P}_0$ has no time dependence, including no ϕ_a dependence. Spatial variations acquire two new terms in δL, due to $\mathbf{P}_0$ and B. Assuming no dependence on time derivatives of the ϕ_a in L, we have

$$\delta L = \frac{d}{dt}\sum_k (\mathbf{p}^* \cdot (\delta\mathbf{q}))_k + (\delta\mathbf{x}) \cdot \sum_k (\nabla\mathbf{p}^* \cdot \dot{\mathbf{q}} - M\nabla B)_k \tag{4.138}$$

where the terms written with ∇ enter in the same way as $\delta\Phi$ in the first example. Given the spatial coordinate shift in a particular direction, call it ξ, we have $x \rightarrow \xi$ and $(\delta\mathbf{q}) \rightarrow (-\delta\xi)$, recalling the issue with the minus sign in the variation of $\mathbf{q}$ due to a spatial coordinate shift. This leaves

$$\frac{d}{dt}\sum_k \left(\mathbf{p}^*_\xi\right)_k = \sum_k \left(\frac{\partial \mathbf{p}^*}{\partial \xi} \cdot \dot{\mathbf{q}} - M\frac{\partial B}{\partial \xi}\right)_k \tag{4.139}$$

If ξ is a symmetry coordinate, then the $\partial/\partial\xi$ terms vanish, and then the total canonical momentum is conserved. The one new element is that the conserved momentum still includes the $\mathbf{P}_0$ terms due to the background geometry. The consequences this has will be addressed in Chapter 6.

Further Reading

For the basics of classical field theory using Lagrangians, especially Lagrangian densities, see *Classical Mechanics*, by H Goldstein (Academic Press, 1960), and *The Classical Theory of Fields*, by L D Landau and E M Lifshitz (4th edn, Pergamon, 1975). The Noether theorem connecting symmetries to conservation laws is covered in Goldstein (Chapter 12) and Landau and Lifshitz (Chapter 4), and the original is E Noether, 'Invariante Variationsprobleme,' *Nachr. Ges. Wiss. Goettingen, Math.-Phys. Kl.* **235** (1918) 235 [English translation by M A Travel, *Transp. Theory Stat. Phys.* **1** (1971) 183].

For a wider variety of texts concerning the methods of classical field theory, see *Classical Field Theory*, by D E Soper (Wiley, 1976, repr. Dover, 2008), and *Electrodynamics and Chassical Theory of Fields and Particles*, by A O Barut (Macmillan, 1964, repr. Dover, 1980), and *Classical Field Theory*, by F E Low (Wiley, 1997).

For statistical physics and the use of Liouville's theorem, see *Principles of Statistical Physics*, by R C Tolman (Oxford, 1938, repr. Dover, 1979).

The use of the Poincaré–Cartan fundamental one-form is from J Cary and R Littlejohn, 'Noncanonical Hamiltonian mechanics and its application to magnetic field line flow,' *Ann. Phys. (N.Y.)* **151** (1983) 1.

For the use of forms, exterior derivatives, and also Lie derivatives, encountered in the next Chapter, see *Geometrical Methods of Mathematical Physics*, by B Schutz (Cambridge, 1980). For more coordinate-based detail, especially on tensors, see *Tensors, Differential Forms, and Variational Principles*, by D Lovelock and H Rund (Wiley, 1975, repr. Dover, 1989). The Levi-Civita pseudotensor is well covered in this reference and also in Landau and Lifshitz *op cit.*

For the use of coordinate-free index tensor notation and some of the concepts, see R Wald, *General Relativity* (U. Chicago Press, 1984).

For the mathematics of functional derivatives, see I M Gelfand and S V Fomin, *Calculus of Variations* (Prentice-Hall, Englewood Cliffs, NJ, 1963).

For the use of functional derivatives in field theoretical approaches to plasma physics, see P J Morrison, 'Hamiltonian and action principle formulations of plasma physics,' *Phys. Plasmas* **12** (2005) 058102.

For more basic material see *The Theoretical Minimum (Classical Mechanics)*, by L Susskind and G Hrabovsky (Basic Books, 2013), and after that *An Introduction to Lagrangian Mechanics*, by A J Brizard (World Scientific, 2008).

IOP Publishing

Turbulence and Instabilities in Magnetised Plasmas, Volume 2
Gyrokinetic theory and gyrofluid turbulence
Bruce Scott

Chapter 5

Introduction to gyrokinetic theory

5.1 Ideas behind the gyrokinetic representation

In the same sense as in a fluid description, we seek a method to separate and then discard the high frequency motion not followed and replace the kinetic description, which follows particles, with an alternative which follows their low frequency drifts in some sense. This used to be done by averaging the equations describing the evolution of the distribution function over the high frequency orbits, basically the gyromotion. While adequate for linear theory this proved difficult when describing a self consistent nonlinear system in which the field variables and the kinetic distribution functions for each species are all treated on the same footing. One can no longer assume motion of single particles in prescribed fields, and then advance the fields according to general principles. Early attempts to systematically average the equations gave way to a method by which the variables and coordinates are transformed, together, into a different representation solving the same problem with somewhat different language. In the end, what came to be called gyrokinetic theory was developed as a type of gauge transformation. No longer a description of coherently identified particles on specific orbits, it is a transformation between one variationally self-consistent representation into another. When we refer to 'gyrocentres' we mean the constituents of the representation, not necessarily a picture based on charged rings, in which the gyromotion of a particle is averaged into the ring. The self-consistent representation is a Lagrangian field theory set up in much the same way as one does initially for Maxwell's equations and charged particles. Along the way, the charged particle motion gives way to a phase space flow using Liouville's theorem, with both steps as shown in the previous Chapter. In the same way, we set up the drift motion of gyrocentres but then replace that with the phase space flow in the gyrokinetic representation. The resulting equations describe drifts through Hamilton/Poisson brackets and is underlain by the same statistical theory as was originally built for systems of particles. What we get from that is a self-consistent theory with energy and momentum conservation laws arising from the symmetries in the field theory system Lagrangian. The representation is built using approximations, but these are made in the Lagrangian, which continues to preserve exact energetic consistency. One of the key concepts is the way in which gyrocenters replace particles and then the way in which gyrocenter charge density, due to the gyrocenter polarisation, represents the role of vorticity in the equations in any setting.

These steps are done in this Chapter and the next, in which the conservation laws, energetics, and gyrocentre drifts are detailed. Then, after obtaining the small scale ordering limit and then exploring the kinetic responses, the gyrofluid model is constructed. Ultimately, it is possible to show that one is actually solving the same problem in the low frequency limit in which both representations are valid. We will do this explicitly for the fluid side by showing the equivalence of the resulting fluid-drift and gyrofluid equations.

5.2 Lagrangian basis of kinetic theory

Summarising the basis of the previous Chapter, we set up a principle of least action according to textbook methods. The Lagrangian for a charged particle has a free motion piece and an interaction piece, whose terms represent the kinetic energy, the spatial coupling to the vector potential, and the temporal coupling to the electrostatic potential. At each step in the transformation of representation we will do, this structure is preserved so that the resulting Euler–Lagrange equations always have the same form. Once the transformed gyrocentre Lagrangian is ready, the free field portion is added to turn it into a system Lagrangian covering both gyrocentres and fields in the same way as done before for particles and fields. The Euler–Lagrange equations of this Lagrangian determine the evolution equations which are eventually used.

The Lagrangian of either particles or gyrocentres has the structure of a differential action, cast in the structure of the Poincaré–Cartan fundamental one-form,

$$L\,dt = \gamma = \gamma_a\,dz^a - H\,dt \tag{5.1}$$

where the phase space coordinates are z^a over the six dimensions and the γ_a are the corresponding canonical momenta. As time is not transformed (the treatment is non-relativistic), the time component is separated out, with H the Hamiltonian. The sign is due to the traditional arrangement following the $p\,dq - H(p, q)\,dt$ of the textbook case. We can think of $(dt, \mathbf{dz})$ as the differential element of a world line. One of the coordinates is a gyrophase angle, whose evolution tracks the gyromotion. Separating out this high frequency motion also involves ensuring independence of the remaining quantities of this gyrophase angle. Coordinates are always transformed such that this one-form is always in canonical form, namely, any dependence on the time dependent field variables is to be transformed out of γ_a and into H at each step. A total differential, dS, can also be added. Taken together, the twin objectives of gyrophase independence and canonical form are preserved. The conserved energy is given by the time translation symmetry. For any spatial symmetry in the magnetic geometry there is a corresponding conserved momentum given by the translation symmetry in one or more of the coordinates (e.g., axisymmetry, with a conserved toroidal angular momentum). This is Noether's theorem, which is the basis for the use of this type of theory and the origin of its conservation laws. One can also consider what is called a symplectic representation, in which γ_a can depend on the field variables, but we will not do that herein.

5.2.1 Euler–Lagrange equations

The equations governing the evolution of the gyrocentre coordinates and field variables are the resulting Euler–Lagrange equations of the system Lagrangian. The conventional method for coordinates (degrees of freedom) is equivalent to setting the functional derivative of the entire L with respect to each of the coordinates to zero, producing

$$(\gamma_{b,a} - \gamma_{a,b})\frac{dz^b}{dt} = \frac{\partial \gamma_a}{\partial t} + H_{,a} \tag{5.2}$$

as our version of Hamilton's equations. The method for the fields is more directly that of the functional derivative, since the field variable quantities are dependent on spatial coordinates $\mathbf{x}$. We set

$$\delta L(\delta\phi_a) = \int d\mathcal{V}\, \delta\phi_a \frac{\delta L}{\delta\phi_a} \qquad \text{then} \qquad \frac{\delta L}{\delta\phi_a} = 0 \tag{5.3}$$

for each field variable $\phi_a(\mathbf{x})$. In doing this we have to have a method to connect gyrocentres in their space $\mathbf{dz}$ to the field variable space $\mathbf{dx}$. This is combined in the general gyroaverage operation which takes the field variable in $\mathbf{x}$ to the gyroaveraged quantity in terms of $\mathbf{z}$. In our case, $\mathbf{dx}$ is a three-dimensional space, while $\mathbf{dz}$ is a six-dimensional phase space. We then set up a heuristic transform of the phase space into a continuous representation in terms of a spatial domain and a velocity space. This is then renamed as a phase space, and then the operations we do to construct equations and conservation laws will all be done in the continuous representation.

The previous Chapter demonstrated the use of canonical form and the fundamental one-form to recover the well known result of charged particle motion and then the initial ideas behind using the background magnetic field as an anchor for drift motion. Now we move on to the change of representation and the consequences of it which constitute gyrokinetic theory.

5.3 The strategy of gyrokinetics

The first step is to build a kinetic Lagrangian for gyrocentres, while defining and/or finding what that means along the way. The second step is to embed that into the continuous phase space while also adding the free-field piece, thereby finishing the construction of the field theory as a whole.

The standard procedure for the first step divides the process into two, first treating gyrocentre drift motion in the background magnetic field and then introducing the field variables as perturbations. This version allows the spatial scale to be arbitrary but keeps the ExB vorticity small compared to the gyrofrequency of any species.

Motion in the background magnetic field is treated through a coordinate transform, $\mathbf{x} = \mathbf{R} + \mathbf{a}$, choosing $\mathbf{a}$ to enforce canonical form. The gauge transform part adds an arbitrary scalar differential $\gamma \to \gamma + dS$, which will be determined by the requirement that no piece depends on the gyrophase angle. Using these constraints the coordinate transform and the gauge transform are done together. This defines the change in representation from particles to gyrocentres. The resulting motion describes drifts in the spatial plane locally perpendicular to the magnetic field and in the parallel phase space plane, with the gyromotion in the other two velocity space coordinates decoupling from the rest.

Addition of a differential to the one-form doesn't change the dynamics because the exterior derivative of a differential vanishes. The lack of gyrophase angle dependence leads to pure gyrocentre motion. Transforming all time dependence into the time component ensures canonical form, with the consequence that the equations contain no extra time derivatives and that the phase space metric and Jacobian are static. The Euler–Lagrange equations for the gyrocentre coordinates are gyrokinetic equations. Use of Liouville's theorem turns this into the evolution equation for the phase space distribution function. The motion of individual gyrocentres is re-interpreted as a flow in phase space, which de-emphasises motion of individual charges in favour of the system dynamics.

Finally, after the transform into the gyrocentre representation, the system is turned into a field theory by embedding the gyrocentre Lagrangian into phase space, with the distribution function as a phase space density, and adding the free-field Lagrangian density terms for the electromagnetic field. Appropriate low-frequency approximations are made in the Lagrangian (e.g., quasineutrality), the system dynamics is now self-consistent, and the model is ready for use.

5.4 The drift-kinetic Lagrangian

Drift kinetics refers to low frequency drift motion of charged particles in an electromagnetic field, with the field specified. The self-consistent field disturbances will only occur later. A strong background magnetic field, **B** with potential **A** is assumed to establish the existence of a magnetised plasma. The particle momentum **p** is assumed to be the perturbed canonical momentum. In the absence of perturbed dynamical fields, it is just the particle momentum. We then have

$$\gamma = (e\mathbf{A} + \mathbf{p}) \cdot \dot{\mathbf{x}} - H\, dt, \qquad H = \frac{p^2}{2m} \tag{5.4}$$

as a starting point. The background **A** will serve as an anchor for the drift motion. This one-form will yield the equations of motion for a charged particle, with $\mathbf{E} = 0$ and $\mathbf{B} = \nabla \times \mathbf{A}$. Part of that is the gyromotion with frequency $\Omega = eB/m$, namely, the species gyrofrequency. Turbulence that concerns us is well slower than this, which motivates this class of theories.

Starting with the one-form as given, we expand spatial quantities about a position **R** which differs from the particle position **x** by an unspecified amount **a**,

$$\mathbf{x} = \mathbf{R} + \mathbf{a}, \qquad |\mathbf{a}| = a \ll L_B \tag{5.5}$$

i.e., assumed to be small compared to the scale L_B on which **B** varies. The smallness parameter is loosely identified with a/L_B so the expansion proceeds in powers of **a**.

The smallness of quantities scaling with $\mathbf{v} \cdot \nabla$, with $\mathbf{v} = \mathbf{p}/m$, as compared to Ω also specifies that velocity terms enter at first order, not lowest order. Hence, order by order we have

$$\gamma_0 = e\mathbf{A} \cdot \mathbf{dR} \tag{5.6}$$

$$\gamma_1 = e\mathbf{a} \cdot (\nabla\mathbf{A}) \cdot \mathbf{dR} + \mathbf{p} \cdot \mathbf{dR} + e\mathbf{A} \cdot \mathbf{da} - H\, dt \tag{5.7}$$

$$\gamma_2 = e\mathbf{a} \cdot (\nabla\mathbf{A}) \cdot \mathbf{da} + \mathbf{p} \cdot \mathbf{da} \tag{5.8}$$

in all of which the gradient symbol denotes derivatives with respect to $\mathbf{R}$ not $\mathbf{x}$. We then split the particle momentum into

$$\mathbf{p} = z\mathbf{b} + m\mathbf{w}, \qquad \mathbf{B} = B\mathbf{b} \tag{5.9}$$

in which z becomes the parallel velocity space coordinate and $m\mathbf{w}$ is the perpendicular momentum. Some work also expands $\mathbf{b}$ using $\mathbf{a} \cdot \nabla\mathbf{b}$ but such a level of extra complication is unnecessary here.

The set of first order terms is now

$$\gamma_1 = e\mathbf{a} \cdot (\nabla\mathbf{A}) \cdot \mathbf{dR} + m\mathbf{w} \cdot \mathbf{dR} + z\mathbf{b} \cdot \mathbf{dR} + e\mathbf{A} \cdot \mathbf{da} - \left(\frac{z^2}{2m} + m\frac{w^2}{2}\right)dt \tag{5.10}$$

We subtract a differential

$$dS_1 = e\, d(\mathbf{A} \cdot \mathbf{a}) \tag{5.11}$$

from this to obtain

$$\gamma_1 = -e\mathbf{a} \times (\nabla \times \mathbf{A}) \cdot \mathbf{dR} + m\mathbf{w} \cdot \mathbf{dR} + z\mathbf{b} \cdot \mathbf{dR} - \left(\frac{z^2}{2m} + m\frac{w^2}{2}\right)dt \tag{5.12}$$

where we have used the vector identity

$$\mathbf{a} \times (\nabla \times \mathbf{A}) = (\nabla\mathbf{A}) \cdot \mathbf{a} - \mathbf{a} \cdot (\nabla\mathbf{A}) \tag{5.13}$$

Now we specify $\mathbf{a}$ as that quantity which makes the first two terms in γ_1 vanish,

$$-e\mathbf{a} \times \mathbf{B} + m\mathbf{w} = 0 \qquad \text{hence} \quad \mathbf{w} = \Omega\mathbf{a} \times \mathbf{b} \quad \text{or} \quad \mathbf{a} = \frac{\mathbf{b} \times \mathbf{w}}{\Omega} \tag{5.14}$$

This rewrites γ_1 as

$$\gamma_1 = z\mathbf{b} \cdot \mathbf{dR} - \left(\frac{z^2}{2m} + m\frac{w^2}{2}\right)dt \tag{5.15}$$

which adds to γ_0. We don't stop here because the model is incomplete: as yet there is no way to specify the dynamics under which $\mathbf{w}$ or $\mathbf{a}$ operate.

The set of second order terms contains $\mathbf{a}$ twice but in an angle-dependent form. We therefore subtract the differential

$$dS_2 = \frac{e}{2}\, d(\mathbf{a} \cdot \nabla\mathbf{A} \cdot \mathbf{a}) \tag{5.16}$$

from γ_2 to obtain

$$\gamma_2 = -\frac{e}{2}\mathbf{a} \times (\nabla \times \mathbf{A}) \cdot \mathbf{da} + m\mathbf{w} \cdot \mathbf{da} \tag{5.17}$$

in which terms other than arising from **da** are one order down. Using the specifications of **B**, **w** and **a**, this becomes

$$\gamma_2 = \frac{m\Omega}{2}\mathbf{a} \times \mathbf{b} \cdot \mathbf{da} \tag{5.18}$$

To evaluate this we have to introduce the third velocity space coordinate, the gyrophase angle ϑ, by setting the vector **a** into the gyromotion plane perpendicular to **b**, choosing the sense of ϑ as positive for ions. Two arbitrary vectors $\mathbf{e}_1$ and $\mathbf{e}_2$ span the plane locally perpendicular to **b**, forming a right-hand basis and defining **a** such that

$$\mathbf{e}_1 \times \mathbf{e}_2 \cdot \mathbf{b} = 1, \qquad \mathbf{a} = a(\cos\vartheta\mathbf{e}_1 - \sin\vartheta\mathbf{e}_2) \tag{5.19}$$

where $a = w/\Omega$ is the magnitude. This defines the local orientation with respect to the magnetic field. The differential **da** is then

$$\mathbf{da} = \mathbf{a}(d\log a) + a(\cos\vartheta\mathbf{de}_1 - \sin\vartheta\mathbf{de}_2) - a(\sin\vartheta\mathbf{e}_1 + \cos\vartheta\mathbf{e}_2)d\vartheta \tag{5.20}$$

The first term doesn't contribute since $\mathbf{a} \times \mathbf{b} \cdot \mathbf{a}$ vanishes. This leaves ϑ and the unit vectors, so that

$$\gamma_2 = \frac{mw^2}{2\Omega}\left(d\vartheta - \mathbf{dR} \cdot \left(\nabla\mathbf{e}_1\right) \cdot \mathbf{e}_2\right) \tag{5.21}$$

In the resulting $\gamma_{(0,\,1,\,2)}$ all quantities are independent of ϑ, which only appears in the one place in γ_2. The quantity

$$M = \frac{mw^2}{2\Omega} \tag{5.22}$$

is now a conserved quantity due to the symmetry in ϑ. It is the effective drift kinetic magnetic moment and can be used as a coordinate.

The entire drift kinetic one-form is now defined in the coordinates $\{\mathbf{R}, z, M, \vartheta\}$ and is given by

$$\gamma = (e\mathbf{A} + z\mathbf{b} - M\mathbf{W}) \cdot \mathbf{dR} + M\, d\vartheta - H\, dt \tag{5.23}$$

where

$$H = \frac{z^2}{2m} + M\Omega, \qquad \mathbf{W} = (\nabla\mathbf{e}_1) \cdot \mathbf{e}_2 \tag{5.24}$$

We usually write this in the compact form

$$\gamma = e\mathbf{A}^* \cdot \mathbf{dR} + M\, d\vartheta - H\, dt \tag{5.25}$$

where

$$e\mathbf{A}^* = e\mathbf{A} + z\mathbf{b} - M\mathbf{W}, \qquad \mathbf{B}^* = \nabla \times \mathbf{A}^* \tag{5.26}$$

in the notation due to Morozov and Solov'ev (1966). This derivation follows Littlejohn (1983). All Lagrangians we consider will be based on a gyrocentre one-form given by Eq. (5.25), with any contributions due to dynamical fields appearing only in H. This is the definition of canonical form for a drift kinetic or gyrokinetic Lagrangian.

One should review the above to note the points at which the coordinate choices are made. We choose dS_1 to combine the **da** terms into a form containing $\mathbf{B} = \nabla \times \mathbf{A}$, and then choose **a** to move gyrophase dependence out of γ_1. Here as well, the parallel velocity space coordinate z is separated from the gyromotion. Then, at the next order we choose dS_2 again to combine the **da** terms properly, and then finally set the coordinates M and ϑ according to the symmetry we find once the dependences on ϑ have all been removed. In this type of transform the steps at first order are applied according to choice, and those at second order are what is needed to restore the desired symmetries. All hanging pieces are indeed of third or higher order.

5.5 The field variables as perturbations

Having established the methodology on the simpler case of particle motion in the background magnetic field, we now use a similar strategy but with a more general method to treat the perturbed fields. The smallness is simply a requirement that any physical process introduced by the field dynamics remains slow compared to Ω, so that the time scale ratio and anything which results from it can be loosely used as a small parameter. In this case, the transformation order is set up by powers of the field quantities (linear at first order, quadratic at second order, etc). Including second order in this is necessary to obtain field perturbation energies, especially the ExB energy and the polarisation charge density that arises from it. This allows neglect of the electric field energy as small in comparison, the manifestation of quasineutrality in gyrokinetics.

Once the result is found, some pieces that are small by gradient scale ratios (e.g., parallel versus perpendicular) may be dropped in the Lagrangian with no damage to energetic consistency. If one tries this type of method by averaging the full kinetic equations, the result will almost always break energetic consistency. The physical picture of drifting charged rings is only valid at first order, and use of it beyond that causes problems. Energetic consistency in the resulting field theory is supported by Noether's theorem if it is set up variationally as we are doing.

In this case we will show the simplest version of the transform by including the electromagnetic field perturbation solely as $A_\|$ (shear Alfvén dynamics only) and incorporating it into the perturbed canonical momentum as a coordinate at the very beginning. The perturbed fields are ϕ and $A_\|$, functions of **x** and t and therefore, within the transformed coordinates,

$$(\phi, A_\|) = \text{(functions of)}\ (\mathbf{R}, M, \vartheta, t) \tag{5.27}$$

that is, they re-introduce gyrophase angle dependence, which we have to transform away. In addition, the time dependence has to be transformed solely into H to preserve canonical form. So we will start by defining the coordinate z and then the parallel velocity U as a function of other coordinates through

$$z = mU + eA_\parallel, \qquad U = U(\mathbf{R}, z, A_\parallel) \tag{5.28}$$

Then, the perturbed drift-kinetic one-form is

$$\gamma = e\mathbf{A}^* \cdot \mathbf{dR} + M\, d\vartheta - H\, dt \tag{5.29}$$

with $\mathbf{A}^*$ defined as in Eq. (5.26) but in terms of the new z, and Hamiltonian perturbed accordingly,

$$H = m\frac{U^2}{2} + M\Omega + e\phi \tag{5.30}$$

with $A_\parallel$ hidden in the definition of U above. Done this way, all the perturbations are already solely in H, and the transforming we do is facilitated. One can start with symplectic form with $A_\parallel\mathbf{b}$ added to $\mathbf{A}^*$ as a perturbation. The same answer is obtained at the relevant order, but at the cost of a lot more work, which can be left to treatments in which gyrokinetic theory and not turbulence or instabilities is the central point.

5.6 The Lie transform

Work in this direction up to the early 1980s developed the Lie transform as the most solid way to do the gyrokinetic transformations, with the parallel experience of early gyrokinetic work, using averaging methods directly on the equations, essentially serving as a signpost for the difficulties in the theory. The transform was to be done on the Lagrangian following the experience of drift kinetic theory from the 1960s, and the methods grew out of that (see Further Reading, below).

The Lie transform is a near-identity transformation with an inverse,

$$T = \exp[\varepsilon L], \qquad T^{-1} = \exp[-\varepsilon L] \tag{5.31}$$

The action on a scalar and on a one-form is given by

$$F = T^{-1}f, \qquad \Gamma = T^{-1}\gamma + dS \tag{5.32}$$

It is manifestly covariant, so that a transformation from representation $\mathbf{z}$ to $\mathbf{y}$ preserves the same structure,

$$\Gamma = \Gamma_\alpha(\mathbf{y})\, dy^\alpha \qquad \text{corresp to} \qquad \gamma = \gamma_\alpha(\mathbf{z})\, dz^\alpha \tag{5.33}$$

which is how energetic consistency is maintained. The coordinates and the resulting distribution function are transformed together, consistently. Manifest covariance is the reason we don't have to explicitly transform the distribution function or its evolution equation, since we eventually get the field theoretical Lagrangian and so the new distribution function is the only one we will write down. Moreover, the coordinate transformation will not have to be explicitly written. Philosophical discussions on the meaning of the coordinates are thereby obviated.

The layout of the Lie transform is with a Lie derivative and a differential giving gauge freedom, order by order,

$$\Gamma_0 = \gamma_0 + dS_0 \tag{5.34}$$

$$\Gamma_1 = \gamma_1 - L_1\gamma_0 + dS_1 \tag{5.35}$$

$$\Gamma_2 = \gamma_2 - L_1\gamma_1 + \left(\frac{1}{2}L_1^2 - L_2\right)\gamma_0 + dS_2 \tag{5.36}$$

Hence, the S_n is often referred to as the gauge function at each order. The generality is in the Lie derivative, whose generating functions can be chosen arbitrarily. Used together with the gauge freedom in choosing the S, any field dependence is moved into or kept within H, while simultaneously maintaining gyrophase independence. The form at second order is simplified by using the first order equation to substitute for $L_1\gamma_0$, so that

$$\Gamma_2 = \gamma_2 - L_1\frac{\gamma_1 + \Gamma_1}{2} - L_2\gamma_0 + dS_2 \tag{5.37}$$

noting that any operation $L(dS)$ of a Lie derivative upon a scalar differential vanishes. The math is facilitated by noting that any L_n acting on γ_0 produces the same terms at n-th order that $L_1\gamma_0$ does at first order. Note that the same structure, with known quantities, then $L_n\gamma_0$, and then dS_n, occurs at each order, so the operations involving L_n and dS_n are always the same.

5.6.1 The Lie derivative

The Lie derivative is defined by its generating functions, which express the Lie derivative along a vector, here labelled $\mathbf{g}$. The Lie derivative of a scalar along $\mathbf{g}$ in terms of coordinates $\mathbf{z}$ is

$$\pounds_{\mathbf{g}} f = g^b \frac{\partial f}{\partial z^b} = g^b f_{,b} \tag{5.38}$$

The Lie derivative of a vector field along another is the commutator, or in a coordinate basis

$$\pounds_{\mathbf{g}} \mathbf{V} = [\mathbf{g}, \mathbf{V}] = [g^b V^a{}_{,b} - V^b g^a{}_{,b}]\frac{\partial}{\partial z^a} \tag{5.39}$$

These two are used to derive the Lie derivative of a one-form through the contraction of its covector with another vector, or

$$\pounds_{\mathbf{g}}(V^a\gamma_a) = g^b(V^a\gamma_a)_{,b} \tag{5.40}$$

which leads to

$$\pounds_{\mathbf{g}}\gamma = g^b(\gamma_{a,b} - \gamma_{b,a})dz^a + d(g^a\gamma_a) \tag{5.41}$$

which is the Cartan 'magic formula' for this case (for more background see the text by Schutz, in Further Reading, in the previous Chapter). The differential is absorbed into the gauge function dS at each level, so we simply need to calculate the structure in the first term (it is an exterior derivative). This derivative has the same form as the Euler–Lagrange equations, which use the exterior derivative along the world line instead of along $\mathbf{g}$. Therefore, at order n we have the Lie transform as in Eq. (5.36), with the L_n on any γ and the differential dS_n defined as

$$L_n\gamma = g_n^b(\gamma_{a,b} - \gamma_{b,a})dz^a, \qquad dS_n = (S_n)_{,a}\, dz^a \tag{5.42}$$

Note that we will not transform time, so that

$$g_n^t = 0 \tag{5.43}$$

Relativistic versions of this proceed differently (e.g., Brizard 1999).

5.6.2 Initial arrangement

The one-form arrangement starts with the guiding centre one-form, since the fields first enter at first order. At zeroth order we have

$$\gamma_0 = e\mathbf{A}^* \cdot \mathbf{dR} + M\, d\vartheta - H_0\, dt \tag{5.44}$$

with $\mathbf{A}^*$ and H_0 given as in Eqs (5.24), (5.26),

$$e\mathbf{A}^* = e\mathbf{A} + z\mathbf{b} - M\mathbf{W}, \qquad H_0 = \frac{z^2}{2m} + M\Omega \tag{5.45}$$

The dynamical fields go into the perturbed Hamiltonian, in the canonical case in both

$$\gamma_1 = -e\left(\phi - \frac{z}{m}A_\parallel\right) dt, \qquad \gamma_2 = -\frac{e^2}{2m}A_\parallel^2\, dt \tag{5.46}$$

In addition, the second term in Γ_2 (Eq. 5.37) serves as an additional piece in the time component. In our case all terms except those in dS_n will appear only in the time component since we have already set up $\gamma_{(0,\,1,\,2)}$ in canonical form.

5.6.3 First order

The Γ_0 is already done since γ_0 is already in canonical form so we can take $S_0 = 0$. The next step is to apply L_1 to γ_0. Put in the minus sign and add the components of dS_1, and then set all the non-time components to zero, so that

$$dS_a = (L_1\gamma_0)_a, \qquad a \in (1, 2, 3, 4, 5, 6) \tag{5.47}$$

This gives

$$\nabla S_1 = e\mathbf{B}^* \times g_1^{\mathbf{R}} + g_1^z\mathbf{b}, \qquad \frac{\partial S_1}{\partial z} = -e\frac{\partial \mathbf{A}^*}{\partial z} \cdot g_1^{\mathbf{R}} \tag{5.48}$$

$$\frac{\partial S_1}{\partial M} = -g_1^{\vartheta}, \qquad \frac{\partial S_1}{\partial \vartheta} = g_1^{M} \tag{5.49}$$

using the same cross-curl vector identity as in Eq. (5.13) to get ∇S_1 (the space–space components are the ones which produce this anti-symmetric form). To solve these, only the parallel phase space planes (between **b** and z) provide difficulty. Contract ∇S_1 with $\mathbf{B}^*$ to isolate g_1^z and then cross it with $\partial \mathbf{A}^*/\partial z$ to eliminate g_1^z and isolate $g_1^{\mathbf{R}}$, using $\partial S_1/\partial z$ to evaluate the parallel component. The gyromotion components are simply read off. This produces

$$B_\parallel^* g_1^{\mathbf{R}} = -\frac{\partial S_1}{\partial z}\mathbf{B}^* - \frac{\partial \mathbf{A}^*}{\partial z} \times \nabla S_1, \qquad B_\parallel^* g_1^z = \mathbf{B}^* \cdot \nabla S_1 \tag{5.50}$$

$$g_1^{M} = \frac{\partial S_1}{\partial \vartheta} \qquad g_1^{\vartheta} = -\frac{\partial S_1}{\partial M} \tag{5.51}$$

where

$$B_\parallel^* = e\frac{\partial \mathbf{A}^*}{\partial z} \cdot \mathbf{B}^* \tag{5.52}$$

is a scalar quantity serving as the Jacobian of the phase space transformation.

The time component of $L_1\gamma_0$ is simply given by derivatives of H_0 along $\mathbf{g}_1$ in phase space, since time is not transformed and there is no time dependence in the phase space components of γ_0. Putting these with the perturbed Hamiltonian, we have

$$(\Gamma_1)_t = \frac{dS_1}{dt} - e\left(\phi - \frac{z}{m}A_\parallel\right) \tag{5.53}$$

where the total time derivative is with the derivatives of H_0,

$$\begin{aligned} \frac{dS_1}{dt} &= \frac{\partial S_1}{\partial t} + \frac{1}{B_\parallel^*}\frac{\partial \mathbf{A}^*}{\partial z} \times \nabla H_0 \cdot \nabla S_1 \\ &+ \frac{\mathbf{B}^*}{B_\parallel^*} \cdot \left(\frac{\partial H_0}{\partial z}\nabla S_1 - \frac{\partial S_1}{\partial z}\nabla H_0\right) + \left(\frac{\partial H_0}{\partial M}\frac{\partial S_1}{\partial \vartheta} - \frac{\partial S_1}{\partial M}\frac{\partial H_0}{\partial \vartheta}\right) \end{aligned} \tag{5.54}$$

We no longer have freedom to set the components of dS_1 but we still have the freedom to set this time component to anything we want. We simply set it equal to the first-order gyroaverage of the field perturbation:

$$(\Gamma_1)_t = -e\langle\Phi\rangle_{\mathbf{R}} \qquad \text{where} \quad \Phi = \phi - \frac{z}{m}A_\parallel \tag{5.55}$$

The gyroaverage is the operation which at finite gyroradius takes a function defined in $\mathbf{x}$ to one defined in the gyrocentre variables $\{\mathbf{R}, z, M\}$, independent of ϑ, the gyrophase angle. Spatially, it is positioned at $\mathbf{R}$. Mathematically, we have to define it as the operator necessary to set the gyroaverage of the defect to zero according to

$$\tilde{\Phi} = \Phi - \langle\Phi\rangle_{\mathbf{R}}, \qquad \langle\tilde{\Phi}\rangle_{\mathbf{R}} = 0 \tag{5.56}$$

It therefore has to be the annihilation operator for whatever we have on the left hand side. In practice the background is taken to be homogeneous on this scale, so that the average is a simple one over the angle. Representations in the limits commonly taken are given below. In any case, it is defined to be independent of ϑ, which together with preservation of canonical form the transform has to maintain.

The choice for the perturbed Hamiltonian (Eq. 5.55) produces the equation for what we call the gauge function S_1, which satisfies

$$\frac{dS_1}{dt} = e\tilde{\Phi} \tag{5.57}$$

In principle, all the pieces of d/dt apply, but in practice we invoke the ordering to say that the gyromotion bracket $(M\vartheta)$ is the largest term. Since H_0 is independent of ϑ and $\partial H_0/\partial M = \Omega$, we have

$$\Omega\frac{\partial S_1}{\partial \vartheta} = e\tilde{\Phi}, \qquad \langle(\text{both sides})\rangle_{\mathbf{R}} = 0 \tag{5.58}$$

and so the determinations at this order are complete. Note that S_n is not used at order n, but it is specified at order n for use in the Lie derivative upon γ_0 at order $(n + 1)$ and upon higher order pieces after that. But our purpose is satisfied by having S_1 for use at order two, where we will stop.

5.6.4 Second order—polarisation

The motivations for going to second order are as follows. The consequences of the freedom used at first order are still to be resolved. We would not have a closed energy theorem since without term quadratic in the fields we cannot get their equations except as constraints. And there is no role yet for polarisation or quasineutrality, which we have already seen is central to this type of dynamics. The reason for that is that ion inertia has been ordered small, so only at second order do we get the necessary corrections to the charged-ring picture of gyrokinetics and thereby proper energetics. The general result we do learn, however, is that we make free choices at first order but now live with the consequences of those at the next.

The need for a polarisation effect is that the first order H_1 does not lead to any new field terms in the field theory. If the variation of the field potentials is taken, only coefficients survive, and these would have to collectively vanish. But since they are dynamically independent quantities, this would overdetermine the system and eliminate the dynamics (ExB vorticity: turbulence) we want to follow. Some early numerical work simply postulated an ExB field energy term, but it is important to know how to derive these consistently, especially if one would ever want to extend the theory.

The second order Lie transform correction is the Γ_2 line in Eq. (5.36), replaced by Eq. (5.37). We first arrange

$$\frac{\gamma_1 + \Gamma_1}{2} = -\frac{e}{2}\tilde{\Phi}\, dt - e\langle\Phi\rangle_{\mathbf{R}}\, dt \tag{5.59}$$

The Lie derivative on this involves S_1 on every term. We already know S_1 and that its gyroaverage is zero. So only the products with S_1 and $\tilde{\Phi}$ will survive, and those with $\langle\Phi\rangle_{\mathbf{R}}$ will vanish. Hence we need only compute

$$L_1(\tilde{\Phi}\,dt)_t = g_1^{\mathbf{R}}\cdot\nabla\tilde{\Phi} + g_1^{z}\frac{\partial\tilde{\Phi}}{\partial z} + g_1^{M}\frac{\partial\tilde{\Phi}}{\partial M} + g_1^{\vartheta}\frac{\partial\tilde{\Phi}}{\partial\vartheta} \tag{5.60}$$

where the last two derivatives are nonzero because of the dependence upon $\mathbf{x}$ of the gyrocentre coordinates. Setting the phase space components of Γ_2 to zero is the same operation as that for Γ_1 and the same terms result with S_2 in them instead of S_1. This leaves

$$\begin{aligned}(\Gamma_2)_t = {}& \frac{dS_2}{dt} - \frac{e^2}{2m}A_\parallel^2 - \frac{e}{2B_\parallel^*}\frac{\partial\mathbf{A}^*}{\partial z}\times\nabla S_1\cdot\nabla\tilde{\Phi} \\ & + \frac{e}{2}\frac{\mathbf{B}^*}{B_\parallel^*}\cdot\left(\frac{\partial\tilde{\Phi}}{\partial z}\nabla S_1 - \frac{\partial S_1}{\partial z}\nabla\tilde{\Phi}\right) + \frac{e}{2}\left(\frac{\partial\tilde{\Phi}}{\partial M}\frac{\partial S_1}{\partial\vartheta} - \frac{\partial S_1}{\partial M}\frac{\partial\tilde{\Phi}}{\partial\vartheta}\right)\end{aligned} \tag{5.61}$$

In this expression the terms involving $\mathbf{B}^*\cdot\nabla$ are down one order since $k_\parallel \ll k_\perp$. When the last term involving derivatives with respect to $(M\vartheta)$ is gyroaveraged, $\partial/\partial\vartheta$ can be done by parts under the average such that it acts on S_1, whose factor can be replaced using Eq. (5.58). The only role for S_2 is to do the final gyroaverage, as in Eq. (5.58). This leaves

$$(\Gamma_2)_t = -\frac{e^2}{2m}\left\langle A_\parallel^2\right\rangle_{\mathbf{R}} + \frac{e^2}{2\Omega}\frac{\partial}{\partial M}\langle\tilde{\Phi}^2\rangle_{\mathbf{R}} - \frac{e}{2B_\parallel^*}\frac{\partial\mathbf{A}^*}{\partial z}\cdot\langle\nabla S_1\times\nabla\tilde{\Phi}\rangle_{\mathbf{R}} \tag{5.62}$$

The first term completes the square to form the parallel kinetic energy, the second term represents polarisation, and the last term represents what can be called a cross-drift gyrocentre correction. In practice it is small because its largest contribution vanishes due to antisymmetry of the cross operation.

5.6.5 Collected result

Summing the three contributions $\Gamma_{(0,\,1,\,2)}$, we have

$$\Gamma = e\mathbf{A}^*\cdot\mathbf{dR} + M\,d\vartheta - H\,dt \tag{5.63}$$

where the Hamiltonian includes the field perturbations,

$$H = \frac{m}{2}\langle U^2\rangle_{\mathbf{R}} + M\Omega + e\langle\phi\rangle_{\mathbf{R}} - \frac{e^2}{2\Omega}\frac{\partial}{\partial M}\langle\tilde{\Phi}^2\rangle_{\mathbf{R}}, \qquad mU = z - eA_\parallel \tag{5.64}$$

This is the result given by Hahm *et al* (1988), except that we drop the cross-drift gyrocentre correction. Note that the structure is the same canonical form as in Eq. (5.25), just that H has been expanded to include the field contributions. This is the point of gyrokinetic theory. Once we have one version of it, all versions share the same structure, and the steps to build the field theory as in the next two sections are the same for any ordering scheme one may choose.

5.7 The gyroaverage

The gyroaverage is formally defined in Eq. (5.56) as the operation which, when applied to the time dependent equation for S_1 (Eq. 5.58) yields zero. Under this form it is the time average over the orbit of gyration of the perturbed Hamiltonian,

$$\langle\Phi\rangle_{\mathbf{R}} = \Omega_0 \int^{\vartheta} \frac{d\vartheta'}{\Omega} \tilde{\Phi}, \qquad \Omega_0^{-1} = \oint \frac{d\vartheta}{\Omega} \tag{5.65}$$

with $2\pi\Omega_0^{-1}$ the time taken by the motion over one orbit. The average is done with the zeroth order motion, hence $\Omega = \partial H_0/\partial M$. As it must annihilate a gyroaverage defect, its application in Eq. (5.61) must also work properly. Both are done using the same Lie derivative operation (L_0 on the remaining time component at each level), so the approximations which set the form of Eq. (5.58) apply to each. At that level, the gyroaverage is the time average over the orbit of gyration in the unperturbed magnetic field. Moreover, in the calculation of and approximations to the Hamiltonian through second order, only $\partial S_1/\partial\vartheta$ is used, since we drop the last term in Eq. (5.61). So Eq. (5.64) remains even if the time average with variable Ω is needed. In practical terms, the gyroaverage is taken with respect to a homogeneous field, with localised delta-f computations in mind. It is important to note, however, that for computations in global, inhomogeneous systems the issue of the gyroaverage as a time average over a deformed orbit, which might not even be closed in the level of approximation one is using, should be reconsidered.

With Ω taken as constant, the gyroaverage is over the angle by itself. It becomes defined heuristically as an average over the ring described by gyromotion in $\mathbf{x}$ about the gyrocentre position $\mathbf{R}$,

$$\langle\Phi\rangle_{\mathbf{R}} = \frac{1}{2\pi} \oint d\vartheta \Phi(\mathbf{R} + \mathbf{a}) \tag{5.66}$$

where the subscript denotes the dependence on $\mathbf{R}$ not $\mathbf{x}$, although it also has dependence on M.

Due to the dependence of the operation on velocity space variables, we have in general

$$\langle\phi\rangle_{\mathbf{R}} = \int \mathbf{dx}\, G(\mathbf{x}, \mathbf{z})\, \phi(\mathbf{x}) \tag{5.67}$$

taking the field variable $\phi(\mathbf{x})$ into a function of the gyrocentre coordinate space $\mathbf{z}$. To obtain an equation for the field variable $\phi(\mathbf{x})$ in terms of a phase space quantity $f(\mathbf{z})$, we need the conjugate operation

$$\langle f\rangle_{\mathbf{x}} = \int \mathbf{dz}\, G(\mathbf{x}, \mathbf{z}) f(\mathbf{z}) \tag{5.68}$$

This is a 'charge assignment' which places the function $f(\mathbf{z})$ into the physical space $\mathbf{x}$. It is an issue for any Lagrangian coordinate based representation, not just a gyrokinetic one. It allows the gyrocentre variables to be sources for the field equation's operations to solve for the field variable $\phi(\mathbf{x})$. These are conjugate operations with a symmetric kernel $G(\mathbf{x}, \mathbf{z})$. The operation will become Hermitian when we get it into a continuous representation.

In the case of homogeneous Ω, we have $\mathbf{x} = \mathbf{R} + \mathbf{a}$ where $\mathbf{a}$ is the directed gyroradius defined in Eq. (5.19). The set of points in $\mathbf{x}$ which come in question to form the gyroaveraged quantity at a point $\mathbf{R}$ traces out a closed circle, and the set of points in $\mathbf{R}$ which contribute to the source at a point $\mathbf{x}$ is the same circle. This gives

$$\langle\phi\rangle_{\mathbf{R}} = \int \mathbf{dx}\, \delta(\mathbf{x} - \mathbf{R} - \mathbf{a})\, \phi(\mathbf{x}), \qquad \langle f\rangle_{\mathbf{x}} = \int \mathbf{dz}\, \delta(\mathbf{x} - \mathbf{R} - \mathbf{a}) f(\mathbf{z}) \tag{5.69}$$

where the symmetric kernel is just the delta function collecting the points in one space which contribute to the other. The weak statement of symmetry is actually the one we need, namely

$$\int \mathbf{dz} f(\mathbf{z})\langle\phi\rangle_{\mathbf{R}} = \int \mathbf{dx}\, \phi(\mathbf{x})\langle f\rangle_{\mathbf{x}} \tag{5.70}$$

or more explicitly,

$$\int \mathbf{dz} \int \mathbf{dx} f(\mathbf{z})\, G(\mathbf{x}, \mathbf{z})\, \phi(\mathbf{x}) = \int \mathbf{dx} \int \mathbf{dz} \phi(\mathbf{x})\, G(\mathbf{x}, \mathbf{z}) f(\mathbf{z}) \tag{5.71}$$

which is obvious. This is the operation we will need in order to express the interaction Lagrangian and enable the functional derivative of the action with respect to the field variables.

These are essentially filter operations. They are not inverses of each other, since in the averaging some information is lost. Essentially, the gyroaverage defect, given by

$$\tilde{\phi}(\mathbf{x}) = \phi(\mathbf{x}) - \langle\phi\rangle_{\mathbf{R}} \tag{5.72}$$

gets averaged to zero and is lost by the operation, leaving no piece which depends on the gyrophase angle.

The gyroaverage defect appears in the second order field term in Eq. (5.64), under another gyroaverage. As such, there is no ambiguity since the resulting quantity is properly a function of $\mathbf{z}$. But if the defect appears by itself there is a piece which depends on $\mathbf{x}$ and another one which depends on $\mathbf{R}$. Mathematically, this is not kosher since one should not be able to add two quantities which belong to different spaces. This is why we take the steps to redefine the gyroaverage and its defect, essentially to make a simple approximation to the Hamiltonian, so that this ambiguity is eliminated and we can do practical computations.

The first step is to notice that

$$\langle\tilde{\phi}^2\rangle_{\mathbf{R}} = \langle(\phi - \langle\phi\rangle_{\mathbf{R}})^2\rangle_{\mathbf{R}} = \langle\phi^2 - 2\phi\langle\phi\rangle_{\mathbf{R}} + \langle\phi\rangle_{\mathbf{R}}^2\rangle_{\mathbf{R}} \tag{5.73}$$

The second term combines with the last so that we have

$$\langle\tilde{\phi}^2\rangle_{\mathbf{R}} = \langle\phi^2\rangle_{\mathbf{R}} - \langle\phi\rangle_{\mathbf{R}}^2 \tag{5.74}$$

At this point we no longer have the defect as part of the mathematical description. In each of the terms involving the field variables, the gyroaverage unambiguously takes a quantity defined in **x** to one defined in **z**. We can now take the step to regard **x** and **R** as a set of points within the same space, and then assign the gyroaveraging operation to an operator which depends on M (or z, though usually not) as well as spatial quantities.

5.7.1 Continuous basis

We redefine the Lagrangian in terms of continuous operations in a phase space covering the gyrocentre coordinates, and identify points in **R** with the same points in **x**. The gyroaveraging operator becomes a filtering kernel standing between a distribution function, $f(\mathbf{R}, z, M)$ in the gyrocentre variables, noting that it is independent of the gyrophase angle, and the fields $\phi_a(\mathbf{x})$. This replaces the gyroaveraging operator between two spaces with a similar operator with the same functionality acting within the same space, according to

$$\langle f \rangle_{\mathbf{x}} \rightarrow J_0(f), \qquad \langle \phi \rangle_{\mathbf{R}} \rightarrow J_0(\phi) \tag{5.75}$$

It is labelled in this way because in a local domain with a Fourier decomposition we have

$$\phi(\mathbf{x}) = \sum_{\mathbf{k}} \phi_{\mathbf{k}}\, e^{i\mathbf{k}\cdot\mathbf{x}} \tag{5.76}$$

and therefore using

$$\mathbf{x} = \mathbf{R} + \mathbf{a} \qquad \text{and} \qquad \mathbf{k}\cdot\mathbf{a} = a\cos\theta \qquad a^2 = \rho_L^2 = \frac{2M}{m\Omega} \tag{5.77}$$

with the integral definition

$$\frac{1}{2\pi}\oint d\theta\, \exp(i\alpha\cos\theta) = J_0(\alpha) \tag{5.78}$$

where $J_0(\alpha)$ is the Bessel function with argument α, we have

$$\langle \phi \rangle_{\mathbf{R}} = \sum_{\mathbf{k}} \phi_{\mathbf{k}}\, e^{i\mathbf{k}\cdot\mathbf{R}} J_0(k_\perp \rho_L) \tag{5.79}$$

where $\rho_L = 2M/m\Omega$ is the gyrocentre Larmor radius. The gyroaveraging operation is therefore equivalent to multiplication of the Fourier amplitudes by the filter function $J_0(k_\perp \rho_L)$. In this form the gyroaveraging as an operator is replaced by one which is spatially Hermitian, since using the definitions leading up to Eq. (5.71) we now have

$$\int d\Lambda f\, J_0(\phi) = \int d\Lambda\, \phi\, J_0(f) \tag{5.80}$$

with all quantities defined on the same space, even though ϕ does not depend on z or M, and none of the quantities dependent on ϑ. This demonstrates that the continuous representation of the gyroaveraging operator is Hermitian and it is the one actually used in computations.

We now have for the gyrokinetic Hamiltonian (Eq. 5.64), the expression

$$H = m\frac{U^2}{2} + M\Omega + e(J_0\phi) - \frac{e^2}{2\Omega}\frac{\partial}{\partial M}(J_0(\Phi^2) - (J_0\Phi)^2) \tag{5.81}$$

and for the parallel velocity function

$$mU = z - e(J_0A_\parallel) \tag{5.82}$$

and we still have Φ as defined in Eq. (5.55). In this form $A_\parallel$ as well as ϕ contributes to polarisation, which is reflected in the field equations and in the energetics.

5.7.2 Integration space

The field variables actually exist in the spatial domain defined by spatial coordinates **x**, essentially a manifold with a metric to define distances. Scalar and vector fields and operations on them are defined according to textbook methods. A spatial integration is over the spatial volume element **dx**, while a phase space integration is over the volume element **dz** given by its span of coordinates. Having done the work to connect **R** to **x** in a continuous representation, $d\mathbf{x}$ and $d\mathbf{R}$ become interchangeable, both replaced by $d\mathcal{V}$ and $G \rightarrow J_0$ is just a spatial operator. In simple models $J_0 \rightarrow 1$ if gyroaveraging is neglected. The description as a gyrokinetic representation, however, remains.

The phase space, spatial domain, and velocity space integration operations are now defined by

$$d\Lambda = d\mathcal{V} \otimes d\mathcal{W}, \quad d\mathcal{V} = \sqrt{g}\, dx^1\, dx^2\, dx^3, \quad d\mathcal{W} = \frac{2\pi}{m^2} B_\parallel^*\, dz\, (e/m) dM \tag{5.83}$$

where the spatial coordinates are $\{x^1, x^2, x^3\}$, the coordinate volume element is $\sqrt{g}$ with g the determinant of metric coefficients g_{ij}, the 2π is for the gyrophase angle and the factors of m and e/m are for units and the sign (positive definite phase space volume element). We now redefine the phase space flow and the system Lagrangian in terms of this continuous phase space.

5.8 The gyrocentre phase space density and flow

In building the gyrocentre one-form we have been considering the world line of one gyrocentre in the background magnetic field perturbed by the two fields ϕ and $A_\parallel$ which represent the time dependent dynamics. However, that is not the point of the entire theory. This is merely a step towards it, basically step one of three. Step two, in this section, is the conversion from world line motion to a conservative phase space flow according to which the phase space gyrocentre density changes. Step three, in the next section, will then complete the field theory.

Given the gyrocentre one-form, we find the Euler–Lagrange equations according to Eq. (5.2). The operations to do the derivatives and isolate the individual time derivatives are the same as those used to find the generating functions of the Lie transform, in Eqs (5.51). This yields

$$B_\parallel^* \dot{\mathbf{R}} = \frac{\partial H}{\partial z}\mathbf{B}^* + \frac{\partial \mathbf{A}^*}{\partial z} \times \nabla H, \qquad B_\parallel^* \dot{z} = -\mathbf{B}^* \cdot \nabla H \tag{5.84}$$

for the drift motion and

$$\dot{M} = 0, \qquad \dot{\vartheta} = \frac{\partial H}{\partial M} - e\dot{\mathbf{R}} \cdot \frac{\partial \mathbf{A}^*}{\partial M} \tag{5.85}$$

for the gyromotion. Since Γ is in canonical form the $\mathbf{A}^*$, $\mathbf{B}^*$, and $B_\parallel^*$ are the same as in Eqs (5.26), (5.52).

The distribution function f is a phase space density depending on $\{\mathbf{R}, z, M, t\}$ but not ϑ. We state that during the motion f is constant along characteristics such that

$$\frac{df}{dt} = \frac{\partial f}{\partial t} + \dot{\mathbf{R}} \cdot \nabla f + \dot{z}\frac{\partial f}{\partial z} + \dot{M}\frac{\partial f}{\partial M} + \dot{\vartheta}\frac{\partial f}{\partial \vartheta} = 0 \tag{5.86}$$

in which the last two terms drop because $\dot{M}$ and $\partial f/\partial\vartheta$ vanish. We now replace the world line time derivatives with a corresponding flow which has the same form,

$$B_\parallel^* Z^{\mathbf{R}} = \frac{\partial H}{\partial z}\mathbf{B}^* + \frac{\partial \mathbf{A}^*}{\partial z} \times \nabla H, \qquad B_\parallel^* Z^z = -\mathbf{B}^* \cdot \nabla H \tag{5.87}$$

The phase space flow actually occurs in the space defined by the gyrocentre coordinates $\mathbf{R}$ and z, with M conserved (i.e., one 4-space for each position in M-space) and no gyrophase angle dependence. There is no discreteness and no difference between particle and phase space coordinates, i.e., only the latter are relevant now. We can use $\mathbf{R}$ and $\mathbf{x}$ interchangeably since the gyroaverage is replaced by a continuous Hermitian operation (Padé approximants involving Laplacians are usually applied). It is easily verified that

$$\nabla \cdot \left(B_\parallel^* Z^{\mathbf{R}}\right) + \frac{\partial}{\partial z}\left(B_\parallel^* Z^z\right) = 0 \tag{5.88}$$

Since all the geometric information in $B_\parallel^*$ is static we also have

$$\frac{\partial}{\partial t} B_\parallel^* = 0 \qquad \text{and hence} \qquad \frac{d}{dt} B_\parallel^* = 0 \tag{5.89}$$

This means that $B_\parallel^*$ can be understood as a phase space volume element, from which follows that phase space conservation is automatically satisfied. Hence, for the gyrokinetic evolution equation we can write either

$$\frac{\partial f}{\partial t} + Z^{\mathbf{R}} \cdot \nabla f + Z^z \frac{\partial f}{\partial z} = 0 \quad \text{or} \quad B_\parallel^* \frac{\partial f}{\partial t} + \nabla \cdot \left(f B_\parallel^* Z^{\mathbf{R}}\right) + \frac{\partial}{\partial z}\left(f B_\parallel^* Z^z\right) = 0 \tag{5.90}$$

the former as $df/dt = 0$ and the latter as a continuity equation. In either form, a conservative collision operator can be added (we will consider only simple forms for that). This is the version of Liouville's theorem which we are using.

5.9 The gyrokinetic field Lagrangian

The self-consistent dynamics results from an action of the form

$$L = \sum_{sp} \int d\Lambda f\, L_p + \int d\mathcal{V}\, \mathcal{L}_f \tag{5.91}$$

where the sum is over species (e.g., electrons and ions), f is the distribution function for one species with gyrocentre Lagrangian L_p, the free field Lagrangian density is $\mathcal{L}_f$, and the integrations are over phase space $d\Lambda$ and the spatial domain $d\mathcal{V}$, respectively. This L_p is simply d/dt on the one-form, so for the coordinates we get time derivatives rather than differentials and H appears by itself,

$$L_p = e\mathbf{A}^{*}\cdot\dot{\mathbf{R}} + M\dot{\vartheta} - H \tag{5.92}$$

In the free field piece we would start with $\varepsilon_0 E^2/2 - B^2/2\mu_0$ but then take

$$\varepsilon_0 \to 0, \qquad B^2 \to \left|\nabla_\perp A_\parallel\right|^2 \tag{5.93}$$

to effect quasineutrality and restrict to shear Alfvén dynamics, respectively. Under these limits we have

$$\mathcal{L}_f = -\frac{1}{2\mu_0}\left|\nabla_\perp A_\parallel\right|^2 \tag{5.94}$$

assuming a treatment in simplified geometry in which $\nabla_\perp$ is the perpendicular gradient (factors of the radius of magnetic field line curvature need to be incorporated in more general instances).

5.9.1 Field equations

The Euler–Lagrange equations for the gyrocentre world lines are found and the dynamics for f constructed as in the last section. We now turn to the field equations. These are the Euler–Lagrange equations for ϕ and $A_\parallel$, respectively. In this context, the entire L is a functional since it takes functions of the dependent variables and maps them to numbers. In general one sets the functional derivative of L with respect to each field to zero as the minimisation principle in each case. In practice, we can still give a very general form using the previous effort to maintain canonical form, which now means the dynamical fields are solely in the time component (H and $\mathcal{L}_f$). Since ϕ appears in H only we can write

$$\delta L(\delta\phi) = -\sum_{sp} \int d\Lambda \left[f\, eJ_0(\delta\phi) - f\, \frac{e^2}{\Omega}\frac{\partial}{\partial M}\left(J_0(\Phi\,\delta\phi) - (J_0\Phi)(J_0\delta\phi)\right)\right] \tag{5.95}$$

We integrate the partial derivative $\partial/\partial M$ by parts to define the polarisability for each species as

$$\mathcal{M} = -\frac{e^2}{\Omega}\frac{\partial f}{\partial M} \tag{5.96}$$

Using the Hermitian property of the J_0 operator, we integrate by parts to isolate the $\delta\phi$ with all operations acting on the rest of the integrand. Finally, we use the fact that ϕ depends only on the spatial variables and separate out the volume integration, yielding

$$\delta L(\delta\phi) = \int d\mathcal{V}(-\delta\phi)\sum_{sp}\int d\mathcal{W}\left[eJ_0(f) + \left(J_0\mathcal{M}\,J_0 - (J_0\mathcal{M})\right)\Phi\right] \tag{5.97}$$

Setting the coefficient of $\delta\phi$ to zero we find the polarisation equation,

$$\sum_{sp}\int d\mathcal{W}\left[eJ_0(f) + \left(J_0\mathcal{M}\,J_0 - (J_0\mathcal{M})\right)\Phi\right] = 0 \tag{5.98}$$

which is also called the gyrokinetic Poisson equation as it balances a polarisation density with the gyroaveraged gyrocentre charge density. However, the system is by definition charge-neutral. This is the consequence of leaving ε_0 terms out of the Lagrangian, namely, the gyrokinetic version of quasineutrality.

The magnetic field perturbation due to $A_\parallel$ is correspondingly found by taking its variation in L. Here also, $A_\parallel$ appears in the time component only, but in both H and $\mathcal{L}_f$. In H we have the form with U as well as the one with Φ, so we use

$$\delta(mU^2/2)(\delta A_\parallel) = -eUJ_0(\delta A_\parallel) \tag{5.99}$$

and for the polarisation the same operations as for $\delta\phi$ above, with a multiplier of $-(z/m)$ due to the definition of Φ in Eq. (5.55). The field term is a straightforward operation on the squared perpendicular gradient (Eq. 5.94),

$$\delta\mathcal{L}_f = -\frac{1}{\mu_0}(\nabla_\perp A_\parallel)\cdot(\nabla_\perp\delta A_\parallel) \tag{5.100}$$

Putting these together and doing the $\nabla_\perp$ and the gyroaverage operations by parts, the result is

$$\begin{aligned}\delta L(\delta A_\parallel) = &\int d\mathcal{V}(\delta A_\parallel)\frac{1}{\mu_0}\nabla_\perp^2 A_\parallel \\ &+ \int d\mathcal{V}(\delta A_\parallel)\sum_{sp}\int d\mathcal{W} \\ &\left[J_0(feU) - \frac{z}{m}\left(J_0\mathcal{M}\,J_0 - (J_0\mathcal{M})\right)\Phi\right]\end{aligned} \tag{5.101}$$

Setting the coefficient of $\delta A_\parallel$ to zero, we find the induction equation,

$$-\frac{1}{\mu_0}\nabla_\perp^2 A_\parallel = \sum_{sp}\int d\mathcal{W}\left[J_0(feU) - \frac{z}{m}\left(J_0\mathcal{M}\,J_0 - (J_0\mathcal{M})\right)\Phi\right] \tag{5.102}$$

which is the gyrokinetic version of Ampère's law. We note that in this representation of velocity space, U as well as the Φ terms depends on $A_\parallel$. In the induction equation we note that the sign of $A_{\|}$ in Φ is negative, so that these polarisation terms contributing to the gyrokinetic current essentially add their $A_{\|}$ contributions to those in the Laplacian.

5.10 Simplified limits

As it stands, the field equations are mixed, i.e., the equation for ϕ depends on both ϕ and $A_{\|}$, and the equation for $A_{\|}$ depends on both $A_{\|}$ and ϕ. But solution of these equations in a computation is still fairly straightforward, even at this level of complexity. In practical cases, however, simpler expressions are desired, and the way to get them is to make the corresponding approximations in the Lagrangian itself, so as not to damage the symmetry and consistency of the equations in any way.

5.10.1 Electrostatic polarisation

The first approximation to make is the low-beta one, namely, that $\beta_e = \mu_0 p_e / B^2 \ll 1$ is usually taken for magnetised plasmas. In this case, we note that the ions contribute most to polarisation since it is fundamentally a finite gyroradius effect and the ions are stronger in that their mass is larger. In the $A_{\|}$ contributions to Φ we have the z/m multiplier. The combination of squared gyroradius (mass) and the thermal velocity (inverse square root of the mass) is larger for the ions by the square root of the mass ratio. In the ions z/m scales like c_s, while the magnetic energy of the disturbances is not larger than the ExB energy (cf. Chapters 3 and 5 of Volume One). The scaling is

$$\frac{\rho}{B^2} |\nabla_\perp \phi|^2 > \frac{1}{\mu_0} |\nabla_\perp A_{\|}|^2 \tag{5.103}$$

which means that for disturbance amplitudes

$$\phi > v_A A_{\|} \gg c_s A_{\|} \sim \frac{z}{m} A_{\|} \tag{5.104}$$

recalling $v_A^2 = B^2/\mu_0 \rho = \beta_e^{-1} c_s^2$, assuming that c_s, the ion thermal speed V_i, and z/m all scale together for ions. In a similar vein, the squared $A_{\|}$ terms in the polarisation are small compared to the magnetic energy by

$$\sum_{sp} \int d\mathcal{W} f \frac{z^2 e^2}{2m^2 \Omega} \frac{\partial}{\partial M} \left[J_0\left(A_{\|}^2\right) - (J_0 A_{\|})^2 \right] \sim \frac{nmV^2}{B^2} |\nabla_\perp A_{\|}|^2 \ll \frac{1}{\mu_0} |\nabla_\perp A_{\|}|^2 \tag{5.105}$$

with mV^2 larger for ions and $V_i^2 \sim c_s^2 = \beta_e v_A^2$. This is a similar measure generally as that by which $\epsilon_0 E^2/2$ is small compared to ExB energy, which is the basis for quasineutrality. Under this limit, then, the polarisation is taken to be electrostatic, and we replace Φ by ϕ in Eqs (5.81), (5.98), and then since the polarisation term no longer depends on $A_{\|}$ the induction equation (Eq. 5.102) becomes simply

$$-\frac{1}{\mu_0} \nabla_\perp^2 A_{\|} = \sum_{sp} \int d\mathcal{W}\, J_0(feU) \tag{5.106}$$

which we can reduce to a skin depth operator form by placing the contribution in U due to $A_\parallel$ on the left side, yielding

$$\left(K_d^2 - \nabla_\perp^2\right)A_\parallel = \mu_0 \sum_{sp} \int d\mathcal{W} \frac{ez}{m}(J_0 f) \tag{5.107}$$

where with finite gyroaveraging the skin depth appears through an operator given by

$$K_d^2 A_\parallel = \sum_{sp} \int d\mathcal{W} \frac{\mu_0 e^2}{m} J_0(f\ J_0 A_\parallel) \tag{5.108}$$

summed over species after the integration over velocity space. This is called the screened form of Ampère's law.

5.10.2 Linearised polarisation

For many models it is computationally problematic to keep the density variability of the f appearing in the polarisation term, even if it has been made electrostatic. Even if the dynamical f includes the evolution of the background, it is admissible in many applications to replace the f in the polarisation term of fH, so that it appears as a field energy only. This is also done with the skin depth term. Then, H is first order (only) in both fields, and the H_2 contributions appear as field energy terms, part of $\mathcal{L}_f$ in the spatially dependent part. The term fH_2 is replaced by F_0H_2, where F_0 is a background distribution function usually taken to be a Maxwellian with density and temperature, possibly spatially dependent, as parameters. The Hamiltonian now appears as $H = H_0 + H_1$ only, or

$$H = \frac{z^2}{2m} + M\Omega + eJ_0(\Phi), \qquad \Phi = \phi - \frac{z}{m}A_\parallel \tag{5.109}$$

and the field Lagrangian acquires the second order pieces,

$$\mathcal{L}_f = -\frac{1}{2\mu_0}\left|\nabla_\perp A_\parallel\right| - \sum_{sp} \int d\mathcal{W}\, F_0 H_2 \tag{5.110}$$

where the second order part of H is

$$H_2 = \frac{e^2}{2m}(J_0 A_\parallel)^2 - \frac{e^2}{2\Omega}\frac{\partial}{\partial M}\left(J_0(\phi^2) - (J_0\phi)^2\right) \tag{5.111}$$

under the electrostatic polarisation limit we have taken above.

The polarisation equation (Eq. 5.98) is more transparent if we have a situation in which the f in $\mathcal{M}$ can be replaced by a Maxwellian for which

$$f \to f^M = a_0(n)\exp[-H_0/T], \quad \int d\mathcal{W} f^M = n, \quad H_0 = m\frac{U^2}{2} + M\Omega \tag{5.112}$$

where n and T are system constants for each species. Then we have

$$\mathcal{M} = \frac{e^2}{T} f^M \tag{5.113}$$

from Eq. (5.96). This yields the well known formula

$$\int d\mathcal{W}\, J_0 \mathcal{M} J_0 = n\frac{e^2}{T}\Gamma_0 \tag{5.114}$$

where Γ_0 is a special function which in k-space is given by

$$\Gamma_0(\mathbf{k}) = e^{-b} I_0(b), \qquad b = k_\perp^2 \rho_{th}^2 \tag{5.115}$$

where I_0 is the zeroth modified Bessel function, and ρ_{th} is the thermal gyroradius whose square is given by mT/e^2B^2 for the species. Then we have

$$ne^2 \frac{1 - \Gamma_0}{T}\phi = \sum_{sp} \int d\mathcal{W}\, eJ_0(f) \tag{5.116}$$

for the polarisation equation. Similarly, we have

$$K_d^2 = \sum_{sp} \frac{\mu_0 n e^2}{m}\Gamma_0 \tag{5.117}$$

as the skin depth operator in the screened form of Ampère's law.

With this linearised polarisation approximation having been taken in the Lagrangian, the evolution equation for f now contains the simplified H but is otherwise unchanged. The field equations also appear with the same structure, except that F_0 instead of f appears in all the field terms, with f appearing only as $eJ_0(f)$ in the polarisation equation (charge density) and as $J_0(feU)$ in the induction equation (current). The computational advantage of doing this is that any of the dynamical speeds (i.e., of waves) which appear on these terms, especially the Alfvén velocity, are determined by the parameterisation of F_0 and not by the fluctuations in f. This helps computational stability.

5.10.3 No gyroaveraging in induction

In practice, also because of the mass ratio, the electron contribution is the largest in both remaining terms of the induction equation, not only the current but also the skin depth term. This leads to forms in which $A_\parallel$ is not gyroaveraged at all, or

$$\left(K_d^2 - \nabla_\perp^2\right) A_\parallel = \mu_0 \sum_{sp} \int d\mathcal{W} \frac{ez}{m} f, \qquad K_d^2 = \sum_{sp} \frac{\mu_0 e^2}{m} \int d\mathcal{W} f \tag{5.118}$$

This is a simple Helmholtz equation for $A_\parallel$ with f known. Note that the J_0 also drops on the right hand side, since $J_0(feU)$ by itself implies gyroaveraging upon $A_\parallel$.

This simplification is not generally used in gyrokinetic models, for which the consistency between $J_0(\phi)$ and $J_0(A_\parallel)$ is usually kept at the kinetic level. But we include it here since it is often used in the gyrofluid models which we will introduce in later Chapters.

5.10.4 Long wavelengths

The long wavelength limit of these expressions is interesting for two reasons. One is to capture the physics of the relaxation of any system into equilibrium, under conditions in which conventional ordering in a magnetised plasma holds, namely, small gyroradius compared to the scale of any thermal gradient. Values of several tens to hundreds, in terms of ion gyroradii per scale length, are common in warm or hot cases appropriate to laboratory plasmas. Much larger values are the case in magnetised plasmas in nature, in planetary or astrophysical situations. The other reason for long wavelength analysis is to show that the gyrokinetic description corresponds, i.e., reduces to the one in fluid drift or magnetohydrodynamical (MHD) descriptions in their respective limits.

The long wavelength limit of the gyroaverage follows from the expansion of the Bessel function J_0, given by

$$J_0(\alpha) = 1 - \frac{\alpha^2}{4} + \frac{\alpha^4}{64} + \cdots \tag{5.119}$$

The leading term is taken everywhere except where it cancels. So we have $J_0 \to 1$ except where $1 - J_0^2 \to \alpha^2/2$ is needed. In the polarisation term we have

$$\lim_{k_\perp \to 0} \frac{e^2}{2\Omega}\frac{\partial}{\partial M}\left(J_0(\phi)^2 - (J_0\phi)^2\right) = \frac{m}{2B^2}\,|\nabla_\perp \phi|^2 \tag{5.120}$$

In terms of the Taylor expansion of $\phi(\mathbf{x})$ around $\mathbf{x} = \mathbf{R}$ we have

$$\phi(\mathbf{x}) = \phi(\mathbf{R}) + \mathbf{a}\cdot\nabla\phi(\mathbf{R}) + \frac{1}{2}\mathbf{aa}:\nabla\nabla\phi(\mathbf{R}) \tag{5.121}$$

through two orders in the smallness of the directed gyroradius $\mathbf{a}$, which is defined in terms of the $\mathbf{e}_{(1,\,2)}$ vectors in Eq. (5.19). Hence for the gyroaverage and gyroaverage defect we have

$$\langle\phi\rangle_{\mathbf{R}} = \phi + \frac{a^2}{4}\nabla_\perp^2\phi, \qquad \tilde{\phi} = \mathbf{a}\cdot\nabla\phi(\mathbf{R}), \qquad \langle\tilde{\phi}^2\rangle_{\mathbf{R}} = \frac{a^2}{2}\,|\nabla_\perp\phi|^2 \tag{5.122}$$

where $a^2 = 2M/m\Omega$ as before. In the gyroaverage, the 1/4 term is the first term past unity in the expansion of J_0, and in the defect, only the first term is kept since it always enters squared and by itself. As a result both sets of terms agree with the continuous representation using Bessel functions. The second order term becomes

$$\lim_{k_\perp \to 0} \frac{e^2}{2\Omega}\frac{\partial}{\partial M}\langle\tilde{\phi}^2\rangle_{\mathbf{R}} = \frac{e^2}{2m\Omega^2}\,|\nabla_\perp\phi|^2 = \frac{m}{2B^2}\,|\nabla_\perp\phi|^2 \tag{5.123}$$

and so it can be reinterpreted as an inertial energy term.

In practical cases, polarisation arises mostly from the ions, while the electrons contribute mostly to the feU term (the main parallel current). Therefore, assuming the electrostatic polarisation case, the Hamiltonian becomes

$$H = m\frac{U^2}{2} + M\Omega + e\phi - m\frac{u_E^2}{2} \tag{5.124}$$

where

$$mU = z - eA_\parallel, \qquad u_E^2 = \frac{1}{B^2}\,|\nabla_\perp\phi|^2 \tag{5.125}$$

The polarisation equation becomes

$$-\nabla\cdot\frac{\rho}{B^2}\nabla_\perp\phi = \rho_G, \quad \rho = \sum_{sp}\int d\mathcal{W}\,fm, \quad \rho_G = \sum_{sp}\int d\mathcal{W}\,fe \tag{5.126}$$

where ρ and ρ_G are the gyrocentre mass and charge densities, respectively. The induction equation becomes

$$-\nabla_\perp^2 A_\parallel = \mu_0 J_\parallel, \qquad J_\parallel = \sum_{sp}\int d\mathcal{W}\,feU \tag{5.127}$$

or equivalently

$$\left(d_e^2 - \nabla_\perp^2\right)A_\parallel = \mu_0\sum_{sp}\int d\mathcal{W}\,f\frac{ez}{m}, \qquad d_e^2 = \sum_{sp}\int d\mathcal{W}\,\frac{\mu_0 fe^2}{m} \tag{5.128}$$

in the screened form, with d_e, a scalar giving the skin depth, replacing K_d, the screening operator. The evolution equation is, again, unchanged except for the use of the long wavelength form of H. This form of the gyrokinetic model is the one used to recover the MHD forms of the conservation laws, as we will do in the next Chapter. When the $\alpha^2/4$ first order correction in the expansion of J_0 is kept, the resulting finite gyroradius corrections will be found to represent the diamagnetic corrections to the vorticity. We will show this after developing the gyrofluid model, in which these effects are kept.

5.10.5 Importance of energetic consistency

In the fluid drift turbulence portion of this work, in Volume One and the first two Chapters of Volume Two, we have emphasised exact energetic consistency with regard to turbulence computations. If the equations at hand support an exact conservation (of any sort, but in this case the emphasis is on energylike quantities, or a set of quantities quadratic in the dependent variables), then the nonlinearities belonging to the turbulence can be run essentially forever without breaking down (i.e., slow wandering of the statistical equilibrium towards different states, or rapid breakdown of convergence). This is the significance of the derivation path of the nonlinear gyrokinetic equations from a variational principle. It gives us a procedure by which we can readily find the quantities which are conserved by any interactions within the theory. In this type of turbulence we have flows and eddies in energetic coupling to thermal fluctuations, with the drive mechanism in that component and not the ExB one. That is the reason we have to have a field theory, as it is not enough to get a gyrokinetic equation. We must also have strictly consistent equations to obtain the field quantities which result from the gyrokinetic plasma component. Approximations are all taken in the Lagrangian, sometimes solely for computational tractability or convenience, but the structure and the derivation path remain intact and so, therefore, does exact energetic consistency. The procedure with which the conservation laws are guaranteed is Noether's theorem, by which the symmetries inherent in the Lagrangian are shown to correspond to quantities which are conserved by the resulting Euler–Lagrange equations. As we have Liouville's theorem as a procedure to obtain the gyrokinetic evolution equation, we prove again that the quantities indicated by Noether's theorem are actually conserved. Then we explore how the part of the system defined by fields, electric and magnetic, exchange energy with the part defined by the gyrocentres, thermal and kinetic. These considerations form the content of the next Chapter.

Further Reading

Background references on the field theoretical methods are given in the Further Reading of the previous Chapter.

The view of drift-kinetic dynamics as corrections to a homogeneous plasma in a background magnetic field is due to A I Morozov and L S Solov'ev 'Motion of charged particles in electromagnetic fields,' *Rev. Plasma Phys.* **2** (1966) 201.

The standard form of gyrokinetic ordering is from J B Taylor and R J Hastie, 'Stability of general plasma equilibria—I formal theory,' *Plasma Phys.* **10** (1968) 479, and from P Rutherford and E A Frieman, 'Drift instabilities in general magnetic field configurations,' *Phys. Fluids* **11** (1968) 569. The nonlinear gyrokinetic equation obtained by averaging is from E A Frieman and L Chen, 'Nonlinear gyrokinetic equations for low-frequency electromagnetic waves in general plasma equilibria,' *Phys. Fluids* **25** (1982) 502. The view of the gyrokinetic description as a change of representation is given by W W Lee, 'Gyrokinetic approach in particle simulation,' *Phys. Fluids* **26** (1983) 556. The drift kinetic gauge transform is from R Littlejohn, 'Variational principles of guiding center motion,' *J. Plasma Phys.* **29** (1983) 111.

The foundation of the Lie transform approach to gyrokinetics is given by J Cary and R Littlejohn, 'Noncanonical Hamiltonian mechanics and its application to magnetic field line flow,' *Ann. Phys. (N.Y.)* **151** (1983) 1. The use of the Poincaré–Cartan fundamental one-form is from this overview. A review of gyrokinetic theory is given by A Brizard and T S Hahm, 'Foundations of nonlinear gyrokinetic theory,' *Rev. Mod. Phys.* **79** (2007) 421. The application of the Lie transform to gyrokinetics was given by D H E Dubin and J A Krommes and C Oberman and W W Lee, 'Nonlinear Gyrokinetic equations,' *Physics of Fluids* **26** (1983) 3524; and T S Hahm (1988), 'Nonlinear gyrokinetic equations for Tokamak microturbulence,' *Physics* *of Fluids* **31** (1988) 2670, whose arrangement of the Lie transform is the one used herein. The electromagnetic Lagrangian using related methods is given by T S Hahm, W W Lee, and A Brizard, 'Nonlinear gyrokinetic theory for finite-beta plasmas,' *Physics of Fluids* **31** (1988) 1940, whose resulting Lagrangian is the one given herein. A relativistic generalisation is given by A Brizard, 'Nonlinear relativistic gyrokinetic Vlasov–Maxwell equations,' *Physics of Plasmas* **6** (1999) 4548.

The field theoretical formulation of gyrokinetic theory is given by A Brizard, 'Variational principle for nonlinear gyrokinetic Vlasov–Maxwell equations,' *Physics of Plasmas* **7** (2000) 4816, and by H Sugama, 'Gyrokinetic field theory,' *Physics of Plasmas* **7** (2000) 466. A review up to 2007 is given by A Brizard and T S Hahm, 'Foundations of nonlinear gyrokinetic theory,' *Rev. Mod. Phys.* **79** (2007) 412. Sugama's paper is particularly elucidating with respect to the way the simplified limits fit into the theory without loss of consistency. It also gives high frequency versions of the theory.

IOP Publishing

Turbulence and Instabilities in Magnetised Plasmas, Volume 2
Gyrokinetic theory and gyrofluid turbulence
Bruce Scott

Chapter 6

Phase space and energetic consistency

6.1 Summary of ideas

The basic ideas of the gyrokinetic formulation covered in the previous Chapter are: (1) the change of coordinates in going from the particle formulation to the drift kinetic model, in which the magnetic moment is conserved, no quantity in the equation describing the evolution of the phase space distribution function depends on the gyrophase angle, and the equilibrium magnetic field is kept but the perturbations, including the entirety of the electric field, are left to the subsequent expansion. (2) The Lie transform which takes this to the gyrokinetic description, by shifting any appearance of the perturbed fields into the time component of the Lagrangian, which now describes evolution of gyrocentres. (3) Polarisation and induction appear separately in the equations for the field perturbations, not in the gyrokinetic evolution equations. Polarisation is specifically removed from the drifts and kept instead as part of the charge density. The overall charge density consists of contributions from all the gyrocentres across species, and the polarisation as well, such that the total is zero but the polarisation determines the electric field in terms of the gyrocentre charge density distribution. Similarly, induction does not appear directly in the evolution equations but appears instead in the field equation for the magnetic perturbation. One still recovers the same kinetic shear Alfvén waves, as we will address in Chapter 8. (4) The incorporation of the fields into the overall system Lagrangian, formally arranging the model within classical field theory, guarantees conservation laws which belong to any particular representation. Exact energetic consistency is thereby guaranteed. These considerations support understanding of the phase space symmetry which underlies the gyrokinetic description. This Chapter covers these topics in detail, before we turn to local descriptions and then use the theory to build the gyrofluid model with which to address the turbulence specifically within the combination of temperature gradient drive and weakly collisional parameter regimes.

The main point of the phase space structure is the incompressible nature of the phase space motion, a fundamental consequence of Hamiltonian dynamics. The Euler–Lagrange equations for the gyrocentres are used in this fashion to determine the phase space volume element, or Jacobian. Particle and entropy conservation follows trivially from the linearity of the evolution equation in the distribution function and the Hamiltonian, regardless of how complicated the dependence of the Hamiltonian on the perturbed fields. Energy conservation follows from the time symmetry of the background, and momentum conservation follows from spatial symmetry of the background. That is, the background magnetic field geometry is static and is independent of one or more coordinates. If there is structure in general but symmetry in one coordinate, then the momentum associated with that coordinate is conserved. These are results of the application of Noether's theorem to this system. The conservation laws for energy and momentum are not preserved within the single gyrocentre motion or even within a single species, but only in terms of entire system across species and also including the fields. This was a lesson of Maxwell's equations, and the same lesson applies here. Both forms of conservation are mediated through interactions between each species and the fields, exchanging thermal and kinetic energy with the field energy components. Approximations are made, but their consistency is kept exact. In this case the guarantee is given by approximating terms in the Lagrangian but never directly in the equations. The Lagrangian can be approximated, but then exactly consistent equations are then found anew using the Euler–Lagrange procedure. This will elucidate how the forms of the phase space Jacobian $B_{\|}^{*}$, the form of the unperturbed canonical momentum, $\mathbf{p}^{*} = e\mathbf{A}^{*}$, and of the drift tensor (counterpart of $B^{-2}\mathbf{B}\times$), all have to go together consistently. Approximate forms can be found according to which these quantities recover familiar appearance, but they work only if they are done properly together. There is a way to guarantee this and we will explore it. The main point is that approximations are made only in the system Lagrangian, so that a model with exact energetic consistency is guaranteed at any level of approximate treatment.

6.2 Basic structure of the model

We have a Lagrangian for individual gyrocentres, of the form

$$L_p = \mathbf{p}^{*} \cdot \dot{\mathbf{R}} + M\dot{\vartheta} - H \tag{6.1}$$

as in Eq. (5.92) from the previous Chapter, where $\mathbf{p}^{*} = e\mathbf{A}^{*}$ is the canonical momentum, $\mathbf{R}$ is the gyrocentre position, M is the magnetic moment coordinate, and H is the Hamiltonian. Only H has any dependence on the fluctuating perturbed fields, which can be of arbitrary order, and the other quantities are static. Both $\mathbf{p}^{*}$ and M are signed due to the charge, and $\mathbf{p}^{*}$ can be spatially dependent and also dependent on the parallel canonical momentum coordinate z and even on M. But no quantity is dependent on the gyrophase angle ϑ. The Hamiltonian has the structure

$$H = m\frac{U^2}{2} + M\Omega + e\phi_E \tag{6.2}$$

where $\Omega = eB/m$ is the species gyrofrequency, $U = U(\mathbf{R}, z, M, A_\parallel)$ is a parallel velocity dependent in general on all the coordinates except ϑ, but depends on time and any symmetry coordinate of the equilibrium only through its dependence on the dynamical field $A_\parallel$, and similarly for ϕ_E, which is a generalised potential dependent on the dynamical fields in an arbitrary manner.

Under the Lie transform theory we did in the previous Chapter, these quantities are

$$\mathbf{p}^* = e\mathbf{A}^* = e\mathbf{A} + z\mathbf{b} - M\mathbf{W}, \qquad mU = z - e(J_0 A_\parallel) \tag{6.3}$$

as in Eqs (5.26), (5.82) from the previous Chapter. The Hamiltonian, therefore, contains $A_\parallel$ to second order in the parallel energy, $mU^2/2$. Additionally, the generalised potential is given by

$$\phi_E = (J_0\phi) - \frac{e}{2\Omega}\frac{\partial}{\partial M}\left(J_0(\Phi^2) - (J_0\Phi)^2\right) \tag{6.4}$$

with $\Phi = \phi - (z/m)A_\parallel$. The $A_\parallel$ part of the first term is arranged with the parallel energy $mU^2/2$. The arrangement of the general gyroaverage in terms of the J_0 operators based on the $J_0(k_\perp\rho_L)$ Bessel functions was done in the previous Chapter.

If an electrostatic model is constructed, simply set $A_\parallel$ to zero. The parallel energy reduces to $z^2/2m$ and the generalised potential reduces to

$$\phi_E = J_0(\phi) - \frac{e}{2\Omega}\frac{\partial}{\partial M}\left(J_0(\phi^2) - (J_0\phi)^2\right) \tag{6.5}$$

This recovers the model of Hahm (1988) and eventually also the equations of Lee (1983).

For an electromagnetic model in the long wavelength approximation without finite gyroradius corrections, namely, the MHD limit, the parallel velocity and generalised potential reduce to

$$mU = z - eA_\parallel, \qquad \phi_E = \phi - \frac{m}{e}\frac{u_E^2}{2}, \qquad u_E^2 = \frac{1}{B^2}\,|\nabla_\perp\phi|^2 \tag{6.6}$$

This is the MHD shear Alfvén Lagrangian one can study equilibria and MHD stability with, in a confined plasma.

6.3 The Euler–Lagrange equations for gyrocentres

From the individual gyrocentre Lagrangian L_p we find the Euler–Lagrange equations by variation of each of the phase space coordinates. The methods are given in the previous two Chapters. The variation of $\mathbf{R}$ yields

$$(\nabla\mathbf{p}^*)\cdot\dot{\mathbf{R}} - \dot{\mathbf{R}}\cdot(\nabla\mathbf{p}^*) - \dot{z}\frac{\partial\mathbf{p}^*}{\partial z} - \nabla H = 0 \tag{6.7}$$

where ∇ is the same as $\partial/\partial\mathbf{R}$ and when taking derivatives $\mathbf{R}$ and the spatial coordinates $\mathbf{x}$ are used interchangeably once the gyroaveraging has been recast in terms of J_0 operations. The variation of z yields

$$\frac{\partial \mathbf{p}^*}{\partial z} \cdot \dot{\mathbf{R}} - \frac{\partial H}{\partial z} = 0 \tag{6.8}$$

The variations of M and ϑ yield

$$\dot{\vartheta} + \frac{\partial \mathbf{p}^*}{\partial M} \cdot \dot{\mathbf{R}} - \frac{\partial H}{\partial M} = 0, \qquad \dot{M} = 0 \tag{6.9}$$

just the gyromotion and the constancy of M. This part is now decoupled from the rest of the dynamics. The first two equations are solved together, yielding

$$B_\parallel^* \dot{\mathbf{R}} = \frac{\partial H}{\partial z}\mathbf{B}^* + \frac{1}{e}\frac{\partial \mathbf{p}^*}{\partial z} \times \nabla H, \qquad B_\parallel^* \dot{z} = -\mathbf{B}^* \cdot \nabla H \tag{6.10}$$

The basic identity used is

$$(\nabla \mathbf{A}) \cdot \mathbf{B} - \mathbf{B} \cdot (\nabla \mathbf{A}) = \mathbf{B} \times (\nabla \times \mathbf{A}) \tag{6.11}$$

and the definitions of $\mathbf{B}^*$ and $B_\parallel^*$ are

$$\mathbf{B}^* = \nabla \times \mathbf{A}^*, \qquad B_\parallel^* = \mathbf{B}^* \cdot \frac{\partial \mathbf{p}^*}{\partial z} \tag{6.12}$$

To find the equation for $\dot{\mathbf{R}}$, cross the variation equation for $\mathbf{R}$ with $\partial \mathbf{p}^*/\partial z$, and use the variation equation for z. To find the equation for $\dot{z}$, dot the variation equation for $\mathbf{R}$ with $\nabla \times \mathbf{p}^*$. These operations also define the phase space Jacobian $B_\parallel^*$ as well as the generalised field vector $\mathbf{B}^*$. This is the same derivation path as for the original drift kinetics of particle motion in the background magnetic field, originally given by Morozov and Solov'ev (1966). The equations can be formulated entirely in terms of $\nabla \times \mathbf{p}^*$ and $\partial \mathbf{p}^*/\partial z$, but it is traditional to divide through by e thereby using $\mathbf{B}^*$ instead of $\nabla \times \mathbf{p}^*$. The largest contribution to $\mathbf{p}^*$ is always $e\mathbf{A}$, so the resulting $B_\parallel^*$ is B plus corrections; in other words, always positive. The traditional forms descend from the choice

$$\mathbf{p}^* = e\mathbf{A} + z\mathbf{b} \qquad \text{hence} \quad B_\parallel^* = B + \frac{z}{e}\mathbf{b} \cdot \nabla \times \mathbf{b} \tag{6.13}$$

but the forms using $\mathbf{p}^*$ and H abstractly are general. The correction to $B_\parallel^* = B$ is always of order gyroradius divided by magnetic field scale length.

The main point of this is that the derivation path and the resulting Euler–Lagrangian equations in the form we will write them are the same for any Lagrangian which has been put into canonical form. All differences between versions of the theory are due to the differences in the definitions of $\mathbf{B}^*$ and $B_\parallel^*$ and eventually the scalar B (through Ω) in terms of coordinates, and in the dependence of H on (any and all) dynamical field variables.

6.4 Symmetry in gyrocentre dynamics

The description is kept general in terms of the explicit factors of $B_\parallel^*$ and the derivatives of H as given above. The equivalent phase space model is given in terms of a distribution function $f(\mathbf{R}, z, M, t)$, dependent on all coordinates including time, except the gyrophase angle. The evolution equation is set by the method of characteristics, with the statement that f is advected without change along the world lines of the gyrocentres. By the chain rule we then have

$$\frac{df}{dt} = \frac{\partial f}{\partial t} + \dot{\mathbf{R}} \cdot \nabla f + \dot{z}\frac{\partial f}{\partial z} = 0 \tag{6.14}$$

The phase space coordinates $\{M, \vartheta\}$ do not appear, because dM/dt and $\partial f/\partial\vartheta$ both vanish. Hence, the actual motion is in a 4D phase space, not a 6D one. Any evolution in M is due to collisions.

6.4.1 Poisson bracket form

When the Euler–Lagrange equations are substituted into Eq. (6.14), the characteristic form of the evolution equation, the result is

$$B_{\parallel}^{*}\frac{\partial f}{\partial t} + \frac{1}{e}\frac{\partial \mathbf{p}^{*}}{\partial z} \cdot (\nabla H \times \nabla f) + \mathbf{B}^{*} \cdot \left(\frac{\partial H}{\partial z}\nabla f - \frac{\partial f}{\partial z}\nabla H\right) = 0 \tag{6.15}$$

where we have multiplied through by $B_{\parallel}^{*}$ to obtain a clear antisymmetric structure in all the coordinate derivatives.

The structure of the evolution equation, following that of the phase space flow equations themselves, is that of a Poisson bracket. The simplest form of that is of a dynamical system with one space (q) and one canonical momentum (p) serving as coordinates, with Hamilton's equations

$$\frac{dq}{dt} = \frac{\partial H}{\partial p}, \qquad \frac{dp}{dt} = -\frac{\partial H}{\partial q}, \qquad H = H(p, q) \tag{6.16}$$

giving the flow of material, described by $f = f(p, q)$, in phase space. The evolution equation is given by

$$\frac{\partial f}{\partial t} + \{H, f\} = 0, \qquad \{H, f\} = \frac{\partial H}{\partial p}\frac{\partial f}{\partial q} - \frac{\partial H}{\partial q}\frac{\partial f}{\partial p} \tag{6.17}$$

The structure, taking two scalars as arguments and returning another scalar, satisfies antisymmetry and the Leibnitz property,

$$\{a, b\} = -\{b, a\}, \qquad c\{a, b\} + b\{a, c\} = \{a, bc\} \tag{6.18}$$

and also the Jacobi identity,

$$\{a, \{b, c\}\} + \{b, \{c, a\}\} + \{c, \{a, b\}\} = 0 \tag{6.19}$$

for scalars abc. For one bracket structure like this, the derivatives are with respect to the two conjugate coordinates. In general (more than two dimensions), it is a set of such structures in which the derivatives are with respect to pairs of conjugate coordinates.

One can always write the evolution equation in this way, since following Hamilton's equations the phase space flow always has this structure. So the notation

$$\frac{df}{dt} = \frac{\partial f}{\partial t} + \{H, f\} = 0 \tag{6.20}$$

is general, but the definition of the bracket itself is dependent on the problem. For a Lagrangian in canonical form, Eq. (6.15) is general.

Given the phase space flow equations in gyrokinetic theory, with $\mathbf{p}^*$ dependent on the coordinates, the bracket pieces still enter in conjugate pairs. The drifts combine spatial coordinates, which only reduces to two perpendicular coordinates if the spatial coordinate geometry is chosen accordingly. The parallel phase space enters in the same way the $\{pq\}$ one does in the simplest system. The perpendicular spatial coordinates act as conjugate pairs (three different ones, in three dimensions), which represents the main difference in a gyrokinetic theory to the textbook Poisson bracket. Moreover, since $\mathbf{p}^*$ has coordinate dependence, the coefficients of the conjugate derivative pairs are not unity, in general not even constant. The dependence of $\mathbf{B}^*$ and $B_\parallel^*$ on coordinates is just such as to preserve the divergence free property of the flow, through the divergence property of the bracket (more on this in Chapter 17).

6.4.2 Phase space incompressibility

We note that the simple derivatives of $\dot{\mathbf{Z}}$ with respect to $\mathbf{Z}$, where $\mathbf{Z} = \{\mathbf{R}, z\}$ is the 4D phase-space vector of coordinates, does not produce a divergence free form since the contraction does not vanish. However, with $B_\parallel^*$ multiplying $\dot{\mathbf{Z}}$, the product is divergence free. We require that

$$\nabla \cdot \mathbf{B}^* = 0, \qquad \frac{1}{e}\nabla \times \frac{\partial \mathbf{p}^*}{\partial z} + \frac{\partial \mathbf{B}^*}{\partial z} = 0 \tag{6.21}$$

which is trivial since it descends from

$$\nabla \cdot \nabla \times \mathbf{p}^* = 0, \qquad \nabla \times \frac{\partial \mathbf{p}^*}{\partial z} + \frac{\partial}{\partial z}(\nabla \times \mathbf{p}^*) = 0 \tag{6.22}$$

To have this lead to the expression $\nabla \cdot \dot{\mathbf{Z}} = 0$ one has to have the coefficient $B_\parallel^*$ in the equations for $\dot{\mathbf{R}}$ and $\dot{z}$ be one and the same with the phase space Jacobian, so that

$$\nabla_Z \cdot \dot{\mathbf{Z}} = \frac{1}{B_\parallel^*}\nabla \cdot B_\parallel^* \dot{\mathbf{R}} + \frac{1}{B_\parallel^*}\frac{\partial}{\partial z}B_\parallel^* \dot{z} = 0 \tag{6.23}$$

expresses the entire phase space divergence, which vanishes. With this to hand, we can convert the evolution equation from the advective form in Eq. (6.14) to

$$B_\parallel^* \frac{\partial f}{\partial t} + \nabla \cdot \left(f B_\parallel^* \dot{\mathbf{R}}\right) + \frac{\partial}{\partial z}\left(f B_\parallel^* \dot{z}\right) = 0 \tag{6.24}$$

which is the divergence form.

This shows that the incompressibility of the phase space flow is a simple consequence of the structure of the geometry, which is a consequence of setting up the Lagrangian with $\mathbf{p}^* \cdot \dot{\mathbf{R}}$ and the rest of the drift dynamics in the time component (H), namely, canonical form. The operations establishing $\nabla_Z \dot{\mathbf{Z}} = 0$ and Eq. (6.24) are what defines $B_\parallel^*$ as the phase space Jacobian. Consistency in a computation is maintained only if $\partial \mathbf{p}^*/\partial z$ and $\mathbf{B}^*$ are consistently defined. This requirement is trivial if the problem has been set up in the manner shown, but not trivial if one proceeds with the pre-1980 operations of applying expansion schemes to the kinetic evolution equation.

6.4.3 Phase space conservation

The divergence form of the evolution equation, Eq. (6.24), is behind all the conservation laws in the gyrokinetic description. At this point, before discussing the role of the fields, there is no essential difference in structure between kinetic and gyrokinetic theory. It follows that any result of the structure of phase space or phase space flow in kinetic theory carries over into gyrokinetic theory. This fundamental basis is the reason we can build models with exact energetic consistency.

Phase space ('particle' or 'gyrocentre') conservation is a consequence of integrating Eq. (6.24) over velocity space to produce a continuity equation in space. To proceed we have to define the properties of the phase space as we did in the previous Chapter. We have

$$d\mathcal{V} = \sqrt{g}\, dx\, dy\, ds, \qquad d\mathcal{W} = \frac{2\pi}{m^2} B_\parallel^*\, dz\, d\mu, \qquad d\Lambda = d\mathcal{V} \otimes d\mathcal{W} \tag{6.25}$$

for space, velocity space, and the entire phase space, where $\mu = (e/m)M$ defines a positive magnetic moment coordinate. The integrations over space and velocity space are separate, with $B_\parallel^*$ dependent on all the coordinates except the gyrophase angle. The factor of 2π is from the integration over that angle. The factor of m^{-2} is for the units of z and μ in terms of mass. The important thing to know is that integration over $d\mathcal{W}/B_\parallel^*$ annihilates $\partial/\partial z$ and commutes past ∇, so that the factors of $B_\parallel^*$ appearing in a spatial divergence in the phase space are properly integrated away by integration over $d\mathcal{W}$. So we have upon integration of Eq. (6.24) over velocity space,

$$\int d\mathcal{W} \left[\frac{\partial f}{\partial t} + \frac{1}{B_\parallel^*} \nabla \cdot \left(f B_\parallel^* \dot{\mathbf{R}} \right) + \frac{1}{B_\parallel^*} \frac{\partial}{\partial z} \left(f B_\parallel^* \dot{z} \right) = 0 \right] \tag{6.26}$$

the spatial continuity equation,

$$\frac{\partial n}{\partial t} + \nabla \cdot \mathbf{F}_G = 0, \qquad \mathbf{F}_G = \int d\mathcal{W} \left(f B_\parallel^* \dot{\mathbf{R}} \right) \tag{6.27}$$

in which n and $\mathbf{F}_G$ define the gyrocentre density (note: not the particle density) and the gyrocentre flux (strictly: flux density), which is not the same thing as the particle flux.

Any function of f can be integrated in the same way. Suppose we have $G = G(f)$. Then define $G' = \partial G/\partial f$. Multiplying Eq. (6.14) by G' produces the phase space advection equation for G, and then integration over velocity space produces

$$\frac{\partial}{\partial t}\int d\mathcal{W}\, G + \nabla \cdot \int d\mathcal{W}\left(GB_{\parallel}^{*}\dot{\mathbf{R}}\right) = 0 \tag{6.28}$$

which serves as the continuity equation for G, hence the conservation of the phase space integral of G. The entropy conservation equation follows from setting $G = -T_0 f \log f$ whereupon

$$\frac{\partial S}{\partial t} = 0, \qquad S = -T_0 \int d\Lambda f \log f \tag{6.29}$$

Such a quantity, $S(f)$, is a functional of f, a real number quantity taking a phase space density as its argument. Any functional of f whose phase space density is differentiable is conserved. These quantities are called Casimirs, after the author of the paper which introduced them. More on this will be discussed in Chapter 17.

6.5 Application of Noether's theorem

The use of Noether's theorem for field theoretical problems was addressed in Chapter 4. The basic statement is that for each symmetry of the Lagrangian against a transformation, there is an associated quantity which is conserved by the dynamics. In general, a variation of the total Lagrangian, after the Euler–Lagrange equations have been applied to constrain the variables, is given by a time derivative of an expression containing products of a dependent variable's variation multiplied by the derivative of the Lagrangian (for discrete variables) or the Lagrangian density (for field variables) with respect to that variable's time derivative. The easiest is the magnetic moment variable M, whose conservation is given by the symmetry with respect to the gyrophase angle ϑ. A shift of the coordinate representing ϑ in the velocity space gives conservation of the derivative of L with respect to the time derivative of the ϑ of the gyrocentre, namely M. A shift of the time coordinate gives energy conservation and a shift of a spatial symmetry coordinate ξ gives a corresponding momentum conservation P_{ξ}^{*}. In the case of axisymmetry, this ξ is φ, the geometric toroidal angle, and the conserved momentum is toroidal angular momentum. We will highlight energy and momentum conservation, especially the latter, as we also have to show how canonical toroidal momentum translates to plasma momentum, as briefly encountered in the discussion of momentum in the Maxwell-plasma system in Chapter 4.

In the case of gyrokinetics, we have no terms in L which depend directly on the time derivative of any field variable, so we have the simpler version of the theorem. With no explicit time dependence in the geometry, L as a functional is symmetric to time translations and therefore the associated conserved energy is given by

$$\frac{d\mathcal{E}}{dt} = 0 \qquad \text{where} \qquad \mathcal{E} = -L + \sum_{k} \frac{\partial L}{\partial \dot{\mathbf{z}}_k} \cdot \dot{\mathbf{z}}_k \tag{6.30}$$

where the sum is over individual gyrocentres and $\mathbf{z}_k$ are the Lagrangian coordinates as discrete dependent variables for the k-th gyrocentre. Due to the structure of L with a symplectic part and a time component L_t, due to H for the gyrocentres and $\mathcal{L}_f$ for the fields, the negative of this time component gives the conserved energy. With the canonical momenta given by coordinate geometry independent of time, the conserved energy is always this one, $\mathcal{E} = -L_t$.

In the absence of dependence of the geometric quantities on a symmetry coordinate ξ, the spatial shift in ξ yields conservation of the total canonical momentum in that coordinate, P_ξ^*, specifically the sum of all the discrete p_ξ^* variables. Re-expressed as a phase space density fp_ξ^* integrated over phase space, this quantity is the one Noether's theorem for momentum says is conserved,

$$\frac{dP_\xi^*}{dt} = 0 \qquad P_\xi^* = \sum_k \frac{\partial L}{\partial \dot{\xi}_k} = \sum_{sp} \int d\Lambda\, fp_\xi^* \tag{6.31}$$

where the first sum is over individual gyrocentres, the second sum is over species, ξ is the coordinate and ξ_k is the Lagrangian coordinate for the k-th gyrocentre. This says that the sum of all gyrocentre canonical momenta is conserved, even though they may exchange momenta. The local formulation for this is constructed below, including the wave flux terms coming from $\partial H/\partial\xi$, and as well an important cancellation.

There remain two issues. First, since we are using the distribution function f for each species as a variable rather than the $\mathbf{z}_k$ for each gyrocentre, we have invoked Liouville's theorem to construct the evolution equation for the distribution function. Liouville's theorem was separately applied—it does not come from the Lagrangian. So we have Noether's theorem to tell us what should be conserved, but then we have to actually show it is conserved using the evolution equation. So we must confirm conservation of the Noether energy and momentum independently.

The second issue is that Noether's theorem tells us that the canonical momentum associated with a symmetry coordinate is conserved. We actually need the plasma momentum. We saw how that worked with Maxwell's equations, where the conserved Noether momentum only became the Poynting momentum plus the plasma momentum after we again applied the definitions of $\mathbf{E}$ and $\mathbf{B}$ in terms of the potentials ϕ and $\mathbf{A}$ and the Maxwell equations for the fields. In the gyrokinetic case we don't have the field momentum directly, but what we will see is that the canonical momentum becomes the plasma momentum after an important cancellation involving the polarisation current. Moreover, the ExB momentum does not enter directly from the ExB drift but through this operation involving the field equations, in a very similar manner as the use of $\nabla \cdot \mathbf{E}$ in the Maxwell-plasma case. There is a strong correspondence in how it works: there, in the Maxwell plasma case, the displacement current, here, in the gyrokinetic case, the polarisation current. There, the charge density, here, the polarisation density. There, the Poynting momentum, here, the ExB momentum. This correspondence is central to the understanding of the physics.

6.6 Energy conservation

Phase space and entropy conservation work separately for each species, since any functional of f is conserved by itself. This is a consequence of the incompressibility of phase space flow and the structure of it in terms of a Poisson bracket. In the case of energy, we multiply the evolution equation for f by H, to obtain

$$H\frac{\partial f}{\partial t} + \{H, fH\} = 0 \tag{6.32}$$

since $\{H, H\}$ vanishes. The Noether energy in the gyrokinetic field theory is not given by the species-summed phase space integral of fH, but that plus the field energy piece $\mathcal{L}_f$ (Eq. (5.91)). In general we have

$$\frac{\partial}{\partial t}fH + \{H, fH\} = f\frac{\partial H}{\partial t} \tag{6.33}$$

by simple algebra. Now we encounter the issue that the right hand side does not vanish by itself; further operations are needed. The time derivative of H can only be evaluated using the field equations. It is given in turn by the field variable time derivatives and the derivatives of H with respect to the fields, complicated by chain rule operations involving their derivatives with respect to the phase space coordinates. Then, the field equations always involve the sum over species. So any quantity like this is only conserved as a whole, across species. In this case, it is the total energy which is exchanged between species mediated by the fields. We have seen this before, in the fluid drift model. Now we see it with the gyrokinetic model.

If only the global conservation law is desired, we can use the properties of functional derivatives to evaluate the $\partial H/\partial t$ term. Since f itself does not depend on the field variables, and the only time dependence of H is through the field variables, the phase space integral over $f\partial H/\partial t$ can be written in terms of functional derivatives. We define the two functionals L_f and L_H as

$$L_f = \int d\mathcal{V}\,\mathcal{L}_f, \qquad L_H = \sum_{sp}\int d\Lambda\, fH \tag{6.34}$$

giving the free-field contribution and the overall set of the Hamiltonian contributions to the overall Lagrangian L, respectively. Recall the variational derivative of a functional, called a functional derivative, with respect to ϕ_a, one of the set of field variables, is given by

$$\delta L_f(\delta\phi_a) = \int d\mathcal{V}\,\frac{\delta L_f}{\delta\phi_a}\,\delta\phi_a \tag{6.35}$$

using L_f as an example. This represents all the necessary integrations by parts as needed to isolate $\delta\phi_a$ under the volume integral. More detail was given in Chapter 4 and the Gelfand/Fomin reference cited therein. For the phase space quantity L_H, it is given by varying each ϕ_a, doing the necessary integrations by parts, and then doing the velocity space integral and the species sum. The chain rule of derivatives says we can also do this with the derivatives of H with respect to space or time, through the functional derivatives of H with respect to each of the ϕ_a then multiplied by the derivatives of ϕ_a with respect to space or time. So we have

$$\sum_{sp} \int d\Lambda\, f \frac{\partial H}{\partial t} = \int d\Lambda\, f \frac{\partial}{\partial t} H(\phi_a) = \int d\mathcal{V} \frac{\delta L_H}{\delta \phi_a} \frac{\partial \phi_a}{\partial t} \tag{6.36}$$

where the repeated index of a is summed since each of the ϕ_a is differentiated independently. The field equations themselves are given by the vanishing of the functional derivative of L, the entire Lagrangian, with respect to each of the field variables,

$$\frac{\delta L}{\delta \phi_a} = \frac{\delta L_f}{\delta \phi_a} - \frac{\delta L_H}{\delta \phi_a} = 0 \tag{6.37}$$

Therefore, we can replace the functional derivative involving fH with the one involving L_f,

$$\sum_{sp} \int d\Lambda\, f \frac{\partial H}{\partial t} = \int d\Lambda \frac{\delta L_f}{\delta \phi_a} \frac{\partial \phi_a}{\partial t} \tag{6.38}$$

and then rebuild the expression as

$$\sum_{sp} \int d\Lambda\, f \frac{\partial H}{\partial t} = \frac{\partial L_f}{\partial t} = \frac{\partial}{\partial t} \int d\Lambda\, \mathcal{L}_f \tag{6.39}$$

Finally, we have the conservation law for total energy,

$$\frac{\partial}{\partial t}(L_H - L_f) = \frac{\partial}{\partial t}\left(\sum_{sp} \int d\Lambda\, fH - \int d\mathcal{V}\, \mathcal{L}_f \right) = 0 \tag{6.40}$$

or defining it directly,

$$\frac{\partial \mathcal{E}}{\partial t} = 0, \qquad \mathcal{E} = \sum_{sp} \int d\Lambda\, fH - \int d\mathcal{V}\, \mathcal{L}_f \tag{6.41}$$

thereby proving conservation of the Noether energy independently. The important thing to note is that we are using the result of having cast L in canonical form, namely, all the dependence on the ϕ_a is solely within the time component of L, in this case, $L_t = L_f - L_H$. And the conserved energy is simply the negative of this L_t, or $\mathcal{E} = L_H - L_f$, namely all the Hamiltonian pieces minus the field piece.

6.7 Momentum conservation

The same applies to the conservation of momentum. Since the gyrocentre canonical momentum is spatially dependent, we don't actually have momentum conservation in all directions but only in directions associated with symmetry coordinates. If ξ is a spatial coordinate for which $\partial \mathbf{p}^*/\partial\xi$ vanishes, and the same applies to the geometry factors in H, then ξ is a symmetry coordinate. This usually only applies to the geometric toroidal angle in axisymmetric systems or the axial coordinate in the case of a straight magnetic field. In each case the magnetic field can be sheared, but as long as $\partial/\partial\xi$ vanishes for any geometric quantity the associated canonical momentum is conserved[1].

At the level of an individual gyrocentre, we have a Lagrangian given by

$$L_p = \cdots + p_\xi^* \frac{d\xi}{dt} - H \tag{6.42}$$

where ξ is the symmetry coordinate and the rest of L_p is independent of ξ or $\dot{\xi} = d\xi/dt$. The canonical momentum associated with ξ is defined by

$$p_\xi^* = \frac{\partial L_p}{\partial \dot{\xi}} = eA_\xi^* \tag{6.43}$$

The Euler–Lagrange equation for the toroidal coordinate of the gyrocentre is given by

$$\frac{dp_\xi^*}{dt} + \frac{\partial H}{\partial \xi} = 0 \tag{6.44}$$

showing how the gyrocentre can gain or lose momentum by moving up or down the slope given by $\partial H/\partial\xi$, which depends on the derivatives of the field variables. So p_ξ^* is not conserved by itself.

At the level of the entire system, the momentum which is conserved is given by the canonical momentum defined by P_ξ^* given in Eq. (6.31), where now all gyrocentres, across species, are involved, hence the phase space integral with f. This is shown by using the same set of steps used to reduce $f(\partial H/\partial t)$ in the demonstration of energy conservation. Simply replace $\partial/\partial t$ by $\partial/\partial\xi$ in all the steps. The difference is that now at the end we have the spatial derivative and not the time derivative under the final integral, namely

$$\int d\mathcal{V} \frac{\partial \mathcal{L}_f}{\partial \xi} = 0 \tag{6.45}$$

This results in P_ξ^* being conserved as a system quantity, another functional. It is important to note that only holds under the sum across species, since that step is used to replace the fH terms with the $\mathcal{L}_f$ ones.

At the level of the Poisson brackets, we have

$$p_\xi^* \frac{df}{dt} = \frac{d}{dt} f p_\xi^* - f\frac{dp_\xi^*}{dt} = \frac{d}{dt} f p_\xi^* + f\frac{\partial H}{\partial \xi} = 0 \tag{6.46}$$

using Eqs (6.14), (6.44). We then have

[1] Note that $\partial/\partial\xi$ is a partial, not a covariant derivative—just because the direction of the geometric toroidal angle unit vector changes with rotation about the symmetry axis does not mean the symmetry is broken.

$$\frac{\partial}{\partial t} f p_\xi^* + \{H, f p_\xi^*\} + f \frac{\partial H}{\partial \xi} = 0 \tag{6.47}$$

using Eq. (6.20). This is the phase space continuity equation for the canonical momentum associated with ξ. Integrating over phase space and summing over species, we have

$$\sum_{sp} \int d\Lambda \left(\frac{\partial}{\partial t} f p_\xi^* + f \frac{\partial H}{\partial \xi} \right) = 0 \tag{6.48}$$

since the integration annihilates the bracket term. The last term is treated in the same way as for energy conservation. Start with the chain rule,

$$\sum_{sp} \int d\Lambda\, f \frac{\partial H}{\partial \xi} = \int d\mathcal{V}\, \frac{\delta L_H}{\delta \phi_a} \frac{\partial \phi_a}{\partial \xi} \tag{6.49}$$

Then use the field equation for each ϕ_a to replace L_H with L_f, and then reverse the chain rule, showing that

$$\sum_{sp} \int d\Lambda\, f \frac{\partial H}{\partial \xi} = \int d\mathcal{V}\, \frac{\delta L_f}{\delta \phi_a} \frac{\partial \phi_a}{\partial \xi} = \int d\mathcal{V}\, \frac{\partial \mathcal{L}_f}{\partial \xi} = 0 \tag{6.50}$$

This proves the conservation of canonical momentum given by

$$\frac{\partial P_\xi^*}{\partial t} = 0, \qquad P_\xi^* = \sum_{sp} \int d\Lambda\, f p_\xi^* = \sum_{sp} \int d\Lambda\, f e A_\xi^* \tag{6.51}$$

writing Eq. (6.31) again for the gyrokinetic case, in terms of $\mathbf{A}^*$ using Eq. (6.43).

6.7.1 Plasma versus canonical momentum

The foregoing shows the conservation of canonical momentum associated with a symmetry coordinate, but what about plasma momentum, or, what is the role of the ExB momentum, which does not at first sight appear to be involved in p_ξ^* or P_ξ^*? The answer involves noticing that in any practical model the largest term in p_ξ^* is eA_ξ, a spatial quantity multiplied by the charge. So the largest term in P_ξ^* is

$$\sum_{sp} \int d\Lambda\, f e A_\xi = \int d\mathcal{V}\, \rho_G A_\xi, \qquad \rho_G = \sum_{sp} \int d\mathcal{W}\, f e \tag{6.52}$$

where ρ_G is the gyrocentre charge density. This is just as in the Maxwell-plasma system, except we now have ρ_G rather than ρ_{ch}, which in the gyrokinetic case is zero by construction. Nevertheless, we treat this $\rho_G A_\xi$ piece the same way here as we did there for $\rho_{ch}\mathbf{A}$, namely, using the field equation for ρ_G. This is none other than the polarisation equation. So let us do charge conservation first.

The gyrocentre charge is just another functional, but one which involves f across species. All we have to do is multiply the evolution equation in divergence form

(Eq. 6.24), by the species charge, sum over species, and integrate over velocity space, which annihilates the $\partial/\partial z$ divergence to leave the spatial one only,

$$\sum_{sp} \int d\mathcal{W} \left[\frac{\partial}{\partial t} fe + \frac{1}{B_\parallel^*} \nabla \cdot \left(feB_\parallel^* \dot{\mathbf{R}} \right) = 0 \right] \tag{6.53}$$

or

$$\frac{\partial \rho_G}{\partial t} + \nabla \cdot \mathbf{J}_G = 0 \tag{6.54}$$

where

$$\rho_G = \sum_{sp} \int d\mathcal{W} fe, \qquad \mathbf{J}_G = \sum_{sp} \int d\mathcal{W} feZ^{\mathbf{R}} \tag{6.55}$$

are the gyrocentre density and current (density). Now, we put in the polarisation equation,

$$\frac{\delta L}{\delta \phi} = 0, \qquad \nabla \cdot \mathbf{P} = \rho_G, \qquad \mathbf{P} = -\frac{\delta L}{\delta \nabla \phi} \tag{6.56}$$

where $\mathbf{P}$ is the gyrocentre polarisation vector. We these definitions we have the continuity of the total current,

$$\nabla \cdot \mathbf{J} = 0, \qquad \mathbf{J} = \frac{\partial \mathbf{P}}{\partial t} + \mathbf{J}_G \tag{6.57}$$

The time derivative of the polarisation vector is the time dependent polarisation current, and the (minus) divergence of the polarisation adds to the charge densities in general, and in this case adds to the gyrocentre charge density to yield zero for a total.

We now know what to do with the momentum. The contribution due to A_ξ in the canonical momentum is reduced according to

$$\sum_{sp} \int d\Lambda\, feA_\xi = \int d\mathcal{V}\, \rho_G A_\xi = \int d\mathcal{V}\, \nabla \cdot \mathbf{P} A_\xi = -\int d\mathcal{V}\, \mathbf{P} \cdot \nabla A_\xi \tag{6.58}$$

This is none other than the ExB momentum, namely its covariant component in the symmetry coordinate. Now we have the conservation of the plasma momentum given by

$$\frac{\partial M_\xi}{\partial t} = 0, \qquad M_\xi = \int d\mathcal{V} \left(-\mathbf{P} \cdot \nabla A_\xi + \sum_{sp} \int d\mathcal{W} fe(A_\xi^* - A_\xi) \right) \tag{6.59}$$

In a traditional model with

$$e\mathbf{A}^* = e\mathbf{A} + z\mathbf{b}, \qquad \mathbf{P} = -\frac{\rho}{B^2} \nabla_\perp \phi \tag{6.60}$$

this reduces to

$$M_\xi = \int d\mathcal{V}\, \frac{\rho}{B^2} \nabla_\perp \phi \cdot \nabla A_\xi + \sum_{sp} \int d\Lambda f z\, b_\xi \tag{6.61}$$

In an MHD equilibrium with a symmetry coordinate the background magnetic field is

$$\mathbf{B} = B_\xi \nabla\xi + \nabla A_\xi \times \nabla\xi \tag{6.62}$$

so that

$$\nabla_\perp \phi \cdot \nabla A_\xi = \nabla_\perp \phi \cdot \left(|\nabla\xi|^{-2}\, \nabla\xi \times (\nabla A_\xi \times \xi) \right) \tag{6.63}$$

since A_ξ is independent of ξ. We can add the $B_\xi \nabla\xi$ piece into the inner parentheses since its cross product with $\nabla\xi$ vanishes. This gives

$$\nabla_\perp \phi \cdot \nabla A_\xi = \nabla_\perp \phi \cdot (|\nabla\xi|^{-2}\, \nabla\xi \times \mathbf{B}) \tag{6.64}$$

and finally

$$\nabla_\perp \phi \cdot \nabla A_\xi = |\nabla\xi|^{-2}\, \nabla\xi \cdot (\mathbf{B} \times \nabla\phi) \tag{6.65}$$

so that

$$M_\xi = \int d\mathcal{V}\, \rho \mathbf{v}_E \cdot \nabla\xi \left(|\nabla\xi|^{-2} \right) + \sum_{sp} \int d\Lambda f z\, b_\xi \tag{6.66}$$

where

$$\mathbf{v}_E = \frac{1}{B^2} (\mathbf{B} \times \nabla\phi) \tag{6.67}$$

We see that M_ξ is given by the covariant component of the mass density times the ExB velocity, plus the kinetic parallel perturbed canonical momentum times the covariant component of the magnetic field unit vector, both in the ξ direction. So, the plasma momentum which is conserved is the ExB together with the parallel momentum, both in the covariant ξ-component. For general polarisation $\mathbf{P}$ and canonical momentum A_ξ^*, the plasma momentum is merely the associated generalisation of the ExB and parallel momentum. If the symmetry coordinate is the geometric toroidal angle (axisymmetric system), then the conserved quantities just derived are the toroidal angular canonical and plasma momenta, respectively.

6.7.2 The polarisation cancellation

It is worth documenting a result from the local form of this conservation law. If the integration done is the one over all phase space, then we have the global conservation law, namely, the constancy of a functional. If the integration done is just the one over velocity space, then we have a continuity equation for the density of the functional. Doing this operation on Eq. (6.47) produces

$$\sum_{sp} \int d\mathcal{W} \left[\frac{\partial}{\partial t} f p_\xi^* + \{H, f p_\xi^*\} + f \frac{\partial H}{\partial \xi} \right] = 0 \tag{6.68}$$

Writing this in divergence form, we have

$$\sum_{sp} \int d\mathcal{W} \left[\frac{\partial}{\partial t} f p_\xi^* + \frac{1}{B_\parallel^*} \nabla \cdot \left(f p_\xi^* B_\parallel^* \dot{\mathbf{R}} \right) + \frac{1}{B_\parallel^*} \frac{\partial}{\partial z} \left(f p_\xi^* B_\parallel^* \dot{z} \right) + f \frac{\partial H}{\partial \xi} \right] = 0 \tag{6.69}$$

The third term is annihilated by the integration, and in the second term we commute $d\mathcal{W}/B_\parallel^*$ past the divergence to obtain

$$\frac{\partial}{\partial t} \sum_{sp} \int d\mathcal{W} f p_\xi^* + \nabla \cdot \left(\sum_{sp} \int d\mathcal{W} f p_\xi^* \dot{\mathbf{R}} \right) + \sum_{sp} \int d\mathcal{W} f \frac{\partial H}{\partial \xi} = 0 \tag{6.70}$$

We have already done the work to show that the last term must be a spatial divergence as well, since the volume integral over it was shown to vanish in Eq. (6.50). Let us examine the contribution due to the largest term in p_ξ^*, namely eA_ξ, to the first two terms. We have

$$\frac{\partial}{\partial t} \sum_{sp} \int d\mathcal{W} f e A_\xi + \nabla \cdot \left(\sum_{sp} \int d\mathcal{W} f e A_\xi \dot{\mathbf{R}} \right) = \frac{\partial}{\partial t} A_\xi \rho_G + \nabla \cdot \left(A_\xi \mathbf{J}_G \right) \tag{6.71}$$

since A_ξ is pulled out of the species sum and the velocity space integration. The two terms on the right hand side involve the gyrocentre charge density and current, respectively. Putting in the polarisation equation to eliminate ρ_G in favour of $\mathbf{P}$, and then the current continuity Eq. (6.57), we have

$$\frac{\partial}{\partial t} A_\xi \rho_G + \nabla \cdot \left(A_\xi \mathbf{J}_G \right) = \frac{\partial}{\partial t} \left(-\mathbf{P} \cdot \nabla A_\xi \right) + \nabla \cdot \left(A_\xi \mathbf{J} \right) \tag{6.72}$$

noting that we can slip A_ξ into or out of the time derivative as it is time independent. So the two terms on the left side, the parts of the left side of Eq. (6.70) due to A_ξ, are replaced by those on the right side. This result is the 'polarisation cancellation' since it has used polarisation to exactly cancel any transport term (under the divergence) dependent on A_ξ times drifts. One just has A_ξ times the current, whose divergence is zero. In the structure of the magnetic field with symmetry coordinate A_ξ, the magnetic field lies in surfaces of constant A_ξ, which are called flux surfaces. Recall (Chapter 16 of Volume One) that the property of the flux surface average over a divergence implies

$$\langle \mathbf{J} \cdot \nabla V \rangle = 0, \qquad \text{if} \quad \nabla \cdot \mathbf{J} = 0 \tag{6.73}$$

where V as a coordinate is the volume enclosed by the flux surface. In other words, the flux surface of the radial current vanishes. Since $\nabla \cdot \mathbf{J} = 0$ we have

$$\nabla \cdot (A_\xi \mathbf{J}) = \mathbf{J} \cdot \nabla A_\xi = \frac{\partial A_\xi}{\partial V} \mathbf{J} \cdot \nabla V \qquad \text{hence} \quad \left\langle \nabla \cdot (A_\xi \mathbf{J}) \right\rangle = 0 \tag{6.74}$$

since the meaning of a flux function is that it is constant on a flux surface. In an MHD equilibrium with closed flux surfaces V and A_ξ are functions of each other by definition. It follows that in the flux surface average of the canonical momentum transport equation,

$$\frac{\partial}{\partial t}\langle \rho_G A_\xi \rangle + \langle \nabla \cdot (A_\xi \mathbf{J}_G) \rangle = \frac{\partial}{\partial t}\langle -\mathbf{P} \cdot \nabla A_\xi \rangle \tag{6.75}$$

the term involving $\mathbf{J}$ vanishes and hence there is no term directly involving A_ξ as a coefficient in the resulting transport flux. This version of the polarisation cancellation appears in the 2010 reference given below.

6.7.3 Wave fluxes

This leaves the wave fluxes, arising from the quantity

$$M_H = \sum_{sp} \int d\Lambda f \frac{\partial H}{\partial \xi} = 0 \tag{6.76}$$

The integral over it vanishes according to the procedure given above, namely, (1) the functional derivative of L_H with respect to each ϕ_a is replaced by the one of $-L_f$, then (2) the latter is used to rebuild the integral into one over a differential of $\mathcal{L}_f$, which vanishes. Hence we can always write

$$M_H = \nabla \cdot \mathbf{\Pi}_H, \qquad \mathbf{\Pi}_H = \nabla \times \mathbf{M} - \nabla \chi \tag{6.77}$$

The latter expression is the Helmholtz decomposition of the wave flux $\mathbf{\Pi}_H$ in terms of a solenoidal vector $\mathbf{M}$ and a wave flux potential χ. This potential is determined by

$$-\nabla^2 \chi = \sum_{sp} \int d\mathcal{W} f \frac{\partial H}{\partial \xi} \tag{6.78}$$

When the radial component of this wave flux is flux surface averaged, the curl drops since it is divergence free, leaving

$$\langle \mathbf{\Pi}_H \cdot \nabla V \rangle = -\langle \nabla \chi \cdot \nabla V \rangle \tag{6.79}$$

Formally, this operation determines the overall wave flux due to all of the $f(\partial H/\partial \xi)$, and it can be separated into pieces due to any desired separation of the terms in H.

The simpler form of this is found under the long wavelength approximation, Eq. (6.6). In that case we have

$$H = H(\phi, \nabla \phi, \nabla A_\parallel) \tag{6.80}$$

The derivative with respect to ξ gives

$$M_H = \sum_{sp} \int d\Lambda \left[f\left(e\frac{\partial \phi}{\partial \xi} + \frac{\partial H}{\partial \nabla \phi} \nabla \frac{\partial \phi}{\partial \xi} \right) + feU\frac{\partial A_\parallel}{\partial \xi} \right] \tag{6.81}$$

Doing the ∇ on the second term by parts, we have

$$M_H = \sum_{sp} \int d\Lambda \left[f\left(e\frac{\partial\phi}{\partial\xi} - \nabla\frac{\partial H}{\partial\nabla\phi}\frac{\partial\phi}{\partial\xi} \right) + \nabla\cdot\left(f\frac{\partial H}{\partial\nabla\phi}\frac{\partial\phi}{\partial\xi} \right) + feU\frac{\partial A_\parallel}{\partial\xi} \right] \tag{6.82}$$

Using the field equations (Eqs (5.126) and (5.127)), the functional derivative with respect to ϕ vanishes, and feU is replaced accordingly,

$$M_H = \int d\mathcal{V}\,\nabla\cdot\left[\sum_{sp}\int d\mathcal{W}\left(f\frac{\partial H}{\partial\nabla\phi}\frac{\partial\phi}{\partial\xi} \right) - \frac{1}{\mu_0}\nabla_\perp^2 A_\parallel\frac{\partial A_\parallel}{\partial\xi} \right] \tag{6.83}$$

Finally, the divergence in the Laplacian is integrated by parts with a pure ξ-derivative vanishing, in the last term, and the dependence of H on $\nabla\phi$ is inserted, and the result is Eq. (6.77) with

$$\mathbf{\Pi}_H = -\left(\frac{\rho}{B^2}\nabla_\perp\phi \right)\frac{\partial\phi}{\partial\xi} - \left(\frac{1}{\mu_0}\nabla_\perp A_\parallel \right)\frac{\partial A_\parallel}{\partial\xi} \tag{6.84}$$

This expression gives the contributions of the ExB Reynolds and magnetic Maxwell stresses to the covariant ξ-component of the momentum flux. For any dependence of H on ϕ and $A_\parallel$, the wave fluxes due to the field variables are found to be the appropriate generalisations of these expressions.

In general, the fH term in the momentum conservation law gives rise to a set of wave fluxes, which under the divergence preserve the structure of the (canonical or plasma) momentum conservation law. This will work in any field theory, since the exactly consistent field equations which are required to make it work are actually the ones produced by the field theory. In any other procedure to obtain a set of gyrokinetic equations wherein the field equations are found or approximated due to other criteria, this will not work properly and there will be problems with momentum conservation.

6.7.4 Local momentum conservation

With these operations taken care of, we have

$$\begin{aligned} &\frac{\partial}{\partial t}\left\langle -\mathbf{P}\cdot\nabla A_\xi + \sum_{sp}\int d\mathcal{W} fe(A_\xi^* - A_\xi) \right\rangle \\ &+ \frac{\partial}{\partial V}\left\langle \sum_{sp}\int d\mathcal{W} fe(A_\xi^* - A_\xi)Z^V \right\rangle + \frac{\partial}{\partial V}\langle -\nabla\chi\cdot\nabla V\rangle = 0 \end{aligned} \tag{6.85}$$

as the radial transport equation of the plasma momentum. The transport term with Z^V (due to drifts) reduces to $v_\parallel b_\xi Z^V$ in a conventional electrostatic model. The last term is the general expression of the radial wave fluxes.

6.7.5 MHD correspondence

In the ideal MHD model the momentum equation can be written

$$\frac{\partial}{\partial t}\rho\mathbf{v} + \nabla\cdot\left[\rho\mathbf{vv} + \left(p + \frac{B^2}{2\mu_0}\right)\mathbf{g} - \frac{1}{\mu_0}\mathbf{BB}\right] = 0 \tag{6.86}$$

where $\mathbf{g}$ is the metric tensor ($\nabla\cdot a\mathbf{g} = \nabla a$ for any scalar a). We have used the continuity equation to put ρ under the time derivative and $\rho\mathbf{vv}$ under the divergence, and Ampère's law $\nabla\times\mathbf{B} = \mu_0\mathbf{J}$ to convert $\mathbf{J}\times\mathbf{B}$ into the two terms involving $\mathbf{B}$. The velocity is the ExB velocity plus the parallel one, so that

$$\mathbf{v} = \mathbf{v}_E + u_\|\mathbf{b} \tag{6.87}$$

The toroidal angular momentum transport equation for closed magnetic surfaces in an axisymmetric MHD equilibrium is formed by contraction with $R^2\nabla\varphi$ (to form the covariant toroidal component) and taking the flux surface average. The main p and $\mathbf{B}$ terms satisfy the equilibrium condition, so only fluctuating disturbances remain. The gradient term drops because the flux surface average of $\partial/\partial\varphi$ vanishes. This leaves

$$\frac{\partial}{\partial t}\langle\rho\mathbf{v}_\varphi\rangle + \left\langle\nabla V\cdot\left(\rho\mathbf{v}\mathbf{v}_\varphi - \mu_0^{-1}\mathbf{BB}_\varphi\right)\right\rangle = 0 \tag{6.88}$$

where the subscript φ denotes the covariant toroidal component. The combination $R^2\nabla\varphi$ commutes past the divergence because it is the covariant component operation and the covariant derivative of the metric tensor in all its components vanishes by definition.

If we separate $\mathbf{v}_\varphi$ into two components, $\mathbf{v}_{E\varphi} + u_\|\mathbf{b}_\varphi$, we have

$$\frac{\partial}{\partial t}\langle\rho\mathbf{v}_{E\varphi} + \rho u_\|\mathbf{b}_\varphi\rangle + \left\langle\nabla V\cdot\left(\rho\mathbf{v}u_\|\mathbf{b}_\varphi + \rho\mathbf{v}\mathbf{v}_{E\varphi} - \mu_0^{-1}\mathbf{BB}_\varphi\right)\right\rangle = 0 \tag{6.89}$$

Note that $\mathbf{B}_\varphi$ and $\mathbf{B}$ here refer to fluctuating disturbances.

This corresponds one to one with the long wavelength form of the plasma toroidal angular momentum. Here, the long wavelength Hamiltonian, done in the previous Chapter, is

$$H = m\frac{U^2}{2} + M\Omega + e\phi - m\frac{u_E^2}{2} \tag{6.90}$$

and the field Lagrangian is still with $\nabla_\perp A_\|$ only. So we have

$$f\frac{\partial H}{\partial\varphi} = -feU\frac{\partial A_\|}{\partial\varphi} + fe\frac{\partial\phi}{\partial\varphi} - \frac{fm}{B^2}\nabla_\perp\phi\cdot\nabla_\perp\frac{\partial\phi}{\partial\varphi} \tag{6.91}$$

where the first term comes from differentiation of the $A_\|$ contribution to $mU^2/2$, the parallel energy. Summing over species and integrating over velocity space, the feU

term is eliminated using the induction equation and the next two are combined with the polarisation equation. The result is

$$\begin{aligned}\sum_{sp}\int d\mathcal{W}f\frac{\partial H}{\partial\varphi} &= \nabla\cdot\left(\frac{1}{\mu_0}\nabla_\perp A_\parallel\frac{\partial A_\parallel}{\partial\varphi}\right) - \frac{\partial}{\partial\varphi}\frac{1}{\mu_0}\left|\nabla_\perp A_\parallel\right|^2 \\ &- \nabla\cdot\left(B_\parallel^*\frac{\rho}{B^2}\nabla_\perp\phi\frac{\partial\phi}{\partial\varphi}\right)\end{aligned} \tag{6.92}$$

Finally, taking the flux surface average the second term drops, and the rest becomes a radial derivative multiplied by the toroidal one in each term, or

$$\left\langle\sum_{sp}\int d\mathcal{W}f\frac{\partial H}{\partial\varphi}\right\rangle = \left\langle\nabla V\cdot\left(-\frac{\rho}{B^2}\nabla_\perp\phi\frac{\partial\phi}{\partial\varphi} + \frac{1}{\mu_0}\nabla_\perp A_\parallel\frac{\partial A_\parallel}{\partial\varphi}\right)\right\rangle \tag{6.93}$$

These give the ExB Reynolds and magnetic Maxwell stresses, respectively.

For an axisymmetric MHD equilibrium we have

$$\mathbf{B} = I\nabla\varphi + \nabla\psi\times\nabla\varphi \tag{6.94}$$

We therefore identify ξ with φ and A_ξ with ψ for this purpose. The long wavelength form of the gyrokinetic version of the conservation law is then

$$\begin{aligned}&\frac{\partial}{\partial t}\left\langle -\mathbf{P}\cdot\nabla\psi + \sum_{sp}\int d\mathcal{W}fz\mathbf{b}_\varphi\right\rangle + \frac{\partial}{\partial V}\left\langle\sum_{sp}\int d\mathcal{W}fz\mathbf{b}_\varphi Z^V\right\rangle \\ &+ \frac{\partial}{\partial V}\left\langle\nabla V\cdot\left(-\frac{\rho}{B^2}\nabla_\perp\phi\frac{\partial\phi}{\partial\varphi} + \frac{1}{\mu_0}\nabla_\perp A_\parallel\frac{\partial A_\parallel}{\partial\varphi}\right)\right\rangle = 0\end{aligned} \tag{6.95}$$

The first term is $\rho\mathbf{v}_{E\varphi}$, the second is $\rho u_\parallel\mathbf{b}_\varphi$, the third is $\rho u_\parallel\mathbf{b}_\varphi$, the fourth is the ExB Reynolds stress, and the fifth is the magnetic Maxwell stress. The identification of the flux terms is subtle: the radial component v_E^V or B^V corresponds to the $\partial/\partial\varphi$ derivative, and the covariant toroidal component corresponds to the $\nabla V\cdot\nabla$ derivative. What we see by doing this is that the gyrokinetic analysis reduces to the MHD one in the MHD limit, so that the steps getting there, which may be somewhat unfamiliar, are justified. The method of obtaining plasma momentum from canonical momentum using the polarisation cancellation is supported by the fulfilment of the principle of correspondence in the MHD limit.

6.7.6 Computational consistency

All of this holds abstractly, but in a computation one would have to show that Eq. (6.47) actually results from multiplying the evolution equation by p_ξ^* in the formulation used there. This is very much nontrivial, since it involves proving that

$$\{H, p_\xi^*\} = -\frac{\partial H}{\partial\xi} \tag{6.96}$$

actually holds in the computation. Then, one must also prove that Eq. (6.50) also actually holds. This is also nontrivial, because it involves exact consistency between the evolution equation and the field equations in the formulation of the computation, within its coordinate system.

Each computation will have to show these two results explicitly, before any claim of viability in this total-f regime can carry weight. The delta-f regime is addressed in Chapter 7.

6.8 Gyrokinetic drifts

The equation for the spatial part of the phase space flow is given by $\dot{\mathbf{R}}$ for an individual gyrocentre having the same coordinate values as the given point in phase space. The set of all $\dot{\mathbf{R}}$ and $\dot{z}$ at all points in phase space gives the flow, $Z^{\mathbf{R}}$ and Z^{z}, respectively. The Hamiltonian can be written as

$$H = m\frac{U^2}{2} + M\Omega + e\phi_E \tag{6.97}$$

where the generalised ExB potential are given by

$$e(J_0\phi) - \frac{e^2}{2\Omega}\frac{\partial}{\partial M}\left(J_0(\Phi^2) - (J_0\Phi)^2\right) \tag{6.98}$$

and the parallel velocity and generalised perturbed Hamiltonian are

$$mU = z - e(J_0A_\parallel), \qquad \Phi = \phi - \frac{z}{m}A_\parallel \tag{6.99}$$

Writing these in terms of U and ϕ_E allows the resulting expressions to stand at any level of approximation. The gyrocentre canonical momentum, resulting field, and velocity space Jacobian are given by

$$\mathbf{p}^* = e\mathbf{A}^* = e\mathbf{A} + z\mathbf{b} - M\mathbf{W}, \quad \mathbf{B}^* = \nabla \times \mathbf{A}^*, \quad B_\parallel^* = \frac{\partial \mathbf{p}^*}{\partial z}\cdot\mathbf{B}^* \tag{6.100}$$

The phase space flow is given by

$$B_\parallel^*\, Z^{\mathbf{R}} = \frac{1}{e}\mathbf{b}\times\nabla H + U\mathbf{B}^*, \qquad B_\parallel^*\, Z^{z} = -\mathbf{B}^*\cdot\nabla H \tag{6.101}$$

The spatial drift motion is in the first quantity, while the parallel force is in the second.

The spatial drifts include the ExB drift, the magnetic gradient (grad-B) and curvature drifts, and the parallel motion. The first two are given by $\nabla\phi_E$ and $M\nabla\Omega$ in ∇H. The parallel motion along the background magnetic field and the curvature drift are given in the $U\mathbf{B}^*$ term. Splitting $\mathbf{B}^*$, we have

$$U\mathbf{B}^* = U\mathbf{B} + U\frac{z}{e}\nabla\times\mathbf{b} - U\frac{M}{e}\nabla\times\mathbf{W} \tag{6.102}$$

The z in the second term splits into $mU + eA_{\|}$, so that

$$UB^* = UB + \frac{1}{e}mU^2\nabla \times \mathbf{b} - U\frac{M}{e}\nabla \times \mathbf{W} + UA_{\|}\nabla \times \mathbf{b} \tag{6.103}$$

The last term combines with the $A_{\|}$ part of ∇H through

$$\frac{1}{e}\mathbf{b} \times \nabla H = \cdots - U\mathbf{b} \times \nabla A_{\|}$$
$$\text{then} \qquad UA_{\|}\nabla \times \mathbf{b} - U\mathbf{b} \times \nabla A_{\|} = U\nabla \times (A_{\|}\mathbf{b}) \tag{6.104}$$

This combination adds to the $U\mathbf{B}$ term using

$$\mathbf{B}_T = \mathbf{B} + \nabla \times (A_{\|}\mathbf{b}) \tag{6.105}$$

as the entire magnetic field including the disturbances. With the ExB and grad-B drifts, we have the combination

$$B_{\|}^* Z^{\mathrm{R}} = U\mathbf{B}_T + \mathbf{b} \times \nabla\phi_E + \frac{1}{e}(M\mathbf{b} \times \nabla\Omega + mU^2\nabla \times \mathbf{b} - UM\nabla \times \mathbf{W}) \tag{6.106}$$

This expression for the drifts is valid at any level of approximation (note that throughout, $\mathbf{b}$ generalises back to $\partial\mathbf{p}^*/\partial z$). Common approximations include the neglect of $M\mathbf{W}$ as a small gyroradius correction to the curvature drift and the inertial correction to the ExB drift due to the polarisation term in ϕ_E. Charge separation in the drifts is provided by the parallel current due to $U\mathbf{B}_T$, and the curvature and grad-B drifts due to the factor of $1/e$. The $M\mathbf{W}$ term is smaller by a factor of $(V/\Omega)/L_B$, where L_B is the scale of variation of $\mathbf{B}$, and V is the thermal velocity, and therefore V/Ω scales as a gyroradius.

If the large scale approximation is used, $J_0 \to 1$ is taken, and $M\mathbf{W}$ is neglected, we have

$$\phi_E = e\phi - m\frac{u_E^2}{2} \qquad mU = z - eA_{\|} \tag{6.107}$$

and the result for the drift motion is

$$B_{\|}^* Z^{\mathrm{R}} = U\mathbf{B}_T + \mathbf{b} \times \nabla\phi_E + \frac{1}{e}(M\mathbf{b} \times \nabla\Omega + mU^2\nabla \times \mathbf{b}) \tag{6.108}$$

just Eq. (6.106) without the last term. The general structure is maintained, but now all the drifts have their traditional forms except for the inertial correction to $e\phi$. If linearised polarisation is taken then even this piece drops. But we always have $B_{\|}^*$ instead of B as a factor, and $B_{\|}^*$ can only be replaced by B if the geometric simplifications in the Lagrangian (i.e., in $\mathbf{p}^*$) allow it. We also have $\mathbf{B}^*\cdot\nabla H$ in the effective force, with $\mathbf{B}^*$ instead of $\mathbf{B}$ including some higher order drift effects, but that is always necessary to preserve the phase space structure.

The difference between $B_{\|}^*$ and B is a small correction to all the drifts. The scaling of $(z/e)\mathbf{b} \cdot \nabla \times \mathbf{b}$ against $\mathbf{B}$ is $(V/\Omega)/L_B$. In simplified models, $B_{\|}^* \to B$ is taken and the inertial ExB correction are dropped, but doing this arbitrarily actually breaks the structure of the phase space flow. Such approximations are well taken only if one can simplify the Lagrangian in order to deliver them as a result. Only then will exact consistency not only in the energetics but also in the structure be maintained. This was a main point in the Littlejohn (1983) paper referenced in Further Reading of the previous Chapter.

There is no diamagnetic drift or polarisation drift. This is because we are working with gyrocentre flows rather than fluid flows. They differ by a magnetisation flow given by a curl, so that the resulting fluxes and energetics are basically the same. Once we take the divergence of the diamagnetic drift in a fluid drift model, we obtain an expression very similar to the gyrocentre grad-B and curvature drifts. We will see this expression in the gyrocentre current in the next section. For polarisation the details are much more complicated, since part of the result comes from time dependence, so only in the continuity equation do the models converge. We will encounter this in the proof of correspondence between the gyrofluid and fluid drift models at the delta-f level in Chapter 10.

6.9 Gyrokinetic energetics

The overall conservation laws are all well and good, but to understand the results we have to look at the pieces. This is the lesson learned using the fluid drift model in Volume One. Examining the energy theorem as a set of equations which are only closed in total allowed us to understand how the system passes energy from piece to piece. With the gradient drive, either as a source or a transfer, in the thermal energy components, the ways in which the energy can be passed into the ExB component show how the vorticity, which is the basis of the turbulence, is maintained.

6.9.1 Long-wave electrostatic systems

It is helpful to examine the situation with large scale relaxation dynamics in a confined plasma with an equilibrium state first, to gain understanding in how the energetics works, before the complications of MHD or drift-Alfvén (electromagnetic adiabatic response), or finite gyroradius corrections, are faced. The physical situation is one in which the MHD has relaxed into equilibrium through decaying Alfvénic activity, and then on slower time scales the flows relax onto the thermal background. This can include turbulence deep into the electrostatic regime.

Under the long wavelength electrostatic approximations, there is only one field, ϕ, since the magnetic potential is being neglected. There is also no free field piece for ϕ, which is contained entirely within H. There are two pieces to this, the main one being the gyrocentre charge potential energy and the second being the ExB inertial correction (ϕ_E in Eq. (6.6)). The total energy consists of E_f, the thermal and kinetic energy given by the $mU^2/2$ and wB terms in H, and E_E, the ExB flow energy contained within ϕ_E. These pieces are given by

$$E_f = \sum_{sp} \int d\Lambda f H_0, \quad H_0 = m\frac{U^2}{2} + wB, \quad E_E = \sum_{sp} \int d\Lambda f e\phi_E \tag{6.109}$$

where H_0 is the part of H due solely to gyrocentre kinetic energy. To show that E_E, which looks like a potential energy, is really the ExB kinetic energy, we eliminate the main $e\phi$ term the same way as was done for $\rho_{ch}\phi$ in Maxwell's equations. Multiplying the field equation (Eq. (5.126)) by ϕ and integrating over space, we have

$$\int d\mathcal{V}\,\phi \sum_{sp} \int d\mathcal{W}\left[fe + \frac{1}{B_\parallel^*}\nabla \cdot B_\parallel^* \frac{fm}{B^2}\nabla_\perp \phi\right] = 0 \tag{6.110}$$

Commute $d\mathcal{W}/B_\parallel^*$ past the divergence and then do it by parts to find

$$\int d\mathcal{V}\,\rho_G \phi = \int d\mathcal{V}\,\rho u_E^2 \tag{6.111}$$

with ρ_G and ρ the species sum of fe and fm, respectively, integrated over velocity space. Therefore, we have

$$\sum_{sp} \int d\Lambda\, fe\phi_E = \int d\mathcal{V}\,\rho\frac{u_E^2}{2} = E_E \tag{6.112}$$

namely, the ExB flow (kinetic) energy.

In a similar manner, the time derivative of the ExB energy is found to be

$$\frac{\partial E_E}{\partial t} = \sum_{sp} \int d\Lambda\, e\phi_E \frac{\partial f}{\partial t} \tag{6.113}$$

where a second term due to $\partial\phi_E/\partial t$ vanishes according to

$$\sum_{sp} \int d\Lambda\, fe\frac{\partial \phi_E}{\partial t} = \int d\mathcal{V}\,\frac{\delta E_E}{\delta \phi}\frac{\partial \phi}{\partial t} = 0 \tag{6.114}$$

since ϕ_E is the only term in H, therefore in L, containing ϕ. For the thermal energy we just have

$$\frac{\partial E_f}{\partial t} = \sum_{sp} \int d\Lambda\, H_0 \frac{\partial f}{\partial t} \tag{6.115}$$

since with $A_\parallel \to 0$ the content of H_0 is just coordinates and geometry (through B).

These pieces add up to the $H(\partial f/\partial t)$ combination when summed, but the remainder of each gives the transfer. For the ExB energetics, multiply the evolution equation by $e\phi_E$ and do the species summed phase space integration to find,

$$\frac{\partial E_E}{\partial t} = \int d\mathcal{V}\,\mathbf{J}_G \cdot \nabla\phi_E \tag{6.116}$$

where J_G is the gyrocentre current. To get this, put the divergence in Eq. (6.24) in for the time derivative and integrate by parts, noting that ϕ_E is independent of z. Having already done the gyrocentre drifts in Eq. (6.106), we can simply evaluate the terms by integration. The gyrocentre current is

$$\mathbf{J}_G = \sum_{sp} \int d\mathcal{W} feZ^{\mathbf{R}} \tag{6.117}$$

In a simplified geometry in which $\mathbf{b} \cdot \nabla \times \mathbf{b} = 0$ and hence $B_{\parallel}^* \to B$, and a situation with only one ion species, it reduces to

$$\mathbf{J}_G = J_{\parallel}\mathbf{b} + \rho_G \mathbf{v}_E - nm_i \mathbf{U}_E + p_{\perp}\mathbf{u}_{\nabla B} + P_{\parallel}\mathbf{u}_C \tag{6.118}$$

where the fluid moment quantities are

$$J_{\parallel} = \sum_{sp} \int d\mathcal{W} feU, \quad p_{\perp} = \sum_{sp} \int d\mathcal{W} fM\Omega, \quad P_{\parallel} = \sum_{sp} \int d\mathcal{W} fmU^2 \tag{6.119}$$

and for each species

$$n = \int d\mathcal{W} f, \qquad nu_{\parallel} = \int d\mathcal{W} fU \tag{6.120}$$

Note $P_{\parallel}$, the species sum of $p_{\parallel} + nmu_{\parallel}^2$, includes the parallel Mach correction. The velocity auxiliaries are

$$\mathbf{v}_E = \frac{1}{B}\mathbf{b} \times \nabla\phi, \qquad \mathbf{U}_E = \frac{1}{B}\mathbf{b} \times \nabla(u_E^2/2) \tag{6.121}$$

where $\mathbf{U}_E$ is the inertial correction to the ExB velocity, and

$$\mathbf{u}_{\nabla B} = \frac{1}{B}\mathbf{b} \times \nabla \log B, \qquad \mathbf{u}_C = \frac{1}{B}\mathbf{b} \times (\mathbf{b} \cdot \nabla\mathbf{b}) \tag{6.122}$$

The contributions to $\mathbf{J}_G$ are the parallel current, the gyrocentre ExB current, the inertial correction, and the grad-B and curvature drift currents. The gyrocentre ExB current drops out of the transfer because it represents the ExB vorticity nonlinear dynamics, which keeps energy within E_E. The last two terms represent ExB compression, with the two velocity auxiliaries $\mathbf{u}_{\nabla B}$ and $\mathbf{u}_C$ giving the grad-B and curvature drifts. These contracted with $\nabla\phi$ give the curvature terms $\mathcal{K}(\phi)$ that are present in the fluid model. This shows the basic correspondence to the same energetics we have studied in the fluid drift model. The new contributions are the inertial corrections due to $\mathbf{U}_E$ which contributes through ϕ_E in $\mathbf{J}_G$. This represents a higher order correction not found in the fluid drift model and indeed is usually small. It won't be found in the delta-f gyrofluid equations in subsequent Chapters, either. But this nonlinear inertia drift does represent something interesting for future investigations in a total-f setting.

The electrostatic energetics is interesting in its own right, since it governs how the flows present in any system with thermal gradients relax towards their equilibrium balances. It also governs how turbulence passes energy from gradients to ExB eddies whose vorticity represents the turbulent dynamics. Except for magnetic flutter contributions to transfer in the latter model, these are much the same effects which appear in more generality within an electromagnetic setting. Once one has done the analysis of this simpler large scale electrostatic case, the necessary familiarity is present to turn to the general electromagnetic case.

6.9.2 Electromagnetic energetics

The new ingredients when going back to the general gyrokinetic setting are the magnetic energy, the magnetic corrections in the perturbed canonical momentum, and the polarisation. We still have the structure present in the description with mU and ϕ_E, so much of the analysis is the same, just that the gyrocentre current $\mathbf{J}_G$ has to be left in its more native form. The fluid moments become problematic when $B_\|^*$ is not B, and there is both species and velocity space dependence in all the corrections. Nevertheless, the various energy transfer pathways are the same.

First, since it remains true that ϕ appears only in H due to the quasineutrality approximation $\varepsilon_0 \ll \rho/B^2$, the ExB energy is still given by the $fe\phi_E$ expression in Eq. (6.109), just with the full expression for ϕ_E in Eq. (6.4). Multiplying the full polarisation equation (Eq. (5.98)) by ϕ and integrating over the volume, one uses the Hermitian property of the J_0 operators and also does the $\partial/\partial M$ in $\mathcal{M}$ by parts to recover the equality

$$\sum_{sp} \int d\Lambda\, fe\phi_E = \sum_{sp} \int d\Lambda\, \frac{fe}{2\Omega}\frac{\partial}{\partial M}(J_0(\Phi^2) - (J_0\Phi)^2) \tag{6.123}$$

The right hand side is the more general expression of the ExB energy, with the left hand side remaining valid. Moreover, one still has the time derivative in Eq. (6.116), just with the more general ϕ_E and $\mathbf{J}_G$, the latter always described by Eq. (6.117) with the drifts given in Eq. (6.106). The identity of the first four of the five terms remains the same, with the inertial correction in ϕ_E now due to the full H_2 expression involving Φ. The last term gives a heat flux correction multiplying a further drift due to $\nabla \times \mathbf{W}$, which is usually small.

The parallel coupling due to $feU\mathbf{B}_T$ recovers its Alfvénic form with the magnetic flutter due to $A_\|$ in $\mathbf{B}_T$. The energy it couples to is, as in the fluid models, the parallel energy wherein $eA_\|$ and mU are combined in p_z. The magnetic energy and parallel thermal/kinetic energy,

$$E_M = \frac{1}{2\mu_0}\left|\nabla_\perp A_\|\right|^2, \qquad E_{f\|} = \sum_{sp} \int d\Lambda\, fm\frac{U^2}{2} \tag{6.124}$$

are written separately. But the time derivative one finds by multiplying the evolution equation by $mU^2/2$ covers both pieces. We have

$$\frac{\partial}{\partial t} m\frac{U^2}{2} = -eU\frac{\partial}{\partial t}(J_0 A_\|) \tag{6.125}$$

Multiplied by f under the species summed velocity space integration, it is the counterpart to $J_\| A_\|$ familiar in the fluid drift models. Multiplying the Ohm's law, with the induction term due to $\partial A_\|/\partial t$ and the inertial term due to $m_e \partial J_\|/\partial t$, by $J_\|$ produces the sum of magnetic and parallel kinetic energy in the electrons in those models. Here, we have the species sum with mU^2, which is the counterpart of doing the same thing for the ions and electrons separately, and then noting that the electrons dominate the $J_\| A_\|$ expression giving the magnetic energy after volume integration. In the gyrokinetic case we have

$$m\frac{U^2}{2}\frac{\partial f}{\partial t} = \frac{\partial}{\partial t} fm\frac{U^2}{2} + feU\frac{\partial}{\partial t}(J_0 A_\|) \tag{6.126}$$

within velocity space within a species. Summing over both and then integration over volume gives

$$\begin{aligned}\sum_{sp}\int d\mathcal{V}\,(feU)\frac{\partial}{\partial t}(J_0 A_\|) &= \sum_{sp}\int d\mathcal{V}\,(J_0 feU)\frac{\partial A_\|}{\partial t}\\ &= \int d\mathcal{V}\,\frac{1}{\mu_0}(-\nabla_\perp^2 A_\|)\frac{\partial A_\|}{\partial t} = \frac{\partial E_M}{\partial t}\end{aligned} \tag{6.127}$$

Therefore, we have both in combination,

$$\frac{\partial}{\partial t}\left(E_{f\|} + E_M\right) = \sum_{sp}\int d\Lambda\, m\frac{U^2}{2}\frac{\partial f}{\partial t} = \sum_{sp}\int d\Lambda\left(\frac{\partial}{\partial t} fm\frac{U^2}{2} + feU\frac{\partial}{\partial t}(J_0 A_\|)\right) \tag{6.128}$$

even though Alfvénic processes are arranged with E_M and thermal ones with E_f.

The parallel transfer between E_E and E_M is the Alfvénic transfer from $feU\nabla_\|\phi_E$, part of which is the nonlinear magnetic flutter effect in $\nabla_\|$ due to $[-A_\|,\ ]$, just as in the fluid drift model. The perpendicular transfer is through the curvature drift from $mU^2\mathbf{v}_C$ giving the parallel energy part of ExB compression, due to $\mathbf{v}_C\cdot\nabla\phi$ plus corrections. The parallel Mach correction is neglected in a delta-f model but would enter in a total-f one, prominently for sonic flows (total-f models will be discussed again in Chapter 16).

So all these terms give the generalisations of the fluid drift energy transfer effects—Alfvénic, ExB compressional, parallel flow, diamagnetic, etc—due to the four main drifts in Eq. (6.106), to the gyrokinetic setting. As such, the exact energetic consistency renders them valid at any level of approximation.

The foregoing is a sketch of the structure of the energy theorem. The interested reader can flesh this out with the details, culminating in a set of equations starting with the one for which $e\phi_E(\partial f/\partial t)$ is the left hand side, with divergence terms on the left hand side and transfer terms on the right in each equation.

6.10 Simplified geometry and the form of the Jacobian

It is often asked whether $\mathbf{B}^*$ or $B_\|^*$ can be replaced by $\mathbf{B}$ or B, respectively. One sees that if the same expression for $B_\|^*$ is used everywhere, then there is no damage to energy conservation. The same is true for $\mathbf{b}$ and B. These descend from

$$B_\|^* = \mathbf{B}^*\cdot\frac{\partial\mathbf{p}^*}{\partial z}, \qquad \mathbf{b} = \frac{\partial\mathbf{p}^*}{\partial z} \tag{6.129}$$

and the $M\Omega$ term in H, respectively. To obtain momentum conservation, and phase space conservation in general, then

$$\frac{1}{e}\nabla\times\frac{\partial\mathbf{p}^*}{\partial z}+\frac{\partial\mathbf{B}^*}{\partial z}=0 \tag{6.130}$$

must be maintained, which constrains the freedom of $\mathbf{B}^*$ and $B_\parallel^*$ further. For this reason $\mathbf{B}^*$ cannot be replaced by $\mathbf{B}$ unless the geometry and coordinate representation yields it exactly. Only if the representation has something like

$$\mathbf{p}^*=e\mathbf{A}+z\nabla\xi \tag{6.131}$$

can $\mathbf{B}^*$ be replaced by $\mathbf{B}$.

The subtlety with $B_\parallel^*$ involves the unit vector, $\mathbf{b}$ itself following from $\partial\mathbf{p}^*/\partial z$, according to

$$\mathbf{B}^*\cdot\mathbf{b}=B_\parallel^*=B+\frac{z}{e}\mathbf{b}\cdot\nabla\times\mathbf{b}=B\quad\text{if and only if}\quad\mathbf{b}\cdot\nabla\times\mathbf{b}=0 \tag{6.132}$$

This relation indicates that we can use B for $B_\parallel^*$ only if the model chosen for the geometry allows it exactly, namely, if $\mathbf{b}$ is approximated in L_p such that $\mathbf{b}\cdot\nabla\times\mathbf{b}$ vanishes identically.

In a tokamak like geometry, the model for the curvature and grad-B drifts often approximates the curvature according to

$$\mathbf{b}\cdot\nabla\mathbf{b}\to-\nabla\log R,\qquad\nabla\log B\to-\nabla\log R \tag{6.133}$$

This can be done in the Lagrangian by assuming

$$\mathbf{b}\to R\nabla\varphi\qquad\text{via}\qquad\mathbf{p}^*=e\mathbf{A}+zR\nabla\varphi \tag{6.134}$$

while keeping $\mathbf{B}=I\nabla\varphi+\nabla\psi\times\nabla\varphi$, whereupon

$$\mathbf{B}^*=\mathbf{B}+\frac{z}{e}\nabla R\times\nabla\varphi\qquad B_\parallel^*=R\nabla\varphi\cdot\mathbf{B}=\frac{I}{R} \tag{6.135}$$

since $\nabla\varphi\cdot(\nabla R\times\nabla\varphi)$ vanishes. So if the magnitude B is also identified with I/R, we would have

$$B_\parallel^*=B=\frac{I}{R} \tag{6.136}$$

If further zR is re-defined as the parallel momentum coordinate, then $\mathbf{B}^*$ could be replaced by $\mathbf{B}$. All of this assumes that magnetic field is mostly toroidal, with $I\nabla\varphi$ the dominant component. It is a simplification whose validity is specific to magnetic fields which are mostly toroidal and also axisymmetric, or in straight field line models with $\mathbf{b}=\nabla z$, in which there is no curvature anyway.

One common mistake occurring in some older literature was to use the basic form of the curvature drift, with $\nabla\times\mathbf{b}$ replaced by $-\mathbf{b}\times(\mathbf{b}\cdot\nabla\mathbf{b})$ in $\mathbf{B}^*$. This will not conserve anything unless one really has a compatible form of the drifts, but then one notes that $-\mathbf{b} \times (\mathbf{b} \cdot \nabla\mathbf{b})$ is not a proper curl, so phase space conservation is broken. Similar problems occur if $B_\parallel^{*-1}$ were to be expanded such that

$$\frac{1}{B_\parallel^*} \rightarrow \frac{1}{B}\left(1 - \frac{z}{eB}\nabla \times \mathbf{b}\right) \tag{6.137}$$

This would also break the necessary phase space properties.

Further Reading

For background references, see Further Reading of Chapter 4.

For the gyrokinetic references (Morozov and Solov'ev 1966, Lee 1983, and Hahm 1988, and other ones), see Further Reading of Chapter 5.

Energetic consistency and momentum conservation in the gyrokinetic description of tokamak plasmas was given in a paper of that title by B Scott and J Smirnov, *Physics of Plasmas* **17** (2010) 112302. This contains the polarisation cancellation and emphasises how charge conservation and momentum conservation have to go together, and that the time-dependent polarisation current has to be kept as part of the mix, for any parameter situation, as otherwise consistency is violated.

IOP Publishing

Turbulence and Instabilities in Magnetised Plasmas, Volume 2

Gyrokinetic theory and gyrofluid turbulence

Bruce Scott

Chapter 7

Gyrokinetic theory for local dynamics

7.1 Ideas behind delta-f gyrokinetics

The usefulness in a delta-f model, such as the fluid examples we have been using have shown, is that even relatively violent nonlinearity can be studied in the absence of strong evolution of the background, on the one hand the gradients but especially on the other hand the magnetic field. Even when the numerical program is able to handle the latter task, the diagnostics with which we understand the concepts of the underlying nonlinear physics are as clean as we have seen them only when this part of the physics is isolated from whole-system evolution. Therefore, we would like to do the same with a delta-f version of gyrokinetic theory, and in the past that is the system concept for which the theory was first presented. On the other hand, as in the fluid cases, older models were usually not completely energetically consistent. However, studies of turbulence require exact energetic consistency for statistical clarity, namely, the ability to produce arbitrarily long runs without systematic wander in the results. The situation for fluid models was established early on, with some of the first papers including clear energy theorems for the cases they were studying. In gyrokinetic theory this has taken more time and exact energetic consistency in these models is much more recent. We saw some of the issues in the general case in the previous Chapter. Now we show how the same level of rigor is maintained for the delta-f case after the conservation of total energy is replaced by the conservation of local free energy.

In the following, we summarise the total-f system for the Lagrangian actually used to produce the delta-f model. The energy theorem is re-established, as one must do for every new version of the model according to what is kept and what is omitted. The first step towards the delta-f model is the linearisation of the part of the Hamiltonian H which is second order in the field quantities. Here, f is replaced by a background F_0 in those terms, turning them into field energy quantities belonging to $\mathcal{L}_f$, the pure-field component of the Lagrangian. The energy theorem remains the same with minor re-labelling. Then, the thermodynamic free energy is established,

doi:10.1088/978-0-7503-3855-4ch7

and its second variation defines the local free energy. The concept of free energy in delta-f gyrokinetics therefore becomes grounded within the framework of thermodynamics in general physics. Then, the steps in the delta-f ordering are applied to get the delta-f equation. The free energy equation is first sketched and then derived, and conditions it sets on what has to be kept or discarded to maintain exact energetic consistency are established. Some issues concerning the use or non-use of gyroaveraging on the magnetic potential variable and the fate of momentum conservation in the models are addressed.

This is basically the same set of steps used on the fluid models from Chapter 6, using drift ordering to produce the local fluid drift equations in Chapter 7, both from Volume One, now applied to gyrokinetic theory. In the next Chapter, the wave responses and the drift-Alfvén instability will be covered. These are the gyrokinetic counterparts of the same basic responses we did for the fluid model in Chapters 8 and 9 of Volume One. In the gyrokinetic version, the kinetic dissipation mechanism called Landau damping emerges naturally, and we can see the similarities and differences to the resistive damping in the fluid model. We will see the resulting gyrofluid equations and the proof of correspondence to fluid drift equations in later Chapters.

7.2 Total-f Lagrangian and energetics

Since the turbulence in a magnetised plasma is in a regime defined by the low frequency relative to compressional Alfvén ($k_\perp v_A \gg \omega$) and ion cyclotron motion ($\Omega_i \gg \omega$), we restrict the more general Lagrangian of the last Chapter to this situation. This pertains to the gyrocentre Lagrangian L_p, since the field Lagrangian has already been pared down by the assumption of quasineutrality ($\varepsilon_0 \ll \rho/B^2$) and the neglect of $\mathbf{A}_\perp$ due to the above frequency ordering. Lastly, we also have the low-beta ($\beta_e = \mu_0 p_e/B^2 \ll 1$) situation, which will mean that $(z/m)A_\parallel$ is neglected against ϕ in the polarisation terms.

We split the Hamiltonian into three pieces according to the order in which the field variables appear:

$$H = H_0 + H_1 + H_2 \tag{7.1}$$

where

$$H_0 = \frac{z^2}{2m} + M\Omega \tag{7.2}$$

is the lowest order H which will appear in the equilibrium,

$$H_1 = e\langle\Phi\rangle, \qquad \Phi = \phi - \frac{z}{m}A_\parallel \tag{7.3}$$

is the part of H which accounts for the main field terms in the equations of motion and sources the field variables in their equations, and

$$H_2 = \frac{e^2}{2m}\langle A^2\rangle - \frac{e^2}{2\Omega}\frac{\partial}{\partial M}\langle\tilde{\Phi}^2\rangle \tag{7.4}$$

is the part of H which together with the second-order field terms in $\mathcal{L}_f$, in this case

$$\mathcal{L}_f = -\frac{1}{2\mu_0}\left|\nabla_\perp A_\parallel\right|^2 \tag{7.5}$$

control the responses of the field variables to their sources. Varying the first-order field terms given by H_1 leaves only expressions involving f, which serve as sources in the field equations. Varying H_2 and $\mathcal{L}_f$ yield expressions which are linear in the field variables, controlling their responses.

The part of H which controls polarisation is the second term in H_2, which contains ϕ to second order. The approximation step for the low-beta case is to neglect $A_\parallel$ in this expression, making polarisation electrostatic in the same way $\mathbf{E}_\perp = -\nabla_\perp\phi$ is. There are two statements. First, the size of the second order terms due to $A_\parallel$ in H_2 is small by beta compared to the magnetic energy due to $A_\parallel$:

$$\sum_{sp}\int d\Lambda f\,\frac{m}{2B^2}\frac{z^2}{m^2}\left|\nabla_\perp A_\parallel\right|^2 \sim \frac{\mu_0 p_e}{B^2}\int d\mathcal{V}\,\frac{1}{2\mu_0}\left|\nabla_\perp A_\parallel\right|^2 \tag{7.6}$$

noting that mU^2 scales like temperature, which is comparable for ions and electrons. The multiplier is simply

$$\frac{\mu_0 p_e}{B^2} = \beta_e \ll 1 \tag{7.7}$$

whose smallness defines the low-beta case. In many situations, this is roughly the same measure by which the plasma is quasineutral ($\beta_e \sim \epsilon_0 B^2/\rho$), so it is reasonable to neglect these contributions. The second statement is that under the perpendicular derivative, ϕ is at least as large as $v_A A_\parallel$ since the equipartition lesson in Chapters 3 and 5 from Volume One showed that ExB energy is always as large as or larger than perturbed magnetic energy in the turbulence. So we have

$$\rho v_E^2 > \frac{1}{\mu_0}B_\perp^2 \implies |\nabla_\perp\phi|^2 > v_A^2\left|\nabla_\perp A_\parallel\right|^2 \tag{7.8}$$

and therefore in the polarisation expression with $z/m \sim c_s$

$$|\nabla_\perp\phi|^2 > \frac{v_A^2}{c_s^2}\frac{z^2}{m^2}\left|\nabla_\perp A_\parallel\right|^2 = \beta_e^{-1}\frac{z^2}{m^2}\left|\nabla_\perp A_\parallel\right|^2 \tag{7.9}$$

(for the term linear in $A_\parallel$ the ions are dominant; for the $A_\parallel^2$ term they are comparable, with z^2/m scaling like temperature as above). In both cases, the $\nabla_\perp\phi$ term is at least a factor of $\beta_e^{-1/2}$ larger than the $(z/m)\nabla_\perp A_\parallel$ term, so that the latter can be neglected in the polarisation expressions. Hereafter, Φ is replaced by ϕ therein.

We then go to the continuous form of the gyroaveraging operator, as done in Chapter 5. The forms of H_1 and H_2 are replaced by

$$H_1 = e\,J_0(\Phi), \qquad H_2 = \frac{e^2}{2m}(J_0A)^2 - \frac{e^2}{2\Omega}\frac{\partial}{\partial M}\left(\phi^2 - (J_0\phi)^2\right) \tag{7.10}$$

under the low-beta approximations just given. The action of J_0 as an operator on a scalar quantity represents

$$J_0(\phi) = \sum_{\mathbf{k}} \phi_{\mathbf{k}}\, e^{i\mathbf{k}\cdot\mathbf{x}} J_0(k_\perp \rho_L) \qquad \rho_L = \frac{2M}{m\Omega} \tag{7.11}$$

where in the continuous form we use $\mathbf{x}$ as the coordinate, i.e., the operation is taken in the continuous spatial domain. In the variations and energetic analysis, we need only use the property that this J_0, no matter how it is implemented computationally, is an Hermitian operator, which means that

$$\int d\Lambda\, f(J_0\phi) = \int d\Lambda\; \phi(J_0 f) \tag{7.12}$$

since all quantities are now defined on the same space. Note that J_0 as an operator can be written either as $J_0(\phi)$ or as $(J_0\phi)$, whichever is most clearly readable. The J_0 acting on the square of ϕ is dropped since the quantities to the left of it are mostly due to the equilibrium. This gives the implementation of the gyroaveraging operators in the continuous case.

With the above considerations we start with the Lagrangian

$$L = \sum_{sp} \int d\Lambda f\; L_p + \int d\mathcal{V}\, \mathcal{L}_f \tag{7.13}$$

where the gyrocentre Lagrangian is

$$L_p = e\mathbf{A}^* \cdot \dot{\mathbf{R}} + M\dot{\vartheta} - H \tag{7.14}$$

The free-field Lagrangians is given in Eq. (7.5). The Hamiltonian is now given by

$$H = \frac{m}{2}U^2 + M\Omega + e\phi_E \tag{7.15}$$

where

$$mU = z - eJ_0(A_\parallel), \qquad \phi_E = J_0(\phi) - \frac{e}{2\Omega}\frac{\partial}{\partial M}(\phi^2 - (J_0\phi)^2) \tag{7.16}$$

This completes the description of the system Lagrangian.

The equation for f using Liouville's theorem and the gyrocentre Euler–Lagrange equations for $\mathbf{R}$ and z is

$$B_\parallel^* \frac{\partial f}{\partial t} + \frac{1}{e}\mathbf{b} \times \nabla H \cdot \nabla f + \mathbf{B}^* \cdot \left(\frac{\partial H}{\partial z}\nabla f - \frac{\partial f}{\partial z}\nabla H\right) = 0 \tag{7.17}$$

where the background magnetic field quantities are

$$\mathbf{B}^* = \nabla \times \mathbf{A}^*, \qquad \mathbf{b} = e\frac{\partial \mathbf{A}^*}{\partial z}, \qquad B_\parallel^* = \mathbf{b} \cdot \mathbf{B}^* \tag{7.18}$$

with the second expression giving the **b** actually used (i.e., in this case **B** and **b** and B in Ω are kept separate for consistency purposes). The equation for ϕ using the functional derivative condition $\delta L/\delta\phi = 0$ is

$$\sum_{sp} \int d\mathcal{W}\left(eJ_0(f) + J_0\mathcal{M}J_0(\phi) - \mathcal{M}\phi\right) = 0 \tag{7.19}$$

where

$$\mathcal{M} = -\frac{e^2}{\Omega}\frac{\partial f}{\partial M} \tag{7.20}$$

defines the polarisability of each species. The equation for $A_\parallel$ using the functional derivative condition $\delta L/\delta A_\parallel = 0$ is

$$-\nabla_\perp^2 A_\parallel = \mu_0 \sum_{sp} \int d\mathcal{W}\, J_0(efU) \tag{7.21}$$

In this context, L is a functional since it takes functions of the dependent variables and maps them to numbers.

The conserved energy is given by the Noether theorem for time translation symmetry. With L set up such that all the dependent variables are in the time component, the Noether energy is given by

$$\mathcal{E} = \sum_{sp} \int d\Lambda f\, H - \int d\mathcal{V}\, \mathcal{L}_f \tag{7.22}$$

which is the negative of this time component. Using the equations for the field variables, it can be split into pieces according to

$$\mathcal{E} = E_f + E_E + E_B \tag{7.23}$$

where the thermal/kinetic, ExB, and magnetic energies are, respectively, given by

$$E_f = \sum_{sp} \int d\Lambda f\left(\frac{m}{2}U^2 + M\Omega\right) \tag{7.24}$$

$$E_E = \sum_{sp} \int d\Lambda f \frac{e}{2\Omega}\frac{\partial}{\partial M}(\phi^2 - (J_0\phi)^2) \tag{7.25}$$

$$E_B = \int d\mathcal{V} \frac{1}{2\mu_0} \left|\nabla_\perp A_\parallel\right|^2 \tag{7.26}$$

Note that E_f is not defined with H_0 but with the complete expression for parallel and perpendicular kinetic energy, since U is the actual parallel velocity as a function of the coordinates and the dependent variable $A_{\|}$. This summarises the results of the previous Chapter applied to this parameter regime.

7.3 Linearised polarisation

A version of the equations which is still total-f but yields more manageable equations makes the additional approximation that, although the distribution function f is left general in other respects (including carrying the equilibrium current in addition to representing the parameter gradients), the terms in H which are second-order in the field quantities are replaced by field terms having the same role. That is, the f multiplying H_2 in L_p is replaced by a background F_0, usually a Maxwellian parameterised by n_0 and T_0, the density and temperature of that species, which may or may not be spatially dependent. This approximation is called 'linearised polarisation,' from its original application in electrostatic models. For example, if we start from the simplified electrostatic Lagrangian,

$$L = \sum_{sp} \int d\Lambda f\, L_p \qquad \text{with} \quad H = \frac{z^2}{2m} + M\Omega + e\phi - \frac{m}{2B^2}\,|\nabla_{\perp}\phi|^2 \tag{7.27}$$

the second order term has f replaced by F_0, so that

$$L = \sum_{sp} \int d\Lambda f\, L_p + \int d\mathcal{V}\, \mathcal{L}_f \tag{7.28}$$

$$\text{with} \quad H = \frac{z^2}{2m} + M\Omega + e\phi \quad \text{and} \quad \mathcal{L}_f = \frac{\rho}{2B^2}\,|\nabla_{\perp}\phi|^2 \tag{7.29}$$

where the mass density ρ is a parameter which may be constant or spatially dependent across the magnetic field. This is how it has been applied in most computations. The only change in the field equation for ϕ is to have

$$\sum_{sp} \int d\mathcal{W} fm \qquad \text{replaced by} \quad \rho \tag{7.30}$$

in the polarisation term, with ρ parameterised instead of arising from the dependent variables. The only change in the conserved energy is to have the fH_2 terms replaced by corresponding forms in $\mathcal{L}_f$ with $f \rightarrow F_0$ so that the ExB energy has the same replacement as in the polarisation term just stated above. All the remaining energetics is the same.

Doing this approximation in our Lagrangian, and using constant values for the n_0 and T_0 in F_0 (since we will be dealing with local equations), we have the same as before except with

$$H = H_0 + H_1 \tag{7.31}$$

in the gyrocentre Lagrangian, and

$$\mathcal{L}_f = \sum_{sp} \int d\mathcal{W} \frac{1}{2} \mathcal{M}\left(\phi^2 - (J_0\phi)^2\right) - \sum_{sp} \int d\mathcal{W} \, F_0 \frac{\mu_0 e^2}{2m} (J_0 A_\|)^2 - \frac{1}{2\mu_0} \left|\nabla_\perp A_\|\right|^2 \quad (7.32)$$

where

$$\mathcal{M} = F_0 \frac{e^2}{T_0} \quad (7.33)$$

in the field Lagrangian. The polarisability is the same as before except for the replacement $f \rightarrow F_0$ in the corresponding term in the Lagrangian. The first term in $\mathcal{L}_f$ is the finite gyroradius finite Larmor radius (FLR) generalisation of the ExB energy,

$$\lim_{k_\perp \rightarrow 0} \sum_{sp} \int d\mathcal{W} \frac{1}{2} \mathcal{M}\left(\phi^2 - (J_0\phi)^2\right) = \frac{\rho}{2B^2} |\nabla_\perp \phi|^2 \quad (7.34)$$

The second term in the field Lagrangian density arises from the term in U^2 due to $(J_0 A_\|)^2$,

$$\int d\mathcal{W} \, F_0 \frac{m}{2} U_A^2 = \int d\mathcal{W} \, F_0 \frac{m}{2} \frac{e^2}{m^2} (J_0 A_\|)^2 = \frac{e^2}{2m} \int d\mathcal{W} \, F_0 (J_0 A_\|)^2 \quad (7.35)$$

Each species contributes, though the electrons do the most due to the small mass ratio.

7.3.1 Conserved energy

Although the structure with a completed square for U^2 is broken, there is no real effect on energetic consistency, so long as the $(J_0 A_\|)^2$ term is arranged with z^2 and $z J_0(A_\|)$ to form the parallel thermal/kinetic energy. The conserved energy is

$$\mathcal{E} = E_f + E_E + E_B \quad (7.36)$$

as before, where the pieces are now given by

$$E_f = \sum_{sp} \int d\Lambda \left[\frac{z^2}{2m} - e\frac{z}{m} (J_0 A_\|) + M\Omega \right] f + \sum_{sp} \int d\Lambda \, \frac{e^2}{2m} F_0 (J_0 A_\|)^2 \quad (7.37)$$

for the thermal/kinetic energy

$$E_E = \sum_{sp} \int d\Lambda \, \frac{1}{2} \mathcal{M}\left(\phi^2 - (J_0\phi)^2\right) \quad (7.38)$$

for the ExB energy, and

$$E_B = \int d\mathcal{V} \frac{1}{2\mu_0} \left|\nabla_\perp A_\|\right|^2 \quad (7.39)$$

for the magnetic energy. Once more, the operations are those necessary to convert Hf into these quantities using the field equations (that is, it is still the usual combination involving Hf and $\mathcal{L}_f$, which is general). The ExB energy is still given in terms of a sum over species, consistent with the form of L written above, but in many applications only one ion species is considered, and the electron contribution to E_E is neglected.

7.3.2 Field equations

The field equations are found in the usual manner, taking the functional derivative of the Lagrangian with respect to each field quantity (ϕ and $A_\|$), and setting it to zero. Since the fields appear only in the time component, only the terms due to fH and $\mathcal{L}_f$ are involved. The polarisation equation for ϕ is Eq. (7.19) with $\mathcal{M}$ linearised as in Eq. (7.33), or

$$\sum_{sp} \int d\mathcal{W}\left[e(J_0 f) + J_0\mathcal{M}(J_0\phi) - \mathcal{M}\phi\right] = 0, \qquad \mathcal{M} = F_0\frac{e^2}{T_0} \tag{7.40}$$

The induction equation for $A_\|$ is Eq. (7.21) with the factor of f multiplying the $A_\|$ term in U replaced by F_0 (see Eq. 7.16), is

$$\sum_{sp} \int d\mathcal{W}\left[J_0\frac{\mu_0 e^2}{m}F_0(J_0 A_\|)\right] - \nabla_\perp^2 A_\| = \mu_0 \sum_{sp} \int d\mathcal{W}\, e\frac{z}{m}(J_0 f) \tag{7.41}$$

where as usual the factor of μ_0 has been multiplied through after calculating the functional derivative of L with respect to $A_\|$. The quantities multiplying and acting upon $A_\|$ in the first term may be combined into an operator K_d^2 in

$$\left(K_d^2 - \nabla_\perp^2\right)A_\| = \mu_0 \sum_{sp} \int d\mathcal{W}\, e\frac{z}{m}(J_0 f) \tag{7.42}$$

where

$$K_d^2(A_\|) = \sum_{sp} \int d\mathcal{W}\left[J_0\frac{\mu_0 e^2}{m}F_0\, J_0\right](A_\|) \tag{7.43}$$

has the role of a skin depth operator. Note that the J_0 acts twice, with the F_0 between them if it is spatially dependent. In the case that $A_\|$ is not gyroaveraged because most of it comes from the electrons (see Chapter 9), this operator reduces to

$$K_d^2 = \sum_{sp} \int d\mathcal{W}\,\frac{\mu_0 e^2}{m}F_0 = d_e^{-2} \tag{7.44}$$

the inverse-square of the skin depth. There is an issue of consistency as f is the only dependent variable and should be treated the same way everywhere. Here, we have slight differences in how we represent it in the field sources of the equations for polarisation (only f appears) and induction (zf and $F_0A_\parallel$ appear together). Nevertheless, this system remains exactly consistent energetically.

In both field equations, we leave the integration over $J_0F_0J_0$ as it is rather than evaluating it analytically. This produces better computations.

7.4 The free energy

The concept of free energy as opposed to total energy, that is, the portion of energy which is accessible to the dynamics, is taken from thermodynamics. In general, the self-consistent system described by the fields and particles has an energy, given by the Noether theorem, and a set of Casimir Invariants, which in our case are functionals of f which are conserved by the dynamics.

The total-f gyrokinetic equation may be written schematically as

$$\frac{\partial f}{\partial t} + \{H, f\} = 0 \tag{7.45}$$

where the combined Poisson bracket stands for the combination of terms in Eq. (7.17). For this purpose we need only know that the phase space integral over the combined bracket vanishes, namely the divergence property. A Casimir Invariant is defined as any functional whose Lie–Poisson functional bracket with any functional of the dependent variables vanishes[1]. In our case it is a functional of f whose density is a function of f and is therefore conserved according to Eq. (7.45). In the absence of collisions, any function of f can be used as the density of a Casimir, since

$$C = \int d\Lambda\, G(f), \qquad \frac{\partial C}{\partial t} = \int d\Lambda\, G'(f)\{f, H\} = 0 \tag{7.46}$$

So the combination

$$\mathcal{U} = \mathcal{E} - \sum_i \lambda_i C_i \tag{7.47}$$

defining a free energy in terms of the Noether energy $\mathcal{E}$ and a set of Casimirs can be used to set up a general minimisation principle.

In our case, we are interested in fluctuations on a background, under conditions where the background evolution is slow enough such that it is stationary over the statistical equilibration time of the turbulence, which in turn is several saturation times. In conventional transport analysis of confined plasmas, one has an equilibrated background on the scale of the ion collision time, which is assumed to be short compared to the transport time but long compared to any dynamical relaxation time. Hence, all the disequilibrium can be represented by the dependent variables in the system. Then, it is physically reasonable to use the Boltzmann entropy for the Casimirs (one separately for each species, possibly with different temperatures but usually with the same one). In practice, the ion collision time is

[1] Lie–Poisson functional brackets are treated in Chapter 17. See also the references in Further Reading.

longer than the turbulence saturation time, and the transport time scale is much longer. Even for reasonable situations with steep gradients near the edge of a confined plasma, the transport time is long enough that time scale separation between equilibration and the relaxation of gradients in the absence of sources can be demonstrated.

Using the Boltzmann entropy, the combination

$$\mathcal{U} = E - TS \tag{7.48}$$

where the letters stand for free energy, total energy, system temperature and entropy, gives the relation of free energy to energy. In our case we can allow the system temperature to be different among species, so we replace the TS term with a sum over species of the same thing. For the energy we use the Noether energy

$$\mathcal{E} = \sum_{sp} \int d\Lambda f\, H - \int d\mathcal{V}\, \mathcal{L}_f \tag{7.49}$$

and for the entropy we use the Boltzmann entropy for a kinetic system

$$TS = -\sum_{sp} T_0 \int d\Lambda f\, \log f \tag{7.50}$$

where since temperature is in units of energy the Boltzmann constant is incorporated into the temperature T_0 which then just multiplies $f\,\log f$. This entropy piece is separate for each species, with separate temperatures.

The extremum of the free energy gives the equilibrium state. There are expected to be no gradients in this state, so anywhere a species density n_0 and temperature T_0 appear, they are taken to be constant. The idea is that the rest state is an homogeneous equilibrium at a set of constant parameters, and also at rest, so that gradients, flows, and currents are all part of the disturbances, i.e., departures from equilibrium. Since $\mathcal{E}$ is the negative of the time component of the Lagrangian, which is set up in canonical form, and S does not involve the field quantities, variation of $\mathcal{U}$ with respect to the field quantities yields the same field equations as we already have. So with no loss of generality we can set the field quantities to zero in equilibrium, consistent with (1) a state at rest with no flows or currents, (2) zero gyrocentre charge density hence also vorticity, hence also ϕ, and (3) homogeneous n_0 and T_0 for each species, hence also no forces. Only f has any structure, and that is in velocity space. So we take the variation of $\mathcal{U}$ with respect to the set of f's, and set it to zero. In anticipation of the end result, since the representation is in canonical form, we will label the variation of f as δg, not δf, forming

$$f \rightarrow F_0 + \delta g \tag{7.51}$$

We will see why this is, shortly.

In taking the first variation we keep only terms linear in the variations. The δH terms form the field equations and do not contribute. In other words, since $\mathcal{E}$ is up to a sign the entire time component of L, which has canonical form, the first variations with respect to ϕ and $A_\parallel$ are zero. This leaves the part due to the variation of f directly, which is given by

$$\delta\mathcal{U} = \sum_{sp} \int d\Lambda \left[H + T_0(1 + \log F_0) \right] \delta g \tag{7.52}$$

Setting $\delta\mathcal{U} = 0$ we find

$$1 + \log F_0 = -H_0/T_0 \qquad \text{with} \quad H_0 = \frac{z^2}{2m} + M\Omega \tag{7.53}$$

as the condition for equilibrium $f \to F_0$ for each species, with $H \to H_0$, i.e., with the field potentials set to zero. To get the constant multiplier for F_0 we can easily go back and add a Lagrange multiplier term $\lambda(n - n_0)$ to the Lagrangian density for each species, where

$$n = \int d\mathcal{W} f \tag{7.54}$$

is the species density. This forces $n \to n_0$, and then we have

$$F_0 = a_0(n_0) \exp(-H_0/T_0) \qquad \text{such that} \quad \int d\mathcal{W}\, F_0 = n_0 \tag{7.55}$$

as the equilibrium state for each species. It is nothing more than a Maxwellian with constant parameters. Any form of $F_0 = F_0(H_0)$ will satisfy equilibrium in the equation for f, but the expression involving $f \log f$ in the entropy leads to the Maxwellian just as in the classical theory.

Now we take the second variation. This includes variations in the field quantities ϕ and $A_\parallel$ in addition to f, since the terms which are quadratic in the variations of the dependent variables form the second variation of the free energy. Note that with the linearised polarisation all the second-order field terms are in $\mathcal{L}_f$. Even if that were not done, however, the quadratic field-variable variation terms would still appear solely with F_0 as those with δg would be third order. So δg only appears with first-order variations in H in any case. One must watch out for two things. First, $f \log f$ contains a finite equilibrium piece, $F_0 \log F_0$, so it has to be expanded to second order.

$$f \log f \to F_0 \log F_0 + \frac{(\delta g)}{F_0} + \frac{1}{2}\frac{(\delta g)^2}{F_0^2} \tag{7.56}$$

which eventually gives the factor of 1/2 multiplying the square of (δg). Second, the Noether energy is linear in f but not in the field variables. In the quadratic field terms f therefore appears as F_0. But H is nonlinear in the field variables in general, and quadratic in our case. Expansion of the quadratic field terms follows from functional derivatives, but one obtains the factor of 1/2 in the same way as a simple Taylor expansion of a function. One has to use the functional derivative procedure, because of the spatial operators acting on the field variables. The field equations were already applied at first order (with the trivial solution). The second variation of $\phi P(\phi)$, where P is an operator independent of ϕ, is $(\delta\phi)P(\delta\phi)$. When the field equations are now applied, with the factor of $P(\delta\phi)$ being replaced by a simple expression involving (δg), this cancels half of the $(\delta g)(\delta H)$ term, not all of it.

This gives in our case

$$\delta^2\mathcal{U} = \sum_{sp} \int d\Lambda \left[(\delta H_1)(\delta g) + T_0 \frac{(\delta g)^2}{2F_0} \right] \tag{7.57}$$

$$+\sum_{sp} \int d\Lambda \frac{1}{2}\left[(\delta\phi)(\mathcal{M} - J_0\mathcal{M}J_0)(\delta\phi) + (\delta A_\parallel)\frac{1}{2\mu_0}(K_d^2 - \nabla_\perp^2)(\delta A_\parallel) \right] \tag{7.58}$$

as the second variation of the free energy. Integration by parts and the use of J_0 as Hermitian have been done. Applying the field equations on the last two terms (with $\delta\phi$ and $\delta A_\parallel$) cancels half of the first term involving $(\delta H_1)(\delta g)$. This then yields

$$\delta^2\mathcal{U} = \sum_{sp} \int d\Lambda \frac{1}{2}\left[(\delta g)(\delta H_1) + T_0 \frac{(\delta g)^2}{F_0} \right] \tag{7.59}$$

which combines into

$$\delta^2\mathcal{U} = \sum_{sp} \int d\Lambda \frac{T_0}{2F_0}(\delta g)(\delta h) \tag{7.60}$$

where (δh) is the nonadiabatic part given by

$$(\delta h) = (\delta g) + \frac{F_0}{T_0}(\delta H_1) \tag{7.61}$$

This expression results for any Lagrangian in canonical form given the choice of the equilibrium state as one with no flows and no excitation of the field variables. All dynamics arises from the departures from equilibrium, and the free energy describes how far from equilibrium the system has departed. This exercise shows that what we have been calling free energy is actually the second variation of the thermodynamic free energy away from equilibrium, which in a delta-f model means the same thing. Hence

$$E_T = \delta^2\mathcal{U} = \sum_{sp} \int d\Lambda \frac{T_0}{2F_0}(\delta g)(\delta h) \tag{7.62}$$

and we have the total free energy for this system.

Now we show that E_T can be transformed into a positive-definite quantity. This will explain why we have used δg instead of δf in the notation. Recall the relationship between δg in Eq. (7.51) and δh in Eq. (7.61). Then we shift the factor involving $\delta A_\parallel$ between them (from H_1) and re-define

$$\delta g = \delta f + \frac{e}{T_0}\frac{z}{m}F_0(J_0 A_\parallel), \qquad \delta h = \delta f + \frac{e}{T_0}F_0(J_0\phi) \tag{7.63}$$

so that E_T in Eq. (7.62) becomes

$$E_T = \sum_{sp} \int d\Lambda \, \frac{1}{2}\left[\frac{T_0}{F_0}(\delta f)^2 + e(J_0\phi)(\delta f) + e\frac{z}{m}(J_0A_\parallel)(\delta f) \right] \tag{7.64}$$

where the fourth piece due to $(J_0\phi)(J_0A_\parallel)$ cancels due to the antisymmetry of zF_0 in velocity space. Now we can again apply the field equations to these to eliminate δf in the last two terms.

In this case one must note that the field equations derived above apply to δg, not δf. While there is no distinction in the polarisation equation for ϕ, since the zF_0 term vanishes under velocity space integration, this matters for the induction equation for $A_\parallel$, due to the skin depth term. So we have

$$\sum_{sp} \int d\mathcal{W} \, eJ_0(\delta g) = \sum_{sp} \int d\mathcal{W} \, eJ_0(\delta f) \tag{7.65}$$

and hence Eq. (7.40) in terms of δg is still

$$\sum_{sp} \int d\mathcal{W} \, (\mathcal{M} - J_0\mathcal{M}J_0)\phi = \sum_{sp} \int d\mathcal{W} \, e(J_0\delta g) \tag{7.66}$$

for polarisation. But in the induction equation we have

$$\sum_{sp} \int d\mathcal{W} \, e\frac{z}{m}J_0(\delta g) = K_d^2 A_\parallel + \sum_{sp} \int d\mathcal{W} \, e\frac{z}{m}J_0(\delta f) \tag{7.67}$$

and Eq. (7.42) in terms of δg is

$$(K_d^2 - \nabla_\perp^2)A_\parallel = \mu_0 \sum_{sp} \int d\mathcal{W} \, e\frac{z}{m}J_0(\delta g) \tag{7.68}$$

with K_d given in Eq. (7.43) with the J_0 factors. Hence, with the right hand side in terms of δf, the K_d^2 term cancels the same term on the left hand side, leaving the simple Ampère's law

$$-\nabla_\perp^2 A_\parallel = \mu_0 \sum_{sp} \int d\mathcal{W} \, e\frac{z}{m}J_0(\delta f) \tag{7.69}$$

for induction. This is the 'Ampère cancellation' which is especially significant for computations at large scale ($\gg d_e$).

Using these field equations to eliminate $(ez/m)(\delta f)$ and $eJ_0(\delta f)$ in the last two terms of Eq. (7.64), in which the Hermitian property of J_0 is used to integrate it by parts, we obtain

$$E_T = E_f + E_E + E_B \tag{7.70}$$

where the pieces are re-defined in the delta-f setting as

$$E_f = \sum_{sp} \int d\Lambda \, \frac{1}{2}\frac{T_0}{F_0}(\delta f)^2 \tag{7.71}$$

$$E_E = \sum_{sp} \int d\Lambda \, \frac{1}{2} \mathcal{M}\left(\phi^2 - (J_0\phi)^2\right) \tag{7.72}$$

$$E_B = \int d\mathcal{V} \, \frac{1}{2\mu_0} \left|\nabla_\perp A_\parallel\right|^2 \tag{7.73}$$

for the thermal/kinetic, ExB, and magnetic free energy pieces, respectively (as before, in a computation these are understood as volume averages). This demonstrates E_T as a positive definite quantity. It also shows that the extremal value of $\mathcal{U}$, having set the first variation $\delta\mathcal{U}$ to zero, is indeed a minimum. So we have alternatively minimised the energy for a given entropy, or maximised the entropy for a given energy. This is what minimising the free energy for an equilibrium does. The departure of free energy away from equilibrium is E_T, which is always positive unless all dependent variables are zero.

A similar result involves the time derivative. In a kinetic system E_T and its time derivative may be checked using the forms with δg and δh. But in an energetic analysis it is always useful to separate the three components, thermal/kinetic free energy, ExB energy, and magnetic energy, and the transfer effects between them. It is easiest to start with the positive-definite form of E_T using δf just given. The time derivative is taken to write

$$\begin{aligned} \frac{\partial E_T}{\partial t} &= \sum_{sp} \int d\Lambda \, \frac{T_0}{F_0}(\delta f)\frac{\partial(\delta f)}{\partial t} + \int d\mathcal{V} \, \frac{1}{\mu_0}(\nabla_\perp A_\parallel) \cdot \nabla_\perp \frac{\partial A_\parallel}{\partial t} \\ &+ \sum_{sp} \int d\Lambda \, \mathcal{M}\left(\phi\frac{\partial \phi}{\partial t} - (J_0\phi)\frac{\partial (J_0\phi)}{\partial t}\right) \end{aligned} \tag{7.74}$$

The Hermitian property of J_0 is used, $\nabla_\perp A_\parallel$ is integrated by parts to produce

$$\begin{aligned} \frac{\partial E_T}{\partial t} &= \sum_{sp} \int d\Lambda \, \frac{T_0}{F_0}(\delta f)\frac{\partial(\delta f)}{\partial t} - \int d\mathcal{V} \, \frac{1}{\mu_0}(\nabla_\perp^2 A_\parallel)\frac{\partial A_\parallel}{\partial t} \\ &+ \sum_{sp} \int d\Lambda \, \phi\left[\mathcal{M}\frac{\partial \phi}{\partial t} - \left(J_0 \mathcal{M} J_0 \frac{\partial \phi}{\partial t}\right)\right] \end{aligned} \tag{7.75}$$

The field equations are applied to eliminate the pieces involving $\partial\phi/\partial t$ and $\nabla_\perp^2 A_\parallel$, and then the Hermitian property of J_0 is used to arrange the terms such that J_0 always operates on a field quantity. This yields,

$$\frac{\partial E_T}{\partial t} = \sum_{sp} \int d\Lambda \left[\frac{T_0}{F_0}(\delta f)\frac{\partial(\delta f)}{\partial t} + e\frac{z}{m}(\delta f)\frac{\partial(J_0 A_\parallel)}{\partial t} + e(J_0\phi)\frac{\partial(\delta f)}{\partial t}\right] \tag{7.76}$$

which is equivalent to

$$\frac{\partial E_T}{\partial t} = \sum_{sp} \int d\Lambda \, \frac{T_0}{F_0}(\delta h)\frac{\partial(\delta g)}{\partial t} \tag{7.77}$$

since the additional term we need to add with both $(J_0\phi)$ and $(ez/m)F_0\partial(J_0A_\parallel)/\partial t$ vanishes under the integral due to antisymmetry. Any alternative among these forms can be used for diagnosis as needed. This is the gyrokinetic equivalent to the energetic analysis we have been doing up to now in all of the turbulence examples from the drift-fluid cases.

7.5 Sketch of the delta-f approach

The delta-f version of the gyrokinetic equation itself follows a hierarchical procedure which may be outlined in the abstract before we apply it to specific cases. Starting with the schematic total-f gyrokinetic equation in Eq. (7.45), we set the following separation between equilibrium and disturbance terms, as in the free energy discussion,

$$f \to F_0 + g, \qquad H = H_0 + H_1 \tag{7.78}$$

where we dispense with the δ symbols. The lowest order equation is

$$\frac{\partial F_0}{\partial t} + \{H_0, F_0\} = 0 \tag{7.79}$$

which is satisfied by having

$$\frac{\partial F_0}{\partial t} = 0, \qquad F_0 = F_0(H_0) \implies \{H_0, F_0\} = 0 \tag{7.80}$$

Any function satisfies this, but the free energy discussion motivates the Maxwellian since we assume a thermal plasma with an equilibrium state. The quantities H_0 and F_0 are as given in Eqs (7.53) and (7.55), respectively.

The next level has g satisfying

$$\frac{\partial g}{\partial t} + \{H_0, g\} + \{H_1, F_0\} = 0 \tag{7.81}$$

with H_1 given in Eq. (7.3). Since $F_0 = F_0(H_0)$ the second term combines with the first into

$$\frac{\partial g}{\partial t} + \{H_0, h\} = 0 \tag{7.82}$$

This is the linear equation, with g and h as given in Eq. (7.63). It satisfies the energy theorem only if the right hand side in

$$\int d\Lambda \frac{T_0}{F_0} h \frac{\partial g}{\partial t} + \frac{1}{2}\left\{H_0, \frac{T_0}{F_0} h^2\right\} = \int d\Lambda\, h^2 \left\{H_0, \frac{T_0}{F_0}\right\} \tag{7.83}$$

vanishes. With $F_0 = F_0(H_0)$ and T_0 constant, the linear terms satisfy this in the continuous limit.

In keeping with the nonlinear delta-f rules, with $g/F_0 \sim \Delta_\perp/L_\perp$, but $L_\perp\nabla \sim L_\perp/\Delta_\perp$ on a fluctuation, the nonlinear advection part of the bracket from the second order level is promoted by one order into this one, so that we have

$$\{H, f\} \to \{H_0, h\} + \frac{1}{e}[H_1, g] \tag{7.84}$$

since the equilibrium part of the bracket vanishes. Once this is set then the energy theorem has to be checked again. It depends on the vanishing of the right hand side in

$$\begin{aligned} &\int d\Lambda\left(\frac{T_0}{F_0}h\frac{\partial g}{\partial t} + \frac{1}{2}\left\{H_0, \frac{T_0}{F_0}h^2\right\} + \frac{1}{2B}\left[H_1, \frac{T_0}{F_0}h^2\right]\right) \\ &= \int d\Lambda\, h^2\left(\left\{H_0, \frac{T_0}{F_0}\right\} + \frac{1}{B}h^2\left[H_1, \frac{T_0}{F_0}\right]\right) \end{aligned} \tag{7.85}$$

which becomes the condition for energetic consistency to hold. Any treatment setting the pieces independently (geometry, Maxwellian F_0 or not, or approximations of derivatives) should be checked as to whether this is the case. It is important to note that global models with structure in F_0 such that the nonlinearity on it produces a nonzero $[H_1, F_0]$ will violate this. Moreover, the factor of B itself in the nonlinearity has to be replaced by B_0, or the coordinates have to be set such that [,] involves derivatives in xy only, while $B = B(s)$, as in the fluxtube fluid models. This situation cannot be remedied if F_0 has structure across magnetic field line flux surfaces. Later, a way will be found to remedy this for local gyrofluid models made global. But the procedure used there is not available to kinetic models, because while we can save the fluid situation by adjusting the treatment of the magnetic geometry, we cannot save the kinetic situation by doing similar things with the structure of F_0, which must simultaneously satisfy the equilibrium condition (Eq. 7.80) and the condition that both brackets on the right hand side of Eq. (7.85) vanish.

7.6 Systematics of the delta-f equations

As in the fluid fluxtube models we will set the geometry up such that the fluctuations vary in all three coordinates, xys, but the nonlinear spatial bracket only involves the first two, xy, while the magnetic geometry local to a reference flux surface varies only in the s-direction. This solves the problems of energetic consistency, since F_0 and any geometric quantity commute past the xy-derivatives, which after construction are the only ones present, except for the parallel phase space zs-bracket, which is linearised and therefore commutes via $F_0 = F_0(H_0)$ with the equilibrium. As long as the model is arranged to do this, energetic consistency is preserved and turbulence computations are viable. How the energetics works in detail, especially with the parallel phase space bracket, will be addressed in the next section.

Field aligning means

$$\mathbf{B}\cdot\nabla = B^i\nabla_i, \qquad B^x = B^y = 0, \qquad B^s \neq 0 \tag{7.86}$$

and the divergence condition $\nabla \cdot \mathbf{B} = 0$ means

$$\frac{\partial}{\partial s}\left(\sqrt{g}\, B^s\right) = \frac{\partial}{\partial s}\left(\sqrt{g}\, B b^s\right) = 0 \tag{7.87}$$

Hence, if B^s or B are given structure the coordinate element $\sqrt{g}$ must set consistently since $\nabla \cdot \mathbf{B} = 0$ means that the field lines trace out constant volumes. Due to the flute mode ordering $k_\parallel \ll k_\perp$ we have

$$\frac{\partial}{\partial x} \sim \frac{\partial}{\partial y} \gg \frac{\partial}{\partial s} \tag{7.88}$$

so that only xy derivatives appear in any of the drift terms in the final form of the model (cf. Chapters 12 and 14 of Volume One)

There is some preparation to go through first. The total-f gyrokinetic equation is given in Eq. (7.17). Setting up in terms of g and h, and H_0 and H_1 as in Eq. (7.84), we have

$$B_\parallel^* \frac{\partial g}{\partial t} + \frac{1}{e}\mathbf{b} \times \nabla H_1 \cdot \nabla g + \frac{1}{e}\mathbf{b} \times \nabla H_0 \cdot \nabla h + \mathbf{B}^* \cdot \left(\frac{\partial H_0}{\partial z}\nabla h - \frac{\partial h}{\partial z}\nabla H_0\right) = 0 \tag{7.89}$$

Under the flute mode ordering and also with $\rho_L / L_\perp$ ordered small, we simplify according to

$$B_\parallel^* \to B \qquad \mathbf{B}^* = \mathbf{B} + \frac{z}{e}\nabla \times \mathbf{b}, \qquad \mathbf{b} = \frac{\mathbf{B}}{B} \tag{7.90}$$

and divide through by B to write

$$\begin{aligned} &\frac{\partial g}{\partial t} + \frac{1}{eB}\mathbf{b} \times \nabla H_1 \cdot \nabla g + \mathbf{b} \cdot \left(\frac{\partial H_0}{\partial z}\nabla h - \frac{\partial h}{\partial z}\nabla H_0\right) \\ &\quad + \frac{1}{eB}\mathbf{b} \times \nabla H_0 \cdot \nabla h + \frac{z}{eB}\nabla \times \mathbf{b} \cdot \left(\frac{\partial H_0}{\partial z}\nabla h - \frac{\partial h}{\partial z}\nabla H_0\right) = 0 \end{aligned} \tag{7.91}$$

with the parallel phase space dynamics split between the parallel phase space bracket (top line) and the curvature drift bracket (bottom, next to the grad-B drift). In this model we have

$$\nabla \times \mathbf{b} \approx \nabla \frac{1}{B} \times B = \mathbf{b} \times \nabla \log B \tag{7.92}$$

so that $\nabla \times \mathbf{b} \cdot \nabla H_0$ vanishes. Finally, we apply $\partial H_0 / \partial z = z/m$ and $\mathbf{b} \cdot \nabla = b^s \partial / \partial s$, and replace $\nabla \log B$ with $-\nabla \log R$, to obtain

$$\begin{aligned} &\frac{\partial g}{\partial t} + \frac{1}{eB}\mathbf{b} \times \nabla H_1 \cdot \nabla g + b^s \left(\frac{\partial H_0}{\partial z}\frac{\partial h}{\partial s} - \frac{\partial h}{\partial z}\frac{\partial H_0}{\partial s}\right) \\ &\quad = \frac{z^2/m + M\Omega}{eB}\mathbf{b} \times \nabla \log R \cdot \nabla h \end{aligned} \tag{7.93}$$

We now have the nonlinearity, the parallel phase space bracket, and the grad-**B** drift and curvature drift terms which have combined into one. Note that the $e\phi$ part of H_1 in the first term gives rise to the ExB advection and the $A_{\|}$ part of it contributes to the parallel dynamics. It remains only to recast the terms involving the perpendicular drifts ($\mathbf{b} \times \nabla$ terms) such that energetic consistency is preserved.

In a straight magnetic field terms due to ∇B or $\mathbf{B}^* \neq \mathbf{B}$ are absent, $B_{\|}^*$ and B^s reduce to $B = B_0$, the spatial gradient of H_0 vanishes, all geometric quantities are independent of xy, and the operations become straightforward. In this case the equation for f is given with the spatial drifts bracket and the parallel phase space bracket,

$$\frac{\partial g}{\partial t} + \frac{1}{eB_0}[H_1, h]_{xy} + [H_0, h]_{zs} = 0 \tag{7.94}$$

with spatial coordinates xys in units of distance, z in units of momentum, and H_0, H_1, and $M\Omega$ (involved in integration and in J_0) in units of energy. In this case with a kinetic phase space a Poisson bracket is written with the coordinates given explicitly, e.g.,

$$[a, b]_{zs} = \frac{\partial a}{\partial z}\frac{\partial b}{\partial s} - \frac{\partial b}{\partial z}\frac{\partial a}{\partial s} \tag{7.95}$$

for scalars a and b and coordinates z and s.

In a magnetic field with structure, we have to calculate the same geometric coefficients encountered in the fluid models. These are the metric coefficients involved in $\nabla_\perp^2$, the magnetic field strength B, the magnetic contravariant coefficient b^s, and the coefficients in the curvature operator which are calculated from the derivatives of $\log R$ in terms of the coordinates and the other coefficients.

In field aligned coordinates we can always re-cast the magnetic field as

$$\mathbf{B} = \nabla\chi \times \nabla y = \chi_x \nabla x \times \nabla y \tag{7.96}$$

where $\chi = \chi(x)$ and $\chi_x = \partial\chi/\partial x$, since $\nabla \cdot \mathbf{B} = 0$ and B^x, B^y vanish. This is the Clebsch representation for the magnetic field. It then follows that

$$\frac{1}{B}\mathbf{b} \cdot \nabla a \times \nabla b = \frac{\chi_x}{B^2} |\nabla x \times \nabla y|^2 \, [a, b]_{xy} = \frac{1}{\chi_x}[a, b]_{xy} \tag{7.97}$$

using Eq. (7.96). The coefficient χ_x becomes constant with the locality in x and can be normalised to unity or $\delta = \rho_s/L_\perp$, the drift parameter.

For the curvature terms due to $\nabla \log B \to -\nabla \log R$ we cannot exploit the Clebsch representation since $\nabla \log R$ does not satisfy flute mode ordering. With a magnetic field of arbitrary structure one has to recast

$$\frac{1}{2}\mathcal{K} = \frac{1}{B}\mathbf{b} \times \nabla(\log R) \cdot \nabla - \to \nabla \cdot () \tag{7.98}$$

namely, as a pure divergence, and also follow the detailed structure of the coefficients according to the coordinates used in a global model, and then restrict to the local model with coordinates xys in an exact way. This is done in the references given below, in Further Reading. For our purposes it is sufficient to use the model described in Chapter 2, which is

$$\frac{1}{2}\mathcal{K} = \frac{1}{BR}\left(\sin s\frac{\partial}{\partial x} + (\cos s + S(s - s_k)\sin s)\frac{\partial}{\partial y}\right) \tag{7.99}$$

In normalised units $\omega_B = 2L_\perp/R_0$ is the coefficient giving the strength of the coupling. In the gyrokinetic model the factors of B and R are left with structure in s, since B serves as the velocity space volume element, and if B varies BR is still nearly constant in a tokamak. To keep $\mathcal{K}$ as a divergence we take $BR \to I_0$ here.

Operations involving Bessel functions or the perpendicular Laplacian all involve evaluating $k_\perp^2$ in xy space with coefficients dependent on the magnetic field strength $B = B(s)$ arising from the dependence of gyroradii upon B. We have

$$\frac{\rho}{B^2}k_\perp^2 \to -\nabla\cdot\frac{1}{B^2}\nabla_\perp \tag{7.100}$$

where in the delta-f model any other coefficient is constant. The perpendicular Laplacian in the Ampère's law involves the metric but not the factors of B.

We then have

$$\frac{\partial g}{\partial t} + \frac{1/e}{\chi_x}[H_1, h]_{xy} + b^s[H_0, h]_{zs} = \frac{z^2/m + M\Omega}{2e}\mathcal{K}(h) \tag{7.101}$$

as the delta-f gyrokinetic equation in an axisymmetric toroidal magnetic field under fluxtube ordering. It has to be axisymmetric because of the way y is used as a symmetry coordinate. The use of $\nabla(\log B + \log R) \approx 0$ is specific to tokamaks. In magnetic fields with different structure, especially those which are mostly poloidal, one has to re-do the curvature terms. One might use the zero-beta equilibrium version of $-\nabla\log B$ to do them instead of $\nabla\log R$. The Clebsch property used to derive the nonlinear drifts term with $[H_1, h]$ is valid in any axisymmetric geometry with flute mode ordering so long as $\rho_s \ll L_\perp$ is satisfied, as that is the extent of a local spectrum.

The simple toroidal model takes circular geometry with $B \to B_0$ except for the existence of the curvature terms. Study of gyrokinetic dynamics requires one more step of complexity, namely, the variation $B = B(s)$, which enters the parallel phase space bracket. In the delta-f model, it is possible to take $B = B(s)$, or not, independently in $k_\perp\rho_L$ as the argument of J_0 (Eq. 7.11), or in H_0, specifically $M\Omega$, in the parallel phase space bracket, or in the velocity space volume element. It is best to keep all three, or not, for consistency although one can vary them independently as control cases in a sensitivity study. In the simplest case, the metric coefficients from Chapter 2 can be modified such that

$$B = B_0(1 + \varepsilon\cos s)^{-1}, \qquad \sqrt{g} = r_0R_0(1 + \varepsilon\cos s), \qquad B^s = Bb^s = \frac{B_0}{qR_0} \tag{7.102}$$

with $\varepsilon = r_0/R_0$ the toroidal aspect ratio with its nominal value, which is sufficient to study the basic physics of the turbulence in a curved magnetic field with mirror trapping due to $\nabla_\parallel \log B$. For this purpose, g^{xx} and g^{yy} and $\mathcal{K}$ can remain unchanged.

7.6.1 Field equations

The field equations carry over from the linearised polarisation model, Eqs (7.40), (7.42), with g replacing the total f since we have already set up such that F_0 placed into the field equations yields zero for the field variables (we see again why to choose the n_0 so as to sum up to zero charge density). They belong solely to the delta-f disequilibrium component. So we have for polarisation,

$$\sum_{sp} \int d\mathcal{W}\left[e(J_0 g) + J_0\mathcal{M}(J_0\phi) - \mathcal{M}\phi\right] = 0, \qquad \mathcal{M} = F_0\frac{e^2}{T_0} \tag{7.103}$$

and for induction,

$$(K_d^2 - \nabla_\perp^2)A_\parallel = \mu_0 \sum_{sp} \int d\mathcal{W}\, e\frac{z}{m}(J_0 g) \tag{7.104}$$

which are the same as Eqs (7.66), (7.68). The field equations are relatively insensitive to whether the model is total-f or delta-f because of the arrangement to place the field variables solely within the disequilibrium component. The price to pay for this, however, is that it is no longer possible to have the main toroidal current in a tokamak within the dependent variable. One has to work in a prescribed equilibrium with the structure of ψ given. But in the use of field aligned coordinates, we have already made this restriction.

7.7 Delta-f energetics and correspondence

The energetics of the delta-f system involve both transfer between free energy components and transfer within components between scales. Here, we are concerned with the consistency of the model and whether we have lost any processes while descending from the total-f to the delta-f model. The time derivative of the free energy form with hg explicitly is given in Eq. (7.76). We have to keep the first two pieces together since (δf) and $A_\parallel$ combine into (δg). Dropping the parentheses on the dependent variables as we did for the derivation of Eq. (7.101), we have

$$\frac{\partial}{\partial t}\left(E_f + E_M\right) = \sum_{sp} \int d\Lambda\, \frac{T_0}{F_0} f \frac{\partial g}{\partial t}, \qquad \frac{\partial E_E}{\partial t} = \sum_{sp} \int d\Lambda\, e(J_0\phi)\frac{\partial g}{\partial t} \tag{7.105}$$

where f is as defined in Eq. (7.63), and we can replace f by g in the second term due to symmetry of F^M. These two sum to zero, so that if we have a right hand side in one it must be equal and opposite to the other. We can therefore obtain the transfer between ExB energy and the rest of the system using the second equivalence.

Using Eq. (7.101) we find

$$\frac{\partial E_E}{\partial t} = \sum_{sp} \int d\Lambda\, eJ_0(\phi)\frac{\partial g}{\partial t}$$
$$= -\sum_{sp} \int d\Lambda\, eJ_0(\phi)\left[b^s[H_0, h]_{zs} + \frac{1/e}{\chi_x}[H_1, h]_{xy} - \frac{z^2/m + M\Omega}{2e}\mathcal{K}(h)\right] \tag{7.106}$$

We can express the phase space bracket as

$$[H_0, h]_{zs} = \frac{\partial}{\partial s}\left(\frac{\partial H_0}{\partial z}h\right) - \frac{\partial}{\partial z}\left(\frac{\partial H_0}{\partial s}h\right) \tag{7.107}$$

We use the fact that integration over dW/B annihilates the $\partial/\partial z$ term and commutes with $\partial/\partial s$, and integrate the latter by parts, to obtain

$$\sum_{sp} \int d\Lambda\, e\frac{\partial H_0}{\partial z}hb^s\frac{\partial J_0(\phi)}{\partial s} = \sum_{sp} \int d\Lambda\, \frac{ez}{m}hb^s\frac{\partial J_0(\phi)}{\partial s} \tag{7.108}$$

The drifts bracket yields zero for the $e(J_0\phi)$ portion of H_1, so that only the $A_\parallel$ part is left,

$$-\sum_{sp} \int d\Lambda\, eJ_0(\phi)\frac{1/e}{\chi_x}[H_1, h]_{xy} = \sum_{sp} \int d\Lambda\, \frac{ez}{m}h\frac{1}{\chi_x}[-A_\parallel, (J_0\phi)] \tag{7.109}$$

The $b^s\partial/\partial s$ term and the $[-A_\parallel,\ ]$ term combine to form

$$\nabla_\parallel = b^s\frac{\partial}{\partial s} + \frac{1}{\chi_x}[-A_\parallel] \tag{7.110}$$

so that we have for these two pieces

$$\frac{\partial E_E}{\partial t} = \cdots + \sum_{sp} \int d\Lambda\, \frac{ez}{m}h\nabla_\parallel(J_0\phi) \tag{7.111}$$

In the absence of gyroaveraging this is just $J_\parallel \nabla_\parallel \phi$, so we can see the correspondence to the fluid drift models, in both the linear and nonlinear parts of the parallel gradient.

The curvature term separates out, so that

$$\frac{\partial E_E}{\partial t} = \cdots - \sum_{sp} \int d\Lambda\left(\frac{z^2}{m} + M\Omega\right)h\mathcal{K}(J_0\phi) \tag{7.112}$$

The velocity space integration yields a pressure times the divergence of the ExB velocity for each species, transferring energy to the thermal component of each. The kinetic version includes all the FLR corrections to what is basically the same set of energy transfer mechanisms between ExB energy and thermal free energy we have in the fluid drift model, so we know we are treating the same physics with a more general model. This is how the principle of correspondence works for our system. The energy transfer mechanisms in magnetised plasma turbulence are the same in both models.

7.7.1 Field equations

The field equations correspond closely both between the delta-f and total-f versions of the gyrokinetic theory and the delta-f gyrokinetic and fluid drift models. What we just derived is basically a recapture of the delta-f energy transfer effects we already studied in the fluid drift model. It is interesting to note that the field equations themselves are basically the same in the delta-f and total-f gyrokinetic models. To show this we revert to writing (δg) and (δf) for the delta-f model to avoid confusion with the f of the total-f model, whose perturbed correction is (δg), not (δf).

We have for the total-f version,

$$-\nabla_\perp^2 A_\parallel = \mu_0 \sum_{sp} \int d\mathcal{W}\, eUf, \qquad \sum_{sp} \int d\mathcal{W} \left[ef + \nabla \cdot \frac{fm}{B^2} \nabla_\perp \phi \right] = 0 \tag{7.113}$$

given the long wavelength Hamiltonian from Section 5.10 of Chapter 5. And for the delta-f version we have Eqs (7.68) or (7.69) for $A_\parallel$ and Eq. (7.66) for ϕ, which if we express $\mathcal{M}$ for $F_0 = F^M$ is rewritten as

$$\sum_{sp} d\mathcal{W} \left[eJ_0(\delta g) + e^2 F^M \frac{J_0^2 - 1}{T_0} \phi \right] = 0 \tag{7.114}$$

where as usual we can use either (δf) or (δg) in the source term for the field. The long wavelength form of this matches the total-f version.

In the total-f Ampère's law, we can rewrite using

$$mU = z - eA_\parallel, \qquad d_e^{-2} = \mu_0 \sum_{sp} \int d\mathcal{W} \frac{e^2 z^2}{m} f \tag{7.115}$$

to find

$$\left(d_e^{-2} - \nabla_\perp^2\right) A_\parallel = \mu_0 \sum_{sp} d\mathcal{W} \frac{ez}{m} f \tag{7.116}$$

which is the same as Eq. (7.68) if we replace f by (δg), assuming that F_0 vanishes by symmetry under the integration. Again, we have correspondence in the field equations between the two models.

Energetically, we have replaced the total-f thermal energy by the delta-f thermal free energy, while leaving the field energy pieces intact and leaving the energy transfer mechanisms unchanged except for corrections. This shows the basic correspondence that we need to justify use of any of these models against any other.

7.8 On consistency

The main requirement of the model besides that it treats the physics we need is that the energetic consistency remains exact. It does not help to make one or two pieces of the model more accurate in some way, if this breaks consistency. This pertains especially to the use of a local model to treat a stratified physical situation. It also pertains to use of $B_{\|}^*$ versus B in any given situation. The main watchword we need to keep is consistency. If greater accuracy or precision breaks this, then the model is not useful for turbulence because it cannot maintain turbulence in a saturated state indefinitely. We have already shown we can do this with the fluid drift models. We need the same for the kinetic ones.

7.8.1 Parallel phase space consistency

The main issue with parallel phase space consistency is the relationship between B^s and $B_{\|}^*$ in the total-f model, and b^s and B in the delta-f one. In most cases, the replacement $B_{\|}^* \to B$ is done through the geometric approximations, even in a total-f setting. So in the standard formulation,

$$B_{\|}^* = B + \frac{p_z}{e}\mathbf{b} \cdot \nabla \times \mathbf{b} \tag{7.117}$$

we are not arguing that the second term is small by $\rho_s/L_\perp$ but that under the approximations we use for the geometry, that $\mathbf{b} \cdot \nabla \times \mathbf{b}$ vanishes in this expression. The simplest example is the replacement of $\mathbf{b}$ by $R\nabla\varphi$ in a tokamak magnetic field model, in which the difference between the total magnetic field and its vacuum value $I_0\nabla\varphi$ is farily small, so that I_0/R is used for B, whereupon we have the above approximation. If that is used, then

$$\mathbf{b} \cdot \nabla \times \mathbf{b} = R\nabla\varphi \cdot (\nabla R \times \nabla\varphi) = 0 \tag{7.118}$$

and then it follows that $B_{\|}^* \to B$. This can be made automatic under the approximation $\mathbf{A}^* \to \mathbf{A} + (z/e)\nabla\varphi$ where now $z = p_z R$ contains the factor of R, and then we even have $\mathbf{B}^* \to \mathbf{B}$ in the vectors. If we keep the $p_z\mathbf{b}$ form yielding $p_z R\nabla\varphi$ with $z = p_z$, then we still (always) have $\mathbf{B}^* = \nabla \times \mathbf{A}^*$ but with $\mathbf{A}^* \to \mathbf{A} + (p_z/e)R\nabla\varphi$ we still have $B_{\|}^* \to B$. If this approximation is done in the Lagrangian, then $B_{\|}^* \to B$ follows throughout.

In the fluxtube delta-f model we have more leeway because B commutes with both directions of $\nabla_\perp$. On the other hand, the need for this is what precludes use of the delta-f model with global geometry. If the commutation between B and the perpendicular derivatives is lost, then so is energetic consistency, and it cannot be saved. More on that in the next Section.

For the parallel phase space bracket, we have to have $d\mathcal{W}/B_{\|}^*$ commuting with spatial derivatives under any form of approximation we might be using. Even the total-f model depends on this, in every aspect of the proofs of conservation we did. In the general formulation it is automatic, so that under any approximation scheme it has to be preserved. Hence the formulation of Eq. (7.101) with b^s and not B^s.

The phase space integration is with the volume element $\sqrt{g}\,B$, so that $d\mathcal{W}b^s$ leaves B^s outside the bracket with $d\mathcal{W}/B$ commuting through $\partial/\partial s$ and annihilating $\partial/\partial z$. Then the volume integration yields $d\mathcal{V}B^s$ annihilating $\partial/\partial s$ due to the divergence condition in Eq. (7.87). So we either have to write

$$B\frac{\partial g}{\partial t} + B^s[H_0, h]_{zs} = \cdots \tag{7.119}$$

or

$$\frac{\partial g}{\partial t} + b^s[H_0, h]_{zs} = \cdots \tag{7.120}$$

to retain this consistency in any particular formulation of the model.

7.8.2 Difficulty of a global model

These considerations carry over into the case of global models. We have already seen that in the fluid case we only have to ensure that spatial divergences work properly. The curvature terms can be either $\mathcal{K}$ in a fluxtube model, in which case

$$\frac{\partial}{\partial x}\mathcal{K}^x + \frac{\partial}{\partial y}\mathcal{K}^y = 0 \tag{7.121}$$

This is automatically satisfied in a flux tube because the geometry depends on s only. But if we want to extend the geometry to treat global models, we still have to keep this divergence. If we formulate such that the spatial bracket $[f, g]$ always has the form $\nabla s \cdot \nabla f \times \nabla g$ then we always satisfy energetic consistency. So then we can set $\mathcal{K} \rightarrow [\log R^2, \,]$ in a global model and keep consistency, even with delta-f equations, in the fluid case.

What we cannot do with this model is stratification in the physical situation. What that means is not that we have global geometry with global gradients and global sources. In that case there is no problem carrying the delta-f fluid drift equations over. Stratification means that the parameters are spatially dependent, even if they are still not given solely by the dependent variables. However, if we do that in fluid drift equations the consistency breaks down once and for all, because the T_0/e factors cannot be made to commute with the spatial brackets in all three sets of the nonlinearities, the linear parallel gradients and divergences, and the curvature forcing terms No version in which some power of this factor is incorporated into integration volume elements can be made satisfactory, in that energetic consistency is recovered. There is no choice but to reformulate from the starting point of the derivation of the model, using total-f equations, with all the computational difficulty that entails. One can form total-f fluid equations that are energetically consistent. But no really satisfactory numerical scheme has been found, despite efforts to paper the problem over.

In the gyrokinetic world this problem is even worse, because the fluxtube approximations have to be used to get the equations to be maintained energetically consistent. This is due to the requirement that the combination of spatially dependent B and spatially dependent parameters n_0 and T_0 in F^M cannot be made to commute with all three of the nonlinearity bracket, the curvature terms, and the parallel phase space bracket. The commutation of F_0 with $[H_0, h]_{zs}$ depends on F_0 solely dependent upon H_0. A variable B can still work here, perhaps (if BR is constant) also simultaneously in the curvature terms, but not simultaneously in the nonlinear spatial drifts bracket. One will always have hanging terms in the free energy theorem in Eq. (7.85). And there is no way to save this situation. This issue has not been generally faced in the computational literature.

7.9 The gyroaveraged magnetic field

In the basic formulation of the gyroaveraging, the J_0 operation factors are kept on both the ϕ and $A_\parallel$ terms. If done somewhere this must be done everywhere, to preserve energetic consistency. But it is possible to drop gyroaveraging on $A_\parallel$ and keep it on ϕ, if desired. The skin depth phenomenon is still treated, as we merely drop the J_0 in Eq. (7.68). This results from the interaction Hamiltonian for $A_\parallel$ given by $g(eJ_0A_\parallel)$ to be replaced by $g(eA_\parallel)$. So if J_0 is dropped on $A_\parallel$ it is important to drop it also in the Ampère's law, due to this origin of all the $A_\parallel$ terms except for $\nabla_\perp^2 A_\parallel$ from the free-field Lagrangian. So one is free to decide whether to keep this, so long as it is done consistently. We will see more of this when the gyrofluid model is developed in Chapter 9, where the motivation of it is to have a better fluid drift model, better able to treat the gyroradii.

One other issue that sometimes emerges is whether the equilibrium magnetic field itself should be gyroaveraged. This depends on how the underlying Lagrangian is formulated. If we have it in canonical form then whichever form of $\mathbf{A}^*$ we use satisfies the requirements, since $\mathbf{A}^*$ and $B_\parallel^*$ can be treated independently. One can then argue whether the small effects one has added actually improve the physics. It has to be noted that we have neglected several $O(\delta)$ effects to build the delta-f model anyway. The extra differences are at this level or even smaller (i.e., ρ_s/L_B instead of $\rho_s/L_\perp$). At some point, extra features do not bring a meaningful return.

7.10 What happened to momentum

In the total-f version of the gyrokinetic theory we had an exact theorem for conservation of the covariant component of canonical momentum in the direction of a symmetry coordinate. From

$$\frac{d}{dt}feA_\xi^* + f\frac{\partial H}{\partial \xi} = 0 \tag{7.122}$$

in which $\mathbf{A}^*$ is independent of ξ, we get the phase space continuity equation

$$B_\parallel^*\frac{\partial}{\partial t}feA_\xi^* + \nabla\cdot B_\parallel^*\left(feA_\xi^* Z^{\mathbf{R}}\right) + \frac{\partial}{\partial z}B_\parallel^*\left(feA_\xi^* Z^z\right) + fB_\parallel^*\frac{\partial H}{\partial \xi} = 0 \tag{7.123}$$

A conservation law arises if we sum over species and integrate over phase space, since the last term turns into a total divergence if (and only if) the field equations are energetically consistent, i.e., they descend from the same Lagrangian as the continuity equation. If ξ is the geometric toroidal angle in an axisymmetric toroidal system, this becomes the conservation law for canonical toroidal angular momentum. And given the charge conservation law and phase space continuity all simultaneously, we find the polarisation cancellation,

$$\sum_{sp} \int d\Lambda \, feA_\xi = -\int d\mathcal{V} \, \mathbf{P} \cdot \nabla A_\xi \tag{7.124}$$

and then we obtain a conservation law for the corresponding plasma momentum,

$$P_\xi = \sum_{sp} \int d\Lambda f z b_\xi - \int d\mathcal{V} \, \mathbf{P} \cdot \nabla A_\xi \tag{7.125}$$

in which the second term is the ExB contribution. This comes from the field pieces in the Lagrangian (anything quadratic or higher in the field variables), and the mathematics of it is the same as the way in which the Poynting momentum emerges from Maxwell's equations to provide conservation in the general theory of classical electrodynamics.

The question is whether anything corresponding to this survives in the derivation of the delta-f form of the theory. We immediately find the first problem from the $\partial H/\partial\xi$ term in Eq. (7.122). The f which multiplies this becomes F^M in the delta-f case. So in a straight magnetic field the conservation law becomes trivial. We multiply Eq. (7.94) by z to find

$$\frac{\partial}{\partial t} zg + \frac{1}{eB_0}[H_1, zh]_{xy} + [H_0, zh]_{zs} - h[H_0, z]_{zs} = 0 \tag{7.126}$$

and the last term vanishes in a straight magnetic field. The rest is a continuity equation for canonical momentum in s when integrated over velocity space. The term we mentioned has become subsumed into the zs-bracket in this equation. So a form of canonical momentum is being conserved. But the issue of the f multiplying A_ξ under the time derivative has vanished as well, also because the f has become F^M and the gyrocentre charge density due to the various F^M is assumed to be zero. In fact, the J_0 does not affect F^M in this case due to the flute mode ordering on the derivatives. So although we have a delta-f conservation law, the way it operates is physically distinct from the total-f one, and so is the meaning of the system momentum.

In the toroidal case another problem emerges. We have the curvature terms, but z commutes with the derivatives in these to form more terms in a continuity equation, just as in the total-f case. However, in the parallel dynamics we have

$$\int d\mathcal{W} \, zb^s[H_0, h]_{zs} = \frac{\partial}{\partial s} \int d\mathcal{W} \left(\frac{\partial H_0}{\partial z} zhb^s \right) + \int d\mathcal{W} \, hb^s \frac{\partial H_0}{\partial s} \tag{7.127}$$

In the total-f theory this last term is an integral over solely $f\partial H/\partial\xi$, but now we have h and $\partial/\partial s$ instead of f and $\partial/\partial\xi$. It is hard to see how to reconcile this. In the total-f theory all the derivatives of $\mathbf{A}^*$ combine to cancel down to a factor of $B_\parallel^*$ times the combination of f and the ξ-derivative of H. Here we even have H_0 instead of H. So it is impossible to get f multiplying $\partial H/\partial\xi$ which becomes a divergence when the integration over velocity space is done on Eq. (6.47) in Chapter 6. The term $h(\partial H_0/\partial s)$ does not make a divergence since at best we would get pieces of H_1 combining with H_0.

It would appear we don't have the conservation law we would like to have for momentum, after we descend from the total-f theory to its delta-f approximation. This can serve as a cautionary tale for the application of practically any ordering scheme. The model we have built is adequate for studies of the physics of the turbulence, but if we try to form a global model out of it we encounter problems in both the energy and momentum conservation considerations. The energy conservation law is now for free energy, and at this level it does not have a global conservation law since the one it has depends on being able to commute F^M past perpendicular derivatives. The momentum conservation law we have for the straight magnetic field case is relatively trivial, and it seems to break down in the toroidal model for a tokamak like magnetic field. The model is fine for our purposes, for building the corresponding gyrofluid model and understanding the physics of the turbulence. Global transport phenomena, on the other hand, would seem to require the total-f model with all its complication, not least the need to retain the entire self consistent magnetohydrodynamic equilibrium. Such questions, being beyond our scope, are left to future work.

Further Reading

Casimir Invariants were introduced for rigid body motion by H B G Casimir, 'Über die Konstruktion einer zu den irreduzibelen Darstellungen halbeinfacher kontinuierlicher Gruppen gehörigen Differentialgleichung,' *Proc. Roy. Akad. Amst* **34** (1931) 844. A summary of the way they emerge in this type of theory is given by P J Morrison and Eliezer, 'Spontaneous symmetry breaking and neutral stability in the noncanonical Hamiltonian formalism,' *Phys. Rev. A* **33** (1986) 4205.

Time scale separation in self consistent dynamics has been shown in the fluid cases corresponding to the delta-f gyrokinetic model in B Scott, 'Dynamical alignment in three species tokamak edge turbulence,' *Phys. Plasmas* **12** (2005) 082305, and in 'Free Energy Conservation in Local Gyrofluid Models,' *Phys. Plasmas* **12** (2005) 102307.

The general use of the free energy minimisation principle descends from the treatment by M D Kruskal and C R Oberman, 'On the stability of plasma in static equilibrium,' *Phys. Fluids* **1** (1958) 275. Other references on the development of the fluctuation free energy concept are given in Chapter 7 of Volume One.

This derivation path to the delta-f gyrokinetic equation was given in B Scott, 'Derivation via free energy conservation constraints of gyrofluid equations with finite-gyroradius electromagnetic nonlinearities,' *Phys. Plasmas* **17** (2010) 102306.

Toroidal coordinates for fluxtube models with exact global consistency were given in B Scott, 'Global consistency for thin flux tube treatments of toroidal geometry,' *Phys. Plasmas* **5** (1998) 2334; and B Scott, 'Shifted Metric Procedure for Flux Tube Treatments of Toroidal Geometry: Avoiding Grid Deformation,' *Phys. Plasmas* **8** (2001) 334, in which the shift operation between successive perpendicular planes is systematically worked out. A more complete list of references is given in Further Reading in Chapter 12 from Volume One.

IOP Publishing

Turbulence and Instabilities in Magnetised Plasmas, Volume 2
Gyrokinetic theory and gyrofluid turbulence
Bruce Scott

Chapter 8

Gyrokinetic treatment of waves

8.1 Introduction

Herein we will briefly address the structure of kinetic theory for wave oscillations and responses, and simple instabilities as done in Chapters 8 and 9 from Volume One. This will by no means be exhaustive. The point is to make the structure clear, not to belabor the details. One should know what the plasma dispersion function is and how it is derived, and the cancellations one faces in obtaining the basic Alfvén and sound wave responses in a straight magnetic field, where all components of the spatial wavenumber can be treated as parameters. The drift Alfvén wave and ion temperature gradient mode can be accessed using similar methods. For anything else, especially linear modes in a torus, it is best to go directly to the nonlinear situation of saturated turbulence, which we have already seen is different from what one sees in mode structure and energetics from the linear mode situation. So we get the basic interactions without being trapped in details which we would then have to struggle to undo.

8.2 Kinetic responses

The basic structure of this is the response function of the distribution function for each species to the fields (in an electrostatic model, the potential by itself). In this, the field is assumed to be given and the complex frequency ω is undetermined. The field equations are then used to close the model and obtain ω as a function of parameters, in which the components of the vector wavenumber $\mathbf{k}$ are included. A background gradient is assumed to be across the magnetic field lines, which lie in surfaces perpendicular to $\mathbf{B}$, the background guide field. So we have the x direction as down-gradient, the s direction as along $\mathbf{B}$, and the y direction mutually perpendicular so that $\nabla x \times \nabla y \cdot \nabla s$ is constant. The coordinates for this were summarised in Chapter 2, so we can just write them here. With $\sqrt{g} = 1$ in a general magnetic field we have x as a volume-type coordinate and y and s as angle-like. Normalisation of that is arbitrary (i.e., where to put the factors of 2π). For the present purpose, though, we will take a straight magnetic field, with

$$B^x = B^y = 0 \qquad B^s = B = B_0 \qquad \mathbf{B} = B_0 \nabla x \times \nabla y \tag{8.1}$$

where at this stage xys are all distances on a Cartesian grid. A Fourier decomposition is assumed, and due to linearity and the constant magnetic field we can take each $\mathbf{k}$ to be constant and separate. This is given as

$$\phi(x, y, s, t) = \sum_{\mathbf{k}} \phi_{\mathbf{k}} \exp[i(\mathbf{k} \cdot \mathbf{x} - \omega t)]$$
$$\text{with} \quad k_\perp^2 = k_x^2 + k_y^2 \quad \text{and} \quad k_\parallel = k_s \tag{8.2}$$

with k_y, $k_\perp$, and $k_\parallel$ as parameters and ω yet to be solved for. For the present purpose, we take this straight magnetic field case and neglect complications of finite gyroradius effects, everywhere taking the limit $J_0 \to 1$ and simple Poisson or Helmholtz equations for the fields.

The first step is to calculate the response of the gyrocentre distribution function to the fluctuating fields. Following the linearisations $f \to F_0 + g$ and $H \to H_0 + H_1$, this takes the linearised delta-f gyrokinetic equation for g, in abstract form given by

$$\frac{\partial g}{\partial t} + \{H_0, h\} = 0 \tag{8.3}$$

where the brace bracket is a shorthand for all the terms (and $\{H_0, F_0\}$ vanishes), and under the Fourier decomposition the subscript $\mathbf{k}$ on g, h and ϕ, $A_\parallel$ is understood rather than written. The ingredients are given by

$$H_0 = \frac{z^2}{2m} + M\Omega \qquad H_1 = e\phi - \frac{e}{m} z A_\parallel, \qquad h = g + \frac{H_1}{T_0} F_0 \tag{8.4}$$

where e is the charge (with charge state e/e_0) and m the mass of the given species, T_0 is the species temperature, and F_0 is taken to be a Maxwellian as in the previous Chapter, since the equilibrium from which we depart is a maximum-entropy state. The normalisation of F_0 is such that it produces the species density n_0 when integrated over velocity space.

Each gyrocentre perturbation g then gives rise to a gyrocentre charge density and current, species by species. After g is solved for in terms of the field variables, the integrals for the charge density and current are put into the equations for the fields. These are given by polarisation,

$$\frac{\rho}{B^2} k_\perp^2 \phi = \sum_{sp} \int d\mathcal{W}\, eg \tag{8.5}$$

and induction,

$$(d_e^{-2} + k_\perp^2) A_\parallel = \mu_0 \sum_{sp} \int d\mathcal{W}\, e \frac{z}{m} g \tag{8.6}$$

in which the sum is over species and the velocity space integration is given by

$$\int d\mathcal{W} = \frac{2\pi}{m^2}\int_0^\infty dM\,\Omega\int_{-\infty}^\infty dz \tag{8.7}$$

as in the previous Chapter (noting that although M and Ω are signed, $M\Omega$ is always positive). In the induction equation, the inverse skin depth d_e^{-2} is the species sum of $\mu_0 n_0 e^2/m$ which will cancel a contribution in the response g due to the field variable $A_\parallel$ (see the discussion around Eq. (7.69) of the previous Chapter). This is the 'Ampère cancellation' which will be noted when it arises. After the skin depth term is cancelled out, the two equations combine in a natural fashion reflecting the response under the limits of gyrokinetic frequency ordering of the electromagnetic field due to the perturbations in the gyrocentre charge density and current.

This type of deliberation gives the response of the gyrocentre charge density and current for each species to the field perturbations, and in turn the responses of the field variables through polarisation and induction to those gyrocentre quantities. As more content and more geometry is added, the basic idea is the same. Since this is linear theory the two steps in the self-consistent response can be separated in this fashion.

The most basic response is given by the situation in the absence of gyroaveraging, gradients, and drifts,

$$\frac{\partial g}{\partial t} + [H_0, h]_{zs} = 0 \tag{8.8}$$

The simplest case is electrostatic, with $H_1 = e\phi$. Doing the Fourier decomposition, we have

$$-i\omega g + ik_\parallel v_\parallel\left(g + \frac{e}{T_0}\phi\, F_0\right) = 0 \tag{8.9}$$

and hence

$$g = -\frac{e}{T_0}\phi\,\frac{k_\parallel v_\parallel}{k_\parallel v_\parallel - \omega}F_0 \tag{8.10}$$

where $v_\parallel = z/m$ is $\partial H_0/\partial z$. Integrating this expression for g over velocity space for the perturbed density, we have

$$n = \int d\mathcal{W}\, g = -\frac{e}{T_0}\phi\int d\mathcal{W}\,\frac{k_\parallel v_\parallel}{k_\parallel v_\parallel - \omega}\,F_0 \tag{8.11}$$

and then the gyrocentre charge density is found by multiplying by e and summing over species.

In processing the integrals, we take advantage of the fact that unless temperature gradients or toroidicity are treated, there is no dependence of the response upon the magnetic moment. So the part of $d\mathcal{W}$ given by ΩdM is integrated away (Eq. 8.7), and the whole problem can be reformulated around a one-dimensional velocity space. The velocity space integration and the background Maxwellian reduce to

$$\int d\mathcal{W} = \frac{1}{m}\int_{-\infty}^{\infty} dz, \qquad \frac{1}{n_0}F_0 = (2\pi T_0/m)^{-1/2}\exp[-z^2/2mT_0] \tag{8.12}$$

recalling m and n_0, T_0 are species parameters.

We then apply the simplification in Eq. (8.12) to do the integration in Eq. (8.11). The basic integral in this is given by

$$Z(\zeta) \equiv \frac{1}{\sqrt{\pi}}\int_{-\infty}^{\infty} \frac{dv\, e^{-v^2}}{v-\zeta} \tag{8.13}$$

where

$$\zeta = \omega \big/ \sqrt{2} k_\| V \qquad \text{where} \quad V^2 = T_0/m \tag{8.14}$$

is the dimensionless argument with V the species thermal velocity. This is the textbook version of the function Z, which is called the 'plasma dispersion function', and the standard definition folds the factor of two in $z^2/2m$ into the integration variable. The plasma dispersion function arises any time a linear model with parallel responses is integrated over velocity space with a background Maxwellian. The integral is taken over an extended path in the complex plane where the contour lies along the real axis and is closed in the upper plane (values with a positive imaginary part). For $\mathrm{Im}(\zeta) < 0$ the contour is deformed via a procedure known as analytical continuation such that it still surrounds the pole at $v = \zeta$. The details are given in the references cited in Further Reading. For our purposes all we need to know is that it has well defined asymptotic limits and is a well defined function for any ζ in the complex plane.

It is important to note that so long as Z appears multiplied by its argument, there is no issue of whether z is normalised to V or $\sqrt{2}\,V$. This is because the unnormalised but dimensionless version is

$$\frac{1}{n_0}\int d\mathcal{W}\, F_0 \frac{\omega}{k_\| v_\| - \omega} = \zeta\, Z(\zeta) \tag{8.15}$$

The Maxwellian F_0 is set such that its integral is n_0, and the fractional function is dimensionless. So as long as we write it as $\zeta\; Z(\zeta)$ and not Z by itself there is no issue. However, when using the standard expansion formulae, one recalls that the argument is the ζ as given in Eq. (8.14).

Processing the integrand in Eq. (8.11) each factor of $v_\|$ is formed using

$$\frac{k_\| v_\|}{k_\| v_\| - \omega} = 1 + \frac{\omega}{k_\| v_\| - \omega} \tag{8.16}$$

to isolate the simple pole in the integrand. In this version there is only one factor of $v_\|$ but in the induction equation with $v_\| A_\|$ there are as many as three. This deliberation is used each time such a factor appears. In the electrostatic case without drifts we obtain

$$n = \int d\mathcal{W}\, g = -n_0 \frac{e}{T_0}\phi[1 + \zeta\, Z(\zeta)] \tag{8.17}$$

for the gyrocentre density response to the field disturbance.

Also in this case we have simply the low-$k_\perp$ field polarisation balancing the density fluctuations to close the system. This gives

$$\frac{\rho}{B^2}k_\perp^2\phi = \sum_{sp} en = -\sum_{sp} n_0\frac{e^2}{T}\phi[1 + \zeta\, Z(\zeta)] \tag{8.18}$$

Note that the charge enters squared on the right hand side, so that all factors except perhaps Z are positive. The simplest case allows only electron motion with $e = -e_0$, assuming that of the ions is zero. This follows from the asymptotic expansion of the Z-function for large argument,

$$1 + \zeta\, Z(\zeta) = -\frac{1}{2\zeta^2} - \frac{3}{4\zeta^4} - \cdots \tag{8.19}$$

In this case we keep the first term for electrons but not for ions, since $V_i^2 \ll V_e^2$ by $m_e \ll m_i$. Then we have

$$K^2 = -[1 + \zeta_e\, Z(\zeta_e)] \tag{8.20}$$

reflecting the response of the electrons only, where as previously

$$K = k_\perp \rho_s \qquad \rho_s^2 = \frac{\rho T_{e0}}{n_{e0}e_0^2 B^2} \tag{8.21}$$

and $e_0\phi/T_{e0}$, the normalised field amplitude, has dropped out as a common factor. The field variable can be divided out in this fashion since it is a complex number of finite amplitude.

The first two terms in the expansion for electrons can be recast as a small correction $(1 + 3/2\zeta_e^2)$ to $1/2\zeta_e^2$ as long as ζ_e^2 is large. With these, Eq. (8.20) can be solved to lowest order and expanded to next, to obtain

$$\omega^2 = k_\parallel^2 V_e^2 \frac{1 + 3K^2}{K^2} \tag{8.22}$$

If we note the isothermal fluid kinetic shear Alfvén wave dispersion relation,

$$V_a^2 = v_A^2 \frac{1 + k_\perp^2 \rho_s^2}{1 + k_\perp^2 d_e^2} \tag{8.23}$$

(Chapter 8 of Volume One) we see that the kinetic result is the same as the $\beta_e \to 0$ limit of the fluid result. The factor of 3 is due to the thermal response. The assumption of large argument for the Z-function is satisfied for $K \ll 1$, that is, long wavelength. So we are on the right track.

It is interesting to note that the difference between this result and the standard textbook one is due to the replacement of the displacement current by the polarisation current. We can recover the textbook result by replacing

$$\frac{\rho}{B^2}k_\perp^2 \to \varepsilon_0 k_\parallel^2 \tag{8.24}$$

reflecting a return to the one-dimensional Vlasov–Poisson problem. With this, Eq. (8.20) becomes

$$k_\parallel^2\lambda_D^2 = -[1 + \zeta_e\, Z(\zeta_e)] \to \frac{1}{2\zeta_e^2} \tag{8.25}$$

with the latter expression in the $(\mathrm{Re}\,\zeta)^2 \gg 1$ limit. In this case the $k_\parallel^2$ factors in the largest terms cancel, and the result is

$$\omega^2 = \frac{V_e^2}{\lambda_D^2} = \frac{n_0 e_0^2}{\varepsilon_0 m_e} = \omega_{pe}^2 \tag{8.26}$$

namely, the textbook Langmuir oscillation. So in a magnetised plasma, the introduction of ion polarisation replaces the Langmuir oscillation with the kinetic shear Alfvén wave, which is now a real wave since ω is proportional to $k_\parallel$ in the largest factor. The phase velocity is V_e/K, which is the low-K zero-β_e limit of V_a in Eq. (8.23). With $k_\parallel \ll k_\perp$ the appearance of $\nabla_\perp^2$ has the same role as ∇^2, but we recall this occurs only due to the properties of dynamics in a magnetised plasma. In purely one-dimensional motion ($k_\perp = 0$) this consideration, as well as the polarisation drift, does not occur. In a magnetised plasma, the existence of gyromotion always introduces $k_\perp$ as a parameter.

If we allow the ions to respond, then the ion and electron responses go into the polarisation together,

$$\frac{\rho}{B^2}k_\perp^2 = -n_0\frac{e^2}{T_{e0}}[1 + \zeta_e\, Z(\zeta_e)] - n_0\frac{e^2}{T_{i0}}[1 + \zeta_i\, Z(\zeta_i)] \tag{8.27}$$

assuming the ions to be singly charged and n_0 equal for both electrons and ions, and using $\zeta_{e,i} = \omega/\sqrt{2}k_\parallel V_{e,i}$ separately for each species. Looking for the acoustic frequencies assuming $V_i < c_s \ll V_e$, we take the electrons to be adiabatic ($\zeta_e \to 0$), and expand $Z(\zeta_i)$ for large argument as for the electrons leading to Eq. (8.20), we obtain

$$1 + K^2 = \frac{T_{e0}}{T_{i0}}\frac{1}{2\zeta^2} = \frac{T_{e0}}{T_{i0}}\frac{k_\parallel^2 V_i^2}{\omega^2} = \frac{k_\parallel^2 c_s^2}{\omega^2} \tag{8.28}$$

so that

$$\omega^2 = k_\parallel^2 V_s^2 \qquad \text{where} \quad V_s^2 = \frac{c_s^2}{1 + K^2} \tag{8.29}$$

which is the same as the ion acoustic frequency in the fluid result (Chapter 8 of Volume One). Taking $T_{i0} \ll T_{e0}$ saves the use of the $(\mathrm{Re}\,\zeta)^2 \gg 1$ limit of the Z-function. Again, the corresponding textbook result is recovered by replacing $k_\perp^2\rho_s^2$ with $k_\parallel^2\lambda_D^2$, and then

$$V_s^2 \to \frac{c_s^2}{1 + (k_\parallel\lambda_D)^2} \qquad \text{if} \quad k_\perp^2\rho_s^2 \to k_\parallel^2\lambda_D^2 \qquad \text{from} \quad \frac{\rho}{B^2}k_\perp^2 \to \varepsilon_0 k_\parallel^2 \tag{8.30}$$

This is the one-dimensional ion acoustic wave from kinetic theory. The lowest order term gives the oscillation frequency. When the next-order wavenumber correction becomes finite, the asymptotic result breaks down and the Landau damping of the oscillation becomes non-negligible.

8.2.1 Landau damping

The simplest version of Landau damping comes from the imaginary part of the Z-function, which comes from the pole of the integral in any situation in which the frequency would otherwise be real. This part is given by

$$\lim_{\zeta\to 0} \operatorname{Im} Z(\zeta) = i\,\sqrt{\pi}\,e^{-\zeta^2} \quad \text{which is also} \quad \lim_{\zeta\to\infty} \quad \text{if} \quad |\operatorname{Im}\zeta| < |\operatorname{Re}\zeta| \tag{8.31}$$

In the acoustic wave result, this correction yields

$$1 + K^2 = \frac{T_{e0}}{T_{i0}}\frac{1}{2\zeta^2}[1 - i\,\sqrt{\pi}\,2\zeta^3\,e^{-\zeta^2}] \tag{8.32}$$

Solving for ζ^2 first and then applying the correction by expanding the square root of the expression in brackets, we find

$$\omega = \pm k_\parallel V_s[1 - i\,\sqrt{\pi}\,\zeta^3\,e^{-\zeta^2}] \qquad \text{with} \quad \sqrt{2}\zeta = \frac{V_s}{V_i} \tag{8.33}$$

If $\zeta \gg 1$ the picture of a weakly damped sound wave continues to hold. As ζ drops to the vicinity of unity as $T_{i0} \to T_{e0}$ the damping becomes progressively stronger, to the point at which for $T_{i0} \sim T_{e0}$ the damping is strong enough to quench the disturbance essentially after one oscillation period. The textbook picture of electron damping of the ion acoustic wave with $T_i \ll T_e$ essentially never occurs in a magnetised plasma, since $\nabla_\perp^2$ replaces ∇^2 well before the scale of λ_D is reached in $k_\perp$ while $k_\parallel$ is always of macroscopic scales.

8.3 Adiabatic drift acoustic wave

When the background gradient is considered the extra terms are the linear contributions from the ExB drift advecting the background gradient, which is written as

$$\frac{1}{B}[\phi, F_0]_{xy} = -\frac{1}{B}[\phi, x]_{xy}\left[\frac{1}{L_n} + \left(\frac{H_0}{T_0} - \frac{3}{2}\right)\frac{1}{L_T}\right] \tag{8.34}$$

where L_n and L_T are the species density and temperature scale lengths (for L_T, the H_0/T_0 comes from the exponential and the 3/2 from the normalising factor including $n_0\,T_0^{-3/2}$ of the Maxwellian). If only the density gradient is considered then

$$\frac{1}{B}[\phi, F_0]_{xy} = \frac{F_0}{BL_n}\frac{\partial\phi}{\partial y} \tag{8.35}$$

The main point to watch here is that this is the same term for every species, not involving charge or temperature, while the parallel dynamics involves both.

With both terms in, we have

$$\frac{\partial g}{\partial t} + \frac{1}{B}[\phi, F_0]_{xy} + [H_0, h]_{zs} = 0 \tag{8.36}$$

with H_0 and h given as above. When the Fourier decomposition is taken, these terms become

$$-i\omega g + i\tau_z\omega_*\Phi F_0 + ik_\| v_\|(g + \Phi F_0) = 0 \tag{8.37}$$

where

$$\tau_z = \frac{T_0/e}{T_{e0}/e_0}, \qquad \tau_z\omega_* = \frac{T_0}{eBL_n}, \qquad \Phi = \frac{e\phi}{T_0} \tag{8.38}$$

are now species-dependent normalised quantities. Factors of $\omega + \tau_z\omega_*$ now arise, where τ_z is the temperature/charge ratio normalised to the electron temperature and the fundamental charge. Note that τ_z is signed, and negative for electrons.

The perturbed distribution function for each species is

$$g = -\frac{k_\| v_\| + \tau_z\omega_*}{k_\| v_\| - \omega}\Phi F_0 \tag{8.39}$$

Multiplying by the charge and integrating over velocity space, we have

$$en = -\frac{e^2\phi}{T_0}\int d\mathcal{W}\,\frac{k_\| v_\| + \tau_z\omega_*}{k_\| v_\| - \omega}F_0 \tag{8.40}$$

where the only difference to Eq. (8.11) is the contribution due to the diamagnetic frequency.

Expressing this in terms of the Z-function as before yields

$$en = -n_0\frac{e^2\phi}{T_0}[1 + (\zeta + \zeta_*)Z(\zeta)] \tag{8.41}$$

where

$$\zeta_* = \tau_z\omega_*/\sqrt{2}k_\| V \tag{8.42}$$

This is a species normalisation, with τ_z carrying the species T_0/e ratio. Each perturbed density contributes to the overall response of the potential, set up the same way as in Eq. (8.18) but with the additional ζ_* term. If the electrons are taken to be adiabatic and the ion temperature to satisfy $\tau_i \ll 1$, the Z-function is expanded for the ions only. Due to the cancellation in $1 + \zeta Z$ we set that piece to the second term in the expansion for $\zeta \gg 1$, while for the ζ_* piece we need only the first term. This yields

$$\frac{\rho}{B^2}k_\perp^2 = -\frac{n_0 e_0^2}{T_{e0}} + \frac{n_0 e^2}{T_{i0}}\left(\frac{k_\|^2 V_i^2}{\omega^2} + \frac{\tau_i\omega_*}{\omega}\right) \tag{8.43}$$

For singly charged ions we can replace $\tau_i = T_{i0}/T_{e0}$, and then

$$\frac{1}{T_{i0}}\frac{k_\parallel^2 V_i^2}{\omega^2} = \frac{k_\parallel^2 c_s^2}{\omega^2} \qquad \text{and} \qquad \frac{1}{T_{i0}}\frac{\tau_z \omega_*}{\omega} = \frac{1}{T_{e0}}\frac{\omega_*}{\omega} \tag{8.44}$$

Putting these into the response and multiplying through by ω^2, we recover

$$(1 + k_\perp^2 \rho_s^2)\omega^2 - \omega\omega_* - k_\parallel^2 c_s^2 = 0 \tag{8.45}$$

which is the fluid result (Chapter 9 of Volume One).

8.3.1 Drift waves with electron Landau damping

We now apply traditional drift wave ordering,

$$k_\parallel V_e \gg \omega_* \gg k_\parallel c_s \sim k_\parallel V_i \tag{8.46}$$

neglecting the ion parallel response, but allow the electrons a nonadiabatic response. Repeating the path to Eq. (8.43), keeping the electron response due to the pole of the Z-function but neglecting the additional sound wave term for the ions, we have

$$\frac{\rho}{B^2} k_\perp^2 = -n_0 \frac{e_0^2}{T_{e0}}[1 + (\zeta_e - \zeta_{*e})Z(\zeta_e)] + n_0 \frac{e_0^2}{T_{e0}}\frac{\omega_*}{\omega} \tag{8.47}$$

where $\zeta_e = \omega/\sqrt{2}k_\parallel V_e$ and $\zeta_{*e} = \omega_*/\sqrt{2}k_\parallel V_e$, that is, the $\tau_z = -1$ factor for electrons is taken into account to give the signs, and the last term is what remains of the ion response, with the temperature normalisation processed as in Eq. (8.44). Giving the nonadiabatic correction to the electrons through the Z-function, we find

$$1 + K^2 = \frac{\omega_*}{\omega} - (\zeta_e - \zeta_{*e})Z(\zeta_e) \tag{8.48}$$

having multiplied through by $T_{e0}/n_{e0}e_0^2$ to form ρ_s^2 and then K^2 as in Eq. (8.21). Assuming the traditional frequency ordering in Eq. (8.46), we expand

$$Z = i\sqrt{\pi} \qquad \text{for} \quad \zeta \ll 1 \tag{8.49}$$

and find through successive substitution

$$\omega = \frac{\omega_*}{1 + K^2} + i\sqrt{\frac{\pi}{2}}\frac{\omega_*^2}{|k_\parallel V_e|}\frac{K^2}{(1 + K^2)^3} \tag{8.50}$$

If the frequency ordering is satisfied then the neglect of parallel ion motion and the use in the Z-function of $\zeta \to 0$ for electrons and $\zeta \to \infty$ and $\mathrm{Re}\,\zeta \gg \mathrm{Im}\,\zeta$ for ions is justified. The $k_\perp$-dependence of this is very similar to the fluid result (Chapter 9 of Volume One). Given there in normalised units, the result with β_e and μ_e small but collisionality C finite and (the ratio of specific heats) $\Gamma = 1$ is

$$\omega = \frac{\omega_*}{1 + K^2} + i\frac{0.51\nu_e}{k_\parallel V_e}\frac{\omega_*^2}{k_\parallel V_e}\frac{K^2}{(1 + K^2)^3} \tag{8.51}$$

where the collision frequency ν_e comes from the parallel resistivity $\eta_\parallel = 0.51 m_e \nu_e / n_e e^2$ in the collisional fluid model. If we set the effective collision frequency to

$$0.51\nu_e \to \sqrt{\frac{\pi}{2}}\,\left|k_\parallel V_e\right| \tag{8.52}$$

then the kinetic and fluid results are the same. The effect of Landau damping is to replace the overt collisionality with an effective value connecting parallel phase space mixing to a dissipation which at the end of the chain is finite.

8.4 Kinetic shear Alfvén wave

We go back to the original expression and keep both ϕ and $A_\parallel$, and moreover carry along both the electron and ion responses. A charge cancellation combines one of the terms in the polarisation equation, and the perturbed gyrocentre current contains a factor which cancels the skin depth term in the induction equation. This latter will be highlighted when it arises. Without the drifts, we have the kinetic response in each species,

$$g = -\frac{e}{T_0}\frac{k_\parallel v_\parallel}{k_\parallel v_\parallel - \omega} F_0(\phi - v_\parallel A_\parallel) \tag{8.53}$$

The fractional function is split as before to isolate the denominator, with each extra occurrence of $v_\parallel$, to facilitate the calculation. The result for each gyrocentre charge density is

$$en = \int d\mathcal{W}\, eg = -n_0\frac{e^2}{T_0}[1 + \zeta\, Z(\zeta)]\left(\phi - \frac{\omega}{k_\parallel}A_\parallel\right) \tag{8.54}$$

Multiplying each g by an extra factor of $v_\parallel$ obtains the contribution by each species to the gyrocentre current. The first '1' is split off to produce a skin depth contribution and the rest proceeds as in the formulation of en. The result is

$$eUn = \int d\mathcal{W}\, ev_\parallel g = n_0\frac{e^2}{m}A_\parallel - n_0\frac{e^2}{T_0}\frac{\omega}{k_\parallel}[1 + \zeta\, Z(\zeta)]\left(\phi - \frac{\omega}{k_\parallel}A_\parallel\right) \tag{8.55}$$

We can rewrite this in the simple form

$$eUn = n_0\frac{e^2}{m}A_\parallel + \frac{\omega}{k_\parallel}en \tag{8.56}$$

as after the first factor of '1' is split off, the rest is like the gyrocentre charge density with a multiplier of $\omega/k_\parallel$, namely, the phase velocity. Except for the skin depth term $n_0 e^2/m$, the fields always occur in combination, giving the potential for the parallel electric field,

$$-ik_\parallel\left(\phi - \frac{\omega}{k_\parallel}A_\parallel\right) = -ik_\parallel\phi + i\omega A_\parallel = E_\parallel \tag{8.57}$$

This indicates that the field equations are to be combined such that the left hand sides add into the combination $E_\parallel$ times a coefficient.

To close the system for the fields, the polarisation equation is

$$\frac{\rho}{B^2}k_\perp^2\phi = -\sum_{sp} n_0\frac{e^2}{T_0}[1 + \zeta\, Z(\zeta)]\left(\phi - \frac{\omega}{k_\parallel}A_\parallel\right) \tag{8.58}$$

and the induction equation is

$$\frac{1}{\mu_0}\left(d_e^{-2} + k_\parallel^2\right)A_\parallel = \sum_{sp} n_0\frac{e^2}{m}A_\parallel - \sum_{sp} n_0\frac{e^2}{T_0}\frac{\omega}{k_\parallel}[1 + \zeta\, Z(\zeta)]\left(\phi - \frac{\omega}{k_\parallel}A_\parallel\right) \tag{8.59}$$

In the induction equation once more, the $\mu_0^{-1}d_e^{-2}$ term is cancelled by the species sum of n_0e^2/m over the first term on the right hand side, in the Ampère cancellation. This leaves the $k_\perp^2$ term. To form common factors for $E_\parallel$ which can then be divided out, we multiply both equations by $T_{e0}/n_{e0}e_0^2$ to form ρ_s^2 and then multiply the induction equation by $\omega/k_\parallel$, the phase velocity so that there are two factors of it, and by $\mu_0\rho/B^2$, which is v_A^{-2}, the inverse square of the Alfvén velocity. This forms the basic Alfvén response, to which the other effects add. The result is

$$K^2 = -\left(1 - \frac{\omega^2}{k_\parallel^2 v_A^2}\right)\sum_{sp}\frac{a_z}{\tau_z}[1 + \zeta\, Z(\zeta)] \tag{8.60}$$

where the coefficients

$$a_z = \frac{n_0 e}{n_{e0}e_0}, \qquad \tau_z = \frac{T_0 e_0}{T_{e0}e} \tag{8.61}$$

are the species charge density and temperature/charge ratio normalised to the electrons. Invoking the standard ordering in Eq. (8.46) for Alfvén or drift waves, implicitly assuming that $1 \gg \beta_e \gg \mu_e$, we use the small argument limit for the electron Z-function,

$$\lim_{\zeta\to 0}\zeta Z(\zeta) = i\sqrt{\pi}\zeta\, e^{-\zeta^2} \tag{8.62}$$

and the large argument limit for the ions,

$$\lim_{\zeta\to\infty}[1 + \zeta Z(\zeta)] = -\frac{1}{2\zeta^2} = -\frac{k_\parallel^2 V_i^2}{\omega^2} \tag{8.63}$$

and assuming singly charged ions ($a_z = 1$ and $\tau_z = \tau_i = T_i/T_e$), we have

$$K^2 = -\left(1 - \frac{\omega^2}{k_\parallel^2 v_A^2}\right)\left(1 + i\alpha_e - \frac{k_\parallel^2 c_s^2}{\omega^2} + i\alpha_i\right) \tag{8.64}$$

where the pole contributions are

$$\alpha_{e,i} = \sqrt{\frac{\pi}{2}} \frac{\omega}{k_\| V_{e,i}} \exp\left(-\omega^2/2k_\|^2 V_{e,i}^2\right) \tag{8.65}$$

assuming for the ions that $\mathrm{Im}\,\omega < \mathrm{Re}\,\omega$.

The fast frequency branch is the Alfvén wave, for which the Z-function multiplier is just $(1 + i\alpha_e)$. This gives the kinetic low-collision limit of the damping,

$$\omega^2 = k_\|^2 v_A^2[1 + K^2(1 - i\alpha_e)] \tag{8.66}$$

in this case Landau damping by the electrons. The damping is given by the coefficient

$$\alpha_e = \frac{1}{2}\sqrt{\frac{\pi}{2}} \frac{\omega}{k_\| V_e} K^2 k_\| v_A = \frac{1}{2}\sqrt{\frac{\pi}{2}} \frac{m_e/M_i}{\beta_e} K^2 k_\| v_A \tag{8.67}$$

and is an absolute value since the sign of $k_\| v_A$ in the plus or minus frequency cancels that in the argument ζ_e which also multiplies the Z-function. The factor of the mass ratio divided by β_e comes from the ratio v_A/V_e, which is assumed to be small in this limit.

The slow frequency branch is the ion acoustic wave, for which the Z-function multiplier includes the '1' from the electrons and the sound wave correction from the ions. This gives the kinetic low-collision limit of the damping,

$$\omega^2 = \frac{k_\|^2 c_s^2}{1 + K^2}\left(1 - \frac{1}{2} i \frac{\alpha_i}{1 + K^2}\right) \tag{8.68}$$

in this case Landau damping by the ions. In the case of cold ions, with $c_s \gg V_i$, the large-argument limit of the Z-function is still valid, and the damping is weak. The real frequency result is the same as in the fluid model (Eq. 8.29).

8.5 Drift-Alfvén wave

We now re-do this calculation with the drifts back in, as in the drift wave. As before, singly charged ions and the frequency ordering in Eq. (8.46) are assumed. The kinetic response in each species is now

$$g = -\frac{e}{T_0} \frac{k_\| v_\| + \tau_z \omega_*}{k_\| v_\| - \omega} F_0(\phi - v_\| A_\|) \tag{8.69}$$

with τ_z and e/T_0 denoting the species dependent temperature normalisation. The gyrocentre charge density is

$$\sum_{sp} en = -\sum_{sp} \int d\mathcal{W} \frac{e^2}{T_0}\left(1 + \frac{\omega + \tau_z \omega_*}{k_\| v_\| - \omega}\right) F_0(\phi - v_\| A_\|) \tag{8.70}$$

and the gyrocentre current is

$$\sum_{sp} eUn = \int d\mathcal{W}\, ev_\parallel \frac{e^2}{T_0}\left(1 + \frac{\omega + \tau_z\omega_*}{k_\parallel v_\parallel - \omega}\right) F_0(\phi - v_\parallel A_\parallel) \tag{8.71}$$

The split of the '1' in these equations is subtle, due to the presence of the ω_* terms such that ω and $\omega + \tau_z\omega_*$ are not the same and have now to be arranged properly. The first '1' in the induction equation gives the skin depth term n_0e^2/m as before, but a second '1' arises from the multiplication by $ev_\parallel$. With the species $\tau_z\omega_*$ done the way it is, we have

$$\sum_{sp} n_0 \frac{e^2}{T_0}\tau_z\omega_* = \frac{e_0}{T_{e0}}\omega_* \sum_{sp} n_0 e = 0 \tag{8.72}$$

which affects both equations, one $A_\parallel$ term in the polarisation equation, and one ϕ term and two $A_\parallel$ terms in the induction equation. Then, the $\omega + \tau_z\omega_*$ multiplier always combines with a Z-function, not the $A_\parallel$ term separately from ϕ. Working through these, the polarisation equation is

$$\frac{\rho}{B^2}k_\perp^2\phi = -\sum_{sp} n_0\frac{e^2}{T_0}[1 + (\zeta + \zeta_*)Z(\zeta)]\left(\phi - \frac{\omega}{k_\parallel}A_\parallel\right) \tag{8.73}$$

and the induction equation is

$$k_\perp^2 A_\parallel = -\mu_0\frac{\omega}{k_\parallel}\sum_{sp} n_0\frac{e^2}{T_0}[1 + (\zeta + \zeta_*)Z(\zeta)]\left(\phi - \frac{\omega}{k_\parallel}A_\parallel\right) \tag{8.74}$$

after the Ampère cancellation of the skin depth terms The same sum appears in both, with only the coefficients different. Multiplying the induction equation by ρ/B^2 and the phase velocity $\omega/k_\parallel$ the coefficient on the right hand side becomes $\omega^2/k_\parallel^2 v_A^2$, forming the basic Alfvén response as before. The combination is

$$K^2 = -\left(1 - \frac{\omega^2}{k_\parallel^2 v_A^2}\right)\sum_{sp}\frac{a_z}{\tau_z}[1 + (\zeta + \zeta_*)\, Z(\zeta)] \tag{8.75}$$

which is the same as Eq. (8.60) except for the ζ_* term, though more work was required to get it.

We can recover all of the above limits with this equation. If $k_\parallel v_A \ll \omega \ll k_\parallel V_e$, that is, arbitrarily small β_e, we neglect all ζ_* terms, take the ion response $1 + \zeta Z \to 0$ and the electron response $1 + \zeta Z \to -1/2\zeta^2 - 3/4\zeta^4$, and also take the Alfvén factor (the first parentheses in Eq. 8.75) to be unity, we revert to the low-K electrostatic branch of the kinetic shear Alfvén wave in Eq. (8.22). If $\omega^2/k_\parallel^2 v_A^2$ is kept as a correction, the $\beta_e/\mu_e K^2$ correction results and is valid as long as that remains small. If $\omega_* \ll k_\parallel v_A \ll k_\parallel V_e$ we revert to the adiabatic acoustic drift wave, by taking the Alfvén factor and the electron response to be unity, and the ions to have

$1 + \zeta Z \rightarrow -1/2\zeta^2$ and $\zeta_* Z \rightarrow -\omega_*/\omega$, as in Eq. (8.43) to recover Eq. (8.45). If the conventional drift ordering in Eq. (8.46) with $v_A \ll V_e$ is kept, then if $k_\| v_A \gg \omega_*$ the ion response is taken to vanish, the electron response becomes $(1 + i\alpha_e)$ and the kinetic shear Alfvén with Landau damping in Eq. (8.66) results. If the ion response with ω_*/ω is then restored, the nonadiabatic drift wave instability in Eq. (8.50) results.

Landau damping therefore takes over the role of collisionality in providing the dissipation for the parallel response waves, and the access to free energy for the drift wave and drift Alfvén instability. In the fluid cases the simplest model got those effects from the resistivity, even though thermal conduction as seen in Chapter 2 also enters. Landau damping is, however, not the same as a resistive friction, which gets its role from collisional friction between electrons and ions. It does not do momentum exchange, for example. Although it ultimately damps wave activity associated with the current it does not damp the current directly. If we take integral moments over the gyrokinetic equation we don't find dissipation effects at all, since both the underlying equation and anything we derive from it conserve free energy. In the fluid Ohm's law we find terms due to the temperature but no damping effects, which have to result from whatever happens as a result of exciting temperature disturbances. There is a simple calculation we can do which does suggest that Landau damping is a heat flux rather than a viscosity or resistivity effect. It is not a truncated closure model as in the procedures summarised in Chapter 2, but a re-arrangement of the equations. Before turning to gyrofluid closure procedures, we motivate them with this result.

8.6 Landau damping as thermal conduction

We consider kinetic shear Alfvén waves, in such a way as to separate the frequency and damping. The dissipationless parts are those which can be expressed in the field variables only in such a way that the result is a wave equation with positive coefficients. This gives a purely real frequency in terms of the time derivative as a function of the parallel derivative, i.e., the square of the frequency is positive coefficients times the square of the parallel wavenumber. The rest is expressed in terms of the perturbed distribution function and we will prove that this is simply the conductive parallel flux of parallel temperature. We will do this without Fourier or Laplace transforms in the parallel direction and time.

8.6.1 Symmetrisation into even and odd components

Starting with the formulation in Eq. (8.8), we split g and h into even and odd components in z,

$$g_\pm = \frac{1}{2}[g(z) \pm g(-z)], \qquad h_\pm = \frac{1}{2}[h(z) \pm h(-z)] \tag{8.76}$$

Since $z = mv_\|$ itself is antisymmetric, we have

$$\frac{\partial}{\partial t}g_{\pm} + v_{\|}\frac{\partial}{\partial s}h_{\mp} = 0 \tag{8.77}$$

In other words, each component drives the other in a simple couple, similar to the coupling between Elsässer variables in MHD (Chapter 2 of Volume One). Moreover, we have

$$h_{+} = g_{+} + F_0\frac{e}{T}\phi, \qquad h_{-} = g_{-} - F_0\frac{e}{T}v_{\|}A_{\|} \tag{8.78}$$

with e/T_0 factors for each species, so that ϕ and $A_{\|}$ are solely associated with the even and odd components, respectively. The polarisation and induction equations become

$$k_{\perp}^2\phi = \mu_0 v_A^2 \int d\mathcal{W}\, eg_{+}, \qquad (d_e^{-2} + k_{\perp}^2)A_{\|} = \mu_0 \int d\mathcal{W}\, ev_{\|}g_{-} \tag{8.79}$$

which is further confirmation of this splitting.

8.6.2 Evolution of the field variables

To proceed we take the time derivatives of the field equations and put in for the time derivative of the g pieces on the right hand side. For ϕ we have

$$k_{\perp}^2\frac{\partial\phi}{\partial t} = \mu_0 v_A^2 \int d\mathcal{W}\, e\frac{\partial}{\partial t}g_{+} \tag{8.80}$$

Putting in the time derivative for g_{+}, the resulting space derivative in h_{-} is replaced in terms of g_{-} and $A_{\|}$, and then the integral over $v_{\|}^2F_0$ multiplying $A_{\|}$ is evaluated, to find

$$k_{\perp}^2\frac{\partial\phi}{\partial t} = -v_A^2\frac{\partial}{\partial s}\left[(d_e^{-2} + k_{\perp}^2)A_{\|} - d_e^{-2}A_{\|}\right] \tag{8.81}$$

having used the moment definition for pressure (the moment of F_0 over $v_{\|}^2$ is nT/m). The two skin depth terms in the last line cancel, manifesting the Ampère cancellation noted above. We end with

$$\frac{\partial\phi}{\partial t} = -v_A^2\frac{\partial A_{\|}}{\partial s} \tag{8.82}$$

as the two $k_{\perp}^2$ factors cancel the same way as in the fluid result.

For $A_{\|}$ we rewrite the induction equation as

$$\left(1 + k_{\perp}^2 d_e^2\right)A_{\|} = \mu_0 d_e^2 \int d\mathcal{W}\, ev_{\|}g_{-} = \frac{1}{ne}\int d\mathcal{W}\, mv_{\|}g_{-} \tag{8.83}$$

so that the time derivative is

$$\left(1 + k_{\perp}^2 d_e^2\right)\frac{\partial A_{\|}}{\partial t} = \frac{1}{ne}\int d\mathcal{W}\, mv_{\|}\frac{\partial}{\partial t}g_{-} \tag{8.84}$$

which is reduced to

$$\left(1 + k_\perp^2 d_e^2\right)\frac{\partial A_\parallel}{\partial t} = -\frac{\partial \phi}{\partial s} - \frac{1}{ne}\frac{\partial}{\partial s}\int d\mathcal{W}\, m v_\parallel^2 g_+ \tag{8.85}$$

using again the pressure moment definition in the ϕ term after replacing h_+ in terms of g_+ and ϕ. This gives the parallel electric field with skin depth correction, and perturbed parallel pressure gradient term we recall from fluid equations.

Here and below, the factors of n and T are assumed to be n_e and T_e, and e is e_0 (i.e., all pertain to the electrons).

8.6.3 The wave equation and thermal conduction

We have equations for the time derivatives of ϕ and $A_\parallel$, so we now take another time derivative of the latter and substitute the former under the spatial derivative, which yields

$$\left(1 + k_\perp^2 d_e^2\right)\frac{\partial^2 A_\parallel}{\partial t^2} = v_A^2\frac{\partial^2 A_\parallel}{\partial s^2} + \frac{1}{ne}\frac{\partial^2}{\partial s^2}\int d\mathcal{W}\, m v_\parallel^3 h_- \tag{8.86}$$

where in the last term we have used the equation for $\partial g_+/\partial t$ under the spatial derivative. The last term is a heat flux moment, but it is the total parallel flux of parallel energy. We isolate the conductive part using its definition

$$\int d\mathcal{W}\, m v_\parallel^3 h_- = \int d\mathcal{W}\,(m v_\parallel^2 - T) v_\parallel h_- + \frac{T}{\mu_0 e} k_\perp^2 A_\parallel \tag{8.87}$$

where in the last term we have used

$$\int d\mathcal{W}\, e v_\parallel h_- = \frac{1}{\mu_0} k_\perp^2 A_\parallel \tag{8.88}$$

Putting this into the wave equation for $A_\parallel$ we find

$$\left(1 + k_\perp^2 d_e^2\right)\frac{\partial^2 A_\parallel}{\partial t^2} = v_A^2\frac{\partial^2 A_\parallel}{\partial s^2} + \frac{1}{ne}\frac{\partial^2}{\partial s^2}\int d\mathcal{W}\,(m v_\parallel^2 - T) v_\parallel h_- + \frac{T}{\mu_0 n e^2} k_\perp^2 \frac{\partial^2 A_\parallel}{\partial s^2} \tag{8.89}$$

The scale of the last term is evaluated as

$$\frac{T}{\mu_0 n e^2} = V_e^2 d_e^2 = v_A^2 \rho_s^2 \tag{8.90}$$

so that

$$\left(1 + k_\perp^2 d_e^2\right)\frac{\partial^2 A_\parallel}{\partial t^2} = v_A^2\left(1 + k_\perp^2 \rho_s^2\right)\frac{\partial^2 A_\parallel}{\partial s^2} + \frac{1}{ne}\frac{\partial^2}{\partial s^2}\int d\mathcal{W}\,(m v_\parallel^2 - T) v_\parallel h_- \tag{8.91}$$

This is the final result. All terms are pure propagation, with two time or space derivatives and positive coefficients, except for the last term which gives twice the conductive parallel flux of parallel energy. The pure frequency terms give a generalised propagation velocity V_a which is the same one as in the isothermal fluid result (Eq. 8.23). Therefore, all thermal results including damping arise from the thermal conduction term.

8.7 Kinetic resonance—Landau damping

We briefly introduce the outline of kinetic computation and address some of the difficulties. The simplest case is the Alfvén response as in the fluid drift model. Background gradients and magnetic curvature effects are neglected as in the above introductions. Then, the one-dimensional formulation of velocity space in Eq. (8.12) can be used. We will do this for the electron and ion responses in a straight magnetic field, with the usual normalisation ($e\tilde{\phi}/T_e$ and $\tilde{n}_e/n_e$, etc), except that each z-space is normalised to its own species.

The variables are the distribution functions and the fields, whose dependence is

$$g_{e,i} = g_{e,i}(s, z), \qquad \phi = \phi(s) \qquad A = A(s) \tag{8.92}$$

where s is the spatial coordinate along the magnetic field, and sz defines the parallel phase space. Numerically we have

$$-\pi < s < \pi, \; -z_L < z < z_L \tag{8.93}$$

with z_L a boundary sufficiently greater than unity to not have significant spillover. Usually, $z_L = 5$. The phase space grid is equidistant and set to $N_s \times N_z = 128 \times 512$. The grid in each is such that the origin falls halfway between two grid points.

The normalised gyrokinetic equation is

$$\frac{\partial g}{\partial t} + \alpha[H_0, h] = 0 \tag{8.94}$$

for each species, where

$$H_0 = \frac{z^2}{2}, \qquad h = g + e(\phi - \alpha z A)F_M, \qquad F_M = c_0 \exp[-z^2/2] \tag{8.95}$$

The constant parameters are $e = \pm 1$ for ions and electrons, and $\alpha = \mu^{-1/2}(K_\parallel L)^{-1}$ for each, where $\mu = m/m_0$ is the mass ratio, or 2.72×10^{-4} for electrons and unity for ions. The scale ratio parameter $K_\parallel L$ is arbitrary but is chosen for convenience to have α_e close to unity in normalised units, which helps choose the timestep. The constant c_0 is calculated from the exponential factor to have the velocity space integral (as a sum over grid points) of F_M over z at unity numerically.

The field equations for polarisation and induction are

$$K^2\phi = \sum[g_i - g_e], \qquad \left(d_e^{-2} + \hat{\beta}^{-1}K^2\right)A = \sum[z(\alpha_i g_i - \alpha_e g_e)] \tag{8.96}$$

where the skin depth d_e is given by

$$d_e^{-2} = \sum[z^2 F_M(\alpha_i^2 - \alpha_e^2)] \tag{8.97}$$

The sums are over grid points in z.

The base case is given by setting

$$\alpha_e = 1, \quad \alpha_i = (3670)^{-1/2}, \quad \hat{\beta} = 1, \quad K = 0.1 \tag{8.98}$$

which sets $v_A = V_e = 60.6c_s$ with K in the MHD regime. It is sufficient to leave $\alpha_{e,i}$ fixed, with $\hat{\mu} = 1$, and vary $\hat{\beta}$ and K. In the parlance of the drift Alfvén work, $\alpha_e = \hat{\mu}^{-1/2}$ and $\alpha_i = \hat{\epsilon}^{-1/2}$. The Courant limit follows z_L times V_e, not v_A. So we set the timestep to 10^{-3} and run to $t = 100$. This is barely enough to see the start of the sound wave response, while for $\hat{\beta} = 1$ or less the Landau resonance is within the thermal distribution and the damping is rapid. The transition to $\hat{\beta} < \hat{\mu}$ and the shift of the electron Landau resonance into the tail of F_M is what is interesting.

For the case with $\hat{\beta} = \hat{\mu} = 1$ the Landau resonance is at $z = \pm 1$, which is V_e in normalised units. The time traces, in figure 8.1, show the Alfvén oscillation and decay superposed on the first quarter period of the sound wave oscillation. The free energy components are given by

$$E_E = c_1 \sum \left[\phi(f_i - f_e)\right], \qquad E_B = c_1 \sum \left[A(\alpha_i f_i - \alpha_e f_e) z\right] \tag{8.99}$$

$$E_e = c_1 \sum \left[f_e^2 / F_M\right], \qquad E_i = c_1 \sum \left[f_i^2 / F_M\right] \tag{8.100}$$

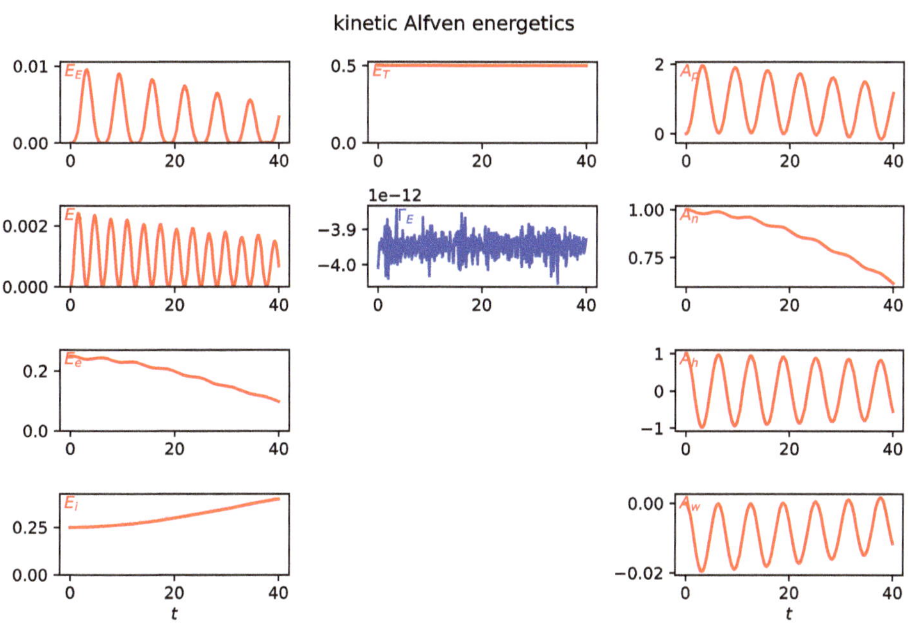

Figure 8.1. Time traces of the free energy components (left), the total free energy and growth rate (center), and four amplitude quantities (right), giving ϕ, n_e, $h = n_e - \phi$, and ϖ at $s = 0$, respectively. This is the base case with $\hat{\beta} = \hat{\mu} = 1$ and $K = 0.1$. The cleanest decay of the Alfvén response is in h, while the sound wave is best followed in n_e. The free energy components from top to bottom are the ExB, thermal electron, magnetic, and thermal ion pieces. The decay is slow because $\hat{\mu}K^2/\hat{\beta} = 10^{-2}$, deep into the inductive regime.

where the sum is over phase space grid points and $c_1 = (2N_sN_z)^{-1}$ does the averaging. Recall that $f = g - e\alpha\, zA\, F_M$ for each species, following the construction of the positive definite quantity corresponding to the thermodynamic free energy (Eqs 7.63, 7.64) from the previous Chapter). The total and its growth rate is

$$E_T = E_E + E_B + E_e + E_i, \qquad \Gamma_T = \frac{1}{2E_T}\frac{\partial E_T}{\partial t} \tag{8.101}$$

Note that due to the absence of dissipation $\Gamma_T = \Gamma_E$ is also the error. The right column of frames shows the $s = 0$ oscillation of amplitudes, each averaged over the two grid points whose interval straddles the origin. The four quantities shown are ϕ, n, $h = n - \phi$ and ϖ denoted by (A_p, A_n, A_h, A_w), respectively. Here, n is the sum of f_e over z-space grid points.

Since there are two wave components active, the kinetic shear Alfvén damping is best measured by the amplitude envelope of h, since this quantity is not involved in the sound wave. This envelope can be seen from the A_h curve to decay symmetrically about the horizontal axis. The sound wave is best visible in A_n. There is a spurious read in the shapes of the E_E and E_B time traces, since ϕ does not oscillate symmetrically. But while this dynamics is proceeding the error is very small indeed.

The phase space morphology at $t = 30$ is shown in figure 8.2. These are the electron and ion perturbed distributions $f_{e,i}$, respectively, in the sz phase space. Even with $\hat{\beta}/\hat{\mu} = 1$ the response is still almost purely inductive, which follows from $\hat{\mu}K^2/\hat{\beta}$ being small. In this regime the electron contours are seen to oscillate back and forth

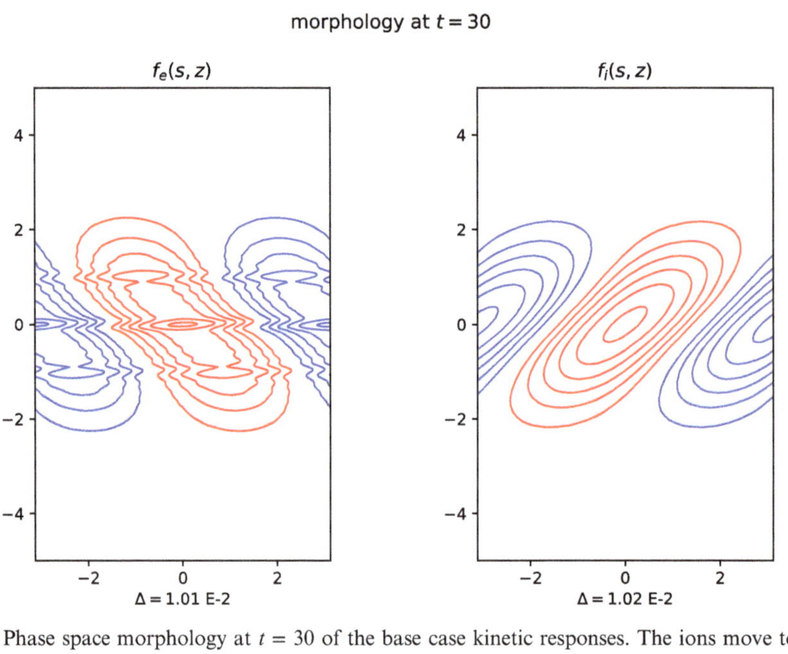

Figure 8.2. Phase space morphology at $t = 30$ of the base case kinetic responses. The ions move to the right (left) for z positive (negative), weakly affected by the waves. The electron response is due to the wave, as this is the inductive regime. This keeps the contours from mixing, except in the very narrow Landau resonance.

in s with a low amplitude, rather than streaming with the kinetic velocity αz like the ions do. At the same time, the resonances form. In the electrons we see two sets, one each for the kinetic and acoustic responses. The latter is at $z = \pm\alpha_i$ in the electron space, essentially zero. The kinetic ones are at $\pm V_a/V_e$ or $z = \pm(\alpha_e\hat{\beta})^{-1/2}$ in the electron space, which is unity here.

The shift of the resonance into the tail of F_M is done by changing v_A relative to V_e, which is being held fixed. So $\hat{\beta}$ is increased to 10 and then also decreased to 0.1, so that the resonance is moved to $z = \pm 0.32$ and then $z = \pm 3.2$, respectively. The latter resonance is not visible in the morphology since it is out of the range. Adding the values $\hat{\beta} = 3.0$ and 0.3 to the run set produces visible resonance layers whose position does follow the expected $\pm\hat{\beta}^{-1/2}$ position. The nature of the response remains the same, since even for $\hat{\beta} = 0.1$ the two factors of $K = 0.1$ lead to a ratio $\hat{\beta}/\hat{\mu}K^2$ of ten. The decay rate also follows $\hat{\beta} - 1$ in this range, also as in the fluid result.

When K is increased we see a transition to a different sort of response. Keeping $\hat{\beta} = 10$, the values are set to $K = 0.3$ and then $K = 1.0$. The latter shows how the kinetic shear Alfvén wave transitions from a pure MHD wave to a more fluidlike response. The time traces for the $K = 1.0$ case are shown in figure 8.3, with the much sharper damping rate reflecting the larger size of $\hat{\mu}K^2/\hat{\beta} = 0.1$ compared to the base case. The sound wave is also sharply damped with this value of K. The phase space morphology at $t = 10$ for this case is shown in figure 8.4. In the $K = 0.1$ cases the electron contours were seen to oscillate back and forth, giving the inductive

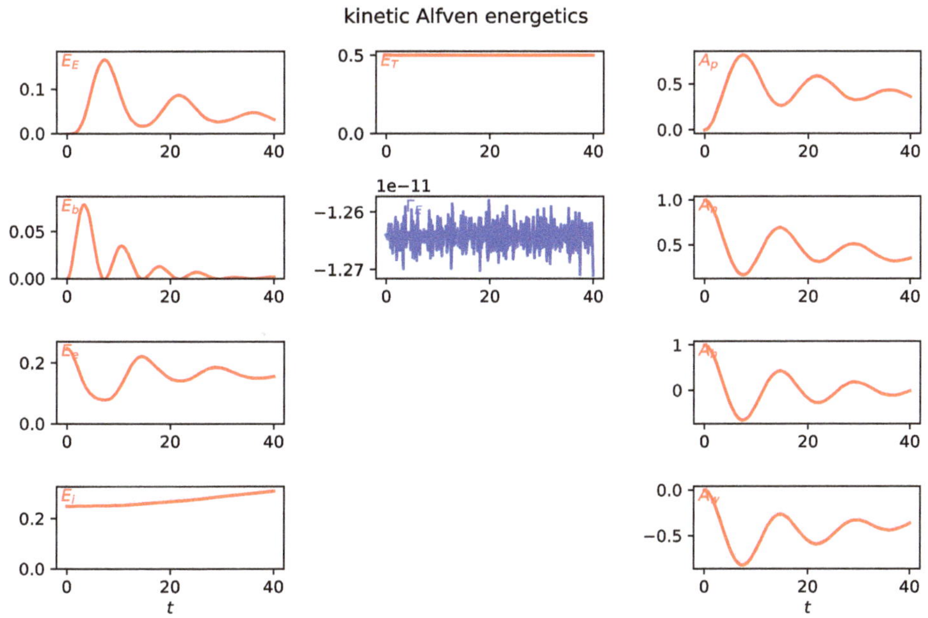

Figure 8.3. Time traces of the free energy components (left), the total free energy and growth rate (center), and four amplitude quantities (right), as in figure 8.1 for the base case. This is for $\hat{\beta} = 10$ and $K = 1$. The decay of both the sound wave and acoustic responses is now very sharp, due mostly to the K value.

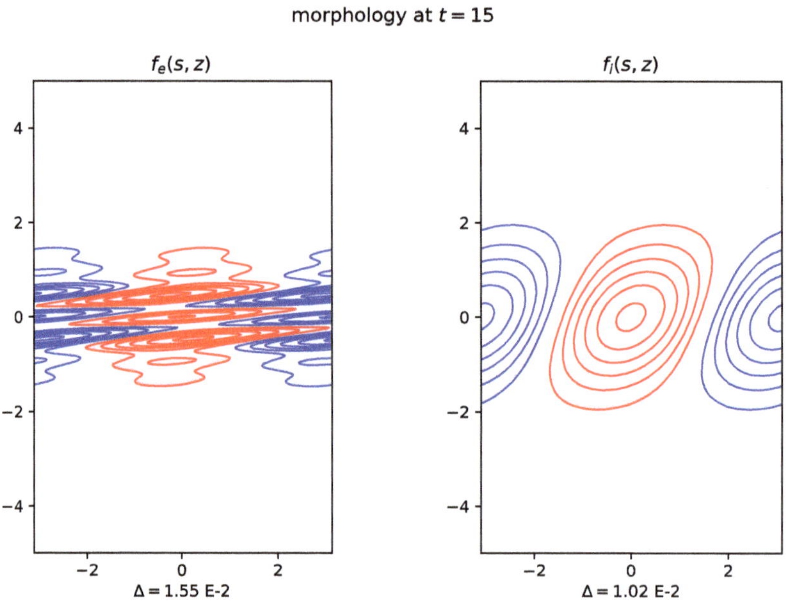

Figure 8.4. Phase space morphology at $t = 15$ of the kinetic responses, for the case with $\hat{\beta} = 10$ and $K = 1$. With the Alfvén speed v_A about one third of the electron speed V_e, the activity is at lower normalised z-values (the scale is still larger than the skin depth). The wave damping is much faster, due to the rapid mixing in the electrons and the polarisation in the ions.

response. For $K = 1.0$, with V_a closer to V_e, the response is like a fluid, namely, the contours move continuously according to which z value they are on, so that the effective shear in phase space tilts the lines more and more to upper right and lower left, much as the ions are doing on their much slower time scale. At this point the Landau damping is not a resonance but a manifestation of the motion of a fluid, which is called phase-space mixing. Spatially, the electron partial charge density becomes mixed as the phase space structure is tilted or sheared in this fashion).

Once again we compare this to the low-K result. The time traces for the $K = 0.1$ case with $\hat{\beta} = 10$ is shown in figure 8.5. With both K and μ_e/β_e small, this case is in the traditional MHD regime. The Alfvén wave is much slower relative to $k_\| V_e$, and on this time scale there is minimal decay in the amplitude A_h, as expected. The phase space morphology is shown in figure 8.6. Now both the acoustic and Alfvén resonances are close to zero in units of V_e, placing maximal pressure on the grid resolution. For kinetic velocities greater than v_A, which is most of the distribution, the electron contours oscillate with the wave, unlike the ions which remain fluidlike.

The main point of this cursory study is the resolution demand on velocity space to resolve this Alfvénic Landau resonance. When the response is mostly inductive, as especially in hot plasma cases, the parameter $\hat{\beta}/\hat{\mu}K^2$ is rather large even for short wavelength. But then in warm plasmas was also have large values for this, because more of the activity is in the long wavelengths. So the inductive response is always the most relevant for the energy containing range. Even if the linear growth phase of an instability is mostly electrostatic, the responses in the turbulence are always electromagnetic.

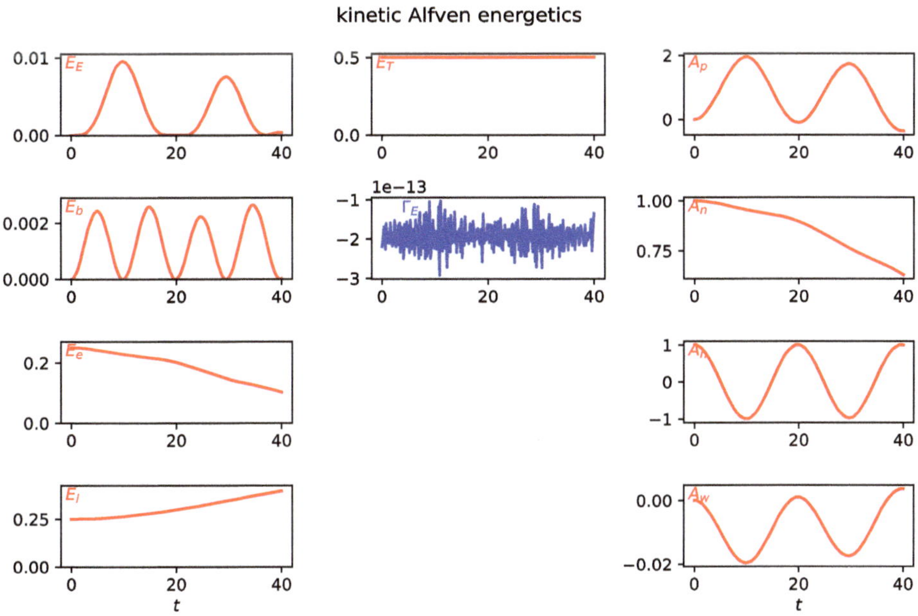

Figure 8.5. Time traces of the free energy components (left), the total free energy and growth rate (center), and four amplitude quantities (right), as in figure 8.1 for the base case. This is for $\hat{\beta} = 10$ and $K = 0.1$. The decay is now minimal, due to the low K and small μ_e/β_e value.

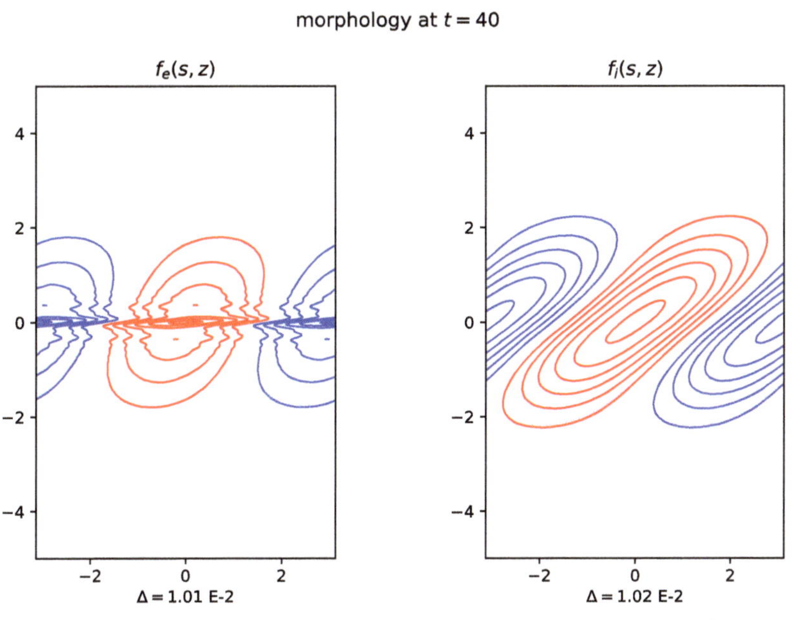

Figure 8.6. Phase space morphology at $t = 40$ of the kinetic responses, for the case with $\hat{\beta} = 10$ and $K = 0.1$. With the Alfvén speed v_A about one third of the electron speed V_e, the activity is at lower normalised z-values. The wave damping is much slower, due to the small $k_\perp d_e$, deeply into the inductive regime.

8.8 Summary

The demands of resolving any possible resonance wherever in velocity space it should occur make well resolved kinetic computation very expensive indeed. We have not discussed collisions beyond what was given in Chapter 6 of Volume One. Landau damping in a kinetic bath does not by itself dissipate free energy, as we see from the small changes to the total free energy in the above cases. In the real world however, the collisions are at the end of the chain. If the simplest pitch angle scattering operator of Section 6.10 therein is used, it is simply added to the gyrokinetic equation of each species, and then it is responsible for dissipation, which reduces to a constant value as the actual size of the coefficient decreases.

Indeed the idea of high Reynolds number convergence with small diffusivities also translates to the velocity space situation with small collisional diffusion in velocity space. Since correct conductive and viscous dissipation rates depend on the correct collision frequency, one cannot artificially make this scale tractable as one does for the spatial situation with subgrid dissipation. A subgrid collision operator would give reasonable results for structure in the distribution function, but the resulting dissipation of resonances would be wrong. The singular nature of the resonance layers is what leads to this difficulty, which is not generally faced in the computational literature.

The result from Section 8.6 equates the Landau damping from the kinetic theory to the parallel heat flux in a description which may be classed as a fluid but nevertheless is just another presentation of the kinetic theory. That is, it is not a truncation but a repackaging. In the collisional regime described by the fluid model, this dissipation is provided by the resistivity and the conductive heat flux, which is given by a collisional model. The same process, in the kinetic regime described by weak collisionality, continuously segues into Landau damping. The resistivity itself goes to zero, as there is no direct damping in the Ohm's law (the parallel momentum equation for electrons). So as a physical process, Landau damping is relegated to the thermal responses and therefore has the same role as thermal conduction. This provides some motivation into the construction of collisionless fluid equations with Landau damping as a process. It is not frictional drag or viscosity, but thermal conduction. So any treatment to model it should appear only in the dynamical equation for the heat flux, in which it should have a similar form to the thermal conduction dissipation as a direct damping of the heat flux. This is what we turn to in the next Chapter.

Further Reading

As a student, I learned this material from the text, *Principles of Plasma Physics*, by N Krall and A Trivelpiece (McGraw-Hill, 1973). Also recommended is *Plasma Physics: An Introduction*, by R Fitzpatrick (CRC Press, 2014). His related lectures are at the University of Texas at Austin, http://farside.ph.utexas.edu/teaching/plasma/lectures/.

The plasma dispersion function is the subject of a book by B Fried and S Conte, *The Plasma Dispersion Function: The Hilbert Transform of the Gaussian* (Academic

Press, 1961). The reference most people use is the pamphlet by J Huba *et al*, 'The NRL Plasma Formulary,' available from https://http://www.nrl.navy.mil/ppd/content/nrl-plasma-formulary.

Landau damping is covered in the above texts, and is due to L D Landau, in 'On the vibrations of the electronic plasma,' *Zh. Eksp. Teor. Fiz.* **10** (1946) 10, in *J. Phys.* **10** (1946) 45, and in *JETP* **16** (1946) 574; reprinted in *Collected papers of L D Landau*, D ter Haar, ed. (Pergamon, Oxford, 1965).

For experimental measurements, see J H Malmberg and C B Wharton, 'Collisionless damping of electrostatic plasma waves,' *Phys. Rev. Lett.* **13** (1964) 184, and H Derfler and T C Simonen, 'Landau waves: an experimental fact,' *Phys. Rev. Lett.* **17** (1966) 172, in two different parameter regimes.

References for the kinetic shear Alfvén wave and Alfvén drift instability were given in Chapter 8 of Volume One.

The question of Landau damping and entropy has been treated by F Winterberg, 'Landau Damping and Entropy,' *Z. Naturforschung* **22a** (1967) 1682. The apparent paradox that the Boltzmann entropy is conserved is resolved by considering macroscopic entropy, which shares features of the delta-f free energy.

The idea of an energy cascade and a velocity space Reynolds number is given by B Coppi and F Santini, in 'Analogy between Turbulent Transfer in Velocity Space and Plasma Collisional Transport in Real Space,' *Phys. Rev. Lett.* **28** (1972) 19.

IOP Publishing

Turbulence and Instabilities in Magnetised Plasmas, Volume 2

Gyrokinetic theory and gyrofluid turbulence

Bruce Scott

Chapter 9

Introduction to gyrofluid theory

9.1 Introduction

It is often the case that gyrokinetic computation is either too expensive or pathological. The former refers to resolution needs and the latter refers to problems whose resolution needs are essentially unobtainable. This can happen either due to the temperature contrast of the problem, requiring a velocity space at least one factor of $\delta = \rho_s/a$ for a perpendicular system size a spanning several gradient scale lengths $L_\perp$, or to the existence of resonances, including simple ones like the Alfvén–Landau resonance for arbitrarily small collisionality, or due to the inability to construct a model which is both energetically consistent and can treat global geometry. It is also often the case that with vast resources a gyrokinetic computation is made tractable, at least according to one's own criteria, but the fact that f contains all available degrees of freedom, some of which have greatly different roles, one cannot discern the governing physics of a situation without creating a narrative which one then simply defends. Especially for learning purposes, one would like something more flexible, and then for research purposes, one which is all of flexible, tractable, energetically consistent, and also treats the underlying physics of the system one is addressing. This Chapter covers the ideas and systematics of gyrofluid models, which capture a necessary subset of the content of the corresponding gyrokinetic theory which describes turbulence in magnetised plasmas. Moreover, this approach provides a direct derivation path via correspondence in the appropriate limit back to the delta-f fluid drift equations we were using prior to Chapter 4. As in the gyrokinetic theory (Chapter 5), a key concept is the way in which vorticity and gyrocenter charge density substitute for each other in the theory, and that in the long wavelength limit they are isomorphic (polarisation and ExB vorticity become proportional).

9.2 Heuristic gyrofluid 2D turbulence

With a simple illustration we can motivate the gyrofluid idea. As with gyrokinetic theory, it is a representation of the dependent variables. Although it facilitates the

doi:10.1088/978-0-7503-3855-4ch9

use of finite Larmor radius (FLR) corrections, it is not natively a result of applying gyroaveraging schemes to the fluid model. Its main point is that the polarisation drift in the velocities is replaced by a polarisation density, such that although the plasma charge density is taken to be arbitrarily small (limit of $\varepsilon_0 \to 0$), the polarisation density can assume the same role in relating the electrostatic potential (i.e., the drift velocity stream function) to the difference between two gyrocentre densities.

We start with the simple 2D hydrodynamic turbulence model with a passive advection equation for a transported scalar (Chapters 3 and 4, Volume One). The quantities advanced by advection equations are the vorticity, ϖ, and the density n, with allowance for a background gradient in the latter. The two ExB advection equations are

$$\frac{\partial \varpi}{\partial t} + [\phi, \varpi] = 0, \qquad \frac{\partial n}{\partial t} + [\phi, n + \log n_0] = 0 \tag{9.1}$$

where $\log n_0(x)$ accounts for a background profile of arbitrary structure. The Poisson bracket is defined as $[a, b] = \hat{z} \times \nabla a \cdot \nabla b$. The definition of the vorticity in terms of the fluid stream function is replaced by a field equation for the electrostatic potential, sourced by the vorticity according to

$$\nabla_\perp^2 \phi = \varpi \tag{9.2}$$

Possibly looking for simply a different computational scheme, we may define an auxiliary variable

$$N = n - \varpi \tag{9.3}$$

and subtract the time dependent equations (Eqs 9.1) to obtain

$$\frac{\partial N}{\partial t} + [\phi, N + \log n_0] = 0 \tag{9.4}$$

What this has done is to re-arrange the variables such that n and N are the dependent variables, and the field equation is now

$$\nabla_\perp^2 \phi = n - N \tag{9.5}$$

This is the simplest gyrofluid model one can construct. At this level it is isomorphic to the 2D fluid equation with a passive scalar transported by the fluid. There is still no FLR correction but we now have the representation: n and N are the dependent variables, each independently advected by ϕ in Eqs (9.1) and (9.4), and the field equation (Eq. 9.5), which is now identified as our gyrofluid polarisation (Poisson) equation. If the electrons are adiabatic, we set $n = \phi$ and then the model is identical to the Hasegawa–Mima model (1977/8).

If we go back and keep $\nabla_\parallel J_\parallel$ on both right hand sides of Eqs (9.1), and furthermore keep gyroaveraging effects through a gyroaveraging operator G, we can reconstruct the model as

$$\frac{\partial n}{\partial t} + [\phi, n + \log n_0] = \nabla_\parallel J_\parallel, \qquad \frac{\partial N}{\partial t} + [G(\phi), N + \log n_0] = 0 \tag{9.6}$$

with field equation

$$\frac{1 - \Gamma_0}{\tau_i}\phi = G(N) - n \tag{9.7}$$

with G in both places due to energetic consistency. The physical interpretation is that $G(\phi)$ gives the gyroaveraged potential to advect the ions and $G(N)$ places the ion gyrocentre charge density into space. The operator G is derived from a background Maxwellian according to

$$G(\mathbf{k}) = \frac{1}{n_0}\int d\mathcal{W}\, F_0 J_0(k_\perp \rho_L) = e^{-b/2} \qquad \text{where} \quad b = k_\perp^2 \rho_i^2 \tag{9.8}$$

where $\rho_i = V_i/\Omega_i$ with $V_i^2 = T_i/m_i$ and $\Omega_i = eB/m_i$ for the ions. The operator G is applied in k-space as

$$G(\phi) = \sum_{\mathbf{k}} G(b)\phi_{\mathbf{k}}\, e^{i\mathbf{k}\cdot\mathbf{x}} \tag{9.9}$$

the same way as for J_0 in Eq. (5.79) of Chapter 5. It is not the same as the Γ_0, defined

$$\Gamma_0(\mathbf{k}) = \frac{1}{n_0}\int d\mathcal{W}\, F_0 J_0^2(k_\perp \rho_L) = e^{-b} I_0(b) \tag{9.10}$$

as in Eq. (5.115) of Chapter 5, since G has one factor of J_0 while Γ_0 has two. The parameter τ_i gives the ion/electron temperature ratio; for most applications it is unity. This model is the one given by Knorr *et al* (1988).

The energetics of the model is the same as in delta-f gyrokinetics. If we multiply the equations for n and N by $n - \phi$ and $\tau_i N + G(\phi)$, respectively, we have

$$\begin{aligned}
&n\frac{\partial n}{\partial t} + \tau_i N\frac{\partial N}{\partial t} + G(\phi)\frac{\partial N}{\partial t} - \phi\frac{\partial n}{\partial t} \\
&+\left[G(\phi), \left(\tau_i\frac{N^2}{2} + G(\phi)(N + \log n_0)\right)\right] + \left[\phi, \left(\frac{n^2}{2} - \phi(n + \log n_0)\right)\right] \\
&+ \nabla_\parallel(\phi - n)J_\parallel \\
&= -\tau_i N[G(\phi), \log n_0] - n[\phi, \log n_0] - J_\parallel \nabla_\parallel(n - \phi)
\end{aligned} \tag{9.11}$$

where the second line gives the ExB free energy fluxes including the nonlinear polarisation current pieces, the third line gives the conservative flux due to the parallel current, and the last line gives the source term due to advection down the background density gradient and the Alfvénic and adiabatic transfer through $J_\parallel$ to the rest of the system. If the Hasegawa–Wakatani closure $\nabla_\parallel J_\parallel \to C^{-1}(n - \phi)$ is used, this becomes $CJ_\parallel^2$, the resistive sink in the fluid drift case. The first line is seen to be analogous to the $h(\partial g/\partial t)$ form in the delta-f free energy theorem in Chapter 7, and using the polarisation equation it may be recast as

$$\frac{\partial E_E}{\partial t} = \mathcal{V}^{-1}\int d\mathcal{V}\left[G(\phi)\frac{\partial N}{\partial t} - \phi\frac{\partial n}{\partial t}\right], \quad \frac{\partial E_f}{\partial t} = \mathcal{V}^{-1}\int d\mathcal{V}\left[\tau_i N\frac{\partial N}{\partial t} + n\frac{\partial n}{\partial t}\right] \tag{9.12}$$

where the ExB energy itself is

$$E_E = \mathcal{V}^{-1} \int d\mathcal{V}\, \phi \frac{1 - \Gamma_0}{2\tau_i} \phi = \mathcal{V}^{-1} \int d\mathcal{V}\, \frac{1}{2}[G(\phi)N - \phi n] \tag{9.13}$$

The two forms are identical under the integral. As in the delta-f gyrokinetic case, we do not write the field wave flux explicitly because it is the solution of an equation of the type

$$aG(b) - bG(a) = \nabla \cdot \mathbf{F} \tag{9.14}$$

which we can state since G is Hermitian. The thermal free energy due to ions and electrons E_f is the same as in the fluid models, except that here we have N for the ions rather than n. This is another FLR effect due to the representation. We return to the significance of this in Section 10.5 on the correspondence to the delta-f drift fluid model under the same level of simplification.

9.3 Heuristic gyrofluid 3D turbulence

This same system works as well for the four-field drift Alfvén model we were using in the latter half of Volume One. If we take the density and vorticity equations

$$\frac{\partial n}{\partial t} + [\phi, n + \log n_0] + \nabla_\parallel(u - J) = \mathcal{K}(n - \phi) \tag{9.15}$$

$$\frac{\partial \varpi}{\partial t} + [\phi, \varpi] + \nabla_\parallel(-J) = \mathcal{K}(-n) \tag{9.16}$$

and subtract them, defining N as before we find

$$\frac{\partial N}{\partial t} + [\phi, N + \log n_0] + \nabla_\parallel(u) = \mathcal{K}(\phi) \tag{9.17}$$

This will serve as the cold-ion equation for the ion gyrocentre density. If we take the Ohm's law and parallel momentum equation without the viscosity term

$$\frac{\partial}{\partial t}\left(A + \mu_e J\right) + \mu_e[\phi, J] + \nabla_\parallel(\phi - n - \log n_0) = CJ \tag{9.18}$$

$$\mu_i \frac{\partial u}{\partial t} + \mu_i[\phi, u] + \nabla_\parallel n = 0 \tag{9.19}$$

and add them, restoring the electron contribution to the parallel momentum equation beforehand (with $v = u - J$), and replacing J with $-v$ in the Ohm's law, we find

$$\frac{\partial}{\partial t}\left(A - \mu_e v\right) - \mu_e[\phi, v] + \nabla_\parallel(\phi - n - \log n_0) = CJ \tag{9.20}$$

$$\frac{\partial}{\partial t}\left(A + \mu_i u\right) - \mu_i[\phi, u] + \nabla_\parallel(\phi) = CJ \tag{9.21}$$

where the latter is a cold-ion equation for the ion parallel momentum (in these, we normalise in terms of β_e rather than $\hat{\beta}$, etc, and $\nabla_\parallel$ is normalised to $L_\perp^{-1}$ rather than $K_\parallel$, and the masses are normalised in terms of a reference ion mass, so that $\mu_i = 1$ is usual). Now, the parallel electric field together with resistive friction is in both equations, as we have decoupled them. Heuristically restoring the ion temperature and gyroaveraging as for the 2D case, we can write the four equations as

$$\frac{\partial n}{\partial t} + [\phi, n + \log n_0] + \nabla_\parallel v = \mathcal{K}(\phi - n) \tag{9.22}$$

$$\frac{\partial}{\partial t}(A - \mu_e v) - \mu_e[\phi, v] + \nabla_\parallel(\phi - n - \log n_0) = C(u - v) \tag{9.23}$$

$$\frac{\partial N}{\partial t} + [\phi_G, N + \log n_0] + \nabla_\parallel u = \mathcal{K}(\phi_G + \tau_i N) \tag{9.24}$$

$$\frac{\partial}{\partial t}(A + \mu_i u) + \mu_i[\phi_G, u] + \nabla_\parallel(\phi_G + \tau_i N + \tau_i \log n_0) = C(u - v) \tag{9.25}$$

where $\phi_G = G(\phi)$ is written as a shorthand. This can serve as a gyrofluid model containing two degrees of freedom per species, density (n, N) and parallel velocity (v, u), in each case for electrons and ions, respectively. The field equations are written analogously to the 2D model,

$$\frac{1 - \Gamma_0}{\tau_i}\phi = G(N) - n, \qquad -\beta_e^{-1}\nabla_\perp^2 A = u - v \tag{9.26}$$

where the first one is the polarisation equation written explicitly as a gyrofluid Poisson equation, and the second one is the induction equation (Ampère's law) written in terms of u and v rather than J. As in the fluid case, we keep background gradient forcing involving $\log n_0$ only in ExB advection or magnetic flutter, but not in curvature so as to avoid doing the full toroidal equilibrium.

The conserved energy for this model is

$$E_T = E_E + E_M + E_e + E_i + E_v + E_u \tag{9.27}$$

where the pieces are the ExB energy,

$$E_E = \mathcal{V}^{-1}\int d\mathcal{V}\,\phi\frac{1 - \Gamma_0}{2\tau_i}\phi = \mathcal{V}^{-1}\int d\mathcal{V}\,\frac{1}{2}[\phi_G N - \phi n] \tag{9.28}$$

the magnetic energy,

$$E_M = \mathcal{V}^{-1}\int d\mathcal{V}\,\frac{1}{2\beta_e}|\nabla_\perp A|^2 = \mathcal{V}^{-1}\int d\mathcal{V}\,\frac{1}{2}(u - v)A \tag{9.29}$$

the electron thermal free energy and parallel kinetic energy

$$E_e = \mathcal{V}^{-1}\int d\mathcal{V}\,\frac{1}{2}n^2, \qquad E_v = \mathcal{V}^{-1}\int d\mathcal{V}\,\frac{1}{2}\mu_e v^2 \tag{9.30}$$

and the ion thermal free energy and parallel kinetic energy

$$E_i = \mathcal{V}^{-1} \int d\mathcal{V} \, \frac{1}{2} \tau_i N^2, \qquad E_u = \mathcal{V}^{-1} \int d\mathcal{V} \, \frac{1}{2} \mu_i u^2 \tag{9.31}$$

In the time dependent equations the magnetic and parallel kinetic energy contributions per species enter as a unit through the parallel canonical momentum, which cannot be split. Therefore, we have to re-arrange them into a parallel energy per species. The time derivatives and energy transfer are given by

$$\begin{aligned} \frac{\partial E_E}{\partial t} &= \mathcal{V}^{-1} \int d\mathcal{V} \left[\phi_G \frac{\partial N}{\partial t} - \phi \frac{\partial n}{\partial t} \right] \\ &= \mathcal{V}^{-1} \int d\mathcal{V} \left[(u - v) \nabla_\parallel \phi + u \nabla_\parallel (\phi_G - \phi) - n \mathcal{K}(\phi) - \tau_i N \mathcal{K}(\phi_G) \right] \end{aligned} \tag{9.32}$$

$$\begin{aligned} \frac{\partial E_{e\parallel}}{\partial t} &= \mathcal{V}^{-1} \int d\mathcal{V} \left[-v \frac{\partial}{\partial t} (A - \mu_e v) \right] \\ &= \mathcal{V}^{-1} \int d\mathcal{V} \left[v \nabla_\parallel \phi - v \nabla_\parallel n + v \, C(u - v) \right] \end{aligned} \tag{9.33}$$

$$\begin{aligned} \frac{\partial E_{i\parallel}}{\partial t} &= \mathcal{V}^{-1} \int d\mathcal{V} \left[u \frac{\partial}{\partial t} (A + \mu_i u) \right] \\ &= \mathcal{V}^{-1} \int d\mathcal{V} \left[-u \nabla_\parallel \phi + u \nabla_\parallel (\tau_i N) - u \, C(u - v) \right] \end{aligned} \tag{9.34}$$

$$\frac{\partial E_e}{\partial t} = \mathcal{V}^{-1} \int d\mathcal{V} \left[n \frac{\partial n}{\partial t} \right] = \mathcal{V}^{-1} \int d\mathcal{V} \left[-v \nabla_\parallel n + n \mathcal{K}(\phi) \right] \tag{9.35}$$

$$\frac{\partial E_i}{\partial t} = \mathcal{V}^{-1} \int d\mathcal{V} \left[\tau_i N \frac{\partial N}{\partial t} \right] = \mathcal{V}^{-1} \int d\mathcal{V} \left[-u \nabla_\parallel (\tau_i N) + \tau_i N \mathcal{K}(\phi_G) \right] \tag{9.36}$$

which shows the way towards investigating energetics as we have done previously.

These equations and their resulting energetics follow directly from the delta-f gyrokinetic model, if we do a moment breakdown according to

$$n_z = \frac{1}{n_0} \int d\mathcal{W} f_z, \qquad u_z = \frac{1}{n_0 V} \int d\mathcal{W} \, v_\parallel f_z \tag{9.37}$$

for each species z, with n_0, m and V the species background density, mass and thermal velocity, respectively (here in this Chapter, we use $v_\parallel$ instead of z/m since we will use z as the corresponding normalised coordinate). In turn, this implies a decomposition of f with two degrees of freedom,

$$f_z = \frac{1}{n_0} \left(n_z + u_z \frac{v_\parallel}{V} \right) F_0 \tag{9.38}$$

in which the moment variables are the coefficients. Both n_z and u_z are normalised quantities. It will be observed that the two pieces in f_z are orthogonal—each moment definition picks off one of them but no part of the other. This tells us in general how to build these models as moment hierarchies upon the foundation of the underlying gyrokinetic equation. We can only go so far with the heuristic method however. To treat temperatures and especially the anisotropy between parallel and perpendicular temperatures, not to mention understanding the reason why temperature anisotropy is a central feature of the model, we will have to turn to the systematics, which we cannot delay any further.

9.4 Gyrofluid systematics

We follow a procedure which for vector-velocity fluid equations with isotropic temperatures but finite viscosities representing off-diagonal pressure tensor components (*i.e.*, $P_{ij} = pg_{ij} + \Pi_{ij}$) is due to Grad (1949). This involved establishing the properties of a set of orthogonal Hermite polynomials over the coordinates of velocity space, and then representing f in terms of the relevant decomposition.

In a magnetised plasma we have different dynamics for z and w, which are normalised coordinates associated with the parallel velocity and the magnetic moment, respectively. This is done separately for each species, with the field equations represented accordingly. All coordinates in velocity space are normalised to V, the species thermal velocity, and densities and temperatures and their products in terms of n_0 and T_0, the background density and temperature of each species. In these terms we have

$$\int d\mathcal{W}\, F_0 = n_0 \int_{-\infty}^{\infty} \frac{dz\; e^{-z^2/2}}{\sqrt{2\pi}} \int_0^{\infty} dw\; e^{-w} = n_0 \tag{9.39}$$

as the representation of velocity space. Spatially, we will leave the equations unnormalised. The integrals give unity in this case without any polynomials in the integrands.

Unlike the case of isotropic velocity space, we have different kernels against which to set up the orthogonal polynomials. With $-z^2/2$ as the argument of the exponential in the parallel part, Hermite polynomials are the clear choice. With $-w$ as the argument of the exponential in the perpendicular part of velocity space, Laguerre polynomials become the choice.

Hermite polynomials $H_m(z)$, with the subscript giving the degree, satisfy the orthogonality relation

$$\int_{-\infty}^{\infty} \frac{dz\; e^{-z^2/2}}{\sqrt{2\pi}}\, H_k(z)\, H_m(z) = \delta_{km} \tag{9.40}$$

In our case, in each polynomial the factor with the highest degree is listed first with a positive sign, and the orthogonality relation gives unity for $k = m$ and zero otherwise—this is the meaning of the Kronecker delta symbol on the right hand side. In other words, the polynomials are orthonormal. These are calculated manually with the Gram–Schmidt process: with $H_0 = 1$ and H_k with $k = 0, 1, \ldots, (m - 1)$ to hand we systematically find H_m by demanding that

$$\int_{-\infty}^{\infty} \frac{dz\, e^{-z^2/2}}{\sqrt{2\pi}}\, z^k\, H_m(z) = \int_{-\infty}^{\infty} \frac{dz\, e^{-z^2/2}}{\sqrt{2\pi}}\, z^m\, H_k(z) = 0 \qquad \text{for} \quad k < m \tag{9.41}$$

and then the case with $k = m$ is decided by the demand that the integral with H_m^2 yields unity. The first few of these are

$$\begin{aligned} &H_0 = 1 \qquad H_1 = z, \qquad H_2 = \frac{z^2 - 1}{\sqrt{2}}, \qquad H_3 = \frac{z^2 - 3}{\sqrt{6}} z, \\ &H_4 = \frac{z^4 - 6z^2 + 3}{\sqrt{24}}, \qquad H_5 = \frac{z^4 - 10z^2 + 15}{\sqrt{120}} z \end{aligned} \tag{9.42}$$

The following sequence and recurrence relations

$$\frac{dH_m}{dz} = \sqrt{m}\, H_{m-1}, \qquad \sqrt{m+1}\, H_{m+1} = z\, H_m - \sqrt{m}\, H_{m-1} \tag{9.43}$$

are used in setting up the eventual moment equations.

Laguerre polynomials $L_n(w)$, with the subscript giving the degree, satisfy the orthogonality relation

$$\int_0^{\infty} dw\, e^{-w}\, L_l(w)\, L_n(w) = \delta_{ln} \tag{9.44}$$

The same order and sign convention is used, with the w^n term in L_n appearing first and the normalisation set such that the L's are orthonormal. Also calculated manually with the Gram–Schmidt process, the first few of these are

$$\begin{aligned} L_0 &= 1, \qquad L_1 = w - 1, \qquad L_2 = \frac{w^2 - 4w + 2}{2}, \\ L_3 &= \frac{w^3 - 9w^2 + 18w - 6}{6}, \qquad L_4 = \frac{w^4 - 16w^3 + 72w^2 - 96w + 24}{24} \end{aligned} \tag{9.45}$$

The sequence and recurrence relations we will use are

$$\frac{dL_n}{dw} = \left(1 - \frac{d}{dw}\right) L_{n-1}, \qquad (n+1)L_{n+1} = (w - 2n - 1)L_n - nL_{n-1} \tag{9.46}$$

The procedure is to multiply the delta-f gyrokinetic equation by H_m or L_n in turn and integrate over velocity space. Before doing that, though, we turn to the representation for f, the free energy, and the field equations, as these will tell us how to arrange the truncation such that exact energetic consistency is maintained.

9.4.1 Representation

With the polynomials established, we define the representation of f, essentially the transform into gyrofluid moment variables and then the back transform which returns f in the finite representation, much as in a Fourier decomposition in space.

Since the velocity space integral separates into individual integrations over z and w, we also have in the orthogonality relations

$$\frac{1}{n_0}\int d\mathcal{W}\,H_m(z)\,H_n(z)\,F_0 = \frac{1}{n_0}\int d\mathcal{W}\,L_m(w)\,L_n(w)\,F_0 = \delta_{mn} \tag{9.47}$$

as well as the combined one

$$\frac{1}{n_0}\int d\mathcal{W}\,H_m(z)\,H_{m'}(z)\,L_n(w)\,L_{n'}(w)\,F_0 = \delta_{mm'}\,\delta_{nn'} \tag{9.48}$$

Therefore, we define

$$f(\mathbf{x}, z, w) = \sum_{mn} a_{mn}(\mathbf{x})\,H_m(z)\,L_n(w)\,F_0(s, z, w) \tag{9.49}$$

in which the dimensional velocity space coordinates are $p_z = mv_\parallel$ and M, and the normalised ones are $z = p_z/mV$ and $w = M\Omega/T_0$. We incorporate the factor of B into the definition of w and remember that it commutes with $[,]$ and $\mathcal{K}$ but not $\nabla_\parallel$ during the derivation until we are finished. The space coordinates are x, y, s with s the parallel coordinate. The geometric coefficients depend on s only and the perpendicular derivatives on the dependent variables involve x, y only such that the commutation rules are maintained.

The reverse of the decomposition of f is the recovery of the moment variables from f. This proceeds according to the orthogonality relations, so that

$$a_{mn}(\mathbf{x}) = \frac{1}{n_0}\int d\mathcal{W}\,H_m(z)\,L_n(w)\,f(\mathbf{x}, z, w) \tag{9.50}$$

Hence, with the a_{mn} to hand we can reconstruct f, and the f is finitely discretised in velocity space according to a finite set of moment variables a_{mn}.

9.4.2 Free energy

The form of the free energy derived from the minimisation procedure is

$$E_T = \sum_{sp}\int d\Lambda\,\frac{T_0}{2F_0}(\delta g)(\delta h) \tag{9.51}$$

where

$$(\delta h) = (\delta g) + \frac{F_0}{T_0}(\delta H), \qquad \delta H = eJ_0(\phi) - e\frac{p_z}{m}J_0(A_\parallel) \tag{9.52}$$

which we derived in the run up to Eq. (7.62) from Chapter 7 (here the notation δH is used rather than H_1, which is reserved for the $m = 1$ Hermite polynomial). The positive-definite form of the free energy is

$$E_T = \sum_{sp}\int d\Lambda\,\frac{1}{2}\left[\frac{T_0}{F_0}(\delta f)^2 + e(J_0\phi)(\delta f) + e\frac{p_z}{m}(J_0A_\parallel)(\delta f)\right] \tag{9.53}$$

such that

$$\delta g = \delta f + \frac{e}{T_0}\frac{p_z}{m}F_0(J_0 A_\parallel), \qquad \delta h = \delta f + \frac{e}{T_0}F_0(J_0\phi) \tag{9.54}$$

as we had in Chapter 7. It is this δf which we call f and subject to the decomposition in terms of a_{mn}.

In terms of the decomposition, the thermal free energy due to f^2 for each species is given by

$$E_f = \mathcal{V}^{-1}\int d\mathcal{V}\int d\mathcal{W}\,\frac{T_0}{2F_0}f^2 = \mathcal{V}^{-1}\int d\mathcal{V}\,\frac{1}{2}n_0 T_0 \sum_{mn} a_{mn}^2 \tag{9.55}$$

now defined as a volume average. One finds this by substituting Eq. (9.49) into each factor of f, with two factors of F_0 cancelling the one in the denominator, and then the orthogonality relations taking care of each product of H's and L's, yielding a diagonal sum with no cross terms. We now see the origin of the free energy expressions we used in the fluid models, and in the heuristic derivations given above: the first two moments a_{00} and a_{10} correspond to n/n_0 and $u_\parallel/mV$ for the species, respectively. Note that the a's are always dimensionless, and that there is one E_f per species.

The ExB energy is the first place we see the need for an FLR closure. It is given by

$$E_E = \mathcal{V}^{-1}\int d\mathcal{V}\int d\mathcal{W}\sum_{sp}\frac{1}{2}e(J_0\phi)f = \mathcal{V}^{-1}\int d\mathcal{V}\sum_{sp}\sum_{n}\frac{1}{2}n_0 e\phi_n\, a_{0n} \tag{9.56}$$

where we now have a hierarchy of gyroaveraged potentials ϕ_n defined by

$$\phi_n(\mathbf{x}) = \frac{1}{n_0}\int d\mathcal{W}\, L_n(w)\, F_0(J_0\phi) \tag{9.57}$$

which in k-space yields

$$\phi_n(\mathbf{x}) = \sum_{\mathbf{k}}\phi_{\mathbf{k}}\, e^{i\mathbf{k}\cdot\mathbf{x}}\,\frac{1}{n_0}\int d\mathcal{W}\, L_n(w)\, F_0 J_0(k_\perp\sqrt{2w}) \tag{9.58}$$

in terms of the normalised velocity space coordinate. The integral over dz has produced unity for $m = 0$ and zero for $m > 0$ according to the orthogonality relation for $H_m(z)$. Had we had no FLR gyroaveraging, namely replaced $J_0 \to 1$, we would have the same for $n = 0$ and $n > 0$ and $\phi_0 = \phi$ and $\phi_{n>0} = 0$ according to the orthogonality relation for $L_n(w)$. With the J_0, we have

$$\phi_n(\mathbf{k}) = G_n(\mathbf{k})\phi_{\mathbf{k}} \tag{9.59}$$

where the coefficient is given by

$$G_n(\mathbf{k}) = \int_0^\infty dw\, e^{-w}\, L_n(w)\, J_0(k_\perp\sqrt{2w}) = \frac{1}{n!}\left(-\frac{b}{2}\right)^n e^{-b/2} \tag{9.60}$$

with $b = k_\perp^2\rho_z^2$ and ρ_z the species thermal gyroradius V/Ω. This is an exact integral, so we do not adorn it with any fitting coefficients. For computability, especially if the Fourier transform in the perpendicular plane is not possible, we use the Padé approximant

$$G_n(\mathbf{k}) = \frac{1}{n!}\frac{(-b/2)^n}{(1+b/2)^{n+1}} \tag{9.61}$$

so that in xy-space we can use

$$G_n(\mathbf{x}) = \frac{1}{n!}\left(1 - \frac{1}{2}\rho_z^2\nabla_\perp^2\right)^{-1}\left[\left(1 - \frac{1}{2}\rho_z^2\nabla_\perp^2\right)^{-1} - 1\right]^n \tag{9.62}$$

in which $\nabla_\perp^2$ contains the spatial volume element in the usual way. The ϕ_n are computed successively from ϕ_0 to ϕ_n so they can be done in sequence, so that the effort scales as n rather than its square.

The magnetic energy is given by

$$E_M = \mathcal{V}^{-1}\int d\mathcal{V}\int d\mathcal{W}\sum_{sp}\frac{1}{2}e\frac{p_z}{m}(J_0A_\parallel)f = \mathcal{V}^{-1}\int d\mathcal{V}\sum_{sp}\sum_n\frac{1}{2}n_0eVA_n\,a_{1n} \tag{9.63}$$

where A_n is defined analogously to ϕ_n as

$$A_n(\mathbf{x}) = \frac{1}{n_0}\int d\mathcal{W}\,L_n(w)\,F_0(J_0A_\parallel) \tag{9.64}$$

The integral over dz has produced unity for $m = 1$ and zero otherwise, according to the orthogonality relation for $H_m(z)$, since $H_1 = z$. These reduce to $A_0 = A_\parallel$ and $A_{n>0} = 0$ if no gyroaveraging is done on $A_\parallel$, namely if $J_0 \to 1$ on these terms. In that case, the magnetic energy reduces to integration over $J_\parallel A_\parallel$ as in the fluid models, where

$$J_\parallel = \sum_{sp} n_0eV\,a_{10} \tag{9.65}$$

is the parallel current in this representation.

The representation of the free energy motivates the choice of f rather than g or h for the decomposition into gyrofluid moment variables. It also defines the FLR representation, which is not a closure but a truncation equivalent to the nature of the model itself, because the integrals defining the G_n operators are exact. The Padé approximants are an approximation, not a closure, since it is not a matter of defining higher moments in terms of those in the set which are kept.

In terms of these definitions, the time derivatives of the free energy pieces are given by

$$\frac{\partial E_f}{\partial t} = \mathcal{V}^{-1}\int d\mathcal{V}\,n_0T_0\sum_{mn}a_{mn}\frac{\partial a_{mn}}{\partial t} \tag{9.66}$$

for the thermal free energy, one per species,

$$\frac{\partial E_E}{\partial t} = \mathcal{V}^{-1} \int d\mathcal{V} \sum_{sp} n_0 e \sum_n \phi_n \frac{\partial a_{0n}}{\partial t} \tag{9.67}$$

for the ExB energy, and

$$\frac{\partial E_M}{\partial t} = \mathcal{V}^{-1} \int d\mathcal{V} \sum_{sp} n_0 e V \sum_n a_{1n} \frac{\partial A_n}{\partial t} \tag{9.68}$$

for the magnetic energy. For the $1n$-moments of each species, the magnetic and kinetic energies are combined as

$$\frac{\partial E_\parallel}{\partial t} = \mathcal{V}^{-1} \int d\mathcal{V} \sum_{sp} n_0 e V \sum_n a_{1n} \frac{\partial}{\partial t}\left(A_n + \frac{T_0}{eV} a_{1n}\right) \tag{9.69}$$

analogously to $J_\parallel \partial A^*/\partial t$ for fluid models, where $A^* = A_\parallel + (m_e/n_e e^2)J_\parallel$ is the combined potential under the time derivative in the Ohm's law.

9.4.3 On not gyroaveraging the magnetic potential

Nevertheless, we have a choice to make about the gyroaveraging of $A_\parallel$—the model will contain terms corresponding to $[\phi, A_\parallel]$ in the fluid models, and these cannot be done exactly. After the orthogonality of the H's is processed, the expression will contain

$$[\phi, A_\parallel] \rightarrow \frac{1}{n_0} \int d\mathcal{W}\, L_n(w)\, F_0[(J_0\phi), (J_0 A_\parallel)] \tag{9.70}$$

This is the opposite problem to the products of L's multiplying $(J_0\phi)$ in the ExB advection. Here, the two J_0's act separately on each of the field variables, so that the velocity space integral cannot be separated out. This nonlinear electric field bracket must therefore be given a closure approximation, with energetic consistency being the only useful guide (this is what I did in 2010).

On the other hand, $A_\parallel$ itself is mostly a result of electron dynamics, with the ions entering behind a factor of the electron/ion mass ratio. In the correspondence to magnetohydrodynamics (MHD) shown in Chapter 6, the Ohm's law was found by evaluating $\partial A_\parallel/\partial t$ in terms of the terms in the gyrokinetic equation for each species. Each species contributes to that with a coefficient given by $n_0 e^2/m$, its contribution to the skin depth. The electrons go in with m_e^{-1} and the ions with m_e/m_i times that. With ion gyroaveraging correcting by $k_\perp^2\rho_i^2$ for low-$k_\perp$, the correction times the mass ratio is actually as large as $k_\perp^2\rho_e^2$, which is very small in most applications. Under these conditions it becomes apparent that the gyroaveraging of $A_\parallel$ is of little consequence for most problems.

The consistency issue with Alfvén responses apparent in the gyrokinetic case does not appear in the gyrofluid case. For the kinetic case all of (δf) must be treated the same way, so that the $J_0(\delta g)$ appearing in both polarisation and induction implies that the right hand sides of the equations for both of the field variables have to be done the same way. In the gyrofluid case, $A_{\|}$ appearing with a_{1n} or a_{10} is separated from ϕ appearing with a_{0n} or a_{00}, and they can be treated separately without loss of consistency. In the physics, the tendency is for $E_{\|}$ to go to zero at long wavelength for higher-beta cases, as it does in MHD. At smaller scale the tendency is more towards a static balance between $\nabla_{\|}\phi$ and $\nabla_{\|}p_e$ in the adiabatic response. Both of these are controlled almost solely by the electrons, where subtleties involving the FLR treatment do not enter. Indeed, one relaxes towards this in the kinetic case as well, since $\rho_e \to 0$ if $k_\perp\rho_s$ does not go too much beyond unity (the correction, entering through the square of $k_\perp\rho_e$, is of order 10^{-3} then).

9.4.4 Field variable equations

Before moving on to the equations for the moment variables a_{mn}, we address the field equations. The hierarchy of gyroaveraged potentials ϕ_n arising from the operators G_n acting upon ϕ has already been introduced. The place in the equations where these occur in isolation is the field equations, polarisation for ϕ and induction for $A_{\|}$, so we treat these first.

The delta-f gyrokinetic polarisation equation is

$$\sum_{sp}\int d\mathcal{W}\,(\mathcal{M} - J_0\mathcal{M}J_0)\phi = \sum_{sp}\int d\mathcal{W}\,e(J_0\delta g), \qquad \mathcal{M} = F_0\frac{e^2}{T_0} \tag{9.71}$$

which is Eq. (7.66) of Chapter 7. With the spatial derivative commutation rules, the F_0/T_0 is pulled out and then the integration over $F_0J_0^2$ yields Γ_0, so that the equation is

$$\sum_{sp}\int d\mathcal{W}\,n_0e^2\frac{1-\Gamma_0}{T_0}\phi = \sum_{sp}\int d\mathcal{W}\,e(J_0\delta f) \tag{9.72}$$

in which we can use δg or δf interchangeably since the part in δg due to $A_{\|}$ drops out of the integration by anti-symmetry of zF_0. Note there is still one $1-\Gamma_0$ term for each species.

Now putting in Eq. (9.49) for $\delta f \to f$, we have for the gyrocentre charge density

$$\rho_G = \sum_{sp}\int d\mathcal{W}\,eJ_0\sum_{mn}a_{mn}H_m(z)\,L_n(w)\,F_0 \tag{9.73}$$

in which the J_0 acts upon the a_{mn} only due to the spatial derivative commutation rules. Separating the velocity-space integration from the rest, we have

$$\rho_G = \sum_{sp}\left[n_0e\sum_{mn}\left(\frac{1}{n_0}\int d\mathcal{W}\,H_m(z)\,L_n(w)\,F_0\,J_0\right)a_{mn}\right] \tag{9.74}$$

The sum over m eliminates all m except $m=0$, so that

$$\rho_G = \sum_{sp}\left[n_0e\sum_{n}\left(\frac{1}{n_0}\int d\mathcal{W}\,L_n(w)\,F_0\,J_0\right)a_{0n}\right] \tag{9.75}$$

The quantity in parentheses is just the operator G_n so we have

$$\rho_G = \sum_{sp} n_0 e \sum_n G_n(a_{0n}) \tag{9.76}$$

For each species we have a series of contributions from all $m = 0$ moments up to the maximum n, which reduces back to a_{00}, namely the normalised density, in the case $J_0 \to 0$ neglecting gyroaveraging (this is normally used for the electrons). The two-moment model listed above is recovered if $n = 0$ only and $m = 0$ or 1, namely the normalised density and parallel velocity for each species. In general, we have the full sum in ρ_G and the species sum over polarisation contributions on ϕ, which then has to be solved for. We then have

$$P(\phi) = \rho_G \tag{9.77}$$

as the gyrofluid polarisation equation, where

$$P = \sum_{sp} n_0 e^2 \frac{1 - \Gamma_0}{T_0} \tag{9.78}$$

is the gyrofluid polarisation operator and ρ_G is the gyrofluid charge density given by Eq. (9.76).

The delta-f gyrokinetic induction equation is

$$(K_d^2 - \nabla_\perp^2)A_\parallel = \mu_0 \sum_{sp} \int d\mathcal{W}\, e\frac{p_z}{m}(J_0 \delta g) \tag{9.79}$$

which is Eq. (7.68) of Chapter 7. If we now put in the definition of δg from Eq. (9.54), the contribution due to $A_\parallel$ cancels the skin depth term $K_d^2(A_\parallel)$ on the left hand side, leaving

$$-\nabla_\perp^2 A_\parallel = \mu_0 \sum_{sp} \int d\mathcal{W}\, e\frac{z}{m} J_0(\delta f) \tag{9.80}$$

which is Eq. (7.69) of Chapter 7. Putting in the representation in Eq. (9.49) for $\delta f \to f$, we have for the gyrocentre current

$$J_G = \sum_{sp}\left[\int d\mathcal{W}\, eVzJ_0\left(\sum_{mn} a_{mn}\, H_m(z)\, L_n(w)\, F_0\right)\right] \tag{9.81}$$

Due to the factor $z = H_1$, the sum over m eliminates all m except $m = 1$, so that

$$J_G = \sum_{sp}\left[n_0 eV \sum_n \left(\frac{1}{n_0}\int d\mathcal{W}\, L_n(w)\, F_0\, J_0\right) a_{1n}\right] \tag{9.82}$$

Again with G_n in the parentheses we have

$$J_G = \sum_{sp} n_0 eV \sum_n G_n(a_{1n}) \tag{9.83}$$

With no gyroaveraging on $A_\|$ the J_0 is replaced by unity, with $G_0 \to 1$ and $G_{n>1} \to 0$, this reduces to

$$J_G = J_\| = \sum_{sp} n_0 e V a_{10} \tag{9.84}$$

so that we end with

$$-\nabla_\perp^2 A_\| = \mu_0 \sum_{sp} n_0 e V a_{10} = \mu_0 J_\| \tag{9.85}$$

that is, the simplest Ampère's law as the gyrofluid induction equation. We can also use the form with

$$\left(d_e^{-2} - \nabla_\perp^2\right) A_\| = \mu_0 \sum_{sp} n_0 e V g_{10} \tag{9.86}$$

with $g_{10} = a_{10} + (eV/T_0)A_\|$ if doing so is more convenient to the numerical scheme.

9.4.5 Moment variable equations

To find the equations for the a_{mn}, we start with the delta-f gyrokinetic equation

$$\frac{\partial g}{\partial t} + \frac{1/e}{\chi_x}[(\delta H), h] + b^s\left(\frac{p_z}{m}\frac{\partial h}{\partial s} - \frac{\partial h}{\partial p_z}\frac{\partial}{\partial s}M\Omega\right) = \frac{p_z^2/m + M\Omega}{2e}\mathcal{K}(h) \tag{9.87}$$

which is Eq. (7.101) from Chapter 7, with the parallel phase space bracket written out and with p_z used as the dimensional coordinate. The spatial bracket is in xy only due to the flute mode ordering,

$$[a, b] = \frac{\partial a}{\partial x}\frac{\partial b}{\partial y} - \frac{\partial b}{\partial x}\frac{\partial a}{\partial y} \tag{9.88}$$

with the coefficient $1/\chi_x$ written explicitly. The factor $1/e$ appears due to the factor e in δH.

In terms of the normalised velocity space coordinates, the equation is

$$\frac{\partial g}{\partial t} + \frac{1}{\chi_x}[J_0(\phi), f] + B\nabla_\|\left(zV\frac{h}{B}\right) - \frac{\partial}{\partial z}\left(b^s Vwh\frac{\partial}{\partial s}\log B\right) = \frac{T_0}{e}\frac{z^2 + w}{2}\mathcal{K}(h) \tag{9.89}$$

where we re-cast the parallel phase space bracket as two divergences and the nonlinear parallel gradient as

$$\nabla_\| = b^s\frac{\partial}{\partial s} + \frac{1}{\chi_x}[-A_\|, \,] \tag{9.90}$$

with no gyroaverage on $A_\|$. The auxiliary variables are

$$g = f + eVzA_\|\frac{F_0}{T_0}, \qquad h = f + eJ_0(\phi)\frac{F_0}{T_0} \tag{9.91}$$

The bracket with $[-A_\parallel, J_0\phi]$ is included in the expression $\nabla_\parallel h$ which will give rise to the term $n_0 e\nabla_\parallel(J_0\phi)$ in each momentum equation. Hence, the bracket with $[J_0\phi, f]$ is written as such and not with g or h.

The equation has four sets of terms: the ExB nonlinear advection ($[J_0\phi, f]$), the parallel gradient ($\nabla_\parallel$) which is alternatively written as such or as a divergence, so-called magnetic pumping terms which arise from $\nabla_\parallel w \to w\nabla_\parallel \log B$, and magnetic curvature terms. Afterwards, there is the closure treatment for Landau damping, which is made systematic according to the natural speeds of the relevant waves. We will address each in turn.

Distinction is made between 'state variables' which are those a_{mn} with odd m, and 'flux variables' which are those a_{mn} with even m. This follows the treatment of ϖ and n on the one hand and J and u on the other, in the four field fluid model. The equations for state variables are conservation laws in which $\partial/\partial t$ is excited by flow/flux imbalances, so that the parallel dynamics is represented by a divergence, e.g., $B\nabla_\parallel(J_\parallel/B)$ in the equation for ϖ. The equations for flux variables are force laws in which the $\partial/\partial t$ is excited by force imbalances, so that the parallel dynamics is represented by a gradient, e.g., $\nabla_\parallel\phi$ in the equation for $J_\parallel$. That is, we keep the same usage as in the toroidal drift Alfvén model. The combination of the parallel gradient or divergence terms and the magnetic pumping terms is therefore slightly different for flux variables, and the generic delta-f gyrofluid which results will reflect this difference according to whether m in mn is odd or even. These state and flux variables always occur in pairs, for example n and $u_\parallel$ or T and $q_\parallel$, and in the gyrofluid case this will always be a_{mn} and $a_{m+1,n}$ with m even. We therefore always truncate the moment hierarchy at the level of flux variables not state variables, for example never close in T but in $q_\parallel$, or one or more pairs of moments higher. When finished, we will address why it is always a bad idea to close in the first flux variable moment (parallel velocity), unless a purely collisional closure is used. If Landau damping is treated via closure, as here, the lowest degree of complexity which can be used closes in the third moment level, with a_{30} and a_{11}, reflecting the two heat flux components under thermal anisotropy. Due to the complexity, we address each set of terms in isolation before recombining them.

9.4.6 Auxiliary gyrofluid variables

We have the auxiliary variables g and h associated with f according to Eqs (9.91). Before doing the moment equation systematics we establish the moment hierarchy for these variables, essentially constructing the g_{mn} and h_{mn} in terms of a_{mn} and the field variables.

Using the definition of g we have

$$\frac{1}{n_0}\int d\mathcal{W}\, H_m(z)\, L_n(w)\, g = a_{mn} + \frac{1}{n_0}\int d\mathcal{W}\, H_m(z)\, L_n(w)\, (zF_0)\, \frac{eV}{T_0}A_\parallel \qquad (9.92)$$

Without the gyroaveraging of $A_\parallel$ this gives

$$g_{mn} = a_{mn} + \delta_{m1}\,\delta_{n0}\,\frac{eV}{T_0}A_\parallel \qquad (9.93)$$

that is, solely $g_{mn} = a_{mn}$ unless $m = 1$ and $n = 0$, in which case $g_{10} = a_{10} + (eV/T_0)A_\|$. We also note that $eV/T_0 = e/mV$.

Using the definition of h we have

$$\frac{1}{n_0}\int d\mathcal{W}\, H_m(z)\, L_n(w)\, h = a_{mn} + \frac{1}{n_0}\int d\mathcal{W}\, H_m(z)\, L_n(w)\, F_0\, \frac{e}{T_0} J_0(\phi) \tag{9.94}$$

The integration over H_m gives δ_{m0} and the integration over $L_n J_0$ gives G_n, so that

$$h_{mn} = a_{mn} + \delta_{m0}\, \frac{e}{T_0}\phi_n \tag{9.95}$$

so that if $m \neq 0$ there is no ϕ_n piece, but every n with $m = 0$ has such a piece.

9.4.7 ExB advection

The first set of terms is given by the ExB advection. There is more than one term in the set because of the way J_0 mixes the moments—all available L_n's are coupled so that all available n's for each m in a_{mn} are coupled.

The equation for a_{mn} is found by applying

$$\frac{1}{n_0}\int d\mathcal{W}\, H_m(z)\, L_n(w)\, \frac{\partial g}{\partial t} + \frac{1}{\chi_x}\frac{1}{n_0}\int d\mathcal{W}\, H_m(z)\, L_n(w)\, [J_0\phi, f] + \cdots = 0 \tag{9.96}$$

Concerning the bracket itself, we have a new complication. Commuting the velocity-space dependence past the spatial derivatives, we have

$$\begin{aligned}&\frac{1}{n_0}\int d\mathcal{W}\, H_m(z)\, L_n(w)\, [J_0\phi, f]\\ &= \sum_{kl}\left[\left(\frac{1}{n_0}\int d\mathcal{W}\, H_k(z)H_m(z)\, L_l(w)L_n(w)\, F_0 J_0\right)\phi,\ a_{kl}\right]\end{aligned} \tag{9.97}$$

explicitly substituting the representation Eq. (9.49) in for f. The quantity within large parentheses is an operator which acts upon ϕ. The sum over k isolates the index m as usual, but the sum over l reflects the moment mixing,

$$\frac{1}{n_0}\int d\mathcal{W}\, H_m(z)\, L_n(w)\, [J_0\phi, f] = \sum_l\left[\left(\frac{1}{n_0}\int d\mathcal{W}\, L_l(w)L_n(w)\, F_0 J_0\right)\phi,\ a_{ml}\right] \tag{9.98}$$

This shows how products of Laguerre polynomials enter the procedure. Fortunately, polynomials are collections of terms with varying degree and so are their products. Unfortunately, there is no simple generation relation for products of Laguerre polynomials, so they have to be done by hand. In general L_n goes up to degree n and so $L_l\, L_n$ goes up to degree $n + l$. In the above list, we have L_n up to $n = 4$ so we are covered up to moment equations for n up to two. The relevant products up to that level are

$$
\begin{aligned}
L_0^2 &= 1, \qquad L_0\, L_1 = L_1, \qquad L_0\, L_2 = L_2, \\
L_1^2 &= 2L_2 + 2L_1 + L_0, \\
L_1\, L_2 &= 3L_3 + 4L_2 + 2L_1, \\
L_2^2 &= 6L_4 + 12L_3 + 10L_2 + 4L_1 + L_0
\end{aligned}
\tag{9.99}
$$

We see that these can be expressed as

$$
L_l(w)\, L_n(w) = \sum_{p=n-l}^{n+l} c_{lnp}\, L_p(w) \tag{9.100}
$$

where n is the larger of the two indices; otherwise use $|n = l|$ as the lowest of the terms in the summation. This is formally correct, but it is more useful to leave the expression in terms of L_l and L_n, since the moment which enters is a_{ml}. We write these in terms of a general operation G_{nl} which is a sum of the G_p with p between $|n - l|$ and $n + l$, given by

$$
G_{nl} = \frac{1}{n_0} \int d\mathcal{W}\, L_n(w)\, L_l(w)\, F_0\, J_0 \tag{9.101}
$$

as for G_n in Eq. (9.57), where the details of the integration are the same as in Eq. (9.60), in the same notation. Then we have

$$
\begin{aligned}
&\frac{1}{n_0} \int d\mathcal{W}\, H_m(z)\, L_n(w)\, [J_0\phi, f] \\
&\quad = \sum_l \left[\left(\frac{1}{n_0} \int d\mathcal{W}\, L_n(w)\, L_l(w)\, F_0\, J_0 \right) \phi,\, a_{ml} \right]
\end{aligned}
\tag{9.102}
$$

which becomes

$$
\frac{1}{n_0} \int d\mathcal{W}\, H_m(z)\, L_n(w)\, [J_0\phi, f] = \sum_l [\phi_{nl}, a_{ml}] \tag{9.103}
$$

At each n, this is a second-rank matrix of coefficients and ϕ_{nl} for each l in the moment list, so that the effective matrix is square and also dense. Fortunately, the combination of $\phi_{n'}$ terms in ϕ_{nl} is linear, so that they can be combined at no additional cost in the computation.

For n up to 1 only (total of six moments), they are

$$
\frac{\partial g_{m0}}{\partial t} + \frac{1}{\chi_x}\left([\phi_0, a_{m0}] + [\phi_1, a_{m1}]\right) + \cdots = 0 \tag{9.104}
$$

$$
\frac{\partial g_{m1}}{\partial t} + \frac{1}{\chi_x}\left([\phi_1, a_{m0}] + [\phi_0 + 2\phi_1, a_{m1}]\right) + \cdots = 0 \tag{9.105}
$$

the latter of which can also be written as

$$\frac{\partial g_{m1}}{\partial t} + \frac{1}{\chi_x}\big([\phi_0, a_{m1}] + [\phi_1, a_{m0} + 2a_{m1}]\big) + \cdots = 0 \qquad (9.106)$$

which is what appears in the models I have used in the past (Scott 2005, 2010). All the off-diagonal pieces have a coefficient of unity, so the demands of energy conservation are clearly met. Note how the first form of the equation for a_{m1} has ϕ_1 in the off diagonal acting on a_{m0}, so that multiplying by a_{m0} and a_{m1}, respectively, gives a combination into ϕ_1 acting on the product, which as a bracket is a pure divergence. Similarly, in the second form, multiplication by ϕ_0 and ϕ_1, respectively, combines on the off diagonal pieces into the product

$$[\phi_0\phi_1, a_{m1}] = [\phi_0, \phi_1 a_{m1}] + [\phi_1, \phi_0 a_{m1}] \qquad (9.107)$$

with either form expressible as the divergence of a transport flux.

For the moment system up to two in factors of w, hence n or l up to two, the main point is to note that only L_1^2 or L_2^2 contains the single L_0 term, that the same factor of two appears on L_2 in L_1^2 and on L_1 in L_1L_2, and that the same factor of four appears on L_2 in L_2^2 and on L_1 in L_1L_2. Then the matrix can be worked out in the same way as for the six-moment case just given. It is also important to recall that the L_3 and L_4 factors appearing in the products are dropped because moments such as a_{03} and a_{04} are not being kept at this level.

For higher moment levels, one finds more non-unity coefficients which have to match up. In each case, as just noted, the moment list for L_n^2 only proceeds up to the level kept, which is the same as the highest n in the sum for the gyrocentre charge density, ρ_G in Eq. (9.76). In the equation for a_{0n} at the highest n, the terms in the product L_n^2 reach up only to n, not n^2, since the moments past n are not kept, and therefore ϕ_n is the highest ϕ-moment appearing. This is required by energy conservation, since ϕ_n couples to a_{0n} at every n, and so the highest n in the set of ϕ_n is the same as that in the set of a_{0n}, etc. Moreover, each product of L's reaches down to $L_0 = 1$ only for L_n^2 and the coefficient there is also unity. Hence, if $J_0 \to 1$ only these latter terms survive. These considerations form useful checks against errors when generating the terms and coefficients by hand. The systematics always works out in the end.

9.4.8 Curvature terms

The terms involving $\mathcal{K}$ are collectively referred to as curvature terms They conserve energy as two sets, coupling within state variables and within flux variables separately. They are found by applying

$$\frac{1}{n_0}\int d\mathcal{W}\, H_m(z)\, L_n(w)\left(\frac{\partial g}{\partial t} - \frac{T_0}{e}\frac{z^2 + w}{2}\mathcal{K}(h)\right) + \cdots = 0 \qquad (9.108)$$

With the commutation rules on spatial derivatives, the velocity space integration and coordinates commute past $\mathcal{K}$ and act upon h, so that

$$\frac{1}{n_0}\int d\mathcal{W}\, H_m(z)\, L_n(w)\frac{z^2 + w}{2}\mathcal{K}(h) = \mathcal{K}\left(\frac{1}{n_0}\int d\mathcal{W}\, H_m(z)\, L_n(w)\frac{z^2 + w}{2}h\right) \qquad (9.109)$$

These integrals involve the recurrence relations for H_m and L_n [Eqs (9.43) and (9.46)], with the one for the H's used twice due to two factors of z. We find

$$
\begin{aligned}
z^2 H_m &= z\left(\sqrt{m+1}\,H_{m+1} + \sqrt{m}\,H_{m-1}\right) \\
&= \sqrt{m+1}\left(\sqrt{m+2}\,H_{m+2} + \sqrt{m+1}\,H_m\right) \\
&\quad + \sqrt{m}\left(\sqrt{m}\,H_m + \sqrt{m-1}\,H_{m-2}\right)
\end{aligned}
\tag{9.110}
$$

$$
w\,L_n = (n+1)L_{n+1} + (2n+1)L_n + nL_{n-1} \tag{9.111}
$$

so that

$$
\begin{aligned}
\frac{\partial g_{mn}}{\partial t} - \frac{T_0}{2e}\mathcal{K}\Big(&\sqrt{(m+1)(m+2)}\,h_{m+2,n} + (2m+1)h_{mn} + \sqrt{m(m-1)}\,h_{m-2,n} \\
&+ (n+1)h_{m,n+1} + (2n+1)h_{mn} + nh_{m,n-1}\Big) + \cdots = 0
\end{aligned}
\tag{9.112}
$$

is the result. This creates combinations

$$
\frac{T_0}{e}h_{mn} = \delta_{m0}\,\phi_n + \frac{T_0}{e}a_{mn} \tag{9.113}
$$

as the effective arguments of the curvature terms Hence, $\phi - n$ and $\phi_G + \tau_i N$ in the heuristic 3D model given above.

9.4.9 Parallel dynamics

The parallel dynamics in the gyrokinetic equation is governed by the parallel phase space bracket with derivatives in s and z. Splitting this into divergences creates a set of terms associated with $\partial/\partial z$, which is annihilated by the velocity space integration. In the equation for g_{mn} there is an extra set of terms arising from taking the $\partial/\partial z$ by parts so that it acts upon the $H_m L_n$ product. We have

$$
\frac{\partial}{\partial z}H_m(z)L_n(w) = \sqrt{m}\,H_{m-1}L_n, \qquad w\frac{\partial}{\partial s}h = \frac{\partial}{\partial s}wh - wh\frac{\partial}{\partial s}\log B \tag{9.114}
$$

The pieces in the $\partial/\partial z$ term and any remnant from commuting w or $L_n(w)$ past $\partial/\partial s$ all contain $w\nabla_\parallel \log B$, so they explicitly become what are called 'magnetic pumping' terms, after the original concept according to which parallel flows and heat fluxes excite thermal anisotropy (in the collisional limit, viscosity) and vice versa. One other place contributing to magnetic pumping is in the parallel divergence term in the kinetic model. State variable equations have the parallel terms arranged such that they are pure divergences, since they represent conservation laws. Flux variable equations have the parallel terms arranged such that they are pure gradients, since they represent force laws. The remaining terms, all of which contain $\nabla_\parallel \log B$, form the set of magnetic pumping terms. An important check on both these sets is that they each conserve free energy within the set, much as like curvature terms within each set of state variables and flux variables. Therefore, we have to do all the parallel bracket terms together.

The simplest way to identify the terms needed in the parallel dynamics, forces and fluxes, is to first derive the equations assuming that $\nabla_\parallel \log B$ vanishes. Then w and B

commute with $\partial/\partial s$ whereever needed and the fundamental set of couplings is easily built. For example, in the reduced MHD (RMHD) model we have

$$\frac{\partial \varpi}{\partial t} = \nabla_{\|} J_{\|}, \qquad \frac{\partial \psi}{\partial t} = -\nabla_{\|} \phi \tag{9.115}$$

with the simple energy conservation found by multiplying by $-\phi$ and $J_{\|}$, respectively. Then, the term in the vorticity equation, which represents charge conservation, is converted into a divergence in the case of inhomogeneous magnetic fields, so that

$$\frac{\partial \varpi}{\partial t} = B\nabla_{\|} \frac{J_{\|}}{B}, \qquad \frac{\partial \psi}{\partial t} = -\nabla_{\|} \phi \tag{9.116}$$

and the same energetic consistency prevails, now with $\phi J_{\|}/B$ rather than $\phi J_{\|}$ behind $B\nabla_{\|}$, which is the parallel divergence in that case. Then, go back and collect all the magnetic pumping terms and verify energetic consistency (this was the procedure I used in the heuristic derivation given in Scott (2005) before doing the real one in 2010). In the case of RMHD there are no magnetic pumping terms since we are not keeping any more dependent variables. But this case gives a simple outline of the basic procedure.

Assuming $\nabla_{\|} \log B = 0$ we have

$$\frac{1}{n_0} \int d\mathcal{W}\, H_m(z) L_n(w) [\varepsilon, h]_{zs} = \frac{1}{n_0} \int d\mathcal{W}\, z H_m(z) L_n(w) \frac{\partial h}{\partial s} \tag{9.117}$$

where $\varepsilon = z^2/2 + w$ is the zeroth Hamiltonian in normalised velocity space (the units of p_z are mV and of H are T_0, so the entire term has dimensionality of $V\nabla_{\|}$ which is the same as that of $\partial/\partial t$). For the purpose of derivation we can use $\partial/\partial s$ or $\nabla_{\|}$ interchangeably until the end.

Using the recursion relation in Eq. (9.43) and doing the velocity-space integration, we have

$$\frac{1}{n_0} \int d\mathcal{W}\, H_m(z) L_n(w) [\varepsilon, h]_{zs} = \frac{\partial}{\partial s} \left(\sqrt{m+1}\, h_{m+1,n} - \sqrt{m}\, h_{m-1,n} \right) \tag{9.118}$$

with the velocity space quantities commuting past $\partial/\partial s$. This gives the basic pattern of the chain of dependences. The equation for g_{mn} contains $h_{m+1,n}$, the original statement of the closure problem. Closure is the art of making a model for the $h_{m+1,n}$ not kept in the set of dependent variables. Truncation is to simply neglect all moments above the maximum moment level $m + 2n$. In our case, we truncate everywhere, with no ambiguity, except for a Landau closure which is set to treat Landau damping. We will delay this until after verifying exact energetic consistency with the dissipation-free part of the model. For now, this chain of dependences shows how the exchanges conserve energy. In the equation for g_{mn} we find $h_{m\pm 1,n}$ and in the equations for $g_{m\pm 1,n}$ we find couplings back to h_{mn} always with the same coefficient. Above the maximum moment level there is no further term. So by multiplying the equation for g_{mn} by h_{mn} and doing

$$\int d\mathcal{V}\, h_{mn}\frac{\partial g_{mn}}{\partial t} = \cdots - \int d\mathcal{V}\; Vh_{mn}\nabla_{\|}\left(\sqrt{m+1}\, h_{m+1,n} - \sqrt{m}\, h_{m-1,n}\right) \tag{9.119}$$

and then the sum over m combines them into

$$\sum_m \int d\mathcal{V}\, h_{mn}\frac{\partial g_{mn}}{\partial t} = \cdots - V\sum_m \int d\mathcal{V}\; \nabla_{\|}\left(\sqrt{m+1}\, h_{mn}h_{m+1,n}\right) \tag{9.120}$$

as can be verified by shifting the index of the sum on the second term and using the fact that h_{mn} does not exist for $m = -1$ or $m + 2n$ past the maximum moment level. Each state/flux variable pair appears together in a conservative transport flux. There is one such chain for each n in the set.

With $\nabla_{\|}\log B$ nonzero the energy theorem for all of the parallel terms should consist of

$$\sum_{mn} \int d\mathcal{V}\, h_{mn}\frac{\partial g_{mn}}{\partial t} = \cdots - V\sum_{mn} \int d\mathcal{V}\; B\nabla_{\|}\left(\sqrt{m+1}\,\frac{h_{mn}h_{m+1,n}}{B}\right) \tag{9.121}$$

This will be our check on energetics at the end.

The task now is to arrange the gyrokinetic equation differently, with $B\nabla_{\|}$ for state variables and with $\nabla_{\|}$ for flux variables. This has to be done together with taking the moment integrals, since the dependence $w = M\Omega$ which includes B means that $L(w)$ has finite parallel derivatives. In processing the zs-bracket, we watch for the fact that $\sqrt{g}\,Bb^s\partial/\partial s$ is the true divergence, while $d\mathcal{W}/B$ commutes past $\partial/\partial s$. With $\nabla_{\|}\log B$ correspondingly finite, we have

$$\begin{aligned} &\frac{1}{n_0}\int d\mathcal{W}\, H_m(z)L_n(w)\; Vb^s[\varepsilon, h]_{zs} \\ &\quad= \frac{1}{n_0}\int d\mathcal{W}\, H_m(z)L_n(w)\; Vb^s\left(\frac{\partial}{\partial s}zh - \frac{\partial}{\partial z}wh\frac{\partial}{\partial s}\log B\right) \end{aligned} \tag{9.122}$$

Combining the polynomials inside $\partial/\partial s$ and integrating the second term by parts we have

$$\begin{aligned} \frac{1}{n_0}\int d\mathcal{W}\, H_m(z)L_n(w)\; Vb^s[\varepsilon, h]_{zs} &= \frac{1}{n_0}\int d\mathcal{W}\; Vb^s\left(\frac{\partial}{\partial s}zH_m\, L_n\, h\right. \\ &\quad \left. - zH_m\, w\frac{dL_n}{dw}\, h\,\frac{\partial}{\partial s}\log B + \sqrt{m}\, H_{m-1}\, wL_n\, h\frac{\partial}{\partial s}\log B\right) \end{aligned} \tag{9.123}$$

Now commute $d\mathcal{W}/B$ past $\partial/\partial s$ in the first term and make it a divergence, using

$$\frac{d\mathcal{W}}{B}b^s\frac{\partial}{\partial s} = \frac{1}{\sqrt{g}}\frac{d\mathcal{W}}{B}\left(\sqrt{g}\,b^s\right)\frac{\partial}{\partial s} = \frac{1}{\sqrt{g}}\frac{\partial}{\partial s}\left(\sqrt{g}\,b^s\right)\frac{d\mathcal{W}}{B} \tag{9.124}$$

Then, process the term $(zH_mL_n\, h)$ as we did above, and combine the terms multiplying $\nabla_{\|}\log B$ into a set. The result is

$$
\begin{aligned}
\frac{1}{n_0}\int d\mathcal{W}\, H_m(z)L_n(w)\; Vb^s[\varepsilon, h]_{zs} \;=\; & VB\nabla_{\parallel}\frac{1}{B}\Big[\sqrt{m+1}\,h_{m+1,n} + \sqrt{m}\,h_{m-1,n}\Big] \\
& - VD\sqrt{m+1}\; n\left(h_{m+1,n} + h_{m+1,n-1}\right) \\
& + VD\sqrt{m}\;(n+1)\left(h_{m-1,n+1} + h_{m-1,n}\right)
\end{aligned}
\tag{9.125}
$$

noting some cancellations between coefficients of the $h_{m-1,n}$ and $h_{m-1,n-1}$ terms, eliminating the latter entirely. The logarithmic magnetic derivative is given by

$$
D = \nabla_{\parallel}\log B = b^s\frac{\partial}{\partial s}\log B \tag{9.126}
$$

with the second form valid in a fluxtube model, since $B = B(s)$ while $[-A_{\parallel},\ ]$ contains only x and y derivatives. This is the version to be used in state variable equations (m even), since it is a parallel divergence.

In the case of a parallel gradient we do not combine the $1/B$ into a divergence, leaving the main contribution as a parallel force. The same steps as before yield

$$
\begin{aligned}
\frac{1}{n_0}\int d\mathcal{W}\, H_m(z)L_n(w)\; Vb^s[\varepsilon, h]_{zs} = & V\nabla_{\parallel}\Big[\sqrt{m+1}\,h_{m+1,n} + \sqrt{m}\,h_{m-1,n}\Big] \\
& - VD\sqrt{m+1}\big((n+1)h_{m+1,n} + nh_{m+1,n-1}\big) \\
& + VD\sqrt{m}\big((n+1)h_{m-1,n+1} + nh_{m-1,n}\big)
\end{aligned}
\tag{9.127}
$$

The difference from the first version is the absence of the factors of B and $1/B$ in the first line and the factors of $[n+1]$ versus n in the last two lines. This second version is used in flux variable equations (m odd), which are force rather than transport equations. Written this way one guarantees energy conservation between a flux divergence and a state variable gradient in the resulting model wherever they occur. In the section on energetics below, we will give the details of how the terms involving D mutually cancel in the free energy theorem and hence serve as transfer effects.

9.4.10 The Landau closure

Up to this point all the terms in the equations are given by exact integrals over velocity space, so there is no ambiguity in any of the coefficients which remain. The model is a truncation, not a closure. Above all, there is no place for any fitting parameters. This leaves the Landau closure. With the free energy conserved, the free energy in an initial perturbation has nowhere to go and oscillations can ring forever. Landau damping of these waves requires a reasonable closure. Models using formulae involving $|k_{\parallel}|$ in Fourier decompositions are impossible to calculate in general geometry, even in a fluxtube in which the s-direction is not amenable to Fourier transforms, due to the phase-shifted boundary condition linking $f(x, y, s + L_{\parallel})$ to $f(x, y, s)$. In this model we use a Landau closure consisting of a simple diagonal damping of the highest-level flux moments according to their natural wave speeds. Specifying this removes any ambiguity—establishing the closure on theoretical grounds eliminates any opportunity or temptation for fitting coefficients.

The wave speed associated with a flux variable a_{mn} and its associated state variable $a_{m-1,n}$, with m odd, is given by the coefficient in Eq. (9.121), namely

$$V_{m-1,m,n} = \sqrt{m}\ V \tag{9.128}$$

namely, the square root of the m of the flux variable (not the state variable) times the thermal speed of the species. Since Landau damping takes place on closed magnetic flux surfaces with a parallel connection length of $L_{\|} = 2\pi/K_{\|}$, the natural choice of a damping rate is $VK_{\|}$ times the numerical factor. With this prescription there is no ambiguity and no place for a fitting coefficient. One simply takes the model as it stands.

We specify that the moments with $m + 2n$ equal to the maximum are damped with the coefficient

$$\frac{\partial g_{mn}}{\partial t} = \cdots - \sqrt{m}K_{\|}V\ a_{mn} \qquad \text{for} \quad m + 2n = \text{maximum} \tag{9.129}$$

In a tokamak $K_{\|} = 1/qR_0$ in simple geometry or $1/q$ times $\langle 1/R\rangle$ in general, where the angle brackets give the flux surface average. Similar methods can be used for a stellarator. With open field lines, the outflow boundary conditions set by setting incoming wave characteristics to zero (Chapter 15 from Volume One) will overwhelm the Landau damping anyway, so one can use the wave-damp model in Eq. (9.129) everywhere, with $K_{\|}$ set appropriately for the given system geometry.

It is not recommended to use a Landau closure at all if one chooses to retain only the first two moments. This is because Landau damping reflects thermal conduction, not resistivity or viscosity (Chapter 8). A Landau closure for α_{10} in a two-moment model will have undesirable consequences resulting from artificial resistivity or flow viscosity. It should be reserved for models whose highest-moment flux variables are heat fluxes, thermal capacity fluxes, or higher. A two-moment model should use a collisional closure, with $\eta_{\|}J_{\|}$ in each equation for $A + (T_0/eV)a_{10}$ and a viscosity term involving derivatives of a_{10} for the ions.

9.4.11 Expression of the moment equations

We have the expressions for each set of terms So the moment equations for the state variables follow

$$\begin{aligned}
&\frac{\partial g_{mn}}{\partial t} + \sum_{l}\sum_{p=n-l}^{n+l} c_{lnp}\left[\phi_p, a_{ml}\right] \\
&- \frac{T_0}{2e}\mathcal{K}\Big(\sqrt{(m+1)(m+2)}\,h_{m+2,n} + (2m+1)h_{mn} + \sqrt{m(m-1)}\,h_{m-2,n} \\
&+ (n+1)h_{m,n+1} + (2n+1)h_{mn} + nh_{m,n-1}\Big) \\
&+ VB\nabla_{\|}\frac{1}{B}\Big(\sqrt{m+1}\,h_{m+1,n} + \sqrt{m}\,h_{m-1,n}\Big) \\
&- VD\sqrt{m+1}\ n\left(h_{m+1,n} + h_{m+1,n-1}\right) \\
&+ VD\sqrt{m}\ (n+1)\left(h_{m-1,n+1} + h_{m-1,n}\right) \\
&= 0
\end{aligned} \tag{9.130}$$

and the moment equations for the flux variables follow

$$
\begin{aligned}
&\frac{\partial g_{mn}}{\partial t} + \sum_l \sum_{p=n-l}^{n+l} c_{lnp}\left[\phi_p, a_{ml}\right] \\
&- \frac{T_0}{2e}\mathcal{K}\Big(\sqrt{(m+1)(m+2)}\,h_{m+2,n} + (2m+1)h_{mn} + \sqrt{m(m-1)}\,h_{m-2,n} \\
&+ (n+1)h_{m,n+1} + (2n+1)h_{mn} + nh_{m,n-1}\Big) \\
&+ V\nabla_{\|}\Big(\sqrt{m+1}\,h_{m+1,n} + \sqrt{m}\,h_{m-1,n}\Big) \\
&- VD\sqrt{m+1}\Big((n+1)h_{m+1,n} + nh_{m+1,n-1}\Big) \\
&+ VD\sqrt{m}\Big((n+1)h_{m-1,n+1} + nh_{m-1,n}\Big) \\
&= -\sqrt{m}K_{\|}V\,a_{mn} \qquad \text{for} \quad m + 2n = \text{maximum}
\end{aligned}
\tag{9.131}
$$

where in the last line the Landau closure term on the right hand side appears only at the highest moment level.

These equations satisfy

$$
\int d\mathcal{V}\, n_0T_0 \sum_{mn} h_{mn}\frac{\partial g_{mn}}{\partial t} = -\int d\mathcal{V}\, n_0T_0 \sum_{m+2n=\max} \sqrt{m}K_{\|}V\, a_{mn}^2 \tag{9.132}
$$

namely, all terms except for Landau damping conserve free energy and the Landau closure represents positive-definite damping (this uses the fact that $h_{mn} = a_{mn}$ for any odd m). The model therefore satisfies exact energetic consistency with thermodynamically correct dissipation. The details will be shown in the section on energetics, below.

9.4.12 System normalisation

The entire system is normalised in terms of a reference T and m which define the sound speed c_s, which is why we obtain species-dependent normalised charge densities and mass/charge and temperature/charge ratios when we are finished. It is almost always preferable to use the electron density and temperature, and an ion mass m_i for this purpose. The species parameters are

$$
a_z = \frac{n_0 e}{n_{e0}e_0}, \qquad \tau_z = \frac{T_0/e}{T_{e0}/e_0}, \qquad \mu_z = \frac{m/e}{m_i/e_0} \tag{9.133}
$$

This is done by multiplying each flux variable equation through by mV/e and using $mV^2 = T_0$, so that mV/e appears in the inertia and ExB advection terms, $A_{\|}$ in g_{10} appears by itself, T_0/e appears with all parallel dynamics terms, and $(mV/e)(T_0/e)$ appears with all curvature terms. In the state variable equations we have V with the parallel dynamics terms and T_0/e with the curvature terms. In all cases, h_{mn} in state variables appears with T_0/e and in flux variables with V only, and g_{mn} in state variables appears by itself and in flux variables with mV/e. Then, when the entire system is normalised in terms of a sound speed c_s and drift scale ρ_s given by

$$c_s^2 = \frac{T_{e0}}{m_i}, \qquad \rho_s^2 = \frac{m_i T_{e0}}{e_0^2 B_0^2} \tag{9.134}$$

we obtain μ_z where mV/e was and τ_z where T_0/e was. In terms of the species parameters $n_0 e$ becomes a_z and $n_0 T_0$ becomes $a_z\tau_z$. Further, $\rho_z^2/\rho_s^2 = \mu_z\tau_z$, the inverse-square skin depth is the species sum of $(a_z/\mu_z)\beta_e$, and the species pressure is $a_z\tau_z$. The field equations appear with a_z multiplying the normalised quantities in the charge density and parallel current.

The example for a two-moment the model is then

$$\frac{\partial a_{00}}{\partial t} + [\phi_0, a_{00}] + B\nabla_{\|}\frac{a_{10}}{B} = \mathcal{K}(\phi_0 + \tau_z a_{00}) \tag{9.135}$$

$$\frac{\partial}{\partial t}(A + \mu_z a_{10}) + \mu_z[\phi_0, a_{10}] + \nabla_{\|}(\phi_0 + \tau_z a_{00}) = \mu_z\tau_z\mathcal{K}(2a_{10}) - CJ \tag{9.136}$$

with field equations

$$\sum_{sp} a_z\frac{1-\Gamma_0}{\tau_z}\phi = \sum_{sp} a_z G_0(a_{00}) \tag{9.137}$$

$$(k_d^2 - \delta_0^2\nabla_\perp^2)A = \beta_e J, \qquad k_d^2 = \sum_{sp}\frac{a_z}{\mu_z}\beta_e \qquad J = \sum_{sp} a_z a_{10} \tag{9.138}$$

where

$$[a, b] = \delta_0\left(\frac{\partial a}{\partial x}\frac{\partial b}{\partial y} - \frac{\partial b}{\partial x}\frac{\partial a}{\partial y}\right), \qquad \delta_0 = \frac{\rho_s}{L_\perp} \tag{9.139}$$

and $L_\perp$ is the length to which the spatial derivatives (in xys) are normalised, with time normalised to $L_\perp/c_s$.

9.5 Gyrofluid energetics

The basics of the energetics was already given in the beginning of this Chapter. We now address the general case. The free energy pieces are given in Eqs (9.55, 9.56, 9.63), and the time derivatives are given in Eqs (9.66–9.69). These are equivalent to

$$E_T = \mathcal{V}^{-1}\int d\mathcal{V}\,\frac{1}{2}n_0T_0\sum_{mn} h_{mn}\, g_{mn},$$
$$\frac{\partial E_T}{\partial t} = \mathcal{V}^{-1}\int d\mathcal{V}\, n_0T_0\sum_{mn} h_{mn}\frac{\partial g_{mn}}{\partial t} \tag{9.140}$$

One can show this by substituting in for h_{mn} using Eq. (9.95) and for g_{mn} using either of Eq. (9.92, 9.93), and noting the fact that there is no moment index mn for which both h_{mn} and g_{mn} are different from a_{mn}. Summed over all mn, one recovers Eq. (9.132).

The interested reader should start with the schematic expression

$$\frac{\partial E_T}{\partial t} = \mathcal{V}^{-1}\int d\mathcal{V}\frac{1}{2}\left[\phi P(\phi) + A_\| S(A_\|) + \sum_{sp} n_0 T_0 \sum_{mn} a_{mn}^2\right] \tag{9.141}$$

where P and S are written as general Hermitian operators representing the ExB and magnetic energies, and the last term is the thermal free energy. Then use Eqs (9.78, 9.85) to re-express the ExB and magnetic energies so that

$$\frac{\partial E_T}{\partial t} = \mathcal{V}^{-1}\int d\mathcal{V}\frac{1}{2}\sum_{sp}\left[n_0 e\phi\sum_n G_n(a_{0n}) + n_0 eVA_\|\sum_n G_n(a_{1n}) + n_0 T_0\sum_{mn} a_{mn}^2\right] \tag{9.142}$$

Then, use the Hermitian property of the G operators to have them act on ϕ and $A_\|$, and combine the ϕ terms into the $0n$-moments and the $A_\|$ terms into the $1n$-moments. The result is the first equation in Eq. (9.140). Similarly, do the same for the time derivatives, which combine in the same way, to recover the second equation in Eq. (9.140). This works whether or not we keep gyroaveraging on $A_\|$.

Doing this for any particular mn for one species, one can find energy transfer effects among the free energy pieces. One has merely to multiply the equation for g_{mn} by $n_0 T_0\, h_{mn}$ and integrate over the volume, and find the terms which 'hang' on the right hand side, i.e., do not combine into pure divergences which vanish under the integration. Then, find the other equation which has the same hanging term with opposite sign. The simplest example is $J_\|\nabla_\|\phi$ in the RMHD model, which gives the Alfvénic transfer between magnetic and ExB energy. The rules are the same: combine into divergences in the state variable equations, and see what hangs, and then find the same piece in a flux variable equation.

The exceptions are the curvature terms, which always act within state variables or flux variables (due to the index shifts by two in m or one in n). We have seen that in the lowest moment, $\mathcal{K}(h_{00})$ in the equation for h_{00} combines into a divergence, but a_{00} times this, representing a piece in E_f, exchanges with E_E by combining with the piece due to $e\phi_0/T_0$. This is the ExB compressional transfer due to magnetic curvature, the interchange effect. The adiabatic compressional transfer corresponding to $J_\|\nabla_\| p_e$ in the fluid drift Alfvén model keeps the energy within the electrons, since it is largely represented by $a_{10}\nabla_\|(a_{00} + \sqrt{2}\,a_{20})$ in the electrons (here note that $h_{10} = a_{10}$), transferring between state variable energy (00 and 20 make up the thermal free energy part) and flux variable energy (a_{10} squared gives parallel kinetic energy). The other piece in h_{00} is covered by $a_{10}\nabla_\|(e\phi_0/T_0)$, which multiplied by $n_0 T_0$ and summed over species gives the FLR-corrected version of $J_\|\nabla_\|\phi$, which is the Alfvénic transfer.

The thermal effects can be more interesting, due to the magnetic pumping effects. The parallel divergences in the state variable equations (with m even),

$$\frac{\partial g_{mn}}{\partial t} = \cdots - VB\nabla_\|\frac{1}{B}\left(\sqrt{m+1}\,h_{m+1,n} + \sqrt{m}\,h_{m-1,n}\right) \tag{9.143}$$

and the parallel gradients in the flux variable equations (with m odd),

$$\frac{\partial g_{mn}}{\partial t} = \cdots - V\nabla_{\parallel}\left(\sqrt{m+1}\,h_{m+1,n} + \sqrt{m}\,h_{m-1,n}\right) \tag{9.144}$$

combine in the energetics into a single expression under the sum over moments,

$$\sum_{mn}\left(h_{mn}\frac{\partial g_{mn}}{\partial t} + h_{m+1,n}\frac{\partial g_{m+1,n}}{\partial t}\right) = \cdots - \sqrt{m+1}\,VB\nabla_{\parallel}\frac{1}{B}h_{m,n}h_{m+1,n} \tag{9.145}$$

which one finds by shifting the sum index accordingly. The $m-1$ piece vanishes for $m = 0$ and the $m + 1$ piece vanishes for the maximum m in the set of moments. Isolating the terms in Eq. (9.143), the first one combines with the second in Eq. (9.144) and the second one combines with the first, respectively.

The magnetic pumping terms follow the same schematic, with the four terms in the state variable energy equation (with m even),

$$\begin{aligned}\sum_{mn} h_{mn}\frac{\partial g_{mn}}{\partial t} = \cdots &- VD\sqrt{m+1}\;n\;h_{mn}\left(h_{m+1,n} + h_{m+1,n-1}\right)\\ &+ VD\sqrt{m}\;(n+1)h_{mn}\left(h_{m-1,n+1} + h_{m-1,n}\right)\end{aligned} \tag{9.146}$$

first, second, third, and fourth, cancelling with the four terms in the flux variable energy equation (with m odd),

$$\begin{aligned}\sum_{mn} h_{mn}\frac{\partial g_{mn}}{\partial t} = \cdots &- VD\sqrt{m+1}\;h_{mn}\left((n+1)h_{m+1,n} + nh_{m+1,n-1}\right)\\ &+ VD\sqrt{m}\;h_{mn}\left((n+1)h_{m-1,n+1} + nh_{m-1,n}\right)\end{aligned} \tag{9.147}$$

fourth, third, second, and first, respectively. Shift the sum indices on each of the four terms separately to get this.

But unlike the two pieces in the divergence combination, which keep the transfer within the n-level moments separate, the magnetic pumping process transfers free energy between these sets. This acts to exchange energy between parallel and perpendicular components within a given moment level, and it is not a dissipation process but something which may or may not cause equilibration (keeping the thermal free energy components at each moment level $m + 2n$ more isotropic). This is situation-dependent and becomes an interesting possibility to watch out for. Due to the gyrokinetic conservation of magnetic moment, it is a process fundamental to the gyrofluid model but absent in a fluid model. We will explore this in more detail within the six-moment model for temperature and heat flux dynamics, in the next Chapter. Collisions are also delayed until then.

9.6 Summary

This Chapter has given the details for setting up a gyrofluid model at any moment level, not just up to velocity or heat flux but beyond. Our purposes will be served by going to six moments to cover the temperatures and heat fluxes, enough for a gyrofluid model which accounts for the electron and ion temperature dynamics we have seen in Chapters 2 and 3, also overcoming the restrictions on the finite Larmor radius implicit in the gyroviscosity expressions of the fluid drift model.

In the next Chapter we will lay out the six-moment gyrofluid model, incorporate collisional dissipation at the same level at which we have seen it, and prove the correspondence back to the fluid drift model in the appropriate limits. These limits are scales larger than the gyroradii, specific heat fluxes smaller than velocities in both perpendicular drifts and parallel streaming, and small thermal anisotropy. Thereafter, the gyrofluid model can serve as a better fluid model than the fluid model it replaces, and we can turn back to the physics of the turbulence, exploring it in regimes where the fluid drift model does not work very well.

Further Reading

The Hasegawa–Mima and Hasegawa–Wakatani models, appearing in A Hasegawa and K Mima, 'Stationary spectrum of strong turbulence in magnetized nonuniform plasma,' *Phys. Rev. Lett.* **39** (1977) 205; 'Pseudo-three-dimensional turbulence in magnetized nonuniform plasma,' *Phys. Fluids* **21** (1978) 87; A Hasegawa and M Wakatani, 'Plasma edge turbulence,' *Phys. Rev. Lett.* **50** (1983) 682; M Wakatani and A Hasegawa, 'A collisional drift wave description of plasma edge turbulence," *Phys. Fluids* **27** (1984) 611; were discussed in Chapter 4 of Volume One.

Gyrofluid equations were given by G Knorr, F R Hansen, J P Lynov, H L Pécseli, and J Juul Rasmussen, in 'Finite larmor radius effects to arbitrary order,' *Physica Scr.* **38** (1988) 829, though anticipated by W W Lee in explaining the meaning of his gyrokinetic equations, in 'Gyrokinetic approach in particle simulation,' *Phys. Fluids* **26** (1983) 556. A systematic approach for six moments was given by W Dorland and G Hammett, in *Phys. Fluids B* **5** (1993) 812, and by M A Beer and G Hammett, in *Phys. Plasmas* **3** (1996) 4046. An electromagnetic version was given by B Scott, in 'ExB shear flows and electromagnetic gyrofluid turbulence,' *Phys. Plasmas* **7** (2000) 1845. An energetically consistent version, fixing the problems found while investigating strongly electromagnetic drift Alfvén turbulence with the pre-2000 models, was given by B Scott, in 'Free energy conservation in local gyrofluid models,' *Phys. Plasmas* **12** (2005) 102307.

The moment approach in terms of Hermite polynomials in fluid plasma equations is due to H Grad, in 'On the kinetic theory of rarefied gases,' *Communications of Pure and Applied Mathematics* **2** (1949) 331. An orthogonal polynomial approach was given for parallel moments by H Sugama and T Watanabe and W Horton, in 'Collisionless kinetic-fluid closure and its application to the three-mode ion temperature gradient driven system,' *Phys. Plasmas* **8** (2001) 2617, and for six moments by B Scott, in 'Derivation via free energy conservation constraints of gyrofluid equations with finite-gyroradius electromagnetic nonlinearities,' *Phys. Plasmas* **17** (2010) 102306.

The product of two Laguerre polynomials is a linear combination of single L's but the expression is nontrivial, although a useful recursion method to calculate these has been worked out. See W N Bailey, 'On the product of two Laguerre polynomials,' *Quart. J. Math.* (Oxford) **os-10** (1939) 60; and J Gillis and G Weiss, 'Products of Laguerre polynomials,' *Math. Comp.* **14** (1960) 60; and P-A Lee, S-H Ong, and H M Srivastava, 'Some Integrals of the products of Laguerre polynomials,' *Int. J. Comput. Math.* **78** (2001) 303.

Magnetic pumping as a generation of thermal anisotropy by parallel flows in a toroidal magnetic field was described by A B Hassam and R M Kulsrud, in 'Convective cells and transport in toroidal plasmas,' *Phys. Fluids* **22** (1979) 2097.

References for the time dependent heat flux are its description and derivation by Hassam (1980), and its use in a fluid drift Alfvén turbulence model by Scott (1997), both *op cit* (Chapter 2). The Landau closure model springs from its use in this latter reference.

IOP Publishing

Turbulence and Instabilities in Magnetised Plasmas, Volume 2
Gyrokinetic theory and gyrofluid turbulence
Bruce Scott

Chapter 10

Gyrofluid equations for thermal dynamics

10.1 Introduction

The previous Chapter covered the ideas and systematics of gyrofluid models in general, showing how to build a model of arbitrary detail in velocity space by keeping an arbitrary number of moments. In practice we wish to truncate this hierarchy at a level commensurate to the fluid drift models with temperature dynamics. Ultimately we have to show that the fluid drift model with collisions and diamagnetic momentum fluxes ('gyroviscosity') from the Braginskii (1965) model is recovered by applying its necessary limiting assumptions to the gyrofluid model. With finite Larmor radius (FLR) corrections entering as an expansion, we show that the set of FLR advection nonlinearities reduce to the gyroviscosity expression we encountered in Volume One and in Chapter 3. In the limit of dominant collisions (collision frequency fast compared to the ExB vorticity or to parallel transit frequencies), the collisional damping of temperature anisotropy is found to recover the fluid model's expression for parallel flow viscosity. This also addresses the Braginskii equations' implicit limit that the specific heat flux (heat flux divided by pressure) be small compared to the velocity, species by species. Finally, under these limits of small anisotropy and dominant collisions the Braginskii expression for parallel heat flux including the thermal force and thermo electric effect is recovered. This renders the gyrofluid model as a necessary superset of the content of the corresponding fluid drift model in the description of turbulence in magnetised plasmas. All of this is a prelude to the gyrofluid computations we will turn to in subsequent Chapters.

10.2 The gyrofluid model with thermal responses

The minimal model under this description for treating temperature gradients and disturbances at the same level as the density, and the heat flux dynamics at the same level as the velocity, is the six-moment model. One might think of four, with a density, parallel velocity, temperature, and parallel heat flux for each species.

doi:10.1088/978-0-7503-3855-4ch10

Since the thermal processes differ according to parallel and perpendicular velocity space weighting ($z^2/2$ or w in the dimensionless coordinates), each of the temperature and parallel heat flux variables are split into components according to parallel or perpendicular energy. This makes six moments, with the highest being the parallel heat fluxes.

This gyrofluid model contains moments up to level $m + 2n = 3$ in the language of the previous Chapter, provides the smallest model with temperature effects, or thermal responses. Temperature gradients, thermal conduction, thermal forces in resistivity, and also temperature anisotropy all enter at this level. Since the maximum n is 1, the FLR corrections to the model are still compact. It is possible to manually add a collisional dissipation model in which the coefficients are set by correspondence to the Braginskii model used in Chapters 2 and 3. The basics of temperature gradient dominated dynamics as one finds in high-performance laboratory plasma confinement devices can be addressed with this model.

10.2.1 List of moment variables

The moments contained in the set of the first six represent density, parallel velocity, parallel temperature, perpendicular temperature, and the parallel and perpendicular components of the parallel heat flux. Written as pairs of state and associated flux variables, they are

$$\begin{aligned} a_{00} &= \frac{n}{n_0}, & a_{10} &= \frac{u_\parallel}{V} \\ a_{20} &= \sqrt{\frac{1}{2}}\frac{T_\parallel}{T_0}, & a_{30} &= \sqrt{\frac{2}{3}}\frac{q_{\parallel\parallel}}{n_0 T_0 V} \\ a_{01} &= \frac{T_\perp}{T_0}, & a_{11} &= \frac{q_{\perp\parallel}}{n_0 T_0 V} \end{aligned} \tag{10.1}$$

with density and parallel and perpendicular temperature on the left, and parallel velocity and parallel and perpendicular energy components of parallel conductive heat flux (i.e., without the pressure times velocity pieces in the total).

The coefficients involving square roots come from the differences between the conventional definitions of the moment variables, and the definitions of a_{mn} in terms of $H_m(z)$ and $L_n(w)$ in equation (9.49) of the previous Chapter. All quantities are understood as one set per species and as perturbed values away from the equilibrium state which has n_0 and T_0 for background density and temperature and all flux and field variables are at zero (i.e., F_0 contains only these two quantities as finite). The meaning of parallel and perpendicular temperature is the thermal component of parallel and perpendicular energy (p_z^2/m and $M\Omega$, respectively), and the associated heat fluxes are the conductive parallel flux of parallel and perpendicular energy, respectively. Isotropic quantities can be constructed according to

$$T = \frac{1}{3}(T_\parallel + 2T_\perp), \qquad q_\parallel = q_{\parallel\parallel} + q_{\perp\parallel} \tag{10.2}$$

We will have more on this below on correspondence and in later Chapters treating the turbulence.

10.2.2 Moment variable equations

The moment variables a_{mn} are normalised quantities, but here we leave the equations un-normalised. The equations for the a_{mn} are given explicitly by

$$\begin{aligned}&\frac{\partial a_{00}}{\partial t} + \frac{1}{\chi_x}\left[\phi_0, a_{00}\right] + \frac{1}{\chi_x}\left[\phi_1, a_{01}\right] + VB\nabla_{\|}\frac{a_{10}}{B}\\&\quad = \frac{1}{2}\frac{T_0}{e}\mathcal{K}\left(2h_{00} + h_{01} + \sqrt{2}\,a_{20}\right)\end{aligned} \tag{10.3}$$

$$\begin{aligned}&\frac{\partial g_{10}}{\partial t} + \frac{1}{\chi_x}\left[\phi_0, a_{10}\right] + \frac{1}{\chi_x}\left[\phi_1, a_{11}\right] + V\nabla_{\|}\left(h_{00} + \sqrt{2}\,a_{20}\right)\\&\quad - V\left(\sqrt{2}\,a_{20} - h_{01}\right)\nabla_{\|}\log B\\&\quad = \frac{1}{2}\frac{T_0}{e}\mathcal{K}\left(4a_{10} + \sqrt{6}\,a_{30} + a_{11}\right) - \frac{e}{mV}R_{ie}\end{aligned} \tag{10.4}$$

for the density/velocity pair,

$$\begin{aligned}&\frac{\partial a_{20}}{\partial t} + \frac{1}{\chi_x}\left[\phi_0, a_{20}\right] + VB\nabla_{\|}\frac{\sqrt{2}\,a_{10} + \sqrt{3}\,a_{30}}{B} + V\sqrt{2}\,(a_{10} + a_{11})\nabla_{\|}\log B\\&\quad = \frac{1}{2}\frac{T_0}{e}\mathcal{K}\left(\sqrt{2}\,h_{00} + 6a_{20}\right) - \sqrt{2}\,D_T\end{aligned} \tag{10.5}$$

$$\begin{aligned}&\frac{\partial a_{30}}{\partial t} + \frac{1}{\chi_x}\left[\phi_0, a_{30}\right] + V\nabla_{\|}\left(\sqrt{3}\,a_{20}\right)\\&\quad = \frac{1}{2}\frac{T_0}{e}\mathcal{K}\left(\sqrt{6}\,a_{10} + 8a_{30}\right) - \sqrt{3}\,K_{\|}Va_{30} - \frac{e}{mV}R_{q\|}\end{aligned} \tag{10.6}$$

for the parallel temperature/heat flux pair, and

$$\begin{aligned}&\frac{\partial a_{01}}{\partial t} + \frac{1}{\chi_x}[\phi_0, a_{01}] + \frac{1}{\chi_x}[\phi_1, (a_{00} + 2a_{01})] + VB\nabla_{\|}\frac{a_{11}}{B} - V(a_{10} + a_{11})\nabla_{\|}\log B\\&\quad = \frac{1}{2}\frac{T_0}{e}\mathcal{K}(h_{00} + 4h_{01}) + D_T\end{aligned} \tag{10.7}$$

$$\begin{aligned}&\frac{\partial a_{11}}{\partial t} + \frac{1}{\chi_x}[\phi_0, a_{11}] + \frac{1}{\chi_x}[\phi_1, (a_{10} + 2a_{11})] + V\nabla_{\|}\left(h_{01}\right) - V\left(\sqrt{2}\,a_{20} - h_{01}\right)\nabla_{\|}\log B\\&\quad = \frac{1}{2}\frac{T_0}{e}\mathcal{K}(a_{10} + 6a_{11}) - K_{\|}Va_{11} - \frac{e}{mV}R_{q\perp}\end{aligned} \tag{10.8}$$

for the perpendicular temperature/heat flux pair.

The auxiliary variables are

$$h_{00} = a_{00} + \frac{e\phi_0}{T_0}, \qquad h_{01} = a_{01} + \frac{e\phi_1}{T_0}, \qquad g_{10} = a_{10} + \frac{eA_\parallel}{mV} \tag{10.9}$$

and the h_{mn} and g_{mn} not listed are left as a_{mn}, which is what they are.

The gyroaveraging operators are in the Padé approximants,

$$G_0 = \frac{1}{1 + b/2}, \qquad G_1 = G_0 \frac{-b/2}{1 + b/2} \tag{10.10}$$

The operators are implemented via

$$\frac{1}{1 + b/2} = \left(1 - \frac{1}{2}\rho_z^2 \nabla_\perp^2\right)^{-1}, \qquad \frac{-b/2}{1 + b/2} = \frac{1}{1 + b/2} - 1 \tag{10.11}$$

For electrons, $b \to 0$ hence $G_0 \to 1$ and $G_1 \to 0$. The exponent -1 on an operator denotes solution of a Helmholtz equation.

The collisional dissipation terms are given by

$$R_{ie} = \frac{m_e \nu_e}{n_{e0} e_0^2}\left(0.51 J_\parallel + \frac{\alpha_e}{\kappa_e}\frac{1}{T_{e0}}\left(e_0 q_{e\parallel} + \alpha_e T_{e0} J_\parallel\right)\right) \tag{10.12}$$

$$D_T = \frac{\nu_z}{3\pi_z}\left(\sqrt{2}\, a_{20} - a_{01}\right) \tag{10.13}$$

$$\frac{e}{mV} R_{q\parallel} = \nu_z \frac{5/2}{\kappa_z}\sqrt{\frac{2}{3}}\left(\sqrt{\frac{3}{2}}\, a_{30} + \Delta_q + 0.6\alpha_z J\right) \tag{10.14}$$

$$\frac{e}{mV} R_{q\perp} = \nu_z \frac{5/2}{\kappa_z}\left(a_{11} - \Delta_q + 0.4\alpha_z J\right) \tag{10.15}$$

$$\Delta_q = 1.92\left(\sqrt{\frac{2}{3}}\, a_{30} - a_{11}\right) \tag{10.16}$$

with auxiliary variables

$$J_\parallel = \sum_{sp} n_0 e V\, a_{10}, \qquad J = \frac{J_\parallel}{n_{e0} e_0 V_e}, \qquad q_{e\parallel} = \left[n_0 T_0 V\left(\sqrt{\frac{3}{2}}\, a_{30} + a_{11}\right)\right]_{\text{electrons}} \tag{10.17}$$

and coefficients

$$\alpha_e = 0.71, \qquad \kappa_e = 3.2, \qquad \pi_e = 0.73, \tag{10.18}$$

$$\alpha_i = 0, \qquad \kappa_i = 3.9, \qquad \pi_i = 0.96 \tag{10.19}$$

and species collision frequencies given by the Braginskii model. The factor e/mV in the resistive friction R_{ie} combines it with $\partial/\partial t$ on $eA_\parallel/mV$, since induction and resistivity combine in this way (as in $E_\parallel - \eta_\parallel J_\parallel$ in an isothermal model). The factor

$\sqrt{2/3}$ in $R_{q\parallel}$ quarantees positive-definite damping in the combination of heat flux dissipation. The factor of e/mV in front of $R_{q\parallel}$ and $R_{q\perp}$ in Eqs (10.14) and (10.15) puts the resistive and conductive dissipation effects into the same units. The α_z arrange that the thermal force for electrons, always present in R_{ie}, affects only the electrons in the heat flux dissipation. We will return to this in Section 10.5 on the correspondence to the fluid drift equations with Braginskii dissipation.

10.2.3 Field equations

The field equations for six moments are

$$P(\phi) = \sum_{sp} n_0 e[G_0(a_{00}) + G_1(a_{01})] \tag{10.20}$$

for polarisation [Eqs (9.76) and (9.77)] and

$$-\nabla_\perp^2 A = \mu_0 \sum_{sp} n_0 e V a_{10} = \mu_0 J_\parallel \tag{10.21}$$

for induction (Eqs 9.85). The polarisation operator is (Eq. 9.78)

$$P = \sum_{sp} n_0 e^2 \frac{1 - \Gamma_0}{T_0} \tag{10.22}$$

where under Padé approximant the operator Γ_0 is implemented as

$$\Gamma_0 = \frac{1}{1+b} = \left(1 - \rho_z^2 \nabla_\perp^2\right)^{-1} \tag{10.23}$$

as with the gyroaveraging operators. With the definition of g_{10} we use the alternative for the Ampère's law with the skin depth term (Eq. 9.86),

$$\left(d_e^{-2} - \nabla_\perp^2\right) A_\parallel = \mu_0 \sum_{sp} n_0 e V g_{10} \tag{10.24}$$

since it is g_{10} which is advanced under the time derivative.

The electrons and ions are advanced separately according to their time-dependent equations, after which the field variables are solved for, constituting one time step. The same procedures as in the fluid drift model are used.

10.3 Collisions in general

Collisional processes at the resistive fluid level lead to the expressions encountered in Volume One, in which the expansions for large collisionality, $\nu \gg \omega$, were given. The model used assumed both these are below the gyrofrequency, $\Omega \gg \nu \gg \omega$, which justifies both the use of drifts and of the collisional expressions for parallel dynamics (only) which are still in the collision dominated regime. We did see, however, in Chapters 2 and 3, that the limit $\nu \gg \omega$ is basically never found in the warm and hot plasma cases which are of the most interest in magnetised plasma turbulence, with the large scale strongly collision dominated regime mostly well described by resistive MHD. In those

Chapters we set up time dependent equations for the heat fluxes but still didn't have the Landau damping model to cover dissipation. One needs a model which serves the weakly collisional regime, in which the Landau damping is practically the only physical dissipation mechanism of interest, across the trans-collisional regime ($\nu \sim \omega$) which we have treated so far, and into the collision dominated regime, in which the Braginskii expressions, for thermal forces and parallel viscosity as well as the Spitzer ones for resistivity and thermal conductivity, are recovered. We get that in this six-moment model by having separate equations for heat fluxes and 'parallel viscosity' with collisions providing damping for these. When the collisions are very weak, Landau damping remains present to cover the dissipation. When the collisions are very strong, they eclipse the Landau damping, and so the Braginskii expressions for the damping are what is needed. Parallel viscosity is put into quotes here, since it is not explicitly among the dependent variables. However, the thermal anisotropy accounts for it naturally, including the way it corrects the pressure effects in both parallel dynamics and diamagnetic compressibility, as we will see below. For now, the damping term needed is simply a damping of thermal anisotropy in each temperature equation (parallel and perpendicular) satisfying the H-theorem, namely, it must display positive definite damping of free energy. We get this by setting up a coefficient multiplying $T_\| - T_\perp$ which yields the Braginskii coefficient for parallel viscosity in the collision dominated limit.

Done in this way, the coefficients can be merely chosen, but for one exception. Thermal conduction anisotropy is also damped by collisions, and there is no counterpart for it in the Braginskii model. We set this up by doing the drift kinetic Chapman–Enskog expansion as in Chapter 6 from Volume One, identifying all the terms with their Braginskii counterparts, and then using that as guidance for setting the coefficient we will need in front of the heat flux anisotropy. The correspondence to the Braginskii collisional model will be shown in Section 10.5.

To do this we briefly describe the Chapman–Enskog procedure for parallel dynamics using a bi-Maxwellian, so that the thermal anisotropy is part of the model. The example is given for the electrons. The zeroth order f_e is given by

$$f_e^{(0)} = a_0 \exp(-m_e w_\perp^2/2T_e) \exp(-m_e w_\|^2/2T_e) \tag{10.25}$$

with $w_\|$ and $w_\perp$ the kinetic velocity components of the electrons in the rest frame of the ions, where a_0 sets the density to n_e and the velocities are in physical units. The $T_\|$ and $T_\perp$ are assumed to have a small anisotropy. The first order equation is

$$\begin{aligned} C[f_e^{(1)}] &= \left(\frac{\partial}{\partial t} + w_\|\nabla_\| - \frac{e}{m_e}E_\|\frac{\partial}{\partial w_\|}\right) f_e^{(0)} - C\,[f_e^{(0)}] \\ C &= \frac{\nu}{w^3}\frac{\partial}{\partial\zeta}(1 - \zeta^2)\frac{\partial}{\partial\zeta} \end{aligned} \tag{10.26}$$

an equation for parallel drift kinetic dynamics, with $-e/m_e$ reflecting electrons and $E_\|$ the parallel electric field. The last term on the right hand side is the anisotropy correction. The second expression defines the Lorentz pitch angle angle collision operator with arbitrary coefficient ν. The pitch angle $\zeta = w_\parallel/w$ is an eigenfunction of $\mathcal{C}$ with eigenvalue -2ν. Taking the moments of 1 and $m_e w_\parallel^2/2$ and $m_e w_\perp^2/2$ produces the consistency conditions at zeroth order

$$\frac{\partial n_e}{\partial t} = 0, \qquad \frac{1}{2} n_e \frac{\partial T_\parallel}{\partial t} = -\frac{\nu}{3}\Delta T, \qquad n_e \frac{\partial T_\perp}{\partial t} = \frac{\nu}{3}\Delta T \tag{10.27}$$

This yields the dissipation of the thermal anisotropy, which is

$$\Delta T = T_\parallel - T_\perp \tag{10.28}$$

If the temperature equations are added the isotropic version of the original calculation is recovered. The factor of 3 is just geometry. Applying these into the equation for the first order correction and solving for it yields

$$\begin{aligned} -2\frac{\nu}{w^3} f_e^{(1)} = w_\parallel \Bigg[& \frac{\nabla_\parallel n_e}{n_e} + \left(\frac{m_e w_\parallel^2}{2T_{e\parallel}} - \frac{1}{2} \right) \frac{\nabla_\parallel T_{e\parallel}}{T_{e\parallel}} \\ & + \left(\frac{m_e w_\perp^2}{2T_{e\perp}} - 1 \right) \frac{\nabla_\parallel T_{e\perp}}{T_{e\perp}} + \frac{eE_\parallel}{T_{e\parallel}} \Bigg] f_e^{(0)} \end{aligned} \tag{10.29}$$

Collisions in a gyrofluid model can also be found by simply taking the Hermite–Laguerre polynomial moments of a reasonable collision operator. The example used is again the Lorentz operator with z reckoned in the centre of mass frame of the ions. The translation of the velocity space derivatives into the $\{zw\}$ coordinates is

$$v\frac{\partial}{\partial v} = z\frac{\partial}{\partial z} + 2w\frac{\partial}{\partial w}, \qquad \frac{1}{v}\frac{\partial}{\partial \zeta} = \frac{\partial}{\partial z} - z\frac{\partial}{\partial w} \tag{10.30}$$

noting that $\partial/\partial z$ with $A_\parallel$ fixed is the same thing as $\partial/\partial(mU)$ in physical units, noting the use of v for kinetic velocity to distinguish from the w coordinate. With the decomposition of δf through the $H_m(z)\, L_n(w)$ terms with $m + 2n$ up to 6, this gives

$$\begin{aligned} \mathcal{C}[(\delta f)] = & -2\frac{\nu}{v^3} F_0 \Bigg[H_1 a_{10} + H_3 a_{30} + H_1 L_1 a_{11} \\ & + \left(\sqrt{2} H_2 - L_1\right)\left(\sqrt{2} a_{20} - a_{01}\right) \\ & + 3\left(\sqrt{\frac{2}{3}} H_3 - H_1 L_1\right)\left(\sqrt{\frac{2}{3}} a_{30} - a_{11}\right) \Bigg] \end{aligned} \tag{10.31}$$

where $F_0 = F^M(z, w)$ is the background Maxwellian with the given density and temperature. The presence of the factor in the denominator is due to the v^{-3} dependence of the Coulomb collision process. Without it, the collision terms are all diagonal in the moments and polynomials they appear in, namely, ΔT appears only in the equations for $T_\parallel$ and $T_\perp$, etc. With it, the moments are mixed just as in the

Chapman–Enskog expansion, providing the thermal force and thermo-electric effect as was one of the points in the original work by Braginskii. One can do the moments in a rather straightforward manner in the diagonal case. It is much more work in the Coulomb case. This is a useful method to get the relevant terms in a model with an arbitrary deep hierarchy (many more than six moments). We mention it here mainly for future work. Ultimately, one would want to replace the Lorentzian pitch angle scattering model with the full Landau collision integral and then find the coefficients numerically. Since the point herein is physical process and physical insight, we will leave this to other work. For now, we use the Chapman–Enskog based version just given (see Further Reading).

10.4 Thermal gyrofluid energetics

Gyrofluid energetics in general was given in the previous Chapter. We now restrict to the specific case of the six moment model. For the ExB and magnetic energies we have

$$E_E = \mathcal{V}^{-1}\int d\mathcal{V}\,\frac{1}{2}\,\phi P(\phi), \qquad E_M = \mathcal{V}^{-1}\int d\mathcal{V}\,\frac{1}{2\mu_0}\,\left|\nabla_\perp A_\parallel\right|^2 \tag{10.32}$$

Using the field equations to eliminate $P(\phi)$ and $\nabla_\perp^2 A_\parallel$ we can recast these as

$$E_E = \mathcal{V}^{-1}\int d\mathcal{V}\,\sum_{sp}\frac{1}{2}\,n_0 e\left[\phi_0 a_{00} + \phi_1 a_{01}\right] \tag{10.33}$$

and

$$E_M = \mathcal{V}^{-1}\int d\mathcal{V}\,\frac{1}{2}\,J_\parallel A_\parallel = \mathcal{V}^{-1}\int d\mathcal{V}\,\sum_{sp}\frac{1}{2}\,n_0 eV\left[A_\parallel a_{10}\right] \tag{10.34}$$

respectively. The thermal free energy represented by (δf) is separated into thermal pieces from the state variables, given by

$$E_{th} = \mathcal{V}^{-1}\int d\mathcal{V}\,\sum_{sp}\frac{1}{2}\,n_0 T_0\left[a_{00}^2 + a_{20}^2 + a_{01}^2\right] \tag{10.35}$$

and the parallel kinetic pieces from the flux variables, given by

$$E_{fl} = \mathcal{V}^{-1}\int d\mathcal{V}\,\sum_{sp}\frac{1}{2}\,n_0 T_0\left[a_{10}^2 + a_{30}^2 + a_{11}^2\right] \tag{10.36}$$

with the latter including the heat flux contributions. In the normalisation using the a_{mn}'s, each of these enter without numerical coefficients like 1/2 for the parallel temperature.

The thermal effects can be more interesting, due to the magnetic pumping effects. The thermal anisotropy is driven by parallel fluxes of perpendicular energy, in total represented by $a_{10} + a_{11}$. The magnetic pumping terms in the equations for a_{20} and a_{01} are

$$\frac{\partial a_{20}}{\partial t} = \cdots - V\sqrt{2}(a_{10} + a_{11})\nabla_\| \log B \tag{10.37}$$

$$\frac{\partial a_{01}}{\partial t} = \cdots + V(a_{10} + a_{11})\nabla_\| \log B \tag{10.38}$$

and in the equations for g_{10} and a_{11} are

$$\frac{\partial g_{10}}{\partial t} = \cdots + V\left(\sqrt{2}\, a_{20} - h_{01}\right)\nabla_\| \log B \tag{10.39}$$

$$\frac{\partial a_{11}}{\partial t} = \cdots + V\left(\sqrt{2}\, a_{20} - h_{01}\right)\nabla_\| \log B \tag{10.40}$$

The state variable combination in the free energy is

$$a_{20}\frac{\partial a_{20}}{\partial t} + h_{01}\frac{\partial a_{01}}{\partial t} = \cdots - V\left(\sqrt{2}\, a_{20} - h_{01}\right)(a_{10} + a_{11})\nabla_\| \log B \tag{10.41}$$

and the flux variable combination is

$$a_{10}\frac{\partial g_{10}}{\partial t} + a_{11}\frac{\partial a_{11}}{\partial t} = \cdots + V(a_{10} + a_{11})\left(\sqrt{2}\, a_{20} - h_{01}\right)\nabla_\| \log B \tag{10.42}$$

It is now apparent that the combination on the right hand side is a transfer term between flux and state variables, much in the same way that the divergence terms in Eq. (10.41) (within the dieresis) leave a residual on the right hand side,

$$\cdots = \cdots + \left(\sqrt{2}\, a_{10} + \sqrt{3}\, a_{30}\right)\nabla_\| a_{20} + a_{11}\nabla_\| h_{01} \tag{10.43}$$

which provides the balance for the corresponding force terms on the right side of Eq. (10.42) (see also Eqs 9.119 to 9.121 of the previous Chapter). But unlike the two pieces in the divergence combination, which keep the transfer within the 30/20 and 11/10 pairs separate, the magnetic pumping process transfers free energy between these pairs. This acts to exchange energy between parallel and perpendicular temperatures independently from collisions, and it is not a dissipation process but something which may or may not cause equilibration (keeping the thermal free energy more isotropic). This is situation-dependent and becomes an interesting possibility to watch out for. Due to the gyrokinetic conservation of magnetic moment, it is a process fundamental to the gyrofluid model but absent in a fluid model. If we add the temperatures to form an equation for an isotropic temperature,

$$a_T = \sqrt{\frac{1}{2}}\, a_{20} + a_{01} = \frac{T_\| + 2T_\perp}{2T_0} \rightarrow \frac{3}{2}\frac{T}{T_0} \tag{10.44}$$

we find

$$\begin{aligned} &\frac{\partial a_T}{\partial t} + \frac{1}{\chi_x}\left[\phi_0, a_T\right] + (\text{FLR}) + VB\nabla_\| \frac{a_{10} + \sqrt{3/2}\, a_{30} + a_{11}}{B} \\ &\qquad = \frac{T_0}{e}\mathcal{K}\left(h_{00} + 3\sqrt{2}\, a_{20} + 2h_{01}\right) + (\text{diss}) \end{aligned} \tag{10.45}$$

Up to FLR corrections and anisotropy on the right hand side, this is the same as the temperature equation in the fluid drift Alfvén model, and the magnetic pumping effect has cancelled out since it is a conservative exchange between $(1/2)T_{\parallel}$ and $T_{\perp}$. This is the first step in showing the correspondence between the gyrofluid and fluid drift Alfvén models. The next is to consider polarisation and dissipation, to which we turn next.

10.5 Correspondence to the fluid model

When we derived the fluid drift model in Volume One, we described the polarisation drift as the inertial correction which should be formally small by $O(\delta)$, but which still has a physical role because its divergence has a correspondingly small scale. Due to this scale disparity, the polarisation alone among corrections to the force balance represented by the basic drifts is kept. Also, for the same reason, the nonlinear drifts bracket among all nonlinear corrections is kept, with others small in the equations by $O(\delta)$. The same thing carries over to the gyrokinetic representation, wherein gyromotion and the ExB energy, and at the same time in the same breath, polarisation is promoted by one order into the general $O(\delta)$ dynamics below the $O(1)$ equilibrium represented by H_0 and F_0. This step is needed to provide energetic consistency and is, or should be, among the first alarms that blind ordering chains one knows from linear stability theory do not apply to a nonlinear situation in which statistical equilibration over indefinite time—saturation—is a central part of the physical description.

When looking at these equations, though, one might ask where the polarisation drift went. Might the lowest approximation to the (density) continuity equation be missing it? Indeed in the absence of curvature and parallel divergences one only finds what looks like an Eulerian fluid equation, with only $[\phi, n_i]$ left in the equation besides the time derivative. This and other questions about correspondence are answered by deriving the path back to the fluid equations consistently. The main point between gyrokinetic and drift kinetic theory is not the FLR corrections but the representation. It is a good exercise to do the following calculation first with $T_{i0} \to 0$ hence $\tau_i \to 0$ in normalised units. One finds, solely due to the representation of what n_i means, that the polarisation flow divergence is automatically kept in the continuity equation. In fact, it is not possible to compute a model with n_i and ϕ consistently without it.

We also cover the route back to the collisional limit represented by the formulae for dissipation given by Braginskii. This is simple for the parallel heat flux and was already done in the fluid model for the electron one, straightforward because we take the electrons as contributing neither to polarisation nor to FLR corrections. It is more interesting to do that for the parallel viscosity, since we are now taking care not to add any dissipation terms directly to the parallel flow equations other than resistive friction. We will find that this effect results from the thermal anisotropy, $T_{\parallel} - T_{\perp}$, which adds to the reasons why thermal anisotropy is a fundamental ingredient of the model and must always be kept if temperature gradients or dynamics is to be addressed.

10.5.1 Polarisation and vorticity

The look and feel of a gyrofluid model can be very similar to a drift Alfvén fluid model, at least at the level of four dependent variables covering ExB vorticity and velocity, electron density, parallel current and magnetic induction, and parallel velocity. When the ion temperature is brought in, they indeed do look somewhat different.

To make the following algebra as clear as possible, we display it in detail for the isothermal (two-moment) version of the model. With the thermal corrections added, then the six-moment model is assumed. The normalisation and usage is as in the drift Alfvén model. Hence the spatial bracket is written without the factor χ_x, which is normalised to unity, and $[a, b]$ appears by itself. The factors of ρ_s^2 appearing with $\nabla_\perp^2$ are written explicitly, due to the manipulation of factors of τ_i in $p_i = \tau_i n$ and $\rho_i^2 = \tau_i \rho_s^2$ which make the connection between FLR corrections in the gyrofluid model and the contribution to polarisation due to diamagnetic vorticity in the fluid model.

The vorticity equation in the fluid model is

$$\nabla \cdot \left(\frac{d}{dt}\right) \rho_s^2 \nabla_\perp (\phi + p_i) = B\nabla_\| \frac{J_\|}{B} - \mathcal{K}(p_e + p_i) \tag{10.46}$$

while the difference between the electron and ion density equations in the gyrofluid model is

$$\frac{d}{dt}(n_e - n_i) + [(\phi - \phi_G), n_i] = B\nabla_\| \frac{u_{i\|} - u_{e\|}}{B} - \mathcal{K}\big(p_e + p_i - (\phi - \phi_G)\big) \tag{10.47}$$

where

$$\frac{d}{dt} = \frac{\partial}{\partial t} + [\phi, \,], \quad p_e = n, \quad p_i = \tau_i n, \qquad \phi_G = G(\phi), \quad \tau_i = T_{i0}/T_{e0} \tag{10.48}$$

The gyrofluid polarisation equation and gyroaveraging under Padé approximant for this model are

$$P(\phi) = G(n_i) - n_e, \quad G = \left(1 - \frac{1}{2}\rho_i^2 \nabla_\perp^2\right)^{-1}, \quad P = \tau_i^{-1}\left[1 - \left(1 - \rho_i^2 \nabla_\perp^2\right)^{-1}\right] \tag{10.49}$$

In each model there are four variables, $(\varpi, n, J_\|, u_\|)$ in the fluid case and $(n_e, u_{e\|}, n_i, u_{i\|})$ in the gyrofluid case, and one ion with $e = e_0$ is assumed.

These two equations are actually much more correspondent than they look. Recall the fluid model is derived assuming $k_\perp^2 \rho_i^2 \ll 1$, which is not actually true. Though it is energetically consistent, its responses break down when $k_\perp \rho_i \sim 1$ is reached, even if as in our case robust computations with it were shown to be possible in Chapter 3. But it is of interest to know how well these two models correspond when indeed $b_i = k_\perp^2 \rho_i^2 \ll 1$ can be assumed and used as a condition for expansion, order by order. So, taking $b_i \ll 1$, we can expand

$$\phi_G = \phi + \frac{1}{2}\rho_i^2\nabla_\perp^2\phi, \qquad n_i = n_e - \rho_s^2\nabla_\perp^2\phi - \frac{1}{2}\rho_i^2\nabla_\perp^2 n_e \tag{10.50}$$

and we can re-write this using

$$p_e = n_e = n, \qquad p_i = \tau_i n, \qquad \rho_i^2 = \tau_i\rho_s^2 \tag{10.51}$$

so that

$$\phi - \phi_G = -\frac{1}{2}\tau_i\rho_s^2\nabla_\perp^2\phi, \qquad n_e - n_i = \rho_s^2\nabla_\perp^2\left(\phi + \frac{1}{2}p_i\right) \tag{10.52}$$

Moreover, we note that the Padé approximants for the FLR operators in the gyrofluid model coincide with the exact ones in $k_\perp$-space through $O(b_i)$.

Now we insert these forms for $n_e - n_i$ and $\phi - \phi_G$ into Eq. (10.47). Since $\phi - \phi_G$ is already first order in b_i, the n_i it appears with is replaced by n at $O(b_i)$, and its appearance under $\mathcal{K}()$ is also an $O(b_i)$ correction. Since $n_e - n_i$ is already at $O(b_i)$ as well, the d/dt that acts upon it is the $O(1)$ version given in Eq. (10.48). So all the FLR terms appear on the left and the terms on the right are left at $O(1)$, in keeping with the polarisation ordering in general where the lowest-order polarisation term formally at $O(b_i)$ make up the current mismatch imbalance between terms formally at $O(1)$. Evaluating the right hand side at $O(1)$, the two right hand sides are the same, with $J_\| = u_{i\|} - u_{e\|}$. Then, the left side is

$$\begin{aligned}\frac{d}{dt}(n_e - n_i) + [(\phi - \phi_G), n_i] &= \frac{\partial}{\partial t}\rho_s^2\nabla_\perp^2\left(\phi + \frac{1}{2}p_i\right) \\ &+ [\phi, \rho_s^2\nabla_\perp^2\phi] + \frac{1}{2}[\phi, \rho_s^2\nabla_\perp^2 p_i] - \frac{1}{2}[\rho_s^2\nabla_\perp^2\phi, p_i]\end{aligned} \tag{10.53}$$

The operations to be used are the vector calculus and algebra of the brackets,

$$\begin{aligned}[a, \nabla_\perp^2 b] &= \nabla\cdot[a, \nabla_\perp b] - [\nabla_\perp a; \nabla_\perp b] \\ [\nabla_\perp^2 a, b] &= \nabla\cdot[\nabla_\perp a, b] - [\nabla_\perp a; \nabla_\perp b] \\ \nabla\cdot[\nabla_\perp a, b] &= \nabla_\perp^2[a, b] - \nabla\cdot[a, \nabla_\perp b]\end{aligned} \tag{10.54}$$

where the semicolon denotes contraction as well as the usual bracket operation. The result is

$$\begin{aligned}&\frac{d}{dt}(n_e - n_i) + [(\phi - \phi_G), n_i] \\ &= \nabla\cdot\left(\frac{\partial}{\partial t} + [\phi, \,]\right)\rho_s^2\nabla_\perp(\phi + p_i) + \rho_s^2\nabla_\perp^2\left(\frac{\partial p_i}{\partial t} + [\phi, p_i]\right) \\ &= \nabla\cdot\left(\frac{d}{dt}\right)\rho_s^2\nabla_\perp(\phi + p_i) + \rho_s^2\nabla_\perp^2\left(\frac{dp_i}{dt}\right)\end{aligned} \tag{10.55}$$

The last term is replaced by the right hand side of the equation for p_i operated upon by an $O(b_i)$ correction, so it is a correction, as $\nabla_\| J_\|$ is as large as $\nabla_\| u_\|$ and $\mathcal{K}()$ acts upon similar order terms. Up to $O(1)$ and $O(b_i)$, then, the two models are convergent. At low-$k_\perp$, where the fluid model is valid, the two coincide. As $b_i \to 1$, the fluid model breaks down but the gyrofluid model can still be used, in fact for arbitrary order as pointed out by Knorr *et al* (1988). This is how the correspondence works for the polarisation current. The nonlinear FLR corrections account for the diamagnetic component (contributions by p_i) to the vorticity nonlinearities, including gyroviscosity, and the ion inertial polarisation (ρ_s only at low-$k_\perp$) accounts for the ExB vorticity. The basic dynamics is the same, and convergent at large scale.

One other thing to note is that if we recover the ion density by itself we have alternatively

$$\frac{dn}{dt} + B\nabla_\| \frac{u_\| - J_\|}{B} = \mathcal{K}(\phi - p_e) \tag{10.56}$$

for the electrons or

$$\frac{dn}{dt} - \nabla \cdot \left(\frac{d}{dt}\right) \rho_s^2 \nabla_\perp (\phi + p_i) + B\nabla_\| \frac{u_\|}{B} = \mathcal{K}(\phi + p_i) \tag{10.57}$$

to $O(b_i)$ using $n_e = n$. The second equation, for the ion density, is found by subtracting the vorticity equation from the first equation, for the electron density. In the fluid model the two densities are the same, and the vorticity equation is for the divergence of the current, including polarisation. In the gyrofluid model the two densities are different, and the polarisation equation is used to solve for ϕ. This difference is the same one between the use of ϖ and n or N and n as dependent variables in the 2D heuristic model at the start of Chapter 9.

This calculation can be repeated with the temperatures. The additional terms are

$$\phi_1 \to \frac{1}{2} \rho_i^2 \nabla_\perp^2 \phi, \qquad T_\| \to T_i, \qquad T_\perp \to T_i \tag{10.58}$$

the linearised pressures are

$$p_e = n_e + T_e \qquad p_i = \tau_i (n + T_i) \tag{10.59}$$

and the density equation is

$$\frac{dn}{dt} - \nabla \cdot \left(\frac{d}{dt}\right) \rho_s^2 \nabla_\perp \left(\phi + p_i + \frac{5}{2}\tau_i T_i\right) + B\nabla_\| \frac{u_\|}{B} = \mathcal{K}(\phi + p_i) \tag{10.60}$$

and the combined temperature equations are

$$\frac{3}{2}\frac{dT}{dt} - \nabla \cdot \left(\frac{d}{dt}\right) \rho_s^2 \nabla_\perp \left(\phi + p_i + \frac{5}{2}\tau_i T_i\right) + B\nabla_\| \frac{u_\| + q_\|}{B} = \mathcal{K}\left(\phi + p_i + \frac{5}{2}\tau_i T_i\right) \tag{10.61}$$

to $O(b_i)$ using $n_e = n$, after several applications of the algebra in Eqs (10.54). These are the reduced Braginskii equations for the density and temperature, with the polarisation drift and also, with the $(5/2)\tau_i T_i$ term, the polarisation heat flux, which (in units of velocity) one usually orders small compared to the polarisation flow. The main point is the same, except for the fact that the gyrofluid model never orders the heat flux to be small. More on this in the subsection on viscosity, below.

One issue we faced in the ion temperature deliberations in Chapter 3 was whether the model with the magnetohydroynamic (MHD) form of polarisation or the two-fluid one is the more valid, and which among the ion velocity kinetic energies, the ExB one or the two-fluid one also including the diamagnetic velocity, is the more valid. The polarisation and energy density in the MHD version are

$$\nabla \cdot \left(\frac{\partial}{\partial t} + \left[\phi + p_i, \quad \right] \right) \nabla_\perp \phi \quad \text{and} \quad \frac{1}{2} \left| \nabla_\perp \phi \right|^2 \tag{10.62}$$

while in the two-fluid version they are

$$\nabla \cdot \left(\frac{\partial}{\partial t} + [\phi, \quad] \right) \nabla_\perp (\phi + p_i) \quad \text{and} \quad \frac{1}{2} \left| \nabla_\perp (\phi + p_i) \right|^2 \tag{10.63}$$

This question can also be addressed with the gyrofluid model. The free energy density associated with the ion density variable is

$$\begin{aligned} \mathcal{E}_n = \frac{1}{2}(\phi_G + \tau_i n_i) n_i &= \frac{1}{2}(\phi + p_i) n \\ &- \frac{1}{2}\left((\phi - \phi_G) + \tau_i (n - n_i)\right) n - \frac{1}{2}(\phi + p_i)(n - n_i) \end{aligned} \tag{10.64}$$

Again we insert the above forms for $\phi - \phi_G$ and $n - n_i$ into this expression, and upon combining terms and doing some operations of the form

$$-a \nabla_\perp^2 b = (\nabla_\perp a) \cdot (\nabla_\perp b) - \nabla \cdot (a \nabla_\perp b) \tag{10.65}$$

we find

$$\mathcal{E}_n = \frac{1}{2}(\phi_G + \tau_i n_i) n_i = \frac{1}{2}(\phi + p_i) n + \frac{1}{2} \left| \nabla_\perp (\phi + p_i) \right|^2 \tag{10.66}$$

In this form the factor of ϕn cancels between the electrons and ions, the factor $p_i n = \tau_i n^2$ is the ion contribution to the density free energy which adds to the one from the electrons, and the remaining piece proportional is the same two-fluid perpendicular ion kinetic energy in Eq. (10.63) This cements the two-fluid forms as the correct ones, and also shows that with that model we can extend the spectrum through $k_\perp \rho_i \sim 1$ by using the gyrofluid model instead of the reduced Braginskii one for these problems.

10.5.2 Thermal conduction

The conventional picture of the fluid model is of an ideal fluid upon which dissipative processes act as corrections. One starts with a shifted Maxwellian given by a density, a flow, and a temperature, and then gradients in all those parameters then lead to collisional fluxes or friction which effect dissipation. In a neutral fluid by itself one has viscosity (flow gradients) and thermal conduction (temperature gradients), while density gradients at finite temperature lead to pressure gradients, which lead to flows, whose viscosity dissipates the dynamics. In a plasma, we also have collisional friction between electrons and ions, represented by resistivity. Due to the dependence of the Coulomb collision cross section on particle relative velocity, resistive friction and thermal conduction are coupled by thermal forces as detailed in Section 6.10 of Volume One. This is why we have $\alpha_e = 0.71$ in the thermal dynamics herein (Eq. 10.19), since for electrons the background ion velocities cluster near zero and the v^{-3} dependence is felt fully. In general the dissipation model is linearised: there are just the three parameters and then their gradients and the collision frequency determines the dissipation. In the gyrofluid model, we take the same limits to find thermal conduction $q_\parallel$ and parallel viscosity, represented by $\Delta T = T_\parallel - T_\perp$, the thermal anisotropy. Hence, these effects are assumed to be small and the collision frequency much larger than $\partial/\partial t$, so the collision frequency multiplying these corrections balances the parallel and perpendicular fluxes. The lowest-order quantities are

$$n = n_0 a_{00}, \qquad u_\parallel = V a_{10}, \qquad T = T_0 \frac{\sqrt{2}\, a_{20} + 2a_{01}}{3} = \frac{T_\parallel + 2T_\perp}{3} \tag{10.67}$$

and the perturbed quantities are

$$\Delta T = T_0\left(\sqrt{2}\, a_{20} - a_{01}\right), \qquad q_\parallel = n_0 T_0 V\left(\sqrt{\frac{3}{2}}\, a_{30} + a_{11}\right) \tag{10.68}$$

The latter are neglected everywhere except multiplied by ν_z, and the anisotropy in $q_\parallel$ is neglected. Solving for $q_\parallel$ and ΔT, in this limit should yield a match to the fluid equations, in this case the Braginskii model in the drift limit.

We will show this for the electrons, for which FLR corrections are not kept. The equation for $q_\parallel$ is found using those for a_{30} and a_{11} and is given by

$$\frac{5}{2} n_{e0} V_e^2 \nabla_\parallel T = -\nu_e \frac{5/2}{\kappa_e}\left(q_\parallel + \alpha_e \frac{T_{e0}}{e_0} J_\parallel\right) \tag{10.69}$$

Inertia, curvature, and Landau damping are neglected as small compared to the collision dissipation. Solving this for the collisional flux yields

$$q_\parallel = -\kappa_e \frac{V_e^2}{\nu_e} \nabla_\parallel T - \alpha_e \frac{T_{e0}}{e_0} J_\parallel \tag{10.70}$$

which, as also shown in Chapter 2, is the conductive parallel heat flux in the collisional limit, with nonzero α_e giving the thermoelectric effect. For ions we have the same result with $\alpha_i = 0$ and $\kappa_i = 3.9$ a slightly different coefficient. That this result should match the Braginskii limit is what sets the values of the coefficients α and κ within the gyrofluid model in the first place. But it guarantees that thermal conductive responses in the gyrofluid model smoothly become those in the

Braginskii model where the latter is valid. Of course, in many applications the limit $\nu_z \gg K_\| V$ or the ExB advection is far from satisfied. But this motivates the use of extended models.

10.5.3 Thermal anisotropy and parallel viscosity

In the collisional model viscosity is the name used for the part of the pressure tensor which is different from the isotropic pressure multiplied by the spatial metric tensor. Under the drift approximations the part we keep is the diagonal part. The diagonal pressure tensor splits into an isotropic part $p\mathbf{g}$ and another diagonal contribution which is specified to have zero trace. Since the two perpendicular dimensions in a Cartesian system are interchangeable, we write them as equal, and then we have

$$P_{ij} = \mathrm{diag}\{p_\perp, p_\perp, p_\|\} = pg_{ij} + \frac{1}{3}\mathrm{diag}\{-\Delta p, -\Delta p, 2\Delta p\} \tag{10.71}$$

where

$$p = \frac{1}{3}(2p_\perp + p_\|), \qquad \Delta p = p_\| - p_\perp, \qquad \Pi_\| = \frac{2}{3}\Delta p \tag{10.72}$$

and conversely

$$p_\| = p + \frac{2}{3}\Delta p, \qquad p_\perp = p - \frac{1}{3}\Delta p \tag{10.73}$$

noting that $\Delta p = n\Delta T$. In all of these, for p one can read T since the density is always scalar. In terms of the moment variables, T and ΔT are as given above. The parallel viscosity $\Pi_\|$ is given explicitly in terms of the anisotropy.

The equation for Δp is found from Eqs (10.5) and (10.7) and using $\delta T = T_0(\sqrt{2}a_{20} - a_{01})$ to form the time derivative of ΔT, given for any given species by

$$\begin{aligned}&\frac{\partial}{\partial t}(\Delta p) + [\text{NL and FLR}] + B\nabla_\|\frac{2p_0u_\| + 2q_{\|\|\|} - q_{\perp\|}}{B} - 3(p_0u_\| + q_{\perp\|})\nabla_\|\log B \\ &= \frac{1}{2e}\mathcal{K}\left(n_0e\phi + \frac{p_\| + p_\perp}{2} + 4n_0T_\| - 3n_0T_\perp\right) - \frac{\nu_z}{\pi_z}\Delta p\end{aligned} \tag{10.74}$$

where $p_0 = n_0T_0$ is the species partial pressure, with nonlinear and FLR terms not explicitly written. Neglecting these along with the inertia, we can solve for Δp from the dissipation term, assuming the deeply collisional limit. This yields

$$\begin{aligned}&\frac{\nu_z}{\pi_z}\Delta p + B\nabla_\|\frac{2q_{\|\|\|} - q_{\perp\|}}{B} - 3\left(p_0u_\| + q_{\perp\|}\right)\nabla_\|\log B - \frac{8}{3e}\mathcal{K}(\Delta p) \\ &= \frac{1}{2e}\mathcal{K}(n_0e\phi + p + n_0T) - 2p_0B\nabla_\|\frac{u_\|}{B}\end{aligned} \tag{10.75}$$

where the terms the collisional limit assumes to be small are on the left hand side. In terms of flow and heat flux divergences this is

$$\Pi_{\|} = -\frac{2}{3}\pi_z n_0 \frac{V^2}{\nu_z} \nabla \cdot \left(2u_{\|}\mathbf{b} + \frac{1}{2}\mathbf{u}_{\perp} + \frac{4}{5}p_0^{-1}q_{\|}\mathbf{b} + \frac{1}{5}p_0^{-1}\mathbf{q}_{\perp} \right) \tag{10.76}$$

neglecting the magnetic pumping term and the anisotropy except for the finite parallel viscosity. The perpendicular flows and heat fluxes are the diamagnetic ones (derived in Chapter 6 of Volume One). In the Braginskii limit the specific heat fluxes (in units of velocity) are ordered small compared to the velocity, so the terms due to the heat flux divergences should not appear. However, in dynamics driven by temperature gradients this is never true. Even if the inertia effects in the equation for $\Pi_{\|}$ are small, the thermal flux effects are not—for example, the term due to $\mathcal{K}(T)$ is never small compared to the others in these cases (this is the term giving rise to $\mathbf{q}_{\perp}$ in Eq. 10.76). In an isothermal model (two moments per species), the viscosity must be put in manually, as we did with the four field drift Alfvén model. In general, the isotropic limit is never realised, so that if parallel viscosity actually acquires a role in the dynamics besides limiting the cascade dynamics in $k_{\|}$, then this six-moment model with thermal anisotropy is the minimal model. We refer to it as the minimal model due to these features, wherein the actual parallel heat fluxes and viscosities are determined by the dynamics rather than a pre-specified formula.

10.6 On usefulness

Hence, with the gyrofluid basis we solve the same problem as with the fluid drift Alfvén model based on the Braginskii collisional fluid equations set up in the drifts regime ($\omega \ll \Omega_i$). It is done merely in a different representation, ultimately with a better model due to the greater range of applicability and better computational behaviour.

Further Reading

These equations are the same as the ones given in B Scott, 'Free energy conservation in local gyrofluid models,' *Phys. Plasmas* **12** (2005) 102307; and B Scott, 'Derivation via free energy conservation constraints of gyrofluid equations with finite-gyroradius electromagnetic nonlinearities,' *Phys. Plasmas* **17** (2010) 102306.

The standard model for collisional dissipation effects in a magnetised plasma is the review given by Braginskii (1965) *op cit*, Chapter 2. The coefficients for resistivity and thermal conductivity are from the Spitzer *et al* references given in Chapter 2.

The description of the Chapman–Enskog procedure for electrons as used here is given by Hassam (1980) *op cit*, Chapter 2. The anisotropic temperature version is from Scott (2000) *op cit*, Chapter 9.

The polarisation heat flux enters at the same order as the polarisation velocity, just as for the diamagnetic flux case, for dynamics driven by temperature gradients. This was detailed by I O Pogutse and A I Smolyakov and A Hirose, in *J. Plasma*

Phys. **60** (1998) 133. The same point for viscosity was made by A Smolyakov, *Can. J. Phys.* **76** (1998) 321. The transcollisional form for viscosity via damping of thermal anisotropy was also used by M A Beer and G Hammett, in *Phys. Plasmas* **3** (1996) 4046.

The full details of the correspondence and energetic consistency between the gyrofluid and fluid drift equations as outlined above were given by B Scott, in 'Nonlinear polarisation and dissipative correspondence between low frequency fluid and gyrofluid equations,' *Phys. Plasmas* **14** (2007) 102318.

IOP Publishing

Turbulence and Instabilities in Magnetised Plasmas, Volume 2

Gyrokinetic theory and gyrofluid turbulence

Bruce Scott

Chapter 11

Gyrofluid Drift–Alfvén turbulence

11.1 Introduction—gyrofluid turbulence

With the gyrofluid model in hand the first thing to do is to revisit the main types of drift wave and drift Alvfén turbulence with this model. The main points are temperature anisotropy and any changes in the turbulence as one moves to weakly collisional regimes where the Landau damping dissipation model takes over from collisions in processes like resistivity, thermal conductivity, and parallel viscosity. Landau damping has a similar role to thermal conduction, as covered in Chapter 8, but it does not provide resistivity. This is why the closure has to avoid affecting the Ohm's law (parallel electron velocity equation). The minimal model becomes one in which the highest-moment flux variables are the heat fluxes, namely, the one with six moments. Density, parallel velocity, temperature, and heat flux would make four, but in this case the anisotropy between parallel and perpendicular temperature is fundamental to the model. Each of these temperature components has its associated heat flux variable, one factor of the Hermite polynomial index higher than the temperature. This makes six moments. If one closes after two moments, the model is more or less limited to a collisional closure, since Landau damping does not provide resistivity. This latter arises solely from collisions between the species, mostly electrons on ions.

The gyrofluid model is run on a variety of collisionality regimes to show how the Braginskii regime is recovered. This was shown analytically in Chapter 9, but it remains to be seen exactly how strong the collisionality has to be for that to occur. In hot plasma cases the collisionality is very weak. In the warm plasma cases, corresponding to the boundary regions of at least moderately ambitious confined plasmas the collisionality is in the transitional regime with electron collision frequency compared to nominal drift wave frequency, i.e., vorticity. This has always been the regime in which these considerations have been most strenuously argued. And the transcollisional situation is present for electrons, but for ions the collision frequency is two orders of magnitude slower. This is where the Braginskii dissipation

doi:10.1088/978-0-7503-3855-4ch11

model, and even the assumption of temperature anisotropy, is most problematic. The worst failures are for thermal conduction and parallel viscosity, both of which are strongly overestimated by the collisional model.

We will focus mostly on the fusion plasma application because these considerations depend on dominant drive of the turbulence by temperature gradients. In most other situations the pressure gradient is generic and density and temperature are not distinguished. The four-field model used most extensively is sufficient for this. But for confined plasmas the temperature gradient is dominant, not only in the hot plasma core regions but also in the warm plasma edge regions. Measured gradient scale ratios $d \log T / d \log n$ are in the vicinity of two in the edge regions to three or more in the core regions, with the electron temperature nearly always having the steepest gradient among the background profiles. This is the situation of interest for which the gyrofluid model is most appropriate. Of course it is electromagnetic, since mesoscale magnetohydrodynamic (MHD) processes intermediate between the overall size of the plasma and the ion gyroradius and drift scale sizes have to be covered. Whether the turbulence is in fact electromagnetic, adiabatic, or hydrodynamic, or MHD-like, is among the results of any calculations rather than part of the input.

11.2 Electromagnetic gyrofluid equations

The model with six moments consists of equations for the moments 00, 10, 20, 01, 30, and 11 for the density, parallel velocity, parallel temperature, perpendicular temperature, pure parallel heat flux, and perp/parallel heat flux. The two heat fluxes are the parallel conductive flux of parallel and perpendicular energy. These are conductive, as the convective part is separated out. For example, for an isotropic case we have the total parallel heat flux as the moment of energy times parallel velocity, with the flow piece split off, according to

$$Q_\parallel = q_\parallel + \frac{5}{2} p u_\parallel + \frac{1}{2} nm \frac{u^2}{2} u_\parallel \tag{11.1}$$

where the first term on the right hand side is the parallel conductive heat flux. Similarly, for this model we have

$$Q_{\parallel\parallel} = q_{\parallel\parallel} + \frac{3}{2} n T_\parallel u_\parallel + \frac{1}{2} nm \frac{u_\parallel^2}{2} u_\parallel \tag{11.2}$$

for the pure parallel heat flux, and

$$Q_{\perp\parallel} = q_{\perp\parallel} + n T_\perp u_\parallel \tag{11.3}$$

for the perp parallel heat flux (note we often use the shorthand ‘perp’ for perpendicular in this context). Note the differences in the coefficients, and that there is no pure Mach term in the second case. For the delta-f gyrofluid model this split occurs according to the orthonormal polynomial decomposition, so that we have directly

$$Q_{\|\|\|} = nTV\left(\sqrt{\frac{3}{2}}\ a_{30} + \frac{3}{2}\ a_{10}\right) \qquad \text{hence} \quad q_{\|\|\|} = \sqrt{\frac{3}{2}} nTV\ a_{30} \tag{11.4}$$

for the pure parallel piece, and

$$Q_{\perp\|} = nTV(a_{11} + a_{10}) \qquad \text{hence} \quad q_{\perp\|} = nTV\ a_{11} \tag{11.5}$$

for the perp parallel piece. The background is isotropic with no flows, so there is no Mach term in a delta-f model. The factors of n, T, and V, are the background density and temperature for each species, as these terms are separate for each species. The moment variables themselves are dimensionless. This makes the equations very easy to write in either physical or normalised units.

Physical equations have coefficients like V and T/Ze in many places, and the corresponding normalised equations replace these by μ_z and τ_z, where the normalisation is to the temperature of each species but to one velocity. In the field equations one also has the partial charge density a_z. These three species quantities are given by

$$a_z = \frac{n_z Ze}{n_e e}, \qquad \mu_z = \frac{m_z/Z}{m_0}, \qquad \tau_z = \frac{T_z/Z}{T_e} \tag{11.6}$$

where the subscript z refers to species, for which Z is the charge state, m_z is the mass, and n_z and T_z are the background density and temperature. The reference parameters are the n_e and T_e for electrons and a main ion mass for m_0, which in a fusion plasma application is usually deuterium. The mass parameter appears in the equations since one common velocity, usually the sound speed c_s is used rather than the species thermal velocity V in the normalisation. This is made necessary to have one unit of time in a computation, such as $L_\perp/c_s$, and the mass enters through the ratio between velocities. Besides a_z as the partial charge density, the normalised partial pressure, mass density, thermal velocity, and gyroradius are

$$\frac{p_{z0}}{p_{e0}} = a_z\tau_z, \qquad \frac{n_{z0}m_z}{n_{e0}m_0} = a_z\mu_z, \qquad \frac{V^2}{c_s^2} = \frac{\tau_z}{\mu_z}, \qquad \frac{\rho_z^2}{\rho_s^2} = \tau_z\mu_z \tag{11.7}$$

for each species. In the expressions appearing below, n_0 refers to the background density for a given species, and n_e and n_i are the n_0 for electrons and ions, respectively, and the same for the temperatures.

For example, in a simplified two-moment model without FLR corrections, the density equation would be

$$\frac{\partial n}{\partial t} + [\phi, n + n_0] + n_0 B\nabla_\| \frac{u_\|}{B} = n_0\mathcal{K}(\phi) + \frac{T_0}{Ze}\mathcal{K}(n) \tag{11.8}$$

Note we already see where the τ_z will go. The factors of n_0 and T_0 refer to the background, and n_0 within the bracket refers to the background gradient. In terms of the moment variables a_{00} and a_{10}, this is

$$n_0\left(\frac{\partial a_{00}}{\partial t} + [\phi, a_{00} + \log n_0] + BV\nabla_\| \frac{a_{10}}{B}\right) = n_0\mathcal{K}(\phi) + n_0\frac{T_0}{Ze}\mathcal{K}(a_{00}) \tag{11.9}$$

where now we have the logarithmic gradient in the bracket. The subscript zero for the background is kept explicit. The factor of n_0 is already farily redundant. Note that the units of ϕ are T_e/e while the factor on the second curvature term is the species ratio. This is how the τ_z factor emerges, especially with Z signed. To get the normalised equation however, we have to deal with the time. Multiplying by $L_\perp/c_s$ and dividing out the n_0, we have

$$\frac{\partial a_{00}}{\partial t} + [\phi, a_{00} + \log n_0] + \frac{V}{c_s} L_\perp B \nabla_\| \frac{a_{10}}{B} = \mathcal{K}(\phi) + \tau_z \mathcal{K}(a_{00}) \tag{11.10}$$

To get this we first normalise ϕ to T_e/e as $e\phi/T_e$, and then fold the factors of $c_s/L_\perp$ and T_e/e into the definition of the bracket and curvature operator. These are given in both normalised and un-normalised versions in Chapter 2. All perpendicular coordinate distances are normalised to $L_\perp$ and the coefficients are

$$\delta_0 = \frac{\rho_s}{L_\perp}, \qquad \omega_B = \frac{2L_\perp}{R_0} \tag{11.11}$$

These come from the relation $T_e/eB_0 = \rho_s c_s$ in the scales. Finally the factor of V/c_s is now folded into a_{10}, so that the parallel flux variables now follow c_s rather than V in their definition (simply put this factor into the definitions of all moments mn with odd m, namely, the flux variables). The effect of doing this will put the factors of μ_z into the flux variable equations, as we see next.

The simplest version of the parallel velocity equation is

$$n_0 m\left(\frac{\partial u_\|}{\partial t} + [\phi, u_\|]\right) + T_0 \nabla_\| n = m\frac{T_0}{Ze}\mathcal{K}(2u_\|) + n_0 Ze\left(E_\| - \eta_\| J_\|\right) \tag{11.12}$$

in which we keep the curvature term, in contrast to the collisional fluid models. The parallel electric field is

$$E_\| = -\frac{\partial A_\|}{\partial t} - \nabla_\| \phi \tag{11.13}$$

The gyrofluid model naturally combines the induction involving $A_\|$ with the inertia involving $u_\|$ under the time derivative. The parallel current is the sum over species of the charged flows, and $\eta_\|$ is the parallel resistivity,

$$J_\| = \sum_{sp} n_0 Zeu_\|, \qquad \eta_\| = 0.51\frac{m_e \nu_e}{n_e e^2} \tag{11.14}$$

This equation is multiplied through by $L_\perp/c_s$ and divided through by $n_0 Ze$, the partial charge density. With ϕ normalised to T_e/e, the other field variable $A_\|$ is normalised to $B_0\rho_s$, so that

$$E_\| = -\frac{T_e/e}{L_\perp}\left(\frac{\partial}{\partial t}\frac{A_\|}{B_0\rho_s} - L_\perp \nabla_\| \frac{e\phi}{T_e}\right) \tag{11.15}$$

Note how $T_e/eB_0 = \rho_s c_s$ is used repeatedly. The resistive term has coefficients which combine into

$$\frac{n_0 Ze}{n_e e}\frac{c_s^2}{L_\perp}\frac{m_e}{e}\frac{\nu_e L_\perp}{c_s} \tag{11.16}$$

so that with $c_s^2 = T_e/m_i$ we obtain

$$\eta_\| J_\| = 0.51\frac{\nu_e L_\perp}{c_s}\frac{T_e/e}{L_\perp}\left(\frac{m_e}{m_i}\frac{J_\|}{n_e e c_s}\right) \qquad \text{with} \qquad \frac{J_\|}{n_e e c_s} = \sum_{sp} a_z a_{10} \tag{11.17}$$

Then the parallel velocity equation of each species becomes

$$\frac{\partial}{\partial t}\left(A + \mu_z a_{10}\right) + \mu_z[\phi, a_{10}] = \nabla_\|(\phi + \tau_z a_{00}) + \mu_z \tau_z \mathcal{K}(2a_{10}) - \eta\, J \tag{11.18}$$

with induction potential, resistivity, and current given by

$$A = \frac{A_\|}{B_0 \rho_s}, \qquad \eta = 0.51\frac{m_e}{m_i}\frac{\nu_e L_\perp}{c_s}, \qquad J = \frac{J_\|}{n_e e c_s} = \sum_{sp} a_z a_{10} \tag{11.19}$$

in the normalisation. We will have to keep the subscript z to distinguish the species when working with more than one of them.

Usually, computations are done with a single ion species of single charge state, so that these parameters are

$$\text{for } z = e\text{:} \quad a_z = -1, \quad \tau_z = -1, \quad \mu_z = -\frac{m_e}{m_i} \tag{11.20}$$

for the electrons and

$$\text{for } z = i\text{:} \quad a_z = 1, \quad \tau_i = \frac{T_i}{T_e}, \quad \mu_i = 1 \tag{11.21}$$

for the ions. Note that for such a case the reference mass m_0 is the mass m_i of an ion.

It is a matter of taste whether one would rather read through equations in physical units or in a normalisation. There is no problem keeping the T_z/Ze and V factors as it is about the same as keeping the μ_z and τ_z factors. The equations have been presented both ways in this work, to help familiarise with the system. To become an experienced practitioner one should be able to slide back and forth between the physical and normalised presentations with little difficulty anyway, and this takes practice. When one discusses physical processes and consistency issues but is not very interested in computations, it is better to remain with the physical units, especially since with this model the plethora of coefficients is not too cumbersome. But to do computations with self- consistent dynamics, especially with electrons and more than one ion species, the normalisation is essential, and the usefulness of the drift wave normalisation rather than an MHD one becomes obvious when dealing with gyrokinetic or gyrofluid equations at this level of complexity.

11.2.1 Six-moment electromagnetic equations

But for this outline the equations will be presented in physical units. For readability we will fold the background gradient pieces into the moment variables in the ExB advection brackets and the parallel gradients, which contain the nonlinearities. So for the density and temperature moments we have

$$A_{00} = a_{00} + \log n_0, \qquad A_{20} = a_{20} + \frac{1}{\sqrt{2}} \log T_0, \qquad A_{01} = a_{01} + \log T_0 \tag{11.22}$$

where the n_0 and T_0 like the variables themselves are for each species. Combinations of state variables and potentials are

$$h_{00} = a_{00} + Z\frac{e\phi_0}{T_0}, \qquad h_{20} = a_{20}, \qquad h_{01} = a_{01} + Z\frac{e\phi_1}{T_0} \tag{11.23}$$

and those including the gradient drive terms are

$$H_{00} = A_{00} + Z\frac{e\phi_0}{T_0}, \qquad H_{20} = A_{20}, \qquad H_{01} = A_{01} + Z\frac{e\phi_1}{T_0} \tag{11.24}$$

For the variables with nonzero moment index m no factors of the ϕ terms appear in h_{mn}, which then reduces back to a_{mn}.

The gradient terms in A_{mn} appear in the ExB advection brackets, leading to the ExB gradient drive. The gradient terms in H_{mn} appear under the parallel gradient, leading to the magnetic flutter gradient drive. In the curvature coupling the gradient drive is left out because otherwise we would have to do the global equilibrium. That level of complexity is left to global tokamak computations which are not the focus of this work. The present methodology is sufficient to treat turbulence driven by the overall thermal gradient.

The ϕ_n are the gyroaveraged potentials, given by

$$\phi_n = G_n(\phi), \qquad G_n = G_n(b_z), \qquad b_z = k_\perp^2 \rho_z^2 \tag{11.25}$$

where ρ_z is the species gyroradius, so that the ϕ_n become species quantities. The details of the gyroaveraging operators are given in Chapter 9, and given below, after the field equations.

The density equation is

$$\begin{aligned} &\frac{\partial a_{00}}{\partial t} + [\phi_0, A_{00}] + [\phi_1, A_{01}] + VB\nabla_\parallel \frac{a_{10}}{B} \\ &= \frac{1}{2}\frac{T_0}{Ze}\mathcal{K}\left(2h_{00} + \sqrt{2}h_{20} + h_{01}\right) \end{aligned} \tag{11.26}$$

The parallel velocity equation is

$$
\begin{aligned}
&\frac{\partial A_\|}{\partial t} + \frac{m_z}{Ze}V\left(\frac{\partial a_{10}}{\partial t} + [\phi_0, a_{10}] + [\phi_1, a_{11}]\right) \\
&\quad + \frac{T_0}{Ze}\nabla_\|\left(H_{00} + \sqrt{2}\,H_{20}\right) - \frac{T_0}{Ze}\left(\sqrt{2}\,h_{20} - h_{01}\right)\nabla_\| \log B \\
&\quad = \frac{1}{2}\frac{m_z}{Ze}\frac{T_0}{Ze}\mathcal{K}\left(4a_{10} + \sqrt{6}\,a_{30} + a_{11}\right) - R_{ie}
\end{aligned}
\tag{11.27}
$$

including the resistive dissipation R_{ie}, given with the rest of the collisional effects below. The parallel temperature equation is

$$
\begin{aligned}
&\frac{\partial a_{20}}{\partial t} + [\phi_0, A_{20}] \\
&\quad + VB\nabla_\| \frac{\sqrt{2}\,a_{10} + \sqrt{3}\,a_{30}}{B} + V\sqrt{2}(a_{10} + a_{11})\nabla_\| \log B \\
&\quad = \frac{1}{2}\frac{T_0}{Ze}\mathcal{K}\left(\sqrt{2}\,h_{00} + 6h_{20}\right) - \sqrt{2}\,D_T
\end{aligned}
\tag{11.28}
$$

including the dissipation of thermal anisotropy D_T, given below. The perpendicular temperature equation is

$$
\begin{aligned}
&\frac{\partial a_{01}}{\partial t} + [\phi_0, A_{01}] + [\phi_1, (A_{00} + 2A_{01})] \\
&\quad + VB\nabla_\| \frac{a_{11}}{B} - V(a_{10} + a_{11})\nabla_\| \log B \\
&\quad = \frac{1}{2}\frac{T_0}{Ze}\mathcal{K}\left(h_{00} + 4h_{01}\right) + D_T
\end{aligned}
\tag{11.29}
$$

also including the dissipation of thermal anisotropy. The pure parallel heat flux equation is

$$
\begin{aligned}
&\frac{m_z}{Ze}V\left(\frac{\partial a_{30}}{\partial t} + [\phi_0, a_{30}]\right) + \frac{T_0}{Ze}\nabla_\|\left(\sqrt{3}\,H_{20}\right) \\
&\quad = \frac{1}{2}\frac{m_z}{Ze}\frac{T_0}{Ze}\mathcal{K}\left(\sqrt{6}\,a_{10} + 8a_{30}\right) - R_{q\|} - \sqrt{3}\,VK_\|\frac{m_z}{Ze}Va_{30}
\end{aligned}
\tag{11.30}
$$

including the parallel part of the thermal conduction dissipation. The last term is the Landau damping closure dissipation in this channel. The perp parallel heat flux equation is

$$
\begin{aligned}
&\frac{m_z}{Ze}V\left(\frac{\partial a_{11}}{\partial t} + [\phi_0, a_{11}] + [\phi_1, a_{10} + 2a_{11}]\right) \\
&\quad + \frac{T_0}{Ze}\nabla_\|(H_{01}) - \frac{T_0}{Ze}\left(\sqrt{2}\,h_{20} - h_{01}\right)\nabla_\| \log B \\
&\quad = \frac{1}{2}\frac{m_z}{Ze}\frac{T_0}{Ze}\mathcal{K}\left(a_{10} + 6a_{11}\right) - R_{q\perp} - VK_\|\frac{m_z}{Ze}Va_{11}
\end{aligned}
\tag{11.31}
$$

including the perp part of the thermal conduction dissipation. The last term is the Landau damping closure dissipation in this channel. In all of these n_0 and T_0 are the background density and temperature and V is the thermal velocity given by $V^2 = T_0/m_z$, and all are species quantities. The Landau damping rates are $\sqrt{3}\,VK_\parallel$ and $VK_\parallel$ for flux variable moments 30 and 11, respectively, corresponding to the native wave speeds of their coupling to the state variable moments 20 and 01, respectively.

11.2.2 Collisional dissipation

The overall collisional dissipation consists of three groups of effects, the dissipation due to resistivity, due to thermal conduction, and the parallel viscosity effect due to thermal anisotropy. The electron–ion frictional momentum transfer is $n_e e R_{ie}$, with R_{ie} the resistive dissipation given by

$$R_{ie} = \eta_\parallel \left[J_\parallel + \frac{\alpha_e}{\kappa_e}\left(n_e e V_{qe} + \alpha_e J_\parallel\right) \right] \tag{11.32}$$

where

$$\eta_\parallel = 0.51\frac{m_e \nu_e}{n_e e^2}, \qquad J_\parallel = \sum_{sp} n_0 Ze V a_{10}, \qquad V_{qe} = V_e\left(\sqrt{\frac{3}{2}}a_{30} + a_{11}\right)_e \tag{11.33}$$

give the resistivity, the parallel current, and the specific electron parallel conductive heat flux (i.e., as a velocity). Note that this quantity is determined across species and always appears together with the inductive parallel electric field.

In the other dissipation expressions we have the anisotropy effects, which appear in all the thermal dissipation channels. The temperature splits according to

$$T = \frac{1}{3}\left(T_\parallel + 2T_\perp\right), \qquad \Delta T = T_\parallel - T_\perp \tag{11.34}$$

and the parallel heat flux splits according to

$$q_\parallel = q_{\parallel\parallel} + q_{\perp\parallel}, \qquad \Delta q_\parallel = q_{\parallel\parallel} - \frac{3}{2}q_{\perp\parallel} \tag{11.35}$$

In the language of the moment variables we have

$$\Delta T = T_0\,\Delta_{20}, \qquad \Delta q_\parallel = n_0 T_0 V\,\Delta_{30} \tag{11.36}$$

with

$$\Delta_{20} = \sqrt{2}\,a_{20} - a_{01}, \qquad \Delta_{30} = \sqrt{\frac{3}{2}}a_{30} - \frac{3}{2}a_{11} \tag{11.37}$$

giving the moment variable expressions which are at once the normalised temperature and heat flux anisotropy. The isotropic part T does not contribute to dissipation because the collision operator conserves energy. The other three pieces, ΔT, $q_\parallel$, and $\Delta q_\parallel$ all contribute.

In the equations all the heat flux dissipation is in the R_q terms. For an isotropic case, the pure parallel conductive heat flux is 3/5 of the total, and the perp parallel part is 2/5 of the total. So in the thermal force effect the term proportional to $\alpha_e J_\parallel$ is split accordingly. There is also the anisotropy piece in the heat flux itself. The parallel part is given by

$$R_{q\|} = \sqrt{\frac{2}{3}}\frac{5/2}{\kappa_z}\nu_z\frac{m_z}{Ze}\left(\sqrt{\frac{3}{2}}\,Va_{30} + \frac{3}{5}\alpha_z\frac{1}{n_e e}J_\| + \sqrt{\frac{2}{3}}\,\Delta V_q\right) \tag{11.38}$$

and the perpendicular part is given by

$$R_{q\perp} = \frac{5/2}{\kappa_z}\nu_z\frac{m_z}{Ze}\left(Va_{11} + \frac{2}{5}\alpha_z\frac{1}{n_e e}J_\| - \Delta V_q\right) \tag{11.39}$$

where the specific heat flux contribution is given by

$$\Delta V_q = 1.28\ V\ \Delta_{30} \tag{11.40}$$

These are species specific, and are cast in the same units as the thermal flow inertia term in each equation. The thermal force effect appears only for the electrons, which is controlled by $\alpha_z = \alpha_e$ for them and zero for the ions.

Parallel viscosity appears for each species and is manifested through the collisional dissipation of thermal anisotropy. The temperature dissipation contribution appears in

$$D_T = \frac{\nu_z}{3\pi_z}\Delta_{20} \tag{11.41}$$

This corresponds to a diffusion term for parallel viscosity only in the limit of arbitrarily large collisionality. In general this dissipation is a clamp on the thermal anisotropy which has to compete with all the other time dependent effects in the equations. A small collision frequency does not lead to a large parallel viscosity diffusion coefficient but just to a small damping rate of part of the overall non-Maxwellian delta-f distribution. The same is true for thermal conductivity.

The dissipation coefficients are

$$\alpha_e = 0.71, \qquad \kappa_e = 1.6, \qquad \pi_e = 0.73 \tag{11.42}$$

for the electrons and

$$\alpha_z = 0, \qquad \kappa_z = 3.9, \qquad \pi_z = 0.96 \tag{11.43}$$

for ions, following the Braginskii collisional dissipation model (these numbers assume singly charged ions). The 1.6 for κ_e is the combination between the original factor 3.2 and the 0.51 in the resistivity. The factor of 1.28 in the heat flux anisotropy arises from applying the Chapman–Enskog expansion to a bi-Maxwellian distribution (separate exponential factors for $T_\|$ and $T_\perp$).

11.2.3 Field equations

The field equations for polarisation and induction are

$$\begin{aligned} P(\phi) = \rho_G &\equiv \sum_{sp} n_0 Ze(G_0(a_{00}) + G_1(a_{01})), \\ -\nabla_\perp^2 A_\| = J_\| &\equiv \mu_0 \sum_{sp} n_0 ZeV(a_{10}) \end{aligned} \tag{11.44}$$

where the polarisation operator is

$$P = \sum_{sp} n_0 Z^2 \frac{e^2}{T_0}\left[1 - \Gamma_0(b_z)\right] \tag{11.45}$$

and the ϕ_n are also dependent upon the b_z through the gyroaveraging operators $G_n(b_z)$, applied separately for each species. These are as in Eqs (9.76)–(9.85) from Chapter 9.

The gyroaveraging model is simplified for computational tractability. The electrons are not gyroaveraged, so that

$$z = e: , \qquad b_e = 0, \qquad G_0 = 1, \qquad G_1 = 0, \qquad \Gamma_0(b_e) = 1 \tag{11.46}$$

and so they do not contribute to $P()$. The ions are done with Padé approximants, so that the gyroaveraging is

$$z = i: , \qquad G_0 = \left(1 - \frac{\rho_i^2}{2}\nabla_\perp^2\right)^{-1}, \qquad G_1 = \left(1 - \frac{\rho_i^2}{2}\nabla_\perp^2\right)^{-1} G_0 - G_0 \tag{11.47}$$

namely, they are done recursively. This follows from

$$G_n = \frac{1}{n!}\left(-\frac{b}{2}\right)^n e^{-b/2} \quad \rightarrow \quad \frac{1}{n!}\frac{(-b/2)^n}{(1 + b/2)^{n+1}} \tag{11.48}$$

at the level of Padé approximants to the operations in $k_\perp$-space. The polarisation, now due solely to the ions, is

$$P = n_i Z^2 \frac{e^2}{T_i}\left[1 - \left(1 - \rho_i^2 \nabla_\perp^2\right)^{-1}\right] \tag{11.49}$$

We usually have $Z = 1$ for the one ion species. Note that ρ_i^2/T_i is ρ_s^2/T_e which is how the $k_\perp \rho_i \ll 1$ limit is recovered.

The way induction is done is to time-advance the specific canonical momentum for each species

$$A_z^* = A_\parallel + \frac{m_z}{Ze} V a_{10} \tag{11.50}$$

and then assemble the Ampère's law according to

$$\left(d_e^{-2} - \nabla_\perp^2\right) A_\parallel = \mu_0 \sum_{sp} \frac{n_0 Z^2 e^2}{m_z} A_z^* \qquad \text{where} \quad d_e^{-2} = \mu_0 \sum_{sp} \frac{n_0 Z^2 e^2}{m_z} \tag{11.51}$$

as in Eq. (9.86) from Chapter 9. The skin depth d_e is dominated by the electrons but it is necessary to keep the entire expression for consistency. After solving this for $A_\parallel$, the a_{10} are found from

$$a_{10} = \frac{Ze}{m_z V}\left(A_z^* - A_\parallel\right) \tag{11.52}$$

Note that these operators are all constructed using $\nabla_\perp^2$ and coefficients, so that they do not depend on the use of Fourier transforms. Both normalised and un-normalised versions of $\nabla_\perp^2$ are given in Chapter 2.

11.3 Energetics

The energetics of this model is very similar to the fluid drift model once all the temperature effects are retained, as in Chapter 3. The time dependent heat flux inertial effect was kept then, but the parallel viscosity was done with a conventional model. Now, the latter is done through the temperature anisotropy. Since the collision frequency never occurs in the denominator of any expression, the gyrofluid model never suffers from the pathology of large coefficients in weakly collisional regimes. It also avoids the use of diffusive collisional dissipation terms, which only result from averaging over kinetic trajectories of the parallel motion which we are resolving anyway. So we simply set the collision frequencies according to the parameters, and then the manifestation of the parallel viscosity and thermal conduction effects occurs naturally. No pathologically large conduction or viscosity coefficients arise.

The conserved free energy and the transfer effects are as in the derivation of the model in Chapter 9. The free energy is

$$E_T = \mathcal{V}^{-1}\int d\mathcal{V}\,\frac{1}{2}\left(\sum_{sp}\left[n_0T_0\sum_{mn}a_{mn}^2\right] + \phi P(\phi) + \mu_0^{-1}\left|\nabla_\perp A_\parallel\right|^2\right) \tag{11.53}$$

where the second term yields the ExB kinetic energy and the last term yields the magnetic energy. Under the integral, this is equivalent to

$$E_T = \mathcal{V}^{-1}\int d\mathcal{V}\,\frac{1}{2}\sum_{sp}\left[n_0T_0\sum_{mn}a_{mn}^2 + n_0Ze\sum_n \phi_n a_{0n} + n_0ZeVa_{10}A_\parallel\right] \tag{11.54}$$

using the field equations. The time derivative of this expression is

$$\frac{\partial E_T}{\partial t} = \mathcal{V}^{-1}\int d\mathcal{V}\sum_{sp}\left[n_0T_0\sum_{mn}a_{mn}\frac{\partial a_{mn}}{\partial t} + n_0Ze\sum_n \phi_n\frac{\partial a_{0n}}{\partial t} + n_0ZeVa_{10}\frac{\partial A_\parallel}{\partial t}\right] \tag{11.55}$$

The ExB and state variable and heat flux inertial energy components can be split accordingly. But now the parallel velocity and magnetic energy are always combined in the energetics, since we get them by multiplying the equation for a_{10} by n_0ZeVa_{10}, which produces

$$\sum_{sp} n_0ZeVa_{10}\frac{\partial}{\partial t}\left(A_\parallel + \frac{m_z}{Ze}Va_{10}\right) = J_\parallel\frac{\partial A_\parallel}{\partial t} + \sum_{sp} n_0T_0a_{10}\frac{\partial a_{10}}{\partial t} \tag{11.56}$$

where the factor of T_0 arises from m_zV^2 obtained from the V factors.

Putting these factors in, the energetic drive and dissipation quantities used for the fluid drift model can be expressed in gyrofluid terms. Special care is taken with any

ExB mechanism. These are not a single quantity found from one equation, but a set of them found by using ϕ_0 with each density (a_{00}) equation and ϕ_1 with each perpendicular temperature (a_{01}) equation. Hence the transfer of energy through the parallel current from the thermal free energies into the ExB kinetic energy is

$$T_J = \mathcal{V}^{-1}\int d\mathcal{V}\sum_{sp} n_0 ZeV\left[a_{10}\nabla_\parallel\phi_0 + a_{11}\nabla_\parallel\phi_1\right] \tag{11.57}$$

Note this can only be evaluated after the sum over species, as the charge dependent cancellations between the individual pieces are necessary. Similarly, the transfer from thermal free energy over the curvature into the ExB kinetic energy is

$$T_E = \mathcal{V}^{-1}\int d\mathcal{V}\sum_{sp}\frac{1}{2}n_0T_0\left[\phi_0\mathcal{K}\left(2h_{00} + \sqrt{2}h_{20} + h_{01}\right) + \phi_1\mathcal{K}(h_{00} + 4h_{01})\right] \tag{11.58}$$

Note the appearance of the h's rather than the a's—this brings in a set of FLR corrections which in practice are not very significant but still contribute to energetic consistency. The terms due to ϕ_1 are themselves FLR corrections. As in the fluid drift model, these can be done as time traces or spectra with respect to k_y or $k_\perp$ or $k_\parallel$ decompositions.

The ExB thermal gradient drive is split into the overall density contributions,

$$\Gamma_n = \frac{1}{2E_T}\mathcal{V}^{-1}\int d\mathcal{V}\sum_{sp} n_0T_0\left(a_{00}[\log n_0, \phi_0] + a_{01}[\log n_0, \phi_1]\right) \tag{11.59}$$

the electron temperature contribution

$$\Gamma_t = \frac{1}{2E_T}\mathcal{V}^{-1}\int d\mathcal{V}\, n_eT_e\left(\sqrt{\frac{1}{2}}a_{20} + a_{01}\right)[\log T_e, \phi] \tag{11.60}$$

and the ion temperature contribution

$$\Gamma_i = \frac{1}{2E_T}\mathcal{V}^{-1}\int d\mathcal{V}\sum_{\text{ions}} n_0T_0\left[\left(\sqrt{\frac{1}{2}}a_{20} + a_{01}\right)[\log T_0, \phi_0] + (a_{00} + 2a_{01})[\log T_0, \phi_1]\right] \tag{11.61}$$

These are written differently because the electrons are not gyroaveraged.

The magnetic flutter drive arises in the flux variable equations. It is summed together as

$$\Gamma_v = \frac{1}{2E_T}\mathcal{V}^{-1}\int d\mathcal{V}\sum_{sp} n_0T_0\left(Va_{10}[\log n_0, -A_\parallel] + \left(Va_{10} + V_q\right)[\log T_0, -A_\parallel]\right) \tag{11.62}$$

where

$$V_q = V\left(\sqrt{\frac{3}{2}}a_{30} + a_{11}\right) \tag{11.63}$$

is the specific parallel conductive heat flux for each species. This is always dominated by the electron parallel heat flux term if it is positive.

The resistive dissipation is fairly straightforward, found from the equations for $J_\parallel$ through all the a_{10} coupled across species, and from the electron heat fluxes. It is a matter of choice how to do the split. One can find the main resistive and thermal conductive sinks from an assumption of isotropy in $q_\parallel$, found by combining the a_{30} and a_{11} pieces into V_q. Then the anisotropy dissipation in both T and $q_\parallel$ are combined into a generalised viscosity sink for each species. Note that we will now have an electron viscosity in addition to the one in the ions. These are all found from damping terms, as there are no longer any diffusion sink terms in the model. Damping of higher moments is a better approximation at low collisionality, because the collision operator itself acts pointwise in space through diffusion in velocity space. It only appears as a spatial diffusion when one averages over trajectories. However, in this model those time scales are being resolved and are forced upon by the turbulence, so that trajectories exist only as infinitesimals. This is the essence of the gyrofluid representation with respect to dissipation.

The resistive dissipation is

$$\Gamma_c = \frac{1}{2E_T}\mathcal{V}^{-1}\int d\mathcal{V}\,\eta_\parallel J_\parallel^2 \tag{11.64}$$

with $\eta_\parallel$ and $J_\parallel$ expressed as above. The electron thermal conduction dissipation is

$$\Gamma_k = \frac{1}{2E_T}\mathcal{V}^{-1}\int d\mathcal{V}\,\kappa_e^{-1}\eta_\parallel\left(n_e e V_{qe} + \alpha_e J_\parallel\right)^2 \tag{11.65}$$

The ion thermal conduction dissipation is

$$\Gamma_K = \frac{1}{2E_T}\mathcal{V}^{-1}\int d\mathcal{V}\,\kappa_i^{-1} n_i m_i \nu_i V_{qi}^2 \tag{11.66}$$

in which there is no thermal force piece.

With the isotropic part of the temperature arranged into the thermal conduction as just expressed, all anisotropy effects are combined into a viscous dissipation. The effects of adding $\alpha_e(T_e/e)J_\parallel$ to the parallel heat flux for electrons cancels out of the anisotropy so that the same expression in

$$\Gamma_s = \frac{1}{2E_T}\mathcal{V}^{-1}\int d\mathcal{V}\sum_{sp} n_0 T_0 \nu_z\left[\frac{1}{3\pi_z}(\Delta_{20})^2 + \left(\frac{2}{3} + \frac{5}{3}\,1.28\right)(\Delta_{30})^2\right] \tag{11.67}$$

using $V^2 = T_0/m_z$, is valid for each species.

The Landau damping dissipation is split between electrons and ions due to the difference in their responses. The electron one is

$$\Gamma_l = \frac{1}{2E_T}\mathcal{V}^{-1}\int d\mathcal{V}\,n_e T_e (K_\parallel V_e)\left(\sqrt{3}\,a_{30}^2 + a_{11}^2\right) \tag{11.68}$$

and the ion one is

$$\Gamma_L = \frac{1}{2E_T}\mathcal{V}^{-1}\int d\mathcal{V}\, n_i T_i (K_\| V_i)\left(\sqrt{3}\, a_{30}^2 + a_{11}^2\right) \tag{11.69}$$

for this model. Always recall that this depends on the model level. A twelve moment model will have three of these due to a_{50}, a_{31} and a_{12}, and a twenty moment model will have four, due to a_{70}, a_{51}, a_{32}, and a_{13}, and so on. In these higher order cases the Landau damping does not act directly on the thermal conductivity velocities through a_{30} and a_{11}, but only indirectly through the moment chain coupled through the $\nabla_\|$ terms, forced upon by the turbulence through the nonlinear advection at every level. The more indirect the better.

The only new pieces in the energetics compared to the fluid drift model with thermal inertia from Chapter 3 is the Landau damping dissipation, Γ_l for electrons and Γ_L for ions. How important this effect is will be learned from comparing all the sinks in a computation. The perpendicular and parallel temperature gradient drives are combined, but the different channels will be compared using the ExB fluxes, especially in their parallel envelopes (averages over the perpendicular plane, as dependent upon the parallel coordinate s). The ExB specific particle flux of each species is given by

$$F_z = \mathcal{V}^{-1}\int d\mathcal{V}\left(a_{00}[\phi_0, x] + a_{01}[\phi_1, x]\right) \tag{11.70}$$

expressed as a velocity. In the volume average, these should be equal for the single component, singly charged ion case. The ExB specific conductive heat flux of each species is given by

$$G_z = \mathcal{V}^{-1}\int d\mathcal{V}\left[\left(\sqrt{\frac{1}{2}}\, a_{20} + a_{01}\right)[\phi_0, x] + (a_{00} + 2a_{01})[\phi_1, x]\right] \tag{11.71}$$

with the parallel temperature contribution due to the a_{20} term and the perpendicular temperature contribution due to all the rest. The total turbulent ExB heat flux for each species is then

$$Q_z = 1.5\, F_z + G_z \tag{11.72}$$

in the same manner as for the fluid drift model. Note the relation to Γ_n and $\Gamma_{t,i}$, in the fluxes and gradient drives. The same exists for the magnetic flutter, for which the specific heat flux is

$$M_z = \mathcal{V}^{-1}\int d\mathcal{V}\; V_{qz}[-A_\|, x] \tag{11.73}$$

for each species. All are measured but the electron one M_e is almost always found to be the only significant one, if indeed it is positive (this discussion was done in Chapter 3).

These fluxes are all expressed as a velocity, so to obtain the physical units simply multiply by the background $n_{z0}T_{z0}$, the species partial pressure. In normalised units they are given as velocities in units of c_s, so that the values one finds are well lower than unity. As in the fluid drift case, the comparison baseline is δ_0^2, which is of order 10^{-4} or a few times that, depending on the case. Order unity indicates an effective velocity of c_s, which would preclude equilibrium, order δ_0 indicates an effective velocity of v_D, which would indicate order unity disturbances, so that order δ_0^2 indicates a smaller effective velocity, commensurate with order δ_0 (one or more percent) disturbances times v_D. This level is consistent with the idea of nonlinear dynamics and turbulence, nevertheless with low enough amplitudes and small enough scale that a given set of background parameters and gradients makes physical sense.

11.4 ITG turbulence in a hot plasma

It is usually assumed that a hot plasma means negligible collisions and adiabatic electrons. This translates into small values for all three of $\hat{\beta}$, $\hat{\mu}$, and C, which control the electron responses (Chapter 2). Recall however, that $\hat{\mu} \ll 1$ is required, which is a scale ratio issue and not a temperature one, for drift wave turbulence. So the epithet hot plasma actually means a case with moderate gradients for which values of $K_\| L_\perp$ are large enough to keep that inequality. For a confined plasma, this means the core regions. Then, hot plasma merely means small C. If $\hat{\beta}$ is arbitrarily small then the case is electrostatic but a hot plasma will also usually indicate that $\hat{\beta}$ is not so small.

As outlined in Chapter 2, we assume a simplified toroidal geometry with minor and major radii a_0 and R_0, respectively, and set parameters at a reference radius r_0, in this case r_0 in the vicinity $0.5 < r_0/a_0 < 0.7$ in rough terms. That sets the scale ratio $L_\perp/a_0$ of order 1/3 to 1/2. Then, small $\hat{\mu}$ depends on a reasonable aspect ratio. If $R_0 \gg a_0$ then this condition might not be met. However, fusion applications in tokamaks have R_0/a_0 in the vicinity of three (or less), and moreover the field line pitch parameter q similar to or less than two in the core regions, so that we have $qR_0/L_\perp$ in the vicinity of ten (not 100). Then, the high temperature guarantees small C, and the low tokamak beta is enough together with $qR_0/L_\perp$ guarantees $\hat{\beta} < 1$, with larger values running into the MHD ballooning boundary, as we have seen. So when one says 'core turbulence' it is to be remembered that it is the geometry of the device that gives us the hot plasma case. Temperatures are of order keV, densities several 10^{19} m^{-3}, and the ratio $\rho_s/L_\perp$ is of order 10^{-2} and if larger then not by much. This is to allow a domain size that actually separates between MHD scales and that of the drift wave turbulence, which covers $0.1 < k_\perp\rho_s < 1$. So if $L_\perp/\rho_s$ is only 64, then the lowest $k_\perp\rho_s$ is 0.05, which does not achieve this separation. At this tokamak size ($a_0/\rho_s \sim 200$), drift wave turbulence and mesoscale MHD ($1 < k_\perp a_0 < 10$) still run together.

The standard case people use for this was defined by a study published in 2000 (see below). The physical parameters are

$$\begin{array}{llll} T_0 = 2\ \text{keV}, & n = 4.5 \times 10^{19}\ \text{m}^{-3}, & B_0 = 1.9\ \text{T}, & \\ a_0 = 0.625\ \text{m}, & R_0 = 1.70\ \text{m}, & r_0 = a_0/2, & q(r_0) = 1.4 \end{array} \tag{11.74}$$

These lead to normalised parameters given by

$$\begin{aligned}\hat{\beta} &= 0.464, \quad \hat{\mu} = 0.0254, \quad \nu = 0.03 \quad \hat{\varepsilon} = 93.3, \quad \omega_B = 0.29 \\ \delta_0 &= 0.0141, \quad \omega_n = 0.321, \quad \omega_t = \tau_i = 1, \quad S = 0.78, \quad \varepsilon_a = 0.18\end{aligned} \tag{11.75}$$

with $\nu = \nu_e L_\perp / c_s$ and $\varepsilon_a = r_0 / R_0$. The other parameters follow the definitions given in Chapter 2. The value of C is below 10^{-3}. Note that C is no longer the sole determinate of collisionality, since although it is still controls the electron parallel dynamics, the anisotropy is directly controlled by ν. For the ions, the normalised collision frequency is a factor of about 90 smaller, $2^{1/2}$ for the ions and $3670^{1/2}$ for the mass ratio (cf. ν_e and ν_i as defined in Chapter 1).

We run these cases with collision frequencies in both species set to zero, exemplifying the collisionless limit. This eliminates the terms R_{ie}, $R_{qz\|}$, $R_{qz\perp}$, and ΔQ in the equations, leaving only Landau damping as a dissipation mechanism. This case differs from the one used generically in Chapter 3 as it can be considered a deep-core case for tokamaks, due to the effective field line length $qR_0/L_\perp < 10$. Nevertheless we use it here to compare to the published study for tokamak cases.

The full domain size for a flux tube model is given by $2\pi r_0 / qL_\perp$ which is about 5.8 for this case. So we set up a perpendicular domain of

$$L_x = L_\perp = 64h_x, \quad L_y = 5.8L_\perp = 256h_y, \quad h_y/h_x = 1.45 \tag{11.76}$$

where $h_{x,y}$ are the step sizes. The parallel domain is as per field line connection, with $-\pi < s < \pi$ covered by 32 grid points, and the timestep is 0.002 due to the small $\hat{\mu}$. All other considerations are as per the cases run in Chapter 2.

In terms of the geometry the only new effect is the existence of magnetic pumping. We set

$$K_\| \frac{\partial}{\partial s} \log B = \varepsilon_a K_\| \sin s \tag{11.77}$$

to capture the effect. This is where ε_a enters as a parameter.

11.4.1 Dependence on inductivity

This turns out to be a relatively extreme inductivity case ($\hat{\beta}$), and it is close to the MHD stability boundary. Taking the stability criterion $\alpha_M < 0.66\,S$ as before, we have

$$\alpha_M = \hat{\beta}\omega_B[\omega_n + \omega_t + \tau_i(\omega_n + \omega_i)] = 0.355 < 0.515 \tag{11.78}$$

Although the 2000 study only considered electrostatic cases, I ran the inductivity scan for this case in 2006, identifying the start of the electromagnetic range (sensitive to $\hat{\beta}$) and the microtearing gap, and the transition to the MHD regime. As in Chapter 3 for the fluid drift model, this is the first scan to be done. As there and before, all times are given in units of $L_\perp / c_s$, perpendicular wavenumbers as $k_y \rho_s$ or $k_\perp \rho_s$, and fluxes as specific velocities (e.g., heat flux divided by $n_e T_e$) in units of c_s, for which the baseline value is δ_0^2, or about 2×10^{-4} for these cases. The initial amplitude is set to $10^{-4}\delta_0$, which is low enough to see if there is an initial linear growth phase.

The time traces of the advective ExB ion heat flux for various values of $\hat{\beta}$ are shown in figure 11.1, together with the averages over $2000 < t < 4000$ of all the turbulent flux channels. There are three groups of curves. For $\hat{\beta}$ up to about 0.1 the ion advective ExB heat flux saturates in the vicinity of the δ_0^2 level, with the finite inductivity mildly stabilising. Then there is a set of curves either wandering at low values, for $0.2 < \hat{\beta} < 0.4$, or completely stabilised for $\hat{\beta} = 0.5$, namely, at the nominal value. The first three of these are nonlinearly stabilised, since they show positive growth rates in the linear stage. The last is stabilised linearly by the magnetic flutter effect. Interestingly, the cases with $\hat{\beta} = 0.2$, 0.3, and 0.4 have a short linear growth phase and reach a finite peak in the ExB advective ion heat flux before dropping to lower values. More on this below.

The last set of curves, for $\hat{\beta} = 0.6$ and higher, sees the turbulence rise to a factor of ten above the level it has in the mostly electrostatic regime. The proportions in the ExB flux channels are comparable to the $\hat{\beta} < 0.1$ cases, but now the electron magnetic flutter flux channel M_e is the second highest. The value of $\hat{\beta}$ at which $\alpha_M = 0.66\ S$ is reached is nominally 0.67, but we find the first case with much higher amplitude and different mode structure is actually $\hat{\beta} = 0.6$. We need some of the mode structure information to interpret this result.

The wavenumber spectra (k_y) are shown for three $\hat{\beta}$ values in figure 11.2. All quantities are in relative units, as $e\phi/T_e$ and with $T_\| = \sqrt{2}\,a_{20}$ and $q_{\|\|\|} = \sqrt{3/2}\,a_{30}$, and the others given by the corresponding moment variable. As before, these are

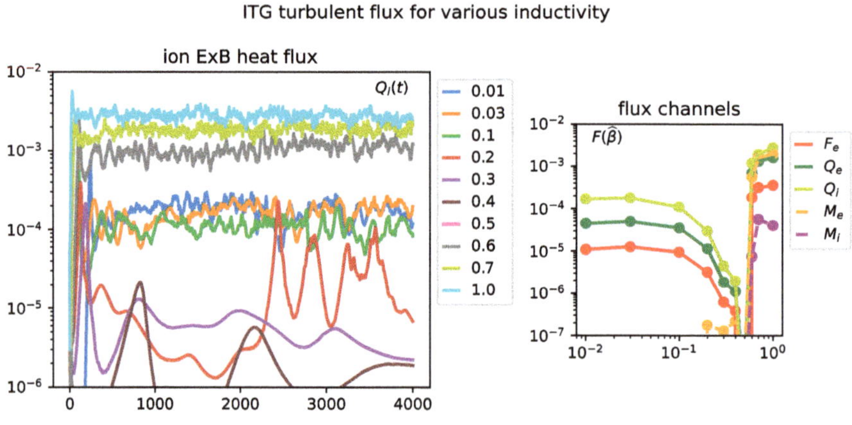

Figure 11.1. Turbulent ExB ion heat flux (Q_i) for various values of inductivity ($\hat{\beta}$), for the standard ITG case given in Eq. (11.75). The mostly electrostatic cases ($\hat{\beta}$ up to 0.1) yield robust turbulence at the δ_0^2 level. For a middle range, about 0.2 and higher, the turbulence is progressively stabilised. Then, just below the nominal MHD boundary at 0.67, the turbulence breaks out again, now to a much higher level. The physical character is different, as the mode structure signatures will show later on. The right frame gives the dependence curves for all the flux channels, showing that the ExB channels all go together while for the higher inductivity states the magnetic flutter transport (M_e) emerges and becomes among the strongest channels (approximately $Q_i > M_e > Q_e$ in the last two cases).

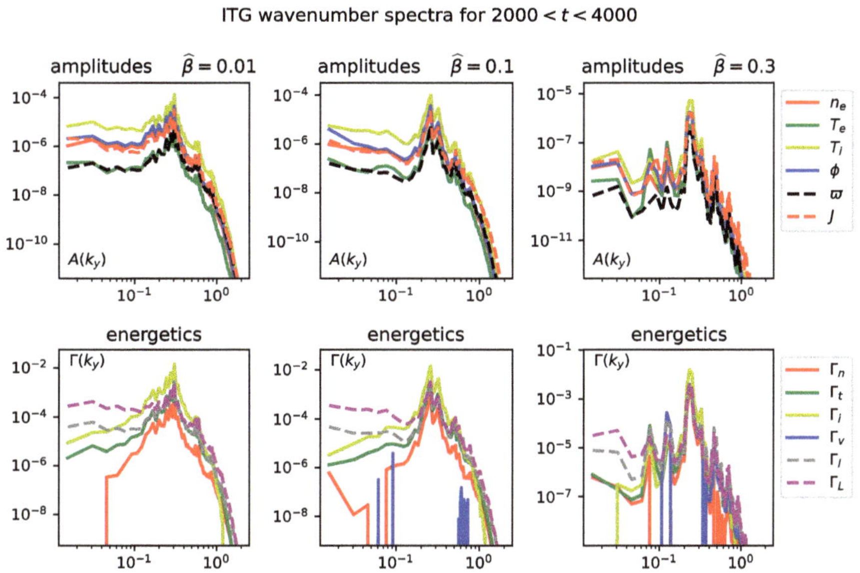

Figure 11.2. Drift-direction wavenumber (k_y) spectra of the amplitudes (top row) and energetics (bottom), for the ITG cases as they change with increasing inductivity into the stabilisation range. The vorticity (ϖ) and current (J) are also shown. The energetics especially show how the spectrum is squeezed as the stabilisation affects the longer wavelengths the most and the nonlinear exchanges become less robust.

quadratic quantities in k_y-space, averaged over s and time, in the interval $2000 < t < 4000$. These are in the mostly electrostatic regime, the beginning of finite beta stabilisation, and near the MHD stability boundary, respectively. The energetics takes place roughly in the decade $0.1 < k_y\rho_s < 1$, with the gradient drive peak in the middle. The nature of the transition toward the MHD side is different than in the more hydrodynamic fluid drift cases. Investigating this further, we next show the three high beta cases in figure 11.3. Here we see drastic changes in the mode structure. The magnetic flutter electron heat flux suddenly becomes positive at all scales (from the completely stabilised $\hat{\beta} = 0.5$ case to the one at 0.6). It is also seen to peak in the longest wavelength. At the same time, the ExB advective fluxes remain peaked in the $k_y \sim 0.3$ range. Electron Landau damping takes over from the ions as the principal sink. The magnetic flutter drive rises to the level of the ExB advective ITG drive, and is about the same level in the last two cases. This is not the same thing as MHD ballooning, which would have the dynamics go mostly to larger scales, lower k_y. So historically it has been called 'kinetic ballooning' (see the Summary, below).

The isotropic perpendicular spectra ($k_\perp$) are shown for three $\hat{\beta}$ values in the top row of figure 11.4, also averaged over the same time interval. These emphasise the scale of motion over the wavelength. The vorticity collects near ρ_s while the energetics peaks in the $k_\perp\rho_s \sim 0.3$ range. As inductivity increases, the width of the spectra reduces slightly. When the kinetic ballooning side is reached, the magnetic flutter drive enters, and this diagnostic shows that the magnetic flutter drive is indeed

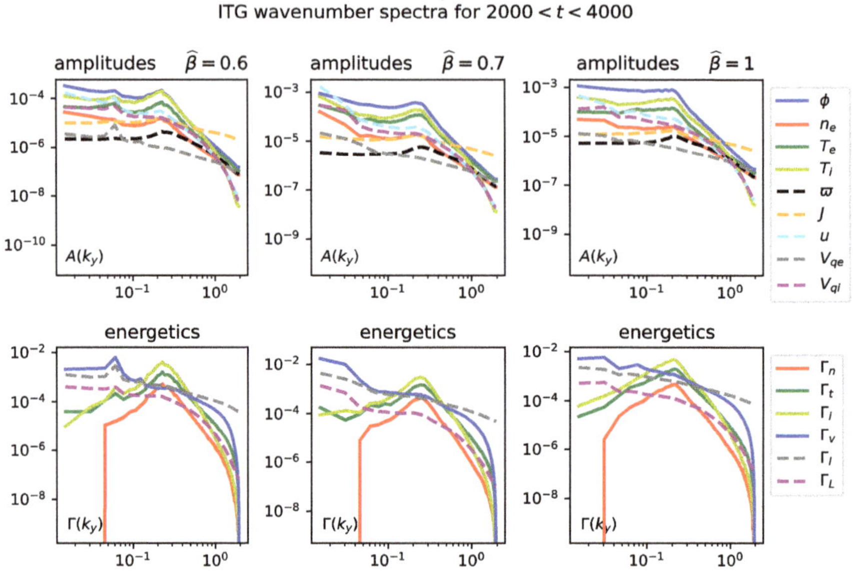

Figure 11.3. Drift-direction wavenumber (k_y) spectra of the amplitudes (top row) and energetics (bottom) as they change with increasing inductivity into the MHD range. The vorticity (ϖ) and current (J) are also shown. The magnetic flutter gradient drive (Γ_v) becomes very strong at long wavelength. The ExB gradient drive remains in the same range it has in the electrostatic regime but is higher as part of the self-consistent mode structure. The collisional sinks follow the magnetic flutter curve and are very flat.

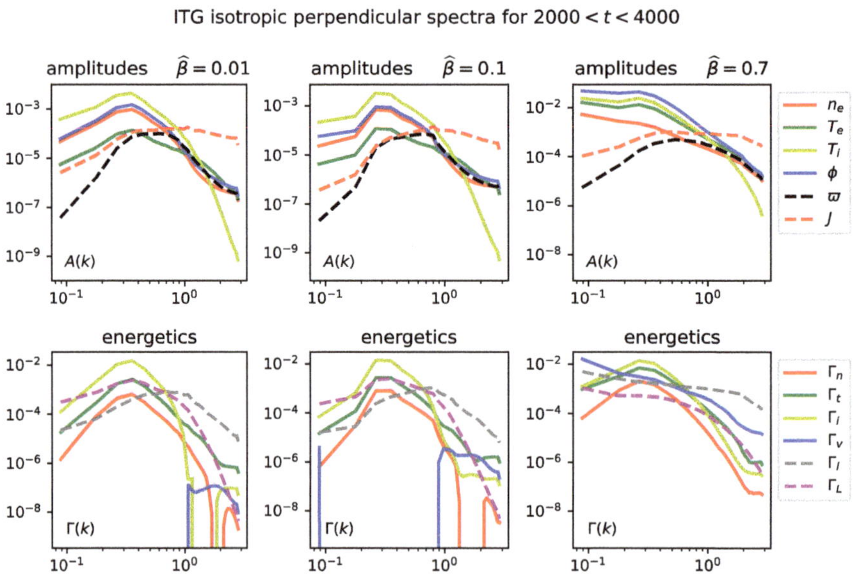

Figure 11.4. Isotropic perpendicular wavenumber ($k_\perp$) spectra of the amplitudes (top row) and energetics (bottom), comparing the electrostatic, electromagnetic, and MHD regimes of the ITG turbulence. The magnetic flutter gradient drive (Γ_v) is negative on the main ITG side but drives the largest scales, as well as the longest wavelengths, on the MHD side. In the middle range the ExB gradient drive is mostly due to the ion temperature gradient (Γ_i) on the ITG side, but the electron temperature gradient (Γ_t) also becomes strong on the MHD side.

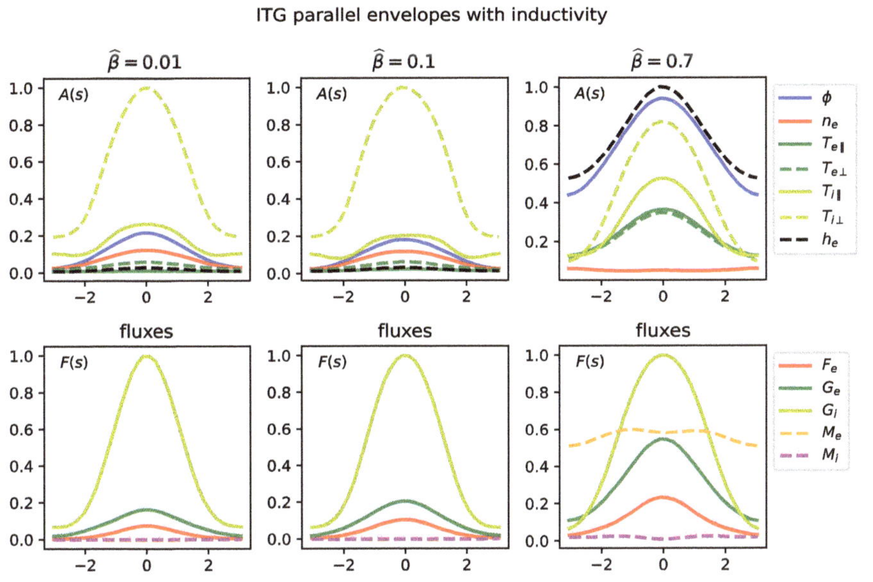

Figure 11.5. Parallel envelopes of the amplitudes (top row) and turbulent fluxes (bottom), comparing the electrostatic, electromagnetic, and MHD regimes of the ITG turbulence. The perpendicular ion temperature disturbances are the largest and most toroidally ballooned (collected near $s = 0$) for the ITG turbulence, with the nonadiabatic density and electron temperature very flat. The resulting ExB ion heat flux is also strongest and the most ballooned. As the MHD regime is reached, the potential is the strongest and the most ballooned, and it carries the electron and parallel ion temperature and the vorticity along with it. The density is very flat even though its associated ExB particle flux is ballooned.

active at large scale as well as long wavelength. It is not all collected in narrow layers at rational surfaces.

The parallel envelopes are shown for three $\hat{\beta}$ values in the top row of figure 11.5. As before, these are the envelopes of the squared amplitudes averaged over xy and over the same time interval. The normalisation is to the maximum value of any of the curves so that the relative amplitudes are what is expressed. Dashed lines are drawn for the perpendicular temperatures, with the electron temperatures in green and the ions in yellow. The electrostatic potential is blue and the electron density is red. The dashed black curve shows h for the electrons (the electron h_{00}, or $n - \phi$ in the parlance of the fluid drift cases). The electron signatures, h and T, are very low. The exemplary signature is the fact that in the ions $T_\perp$ is much larger than $T_\parallel$ and much more ballooned. It does not participate in the sound wave dynamics, having direct coupling only to the perp parallel heat flux. This is why the turbulence can be natively stronger than in the fluid drift case. As the kinetic ballooning side is reached, the other variables become liberated from the parallel dynamics and are therefore larger. The ion temperature is still anisotropic, while the electron one is closer to balanced. Since the collisions have been set to zero, the mechanism behind this is just the set of couplings between all the variables. Note that the electrons are also anisotropic on the ITG side.

The bottom row of the same figure shows the parallel envelopes for the turbulent flux channels. These are all expressed in terms of a specific velocity and arbitrarily normalised to show the relative strengths. The ITG mode structure of the ExB advective fluxes (density, and electron and ion conductive heat) is prominent, and the magnetic flutter channels are small and negative. For $\hat{\beta} = 0.6$ this changes into a ballooning mode type structure, except that here the ion ExB heat flux is still the highest and most strongly ballooned. The electron magnetic flutter flux is now prominent, and is very flat. This shows that sheared slab and ballooning mode effects are very active at the same time.

11.4.2 Temperature gradient threshold

The main point of the 2000 study was the discovery of the nonlinear upshift in the temperature gradient threshold of the ITG turbulence found in this parameter regime. The threshold was discussed in the context of the fluid drift model in Chapter 3. But the kinetic models are the ones in the study which showed the nonlinear upshift. We investigate this closely here and then later give direct comparison to the fluid drift model for this same case set.

The parameters of the baseline case are held fixed, except we set $\hat{\beta} = 0.01$ to artificially recover the electrostatic regime, which is what the study considered. Then, $\omega_t = \omega_i$ are varied together. To find a well defined linear stage, the initial amplitude is set to $A_0 = 10^{-4}\delta_0$, so that the fastest growing eigenmodes can settle out and reliable growth rates can be found.

The time traces of the turbulent ExB ion heat flux for these cases are shown in figure 11.6, together with the averages over $2000 < t < 4000$ of all the turbulent flux channels. The value of the time traces is that the whole run of the case is seen, so the existence of a linear growth phase, an overshoot, and saturation or stabilisation, or an intermediate state of uncertain conclusion, can all be seen if present. These time traces show a pronounced overshoot for any case which shows a linear growth phase, practically all except the first two. The case with $\omega_t = 0.55$ is judged stable if the run length is too short, since the eventual growth rate is so low that it is well past $t = 1000$ before the instability settles out. In rough numbers, the linear threshold is about $R_0/L_\perp = 4$, which agrees with the 2000 study. The overshoot peaks occur earlier as the linear growth rate increases. The case with $\omega_t = 0.9$ is the first one which reaches saturation at a level above 10^{-4}, with 0.8 oscillating well below that. So the cases between $0.55 < \omega_t < 0.8$ are stabilised in effective terms after they become nonlinear. If $\omega_t = 0.9$ or slightly below is taken as the effective nonlinear threshold, it translates to $R_0/L_\perp = 6$, which also agrees with the study. All of the nonlinearly stabilised cases remain at very low amplitude except for 0.8, which tumbles along at an average level of about $0.1\delta_0^2$. This is judged stable if a linear-linear reckoning is used (flux versus control parameter). The measured energetic error Γ_E is about three orders of magnitude less than the ExB ion temperature gradient drive Γ_i, rising to about 10^{-2} the total gradient drive in the last three cases wherein the energetics is much stronger, and some activity reaches the boundaries in the x-direction.

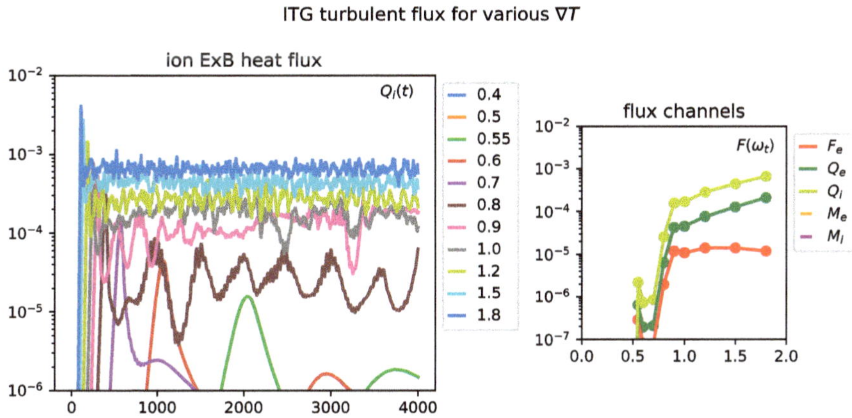

Figure 11.6. Turbulent ExB ion heat flux (Q_i) for various temperature gradient ($\omega_t = \omega_i$), for the near electrostatic ITG case ($\hat{\beta} = 0.01$) with density gradient (ω_n) and other parameters held fixed. The format is the same as in figure 11.1. Very low normalised gradient values are stabilised. Values in the range 0.55, the effective linear threshold, and 0.8 grow initially but are then nonlinearly suppressed. The effective nonlinear threshold is about 0.9. These thresholds are near 4 and 6 in units of R/L_T, respectively, with $R/L_\perp = 6.9$ indicating the $\omega_t = 1.0$ case. The right frame gives the dependence curves for all the flux channels, showing that for $\omega_t = 0.9$ and higher the heat fluxes rise naturally. The magnetic flutter flux is small and negative for both species. The particle flux is flat for the strong cases because ω_n is not changed.

11.4.3 Stabilisation by flows

The finding in the 2000 study on this nonlinear threshold upshift effect as that it is caused by self generated flows, which suppress the instability behind the turbulence when their amplitude is high enough. Their shear, expressed as a rate V', should exceed the growth rate of the instability and then they should remain after the turbulence is suppressed. All of this is measurable if it is true, which we turn to now.

The growth rate of the instability can be measured by watching the free energy grow during the linear stage. If the growth rate is γ, then the dependent variable amplitude grows as $u = u_0 \exp[\gamma t]$. Since the free energy is quadratic in the amplitude, $E \propto u^2$, it grows as $E = E_0 \exp[2\gamma t]$. Then, γ is calculated in the way we have already expressed. Measuring the free energy as a function of time, we find the growth rate using

$$\Gamma_T = \frac{1}{2E_T}\frac{\partial E_T}{\partial t} \tag{11.79}$$

as we have been doing all along. This is the reason we characterise all rates in terms of growth rates, as in the definitions of the Γ's given above and in previous Chapters. The growth rate in the linear stage is then given by $\gamma = \Gamma_T$. It is characterised by the maximum value found in Γ_T in the time phase just prior to the transition to nonlinearity and saturation of the turbulence. During this phase Γ_T is found to plateau and then to rise slowly as the fastest growing more slowly emerges beyond others with slightly lower γ. Then, as the nonlinear phase is entered, Γ_T sharply crashes toward zero, in basically one nonlinear transfer time, which happens to be about the same as the original growth time. This is before the turbulent vorticity becomes fully established. As we have seen before, the nonlinear throughput rate (Γ_+, the total among the drive rates) is not necessarily the same as the linear growth rate.

These cases are characterised by an overshoot of the instability and then the drop into saturation. The time trace of Γ_T therefore shows a negative spike, and then a nonlinear adjustment phase, before saturating such that the standard deviation of the time trace sample set for Γ_T is at least an order of magnitude above the mean. However, in some of these cases the amplitude is seen to drop back to very small values even if not all the way to zero. This is the range of nonlinear stabilisation. To see if it is caused by flows, we measure their shear by the zonal profile of the vorticity, here the actual ExB flow vorticity (Ω_E) and not the polarisation density, and see if it is as large as the linear growth rate. The zonal profile in this case is the average over y and s, leaving a flux function (cf. Section 2.5.4).

The second thing to measure is the quasi equilibrium state of the zonal flow layers. They should be at least close to divergence free. Recall that in toroidal geometry a pure zonal ExB flow is not divergence free. The curvature forcing acts upon the flow to produce a sine sideband in the density. To be in equilibrium this has to be in balance with the corresponding divergence of the diamagnetic flow and the parallel divergence of the parallel flow. This is quantified by measuring the sideband $\langle n_i \sin s \rangle$ by multiplying the ion density (a_{00} moment) equation by $\sin s$ and taking the zonal average. If the three terms given by these three divergence components are in close balance, then the flow is close to a quasisteady equilibrium state. The background is nearly always in this state, but the equilibrium can be perturbed into one with a flow layer by the turbulence. This sideband diagnostic tells us if that has in fact occurred. Does a flow in divergence balance suppress the turbulence, or does the flow divergence balance become established only after the turbulence has been suppresses, with the overshoot and collapse perhaps caused by some other process? Let us find out.

The zonal vorticity is measured as a flux surface averaged profile at each radial point and each time point. Then, the square of it is averaged over the radius, and then the square root is taken. This produces a time series of the square root of the spatial-domain average of the square (RMS) of the zonal vorticity. The mean and standard deviation is taken over the sample of 200 time points over $2000 < t < 4000$. The maximum growth rate for the $\omega_t = 0.8$ case takes place just prior to the overshoot and is measured at 0.037 4 in units of $c_s/L_\perp$. The RMS ExB zonal vorticity is measured at $W_Z = 0.0617 \pm 0.0199$ (Z for the zonal component), averaged over the $2000 < t < 4000$ time interval, in the same units. More relevant than the linear growth rate is the turbulence drive rate. The three ExB gradient drive rates are $\Gamma_n = (0.149 \pm 0.0360) \times 10^{-2}$ for the density, $\Gamma_t = (0.332 \pm 0.0841) \times 10^{-2}$ for the electron temperature, and $\Gamma_i = (2.14 \pm 0.495) \times 10^{-2}$ for the ion temperature, adding up to about 0.026 for a total ExB gradient drive, compared with the above vorticity measurement which is significantly higher. It is indeed plausible that the self generated flow vorticity is able to stabilise the turbulence, which does not drop all the way to zero but tumbles back and forth at very small values.

On the other hand, examination of these numbers for the cases with $\omega_t = 1.0$ and 1.2 tells a different story. For $\omega_t = 1.0$ we have $W_Z = 0.136 \pm 0.0402$ for the zonal vorticity while $\Gamma_n = (0.142 \pm 0.0175) \times 10^{-2}$, and $\Gamma_t = (0.594 \pm 0.0803) \times 10^{-2}$, and $\Gamma_i = (3.12 \pm 0.361) \times 10^{-2}$, for the three gradient drive pieces. The linear growth rate for this case is measured at 0.0617. The RMS turbulent vorticity (not the zonal one) is $\Omega_E = 0.574 \pm 0.0892$, which is larger than W_Z but that is also the case for $\omega_t = 0.8$, for which $\Omega_E = 0.211 \pm 0.0541$ is found. The turbulence is moderated, as we will see shortly, but not suppressed. So while this is indicative, it is not the entire reason for the turbulence becoming suppressed in the one case but not in the other. It is more of a continuous transition from control by the zonal vorticity to control by the turbulence, which the turbulent flux quantities being plotted on a logarithmic scale also shows.

One interpretation is that when the turbulence is weak, then if the zonal vorticity is larger than the drive rate then the turbulence is suppressed. But measurements show that from $\omega_t = 0.8$ to 1.0 the RMS zonal vorticity increases faster than the drive rate, but so does the RMS turbulence vorticity, and increasingly as the dynamics becomes more nonlinear then the zonal vorticity has to overcome the turbulence vorticity, which for self-generated flows it never does (the energies can become comparable but the turbulence, being at slightly smaller scale, has the larger vorticity). It is difficult to see how to prove this quantitatively, however.

The second ingredient of this is whether the flows are in divergence balance or not. The ion density equation (for a_{00}) is multiplied by $\sin s$ and then the flux surface average is taken. The resulting profiles are shown for the ExB piece by itself, given by $\mathcal{K}(\phi)$ without any FLR averaging, the total perpendicular divergence, given by the entire $\mathcal{K}$ term, the parallel divergence given by the $\nabla_\parallel$ term, and the total, given by the sum of the $\nabla_\parallel$ and $\mathcal{K}$ pieces. These give the perturbed equilibrium, represented by all terms except the nonlinearities and the partial time derivative. These are shown in figure 11.7, for the case with $\omega_t = 0.8$ at the $t = 4000$ time point. The six frames are the ExB part of the divergence, the total perp part, the parallel part, the total, and then the zonal potential and the zonal ExB vorticity, left to right and top to bottom. The overall sideband part of the equation is close to equilibrium, not in oscillation and not significantly perturbed by disturbances, as evidenced by the total sideband divergence close to zero. The ExB flow is finite as evidenced by $\phi(x)$, and the vorticity level is significant. The perp and parallel divergences, both finite, are close to balance, with the total sideband divergence profile close to zero with small oscillatory layers on top of it, reflecting the narrow vorticity layers. The total flow divergence will always be close to zero in time average in a saturated state, but this one is shown to be close to zero at a single time point.

We compare these divergences in figure 11.8, for the three cases $\omega_t = 0.8$, 1.0, and 1.2, left to right, showing the ExB part of the divergence (top) and the total (bottom). The ExB divergence is finite for all three cases. For $\omega_t = 0.8$ the total is small, indicating flow divergence balance, but moving to higher values with stronger forcing, the flows increasingly lose control of the dynamics to the turbulence.

We can see that if the turbulence is fairly weak to begin with, the self-generated flows which come and go as a part of it can turn around and stabilise the system if

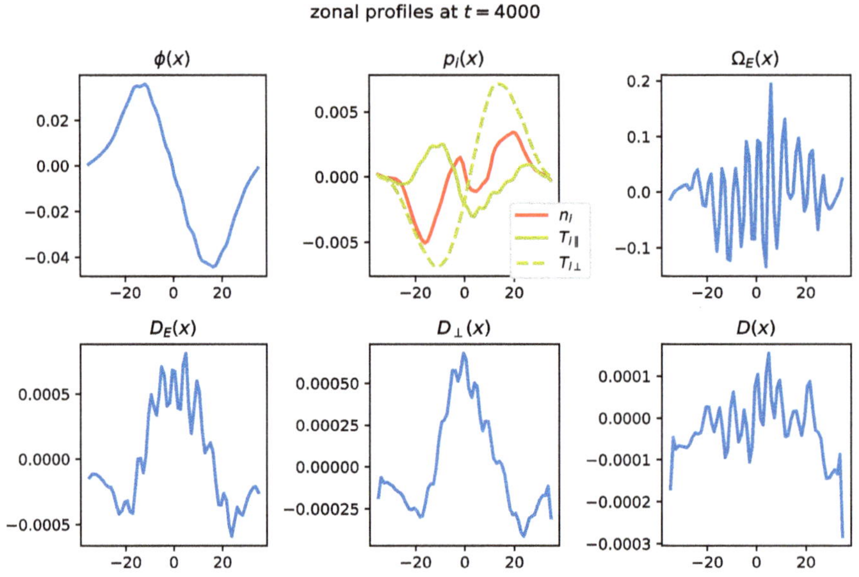

Figure 11.7. Zonal profiles of the state variable sidebands (potential, ion pressure components, and vorticity) and the sideband velocity divergences (ExB D_E, perpendicular $D_\perp$, and total D), for the nonlinearly stabilised ITG case with $\omega_t = \omega_i = 0.8$. Both D_E and D are about two in units of δ_0^2, indicating no cancellation between ExB and diamagnetic velocity components. That D is much lower indicates near cancellation between ExB and parallel velocity components. The diamagnetic component is small because the three pressure components come close to cancelling, a process made possible by the thermal anisotropy (the $T_{i\parallel}$ and $T_{i\perp}$ sidebands are opposite).

they overcome the underlying dynamics. But once the gradient drive is strong enough, the turbulence takes over and while it is mitigated by flows, it is not suppressed. This acts on all the cases, not just the electrostatic ones. We can see in retrospect the signs of nonlinear stabilisation in the $0.2 < \hat{\beta} < 0.4$ cases in the inductivity study above, as well. But it is more subtle than a simple comparison between two of the overall rates. We will obtain more insight in the next Chapter considering adiabatic models of this type of turbulence.

11.4.4 Comparison to the fluid drift model

The baseline case used in this Chapter is different from the more generic one used for the fluid drift model in Chapter 3. So we briefly return to the fluid drift model and thereby obtain a direct comparison between the two models on the same cases. We do this with both scans—the dependence on inductivity and the one on the temperature gradient. Of particular interest is to what extent the same set of phenomenology is seen in both models, and perhaps what goes missing in the fluid drift model.

As with figure 11.1, the time traces of the turbulent ExB ion heat flux are shown in figure 11.9 for various values of $\hat{\beta}$ in the fluid drift case. The lowest case is the $\hat{\beta} = 0.01$ one set as artificially electrostatic. There is a slow, continuous drop in the

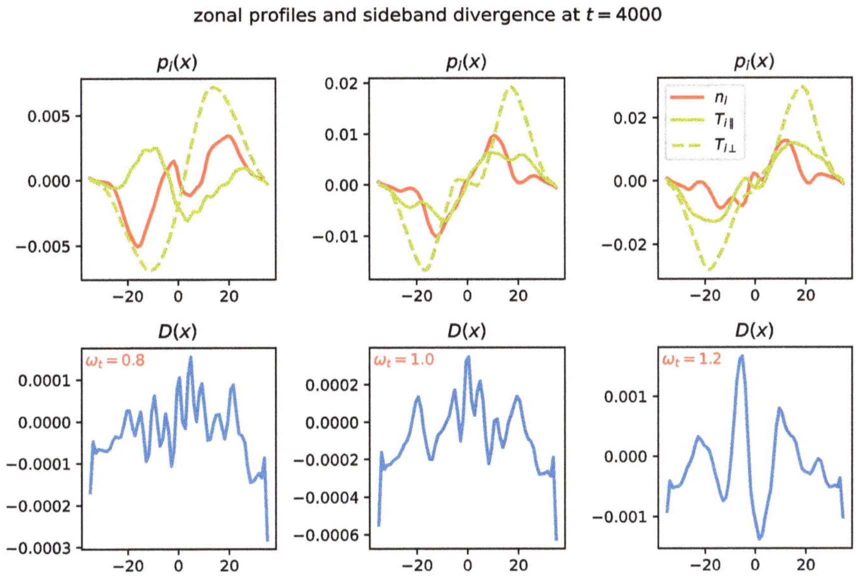

Figure 11.8. Zonal profiles of the sideband ion pressure components and the sideband total velocity divergence, for the three cases showing emergence out of the nonlinearly stabilised gradient range of the ITG turbulence. The temperature components of the sidebands no longer oppose, and the total flow divergence becomes much larger, for the cases with robust turbulence (note the changes in the plot scale, done to show the detail structure in each case).

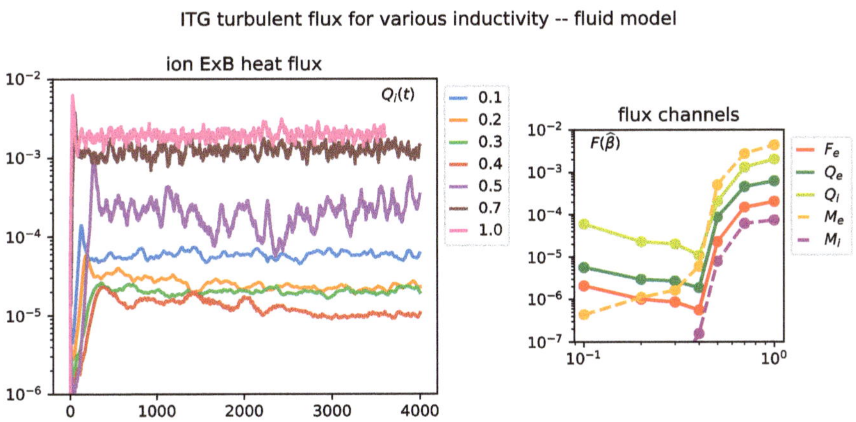

Figure 11.9. As a control case we reproduce the ITG results on inductivity using the fluid drift model for the same cases. Compare with figure 11.1. A similar stabilisation due to finite $\hat{\beta}$ and a transition to the MHD regime is found, but the minimum turbulence level is still finite, and the MHD boundary is at a lower $\hat{\beta}$ value, though the run of fluxes and their ratios is similar in the MHD regime.

intermediate inductivity zone, between $0.1 < \hat{\beta} < 0.4$, but then the 0.5 case is the first one with a stronger result, and the transition into the MHD regime is seen for 0.7 and 1.0. As in the gyrofluid case, the fluid drift case shows a rise in the magnetic flutter electron heat flux before the nominal MHD transition. At $\hat{\beta} = 0.5$

we find M_e is already the largest turbulent flux channel. A similar mode structure to figure 11.2 is found, but the situation is pathological due to the lack of a Landau closure in the model. The collisionality is not strong enough for the thermal conduction sink Γ_k to compensate. In the gyrofluid case Γ_k was zero because the collisions were set to zero. That is not possible in the fluid drift case. In the intermediate inductivity zone, the gyrofluid case sees stabilisation while the fluid drift model sees the gradual decline.

As with figure 11.6, the time traces of the turbulent ExB ion heat flux are shown in figure 11.10 for various values of the temperature gradient forcing $\omega_t = \omega_i$ in the fluid drift case. The first two cases, $\omega_t = 0.5$ and 0.6 are stable, while the 0.7 one grows very slowly to a very low level. This can be taken as the threshold. The cases with $\omega_t = 0.8$ and higher all show a finite overshoot but then saturate at reasonable amplitudes. The effective threshold for the two models is in about the same place, but the details of the phenomenology of the turbulence are different. There is only one ion temperature, not two channels one of which is free of the sound wave dynamics. This matters because of the small size of $qR_0/L_\perp$ for these cases. So there is more constraining on the ion temperature mode structure as there is no freedom to adjust using the extra degrees of freedom.

It is interesting to note that the low beta side yields much stronger turbulence for the gyrofluid model in this comparison. For $\hat{\beta} = 0.01$ we have $Q_e = (0.495 \pm 0.134) \times 10^{-4}$ and $Q_i = (1.85 \pm 0.453) \times 10^{-4}$ in the gyrofluid case, comparing to $\delta_0^2 = 2 \times 10^{-4}$, and $Q_e = (0.044 \pm 0.005) \times 10^{-4}$ and $Q_i = (0.645 \pm 0.064) \times 10^{-4}$ in the fluid drift case (the slight difference in the gyrofluid case to the values given in the analysis of flows above is from the different initialisation, hence different realisation, in this case—such is the nature of turbulence statistics). In the baseline low beta side, then, the ability

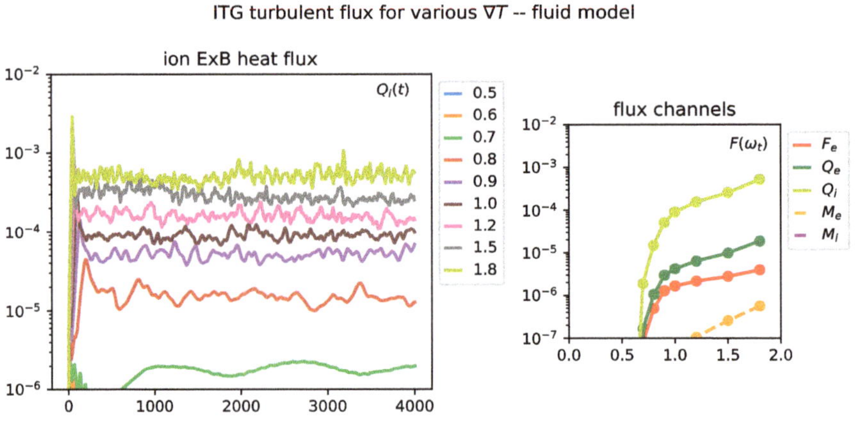

Figure 11.10. As a control case we reproduce the ITG results on ∇T using the fluid drift model for the same cases. Compare with figure 11.6. The linear threshold appears to be near $\omega_t = \omega_i = 0.7$ and the rise to finite level is progressive after that. The 0.8 case saturates at a very low value, but no case is definitively nonlinear suppressed after a linear growth phase. For the cases beyond 0.8 the electron heat flux is a much lower fraction of the ion heat flux, which can be attributed to the assumption of thermal isotropy in the model. The fluxes due to perpendicular temperature are much less free to flourish.

of $T_\perp$ to act independently of $T_\|$ is very significant. So not only is the stabilisation effect more pronounced in the gyrofluid case, it does this on turbulence which is intrinsically stronger to begin with.

The stabilisation effects are in fact rather less pronounced than in the gyrofluid case, and not only in the effect of the inductivity. It is also the case with the dependence on the temperature gradient. No nonlinear upshift of the threshold is seen in the fluid drift cases. It is apparent that the lack of thermal anisotropy in the model does not allow for the generation of flows sufficient to stabilise the turbulence. Not only the flows, but also the rest of the phenomenology.

We return to the issues of thermal anisotropy and flows after considering warm plasma cases, which are finitely collisional and hydrodynamic.

11.5 Drift Alfvén turbulence in a warm plasma

We reflect once more on what constitutes a warm plasma case. The temperatures must be high enough to affect the parallel forces on the electrons, so that we never have a cold plasma response such as in MHD models. The electron pressure is always active in the Ohm's law. It also means that the temperatures are not so strong as to make the adiabatic response so tight that nearly adiabatic electrons result. We have seen that this is a scale ratio question as well. We require a large enough scale ratio $qR_0/L_\perp$ so as to have $\hat{\mu} > 1$. With deuterium ions as in a modern tokamak, this means that $qR_0/L_\perp$ must be larger than about 60. In the outer regions of a tokamak, the temperatures drop from the interior parameter values to something much less, over a few centimeters. Due to the operational constraints we also have q values in the vicinity of 3–5 rather than 1–2. In the generic cases we studied before, we had $qR_0/L_\perp = 100$ so that $\hat{\mu} = 2.72$. That was enough to get into this regime. In Chapter 14 of Volume One, we also touched on a steeper gradient case with $\hat{\mu} = 10$ and $\nu = 1$, arranged from $L_\perp = 3$ cm so that $L_\perp/\rho_s$ is only 52 and $qR_0/L_\perp = 192$. For the present purpose we also require that a warm plasma case have the temperatures as the principal gradient, so that $d \log T/d \log n \sim 2$ in both electrons and ions, as representative tokamak profile data suggests.

A warm plasma case can be stated as follows. Temperature is the main gradient but the density gradient is still significant. The electrons are neither hydrodynamic nor adiabatic. There is enough of an adiabatic response to keep $e\phi/T_e$ and $\log p_e$ similar in the fluctuations, giving ρ_s its sense, but the thermal electrons are nonadiabatic. The thermal fluid electrons are hydrodynamic enough ($\hat{\mu} > 1$) but the parallel wavenumber spectrum needs only spread to a few times $K_\|$ before anything wider would be dominantly adiabatic (measure this using the $k_\|$ spectrum of h_{00} in the electrons). Hence, the electron energetic spectrum is limited to $k_\|/K_\|$ values between one and ten. Larger than that would be MHD like. Smaller would be more of a hot plasma case. Lastly, the case is collisional, but not so much that $\nu_e \gg c_s/L_\perp$, and it is electromagnetic, but not so much that stronger induction can render the parallel electron pressure gradient insignificant in the Ohm's law. In terms of the parameters controlling the electron response we require all of $\hat{\mu}$, C, and $\hat{\beta}$ larger than unity, but not so large as high tens to hundreds.

The base case for warm plasmas that I have used before is determined by the physical parameters

$$\begin{array}{lllll} n_e = 2 \times 10^{19}\,\mathrm{m}^{-3}, & T_e = T_i = 100\,\mathrm{eV}, & B_0 = 2.5\,\mathrm{T}, & & \\ R_0 = 3.3 & a_0 = 1.65\,\mathrm{m}, & q = 3.5, & S = 1.14, & L_\perp = 3.5\,\mathrm{cm} \end{array} \tag{11.80}$$

Normalised parameters are

$$\begin{array}{lllllll} \hat{\beta} & = & 1.75, & \hat{\mu} = 7.41, & \nu = 0.823, & \hat{\varepsilon} = 27\,200, & \omega_B = 0.0424, \\ \delta_0 & = & 0.015, & \omega_n = 0.5, & \omega_t = \omega_i = \tau_i = 1, & S = 1.14, & \varepsilon_a = 0.3 \end{array} \tag{11.81}$$

such that $C = 3.11$ and $R_0/L_\perp = 47$ (cf. Eq. (11.75 for ν)). The closeness of the value of δ_0 to the hot plasma case is a coincidence. For this case $\alpha_M = 0.22$, while the nominal MHD boundary, $\alpha_M = 0.66\,S$, or about 0.75, is met at $\hat{\beta} = 6.0$, which is not very high (we have seen this already in the slightly shallower, in terms of $\hat{\mu}$, cases from the fluid drift Chapters).

The full domain size for a flux tube model is given by $2\pi r_0/qL_\perp$ which is about 25.6 for this case. So we use the 4 × 1 rule as previously,

$$L_x = L_\perp = 64h_x, \qquad L_y = 4L_\perp = 256h_y, \qquad h_y/h_x = 1 \tag{11.82}$$

where $h_{x,y}$ are the step sizes. The timestep is 0.002 and the boundary dissipation is done with $s_f = 0.1$. All other considerations are as above for the hot plasma cases.

We investigate the run of the turbulence against inductivity, varying $\hat{\beta}$ above 0.3. The time traces of the turbulent ExB ion heat flux are shown in figure 11.11, together with the averages over $2000 < t < 4000$ of all the turbulent flux channels. The turbulence strengthens smoothly and monotonically with $\hat{\beta}$, with no indication of stabilisation or even a sharp transition to the MHD regime. In the free energy

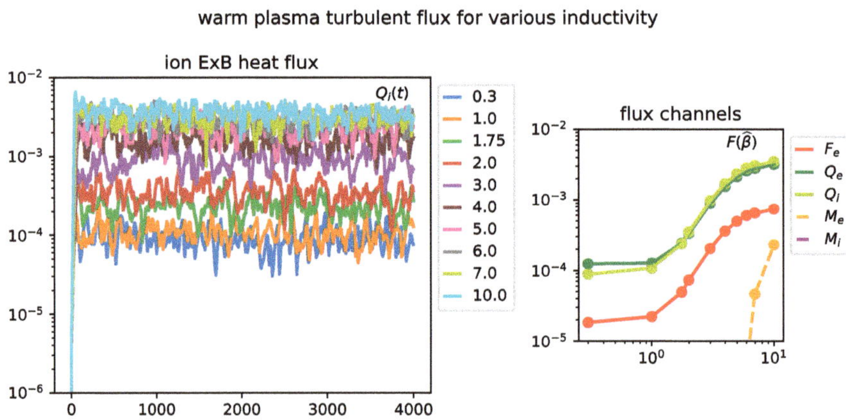

Figure 11.11. Turbulent ExB ion heat flux (Q_i) for various values of inductivity ($\hat{\beta}$), for the warm plasma case given in Eq. (11.81). Below $\hat{\beta} = 1$ there is little dependence, indicating a subdominant role for the electromagnetic component of the electric field. Thereafter, the rise into the MHD regime is steady and gradual. The transition into the MHD regime is not sharp. The turbulence is robust to very robust throughout the range of inductivity. The clutter of the time traces is not an issue as they serve only to show every case is well saturated.

channels not shown, the thermal free energy, ExB and magnetic energy, and enstrophy all rise together in roughly the same proportion. Even more interesting is the relation to the linear growth phase. Linear growth rates fall slightly with $\hat{\beta}$, from a maximum growth of 0.0524 for the first case through 0.0450 for $\hat{\beta} = 3$, which then rises to 0.0866 for $\hat{\beta} = 10$. The increase of the turbulence with $\hat{\beta}$ is much stronger than this, with all energetic channels rising by more than one order of magnitude across this range. This is the nonlinear nature of electromagnetic drift wave turbulence in this parameter regime.

The change in mode structure with rising inductivity is worth following. The k_y-spectra are shown for $\hat{\beta} = 1$, 3, and 5 in figure 11.12. The changes are gradual, with a shift to longer wavelength with higher values. The electron gradient drive is strongest for smaller $\hat{\beta}$, and then the ions take over marginally in the longer wavelengths for larger $\hat{\beta}$. The sink channels are well spread, with the resistivity, electron Landau damping, and electron thermal conductivity in slightly decreasing proportion as the principal channels. The electron thermal anisotropy ('viscosity') is also in this range. The ion parallel dissipation channels are much smaller.

The isotropic perpendicular spectra ($k_\perp$) are shown for $\hat{\beta} = 1$, 3, and 5 in figure 11.13. These are necessary to know that the vorticity remains collected near the ρ_s-scale despite the shift to larger scale for the state variables and flux channels

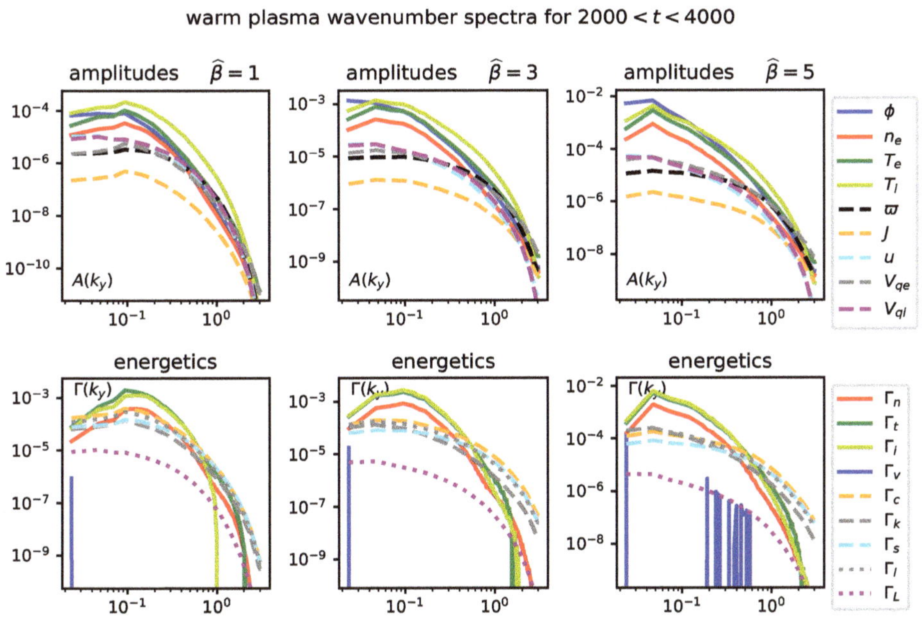

Figure 11.12. Drift-direction wavenumber (k_y) spectra of the amplitudes (top row) and energetics (bottom), for the warm plasma cases up to the nominal MHD boundary at $\hat{\beta} = 6.0$ for these cases. The transition to longer wavelengths is strong and occurs well before the nominal MHD boundary is reached. The rise in vorticity is smaller than the rise in the energy content, though its spectrum becomes flatter. The mesoscale MHD regime, whose nominal boundary $k_y L_\perp = \pi$ occurs at about $k_y \rho_s = 0.05$ is already overlapped at $\hat{\beta} = 1$, due to the steepness of the background gradient ($\hat{\mu} > 1$). Note as well that δ_0 is larger than $K_\parallel L_\perp$ for any of these cases.

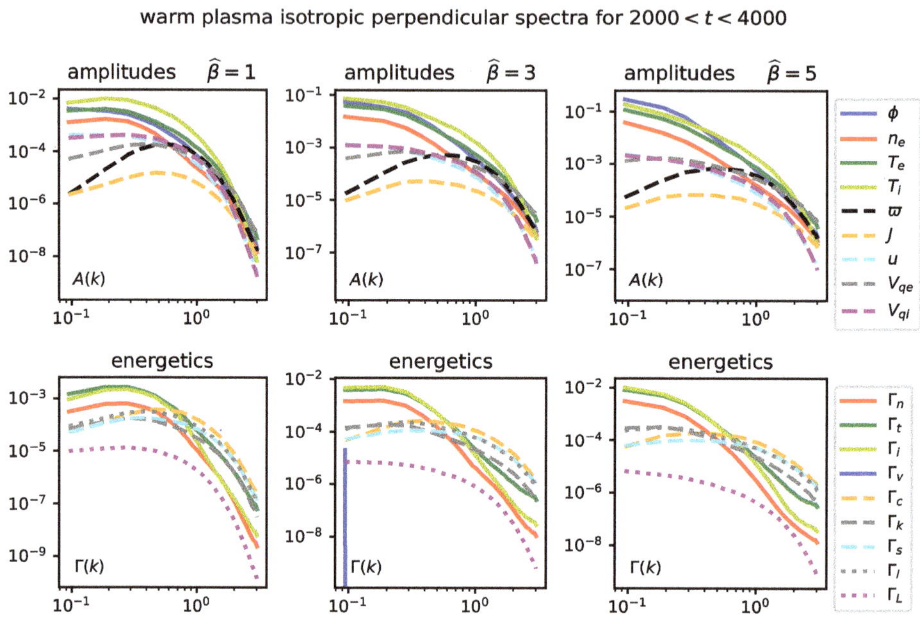

Figure 11.13. Isotropic perpendicular wavenumber ($k_\perp$) spectra of the amplitudes (top row) and energetics (bottom), for the warm plasma cases up to the nominal MHD boundary at $\hat{\beta} = 6.0$ for these cases. The transition to larger scale mirrors the one to longer wavelength in the previous figure. The vorticity peak remains in the vicinity of $k_\perp \rho_s = 1$ as the rest shifts to larger scale. For $\hat{\beta} = 5$ or larger, these spectra peak in the first mode bin for any of the densities or temperatures and their ExB gradient drives. That the magnetic flutter drive (M_e) remains negative indicates the MHD regime is not yet reached.

with increasing inductivity. The sink channels receive the cascade of free energy to smaller scale and so are shifted to smaller scale relative to the rest of the energetics. The parallel wavenumber spectra ($k_\parallel$) are shown for $\hat{\beta} = 1$, 3, and 5 in figure 11.14. These are all of similar character, showing that beyond the longest parallel wavelength represented by the ballooning, the parallel structure does not change very much. Practically the only visible broadening is in the electron temperature, especially from $\hat{\beta} = 1$ to the next case.

The parallel envelopes, in figure 11.15, show basic drift wave mode structure for $\hat{\beta} = 1$, with h very flat and the ion temperature mildly anisotropic. The transition towards MHD mode structure, with ϕ the largest and most ballooned, begins already for $\hat{\beta} = 5$, which is below the nominal boundary. Note the increasing degree of ballooning in the electron temperature and in h for this highest case. The parallel envelopes of the turbulent fluxes show the double-hump structure often seen for low $\hat{\beta}$ values. This follows the run of $\mathcal{K}^y$ going like $\cos s + Ss \sin s$ in the coordinate system with $s_k = 0$, and is seen for strong magnetic shear and electrostatic parameters. Note that the electron ExB heat flux is the strongest channel for $\hat{\beta} = 1$. As the turbulence becomes more electromagnetic, the dip in the envelope at $s = 0$ disappears, and the ions become the strongest channel. This follows the overall shift towards longer wavelength. The magnetic flutter flux is small at these values and only rises for the cases in the nominal MHD regime (figure 11.11).

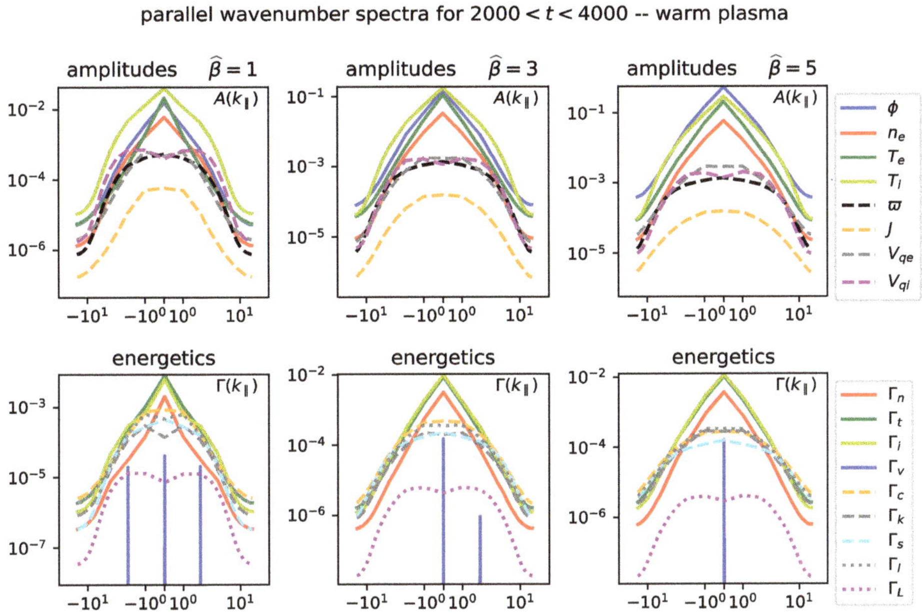

Figure 11.14. Parallel wavenumber ($k_{\|}$) spectra of the amplitudes (top row) and energetics (bottom), for the warm plasma cases up to the nominal MHD boundary at $\hat{\beta} = 6.0$ for these cases. Similar structure is found, with the state variables and ExB gradient drive peaked near the flute mode range ($k_{\|} = 0$), while the flux variables and the vorticity, and the dissipation energetics, are lower and flatter. The ion temperature is the largest state variable at $\hat{\beta} = 1$ but it is overtaken by ϕ as the MHD regime is approached. The electron temperature also becomes relatively stronger, but the gradient drive is closer for ion and electron temperature in all these cases.

All of this mode structure information is qualitatively the same as for drift Alfvén turbulence under the fluid drift model. For the warm plasma cases the differences brought in by the additional physics are not really necessary for fundamental understanding. They are, however, necessary for any comparisons of computations of this type to experimental data, for which the temperature effects define the quantitative situation. They are also necessary to treat the hot plasma cases, so if one wishes to show the transition from warm to hot in the same model, it must be at least this model.

11.6 Thermal anisotropy

The main new ingredient introduced with the gyrofluid model is thermal anisotropy. The results show that the parallel and perpendicular temperatures act independently of each other, with qualitatively different responses to the rest of the turbulence. They only resemble each other in the mode structure when equipartition channels are effective to keep them close. This can either be collisions or it can be turbulent forcing to make the parallel dynamical constraints subdominant. The perpendicular temperature does not participate in parallel force dynamics, only at higher order in the thermal conduction channels. So the coupling to density and potential is indirect. When the parallel/perp scale ratio is not too large, namely $K_{\|}L_{\perp}$ not too small, the

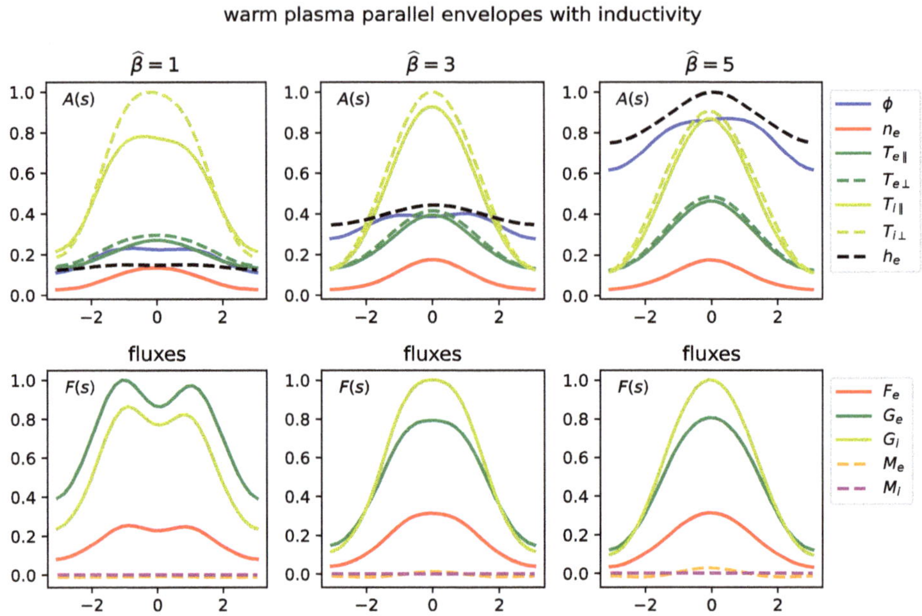

Figure 11.15. Parallel envelopes of the amplitudes (top row) and turbulent fluxes (bottom), for the warm plasma cases up to the nominal MHD boundary at $\hat{\beta} = 6.0$ for these cases. A rise of T_e and even more of ϕ, relative to T_i, with inductivity is found, with ϕ progressively flatter, and Alfvén signature. The $\hat{\beta} = 1$ case is nominally similar to the hot plasma cases (figure 11.5) in that T_i is strongest and most ballooned, with $T_{i\perp} > T_{i\parallel}$, and the nonadiabatic density $h = n_e - \phi$ lowest and flattest, but this signature package is broken up with increasing electromagnetic character. The electron and ion heat fluxes are close despite the greater disparity in their amplitudes. The electron temperature components are kept close by collisions (much less effective for ions).

perpendicular temperature becomes the strongest, especially with toroidal curvature forcing present. The clearest sign in the mode structure is the parallel envelopes, showing that $T_\perp$ is much larger than $T_\parallel$ and more strongly ballooned. In the ions it can carry almost all of the turbulent ExB advective heat flux. This leads not only to stronger instability and turbulence in the ion temperature gradient (ITG) turbulence regimes, but also facilitates self generated flows to exist in a quasi equilibrium while suppressing the turbulence with their zonal vorticity. We have seen these differences in the hot plasma cases, by comparing the gyrofluid and fluid drift models directly on the same cases as we did above. Here, some of the information directly concerning temperature anisotropy is highlighted.

11.6.1 in temperature responses

The temperature anisotropy can be directly measured by comparing $\Delta T = T_\parallel - T_\perp$ to $T = (T_\parallel + 2T_\perp)/3$ for each species. The large degree of anisotropy, $\Delta T/T$ is already obvious from the parallel envelopes in figure 11.5. For the hot plasma cases with varying ω_t they are directly measured by sampling $T_\parallel$ and $T_\perp$ for each species at every grid point and 200 time points from $2000 < t < 4000$, calculating the squares of T and ΔT in volume average, then taking the square root. Then we measure the mean and standard deviation of each RMS quantity, in the same way the statistical values of the Γ's and the fluxes are calculated. Given their definitions, it is often the case that $\Delta T > T$ when $T_\perp > T_\|$, and if $T_\|$ is zero then $\Delta T = 3T/2$ in magnitude. In large measure, we find that ΔT and T are comparable for ions with ΔT a bit larger for electrons. In the last three cases $T > \Delta T$ for ions but by margins less than 30 percent. The standard deviation is in the range of 1/4 to 1/3 for both quantities for both species in each case reaching saturation. When in the inductivity scan the kinetic ballooning regime is reached, the dynamics of the turbulence is stronger compared to the parallel coupling, and the degree of anisotropy decreases.

11.6.2 in flows

The temperature anisotropy in flows can be seen when the diamagnetic flow sideband divergence is separated into the three pieces represented by n, $T_\|$ and $T_\perp$, equal to a_{00}, $\sqrt{2}\,a_{20}$, and a_{01} in normalised units. The curvature forcing represents diamagnetic flows in the fluid drift model, acting upon the density and the isotropic temperature. In the gyrocentre representation these represent the curvature and grad-B drifts, in the gyrofluid model due to $T_\|$ and $T_\perp$, respectively, in equal measure. The effective linearised pressure in this case is $n + T$ in the fluid model, and $n + (T_\| + T_\perp)/2$ in the gyrofluid model. The sideband diagnostic we used in the hot plasma cases characterises the degree to which any self-consistent flows are close to equilibrium, using the sine sideband component of the ion density equation. The details of the divergence balance do indicate a role for the anisotropy in temperature. Returning to the three cases from the hot plasma set above, with $\omega_t = 0.8$, 1.0, and 1.2 these contributions were given in figure 11.8. The components due to the curvature forcing on the zonal components of n, $T_\|/2$, and $T_\perp/2$ are isolated. For the $\omega_t = 0.8$ case, which is nonlinearly stabilised, the flows were shown to be in divergence balance in the analysis above. This shows that the diamagnetic part largely cancels itself out, with n smaller and the temperatures comparable and opposite. The total divergence was already shown to be smaller than the pure ExB one by at least an order of magnitude. So the overall divergence balance is mostly between the ExB part and the parallel part.

11.7 Summary

With the introduction of temperature anisotropy and a model for Landau damping into the dynamical system, we finally have the set of physical ingredients we need to capture the dynamics of turbulence in magnetised plasmas across the entire range of parameters.

If the plasma is dominated by temperature gradients and the collisionality is not too strong, then the temperature in the fluctuations is anisotropic. The parallel and perpendicular temperature disturbances are excited the same way by ExB advection down the gradient, but all the response channels are different. The temperature in any species is only nearly isotropic if its collision frequency is stronger than the RMS vorticity of the turbulence. It can also look approximately isotropic if the ExB advection is so dominant that anisotropy is not strongly excited (e.g., in the MHD regime). The result in such a complicated system is not available in advance, and so the computations have to faithfully allow the responses to provide the result in a consistent manner.

The hot plasma cases are characterised by nearly adiabatic electrons but otherwise robust electromagnetic dynamics driven by both temperature gradients. As seen before, the turbulence occupies the entire scale range between the largest available down to the inertial drift scale, ρ_s. Once more, the ion gyroradius ρ_i is treated by necessity, simply because the electrons and ions are usually at close to the same temperature, which means that ρ_s and ρ_i are close to the same scale.

At any relevant value of plasma beta $\beta = 2\mu_0 p/B_0^2$, the inductivity of the electrons, $\hat{\beta} = \beta_e(K_\parallel L_\perp)^{-2}$ where $\beta_e = \mu_0 p_e/B_0^2$, is sufficient to render the parallel responses to the turbulence electromagnetic. This and the existence of thermal anisotropy in the electrons allow for a finite net electron heat flux across the magnetic field lines. To this is added the parallel heat flux moving down the disturbed field lines which include a radial component. Temperature anisotropy in the ions allows for a dominant response of the perpendicular ion temperature disturbances and consequently a robust net ion heat flux across the magnetic field lines, more than what a simpler model with isotropic temperature would allow. At the same time, the anisotropic temperature facilitates flows whose back reaction upon the turbulence is moderating if not stabilising.

The high beta cases studied herein show a sharp transition into an unstable regime characterised by very strong magnetic flutter gradient drive. This mode type has been given the name 'kinetic ballooning' for the same reason the kinetic shear Alfvén waves got their name: first found with a linear kinetic analysis, it became believed that these were fundamentally kinetic processes. But if we treat the electron and ion fluids on an equal footing, allow for temperature anisotropy, and model the sharp, perhaps not resolvable kinetic Landau resonance with a closure dissipation, then these processes are accessible to capture by gyrofluid computations.

The warm plasma cases, with fluidlike electron responses allowed by the parameter regime $c_s/L_\perp > V_e K_\parallel$, such that thermal electrons are allowed to be hydrodynamic in the spectral range $k_\parallel \sim K_\parallel$, exhibit robustly electromagnetic turbulence at all relevant parameters. This was shown by the choice of a density and temperature in the low range for the edge regions of modern tokamaks. The density is usually larger than $2 \times 10^{19}\ \mathrm{m}^{-3}$ and the temperature is at least 100 eV if not above. The magnetic field is not larger than 3 T in most experiments although there are outliers (often compensated by a higher density and steeper edge gradient). For these parameters the turbulence is already strongly electromagnetic, and its strength increases with the inductivity parameter, even at very moderate temperatures. The response of the long wavelengths is always electromagnetic because (1) the electrons are nonadiabatic therefore there are robust parallel currents, and (2) the scale of the turbulence where most of the gradient drive exists is larger than either the collisional or collisionless skin depth. The collision frequency is comparable to or smaller than the RMS vorticity, and d_e is 1 mm for $n_e = 2.7 \times 10^{19}\ \mathrm{m}^{-3}$ and smaller for higher values. The size of ρ_s is usually between 0.5 and 2 mm depending on T_e and B_0, and the scale of peak energetics is a factor of 5–10 larger.

We recall one does not really need FLR effects, following ρ_i, to treat the region of the peak, but they are necessary to get the entire spectrum, especially the vorticity, and have a converged case with a reasonable dissipation range in which the physics is subsumed by subgrid dissipation. As in the fluid drift cases, the dissipation spectra are flatter than the drive spectra. This is part of the reason we require long-box domains ($L_y > L_x$), even without the considerations of radial flows. In the hot plasma cases above the convergence with respect to L_y is not an issue because this value of L_y is in fact the system size. But in general, though that is not the purpose of the present work, any comparison of computations to experimental observations should contain a test against domain size in the y-direction, perpendicular scale resolution in $k_\perp$-space, and parallel wavelength resolution in $k_\parallel$-space.

In the next Chapter, we will encounter another type of turbulence which lives in the scale range smaller than the ion gyroradius and larger than the electron gyroradius in the same way that the turbulence just treated here occurs above the inertial drift scale. We will have to ascertain how and whether it survives against the dynamics in the drift wave scale range.

Further Reading

This version of electromagnetic gyrofluid equations was developed starting from an electromagnetic, transcollisional extension to the Beer and Hammett reference given in Chapter 9. It is given in B Scott, 'ExB shear flows and electromagnetic gyrofluid turbulence,' *Phys. Plasmas* **7** (2000) 1845. A model with two moment electrons was given in P Snyder and G Hammett, 'Electromagnetic effects on plasma micro-turbulence and transport,' *Phys. Plasmas* **8** (2001) 744. Energetic consistency, problematic in all the early 3D models, was established in B Scott, 'Free energy conservation in local gyrofluid models' *Phys. Plasmas* **12** (2005) 102307. The derivation in terms of orthogonal polynomials was given in B Scott, 'Derivation via free energy conservation constraints of gyrofluid equations with finite-gyroradius electromagnetic nonlinearities,' *Phys. Plasmas* **17** (2010) 102306. Further references were given in Chapter 9.

The hot plasma case is from a study of tokamak core turbulence in A M Dimits *et al*, 'Comparisons and physics basis of tokamak transport models and turbulence simulations,' *Phys. Plasmas* **7** (2000) 969. The temperature gradient threshold upshift was published in this paper. At the time it was stated to result from kinetic processes. The gyrofluid capture of it was made possible by the energetic consistency. It is given in B Scott, 'Computation of turbulence in magnetically confined plasmas,' *Plasma Phys. Contr. Fusion* **48** (2006) B277. It includes the transition to kinetic ballooning discussed above.

The warm plasma case is from a study of tokamak edge turbulence in B Scott, 'Edge turbulence and its interaction with the equilibrium,' *Contrib. Plasma Phys.* **46** (2006) 714.

For representative profile structure in the steep gradient edge regions in tokamaks, see P Gohil, K H Burrell, E J Doyle, R J Groebner, J Kim, and R P Seraydarian, 'The phenomenology of the L-H transition in the DIII-D tokamak,'

Nucl. Fusion **34** (1994) 1057; and W Suttrop, M Kaufmann, H J de. Blank, B Brüsehaber *et al* 'Identification of plasma-edge-related operational regime boundaries and the effect of edge instability on confinement in ASDEX Upgrade,' *Plasma Phys. Contr. Fusion* **39** (1997) 2051; and J Neuhauser, D Coster, H U Fahrbach, J C Fuchs, *et al* 'Transport into and across the scrape-off layer in the ASDEX Upgrade divertor tokamak,' *Plasma Phys. Contr. Fusion* **44** (2002) 855; and L D Horton, A V Chankin, Y P Chen, G D Conway *et al* 'Characterization of the H-mode edge barrier at ASDEX Upgrade,' *Nucl. Fusion* **45** (2005) 856.

IOP Publishing

Turbulence and Instabilities in Magnetised Plasmas, Volume 2

Gyrokinetic theory and gyrofluid turbulence

Bruce Scott

Chapter 12

Electron gyroscale turbulence

12.1 Introduction—the gyroscale

Turbulence driven by temperature gradients under hot plasma conditions tends to have its energy and ExB transport flux near but above the gyroradius scale of each species, in the perpendicular wavenumber spectrum. In general it is possible to have this for electrons as well, not just ions. In the ion temperature gradient (ITG) turbulence, which we cover in Chapter 3 for the fluid model and in Chapter 11 for the gyrofluid model, the electrons are allowed their natural electromagnetic responses self-consistently in the turbulence. This response enters as a restoring force through kinetic shear Alfvén waves excited by the imbalance in the parallel pressure and electrostatic potential in the parallel electron dynamics. If the response parameters controlling the electron dynamics ($\hat{\beta}$, $\hat{\mu}$, and C, for inductivity, inertia, and resistivity, respectively) are all small, then the electrons will hold the electrostatic potential down to the density fluctuations, with the temperature fluctuations kept small by the thermal conduction dynamics. The electrons evolve quasistatically, with the dynamics all resulting from the forcing due to the ion temperature gradient. This is the picture of what we call adiabatic ITG turbulence.

A similar set of dynamics can result from the electron dynamics in the vicinity of the electron gyroradius, which we have largely neglected up to now. Here, the adiabatic response is in the ions, not the electrons. The dynamics is now caused by the electron temperature gradient, which has the same considerations relative to the density gradient that the ion gradient has in the ITG model. The adiabatic ion limit does not result from anything in the dynamics within the model, i.e., similar to the electron adiabatic limit resulting from the parallel dynamics. Rather, it comes from the small scale limit in the ion polarisation. At scales well below the gyroradius, the polarisation is proportional to the potential itself, which adds to the mix of the densities. The ion gyrocentre variables are neglected as their responses below the scale of the ion gyroradius are assumed small, and the only ion response is the small-scale limit of the polarisation. All the dynamics results from the forcing due to the

doi:10.1088/978-0-7503-3855-4ch12

electron temperature gradient. This is the picture of what we call adiabatic electron temperature gradient (ETG) turbulence.

At the level of instabilities, where the ExB and magnetic flutter nonlinearities are absent, the two systems are isomorphic, once they have been normalised to their respective scales. The region of the spectrum occupied by most of the energetics is the decade above the gyroradius, $\rho_i \sim \rho_s$ for the ITG case, and ρ_e for the ETG case. However, the way the adiabatic state for the subdominant species occurs is different, and the nonlinear responses, especially those involving flows but also, for the ETG case, those involving any electromagnetic responses. Indeed, for the ETG case the collisionless skin depth becomes relevant on the long wavelength side. This has previously been called skin depth turbulence, as it was once thought that the skin depth d_e provided an upper limit on the spectral processes. We will now explore these issues in turn.

12.2 Responses below the ion gyroradius

The ion polarisation at the level of the kinetics results from gyroscreening. At long wavelength it merges to the MHD limit, in which

$$P(\phi) \to -\frac{\rho}{B^2}\nabla_\perp^2\phi \qquad \text{where} \quad P(\phi) = \rho_G = \sum_{sp} n_z Ze \tag{12.1}$$

where Z is the charge state of species z. Gyroaveraging is neglected, and the ion inertia accounts for the polarisation response. Any time dependent changes in the gyrocentre charge density lead to corrections to the basic gyromotion plus ExB drift, and at the particle level the polarisation drift results. At the gyrocentre level one has the polarisation charge density. The time dependence of that is the time dependent polarisation current. The two pictures represent the same thing in different language. As the scale becomes smaller, the size of the gyroradius in the polarisation enters to reduce the response so that it does not continue rising with $k_\perp^2$. When the scale is below the gyroradius, or $k_\perp\rho_i \sim 1$ or >1 for the ions, one is still in the large-scale regime for electrons, so that their contribution now also enters. The ion polarisation response caps at unit level times the gyrocentre charge density, while the electron one continues increasing until $k_\perp\rho_e \sim 1$ is reached. The relevant range is then $\rho_i > k_\perp^{-1} > \rho_e$ for this situation.

In the limit of linearised polarisation in a delta-f model, in which the parameters in $P(\phi)$ are all background constants due to the ion density and temperature, one can go to Fourier space to evaluate this response. We did this in Chapter 5 leading up to Eq. (5.116), which gives the polarisation response as

$$P(\phi) = \sum_{sp} n_z Z^2 e^2 \frac{1 - \Gamma_0}{T_z}\phi \tag{12.2}$$

where each species contributes with density n_z, charge state Z, mass m_z, and temperature T_z. This Γ_0 is a gyroscreening operator which cancels the factor of unity at long wavelength but then decreases with $k_\perp$ according to its expression in $k_\perp$-space,

$$\Gamma_0(\mathbf{k}) = e^{-b}I_0(b) \qquad b = k_\perp^2\rho_z^2 \tag{12.3}$$

where I_0 is the zeroth modified Bessel function, and ρ_z is the thermal gyroradius given by m_zT_z and ZeB for the species. Like all the moment integrals we use involving the Bessel function J_0 in velocity space, Γ_0 does not oscillate in sign. Its limiting values are $1 - b$ to the low $k_\perp$ side and zero at high $k_\perp$. Then, the polarisation due solely to the ions is simply

$$\lim_{b\gg 1} P(\phi) = n_iZ_i^2\frac{e^2}{T_i}\phi \tag{12.4}$$

namely, the bare potential times parameters. In the simplest case with no electron gyroaveraging and no ion gyrocentre responses, we would have ϕ proportional to the electron density n_e, with a minus sign,

$$n_i\frac{e^2}{T_i}\tilde{\phi} = -\tilde{n}_e e \tag{12.5}$$

in the fluctuations, for singly charged ions. This is the opposite sign to the adiabatic electron case in drift wave or ITG instabilities. The minus sign is simply due to the charge on the electrons. This is what the ion polarisation response looks like at small scale.

12.3 Heuristic 2D electron gyroscale model

This type of turbulence was introduced with analytical models in the 1970s when it was often referred to as skin depth turbulence. The idea was to have largely electrostatic turbulence based on the electron gyroscale, which would then be limited by a long wavelength cutoff set by the collisionless skin depth, where magnetic fluctuations would enter and change the dynamical character. Between the skin depth d_e and the electron gyroradius ρ_e would be the spectral range of the turbulence. This scale ratio is given by the inverse electron beta in the form

$$\frac{\rho_e^2}{d_e^2} = \frac{m_eT_e}{e^2B^2}\frac{\mu_0n_ee^2}{m_e} = \beta_e \tag{12.6}$$

with the mass and charge cancelling. This is related to the scale ratio we have had before, between the drift scale and the skin depth,

$$\frac{\rho_s^2}{d_e^2} = \frac{m_iT_e}{e^2B^2}\frac{\mu_0n_ee^2}{m_e} = \beta_e\frac{m_i}{m_e} \tag{12.7}$$

with ρ_s and ρ_e differing solely due to the mass ratio. The relevant ratio for screening of currents by magnetic induction is still d_e, but now the scale of motion is several ρ_e rather than several ρ_s, so that the dynamics is expected to occur below the d_e scale, rather than above the ρ_s scale. Here, the parameters n_e, T_e, and T_i refer to the background electron–ion plasma, and we usually assume a single component case with singly charges ions so that the background densities are equal.

With a normalisation in terms of the electron thermal speed V_e rather than the acoustic speed c_s, the relevant mass parameter appears in terms of the scale ratio $K_{\|}L_{\perp}$ as for the sound wave responses in drift wave or ITG turbulence. Overall, the normalisation is in terms of V_e and $L_{\perp}$, the nominal profile scale.

The simplest model in two dimensions (the perpendicular plane), has equations for n_e and T_e for an electron fluid. For simplicity we fold the drift parameter into the normalisation. The normalised variables are

$$n_e = \delta_e^{-1}\frac{\tilde{n}_e}{n_0} \qquad T_e = \delta_e^{-1}\frac{\tilde{T}_e}{T_{e0}} \qquad \phi = \delta_e^{-1}\frac{e\tilde{\phi}}{T_{e0}} \qquad \delta_e = \frac{\rho_e}{L_{\perp}} \tag{12.8}$$

The background densities and temperatures are now given the zero subscript. Going back through the parameters we have used in other models, the only difference is the substitution of m_e for m_i in the parameters. Instead of ρ_s and $c_s/L_{\perp}$, the scales of normalisation are ρ_e for the space scale and $V_e/L_{\perp}$ for the time scale.

The potential is set by invoking ion polarisation below the ρ_s scale, so that the ion polarisation response is

$$P(\phi) = n_0 e^2 \frac{1 - \Gamma_0}{T_{i0}}\phi \qquad \text{limiting to} \quad n_0\frac{e^2}{T_{i0}}\phi \rightarrow \tau_i^{-1}\phi \tag{12.9}$$

where the normalisation is in terms of T_{e0}, with $\tau_i = T_{i0}/T_{e0}$ the usual temperature ratio parameter. This is only the ion term, however, as in this case we will also keep the electron polarisation because we are treating its $k_{\perp}\rho_e$ contribution.

The dynamical equations follow a 2D electron temperature interchange model such as that in Chapter 2,

$$\frac{\partial n_e}{\partial t} + [\phi, n_e + \log n_0] = \mathcal{K}(\phi - (n_e + T_e)) \tag{12.10}$$

$$\frac{3}{2}\frac{\partial T_e}{\partial t} + \frac{3}{2}[\phi, T_e + \log T_0] = \mathcal{K}\left(\phi - (n_e + T_e) - \frac{5}{2}T_e\right) \tag{12.11}$$

where the log terms are the background gradients,

$$-\frac{\partial}{\partial x}\log n_0 = \omega_n \qquad -\frac{\partial}{\partial x}\log T_0 = \omega_t \tag{12.12}$$

The curvature terms follow the fluid model, with the simple interchange form $\omega_B(\partial/\partial y)$ with $\omega_B = 2L_{\perp}/R$ the nominal forcing parameter. The brackets and subgrid dissipation are done as in the fluid case, and the energetics is the same (Chapter 2). The main difference here is the polarisation, which has the factor of τ_i^{-1} adding to the perpendicular Laplacian term. In this early model the FLR corrections following the finite electron gyroadius are considered only in their contribution to the polarisation charge density, so that the polarisation equation is

$$\left(\tau_i^{-1} - \rho_e^2\nabla_{\perp}^2\right)\phi = -n_e \tag{12.13}$$

where the first term is the sub-scale ion polarisation giving the adiabaticity, and the second term is the electron polarisation, which is now kept to control the high-$k_\perp$ responses. This system of equations is basically the same as a Hasegawa–Mima equation added to a passive scalar temperature equation, with curvature forcing added in to provide a coupling mechanism. The instability can be calculated directly, as for the adiabatic-electron ITG case in Chapter 3. If $\tau_i = 1$ the only difference is the sign of ϕ, so that the instability propagates in the electron instead of the ion direction. The main control parameter is $\eta_e = \omega_t/\omega_n$ which is in the same role here as η_i was for the ions in the ITG case.

12.4 ITG and ETG isomorphism

There has been discussion of the isomorphism between the ITG model on the one hand, and the ETG model on the other, once one normalises to the corresponding thermal velocity and leaves $\tau_i = 1$ in the parameters. Only the sign of ϕ is different, but this can be transformed away through a reflection in the coordinates. This isomorphism is only true if one takes the basic model without any FLR corrections beyond the first polarisation term dependent upon $k_\perp^2$, namely, the other species is adiabatic.

However, the origin of an adiabatic limit in the two models is different. In the ETG case it pertains to the entire ion response, since it is due to the high-$k_\perp$ limit of the ion polarisation. Moreover the ion gyrocentre variables are neglected because of the high-$k_\perp$ limit of the gyroaveraging. All degrees of freedom are affected, and the factor of unity in $1 - \nabla_\perp^2$ remains for all components of ϕ, including the zonal one. In the ITG case it pertains only to the non-zonal component of the electron response, since it arises from parallel force balance in the electrons. Due to the closed flux surfaces and finite magnetic shear, the field line connection guarantees a finite $\nabla_\parallel$ for every degree of freedom, except for the zonal component, since the flux surface average of the parallel divergence vanishes and the parallel gradient of a zonal quantity vanishes. This is the only component for which that is true, so to apply the adiabatic response the zonal component has to be subtracted out in the adiabatic term in the polarisation equation.

12.4.1 Flows and the adiabatic response

Flows and flow energetics were covered in Chapter 16 of Volume One. In discussing the zonal ExB flow component of the turbulence, we faced the issue of separating the adiabatic response in any drift wave model, from the simplest 2D cases to the most general 3D cases, from the zonal component of the dependent variables. This allows zonal vorticity and hence flows to evolve naturally, without any artificial adiabatic construction, since the adiabatic response in the electrons does not affect the zonal component, for which $\nabla_\parallel$ vanishes. If we did not do this, a different sort of dynamics enters. One does not get spurious transport in the density, since $[\phi, n]$ vanishes in any case wherein the xy-dependent components of n are adiabatic, namely, proportional to ϕ. However, with the factor of unity present in the Hasegawa–Mima polarisation $(1 - \nabla_\perp^2)$, zonal flows do not effectively arise. With the nonlinearity decreasing in magnitude at low $k_\perp$, any self-consistent large scale response becomes weak. So to apply the adiabatic constraint we had to watch out for the flows, unless for illustrative purposes we simply set the $k_y = 0$ mode to zero for all dependent variables. If not assuming an adiabatic constraint in the electrons, then the adiabatic response through $\nabla_\parallel$ acting upon the state variables proceeds naturally, and the adiabatic state is approached if the parameters and the physical situation indeed delivers that as a result.

The adiabatic response in a drift wave model, by extension the ITG case, allows for zonal flows by subtracting off the zonal component,

$$\tilde{n}_e - \langle \tilde{n}_e \rangle = n_0 \frac{e}{T_{e0}} (\tilde{\phi} - \langle \tilde{\phi} \rangle) \tag{12.14}$$

We already had to do this in some of the cases studied with the 2D model, in which the dissipative coupling term in normalised units,

$$\nabla_\parallel J_\parallel \to D(n - \phi) \tag{12.15}$$

was replaced by its non-zonal component,

$$\nabla_\parallel J_\parallel \to D((n - \phi) - \langle n - \phi \rangle) \tag{12.16}$$

leaving the zonal component unaffected by the coupling.

Now, if applying adiabatic electrons, we have to do the same thing in the polarisation equation for the gyrofluid model, which is

$$\tilde{n}_i e + \frac{\rho}{B^2} \nabla_\perp^2 \phi = \tilde{n}_e e = n_0 \frac{e}{T_{e0}} (\tilde{\phi} - \langle \tilde{\phi} \rangle) \tag{12.17}$$

for a simplified gyrofluid case with singly charged ions and without FLR effects. In the adiabatic case, a zonal component of $\tilde{n}_e$ never arises because the zonal component of the ExB advection term vanishes (integration over y annihilates the xy-Poisson bracket unless both scalar quantities have y-dependence). It is even more challenging in the low-frequency Braginskii equations, where operation on the polarisation has to be folded into the vorticity variable, which for warm ions is given in terms of both ϕ and p_i, making the necessary solve more difficult (see Chapter 3). This is why we didn't go into the adiabatic ITG cases in any detail therein. But in this Chapter we can take advantage of the fact that doing this is relatively straightforward in a simplified gyrofluid model. One simply solves the above equation twice, once for the zonal mode and once for the rest. That works by taking the zonal average,

$$\langle \tilde{n}_i \rangle e + \frac{\rho}{B^2} \nabla_\perp^2 \langle \phi \rangle = 0 \tag{12.18}$$

and subtracting from the rest, obtaining

$$\widehat{n}_i e + \frac{\rho}{B^2} \nabla_\perp^2 \widehat{\phi} = n_0 \frac{e}{T_{e0}} \widehat{\phi} \tag{12.19}$$

where

$$\hat{n}_i = \tilde{n}_i - \langle \tilde{n}_i \rangle \qquad \hat{\phi} = \tilde{n}_i - \langle \tilde{\phi} \rangle \tag{12.20}$$

Each of these is solved separately, and then the total is reconstructed from

$$\tilde{\phi} = \langle \phi \rangle + \hat{\phi} \tag{12.21}$$

which completes the solve. In the simplest case $\tilde{n}_i$ is the dependent variable and $\tilde{\phi}$ a dependent field, as in the Hasegawa–Mima equation under the gyrofluid representation, as we saw in Chapter 9.

12.4.2 Polarisation in the ETG and ITG or EZF models

The adiabatic ETG and ITG models are dynamically the same, then, up to the sign of ϕ, the temperature ratio τ_i, and the different scales ρ_e for ρ_s and $V_e/L_\perp$ for $c_s/L_\perp$, except for this difference in polarisation. To illustrate this effect there is no need to actually build an ETG and an ITG model next to each other. We can run the simplified ETG model by itself and just show what the effect of this difference in the adiabatic response in the polarisation equation actually is, reflected in the results, by applying either version of the polarisation. So we can keep the ETG model intact, and define the EZF case by making that one change in the polarisation. Since we will show the comparison from within the six-moment gyrofluid model (moment variables: a_{mn} with mn = 00, 10, 20, 01, 30, and 11), we illustrate the difference in the polarisation equations for this case. Singly charged ions are assumed throughout.

The ETG case has an adiabatic ion response for all degrees of freedom. Its polarisation equation is

$$n_0 \frac{e^2}{T_{i0}} \tilde{\phi} - n_0 \frac{m_e}{B^2} \nabla_\perp^2 \tilde{\phi} = -n_0 e (G_0(a_{00}) + G_1(a_{01})) \tag{12.22}$$

in which the moment variables a_{mn} represent electron gyrocentres, the gyroaveraging operators G_n (species dependent, defined in Eq. (9.62)) are done with $k_\perp \rho_e$, and the sign is due to the negative electron charge. This equation becomes

$$\left(1 - \nabla_\perp^2\right)\phi = -(G_0(a_{00}) + G_1(a_{01})) \tag{12.23}$$

in normalised units.

An ITG case has an adiabatic electron response for all degrees of freedom except the zonal component. Its polarisation equation is

$$n_0 \frac{e^2}{T_{e0}} (\tilde{\phi} - \langle \tilde{\phi} \rangle) - n_0 \frac{m_i}{B^2} \nabla_\perp^2 \phi = n_0 e (G_0(a_{00}) + G_1(a_{01})) \tag{12.24}$$

in which the moment variables a_{mn} represent ion gyrocentres, the gyroaveraging operators G_n are done with $k_\perp \rho_i$, and the zonal electron density is assumed to remain at zero. We recall that

$$\frac{\rho_i^2}{T_i} = \frac{\rho_s^2}{T_{e0}} \tag{12.25}$$

so that the polarisation charge density for ions appears with ρ_s and T_{e0}. This equation becomes

$$\phi - \langle\phi\rangle - \nabla_\perp^2\phi = G_0(a_{00}) + G_1(a_{01}) \tag{12.26}$$

in normalised units. Note the sign difference.

We can keep the signs and scales of the ETG model while doing the adiabatic species as in the ITG model by simply making the appropriate change to the ETG model, so that Eq. (12.23) is replaced by

$$\phi - \langle\phi\rangle - \nabla_\perp^2\phi = -(G_0(a_{00}) + G_1(a_{01})) \tag{12.27}$$

in what is still an ETG model with the adiabatic response as in the adiabatic drift wave or ITG case. We will call this the EZF model and directly compare it with the ETG model. Again, note the sign on the right hand side of the polarisation equation: plus for the ITG model, minus for the ETG or EZF model.

Simple polarisation is used to illustrate this similarity, and eventually to facilitate the computations. In the ITG model with adiabatic electrons there is no dynamics below ρ_s, and in the ETG model with adiabatic ions there is no dynamics above ρ_s, so that computational issues with scales not covered do not arise. These forms are written for the six-moment gyrofluid model, which will be the one under which we do the comparisons. In the EZF model all the scales and signs are as in the ETG model, except we artificially use the EZF polarisation equation to solve for ϕ, namely Eq. (12.27) instead of Eq. (12.23). It is this comparison in which we are interested, as it provides the most minimal control case for this flow polarisation effect in the adiabatic models.

12.5 Three-dimensional adiabatic ETG turbulence

To treat three-dimensional (3D) toroidal ETG turbulence in an adiabatic model, we use the six-moment gyrofluid model for the electrons with adiabatic ions. This takes the moment level up to three: up to three factors of z for the Hermite polynomials H_m, and one in w for the Laguerre polynomials L_n, with $2m + n = 3$ in maximum. We will have six dependent variables covering the electron gyrocentre dynamics, plus a polarisation equation, either the ETG or EZF one, and also an Ampère's law obtaining the magnetic fluctuations from the parallel current, which is given by variable 10 in the moment list (with a minus sign for the charge). The geometry is the toroidal fluxtube model, and in this case it will be possible to run with a 1×1 spatial domain because neither model produces large-scale flow dynamics, and moreover the turbulence is faster than the nominal transit frequency. In this case the comparison is between $K_\parallel V_e$ and $V_e/L_\perp$, not $c_s/L_\perp$, so that the ratio does not involve the mass ratio. Usual values of $K_\parallel L_\perp$ are rather small, as we have used before.

12.5.1 Normalised gyrofluid moment equations

Expressing

$$n = a_{00} \qquad T_\parallel = \sqrt{2}\, a_{20} \qquad T_\perp = a_{01} \tag{12.28}$$

as state variables, and

$$J = -a_{10} \qquad q_{\|} = \sqrt{\frac{3}{2}}\, a_{30} \qquad q_{\perp} = a_{11} \tag{12.29}$$

as flux variables, in normalised units, we have

$$\begin{aligned} &\frac{\partial n}{\partial t} + [\phi_0, n + \log n_0] + [\phi_1, T_{\perp} + \log T_0] - \nabla_{\|} J \\ &\quad = \frac{1}{2}\mathcal{K}\big(2\phi_0 + \phi_1 - (2n + T_{\|} + T_{\perp})\big) \end{aligned} \tag{12.30}$$

$$\begin{aligned} &\frac{\partial}{\partial t}(A + \mu J) + \mu[\phi_0, J] - \mu[\phi_1, q_{\perp}] + \nabla_{\|}\big(\phi_0 - (n + T_{\|})\big) \\ &\quad + D(T_{\|} - T_{\perp} + \phi_1) = \frac{1}{2}\mu\mathcal{K}\big(4J - 2q_{\|} - q_{\perp}\big) \end{aligned} \tag{12.31}$$

$$\begin{aligned} &\frac{1}{2}\frac{\partial T_{\|}}{\partial t} + \frac{1}{2}[\phi_0, T_{\|} + \log T_0] + \nabla_{\|} q_{\|} \\ &\quad + D(q_{\perp} - J) = \frac{1}{2}\mathcal{K}\big(\phi_0 - (n + 3T_{\|})\big) \end{aligned} \tag{12.32}$$

$$\begin{aligned} &\frac{\partial T_{\perp}}{\partial t} + [\phi_0, T_{\perp} + \log T_0] + [\phi_1, n + \log n_0 + 2(T_{\perp} + \log T_0)] + \nabla_{\|} q_{\perp} \\ &\quad - D(q_{\perp} - J) = \frac{1}{2}\mathcal{K}\big(\phi_0 + 4\phi_1 - (n + 4T_{\perp})\big) \end{aligned} \tag{12.33}$$

$$\begin{aligned} &\mu\frac{\partial q_{\|}}{\partial t} + \mu[\phi_0, q_{\|}] + \frac{3}{2}\nabla_{\|} T_{\|} \\ &\quad = \frac{1}{2}\mu\mathcal{K}\big(3J - 8q_{\|}\big) - \mu^{1/2}\sqrt{3}\, q_{\|} \end{aligned} \tag{12.34}$$

$$\begin{aligned} &\mu\frac{\partial q_{\perp}}{\partial t} + \mu[\phi_0, q_{\perp}] + \mu[\phi_1, 2q_{\perp} - J] + \nabla_{\|}(T_{\perp} - \phi_1) \\ &\quad - D(T_{\|} - T_{\perp} + \phi_1) = \frac{1}{2}\mu\mathcal{K}\big(J - 6q_{\perp}\big) - \mu^{1/2} q_{\perp} \end{aligned} \tag{12.35}$$

In these equations the parallel derivative is nonlinear,

$$\nabla_{\|} = \frac{\partial}{\partial s} + [-A,\ ] \tag{12.36}$$

with $\partial/\partial s$ giving the derivative along the undisturbed magnetic field. The quantity D is the magnetic pumping effect, given by

$$D = \frac{\partial}{\partial s}\log B = \varepsilon_a \sin s \tag{12.37}$$

with $\varepsilon_a = r_0/R_0$ the ratio between the reference minor and major toroidal radii. The parameter μ is the scale ratio,

$$\mu = \left(K_\parallel L_\perp\right)^{-2} \tag{12.38}$$

which now appears without the mass ratio as mentioned above. The gradient drive parameters are ω_n and ω_t given as in the 2D case. The curvature coupling parameter is $\omega_B = 2L_\perp/R_0$ as before. The ion/electron temperature ratio is τ_i which here appears only in the polarisation equation.

The gyroaveraged potentials defined by operators

$$\phi_0 = G_0(\phi) \qquad \phi_1 = G_1(\phi) \tag{12.39}$$

defined by

$$\phi_0 = \left(1 - \frac{1}{2}\nabla_\perp^2\right)^{-1}\phi \qquad \phi_1 = \left(1 - \frac{1}{2}\nabla_\perp^2\right)^{-1}\phi_0 - \phi_0 \tag{12.40}$$

in the Padé approximation established in Chapter 9. Note the use of recursion.

12.5.2 ETG and EZF polarisation and induction

The ETG polarisation equation is

$$\left(1 - \nabla_\perp^2\right)\phi = -(G_0(n) + G_1(T_\perp)) \tag{12.41}$$

following Eq. (12.23). The gyroaveraging operators are those just given for the gyroaveraged potentials.

The EZF polarisation equation is the same except for the one change,

$$(\phi - \langle\phi\rangle) - \nabla_\perp^2\phi = -(G_0(n) + G_1(T_\perp)) \tag{12.42}$$

following Eq. (12.27). This gives us the point of comparison between the ETG and EZF models.

The induction equation is the form of Ampère's law for this model,

$$-\nabla_\perp^2 A = \hat{\beta} J \tag{12.43}$$

since the ion gyrocentre variables are absent. Ampère's law is solved for A according to the procedure in Chapter 2 as in this model it is the same as in the fluid drift model.

The parameter $\hat{\beta}$ is β_e times two scale ratio factors,

$$\hat{\beta} = \beta_e\left(K_\parallel L_\perp\right)^{-2} \qquad \text{with} \quad \frac{\rho_e^2}{d_e^2} = \frac{\hat{\beta}}{\mu} \tag{12.44}$$

as in the drift Alfvén models of previous Chapters. With typical values of β_e in the range of a few 10^{-3} in hot plasma cases, $\hat{\beta}$ remains moderate and rather smaller than μ. This is because ρ_e is smaller than d_e as noted above. Only the very longest wavelengths in an ETG model are electromagnetic.

12.5.3 Comparison within the standard case

We show the result for the ETG and EZF models on a standard case given by

$$\hat{\beta} = 0.1, \quad \mu = 93.3, \quad \omega_B = 0.29, \\ \omega_n = 0.321, \quad \omega_t = 1, \quad S = 0.78, \quad \varepsilon_a = 0.18 \tag{12.45}$$

which is one of the hot plasma cases from Section 11.4 of the previous Chapter. The standard domain size and grid with approximately one point per ρ_e is

$$L_x = L_y = 2\pi/K \quad K = 0.025 \quad \text{with} \qquad N_x \times N_y \times N_s = 256 \times 256 \times 32 \tag{12.46}$$

with the s-domain over $-\pi < s < \pi$ as usual for the toroidal model. The timestep is 10^{-2} and the subgrid dissipation is done with $\nu_0 = 1$ and $s_f = 0.1$ in Eq. (2.95). The boundary conditions are the same as in Eq. (2.97).

We first show the results of the ETG and EZF models, differing by the polarisation equation, on this case. The starting amplitude is $A_0 = 10^{-4}$, giving a well established linear stage before the turbulence saturates. Note that with this normalisation the quoted amplitude level is in terms of δ_e, not unity or δ_0. The time traces comparing the two cases are shown in figure 12.1. The electron ExB turbulent heat flux,

$$Q_e = (1.5n + 0.5T_\parallel + T_\perp)[\phi_0, x] + (n + 2T_\perp)[\phi_1, x] \tag{12.47}$$

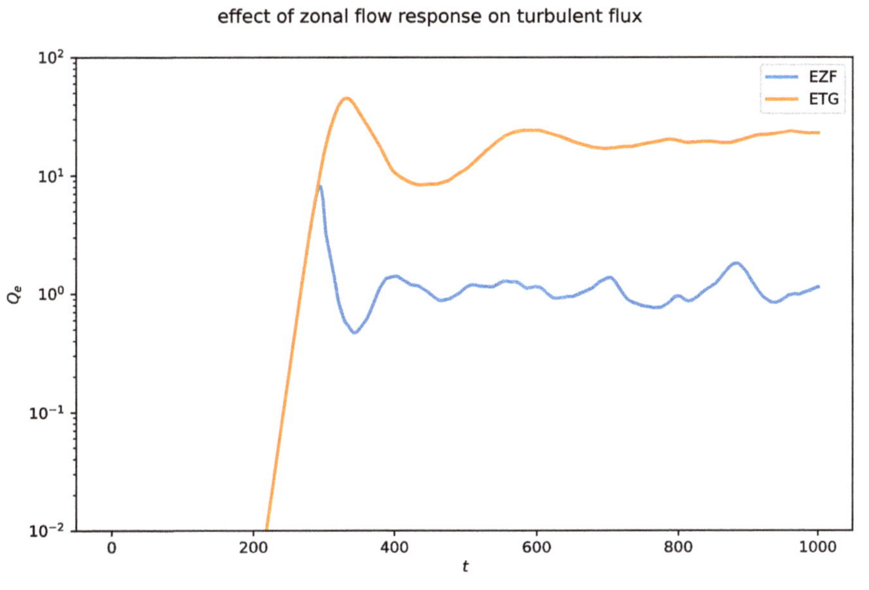

Figure 12.1. Time traces of the ExB electron heat flux Q_e for the ETG and EZF cases, differing only by the way the adiabatic response is applied, as per Eq. (12.23) or Eq. (12.27), respectively. The units are in terms of δ_e, and the time is in units of $L_\perp/V_e$. In the early growth phase, in which the responses are mostly linear, the two cases grow together at the same constant rate. Upon saturation, the difference emerges, and the EZF case saturates at a lower level. The ETG case continues growth and then saturates at a higher level, more than an order of magnitude stronger than the EZF case.

in volume average, rises together in both models identically, but then the EZF one saturates early at reasonable amplitude, while the ETG one continues to much higher amplitude. The type of polarisation is able by itself to produce a difference of well over an order of magnitude in the result. Note that this is Q_e in units of $n_0 T_{e0} V_e \delta_e^2$, so that in the more usual drift wave units of $n_0 T_{e0} c_s \delta_e^0$, with the factor of $(m_e/m_i)^{1/2} \approx 60$ between them, the flux of about 20 in the ETG case is not very large, but not very small either. This is comparable to the ITG transport in the same case. More on this later, below.

The transition to nonlinearity in both the spectra and morphology shows what happens. The k_y-spectra for ϕ and T and the spatial morphology for ϕ are shown at the three times $t = 150$, 330, and 500 for both cases. As seen in the previous figure, these time points are in the linear stage, the transitional stage, and into saturation. The isotropic temperature T is given by $T = (T_\parallel + 2T_\perp)/3$ following the kinetic energy pieces $z^2/2$ and w in normalised units (two degrees of freedom for $T_\perp$). This transition is shown first for the ETG case in figure 12.2. The long streamer structures in the x-direction, with aspect ratios on the order of ten, are what is exemplary of ETG turbulence. These form in the late linear stage and then persist in the saturated stage. The spectra do not change significantly into the turbulent state except to spread out moderately on both sides of the peak. Unlike the drift wave cases we have seen, the peak is in the same place in the early and late stages. The same times and same quantities are shown for the EZF case in

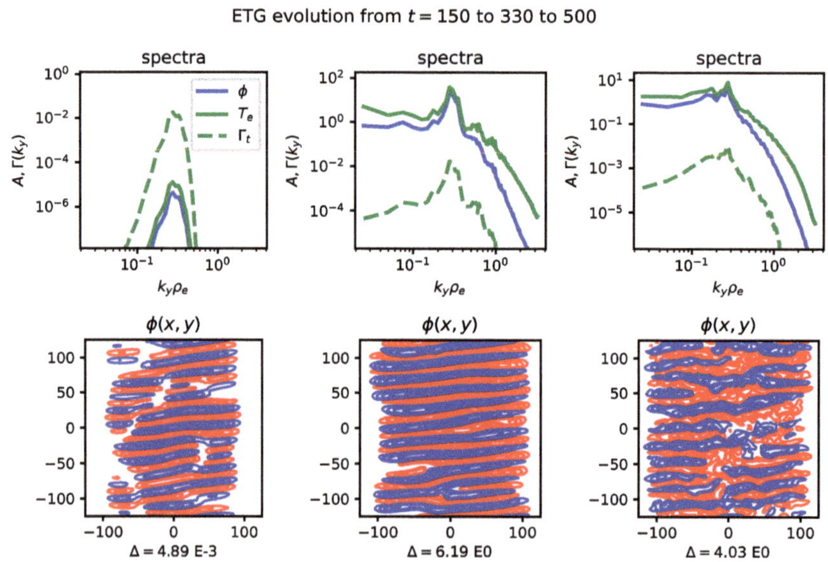

Figure 12.2. Transition to saturation and turbulence, for the ETG case. The k_y-spectra for ϕ and T, and the spatial morphology of ϕ, are shown together at the three time points $t = 150$, 330, and 500, in the linear stage, the transitional stage, and into saturation. The time and space are units of $L_\perp/V_e$ and ρ_e. The peak in the spectra does not move, and the broadening is moderate. The streamer structures (long in the x-direction with aspect ratio of order ten) form and persist.

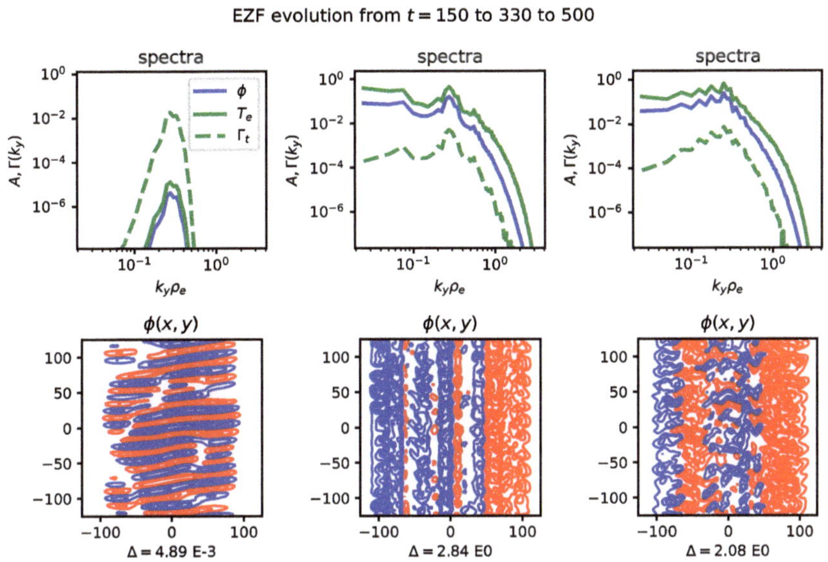

Figure 12.3. Transition to saturation and turbulence, for the EZF case. The k_y-spectra for ϕ and T, and the spatial morphology of ϕ, are shown together at the three time points $t = 150$, 330, and 500, in the linear stage, the transitional stage, and into saturation. The time and space are units of $L_\perp/V_e$ and ρ_e. The spectra are similar to the ETG case in the previous figure, but now the streamer structures are broken up by flows and eddies in the saturated stage. The amplitude is then lower due to more efficient nonlinear processing, as now there exists a component (the zonal one) for which $[\phi, n]$ does not vanish (and for the streamers and eddies, ϕ and n are still proportional).

figure 12.3. We see the same tendency to form streamer structures in the linear stage, but then these become pre-empted by very narrow y-direction flows which break the structures.

The late-time spatial morphology is shown in figure 12.4, at $t = 1000$ for both cases. The ETG case (left frame) shows pronouced streamers extending all the way across the domain into the dissipation buffers on both sides. One sees a roiling structure, as flow eddies at the scale of the streamer thickness in the y-direction form and decay, but the structures keep their identity and extension for comparatively long times. In the EZF case (right frame) these radial flows are broken up, such that the resulting structure is nearly isotropic. The scale in the y-direction is about the same as in the ETG case, but this case with the flow-like polarisation looks much more like the sort of turbulence we have seen in the hot-plasma ITG cases, both fluid and gyrofluid. At late times the turbulence appears mostly as eddies, with the flows in a secondary but still significant role, keeping the amplitude down. This is now seen to be due to the way the adiabatic response works for the non-zonal components but leaves the zonal one (with $\nabla_\parallel$ vanishing) untouched.

The late-time k_y-spectra for ϕ and T as well as the flux Q_e are shown in figure 12.5, at $t = 1000$ for both cases. The ETG case is on the left and the EZF one is on the right. Here we find very little difference. The underlying dynamics is still the same, with just the interaction with the zonal flows different. In the ETG case the streamer

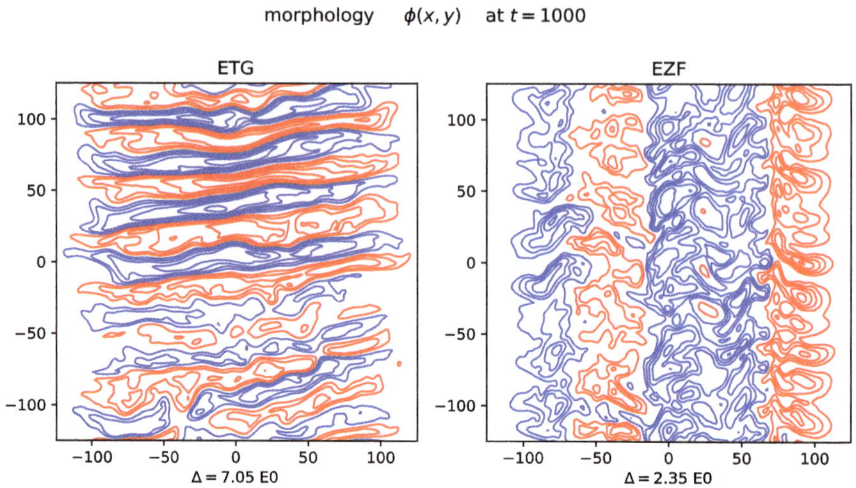

Figure 12.4. Late time ($t = 1000$) morphology of the flows and eddies (through ϕ) for the ETG (left frame) and EZF (right frame) cases. The time and space are units of $L_\perp/V_e$ and ρ_e. The streamer structures in the ETG case persist and cover the entire radial domain, while in the EZF case they are eliminated by the flow dynamics.

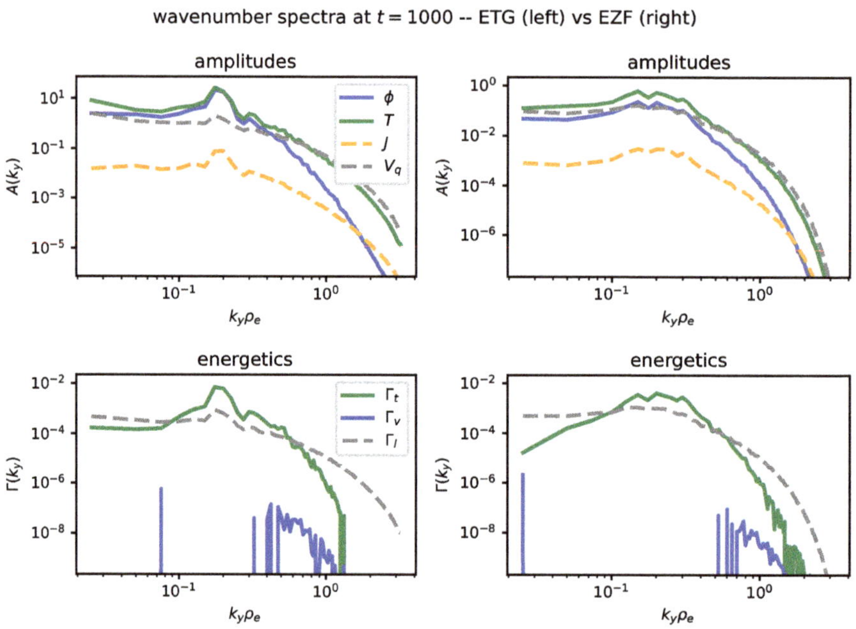

Figure 12.5. Late time ($t = 1000$) k_y-spectra for amplitudes and energetics, for the ETG and EZF cases. Time and space are units of $L_\perp/V_e$ and ρ_e. There is little difference in the shape of the spectra between the two cases, just the amplitude difference (here visible in the differing plot scales). The amplitudes for ϕ, T, and J, the drive rates for magnetic flutter (Γ_v) and ExB temperature gradient advection (Γ_t), and the sink rate for the Landau damping dissipation (Γ_l) are shown.

peak is more obvious in the spectra. Besides this, the difference is in the amplitude of the various spectra, whose relative form (i.e., the ratios between the quantities) remains intact. The drive is solely due to the temperature gradient. The ExB temperature gradient drive is the dominant mechanism, with the magnetic flutter drive small where not negative. The Landau dissipation is the only sink, and as in the case of dissipative drift wave turbulence, this sink due to parallel dynamics has a flatter spectrum than the drive, since the nonlinearities, mainly ExB advection, tends to spread the spectrum. All of this is like ITG turbulence for ion dynamics, but instead of sitting above ρ_s, it sits above ρ_e and for reasonable mass ratio is practically all below ρ_i. The effect of the skin depth is covered below.

12.5.4 Dependence on the temperature gradient

This difference in the flow dynamics in the two cases just studied makes itself felt in the dependence upon the temperature gradient. We now vary ω_t at the fixed ω_n to find the threshold in both models. The cases are carried to $t = 4000$ since growth rates near the threshold are small.

The time traces of the ExB electron heat flux are shown in figure 12.6 for the ETG cases, i.e., without the zonal flow effect in the polarisation. The cases with $\omega_t = 0.5$ is not visible because it decays out of the initial amplitude, which is below the plot limit. The case with $\omega_t = 0.6$ is barely unstable, and is what prompted the long run length. The other cases saturate above unity in terms of δ_e, with $\omega_t = 1.0$ and 1.2

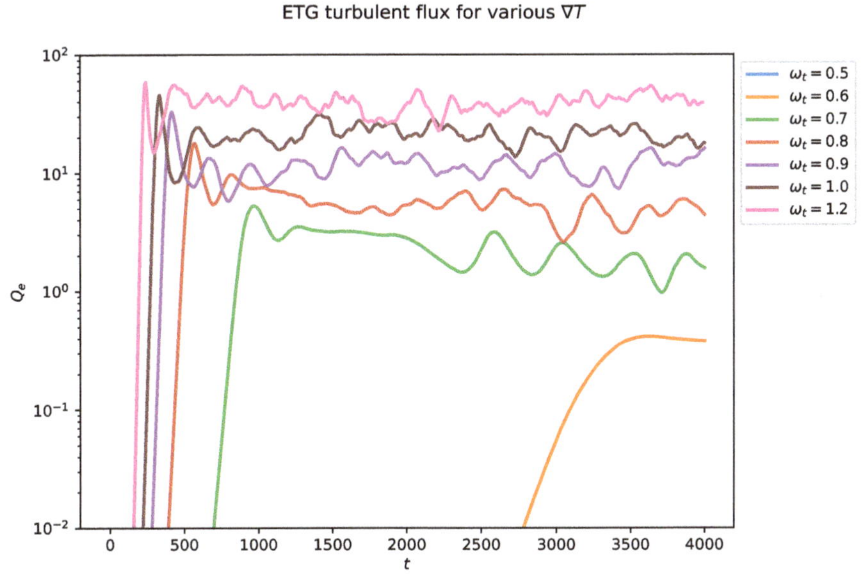

Figure 12.6. Time traces of the ExB electron heat flux Q_e for the ETG cases with various temperature gradient $\omega_t = L_\perp |\nabla \log T_{e0}|$. The units are in terms of δ_e, and the time is in units of $L_\perp / V_e$. All cases but the first two grow robustly in the linear phase (normalised flux below unity), and these ETG ones saturate with little statistical variation.

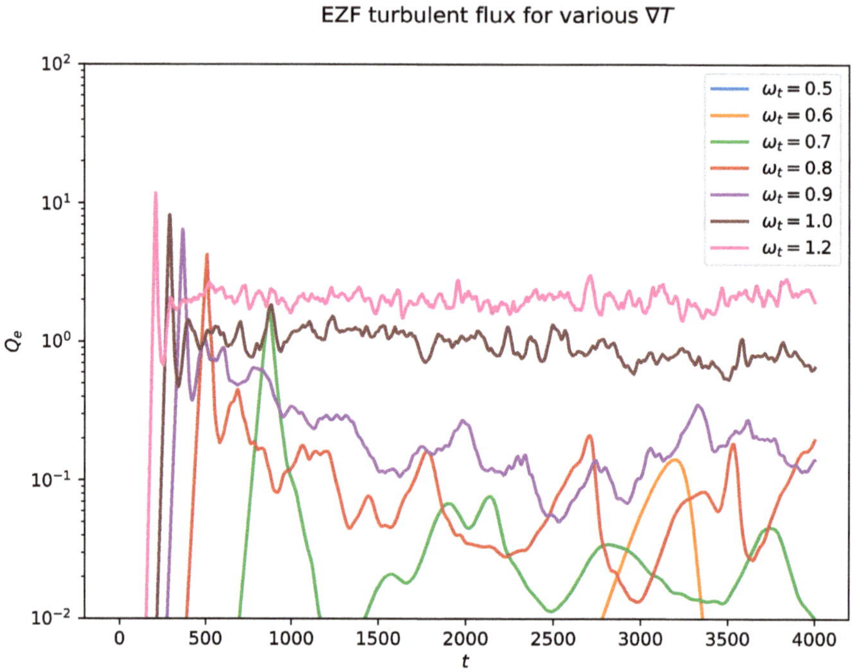

Figure 12.7. Time traces of the ExB electron heat flux Q_e for the EZF cases with various temperature gradient $\omega_t = L_\perp \mid \nabla \log T_{e0} \mid$. The units are in terms of δ_e, and the time is in units of $L_\perp / V_e$. The EZF cases near the threshold overshoot and then remain at low amplitude, essentially stabilised at small though finite amplitude. Only $\omega_t = 1.0$ and greater among the EZF cases reach robust nonlinear amplitude. This stabilisation is due to the zonal flows, which are the only difference between the ETG and EZF models, as in figure 12.1.

reaching high amplitude as shown for 1.0 above. Any case which grows out of the linear stage finds a saturated amplitude without decaying again.

The time traces of the ExB electron heat flux are shown in figure 12.7 for the EZF case with the zonal flow polarisation. The growth phase is as in the ETG cases as detailed above for the $\omega_t = 1.0$ case. The cases with $\omega_t > 0.8$ saturate early in comparison to the corresponding ETG cases, but those with $\omega_t = 0.6$, 0.7, and 0.8 stabilise again after growing. These overshoot into nonlinear amplitude, but now the flows that produce the early saturation defeat the turbulence, which then stabilises again. This is effectively a nonlinear upshift in the threshold, which we treated for more general gyrofluid cases in the previous Chapter. The main finding here is that this upshift is seen when zonal flows are allowed by the adiabatic response term in the polarisation, but not if the system has the ETG adiabatic polarisation.

For any of the unstable cases, the mode structure is the same as in the $\omega_t = 1.0$ cases shown above. The interest is in how the stabilisation by flows works. This comparison study is what finally confirms flows as the reason for the nonlinear stabilisation when it occurs. If the flows are left out, in this case by changing the polarisation to the one actually occurring in the ETG model, the nonlinear stabilisation does not occur.

12.5.5 Dependence on inductivity

The value chosen for $\hat{\beta} = 0.1$ is arbitrary, since the point has been to elucidate the variations in the model closer to the electrostatic cases usually chosen by others. Now we vary $\hat{\beta}$ as we did for the ITG cases within both the fluid and the gyrofluid models. We will find this makes much less of a difference as in the other cases, since most of the dynamics remains below the skin depth even when higher beta values reduces it relative to ρ_e. There is no MHD boundary since large scale dynamics is not being considered, so we can go considerably beyond $\hat{\beta} = 1$. This will allow investigation of the effect of the finite skin depth on the dynamics as it runs into the nominal ETG energetic spectral range.

We can be brief in this segment since there is found to be no significant dependence on $\hat{\beta}$ for these ETG cases. Values of $\hat{\beta} = 0.01$ and 0.1, 0.2, 0.3, 0.5, 0.7, and 1 are taken. The initial amplitude is $A_0 = 10^0$ to shorten the necessary run length. These are taken to $t = 1000$, with saturation occurring after $t = 200$ for all cases. The time traces of Q_e in volume average are shown as before in figure 12.8, for the seven cases. Within the temporal variability these are all roughly the same, while the nominal ratio of d_e/ρ_e is between 100 and 10, from lowest to highest $\hat{\beta}$, respectively. so this is not really skin depth, or 'c/ω_{pe}' turbulence as it was thought perhaps to be in the past.

The one interesting mode structure difference is the relative role of magnetic flutter. We saw in Chapter 3 that especially for cases in which ITG turbulence was weak, the electron temperature might take over and produce microtearing effects

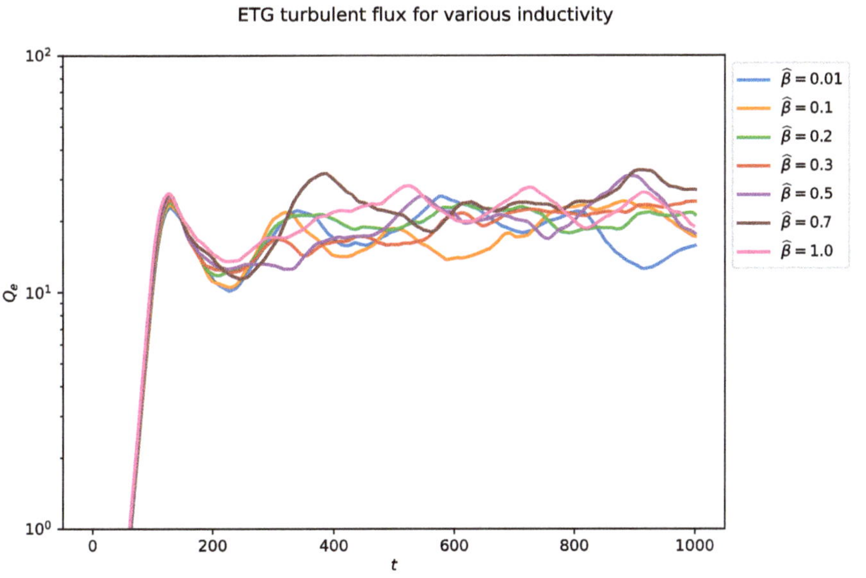

Figure 12.8. Time traces of the ExB electron heat flux Q_e for the ETG cases for the various $\hat{\beta}$ values given in the text and indicated here. The units are in terms of δ_e, and the time is in units of $L_\perp/V_e$. The growth, saturation, and turbulent phases are all roughly convergent, the latter within the temporal variability. Over this range the ratio of skin depth to electron gyroradius varies by 10, down to about 10 for the highest $\hat{\beta}$ value.

which, while not really turbulence in the sense of these studies, still are nonlinear and produce finite if small electron heat flux. The energetic pathway of $q_\parallel$ reacting to the $\nabla_\parallel \log T_{e0}$ resulting from magnetic flutter, to fluctuations in T_e and to currents which then maintain the necessary b^x for the gradient term, is present intact in the ETG model. The spectra of $\tilde{T}_e$ and $\tilde{q}_{e\parallel}$ are much the same for all the cases, magnetic flutter heat flux, given by

$$M_e = (q_\parallel + q_\perp)[-A, x] \tag{12.48}$$

in volume average analogous to Q_e, show the inductivity effect. The spectra for M_e along with Q_e and the Landau damping dissipation rate Γ_L are given in figure 12.9 for the $\hat{\beta} = 0.01$, 0.1 and 1 cases. This magnetic contribution becomes positive and stronger as inductivity increases. It remains a subdominant effect for reasonable $\hat{\beta}$, however. Recall that ITG turbulence changes over into mesoscale MHD at this level of inductivity, with M_e becoming the strongest transport effect. But this takes place in the drift-Alfvén part of the spectrum, which is not covered in an ETG model. It is addressed below.

12.5.6 Dependence on radial domain size

The dependence of the result on the domain size and aspect ratio is briefly checked. From the nominal domain size of $L_x = L_y = 80\pi$ on the 1 × 1 aspect ratio, the y-direction and x-direction are extended by 2 or 4 in separate cases, so that the 1 × 1

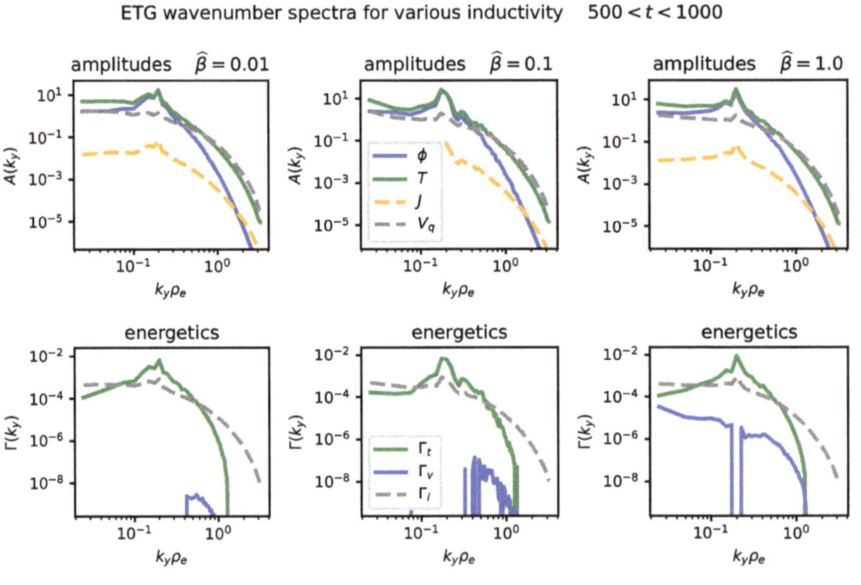

Figure 12.9. Time-averaged ($500 < t < 1000$) k_y-spectra for amplitudes and energetics for the ETG cases with low, medium, and high inductivity ($\hat{\beta}$). Time and space are units of $L_\perp / V_e$ and ρ_e. The amplitudes of ϕ, T, and V_q, and the ExB gradient drive (Γ_t) and Landau damping dissipation (Γ_l), are unaffected, but the magnetic flutter drive (Γ_v) becomes positive for the highest value. At $\hat{\beta} = 1$ the MHD threshold is already passed at these parameters, however.

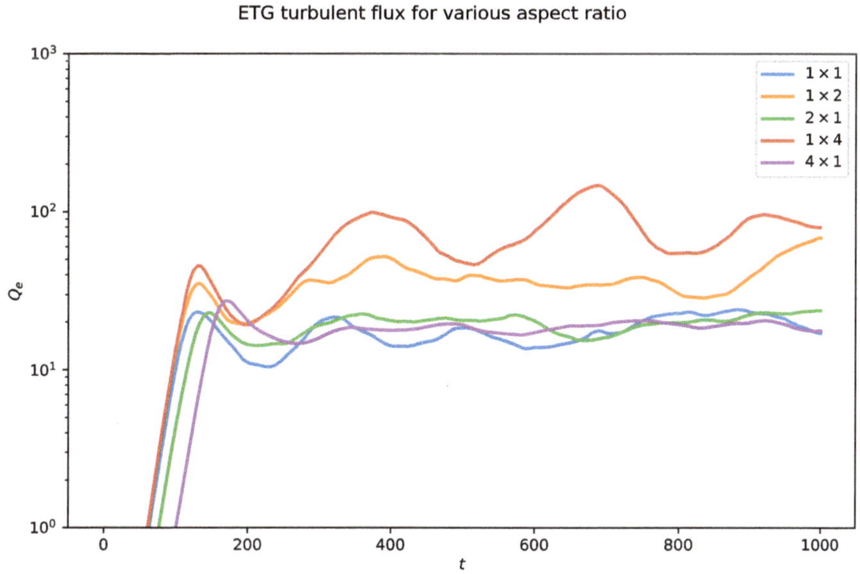

Figure 12.10. Time traces of the ExB electron heat flux Q_e for the ETG cases for the various domain sizes $1 \times [124]$ and $[124] \times 1$ in multiples of $2\pi/K = 80\pi\rho_e$, the nominal size. The units are in terms of δ_e, and the time is in units of $L_\perp/V_e$. The growth, saturation, and turbulent phases are all roughly convergent, the latter within the temporal variability, for extension in y with the same L_x. The turbulence saturates at a flux level roughly proportional to L_x when the extension is in that direction. The streamers persist all the way across the domain, and the flux level is stronger for longer throw length.

case is compared with those with 1×2 or 1×4 for extension in x and 2×1 or 4×1 for extension in y.

The time traces of Q_e in volume average are shown as before in figure 12.10, for the five cases. No significant dependence on L_y is found, as the three curves for $[124] \times 1$ are convergent within the temporal variability. However, the strength of the flux depends roughly linearly on L_x for the $1 \times [124]$ cases. The Q_e values averaged over $500 < t < 1000$ with 2500 samples are 19.0 ± 3.4 and 38.9 ± 9.6 and 83.0 ± 27.7 for those cases. Examination of the spectra and parallel envelopes shows no change in the mode structure. The only difference in any of these cases is the length of the streamers. Up to the longest dimension of $320\pi\ \rho_e$ they are found to extend all the way from the dissipation layer on the left to the one on the right. The late-time spatial morphology is shown in figure 12.11, at $t = 999$ (the timestep limit is briefly hit during this run) for this case. One finds eddies within streamers as before, only with the longer extension of the streamers.

This shows that if the radial flows are not forced to turn around then the turbulence amplitude can rise indefinitely. In contrast to drift wave or ITG cases, the radial forcing does not produce ever-wider flows, and the flows themselves do not break up due to the local Kelvin–Helmholtz instability. These features are due to the factor of unity in $1 - \nabla_\perp^2$ rather than $-\nabla_\perp^2$ in the polarisation, the counterpart to the vorticity in a hydrodynamic fluid model. However, with such long extension, one naturally has to re-consider the model, as streamers extending above the ion

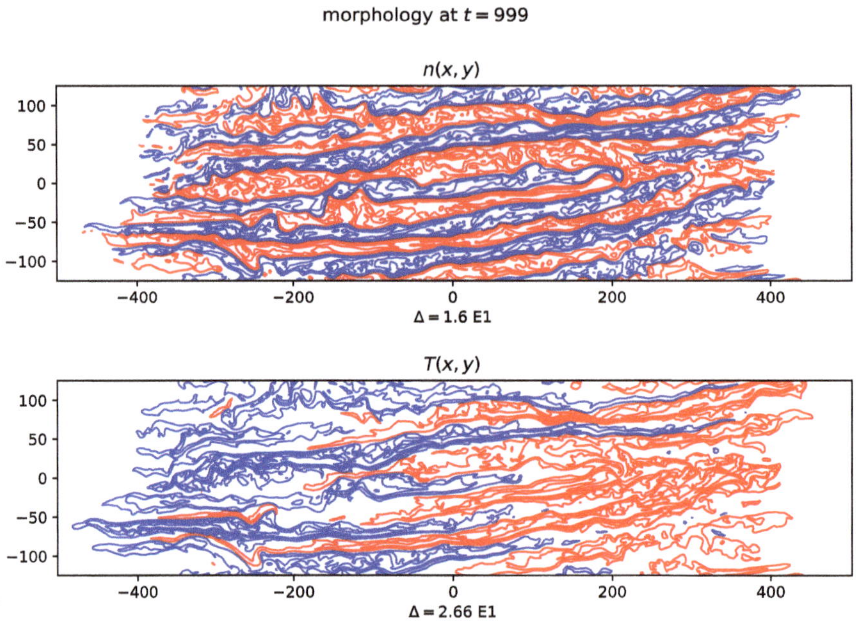

Figure 12.11. Late time (t = 999) morphology of the flows and eddies (through ϕ) for the ETG case with the x-direction domain size extended by a factor of four, the 1 × 4 case of the previous figure. Time and space are units of $L_\perp / V_e$ and ρ_e. The streamer structures in ETG turbulence are seen to extend indefinitely, as long as the domain size allows. This accounts for the larger turbulence level in the longer cases.

gyroradius scale are susceptible to the turbulence there. This can only be considered within a so-called two-scale model, which is covered in the next Section.

12.5.7 Dependence on field line length

A significant dimensionless parameter we have not tested yet is the effective field line length. As with any turbulence of this type, the result is due to a competition between simple Eulerian ExB fluid advection in the perpendicular plane, including the gradient drive, and constraining by parallel coupling in the third dimension. If the magnetic field lines are effectively longer, the turbulence will adjust. Effective field line length follows from the scale ratio $L_\parallel / L_\perp$ between the parallel and perpendicular scales, with the field line length given by the field line connection property and the perpendicular scale length giving the strength of the gradient drive. In the context of ETG turbulence, the competition between $[\phi, \log T]$ and $V_e \nabla_\parallel$ is given by the parameter μ, so that the effective field line length is $(K_\parallel L_\perp)^{-1}$ which is equal to $\mu^{-1/2}$ in these units. So we investigate cases of varying μ around the base ETG case listed above. The inductivity is set to $\hat{\beta} = 1$.

The time traces of the ExB electron heat flux are shown in figure 12.12 for these ETG cases with varying $10^2 < \mu < 3 \times 10^4$, so that the effective field line length varies between 10 and about 170. As in the cases with varying inductivity ($\hat{\beta}$), in figure 12.8, the temporal evolution and saturation of the turbulence and the flux it produces is similar for all the cases.

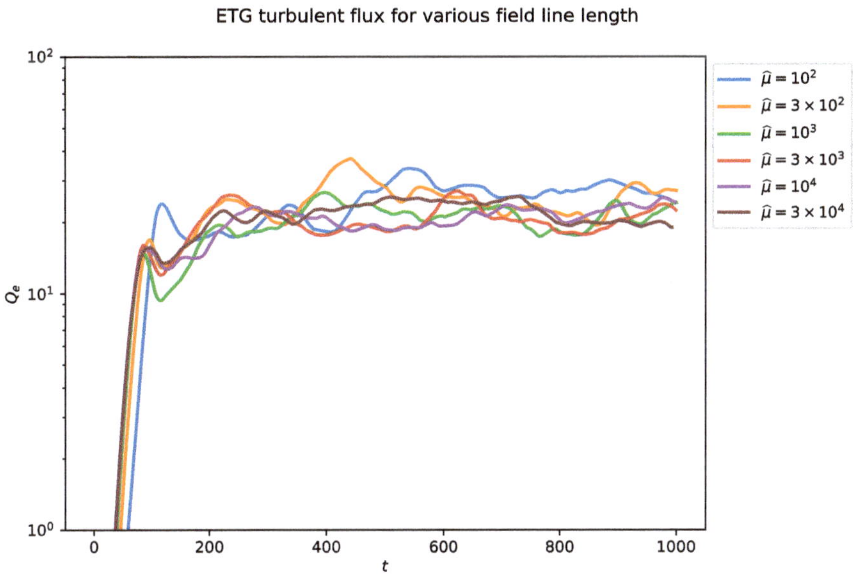

Figure 12.12. Time traces of the ExB electron heat flux Q_e for the ETG cases for the various μ values given in the text and indicated here. The units are in terms of δ_e, and the time is in units of $L_\perp/V_e$. The growth, saturation, and turbulent phases are all roughly convergent, the latter within the temporal variability. Over this range the effective field line length $\mu^{1/2} = (K_\| L_\perp)^{-1}$ ranges from 10 to 170. Mode structure differences are visible however, in the next two figures.

In the inductivity variation the only mode structure effect was the increasing relative strength of the magnetic flutter effect, as the magnetic disturbances become larger for a given parallel current determined by the electron fluid dynamics. Here, we find a scale effect as the field lines become effectively longer. The spectra for M_e, Q_e, and Γ_l are given in figure 12.13 for the three $\mu = 3 \times 10^2$, 3×10^3, and 3×10^4 cases. The structure is similar but the peak in the energetics moves toward longer wavelength for longer field lines. Since $k_y \sim \mu^{-1/4}$ in the variation, the turbulence is seen to preserve the same ratio of wavelength peak to parallel scale.

The increasing scale of motion with effectively longer field lines is seen in the spatial morphology of the streamer flows. The late-time spatial morphology of the three $\mu = 3 \times 10^2$, 3×10^3, and 3×10^4 cases whose spectra were just discussed, is shown in figure 12.14, at $t = 1000$. One finds the same streamer structure, but with widening radial flows as μ increases. The spectra peak is associated with the width of the streamer flows in the y-direction. Together, this is the essential result. The same amplitude is produced by the turbulence for varying μ, with the scale of the streamer flows proportional to the field line length.

12.5.8 The warm plasma case, steeper gradients

Before moving on to the two-scale problem we consider the warm plasma cases studied in the previous Chapter (Section 11.5), within the ETG model. Since the turbulence does not depend on inductivity ($\hat{\beta}$) and the collision frequency is small compared with the ETG time scale ($V_e/L_\perp$ rather than $c_s/L_\perp$), the only salient

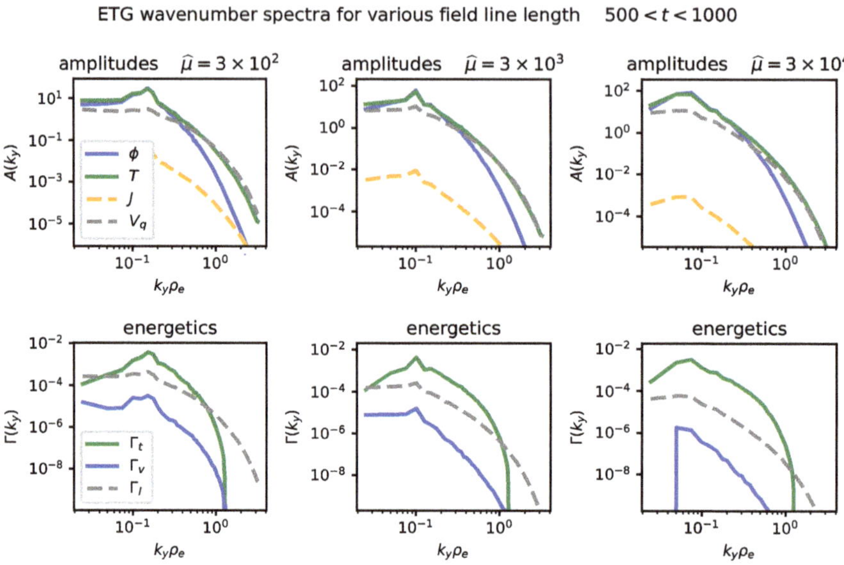

Figure 12.13. Time-averaged ($500 < t < 1000$) k_y-spectra for the energetics for the ETG cases with short, medium, and long effective field line length $\mu^{1/2} = (K_\| L_\perp)^{-1}$, ranging from 10 to 170. Time and space are units of $L_\perp / V_e$ and ρ_e. Although the turbulent flux is about the same for the three cases, the scale of motion shifts longward, as $k_y \sim \mu^{-1/4}$ in rough terms.

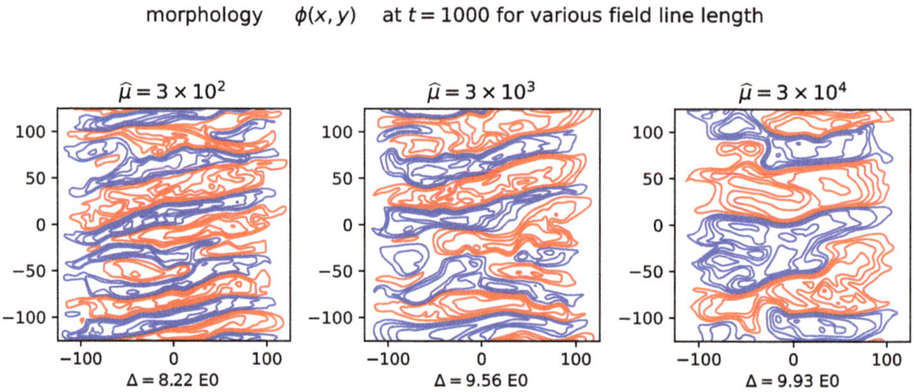

Figure 12.14. Late time ($t = 1000$) morphology of the flows and eddies (through ϕ) for the ETG cases with short, medium, and long effective field line length $\mu^{1/2} = (K_\| L_\perp)^{-1}$, ranging from 10 to 170. Time and space are units of $L_\perp / V_e$ and ρ_e. The shift of the scale of motion longward with field line length indicated by the spectra in the previous figure is visible here in the thickness of the streamers in the y-direction.

difference is the scale ratio. In this case it is both the effective field line length, $K_\| L_\perp$, and the curvature forcing, ω_B. This is the effect of the steepness of the gradient rather than the field line length.

In the warm plasma case we are considering a value of $L_\perp / R_0$ which is smaller by a factor of about seven, so that ω_B drops from 0.29 to 0.0424. At the same time, the pitch parameter q rises so that $K_\| L_\perp$ drops by about ten. The issue of the difference between field line length and gradient steepness, namely ω_B, is why we did the control case represented by the scan of $K_{\|}L_{\perp}$ by itself in the previous set of cases. An interesting finding in ETG turbulence is how the role of the streamers seems to change when the parameters reflect this new situation.

The parameters of the nominal case are given by

$$\begin{aligned} \hat{\beta} = 1.75, \quad \mu = 27\,200, \quad \omega_B = 0.0424, \\ \omega_n = 0.5, \quad \omega_t = 1, \quad S = 1.14, \quad \varepsilon_a = 0.3 \end{aligned} \tag{12.49}$$

The domain size and grid, and the timestep and subgrid and boundary dissipation, are the same as done above for the hot plasma case. The initial amplitude is set to $A_0 = 10^0$ to get into saturation earlier, and the runs are carried to $t = 1000$.

The temperature gradient scale is held fixed while the density gradient scale is varied between $\omega_n = 0.25$, for a scale ratio $\eta_e = L_{ne}/L_{Te} = 4$, and $\omega_n = 0.7$, taking the scale ratio down below two. The time traces of the electron ExB turbulent heat flux, Q_e, in volume average are shown in figure 12.15, with the same format as in the above studies. Only the first two values of ω_n produce a flux amplitude larger than unity in these units. Recall that it is a factor of about 60 lower in units of $c_s/L_{\perp}$, so that the flux level seen for $\omega_n = 0.5$ is very small compared with what the drift Alfvén turbulence models produce for the same case. For the hot plasma case we were dealing with Q_e about 20 by contrast. For η_e about three or less, the turbulence

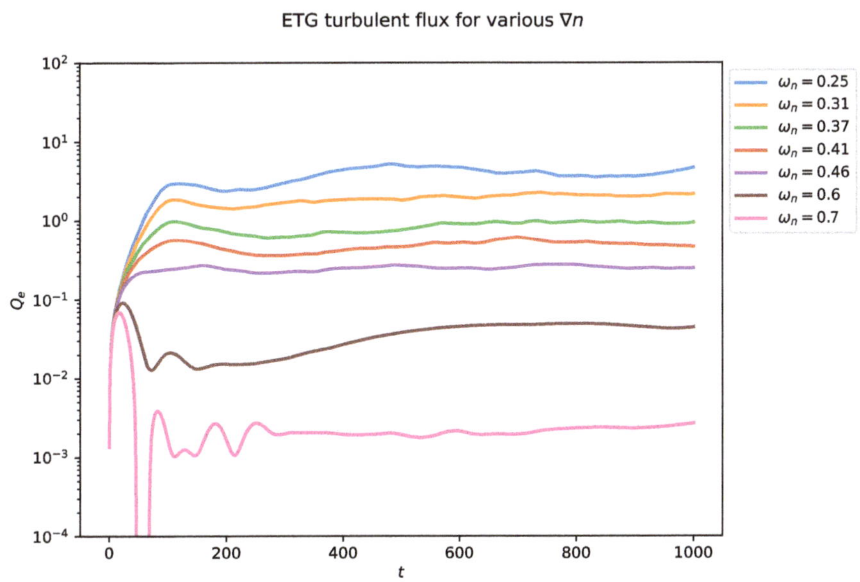

Figure 12.15. Time traces of the ExB electron heat flux Q_e for the warm plasma or steep gradient cases ($K_{\|}L_{\perp} < 10^{-2}$) for the various density gradient forcing parameter ω_n values as indicated. The units are in terms of δ_e, and the time is in units of $L_{\perp}/V_e$. The level is much lower in these units ($V_e\rho_e^2/L_{\perp}^2$) than in the hot plasma cases. Note that unity here is about 0.016 in drift wave turbulence units ($c_s\rho_s^2/L_{\perp}^2$). Only the first two cases are at significant level. Interestingly, the low amplitude cases persist and are not stabilised, i.e., decaying.

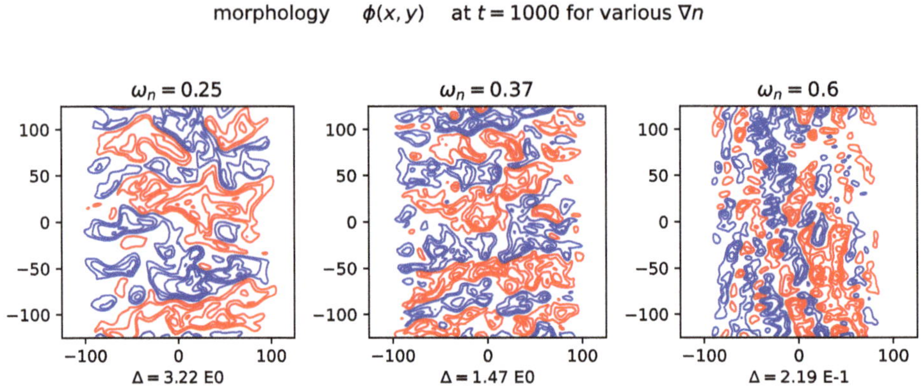

Figure 12.16. Late time ($t = 1000$) morphology of the flows and eddies (through ϕ) for ETG turbulence in the steep gradient cases with various ω_n. Time and space are units of $L_\perp/V_e$ and ρ_e. Streamer structure of the type seen previously is much less clear, and visible only in the first case. The last case, the nominal one, shows nearly isotropic eddies.

amplitude in the pure ETG model is not very significant. Interestingly, the lower cases $\omega_n = 0.6$ and 0.7 are not completely stable, but merely saturate at a very low finite amplitude.

The spatial morphology for three of the cases is shown in figure 12.16, at $t = 1000$ in late saturation. Here we do not see such pronouced streamers as in the hot plasma case with the parallel- to-perpendicular scale ratio $(K_\parallel L_\perp)^{-1}$ only about ten. In these steep gradient cases it is larger than one hundred. There is a hint of streamer flow in the strongest case with $\omega_n = 0.25$ hence $\eta_e = 4$, whose flux level is 4.17 ± 0.46 in these units (V_e and ρ_e) hence less than 10^{-1} in the drift wave normalisation (c_s and ρ_s). But for $\omega_n = 0.37$ one has no obvious streamer structure and for $\omega_n = 0.5$ the eddies are small and close to isotropic.

The parallel mode structure also changes for lower η_e values. The parallel envelope of Q_e is shown in figure 12.17 for three values of ω_n as given. These are averaged over $500 < t < 1000$ with 50 samples. The ballooning structure is present for $\omega_n = 0.25$ but for values above 0.4 the envelope acquires a double-peak structure. Recall for the ballooning model the curvature forcing depends on s as $\cos s + Ss \sin s$, so that for large enough S the dependence becomes concave rather than convex. This structure appears to account for the persistence of ETG turbulence at finite level, even though small, for values of η_e at or below two.

12.6 The two-scale problem

All of these ETG results beg the question of whether such high amplitude turbulence at small scale affects an overall result. After all, for the ETG dynamics above ρ_e instead of above ρ_s, the ETG polarisation without the zonal flow effect is the correct one, not the EZF version which was used as a control case. In the ITG hot plasma turbulence regime, both adiabatic relations, the electron one and the ion one, should emerge naturally in the electromagnetic gyrofluid equations for both species. Up to now we have avoided this by restricting the spectrum to scales near but not too much

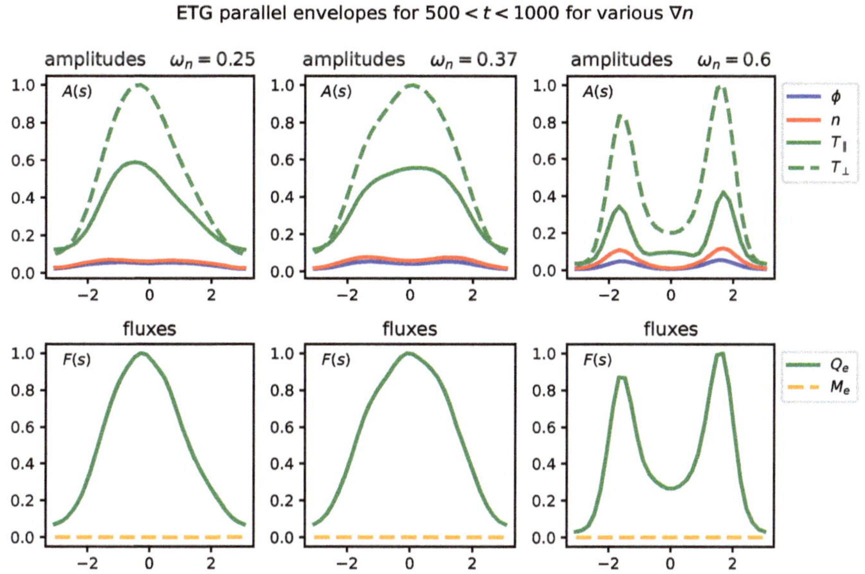

Figure 12.17. Parallel envelopes of amplitudes and fluxes for the steep gradient ETG cases with various ω_n, averaged over $500 < t < 1000$. The time is in units of $L_\perp / V_e$. The perpendicular temperature is largest for all cases. The ballooning structure changes into a double peak structure as $\eta_e = \omega_t/\omega_n$ decreases. This coincides with the weaker turbulence level and more isotropic small scale structure without radially extended streamers at these values of ω_n.

smaller than ρ_s. What happens when we include as much of the smaller range as we can is of definite interest. Will the ETG turbulence remain, influence, and possibly dominate the result, or will the drift Alfvén or ITG turbulence in the scale range between $L_\perp$ and ρ_s overwhelm it?

Conversely, we can ask another question, if by setting the domain size to only a few ρ_s will the drift Alfvén or ITG side be masked with the ETG side remaining dominant. When I first ran ETG turbulence in the electromagnetic gyrofluid model it was like this. The physics and mode structure of the ETG turbulence was shown by isolating its own spectral range, simply by setting the minimum $k_\perp\rho_s$ in a 1×1-square domain to 1.5, so that $L_x = L_y$ is about $4\rho_s$. Since in the ETG dynamics one does not see zonal flows the square domain is appropriate.

To address this we use the same equation set as above for both species, taking care of the normalisation and hence the effective mass and charge parameters. In the conventional drift wave normalisation to c_s and ρ_s, we have $\delta_0 = \rho_s/L_\perp$ as the scale factor as previously. Note that with this normalisation the quoted amplitude level is in terms of δ_0, not unity or δ_e. The domain size is not tied to $L_\perp$, so that realistic parameters are always used. Only the domain size is artificial in restricting the scales of motion which are to be allowed. The relevant species parameters are

$$a_z = \frac{n_{z0}Ze}{n_{e0}e} \qquad \tau_z = \frac{T_{z0}/Ze}{T_{e0}/e} \qquad \mu_z = \frac{m_z/Ze}{m_i/e} \tag{12.50}$$

for species z, with mass m_z, background density and temperature n_{z0} and T_{z0}, and charge state Z. These give the background charge density and temperature/mass-to-charge ratios, relative to the ion mass and electron temperature and fundamental charge. The background parameters are given with the subscript zero since we will use n_e, $T_{e\|}$, and $T_{i\perp}$, for example, as three of the dependent variables. The first parameter, a_z, gives the partial charge of the background. Normalised away in the moment variable equations, it appears in the field equations, which is where the species are coupled together. The temperature/charge ratio, τ_z, gives the sign and speed of the dynamics, and the mass/charge ratio, μ_z, gives that for the inertia. Note that $\tau_z = -1$ for the electrons and $\mu_z = 1$ for the singly charged ions, for which also $\tau_z = \tau_i$. The species gyroradius, and inertia and Landau damping parameters are given by

$$\rho_z^2 = \mu_z \tau_z, \qquad \mu = \mu_z \left(K_\| L_\perp\right)^{-2}, \qquad V_L = \sqrt{\left(\tau_z/\mu_z\right)} \tag{12.51}$$

relative to c_s and ρ_s, respectively. For these purposes we are still neglecting collisions. We keep the electron FLR gyroaveraging but use only the lowest term in $k_\perp \rho_e$ in the electron polarisation, largely for numerical reasons (we don't have the physics which governs things at λ_D, the Debye length).

We have dependent variables

$$n = a_{00} \qquad T_\| = \sqrt{2}\, a_{20} \qquad T_\perp = a_{01} \tag{12.52}$$

$$u = a_{10} \qquad q_\| = \sqrt{\frac{3}{2}}\, a_{30} \qquad q_\perp = a_{11} \tag{12.53}$$

now for both species, not writing the species subscripts, except for the parameters, until they are needed. Eqs (12.30)–(12.35) become

$$\begin{aligned} &\frac{\partial n}{\partial t} + [\phi_0, n + \log n_0] + [\phi_1, T_\perp + \log T_0] + \nabla_\| u \\ &\quad = \frac{1}{2}\mathcal{K}\left(2\phi_0 + \phi_1 + \tau_z(2n + T_\| + T_\perp)\right) \end{aligned} \tag{12.54}$$

$$\begin{aligned} &\frac{\partial}{\partial t}(A + \mu u) + \mu[\phi_0, u] + \mu[\phi_1, q_\perp] + \nabla_\|\left(\phi_0 + \tau_z(n + T_\|)\right) \\ &\quad - D(\tau_z(T_\| - T_\perp) - \phi_1) = \frac{1}{2}\mu\tau_z\mathcal{K}\left(4u + 2q_\| + q_\perp\right) \end{aligned} \tag{12.55}$$

$$\begin{aligned} &\frac{1}{2}\frac{\partial T_\|}{\partial t} + \frac{1}{2}[\phi_0, T_\| + \log T_0] + \nabla_\| q_\| \\ &\quad + D(q_\perp + u) = \frac{1}{2}\mathcal{K}\left(\phi_0 + \tau_z(n + 3T_\|)\right) \end{aligned} \tag{12.56}$$

$$\begin{aligned} &\frac{\partial T_\perp}{\partial t} + [\phi_0, T_\perp + \log T_0] + [\phi_1, n + \log n_0 + 2(T_\perp + \log T_0)] + \nabla_\| q_\perp \\ &\quad - D(q_\perp + u) = \frac{1}{2}\mathcal{K}\left(\phi_0 + 4\phi_1 + \tau_z(n + 4T_\perp)\right) \end{aligned} \tag{12.57}$$

$$\mu \frac{\partial q_{\|}}{\partial t} + \mu[\phi_0, q_{\|}] + \frac{3}{2}\tau_z \nabla_{\|} T_{\|} = \frac{1}{2}\mu\tau_z \mathcal{K}\left(3u + 8q_{\|}\right) - \sqrt{3}\, V_L \mu q_{\|} \tag{12.58}$$

$$\mu \frac{\partial q_{\perp}}{\partial t} + \mu[\phi_0, q_{\perp}] + \mu[\phi_1, u + 2q_{\perp}] + \nabla_{\|}(\phi_1 + \tau_z T_{\perp}) - D(\tau_z(T_{\|} - T_{\perp}) - \phi_1) = \frac{1}{2}\mu\tau_z \mathcal{K}\left(u + 6q_{\perp}\right) - V_L \mu q_{\perp} \tag{12.59}$$

The species are coupled into the field equations, given by

$$P(\phi) = \sum_{sp} a_z\left(G_0(n) + G_1(T_{\perp})\right) \qquad -\nabla_{\perp}^2 A = \hat{\beta} J \tag{12.60}$$

for polarisation and induction, where the polarisation operator and parallel current are

$$P = \tau_i^{-1}\left[1 - \left(1 - \rho_i^2 \nabla_{\perp}^2\right)^{-1}\right] - \rho_e^2 \nabla_{\perp}^2 \qquad J = \sum_{sp} a_z\, u \tag{12.61}$$

respectively. The gyroaveraging operators and gyroaveraged potentials are

$$G_0 = \left(1 - \rho_z^2 \nabla_{\perp}^2\right)^{-1} \qquad G_1 = [G_0 - 1]G_0 \qquad \phi_{0,\,1} = G_{0,\,1}(\phi) \tag{12.62}$$

and it is important to note that these are given separately for each species. We still neglect gyroaveraging upon A since its effect in the ETG scale range is negligible and the gyroaveraging effect scales effectively as $k_{\perp}^2 \rho_e^2$ for the Ohm's law and parallel current dynamics. Ampère's law is solved for A according to the procedure in Chapter 11.

12.6.1 Computational requirements for the two-scale problem

I do not have the resources available to run definitive computations on this problem. It is necessary to take a step back and discuss what one has to do in order to run these cases, down to ρ_e rather than ρ_s, at the same standard as we have been doing for the electromagnetic ITG cases and the general drift Alfvén cases. The mass ratio decides the ratio of the scales found in the ETG case to those found in the ITG cases. If ITG turbulence peaks at $k_y \rho_s \sim 0.3$ and has to resolve ρ_s with $h_x \sim h_y \sim \rho_s$, then ETG turbulence peaks at $k_y \rho_e \sim 0.3$ and has to resolve ρ_e with $h_x \sim h_y \sim \rho_e$. This means that the two-scale problem requires a factor of 60 times more intense in the resolution of the perpendicular plane. This is the start. Now, the timestep is shorter for ETG turbulence in the same measure. So to run the problem properly, one needs the same domain size for the ITG cases and the same timestep and resolution for the ETG cases. Neglecting the complications of the parallel spectrum and finite inductivity, discussed shortly, that translates to a factor of 60^3 more computational effort. If we can relax the timestep requirement slightly, and keep to using powers of two in the grid, this becomes about 50 times 4096, or about two hundred thousand, just in the timestep and the number of grid points, assuming perfect computational scaling.

Now the constraints of the parallel grid. We have just run these ETG cases with 32 grid points over one connection length, as in the ITG cases. However, for electromagnetic turbulence, one must ensure that the magnetic field line disturbances are covered by the resolution in the parallel direction, so that it is always possible to deliver a possible $k_\| = 0$ degree of freedom, in the nonlinear sense, on the disturbed field lines. This means that $h_s B^s < h_x B^x$, or in the ratio, $h_s < h_x b^x$. With differing normalisations we have to be careful about the units, as there is a factor of δ_0 and also $K_\| L_\perp$ to worry about. In practice, b^x is of order up to 10^{-3} so that for smaller $K_\| L_\perp$ more grid points are potentially needed.

This is what we faced for parallel wavenumber convergence in the first place (Chapters 11 and 12 of Volume One), varying N_s from 32 to 512 since values quite larger than 32 would be potentially needed. For low inductivity we were fortunate to see that the field aligned cases with both lower and higher N_s were convergent with the same non-aligned cases. So in principle we can remain with $N_s = 32$ but be aware that if magnetic fluctuations ever set the timestep value then those tests have to be done again. On the two-scale problem, no one can afford that, even with fluid cases let alone kinetic ones.

Completely neglecting the parallel resolution issue, a two-scale computation goes up by 2×10^5 in expense. Increasing the parallel resolution in proportion to $K_\| V_e$ versus $K_\| c_s$ is another factor of 60, or 64 if we are tied to powers of two. That makes the expense factor 10^7, which is why no one has ever done this without relaxing the standard to a point where it is not clear that the results will hold when the problem is done properly. This is also the main problem with using MHD models to do drift wave turblence—one has to resolve ρ_s and d_e. With thirty years of hindsight it is known how to do that, and the results and the underlying theoretical support is the main subject of this work. Nevertheless, awareness of this has not percolated all the way through the scientific community, not even after thirty years (forty, if the formative work in the late 1970s and early 1980s is taken into account).

Most work addresses this by using a less disparate mass ratio. In our view, the mass ratio is the whole point and should not be changed. Note that this is why deuterium has been used throughout, as it is the standard main ion plasma in fusion experiments ever since favourable results were obtained in the 1980s. It is the view here that if the mass ratio (or skin depth, or ρ_s, etc) is changed then one is working on a different problem. So if we cannot do the entire problem properly, we will do the partial result by varying the domain size around a reasonable grid point count, and investigating differing degrees of spectral overlap between the 'ion scale' and 'electron scale' here represented by the gyroradii. It is most important to never forget that we are not doing linear theory, or a set of dueling linear modes. We have a generally electromagnetic turbulence which involves the scale around $L_\perp$ as well as ρ_s, and the turbulence is not properly captured unless both ρ_s and $L_\perp$ are in the system. The best we can do is to take a 1×1 computation for cases in which large-scale radial flows do not develop, fully congnisant of the contingency of the result. Then we run several cases on a large domain, perhaps double the perpendicular ETG grid done above, and with the same resolution in the parallel direction. But if not enough of the $\rho_e < k_\perp^{-1}\rho_s$ side is taken the ETG part is damaged, and if not enough of the $\rho_s < k_\perp^{-1}L_\perp$ side is taken the drift Alfvén part is damaged. The best we can do is investigate the initial onset of vertical motion in the drift Alfvén side and see what effect that has on the development of the ETG streamers.

For strongly electromagnetic warm plasma cases we have the advantage that the streamer formation has been seen to be weak to begin with, and for $K_\| L_\perp \sim 10^{-2}$ their scale is already much larger than ρ_e anyway. So we will do the best we can on both hot plasma and warm plasma cases.

So we are resigned to compromise due to computational limitations, but what we do not do is to relax or compromise the value of any dimensionless parameter. We can neglect scales, but we do not neglect physics.

12.6.2 The hot plasma case

Obviously a widespread parameter scan is beyond our means. So we will investigate the scale competition within two standard cases. Here, we do the hot plasma case with $\hat{\beta} = 0.1$ (Eq. 12.45), which does deliver robust ITG and ETG turbulence separately in their individual cases, and explore variation of the domain size.

When I first showed this six-moment electromagnetic gyrofluid model can recover ETG turbulence natively, it was done by adding the electron FLR corrections to the model, as done here, and running within an ETG domain in which the scale is so small the ITG side is never excited. Then, the length of the run only has to satisfy the ETG standards, namely one or a few $10^3 L_\perp / V_e$, or about 20 to $40 L_\perp / c_s$. As we have seen in the ITG results in Chapters 3 and 11, this is not long enough for the ITG side to develop. The result was very much like setting all the ion gyrocentre variables to zero. That was not done in the equations, just by restricting the domain size to a minimum $k_\perp \rho_s$ of $K = 1.5$, hence $L = L_x = L_y$ about $4\rho_s$, so that the ion gyroaveraging operators do it naturally. Then, due to the resulting near-adiabatic ions and therefore no density cascade, the resolution of 128 × 128 was found to be sufficient. Here we will run 256 × 256 × 32 and vary K. The native ETG case with $K = 1.5$ defines the 'small' case, which we run first to recover the ETG result as a control case.

The ETG growth and dynamics looks impressive in these drift wave (ITG) units of c_s and ρ_s rather than V_e and ρ_e. The initial amplitude is $A_0 = 10^{-6}$ and the subgrid and boundary dissipation is set to $\nu_0 = 20$ and $s_f = 20$, as otherwise the ETG side crashes numerically. The timestep is correspondingly set to 2×10^{-4}, equivalent to about 0.012 in ETG units. The linear phase starts into a growth rate which peaks at about 3 in these units, about 0.05 in ETG units. The nonlinear phase starts at about $t = 5$ and saturation is well before $t = 10$. In figure 12.18 the drive rates are shown in both linear and exponential scales. The electron temperature drive rate Γ_t is about 1.3 while the ion or drift wave channels $\Gamma_{n,i}$ are below 10^{-3}. The magnetic flutter drive Γ_v is small as in the $\hat{\beta} = 0.1$ ETG case above. The dissipation is due to subgrid and Landau dissipation in about a 2/1 ratio, and the error Γ_E is below 10^{-4}.

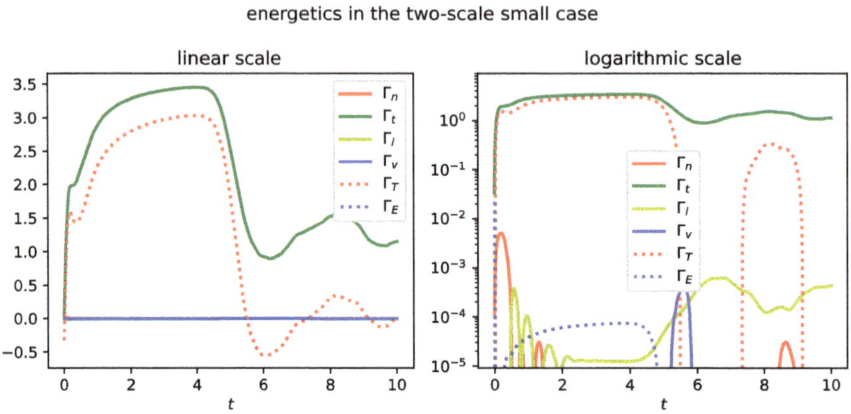

Figure 12.18. Time traces of the drive rates for the 'small' case of the two-scale problem, on a linear scale (left) and a logarithmic scale (right). The units are in terms of δ_0, and the time is in units of $L_\perp/c_s$. The electron temperature drive rate, Γ_t, is several times unity in the linear phase, which corresponds to about 0.05 in ETG units. The logarithmic scale plot shows how much lower the rates of all the other drive mechanisms are. The error is measured at $\Gamma_E = (-1.56 \pm 0.62) \times 10^{-5}$ over the last half of the time interval.

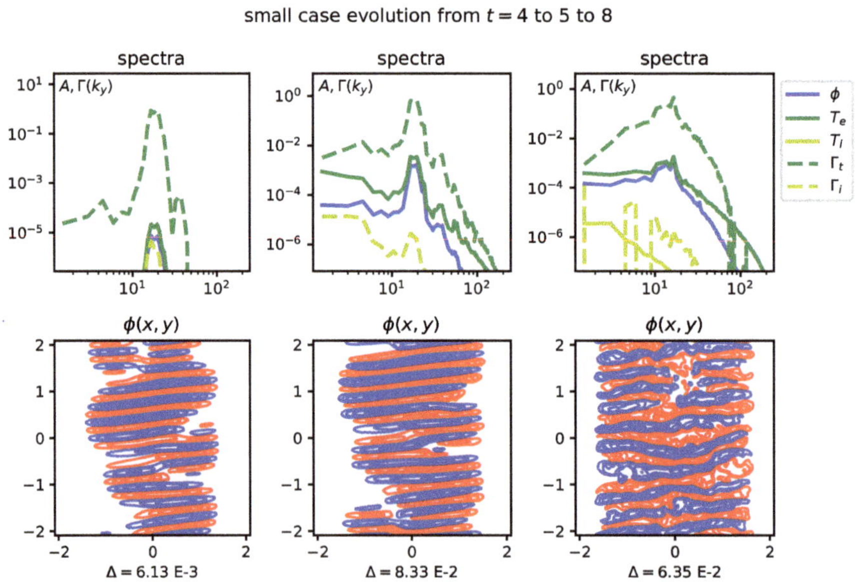

Figure 12.19. Transition to saturation and turbulence, for the 'small' case of the two-scale problem. Time and space are units of $L_\perp/c_s$ and ρ_s. Compare with figures 12.2 and 12.3; the ETG mode structure is closer to the actual case, in which the ion variables are small not by design but due to the ion gyroaveraging at scales below their gyroradius. Note the domain size and spectral scale in units of ρ_s not ρ_e. The ratio of ρ_e/ρ_i is 60.6 for deuterium ions, as here.

The transition to nonlinearity is shown in figure 12.19, which one can compare with figures 12.2 and 12.3. Note that the scales are shown relative to ρ_s not ρ_e here. The development of the ETG streamers through times $t = 4.0$, 5.0, and 8.0 is the principal feature. The spectra and streamer flow morphology is the same as in the ETG case. This also shows that for nonadiabatic turbulence at these scales ETG is the more correct simplified model. Here no form of adiabaticity is assumed, but it is a result here since the parameter situation and allowed scale range are appropriate for it.

This is the basic result of the 2005 work showing that the gyrofluid model could do these cases natively. The transition to ion adiabaticity ($\phi \to -\tau_i n_e$) in the scale range below ρ_s is handled naturally in the ion gyroaveraging, unmodified.

Now we move to a larger domain size, with $K = 0.25$, defining the 'mid' case. The main issue to watch is the emergence of any magnetic flutter problems arising from magnetic disturbances at ever larger scales. The larger the scale, the larger the magnetic response to any current, due to the factors of $k_\perp$ in Ampère's law (two) and in the parallel derivative bracket (one). Nevertheless we set the subgrid and boundary dissipation to $\nu_0 = 4$ and $s_f = 4$, and a hopeful timestep of 10^{-3}, and run to $t = 50$. As the results will show, we get away with this because it is not enough time for the ITG side to reach nonlinearity for itself. It is one of the significant results that there is not enough inverse cascade from the ETG side to excite the ITG side in a preemptory way. That is due to the effective ion adiabaticity for $k_\perp > 1$, and the peak is found at $k_\perp \sim 10$, or 0.15 in ETG units. The lack of resolution puts the dissipation range somewhat close to the actual ETG peak, and the resulting linear growth is slower, about $\gamma_L = 1.3$ rather than the 3 we got before. This is a clear indication that the problem is already under-resolved. Nevertheless, the mode structure formation and saturation is as we saw in the 'small' case before. The streamers look narrower in the spatial morphology because the domain is larger. They extend the entire way across, about $22\rho_s$, from one boundary dissipation layer to the other in the radial direction. The spectrum and morphology at $t = 50$ are shown in figure 12.20.

Now we try to jump-start the ITG side by running exactly the same case but from an initial amplitude set to $A_0 = 10^0$, i.e., marginally nonlinear as for drift wave turblence in the nonlinear instability problem (Chapter 12 of Volume One). The problem is that even with a nonlinear start, the hot plasma cases take longer than $t = 50$ for ITG to form its mode structure and take over even a nonlinear start. So the ETG captures itself first and assumes control of the dynamics anyway, within the first $t = 10$. There is not really enough space in the spectrum for the ITG side to develop quickly. The entire scale range for $k_\perp \sim 1$ and smaller decays over much of this time in the ion variables, which facilitates the ETG takeover. The spectrum at $t = 50$ is not very different from the above case.

This changes at late times. We return to the main 'mid' case initialised at $A_0 = 10^{-6}$, in which the ETG side quickly grows and saturates. Now we run for longer time, up to $t = 1000$. The snapshot interval is 10 so the post process information jumps from time point to time point, not really in an evolution that appears continuous. We find the ETG side in full growth at $t = 10$, largely saturated at $t = 20$, and then after about $t = 70$ there is an ITG peak visible in the second k_y mode, which is $k_y\rho_s = 0.5$. Note that due to the small domain size the k_y spectrum

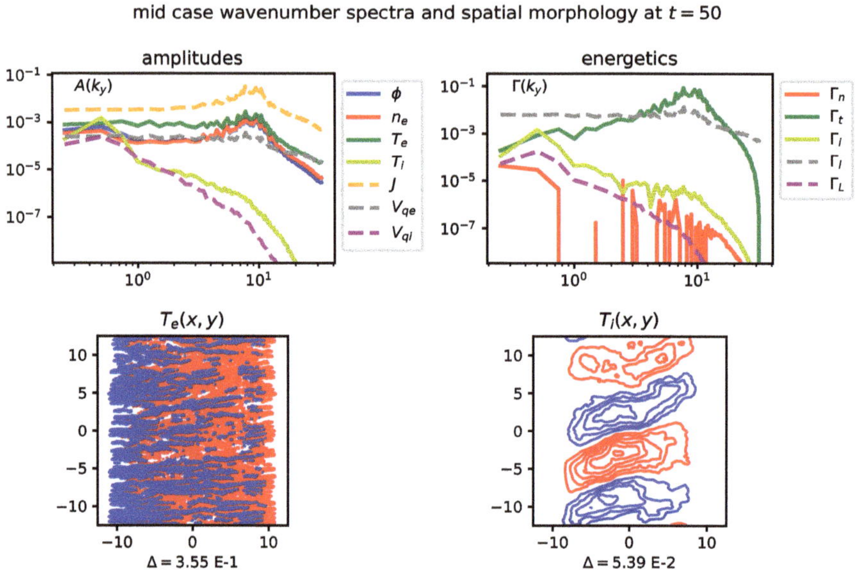

Figure 12.20. Amplitude and energetics (top, left and right) spectra for the 'mid' case of the two-scale problem, at $t = 50$. Time and space are units of $L_\perp/c_s$ and ρ_s. Most variables follow two peaks; the ETG one at short wavelength, and the budding onset of the ITG side at the longest wavelengths, which are not really long enough for that side. The problem is under-resolved both spatially (for ETG) and spectrally (for ITG), with the domain not large enough to converge on the ITG side and the resolution insufficient to converge on the ETG side. The high-amplitude dashed curve is J, which is trying to follow V_e rather than c_s. Note how the spectra for the ion variables, T_i and V_{qi}, show the peak on the ITG side and fall away naturally for $k_y\rho_s > 1$. These features are also reflected in the spatial morphology for T_e and T_i (bottom, left and right).

is too coarsely quantised (which is what is meant by spectral resolution). The first mode is $k_y\rho_s = 0.25$, then 0.5, then 0.75, which is practically the entire active energetic range of ITG turbulence of finite inductivity. Nevertheless, the ITG side looks to assume control during the range $100 < t < 200$. This is shown in figure 12.21, where the structure of ITG vortices superposing upon the ETG streamer flows is visible.

Then, in a slower evolution, the ITG side grows to the largest amplitude, and at the same time its nonlinear activity causes the ETG side to decrease by three orders of magnitude in the energetics. Even with the ITG side too tightly constrained, we still find it definitively taking over the energetics after about $t = 500$, with the ion temperature gradient drive the strongest, the ion $T_\perp$ the largest and most ballooned fluctuation, and the Q_i channel the strongest among the turbulent fluxes. The spectra in terms of both wavelength (k_y) and scale ($k_\perp$) are shown for the late phase in figure 12.22. One can speak of a 50 percent drive point to characterise the rough scale of motion or drive in the turbulence (recall in drift Alfvén turbulence that energy, flux, and vorticity peak in descending order in scale). In either the k_y or $k_\perp$ spectrum of Γ_t, for example, the spectrum is simply summed, entry by entry in the one-dimensional array (note the density of states is already integrated over, to produce this spectrum), until one reaches 50 percent of the overall sum. These measurements

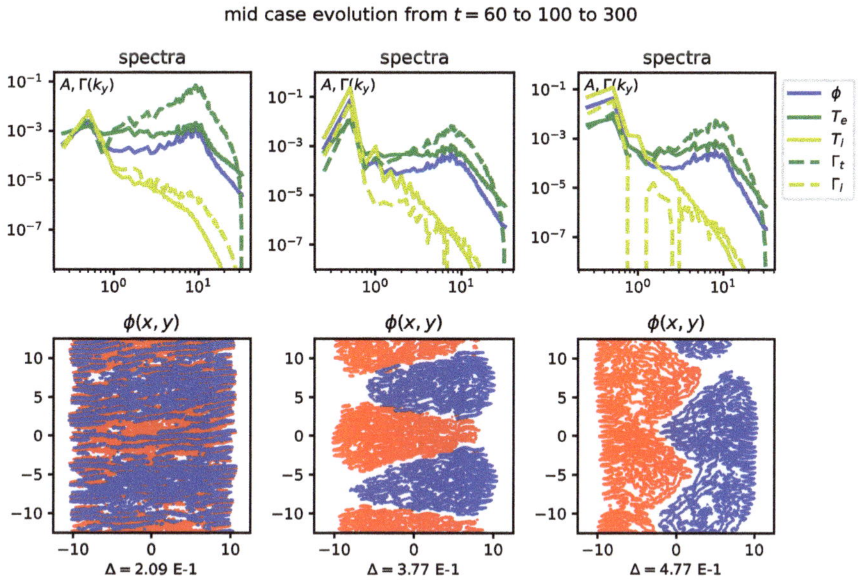

Figure 12.21. Transition between ETG- and ITG-controlled turbulence, for the 'mid' case of the two-scale problem. Time and space are units of $L_\perp/c_s$ and ρ_s. Compare with figure 12.19. The ETG peak remains in this case, but the ITG dynamics assumes control of the energetics. The manifestation of this is the larger scale vortices superposing upon the ETG streamers and reducing their length. The ITG dynamics needs more space in the spectrum than it is given here, namely, the minimum $k_x\rho_s$ of 0.1 and minimum $k_y\rho_s$ of 0.018 it has in the full domain.

are used to evaluate the relevance of any ETG peak or shoulder seen in the spectra of T_e or Γ_t (hence the flux component G_e). For this case we find the 50 and 90 percent drive points are modes 2 and 3 for the k_y spectrum of Γ_t, and modes 40 and 56 for its $k_\perp$ spectrum. The 90 percent drive points are $k_y\rho_s = 10$ and $k_\perp\rho_s = 14$ in real terms, respectively. These are both within the ETG peak, so that is still effecting a reasonable fraction of the energetics and turbulent ExB flux. But the case is under-resolved, in this instance particularly in terms of the spectrum as just noted. One would like to see how this evolution appears in a larger domain.

We try next with a case with $K = 0.1$, so that $L = 62.8$, and set the grid to $512 \times 512 \times 32$, so that the maximum $k_\perp\rho_s$ is 25.6, with $h_x = h_y$ about $8\rho_e$. This is termed the 'big' case. The resolution on the ETG side is less by a factor of 1.25 while the domain size is increased by 2.5, and at $64\rho_s$ it is close to full size. The initial amplitude is $A_0 = 10^{-4}$. The time traces of the drive energetics are shown on two time scales in figure 12.23. Even with the depressed resolution, as in the 'mid' case, the ETG growth is still strong, with Γ_T rising to just below 1.0 with the total drive rate Γ_+ (essentially all Γ_t) just higher. The error Γ_E is small, below 10^{-4} at all times. The difference $\Gamma_+ - \Gamma_T$ is the subgrid dissipation, whose finite size indicates we would do better with more resolution. Nevertheless, the ETG growth proceeds until nonlinearity is reached at $t = 24$ and saturation after $t = 30$. The familiar evolution of linear structures coalescing into streamer flows is what is seen in this

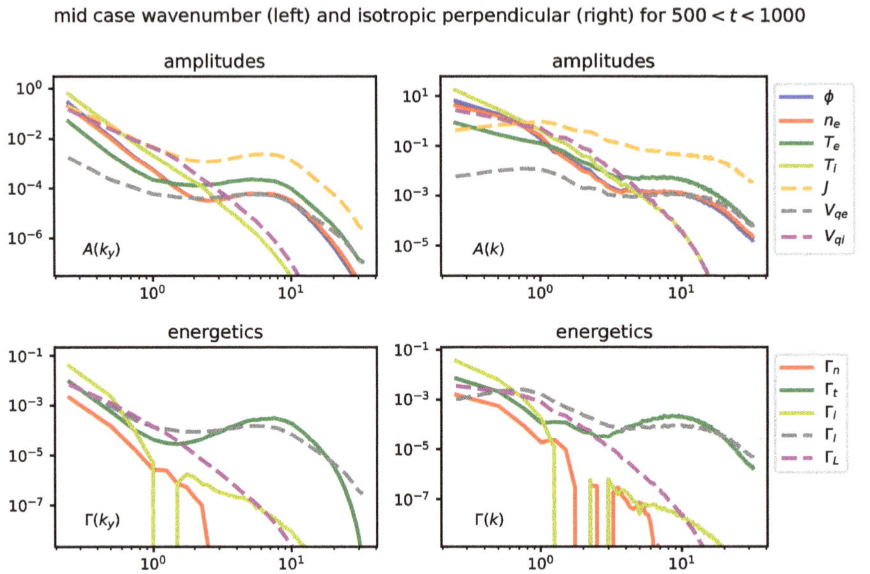

Figure 12.22. Time-averaged ($500 < t < 1000$) k_y-spectra and $k_\perp$-spectra for the 'mid' case of the two-scale problem. Time and space are units of $L_\perp/c_s$ and ρ_s. The ETG 'shoulder' visible especially in $k_\perp$ energetics near $k_\perp\rho_s \sim 10$ does not carry a significant fraction of the energetics. The 50 percent drive points are at $k_y\rho_s = 0.5$ and $k_\perp\rho_s = 0.75$, respectively, but the 90 percent drive points are at $k_y\rho_s = 10$ and $k_\perp\rho_s = 14$, respectively.

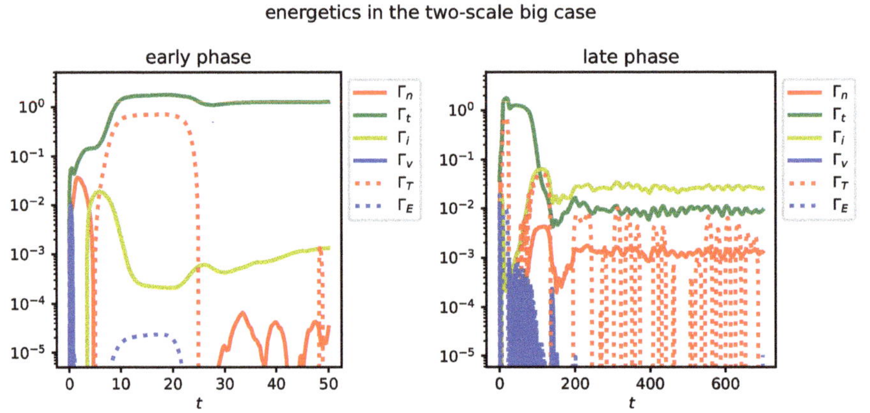

Figure 12.23. Time traces of the drive, total growth, and error rates, for the 'big' case of the two-scale problem, on a logarithmic scale on two time scales. The units are in terms of δ_0, and the time is in units of $L_\perp/c_s$. The two steady growth phases (Γ_T) are the ETG one before $t = 25$ and the ITG one around $t = 110$. The electron temperature drive rate (Γ_t) is above 1.0 in the ETG linear phase, but drops down in the ITG turbulence phase after $t = 200$. The ion temperature drive rate (Γ_i) is small during the ETG phase but becomes the strongest channel at late times. The density gradient drive (Γ_n) is smaller but finite at late times. The magnetic flutter gradient drive (Γ_v) is insignificant at this $\hat{\beta}$ value. The error (Γ_E) is above 10^{-5} in the ETG linear phase but is below 10^{-6} thereafter. The noisiness of this curve for $t < 100$ in the right frame is due to the coarser sampling interval.

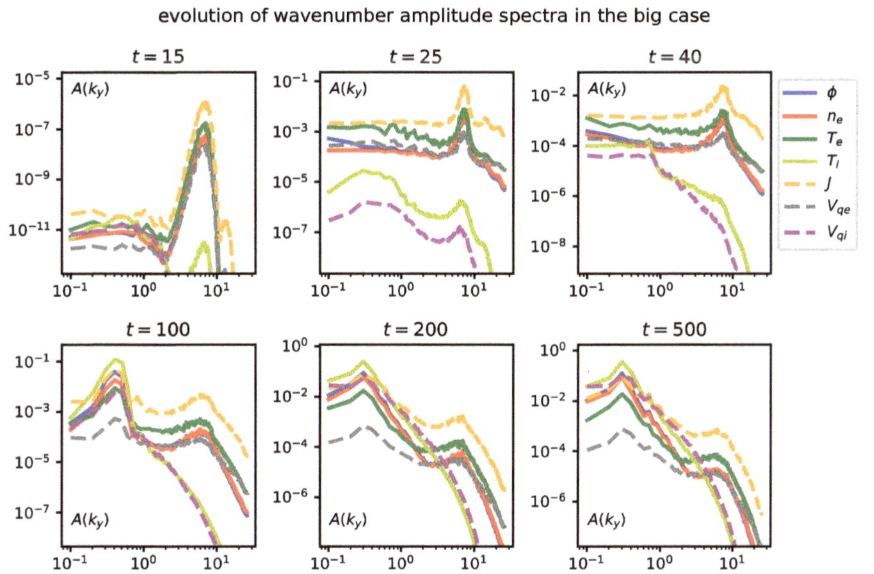

Figure 12.24. Amplitude spectra for the times indicated, during and after the ETG growth phase (top), and then during and after the ITG growth phase (bottom). Time and space are units of $L_\perp/c_s$ and ρ_s. The ETG phase, during which the initial ITG growth phase is visible ($k_y\rho_s < 1$), culminates with development of ExB flow streamers ($k_y\rho_s \sim 10$). The ITG phase culminates in standard hot plasma ITG turbulence mode structure, except the ETG remnant is still visible. In the $400 < t < 600$ phase the 50, 70, and 90 percent drive points for Γ_t are at $k_y\rho_s = 0.4$, 0.8, and 7.0, respectively.

phase. The streamers extend the entire distance, about $3500\rho_e$, between the boundary dissipation layers. The initial development of the k_y spectra is shown in the top row of figure 12.24. This is the same as the establishment of the ETG turbulence as in the standard case (figure 12.2), except that we also see the initial growth of the ITG side before it really emerges.

Then, in a short time interval around $t = 110$ the ITG turbulence grows to nonlinear amplitude. The two main growth phases, ETG early, and ITG later, are visible in the phases where Γ_T remains positive without any dropouts (to negative values, not well displayed on a logarithmic scale) for significant time intervals. The ITG transition to turbulence is around $t = 200$ when the flows begin to enter and then exchange energy with the turbulence. The ITG phase is in saturation until the run ends (due to wall clock time limits). One sees the error Γ_E drop by another order of magnitude, a result of more of the energy being resident at large scales. The later development of the k_y spectra is shown in the bottom row of figure 12.24. The ITG side now assumes control, and after it is established, by $t = 300$ at the latest, the ETG side is relegated to a secondary peak. In the $400 < t < 600$ phase the 50, 70, and 90 percent drive points for Γ_t are modes 4, 8, and 70, corresponding to $k_y\rho_s = 0.4$, 0.8, and 7.0, respectively. This indicates that the ETG scales are carrying 10 to 15 percent of the ExB turbulent electron heat flux.

12.6.3 The warm plasma case

We now run the same 'mid' domain size which we just did, but now with the warm plasma case instead (Eq. 12.49). The subgrid and boundary dissipation and timestep are unchanged. The collisions are restored, with the same dissipation model as in the previous Chapter. The collisional parameters are as given therein. Now, the only difference between these equations and those given therein is the addition of the electron FLR corrections.

The initial amplitude is still $A_0 = 10^{-6}$ as in the 'mid' case above. In figure 12.25 the transition to nonlinearity is shown, as in figure 12.19 for the hot plasma 'small' case. Very early in the run, there is a robust instability on the ETG side but which is not really an ETG streamer instability. Its growth rate reaches 0.45 by $t = 60$, during which the electron temperature drive Γ_t is the only significant drive. The nature changes after $t = 75$, as initial saturation occurs (the total growth rate Γ_T hits zero at $t = 78$) Γ_t drops down but returns to about 0.2 after $t = 90$. As the warm plasma pure ETG cases shown above indicate, the instability still occurs, but with the parameter ω_t/ω_n only 2, and strong streamers do not form. Nevertheless, the strong linear peak at $k_y\rho_s = 3$ and $k_\perp\rho_s = 12$ forms by $t = 40$. It is very short-lived however, and disappears by $t = 250$, as nonlinearity fully develops in the drift wave scales ($k_\perp\rho_s < 1$). Thereafter, the manifestation of the insufficiency 1×1 perpendicular domain presents itself, in the way a small set of vortices forms at the largest allowed

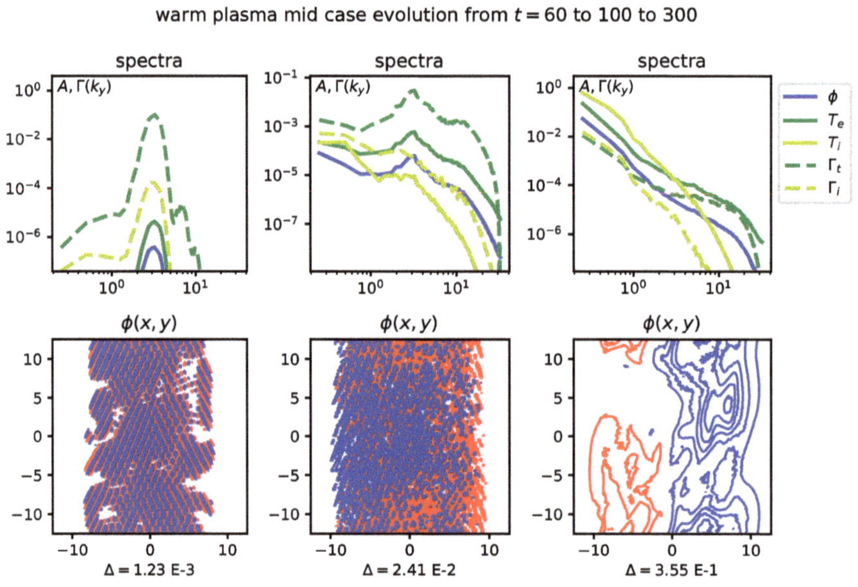

Figure 12.25. Transition to saturation and turbulence, for the warm plasma 'mid' case of the two-scale problem. Time and space are units of $L_\perp/c_s$ and ρ_s. Compare with figures 12.19 The initial instability at scale below ρ_s is later eliminated by the nonlinear cascades coming down from the drift Alfvén dynamics at larger scale. The latter puts most of its energy into the largest or longest mode set ($k_y\rho_s = k_\perp\rho_s = K$) because the domain is not large enough.

scale, which does not efficiently decorrelate via turbulence. This is the same problem affecting the hot plasma cases we saw in the longer run to $t = 1000$ of the 'mid' case above.

In figure 12.26 the resulting k_y wavenumber and isotropic $k_\perp$ perpendicular scale spectra are shown for this case, averaged over $500 < t < 1000$, well after the drift Alfvén side has assumed control of the dynamics. Amplitudes are in the top row, with energetics in the bottom row. For this case we find the 50 and 90 percent drive points are modes 1 and 8 for the k_y spectrum of Γ_t, and modes 1 and 14 for its $k_\perp$ spectrum. The 90 percent drive points are $k_y\rho_s = 2$ and $k_\perp\rho_s = 3.5$ in real terms, respectively. These are both well above where the ETG peak manifests, and at the beginning of the ETG shoulder in this case.

Finally, we start with the standard drift Alfvén setup for this case, and increase the resolution by a factor of four. The domain size is $L_y = 4L_x = 80\pi\rho_s$ with $K_y = 0.025$ and $K_x = 0.1$ giving the fundamental wavenumbers in terms of ρ_s, but the grid is now $256 \times 1024 \times 32$ for $N_x \times N_y \times N_s$. The grid size is about $0.25\rho_s$ or about $16\rho_e$. The initial amplitude is $A_0 = 10^0$, and the subgrid and boundary dissipation parameters are set to $\nu_0 = s_f = 1.0$, anticipating that the cascade will be effective in constraining small scale amplitude. The time traces, in figure 12.27, bear this out. At these parameters the nonlinear start leads quickly enough into turbulence that the ETG growth phase is stunted, with Γ_t never rising above 10^{-1}. The segue into the ITG growth phase and saturation is continuous, up to saturation before $t = 200$. Thereafter the ion and electron heat flux channels are as close as they

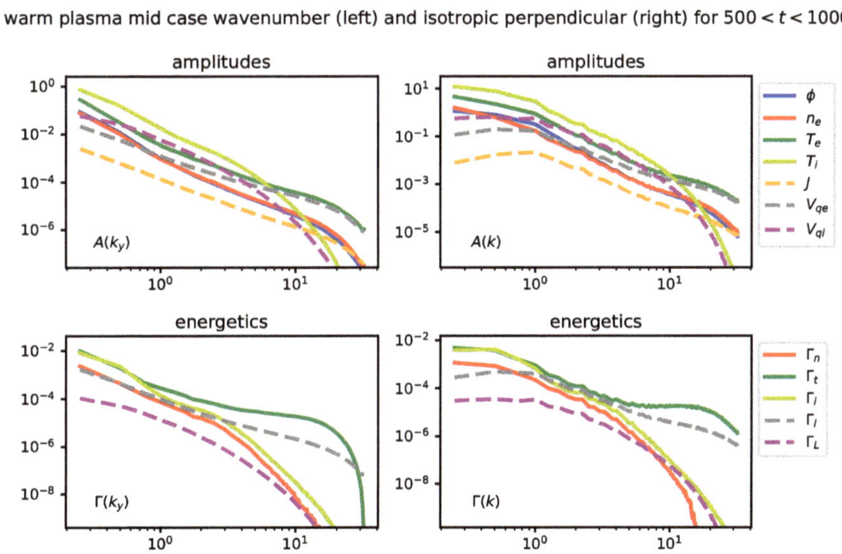

Figure 12.26. Time-averaged ($500 < t < 1000$) k_y-spectra (left side) and $k_\perp$-spectra (right) for the warm plasma 'mid' case of the two-scale problem. Time and space are units of $L_\perp/c_s$ and ρ_s. The ETG 'shoulder' visible especially in $k_\perp$ energetics near $k_\perp\rho_s \sim 10$ does not carry a significant fraction of the energetics. The 90 percent drive points are at $k_y\rho_s = 2$ and $k_\perp\rho_s = 3.5$, respectively.

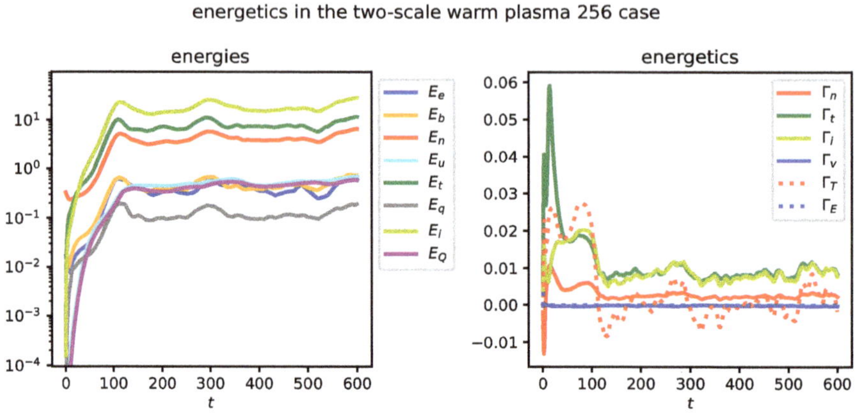

Figure 12.27. Time traces of the drive, total growth, and error rates, for the 256 × 1024 drift Alfvén case of the two-scale problem, on a logarithmic scale. The units are in terms of δ_0, and the time is in units of $L_\perp/c_s$. The initial amplitude is unity, the ETG growth is slower, and the drift wave turbulence growth and saturation is faster, in comparison to hot plasma cases. The saturation is well before $t = 200$. The drive rates are as they are in the usual gyrofluid warm plasma cases, with $\Gamma_t \sim \Gamma_i > \Gamma_n$ following $\omega_t = \omega_i = 2\omega_n$ in the gradient scale ratios. The magnetic flutter is small but positive. The error (Γ_E) is small at all times.

are in the standard drift Alfvén cases of Chapters 3 and 11. All other aspects of the mode structure appear as in those cases. The resulting k_y wavenumber and isotropic $k_\perp$ perpendicular scale spectra, with the same format as in figure 12.26, are shown for this case in figure 12.28, averaged over $400 < t < 600$. In the $400 < t < 600$ phase the 50 and 90 percent drive points for Γ_t are modes 6 and 15, corresponding to $k_y\rho_s = 0.15$ and 0.375, respectively. In the isotropic spectra (note the lowest $k_\perp\rho_s$ is 0.1 not 0.025), these are modes 3 and 6, or $k_\perp\rho_s = 0.3$ and 0.6, respectively. Here, the ETG side is energetically negligible. In these cases with steeper gradients and longer effective magnetic field lines, the ETG side both tries to grow at larger scale and is correspondingly weaker compared with the hot plasma cases, and the drift Alfvén side is also correspondingly stronger. But we would really like to see these cases with about an order of magnitude more resolution.

With the entire set of these results in hand we can conclude that the turbulence cascade coming down from the drift wave scales overwhelms the ETG drive mechanism in the warm plasma drift Alfvén case. But only provisionally, until the definitive cases have been done with future computational resources.

12.6.4 The two-scale problem is work in progress

This is very tantalising, but it remains that one or both sides are under-resolved, the ITG side spectrally and/or the ETG side spatially. Nevertheless, it is a strong indication that the action of the ITG turbulence to form eddies (not coherent vortices) actually does overcome the action of the ETG turbulence to form streamer flows. One does not any longer find streamer flows extending the entire radial domain. The direction of propagation turns around to the ion direction. On the other hand, the insufficient spectral resolution translates into eddy chains which are

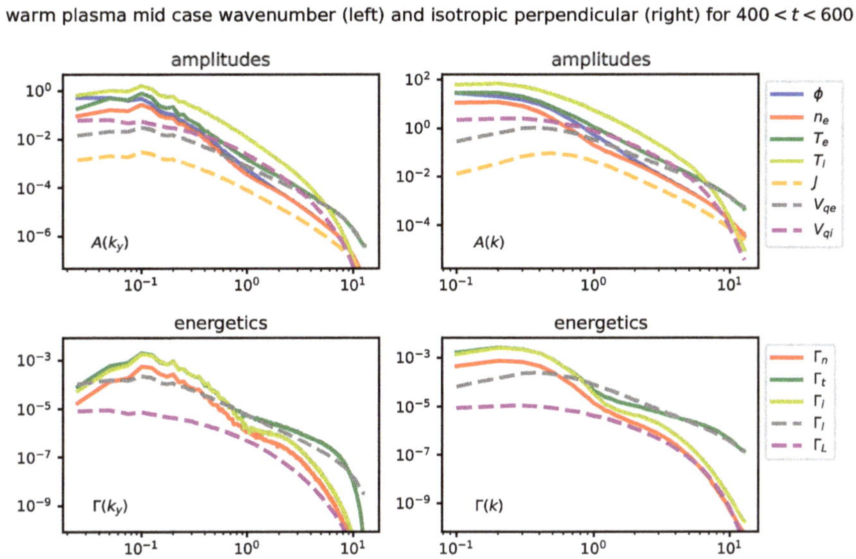

Figure 12.28. Time-averaged ($400 < t < 600$) k_y-spectra (left side) and $k_\perp$-spectra (right) for the 256×1024 drift Alfvén case of the two-scale problem. Time and space are units of $L_\perp / c_s$ and ρ_s. The ETG 'shoulder' is barely visible even in the energetics. With enough spectral space for the turbulence, the 90 percent drive points are now at $k_y \rho_s = 0.375$ and $k_\perp \rho_s = 0.6$, respectively. Compare with figure 12.26. The question of resolution remains open.

not as decorrelated by the turbulence as they are in cases with a large enough domain size, like those of the previous Chapter. But it does indeed appear that streamer flows are prevented from radial extent that leads to the large amplitudes seen in pure ETG turbulence.

Nevertheless, it is important to face the gulf between the largest of the above examples and a case which actually does the full problem. In the largest warm plasma case, the domain size is full but the resolution (hence grid size squared, at least) would have to increase by 16 from $h_x = h_y = \rho_s/4$ down to $h_x = h_y \sim \rho_e$ as we have in the baseline ETG cases. This is 64 times the standard grid for a drift Alfvén case in both of xy, to a size of 4096×16384. This is before the necessary demonstration of convergence in $k_\parallel$ on such a grid. The hot plasma ITG cases are about the same, with similar domain size and the same factor of 60.6 between the species gyroradii. Only the demands on $k_\parallel$ might be less in the hot plasma case. But this is something for the future, despite some of the grandiose claims one can find in the literature.

12.7 Summary

The exemplary features of ETG turbulence are as follows. The underlying dynamics is the same as ITG turbulence in the linear stage and within the non-zonal degrees of freedom in the saturated turbulence. The scale of motion in the y-direction is set by the underlying dynamics. In the ETG case the zonal component of the adiabatic response is present, and in an adiabatic model the polarisation is affected. The state of ETG turbulence at late times remains dominated by the radial streamer structures which extend all the way across the domain. The dynamics depends on the temperature gradient in an adiabatic model, since the drive due to the density gradient vanishes.

In general, however, both sets of dynamics are available to the turbulence, which by filling all available degrees of freedom with its spectral transfer dynamics always finds possible modes of access to free energy. The general prospect is of the thermal free energy cascade descending from perpendicular scales above ρ_s to those below ρ_s, thereby washing out the native dynamics for the region intermediate to the two gyroradii. The only question is whether the vorticity is strong enough. Results will vary in general situations, but to do the problem in a way without predetermined results requires the basic standards for the larger scale region. These are the ones covered for drift wave and ITG turbulence in previous Chapters and in Volume One, namely, the radial direction up to the gradient scale, the drift direction at least four times greater, electromagnetic equations, and the run length long enough to get the longest wavelength ExB and magnetic dynamics and the acoustic and ion thermal channels to reach statistically equilibrated saturation.

Further Reading

It should be noted that the central role of the dynamics in turbulence in a tokamak is the adiabatic response on one side and polarisation on the other. This has been known since the Hasegawa–Mima work of the 1970s and the Hasegawa–Wakatani work of the 1980s. It should also be noted that such as the ion temperature gradient turbulence is impossible to capture in an MHD model, since the mode type does not exist without the adiabatic response. This holds by extension as well for the ETG model. As with the Hasegawa–Mima model, the simplest version with adiabatic response together with polarisation was the first one. These two Hasegawa references were given in Chapter 9.

Early work laying the groundwork for this type of electron-based instability is reviewed by B B Kadomtsev and O P Pogutse, in 'Turbulence in toroidal systems,' *Rev. Plasma Phys.* **5** (1970) 249. A salient refence is B Coppi, 'Fluidlike Electron and Ion Modes in Inhomogeneous Plasmas,' *Phys. Rev. Lett.* **21** (1968) 1055.

The idea of turbulence at the scale of the collisionless skin depth, often called c/ω_{pe} turbulence, was given by T Ohkawa, in 'A transport model for alcator scaling in tokamaks,' *Phys. Lett.* A 67 (1978) 35.

The equations for a fluid electron model of the basic instability together with the adiabatic ion response were given by V A Rozhanskii, in 'Transfer processes in stochastic magnetic fields, hydrodynamic approach,' *JETP Lett.* **34** (1981) 60. His Eqs. (3–5) for the electrons and Eq. (6) for the high-$k_\perp$ limit of the ion polarisation define the basic model. He credits also B B Kadomtsev and O P Pogutse, in 'Voprosy teorii plasmy (Problems of plasma theory),' Atomizdat No. 5 (Moscow 1967), p 209. Finite inductivity (β) modifications in a sheared magnetic geometry were studied in the kinetic version by Y C Lee, J Q Dong, P N Guzdar, and C S Liu, in 'Collisionless electron temperature gradient instability,' *Phys. Fluids* **30** (1987) 1331. Two-dimensional fluid computations were done by J F Drake, P N Guzdar, and A B Hassam, in 'Streamer formation in plasma with a temperature gradient,' *Phys. Rev. Lett.* **61** (1988) 2205. The word 'streamer' may have been coined in this reference. The toroidal version of the mode was given by W Horton, B G Hong, and W M Tang, in 'Toroidal electron temperature gradient driven drift modes,' *Phys. Fluids* **31** (1988) 2971. The toroidal curvature terms including diamagnetic flow and heat flux appear in this reference, which used the tokamak ballooning formulation. These references all note that the electron mode is like the ITG mode, with the electrons and ions reversed in their roles.

ETG turbulence as it is known today was rediscovered in the years around 2000. See F Jenko and W Dorland and M Kotschenreuther and B N Rogers, 'Electron temperature gradient driven turbulence,' *Phys. Plasmas* **7** (2000) 1904; and Y Idomura, M Wakatani, and S Tokuda, 'Gyrokinetic theory of slab electron temperature gradient mode in negative shear tokamaks,' *Phys. Plasmas* **7** (2000) 2456. My capture of it is in the 2005 work cited in the previous Chapter.

The back reaction of the electron-scale range of the spectrum on the main energy containing range was predicted by a statistical model and found to be small, by C Holland and P Diamond, in 'A simple model of interactions between electron temperature gradient and drift-wave turbulence,' *Phys. Plasmas* **11** (2004) 1043.

Initial study of the effect of the ion-scale cascade dynamics on the electron-scale side was done by J Candy, R E Waltz, M R Fahey and C Holland, in 'The effect of ion-scale dynamics on electron-temperature-gradient turbulence,' *Plasma Phys. Contr. Fusion* **49** (2007) 1209. This reference found the result that for hot plasma cases the electron-scale contribution to the turbulent ExB electron heat flux is below 20 percent, and also confirmed the above prediction by Holland and Diamond.

The closest capture of the two scale problem for hot plasma cases in a kinetic model was given by S Maeyama, Y Idomura, T-H. Watanabe, M Nakata, M Yagi, N Miyato, A Ishizawa, and M Nunami, in 'Cross-scale interactions between electron and ion scale turbulence in a tokamak plasma,' *Phys. Rev. Lett.* **114** (2015) 255002. Figure 2 of this reference is close to the 'big' case result of figure 12.24 above.

Chapter 13

Trapped-electron turbulence

13.1 Introduction—magnetic trapping

There is a special class of instabilities one usually associates with kinetic processes but which turns out to be amenable to understanding with simplified fluid models. This is the so-called 'trapped electron' turbulence. It arises from the variation of the strength of the magnetic field in the direction of the field lines. We already saw in turbulence in open field lines that this variation can be very strong in some geometries but that in a model not treating temperature dynamics the variation does not by itself lead to effective parallel localisation (Chapter 15 of Volume One). On the other hand, the development of gyrokinetic theory has given us something else: with the magnetic moment of each gyrocentre fixed, the coefficient B which multiplies it acts as a force potential. The resulting magnetic mirror force can cause the parallel motion of the gyrocentre to turn around. Although this is a single particle picture, its relevance to turbulence of the drift wave type is that these particles do not participate in parallel wave responses, and so for them there is no adiabatic response.

Single particle motion in gyrokinetic theory follows from its Hamiltonian basis. In a system with a symmetry coordinate, the projection of the motion in the plane perpendicular to the symmetry direction follows a contour of constant H, the Hamiltonian. This is the unperturbed motion, as at this point we do not consider the effect of fluctuating fields. In this neglect of fields, the gyrocentre motion actually does represent the particle motion, which is split between the gyromotion and the drift of the centre of the gyro orbit. Without fields there is no polarisation drift. The subtleties involving the gyrocentre representation do not enter if we restrict consideration to this unperturbed motion.

In the motion of each gyrocentre along the level curve of H, the other quantities which are conserved are the magnetic moment and the canonical momentum. These can be tracked by coordinates, so that the fourth coordinate is an angle of the motion. Analogous to the gyrocentre and the gyrophase angle, we then have a 'bounce centre' and a 'bounce angle' if the particle is trapped. A trapped particle in this sense is then one whose orbit is closed in such a way that the entire extent of the parallel coordinate in the plane of motion is not sampled. A passing, or circulating, particle is one which has a motion which is periodic in this angle. A trapped particle is one which has a temporally periodic motion, but the motion is oscillatory in this angle. In an open system, the passing particles are considered to be in the 'loss region' of phase space (energy, magnetic moment, and canonical momentum) since their level curves intersect the open boundary at the ends of the field lines.

If a region of phase space represents trapped orbits, the population of particles occupying that region is considered as the trapped fraction. In a collisionless system, or one whose collisionality is so small as to be negligible, the particles, or better the regions of phase space, interact with each other solely through the self consistent fluctuating fields. The trapped and non-trapped populations therefore are considered to function as separate species and can be treated that way. The trapped particles do not feel the adiabatic response because the parallel motion averages to zero over an orbit, that is, the bounce centre itself is stationary. The non-trapped particles may be taken to be in an adiabatic (force balance) relation with respect to the fields or may be treated self-consistently.

The final consideration involves the time scales. The reason we speak of trapped electron turbulence is due to the conventional time scale ordering. The turbulence is considered fast compared to the parallel ion transit time, so that whether or not the ions are trapped becomes negligible. The electrons, being faster, now separate into the passing or non-trapped fraction, which is subject to parallel dynamics, and a trapped fraction which is free to exhibit drift motion without any effect of field line connection, for closed systems, or exit fluxes, for open ones. Thought of another way, the 'bounce frequency' given by the rate of a cycle through the closed orbit is taken to be fast compared to wave motion or turbulence, so that any parallel forces are averaged out. The integral of a parallel gradient of a force potential around the closed phase-space curve vanishes. So, in simplified models, the trapped particles are prescribed to neglect parallel motion. They are free to participate in interchange motion unconstrained. This is a situation analogous to the ion temperature gradient (ITG) mode (Chapter 3), but in this case the density gradient can drive the trapped fraction and so the concept does not depend on temperature gradients.

Following sections develop the model of the orbit as a closed curve of the Hamiltonian in phase space, then the process of the toroidal precession drift in axisymmetric systems, and then the interaction of trapped and passing electrons within drift Alfvén turbulence. Research into fully self-consistent turbulence on this type of mode is still very active, so as far as three-dimensional cases are concerned we will restrict to models of the process of the fluid type, as these are generally accessible computationally.

13.2 Gyrokinetic Hamiltonian in a system with symmetry

A symmetry coordinate is one in which the components of the geometry in the gyrokinetic Lagrangian has no dependence. This is sometimes called an ignorable coordinate. That means the trajectories in phase space can be drawn in a plane at constant y, the symmetry coordinate. In general, there is still motion in the y-direction, but it is independent of y. The Lagrangian for the gyrocentre is of the form

$$L_p = \mathbf{p}^* \cdot \dot{\mathbf{R}} + M\dot{\vartheta} - H \tag{13.1}$$

where the phase space coordinates are $\{x, y, s\}$ for space and $\{z, w\}$ for velocity space. The coordinate dependences of the pieces are

$$\begin{aligned} \mathbf{p}^* &= \mathbf{p}^*(x, s, z, w), \qquad & M &= M(w), \\ H &= H(x, s, z, w, \phi_a), \qquad & \phi_a &= \phi_a(x, y, s, t) \end{aligned} \tag{13.2}$$

where the ϕ_a form an arbitrary set of time dependent field variables. The coordinate y is assumed to be the symmetry coordinate, which means that the gyrocentre canonical momentum $\mathbf{p}^*$ is independent of y but in general depends on all the other coordinates. The species magnetic moment is M, defined as e/m times the magnetic moment w of the gyrocentre in this representation. As an energy, $M\Omega = wB$, where e and m are the species charge and mass, Ω is the signed gyrofrequency, and B is the strength of the magnetic field. For coordinate dependence, M depends on w only. The Hamiltonian H depends on phase space but can depend on y and time t only through its dependence on the ϕ_a. The ϕ_a in turn can depend on all three space coordinates and time. For the purposes of trapped particle motion in the magnetic field we can assume the traditional picture of test particle motion in a prescribed magnetic field electric fields. The plasma can have a current, but for this purpose it serves only to define the magnetic field.

Under these restrictions the motion of a gyrocentre in phase space depends on four coordinates: $\{x, s\}$ and $\{z, w\}$. This can be defined as the plane with $y = 0$ (any constant is equivalent, zero will suffice). In the absence of the ϕ_a, which are assumed to be zero, there are three conserved quantities. With no time dependence, H, the energy, is conserved. With no y-dependence, the component P_y^*, the canonical momentum associated with y, is conserved. With no ϑ dependence (the gyroangle), M is conserved. With three quantities conserved, the motion is then described by level curves of H in the $y = 0$ plane. These are referred to as trajectories or orbits. In gyrokinetic theory in general, this motion is infinitesimal in time. In this case, the motion describes entire curves, and then the topology of these curves becomes relevant.

13.2.1 Simplest case: 1D mirror field

We consider the case in which $B = B(s)$ in a cylindrical system with axisymmetry, with variation along the axis (s) independent of radius (r). The spatial extent in r is assumed to be small compared to the scale of variation in s. The latter can be taken periodic, with fundamental wavenumber $K_\parallel$. The variation of B can be considered as

$$B = B_0(1 + \varepsilon \cos K_\parallel s)^{-1}, \qquad \varepsilon < 1 \tag{13.3}$$

The canonical momentum associated with s is not conserved since the geometry depends on s. The symmetry coordinate in this case is not the axial one but the azimuthal one, and the momentum in the latter direction can be assumed to be zero. Momentum in r and s is exchanged through the magnetic field. Motion can be considered in the sz phase space plane, with $z = mv_{\parallel}$ the parallel momentum. In turn, the position of $z(s)$ on a contour of constant $H = H_0$ is given by the condition that both the magnetic moment w and the energy H be conserved. So we have

$$H(s, z, w) = \frac{z^2}{2m} + wB_0(1 + \varepsilon \cos K_{\parallel}s)^{-1} = H_0 \tag{13.4}$$

as the condition describing trajectories in sz-space. The parallel motion in the phase space is then given by

$$z = \pm\sqrt{2m(H_0 - wB)} \tag{13.5}$$

with the sign giving the positive or negative sides of the orbit. If wB_0 is a large enough fraction of H_0 then the locations in s given by

$$\varepsilon \cos K_{\parallel}s_0 = \frac{wB_0}{H_0} - 1 \tag{13.6}$$

are the turnaround points at which z vanishes. The effective angle $K_{\parallel}s_0$ is the bounce angle. This is symmetric about $s = 0$ and we note that the right hand side can be negative. This is because the condition on the energies for existence is that the minimum of $wB(s)$ must be less than H_0,

$$\frac{wB_0}{H_0} < 1 + \varepsilon \tag{13.7}$$

not the value at $B = B_0$. For a given H_0 values of wB_0 can range between zero and this limit. The trapping condition is that the maximum of $wB(s)$ be greater than H_0, or

$$\frac{wB_0}{H_0} > 1 - \varepsilon \tag{13.8}$$

Particles satisfying this are trapped between $K_{\parallel}s = \pm\pi$, while particles with $wB_0 = H_0$ are trapped between $K_{\parallel}s = \pm\pi/2$. Then, the curve for $wB_0/H_0 = 1 - \varepsilon$ gives the separatrix orbit and the one for $wB_0/H_0 = 1 + \varepsilon$ gives the point at the origin. The range of wB_0/H_0 for trapped particles is

$$1 + \varepsilon > \frac{wB_0}{H_0} > 1 - \varepsilon \tag{13.9}$$

The bounce centre of a trapped orbit is at the position of the minimum value of B along the field line, labelled $B_{\min}$. So we define an effective pitch angle there,

$$\lambda = \frac{wB_{\min}}{H_0} = 1 - \zeta^2 \tag{13.10}$$

where ζ is the pitch angle cosine. In this case $B_{\min} = B_0(1 + \varepsilon)^{-1}$, not B_0 as in some other cases. The conditions for existence are the λ can vanish but be no higher than unity, the value for which the extent of a trapped orbit goes to zero. Given an initially isotropic distribution, set where $B(s) = B_{\min}$, which is the location of the bounce centres, the fraction of particles which are trapped is given by integration over ζ, or

$$f_t = \sqrt{1 - \frac{B_{\min}}{B_{\max}}} = \sqrt{\frac{2\varepsilon}{1 + \varepsilon}} \tag{13.11}$$

The simple expectation, the one without any plasma physics, is that any passing particles given by a source are lost, and any trapped particles remain confined.

13.2.2 The dipole magnetic field

This is a similar situation to the previous case, now with the metric of the dipole field (Chapter 15 of Volume One). The symmetry coordinate is the toroidal direction which is purely perpendicular, so it is not involved in the parallel motion with finite magnetic moment. The toroidal drift motion is separate from this and will be addressed later on. The magnetic field structure is given in terms of the poloidal angle θ as

$$B^{\theta}(\theta) = B_0 \cos^{-7}\theta, \qquad B = \frac{B_0}{\cos^6\theta}\sqrt{1 + 3\sin^2\theta}, \qquad r = r_0 \cos^2\theta \tag{13.12}$$

where B_0 is the field in the equatorial midplane at a given radius r_0, and r is the radius of the same field line at different angles. Since the field amplitude is singular at $\theta = \pm\pi/2$ all particles can be considered to be trapped. If H_0 and wB_0 are given, then the condition on w is

$$wB_0 < H_0, \qquad \lambda = \frac{wB_0}{H_0} < 1 \tag{13.13}$$

Note that in this case $B_{\min}$ is B_0. The bounce angle is given by setting $wB = H_0$, as

$$\frac{wB}{H_0} = \lambda \cos^{-6}\theta_0\sqrt{1 + 3\sin^2\theta_0} = 1 \tag{13.14}$$

In practical terms the flux tube is considered to be open at some endpoints $\theta = \theta_L$. If we take these at $\theta_L = \pm\pi/4$ as before, then the trapping condition is

$$\lambda > \lambda_0 = \frac{1}{8}\sqrt{5/2} = 0.198 \tag{13.15}$$

The associated trapping fraction for an isotropic initial distribution is given by averaging over the pitch angle cosine ζ, so that

$$f_t = \sqrt{1 - \lambda_0} = 0.896 \tag{13.16}$$

So most of the particles are indeed trapped.

The position at which $B = B_0$ is the outboard midplane in this case. Since this is the minimum value for B, the trapping conditions are different from the axial mirror case, where B_0 was set as the average value. Trapping is actually a matter of the maximum versus the minimum values of B on a field line. For existence we require $wB_{\min} < H_0$, and then given that, for trapping we have $wB_{\max} > H_0$, with equality at the turnaround point or bounce angle. For the dipole, $B_{\max}$ is the value at the declared exit point. If the dipole flux tubes are limited by an atmosphere, then the exit point will depend on the reference radius r_0.

13.2.3 The tokamak magnetic field

So far we have not faced the magnetic drifts because the symmetry coordinate in the two previous cases has no associated component of the magnetic field. With both variation and magnetic field in the poloidal plane, it follows that the drifts are purely toroidal. In the tokamak magnetic field, or anything else with axisymmetry and a finite toroidal magnetic component, this is not the case and then the conserved canonical momentum is part of the mix. The associated drifts are now both radial and poloidal as well as toroidal, and in a mostly toroidal magnetic field these are the largest components of the drifts. We will treat these below. Beforehand we examine trapping in a tokamak magnetic field.

Under conventional tokamak ordering the unit vector $\mathbf{b}$ for $\mathbf{B}$ may be approximated as $R\nabla\varphi$, the position-dependent major radius times the gradient of the geometric toroidal angle, and the magnitude B may be approximated by I/R with I constant. The canonical momentum is then written as

$$\mathbf{p}^* = e\mathbf{A}_{\text{pol}} + z\nabla\varphi, \qquad \nabla \times \mathbf{A}_{\text{pol}} = I\nabla\varphi \qquad z = e\psi + mUR \tag{13.17}$$

where z is the coordinate and U the parallel velocity which depends on the spatial coordinates as well as z. The poloidal magnetic field is represented by ψ, which is incorporated into $\mathbf{p}^*$ so that the toroidal magnetic field is determined by $\mathbf{A}_{\text{pol}}$, which we never have to write explicitly. The Hamiltonian is given by

$$H = \frac{m}{2}U^2 + wB = \frac{1}{2mR^2}(z - e\psi)^2 + wB \tag{13.18}$$

where the second form is used in computing gradients with z and not $v_\parallel = U$ the coordinate. The toroidal canonical momentum p_φ^* is the toroidal covariant component of $\mathbf{p}^*$, which in this representation is just z. So the curve of constant H in the poloidal plane (either $\psi\theta$ or as in this case RZ) is constructed by choosing the three constants of the motion (H, z, and w), thereby producing a 1D curve in the RZ poloidal plane with $H = H_0$. To place the gyrocentre on a particular flux surface $\rho = \rho_0$ where

$$\psi = \psi(\rho), \qquad \psi_0 = \psi(\rho_0) \tag{13.19}$$

we can choose $z = e\psi_0$.

With the experience of the first example above, we can proceed heuristically, with B_0 the value of B at $R = R_0$, the centre of the curve of $\psi = \psi_0$ describing the flux surface. So far, ψ is general. For purposes of illustration we set up the simplest case, which is circular surfaces, neglecting any complications due to finite aspect ratio R_0/ρ_0 or finite pressure and current of the tokamak equilibrium. So we set

$$\rho^2 = (R - R_0)^2 + Z^2, \qquad \frac{1}{\rho}\frac{\partial\psi}{\partial\rho} = -\frac{B_0}{q}, \qquad q(\rho) = q_0 + q_1(\rho/a_0)^2 \tag{13.20}$$

as a model for the flux surfaces, where $\rho = a_0$ is the location where $\psi(a_0)$ vanishes, namely, the overall toroidal minor radius. For B we have

$$B = B_0 R_0/R \tag{13.21}$$

For ψ as a profile following q, we have

$$\psi = \frac{B_0 a_0^2}{2q_1}\log\frac{q_a}{q}, \qquad q_1 = q_a - q_0 \tag{13.22}$$

This is a parabolic q-profile, for which a reasonable case is given by

$$q_0 = 1.5, \qquad q_a = 4.0, \qquad q_1 = 2.5 \tag{13.23}$$

This gives a reasonable model for setting up generic tokamak cases for study.

To illustrate trapped and passing gyrocentre orbits, we can choose parameters

$$\frac{\rho_0}{a_0} = 0.7, \qquad \frac{a_0}{R_0} = 0.3 \tag{13.24}$$

Note that the actual values of B_0 and the plasma parameters are not needed because of the way they cancel out when producing the trajectories. The remaining salient parameters are the local rho-star and the kinetic velocity, given by

$$\rho_* = \frac{v/\Omega}{\rho_0}, \qquad v^2 = 2\frac{H_0}{m} \tag{13.25}$$

where v can correspond to parallel or perpendicular energy, or both. Actual values of ρ_* in a tokamak are of order 10^{-2} or smaller (for $H_0 = 1$ keV and $B_0 = 3$ T one finds $v/\Omega = 0.216$ cm for a deuterium ion). For illustrative purposes one draws a curve for a larger value. The width of a trapped orbit can be anticipated by assuming $wB_0/H_0 = 1/2$ and setting $Z = 0$ while finding the points in R for which $H = H_0$. Taylor expand ψ and then use Eq. (13.20) to evaluate the derivative. This gives

$$\rho - \rho_0 = \pm 2^{-1/2}\frac{v}{\Omega}\frac{qR_0}{\rho_0} \tag{13.26}$$

The shape of the orbit is a rounded crescent which has led to the name 'banana orbit' for the trajectory. The 'banana width' is equal to the quantity on the right hand side, called the 'poloidal gyroradius', for this case. This can be a tenth of a_0 for exceptional cases though it is usually smaller (i.e., much smaller for electrons). So in

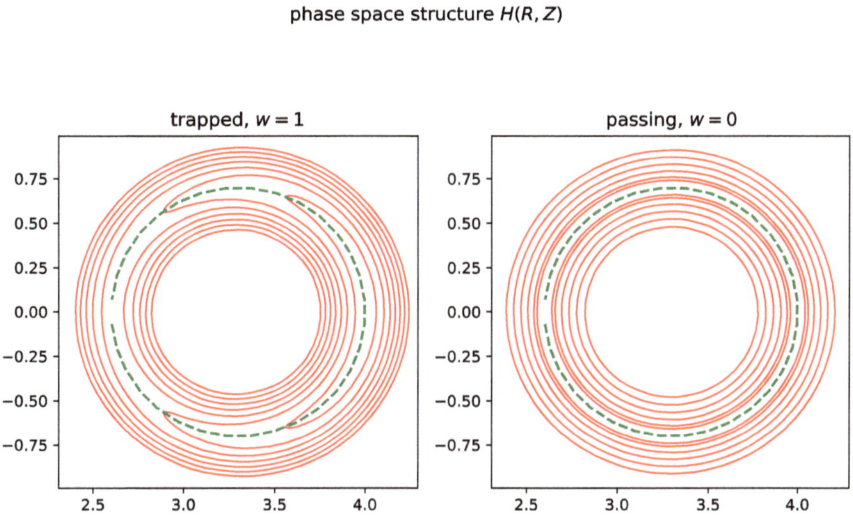

Figure 13.1. Contours of constant undisturbed Hamiltonian H_0 in the poloidal plane of the torus in a circular tokamak model. The magnetic field is lower on the outboard midplane ($Z = 0$ for $R > R_0$) so that the parallel velocity is largest on that side of each contour. The toroidal magnetic field and equilibrium current are both out of the plane on the right of the symmetry axis, so that motion in the direction of the magnetic field has a projection onto the plane of counterclockwise poloidal motion. For thermal perpendicular energy ($\mu B_0 = T$) one finds closed contours which do not contain the magnetic axis ($\rho = 0$), and which do not extend to the inboard midplane. These are the trapped orbits. For zero perpendicular energy all orbits are on the flux surfaces. The reference position is $\psi = \psi_0$ at about $\rho_0/a_0 = 0.7$, set by the choice of canonical toroidal momentum ($p_\varphi^* = e\psi_0$).

displaying curves, we set this width large enough for clarity. In figure 13.1 the contours of constant $H = H_0$ are shown for the simple model

$$H = \frac{(\psi - \psi_0)^2}{2\delta_0^2 R_0^2} + wR_0/R = H_0 \tag{13.27}$$

where H_0 is normalised to the temperature, ψ to $B_0 a_0^2$, and w to T/B_0. The drift parameter is $\delta_0 = \rho_s/a_0$ in the global geometry. The reference flux surface is given by $\psi = \psi_0$ at $\rho_0/a_0 = 0.7$. We set $\delta_0 = 0.01$ which is a factor of two to three larger than it usually is, so that the trapped orbits are visible. In the left frame we have $w = 1$, and can expect that except for very large $\psi - \psi_0$ all the orbits are trapped for this value of δ_0. For a more realistic one the trapped region is much thinner. In the right frame we have the orbits for $w = 0$, such that all are passing orbits. At this value of the canonical momentum, $\psi_0/\delta_0 R_0$, the orbits for $\rho > \rho_0$ are positive, in the counterclockwise sense. For $\rho < \rho_0$ they move the other way. So trapped orbits circulate in the same sense, around the contour. But in doing this they turn around in their poloidal direction at the 'tips' of the 'banana' and thereby do not reach the toroidal inner midplane ($\theta = \pm\pi$ at $Z = 0$ on the left side of the frame). Hence, they are trapped to the outer side of the torus, and the interest for instabilities is that the curvature is unfavourable there (interchange drive is maximum for $\nabla \log B$ aligned with ∇p).

In a fluxtube model described in Chapter 2 we have the geometry along the field line represented by dependence of coefficients in the s-direction, along the magnetic field. The xy-plane is local and homogeneous, namely, no geometric coefficient depends on x or y. Trapping follows from the s-dependence of B, given in normalised units by $1/(1 + \epsilon_a \cos s)$, following that of R/R_0, with $\epsilon_a = r_0/R_0$, where r_0 and R_0 are the reference toroidal minor and major radii. The structure of the undisturbed Hamiltonian H_0 is shown in figure 13.2 in the parallel phase space sz-plane at constant w and the velocity space zw-plane at constant s, where z normalised to mV and w to T/B_0, with V given by $V^2 = T/m$. In the figure, $v_\perp = (2w)^{1/2}$ gives the perpendicular velocity coordinate, so that H_0 is represented as isotropic. The structure in the sz-plane shows the structure of the trapped orbits as the closed contours which do not extend all the way to $s = -\pm\pi$, which is the inner midplane of the torus. The second contour is approximately the separatrix which divides the trapped region from the two passing regions, one positive and the other negative according to the sign of z. In the absence of collisions, the particles move on the sz-contours and are static in zw. If pitch angle scattering collisions are present, the particles diffuse along contours of constant H_0 in zw. If energy scattering is present, then they also diffuse across the contours. In a total-f setting, these diffusion effects conserve energy and momentum pointwise locally in space.

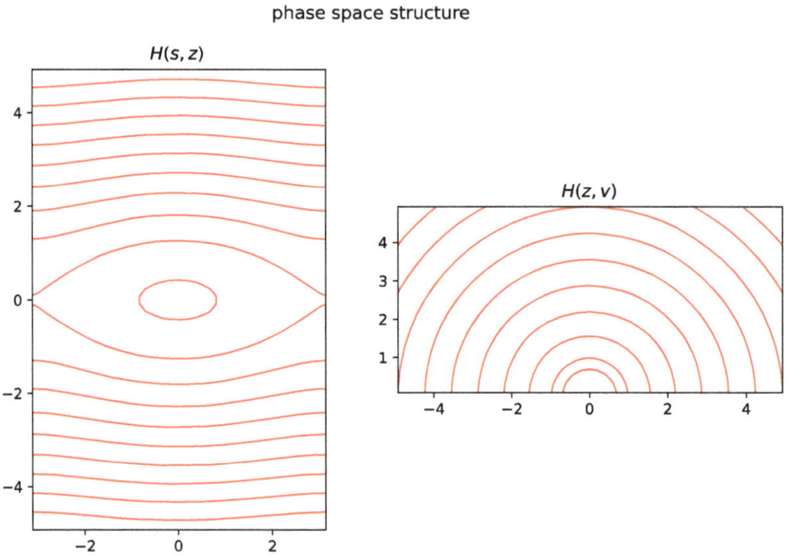

Figure 13.2. Contours of constant undisturbed Hamiltonian H_0 in the sz parallel phase space for $w = 1.9$ (left) and in the zw velocity space for $s = \pi/2$, where $v = (2w)^{1/2}$ (right). The right side shows that velocity space is isotropic in zv, and left side shows the trapped particle region (closed contours) and the passing regions (open contours), with the direction of motion positive (to the right) above and negative (to the left) below the trapped region, respectively. The second contour looking to just close at $z = 0$ and $s = \pm\pi$ is close to the separatrix for this value of w. Particle motion in the absence of collisions and fields is on surfaces of constant H_0, so they follow the contours in the sz-plane. Trapped orbits are those on the closed contours, which turn around to reverse velocity before reaching the $s = \pm\pi$ points which indicate the inner midplane of the torus. Pitch angle scattering by collisions would diffuse in the zw-plane along the contours of H_0.

13.3 The toroidal precession drift

Naïvely, one can propose a heuristic two-dimensional interchange model in which the active species is a trapped electron population, with the circulating population given an adiabatic response. The effective coupling is due to curvature terms as in the interchange forcing models. An ion temperature response can be added to provide a hybrid ion temperature gradient and trapped electron mode (ITG–TEM) turbulence model.

This isn't how trapped electron responses work in toroidal geometry, however. The bounce motion of the trapped orbit sets up a similar picture to a gyrocentre. The centre itself moves solely due to drifts while the fast orbital motion is decoupled from the rest. In a similar way, we can have the bounce centres as the fundamental ingredients and separate out the bounce motion so that the trapped orbit drifts according to a drift motion of its centre. In an axisymmetric background geometry, the bounce centre is found not to move radially, but has a solely toroidal motion with the centre itself remaining at the same radial and poloidal location in the plane perpendicular to the toroidal angle unit vector, namely, the poloidal plane. For a purely poloidal field, the toroidal drift is the simple drift motion averaged over a trapped orbit. In this case there is no radial or poloidal drift (both $\mathbf{B}$ and ∇B or $\mathbf{b} \cdot \nabla \mathbf{b}$ are in the same poloidal plane), just the toroidal one. For a tokamak-like field (mostly toroidal), though, the main drifts are radial and poloidal, and the toroidal drift is of a bounce centre precessing toroidally as a result of the parallel and drift motion in combination. Hence the name toroidal precession drift. We will examine this in the general axisymmetric equilibrium since it becomes possible in the end to use the general result in both the dipole and tokamak field cases.

13.3.1 Equilibrium and Orthogonal Coordinates

Due to the mathematical operations faced, it is helpful to do this calculation in orthogonal coordinates, meaning that all three coordinate gradients are mutually orthogonal (there are no off-diagonal metric coefficients). The axisymmetric magnetic field can be represented as

$$\mathbf{B} = I\nabla\varphi + \nabla\psi \times \nabla\varphi \tag{13.28}$$

and also as

$$\mathbf{B} = I\nabla\varphi + G\nabla\chi \tag{13.29}$$

where φ is the geometric toroidal angle and χ is the poloidal coordinate. The relation between χ and ψ is given by

$$\nabla\psi \times \nabla\varphi = G\nabla\chi \tag{13.30}$$

two equivalent expressions for the poloidal magnetic field. The orthogonality and axisymmetry conditions give

$$\nabla\psi \cdot \nabla\chi = 0, \qquad \nabla\varphi \cdot \nabla f = 0 \tag{13.31}$$

for any scalar quantity f. The quantities G and I are functions of ψ only,

$$G = G(\psi), \qquad I = I(\psi) \tag{13.32}$$

so that $\nabla \cdot \mathbf{B}$ vanishes properly and there is no force in the toroidal direction, respectively.

The strength of the poloidal magnetic field is given by

$$B_\perp = R^{-1}\,|\nabla\psi| = G\,|\nabla\chi| \tag{13.33}$$

It follows that

$$\mathbf{B} \cdot \nabla\chi = \nabla\psi \times \nabla\varphi \cdot \nabla\chi = \frac{B_\perp^2}{G} \tag{13.34}$$

which gives the Jacobian. The sense is that $G > 0$ with

$$(-\nabla\psi), \qquad \nabla\chi, \qquad \nabla\varphi \tag{13.35}$$

forming a right handed coordinate system. The poloidal flux radius is given by

$$\rho^2 = \frac{\psi - \psi_0}{\psi_a - \psi_0} \tag{13.36}$$

where $\psi_a = 0$ and ψ_0 is the axis value. Note that ρ^2 is always positive regardless of the sign of ψ.

The plasma current as given by Ampère's law is

$$\mu_0 \mathbf{J} = \nabla \times \mathbf{B} = \nabla I \times \nabla\varphi + \nabla G \times \nabla\chi \tag{13.37}$$

whose evaluation is assisted by the orthogonality of the coordinate gradients. Note that $\nabla\psi$ and $\nabla\rho$ are opposite if the toroidal components of both $\mathbf{B}$ and $\mathbf{J}$ are positive, with ψ positive and rising from zero at the boundary to a maximum on the magnetic axis.

13.3.2 Drift combinations

What follows is valid for any axisymmetric equilibrium so long as the magnetisation condition

$$\delta_R = \frac{\rho_L}{R} \ll 1 \tag{13.38}$$

is well satisfied everywhere. For drift motion in equilibrium magnetic fields, δ_R is the relevant drift parameter. The drifts in the magnetic field are given by Eq. (6.106), in which the field variables and $O(\delta_R)$ corrections are neglected.

The curvature drift is decomposed according to

$$B\,\nabla \times \mathbf{b} = \mathbf{b} \times \nabla B + \nabla \times \mathbf{B} \tag{13.39}$$

with the first term combining with the grad-$\mathbf{B}$ drift and the second treated separately. The $UM\nabla \times \mathbf{W}$ term is neglected, as it is one order higher in δ_R than anything we keep. This gives

$$B_\parallel^* Z^{\mathbf{R}} = U\mathbf{B} + \frac{mU^2 + M\Omega}{eB^2}\mathbf{B} \times \nabla B + \frac{mU^2}{eB}\nabla \times \mathbf{B} \tag{13.40}$$

as the starting point for the drift motion. We leave this in terms of $M\Omega$, with $\Omega_0/\Omega = B_0/B$.

The magnetic components of the coordinate gradients are given by

$$\mathbf{B} \cdot \nabla\psi = 0, \qquad \mathbf{B} \cdot \nabla\chi = \frac{B_\perp^2}{G}, \qquad \mathbf{B} \cdot \nabla\varphi = \frac{I}{R^2} \tag{13.41}$$

The components of the main drift are

$$\begin{aligned} \mathbf{B} \times \nabla B \cdot \nabla\psi &= I\frac{B_\perp^2}{G}\frac{\partial B}{\partial\chi} \\ \mathbf{B} \times \nabla B \cdot \nabla\chi &= -I\frac{B_\perp^2}{G}\frac{\partial B}{\partial\psi} \\ \mathbf{B} \times \nabla B \cdot \nabla\varphi &= G\frac{B_\perp^2}{G}\frac{\partial B}{\partial\psi} \end{aligned} \tag{13.42}$$

The components of the current term are

$$\begin{aligned} \nabla \times \mathbf{B} \cdot \nabla\psi &= 0 \\ \nabla \times \mathbf{B} \cdot \nabla\chi &= \frac{\partial I}{\partial\psi}\frac{B_\perp^2}{G} \\ \nabla \times \mathbf{B} \cdot \nabla\varphi &= -\frac{\partial G}{\partial\psi}\frac{B_\perp^2}{G} \end{aligned} \tag{13.43}$$

These three contribute to each component of the drift, except that there is no radial current $(\mathbf{J} \cdot \nabla\psi)$ since both I and G are flux functions.

The radial drift is given by the grad-B term, with the parallel motion and the current not contributing,

$$B_\parallel^* Z^\psi = I\frac{B_\perp^2}{G}\frac{mU^2 + M\Omega}{eB^2}\frac{\partial B}{\partial\chi} \tag{13.44}$$

The constants of the motion are ψ, H, M, with

$$H = \frac{m}{2}U^2 + M\Omega \tag{13.45}$$

so we have

$$mU\nabla U = -M\Omega\nabla \log B, \qquad U\nabla\frac{mU}{eB} = -\frac{mU^2 + M\Omega}{eB^2}\nabla B \tag{13.46}$$

in any direction. Using this relation greatly simplifies the algebra. The radial drift is then

$$B_\parallel^* Z^\psi = -I\frac{B_\perp^2}{G}U\frac{\partial}{\partial\chi}\frac{mU}{eB} \tag{13.47}$$

The corresponding poloidal motion is the parallel one, to lowest order in δ_R,

$$B_\parallel^* Z^\chi = \frac{B_\perp^2}{G}U \tag{13.48}$$

To find the radial excursion along the poloidal trajectory, in other words the tangent vector in the poloidal plane, we take the ratio of these, in which $B_\parallel^*$ drops out. We then have

$$\frac{d\psi}{d\chi} = -\frac{\partial}{\partial\chi}I\frac{mU}{eB} \tag{13.49}$$

since I is a flux function. The closed integral over this vanishes, so that there is no radial drift of a bounce centre in an axisymmetric magnetic field. The radial displacement along the trajectory is given by

$$\delta\psi = -I\frac{mU}{eB} \tag{13.50}$$

valid since U is zero at the bounce angle, at which $\delta\psi$ is also zero.

To find the toroidal precession drift, we need the other two components of the drift motion. These are

$$B_\parallel^* Z^\chi = U\frac{B_\perp^2}{G}\left(1 + \frac{\partial}{\partial\psi}I\frac{mU}{eB}\right) \tag{13.51}$$

for the poloidal direction, and

$$B_\parallel^* Z^\varphi = U\frac{I}{R^2}\left(1 - \frac{B_\perp^2/G}{I/R^2}\frac{\partial}{\partial\psi}G\frac{mU}{eB}\right) \tag{13.52}$$

for the toroidal direction, in both cases using Eq. (13.46) and combining in the term from $\nabla\times\mathbf{B}$ involving the derivatives of I or G with respect to ψ. The trajectory in the $\chi\varphi$-plane is given by the ratio, in which the drift terms are expanded as $O(\delta_R)$ corrections. They all add together to the main parallel motion,

$$\frac{d\varphi}{d\chi} = \frac{I/R^2}{B_\perp^2/G}\left(1 - \frac{B_\perp^2/G}{I/R^2}\frac{\partial}{\partial\psi}G\frac{mU}{eB} - \frac{\partial}{\partial\psi}I\frac{mU}{eB}\right) \tag{13.53}$$

Note how the leading factors of U and $B_\parallel^*$ cancel out.

The first term on the right hand side averages to zero to $O(1)$ over the trapped orbit since it is simply the ratio of the field line components, and on that line the orbit traverses the same path up and down. Due to the motion out of the flux surface due to the toroidal magnetic field, though, it has an $O(\delta_R)$ correction which survives and combines with the other terms. The correction coming from the Taylor expansion is

$$\frac{I/R^2}{B_\perp^2/G} \rightarrow \frac{I/R^2}{B_\perp^2/G} + \delta\psi \frac{\partial}{\partial\psi}\frac{I/R^2}{B_\perp^2/G} \tag{13.54}$$

We then have

$$\frac{d\varphi}{d\chi} = O(1) - \left(G + \frac{I/R^2}{B_\perp^2/G}\right)\frac{\partial}{\partial\psi}\frac{mU}{eB} - \frac{mU}{eB}\frac{\partial}{\partial\psi}\left(G + \frac{I/R^2}{B_\perp^2/G}\right) \tag{13.55}$$

in which the second term is the combination of the terms from the radial expansion just done and the $\nabla \times \mathbf{B}$ terms. Since the magnetic field strength is

$$B^2 = \frac{I^2}{R^2} + B_\perp^2 \tag{13.56}$$

the closed integral of the toroidal drift for one poloidal cycle is given by

$$\Delta\varphi = \oint d\varphi = \oint \frac{d\varphi}{d\chi}\, d\chi = -\frac{1}{e}\frac{\partial}{\partial\psi}\oint d\chi\, \frac{GB}{B_\perp^2}\, mU \tag{13.57}$$

The small change of the bounce angle endpoints of the integration with ψ is regarded as an order higher in δ_R and is neglected.

The first factor in the integrand is recognised as the parallel line element,

$$\frac{d\chi}{\mathbf{b}\cdot\nabla\chi} = d\chi\frac{GB}{B_\perp^2} \tag{13.58}$$

The integral of the parallel momentum over this line element is the action angle variable,

$$J_\chi = \oint \frac{d\chi}{\mathbf{b}\cdot\nabla\chi}\, mU \tag{13.59}$$

which is also called the second adiabatic invariant (M or w being the first). Hence the toroidal drift is given by

$$\Delta\varphi = -\frac{1}{e}\frac{\partial J_\chi}{\partial\psi} \tag{13.60}$$

that is, just the radial derivative of J_χ divided by the charge. The minus sign on $d\psi$ results from ψ increasing inward for positive senses in all variables. This result is due to Kadomtsev (1958).

The toroidal precession drift is given by this displacement divided by the time over one orbit, which is given by a similar integral,

$$\tau_\chi = \oint \frac{d\chi}{\mathbf{b}\cdot\nabla\chi}\frac{1}{U} \tag{13.61}$$

With mU in the definition of J_χ and $1/U$ here we recognise this as

$$\tau_\chi = \frac{\partial J_\chi}{\partial H} \qquad \text{with} \quad \frac{\partial H}{\partial U} = mU \tag{13.62}$$

The toroidal precession drift is then given as a contravariant component by

$$V^\varphi = -\frac{1}{e}\frac{\partial J_\chi/\partial \psi}{\partial J_\chi/\partial H} \tag{13.63}$$

It is valid for general axisymmetric equilibria, assuming only that δ_R is small. This result is due to Rosenbluth and Sloan (1971).

13.3.3 The tokamak magnetic field

In the concentric circular model used in Eq. (13.20), the coordinates (r, θ, φ) are already orthogonal, and we have

$$J_\theta = \oint d\theta \; mUqR_0 \tag{13.64}$$

In the analytical calculations in the literature $\varepsilon = r/R_0$ is taken to be the small parameter of an expansion. In this case we have

$$\Omega = \Omega_0(1 - \varepsilon\cos\theta) \qquad \text{hence} \quad mU = \sqrt{2mM\Omega_0\varepsilon}\sqrt{\cos\theta - \cos\theta_0} \tag{13.65}$$

where

$$\cos\theta_0 = -\frac{H - M\Omega_0}{M\Omega_0\varepsilon} \tag{13.66}$$

determines θ_0, the bounce angle. In these terms the bounce time is given by

$$\tau_\theta = \oint \frac{d\theta \, qR_0}{U} = 4\frac{mqR_0}{\sqrt{2mM\Omega_0\varepsilon}}\int_0^{\theta_0}\frac{d\theta}{\sqrt{\cos\theta - \cos\theta_0}} \tag{13.67}$$

These give rise to the results

$$\int_0^{\theta_0}\frac{d\theta}{\sqrt{\cos\theta - \cos\theta_0}} = \sqrt{2}\, K(k) \tag{13.68}$$

$$\int_0^{\theta_0} d\theta\,\sqrt{\cos\theta - \cos\theta_0} = 2\sqrt{2}\left[E(k) - (1 - k^2)\, K(k)\right] \tag{13.69}$$

where

$$K(k) = \int_0^{\pi/2}\frac{d\eta}{\sqrt{1 - k^2\sin^2\eta}}, \qquad E(k) = \int_0^{\pi/2} d\eta\,\sqrt{1 - k^2\sin^2\eta} \tag{13.70}$$

are the complete elliptic integrals of first and second kind, respectively. To get these denote

$$\sin\eta = k^{-1}\sin\frac{\theta}{2}, \qquad k = \sin\frac{\theta_0}{2} \tag{13.71}$$

so that k is ultimately set by the effective pitch angle

$$\frac{H - M\Omega_0(1-\varepsilon)}{M\Omega_0} \to 1 - \lambda = \zeta^2 = 2\varepsilon\, k^2 \tag{13.72}$$

using Eq. (13.66) and the definition of k, and taking the $\varepsilon \ll 1$ limit. The extremes $k = 0$ and $k = 1$ set the bounce points, or banana tips, at the outboard and inboard midplane, $\theta_0 = 0$ and $\theta_0 = \pi$, respectively. If $H = M\Omega_0$, then the tips are on the vertical line $R = R_0$ and the orbit covers the outer half of the torus. In this case $B_{\min} = B_0(1-\varepsilon)$, not B_0, and the trapped fraction is

$$f_t = \sqrt{2\varepsilon} \tag{13.73}$$

following $0 < k < 1$ and $B_{\min}/B_{\max} = 1 - 2\varepsilon$.

The radial gradient in Eq. (13.63) is

$$\frac{\partial J_\theta}{\partial\psi} = -\frac{\partial r}{\partial\psi}\oint d\theta\left(mU\frac{\partial q}{\partial r}R_0 + \frac{qR_0}{U}M\Omega_0\frac{\partial}{\partial r}\varepsilon\cos\theta\right) \tag{13.74}$$

The first term yields J_θ itself times $S = rq'/r$, the magnetic shear parameter, and the radial flux gradient $-\psi'$, with the prime denoting $\partial/\partial r$. The second yields another combination of E, $K(k)$, following

$$\int_0^{\theta_0}\frac{d\theta\cos\theta}{\sqrt{\cos\theta - \cos\theta_0}} = \int_0^{\theta_0} d\theta\sqrt{\cos\theta - \cos\theta_0} + \cos\theta_0\int_0^{\theta_0}\frac{d\theta}{\sqrt{\cos\theta - \cos\theta_0}} \tag{13.75}$$

which are evaluated according to Eqs (13.68), (13.69). Since $-\psi' = rB_0/q$, the result for the toroidal precession drift is

$$V^\varphi = \frac{q}{r}\frac{2M\Omega_0}{eB_0R_0}G(k) \tag{13.76}$$

where the coefficient is given by

$$G = \frac{E(k)}{K(k)} - \frac{1}{2} + 2S\left(\frac{E(k)}{K(k)} + k^2 - 1\right) \tag{13.77}$$

The factor of $M\Omega_0$ is replaced by H since $1 - \lambda$ is proportional to ε. This result is due to Kadomtsev and Pogutse (1967).

In a fluxtube, $y = (r/q)(q\theta - \varphi)$, so that the drift becomes a velocity given by

$$V^y(H, k) = -\frac{2H}{eB_0R_0}G(k) \tag{13.78}$$

So although the toroidal precession drift is given by a combination of processes, which includes the radial excursion from the flux surface and the parallel motion, the effect is similar to that of a curvature term and scales the same way.

The toroidal precession drift averaged over a Maxwellian distribution has two properties of interest. We replaced the factor of $M\Omega_0$ with H to get $V^y(k)$, since $\lambda > 1 - 2\epsilon$. With $M\Omega_0$ left there, an integration over the entire distribution with $-1 < \zeta < 1$ yields the 2/3 factor cancelling the 3/2 factor arising from the velocity (v) integration, yielding unity (the perpendicular pressure is nT). Over the trapped fraction the range of ζ is close to zero, so that the 2/3 factor is replaced by unity, and then the integration over v yields the 3/2 factor. The average of $G(k)$ over $0 < k < 1$ and then over v now yields

$$V^y = -\frac{2T}{eB_0R_0}(0.462 + 0.425\,S) \tag{13.79}$$

which can be used in fluid or gyrofluid models. The curvature operator is reduced to just $-V^y(\partial/\partial y)$, since there is no radial drift of the bounce centre.

13.3.4 The dipole magnetic field

In the dipole magnetic field the calculation leading to Eq. (13.63) is easier. With $I = 0$, there is neither a radial drift nor a perpendicular drift in the poloidal direction. So we have only the parallel motion in the poloidal direction and the native drift in the toroidal direction. However, with $B_\perp = B$ and $\mu_0 p \ll B^2$ the result for the toroidal precession drift is still the one in Eq. (13.63). On the other hand, it is no longer so simple to calculate J_θ or τ_θ in the dipole field in Eq. (13.12). The integrals are evaluated numerically. The native drift is

$$Z^\varphi = \frac{mU^2 + M\Omega}{eB^2}\mathbf{B} \times \nabla B \cdot \nabla\varphi \tag{13.80}$$

from Eq. (13.40). The velocity combination is combined into H and angles according to

$$mU^2 + M\Omega = 2H\left(1 - \frac{B_0}{2B}\right) \tag{13.81}$$

Evaluating the vector combination for the dipole field, we have

$$Z^\varphi = \frac{6H}{er_0B_0}\left(1 - \frac{B_0}{2B}\right)\left[\frac{1 + \sin^2\theta}{(1 + 3\sin^2\theta)^2}\cos^2\theta\right] \tag{13.82}$$

The units of velocity are recovered by multiplying by the local cylindrical radius, $r\cos\theta = r_0\cos^3\theta$, so that

$$V_d(H, \theta) = g_{\varphi\varphi}^{1/2}Z^\varphi = \frac{6H}{er_0B_0}\left(1 - \frac{B_0}{2B}\right)\left[\frac{1 + \sin^2\theta}{(1 + 3\sin^2\theta)^2}\cos^5\theta\right] \tag{13.83}$$

gives the local toroidal velocity (this is not the same thing as V^y since $y = -r_0\vartheta$ here not $r\vartheta$). Fitzpatrick (see Further Reading) gives this result, then on the orbital average

$$\tau_\theta = \frac{r_0}{v}(3.7 - 1.6\lambda^{1/2}), \qquad V_d(H, \lambda) = \frac{6H}{er_0B_0}(0.35 + 0.15\lambda^{1/2}) \tag{13.84}$$

as a useful approximation for the average toroidal drift velocity of the bounce centre. This velocity is proportional to energy as in a diamagnetic drift velocity, with a relatively weak dependence on the pitch angle. Locally however, the kinetic motion is strongly localised to $\theta = 0$, so that most of the current is equatorial, and in space physics is called the toroidal ring current. Fitzpatrick gives simple comparisons to observational data.

13.3.5 Contrast to the tokamak case

In an axisymmetric magnetic field which is purely poloidal, of which the dipole magnetic field is an example, the toroidal drift is direct, not really a precession. The parallel motion is poloidal, there is no radial drift excursion, and the drift is purely toroidal. It does give rise to a current because of the charge separation (proportionality to $1/e$). This current is the counterpart to the diamagnetic current from the fluid motion, and also has for equal temperatures the same magnitude for electrons and singly charged ions.

In a tokamak magnetic field, there is the finite poloidal component and therefore the same pure drift motion in the toroidal direction. This contribution is practically negligible in a tokamak, however. The main process is the mismatch of the positive and negative sides of the orbit. The dominant pieces of this are due to the radial and poloidal drift motion, yielding the largest terms (the two with I/R^2 on the right hand side) in Eq. (13.55). The gradients of I and G and the difference between B^2 and I^2/R^2 are small by $\beta = 2\mu_0 p/B^2$ and ϵ^2/q^2, with ϵ/q giving the angle of the field lines away from purely toroidal in this magnetic field. The directly toroidal pieces of the drift are therefore small corrections (about one percent) in the tokamak case. This is why the drift is called a toroidal precession, and why the parallel motion itself is responsible for it.

The role of the canonical toroidal momentum p_φ^* is present in both cases with the difference that in the purely poloidal field cases there is no parallel velocity component. In those cases p_φ^* is just $e\psi$, and so with p_φ^* conserved by the drift motion the gyrocentre drift cannot have a radial excursion. In the tokamak case p_φ^* includes mUb_φ, so that when U changes so must the local ψ, which leads to the finite radial excursion. Together with the poloidal excursion off the field line, this leads to most of the toroidal precession drift in a magnetic field with these properties.

13.4 Single-centre drifts versus gyrokinetics

In a gyrokinetic setting, we should not think of single-particle motion, but of characteristics of the distribution function f. In general, the motion is only instantaneously along surfaces of constant H (including the fields). Even in a delta-f setting it is only instantaneously along surfaces of constant H_0. Only in linear theory is it really true that 'integration along unperturbed orbits' is a valid step in the calculation. It is only really true for a linear eigenmode that the motion stays on these contours for indefinite time. At all times, it is important to recall we are evolving a self consistent system of arbitrarily many particles and fields, in which the fields are never presented as static to the particles. Nevertheless, when understanding drifts, whose coefficients go into a delta-f fluxtube model, it is convenient to regard them as the motion of the centres of orbits which otherwise do not change in time.

13.5 Trapped electrons as separate species in turbulence

In a fluid model, one can model trapped and passing electrons as separate species, with conservative exchange through a collisional transfer process. Although it is more customary in such models to treat the passing electrons as adiabatic (e.g., Garbet *et al*, 2003), the earlier one proposed by Waltz (1985) treated both trapped electron and collisional drift wave processes in the same model. With the appropriate treatment of the Ohm's law, the passing electron response is allowed to be electromagnetic. Besides trapped/passing collisional exchanges, the species couple through their mutual exchanges with the ExB energy. In a fluid model, the vorticity equation has to be constructed accordingly, and it is hard to get exact energetic consistency in a toroidal model, as Waltz noted. In a gyrofluid model, as we will see, it is possible to (1) treat the species as separate, (2) conserve free energy just as in any other multi-species model, (3) arrange the collisional exchanges so that they conserve particles and satisfy positive-definite damping of free energy, and (4) incorporate the poloidal asymmetry in the distribution of the trapped electron population and its effect on polarisation.

This last issue was not addressed in the early papers, but was a central point of the treatment by Fong and Hahm (1999), one of whose central points was that (1) the potential seen by the bounce centre is a weighted average over trapped orbits in the poloidal angle, and (2) the gyrocentre charge density due to the bounce centres itself has poloidal extent. The main point was to treat the bounce centres in the same way as gyrokinetic theory does a gyrocentre. The bounce centre becomes a two-dimensional object (in ψ and ϑ). Radial motion is due to the ExB velocity while the toroidal precession drift takes the place of the interchange curvature forcing, producing an effect much like interchange curvature but with a different coefficient. The poloidal envelopes work together to provide a free energy theorem for the system.

These poloidal distribution effects can be treated through an envelope function in the interaction Hamiltonian as we will show below. In fact, doing so automatically satisfies the requirement that the poloidal extent of the trapped electron density work together properly with the poloidal averaging of the potential to produce the effective potential seen by the bounce centres. This is analogous to the way that the same gyroaveraging operator used to get the potential seen by a gyrocentre is the same as the one used to get the spatial distribution of the polarisation charge density of a gyrocentre. Energy conservation and exchange eventually works the same way in both cases.

As for the passing electrons as a species, they interact with the potential with an electromagnetic adiabatic response, producing the drift Alfvén physics we have been studying. The poloidal extent of the trapped population is offset by the dynamics of the passing population according to self consistent responses to all the coupling mechanisms, maintaining not only quasineutrality but also the smallness of vorticity compared to density in normalised amplitudes and of the ExB energy compared to the thermal free energy. Although the trapped population is not constrained by an adiabatic response (it has none), its free energy amplitude does not run away to large values in the way exhibited by 3D drift wave turbulence without magnetic shear as seen in Volume One.

13.5.1 Bounce averaged objects

The trapped region is given by the region interior to the separatrix between closed and cyclic open contours of H_0 in the poloidal phase space plane (p_z and θ or as in the computations, z and s). Since the geometry is independent of y in a fluxtube model, which follows $-\varphi$ for fixed θ in the torus, the poloidal extent of this region on average can be used to set a sampling function in the parallel coordinate for the trapped electrons. For this purpose the bounce frequency is assumed to be arbitrarily fast, so that averaging along phase space surfaces (here, the curves) of constant H_0 is similar to averaging over the gyrophase angle.

The position of the centre of this orbit in the outboard midplane is taken as a bounce centre. So, analogous to the gyromotion, the bounce motion is a line contour of averaging in a linearised model (recall the fluxtube model keeps nonlinearity only in the ExB advection and magnetic flutter terms). The bounce centre density is a two-dimensional (2D) object sitting at $s = 0$ in xy-space, but it samples the s-dependence of ϕ according to a prescribed function of s which multiplies the bounce centre density to produce the three-dimensional (3D) trapped electron density. The sampling occurs according to

$$\int dx\, dy\, n_b e \phi_b, \qquad \phi_b(x, y) = \int ds\, p_b(s)\phi(x, y, s) \tag{13.85}$$

This gives the interaction Hamiltonian, where n_b is the bounce centre density in 2D, ϕ_b is the bounce centre field potential, and p_b is a bounce sampling envelope function. The gyrocentre charge density in 3D is given by the functional derivative of the interaction Hamiltonian with respect to ϕ, or

$$\rho_t = e\, n_t, \qquad n_t(x, y, s) = n_b(x, y)\, p_b(s) \tag{13.86}$$

where n_t gives the 3D spatial extent of the trapped electron density. In this way the envelope $p_b(s)$ models the poloidal extent of the trapped orbits taken collectively as the region within the separatrix for each H and then averaged over H. The main complication computationally is that this s must be on the same coordinate system as the $\{xy\}$ in the outboard midplane at $s = 0$. So that if ϕ is being kept at $\{x, y_k, s\}$ at each s_k, those shifts between y_k and y must be undone before finding ϕ_b or ρ_b (recall y is y_k for $s_k = 0$). The choice for the sampling function is constrained by energetic considerations, as detailed below. For now we can set it to vanish smoothly at $s = \pm\pi$, where the extent of the trapped region in velocity space goes to zero. This avoids any geometric issues at the branch cut.

The bounce averaged quantities see a curvature operator given by the toroidal precession drift velocity

$$\mathcal{K}_b = -V^y \frac{\partial}{\partial y} = \lambda_b\, \mathcal{K}^y \frac{\partial}{\partial y}, \qquad \lambda_b = 0.462 + 0.425\, S \tag{13.87}$$

where S is the magnetic shear. It scales the same way as in the usual curvature operator, except that it involves only the y-direction since the bounce centres do not have radial motion. The coefficient λ_b as a dimensionless parameter. In a local model, a form like the one in Eq. (13.79) can be used. Various expressions for λ_b have been taken by various authors, and we can take this one.

13.5.2 Simple exposition of the energetics

To show how the ideas behind the gyrofluid model apply to trapped electrons we can explore it in a simplified two moment fluxtube model without any finite gyroradius or magnetic pumping effects. The moment equations for the ions are

$$\frac{\partial n_i}{\partial t} + [\phi, n_i + \log n_0] + B\nabla_\parallel \frac{u}{B} = \mathcal{K}(\phi + \tau_i n_i) \tag{13.88}$$

$$\frac{\partial}{\partial t}\left(A + \mu_i u\right) + [\phi, \mu_i u] = \mu_i \tau_i \mathcal{K}(2u) - \nabla_\parallel(\phi + \tau_i(n_i + \log n_0)) - C\, J \tag{13.89}$$

and for the passing electrons we have

$$\frac{\partial n_e}{\partial t} + [\phi, n_e + \log n_0] + B\nabla_\parallel \frac{v}{B} = \mathcal{K}(\phi - n_e) \tag{13.90}$$

$$\frac{\partial}{\partial t}\left(A - \mu_e v\right) - [\phi, \mu_e v] = -\mu_e \mathcal{K}(2v) - \nabla_\parallel(\phi - (n_e + \log n_0)) - C\, J \tag{13.91}$$

just as in a usual drift-Alfvén model, while for the trapped electrons

$$\frac{\partial n_b}{\partial t} + [\phi_b, n_b + \log n_0] = \mathcal{K}_b\left(\phi_b - n_b\right) \tag{13.92}$$

is the only one since parallel dynamics is absent for bounce centres. Here, $\mu_i = 1$ and $\mu_e = 1/3670$ are the normalised masses, and $\tau_i = T_i/T_e$ in the background. The electron parameters are not signed in this case, since the minus signs are written explicitly in the equations. The ions and passing electrons respond to ϕ and to $A_\parallel$, while the trapped electrons average the parallel responses away and respond to ϕ_b, the bounce-averaged electrostatic potential. With $\mathcal{K}$ the usual curvature operator, $\mathcal{K}_b$ is the counterpart for the bounce centres, due to the toroidal precession drift.

The resistivity is given by C as before, acting solely on the ion and passing electron parallel velocities, u and v, respectively.

The density variables are normalised to their respective background densities. In total we have n_{e0} as the background value for both ions and electrons, but the trapped and passing background densities are $f_t n_{e0}$ and $f_c n_{e0}$, respectively. The normalised background gradient is then $\log n_0 = -\omega_n x$ for all species, entering in ExB advection and magnetic flutter.

The bounce averaged potential seen by the trapped electrons, $\phi_b(x, y)$, is mediated by the envelope function p_b as in Eq. (13.85). The same envelope becomes a charge assignment function which places the 2D bounce centre density $n_b(x, y)$ in space to produce the 3D trapped electron density $n_t(x, y, s)$.

In the polarisation equation, n_t is the trapped electron density,

$$-\delta_0^2 \nabla_\perp^2 \phi = n_i - \left(f_t n_t + f_c n_c\right) \tag{13.93}$$

The induction equation does not see the trapped population but the passing electrons still enter with their charge density coefficient, such that

$$-\delta_0^2 \nabla_\perp^2 A = \beta_e J, \qquad J = u - f_c v \tag{13.94}$$

The trapped electrons do not appear in the current. The effective charge density for the circulating particles, $-f_c$, appears with the parallel electron velocity. With this form for J, the resistivity satisfies the energetics correctly, since $C J$ gets multiplied by u and $-f_c v$ to produce $C J^2$ as the resistive dissipation.

The form of the equation for n_b suggests that the free energy is formed by multiplying by $n_b - \phi_b$, so that the ExB nonlinearity and the precession drift term conserve energy. The ExB energy is formed by multiplying the polarisation equation by ϕ and integrating, so that

$$E_e = \mathcal{V}^{-1} \int d\mathcal{V} \frac{1}{2} |\delta_0 \nabla_\perp \phi|^2 = \mathcal{V}^{-1} \int d\mathcal{V} \frac{1}{2} \phi\left(n_i - f_c n_e - f_t n_t\right) \tag{13.95}$$

$$\frac{\partial E_e}{\partial t} = \mathcal{V}^{-1} \int d\mathcal{V} \left(\phi \frac{\partial n_i}{\partial t} - f_c \phi \frac{\partial n_e}{\partial t} - f_t \phi \frac{\partial n_t}{\partial t}\right) \tag{13.96}$$

The integration over trapped electrons proceeds in the same way as in the interaction Hamiltonian,

$$\int d\mathcal{V}\, \phi\, n_t = \int dx\, dy\, n_b \oint ds\, \phi\, p_b = \int dx\, dy\, n_b \phi_b \tag{13.97}$$

Since p_b commutes past the time derivative, the same equivalence applies to the energetics,

$$\int d\mathcal{V}\, \phi \frac{\partial n_t}{\partial t} = \int dx\, dy\, \phi_b \frac{\partial n_b}{\partial t} \tag{13.98}$$

The ExB energetics is therefore compatible with either form of density times potential. The thermal free energy should satisfy

$$\int dx\, dy\, n_b^2 = \int d\mathcal{V}\, n_t^2 \qquad \text{hence} \qquad \oint ds\, p_b^2 = 1 \tag{13.99}$$

which becomes the energetic constraint on p_b. With this in hand, the 2D energetics of the bounce averaged quantities works properly in the 3D energetics of the whole, and the collisionless part of the energetics works exactly the same way as in any other multi-species gyrofluid model.

This leaves the trapped/passing collisional exchanges. There are three requirements, (1) the total electron density $f_c n_e + f_t n_t$ is conserved, (2) the equilibrium point of the collisional process is an isotropic state such that $n_e = n_t$, and (3) the effect on free energy is positive definite damping. These are satisfied if we prescribe

$$\frac{\partial n_e}{\partial t} = \cdots + f_c^{-1}\, p_b\, D_t \qquad \frac{\partial n_b}{\partial t} = \cdots - f_t^{-1}\, D_t \tag{13.100}$$

noting that these are 3D and 2D objects, respectively, hence p_b in the one for n_e. The dissipation term, a 2D object, is given by

$$D_t = \nu_t \oint ds\, p_b\, (n_t - n_e) \tag{13.101}$$

where ν_t is the effective de-trapping collisionality. One can verify that D_t satisfies

$$\frac{\partial}{\partial t}\left(f_c n_e + f_t\, p_b\, n_b\right) = \cdots + 0 \tag{13.102}$$

$$D_t = 0 \qquad \text{if} \quad n_t = n_e \tag{13.103}$$

$$\frac{\partial}{\partial t}\int d\mathcal{V}\, \frac{1}{2}\left(f_c\, n_e^2 + f_t\, n_t^2\right) = \cdots - \nu_t \int dx\, dy \left[\oint ds\, p_b\, (n_t - n_e)\right]^2 \tag{13.104}$$

where the ellipsis refers to the collisionless part of the dynamics, the zero denotes the required cancellation, and the last term denotes the positive definite damping of free energy. The collision operator thereby satisfies the required properties.

This details the construction of the equations and free energy theorem for the isothermal model, which is at about the same level of detail as the one by Waltz. Using the fact that the structure and energetics of gyrofluid models with any level of moment truncation (i.e., 2 or 6 or 12 or 20 moments, etc) is the same (indeed, the same as the gyrokinetic one), it is a choice of how much to keep and then a simple matter to preserve the energetics the way it is done in the isothermal model. In the following, we turn to computations using the six-moment model for the passing species, so as to treat hot plasma cases.

13.5.3 Integration into the six moment gyrofluid model

All of the ingredients necessary to construct computations are now present. The separate species treatment for the trapped electrons allows straightforward preservation of energetic consistency for the resulting equations at any level of moment hierarchy truncation. The gyrocentre quantities for each species evolve separately and are then tied together through the field equations (polarisation and induction). The trapped electron variables are state variables only, described in the 2D space of the bounce centres in the outer toroidal midplane. The ions and passing electrons are treated as for electrons and ions in the examples in previous Chapters, with the only difference being the use of *fcr*, different from unity, as the partial charge density coefficient. The equations for ions and passing electrons are as in Section 11.2.1 of Chapter 11, with the FLR corrections keeping G_0 and G_1 as gyroaveraging operators and ϕ_0 and ϕ_1 as the corresponding potentials.

For the trapped electrons, a_{00}, a_{20}, and a_{01} are the nonzero moments. The equations are

$$\frac{\partial}{\partial t}a_{00} + \left[\phi_b,\, a_{00} + \log n_0\right] = \frac{1}{2}\mathcal{K}_b\left(2\phi_b - 2a_{00} - a_{01} - \sqrt{2}\,a_{20}\right) \tag{13.105}$$

$$\frac{\partial}{\partial t}a_{20} + \left[\phi_b,\, a_{20} + 2^{-1/2}\log T_0\right] = \frac{1}{2}\mathcal{K}_b\left(\sqrt{2}\,\phi_b - \sqrt{2}\,a_{00} - 6a_{20}\right) \tag{13.106}$$

$$\frac{\partial}{\partial t}a_{01} + \left[\phi_b,\, a_{01} + \log T_0\right] = \frac{1}{2}\mathcal{K}_b\left(\phi_b - a_{00} - 4a_{01}\right) \tag{13.107}$$

in the same normalisation as above but otherwise in the language of Section 10.2 from Chapter 10. The bounce averaged potential ϕ_b is as given in Eq. (13.85), with the envelope function

$$p_b = \sqrt{\frac{2}{3}}(1 + \cos s) \tag{13.108}$$

To find ϕ_b we shift ϕ from the $\{xy_k s\}$ coordinate system back to $\{xys\}$ and then do the integration over s. The bounce centre curvature operator is as given in Eq. (13.87), or

$$\mathcal{K}_b = \delta_0\omega_b\frac{\partial}{\partial y}, \qquad \omega_b = \lambda_b\omega_B, \qquad \lambda_b = 0.462 + 0.425\,S \tag{13.109}$$

in normalised units.

The field equations are done taking account of the f_c and f_t coefficients for the electrons. The trapped electron density is defined in $\{xy\}$, multiplied by $p_b(s)$ as in Eq. (13.86), and then shifted onto the $\{xy_k s\}$ coordinate system. Then we have

$$P(\phi) = [G_0(a_{00}) + G_1(a_{01})]_i - \left[f_c a_{00}\right]_c - \left[f_t n_t\right] \tag{13.110}$$

for polarisation, with the 3D trapped density n_t given by

$$n_t(x, y, s) = a_{00}(x, y)\,p_b(s) \tag{13.111}$$

and then shifted onto $\{xy_k s\}$. For induction we have

$$-\delta_0^2\nabla_\perp^2 A = \beta_e\left([a_{10}]_i - \left[f_c a_{10}\right]_c\right) \tag{13.112}$$

with the left hand sides unchanged from the previous usage. Note that the trapped electrons do not contribute to the parallel current. The notation including the polarisation operator P is as in Eq. (11.44) of Chapter 11.

For this illustration the collisionless case is taken, so that both the collisional effects we have treated before and the collisional detrapping are set to zero. Landau damping in the ions and passing electrons are the only physical dissipation channels. Subgrid dissipation is done as before.

The ions and passing and trapped electrons are advanced separately according to their time-dependent equations, after which the field variables are solved for, constituting one time step. This is the same structure as with any of the gyrokinetic or gyrofluid models we have used. The free energy theorem is constructed in the same way as in the isothermal model, with the additional moments beyond density and parallel velocity done as in the case without the split of the electrons into trapped and passing components.

13.5.4 Mode structure and effect on the threshold

We take an illustrative case of ITG turbulence in a hot plasma, allowing the self-consistent and electromagnetic adiabatic response of the ions and the passing electrons, together with the density (a_{00}) and parallel (a_{20}) and perpendicular (a_{01}) temperatures of the trapped electrons. If f_t is set to zero, then the model reverts to the one used in Chapter 11 for electromagnetic ITG turbulence. We get a direct comparison by setting f_t to zero, to 0.001, and to 0.4, with the lower value as an additional control test of model consistency. The base case is given by

$$\begin{aligned} &\hat{\beta} = 0.1, \quad \hat{\mu} = 0.0254, \quad \hat{\varepsilon} = 93.3, \\ &\omega_B = 0.29 \quad \delta_0 = 0.0141, \quad \omega_n = 0.321, \\ &\omega_t = \tau_i = 1, \quad S = 0.78, \quad \varepsilon_a = 0.18 \end{aligned} \tag{13.113}$$

with $\nu_e = \nu_i = \nu_t = 0$, chosen somewhat arbitrarily but with anticipation of more of an electromagnetic response. Other details, including grid, domain size and aspect ratio, and subgrid dissipation, are as given for the ITG cases in Chapter 11. We look at the different values of f_t within sweeps of ω_n between zero and 1.5 holding $\omega_t = \omega_i$ fixed to check for a transition between dominance of ion temperature and trapped electron pressure gradient drive of the resulting turbulence.

Time traces of the ion heat flux in each case are shown in figure 13.3, together with the run of the turbulent flux channels with the density gradient parameter ω_n. The format is similar to the fluid drift hot plasma cases in figure 3.12 from Chapter 3. In those cases, the turbulence was robust for low ω_n compared to $\omega_t = 1$ but decreased and then shut off for $\omega_n > 0.7$. The ion heat flux channel was dominant for all cases. With the trapped electron population present, the turbulence is robust for low ω_n and the ion conductive heat flux G_i is the strongest channel, but now also at about $\omega_n = 0.7$ the ion particle flux F_i is the strongest. In terms of a heat flux this translates to $Q_i = 1.5\, F_i + G_i$ as a specific total heat flux in the ions, giving the large values seen in the time traces. It is interesting that it is not the trapped electrons that give the largest values, but by synergy the ions. The main point, though, is that the

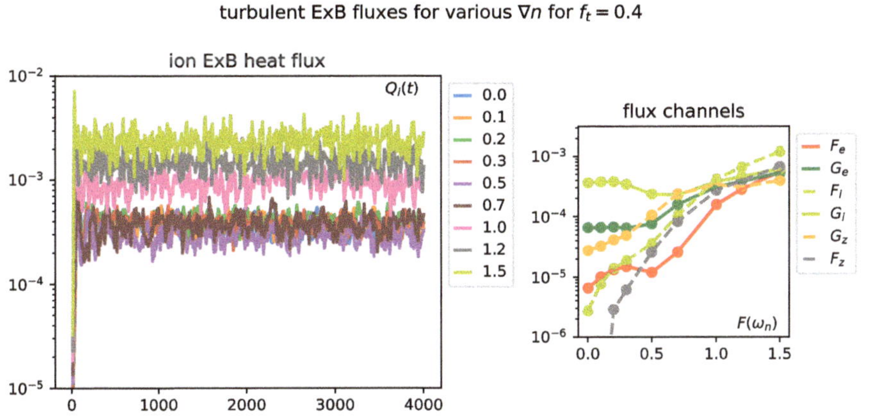

Figure 13.3. (Left) time traces of the ExB advective ion heat flux at various values of density gradient ω_n at fixed temperature gradient $\omega_t = \omega_i = 1$ and other parameters, for the hot plasma case with a trapped fraction $f_t = 0.4$ in the electrons. One finds straightforward evolution, with clear saturation in all cases. The highest three ω_n values (1.0 and higher) produce the strongest turbulence. (Right) time averaged ($2000 < t < 4000$) fluxes at the various ∇T values, with the ExB advective fluxes for particle (F) and conductive heat (G) channels for all species $\{eiz\}$ for passing electrons, ions, and trapped electrons, respectively (given in terms of velocity in units of c_s). The ion heat flux is dominant at low ω_n, and at higher ω_n the ion particle flux, not the trapped electron one, is the strongest. In terms of heat flux, $Q_i = 1.5\, F_i + G_i$, producing the large values seen in the time traces.

threshold in $\eta_i = \omega_i/\omega_n$ is eliminated by the trapped electrons, since now there is ample ability to access the free energy in the density gradient. This occurs because the trapped electron population no longer follows the adiabatic response. Indirectly, the ion density fluctuations are freed from this constraint and then they carry the majority fraction of the energy for the turbulence.

The wavenumber spectra (k_y) are given for the $\omega_n = 0.5$ case, averaged over $2000 < t < 4000$, in figure 13.4. Compare with the $\hat{\beta} = 0.1$ case in figure 11.2. In the spectra, the trapped electron spectra (n_z and T_z), and their density and temperature gradient drives (Γ_{nz} and Γ_{Tz}), are shown separately. The shape of these spectra is much more smoothly rounded near the peak for these cases with trapped electrons, compared to in the pure passing species cases of the gyrofluid and fluid drift models. The ion temperature gradient drive (Γ_i) remains in the intermediate ITG spectral range, while the electron signals are dominated by the trapped component, which as a whole is shifted towards longer wavelength. This is a signature of the trapped electron responses in the turbulence. The Landau damping channels are most prominent at longer wavelength, as in most other cases.

The isotropic perpendicular spectra ($k_\perp$) are shown for the same case in figure 13.5, averaged over the same time interval. These emphasise the scale of motion over the wavelength. Compare with the $\hat{\beta} = 0.1$ case in figure 11.4, where the vorticity was seen to collect near ρ_s as opposed to the rest of the dynamics. With the trapped electron responses present, the vorticity as well as the electron and ion dynamics has shifted towards larger scale. Most of the signal in the electron drive channels is due to the trapped electrons (Γ_{nz} and Γ_{Tz}). The ion temperature gradient drive (Γ_i) remains

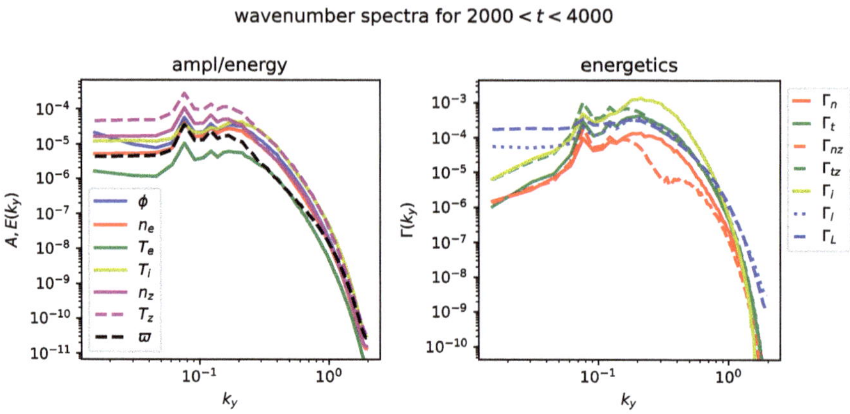

Figure 13.4. Drift-direction wavenumber (k_y) spectra of the amplitudes and energetics, averaged over $2000 < t < 4000$, for the hot plasma case with $\omega_n = 0.5$. The passing (n_e) and trapped (n_z) electron components are given separately. The passing electron and ion density gradient drive is Γ_n, and the trapped electron one is Γ_{nz}. The electron temperature gradient drive is also separated into passing (Γ_t) and trapped (Γ_{tz}) components. The ion temperature drive (Γ_i) is in the intermediate ITG spectral range, while (Γ_{tz}) is very strong and is shifted to longer wavelength. As in the gyrofluid ITG cases, the vorticity peaks where the rest of the dynamics does, but in this case that means longer wavelength. Ion (Γ_L) and electron (Γ_l) Landau damping is strongest at longer wavelength.

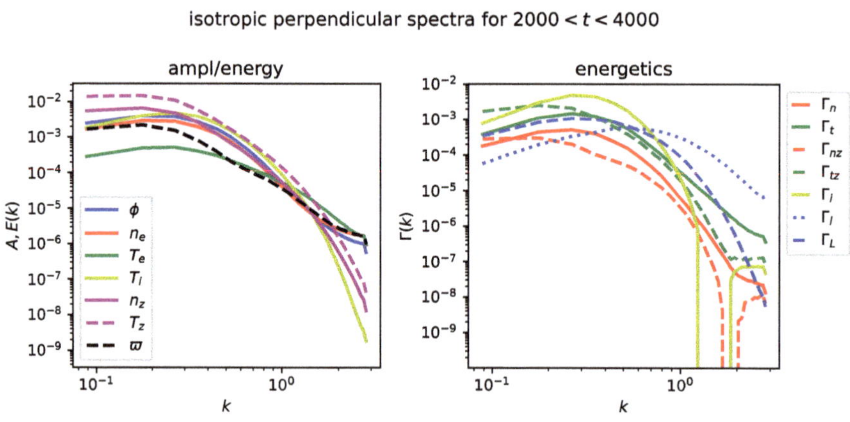

Figure 13.5. Isotropic perpendicular wavenumber ($k_\perp$) spectra of the amplitudes and energetics, averaged over $2000 < t < 4000$, for the hot plasma case with $\omega_n = 0.5$. The format is the same as in the previous figure. The ion temperature drive (Γ_i) is in the intermediate ITG scale range while the trapped electron temperature drive (Γ_{tz}) is strong at large scale. This is reflected in the strength at large scale in both the density (n_z) and temperature (T_z) spectra. As in the gyrofluid ITG cases, the vorticity peaks where the rest of the dynamics does, but in this case that means larger scale. Note the low amplitudes in both the density (n_e) and temperature (T_e) of the passing electrons, as well as their lower drive levels. Note also the strength of ion Landau damping (Γ_L) at large scale and the flatness of the electrons (Γ_l).

in the intermediate scale ITG range, but the trapped electron dynamics collects at larger scale, especially the temperature components. The vorticity now follows the trapped electron density (n_z), with the greater freedom to adjust with three densities present. Recall that ϖ is the difference among the densities, namely, the negative of the gyrocentre charge density. The ExB vorticity itself still collects at small scale (see below). But with three effective species present, there is more room for differences, which is balanced by the FLR corrections to the densities and temperatures as well as the polarisation. In general these features account for the increased strength in the turbulence.

The parallel envelopes of amplitudes and turbulent ExB flux channels are shown for the same case in figure 13.6, averaged over the same time interval. The magnetic flutter flux is small and negative. Each species is normalised to its own background density and temperature, so the trapped component appears higher than it actually is (the fraction factor is $f_t^2 = 0.16$). Nevertheless, the trapped electron conductive heat flux is the second strongest, after the ion one. The degree of ballooning in the trapped electron channels follows p_b^2, and the ion temperature channels adjust to more or less the same structure. This is still a borderline ITG turbulence case, while the trapped electrons take over for larger ω_n/ω_t, especially greater than unity.

The late-time spatial morphology is shown in figure 13.7, at $t = 4000$ for this case. The three densities, trapped (n_z) and passing (n_e) electron and ion (n_i) are shown with ϕ and the ExB vorticity Ω_E (in units of $c_s/L_\perp$), and the parallel current (J). The densities are shown with gradient added to disturbances. The larger scale interchange appearance in n_z stands out. The trapped electron dynamics stamps this large-scale component onto the mix of self-consistent dynamics. However, the actual ExB vorticity, Ω_E, remains in small scale structures much as in all the other cases. So the vorticity spectra seen above are due to the gyrocentre charge density, while the actual ExB vorticity keeps its character.

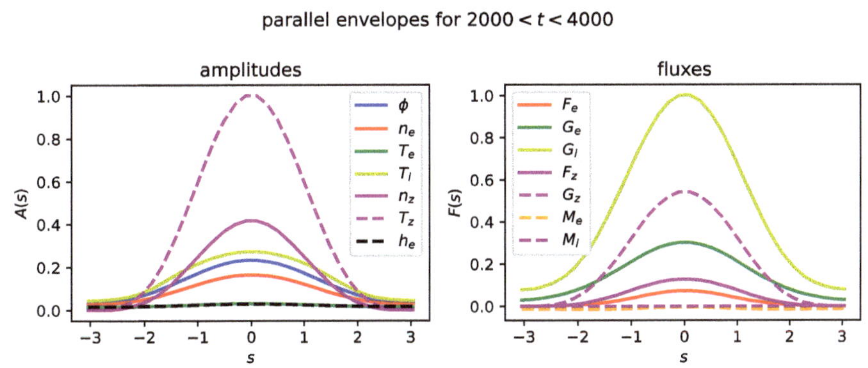

Figure 13.6. Parallel envelopes of the amplitudes and turbulent fluxes, averaged over $2000 < t < 4000$, for the hot plasma case with $\omega_n = 0.5$. Passing and trapped electrons are with the 'e' and 'z' subscripts, respectively. Though the ion conductive heat flux is the largest channel, the trapped electron conductive heat flux is next, and in the normalised fluctuations both n_z and T_z are higher. The square of the $f_t = 0.4$ factor would bring T_z back down below T_i, following the fluxes. The trapped signals follow the square of the bounce envelope $p_b(s)$ in each curve.

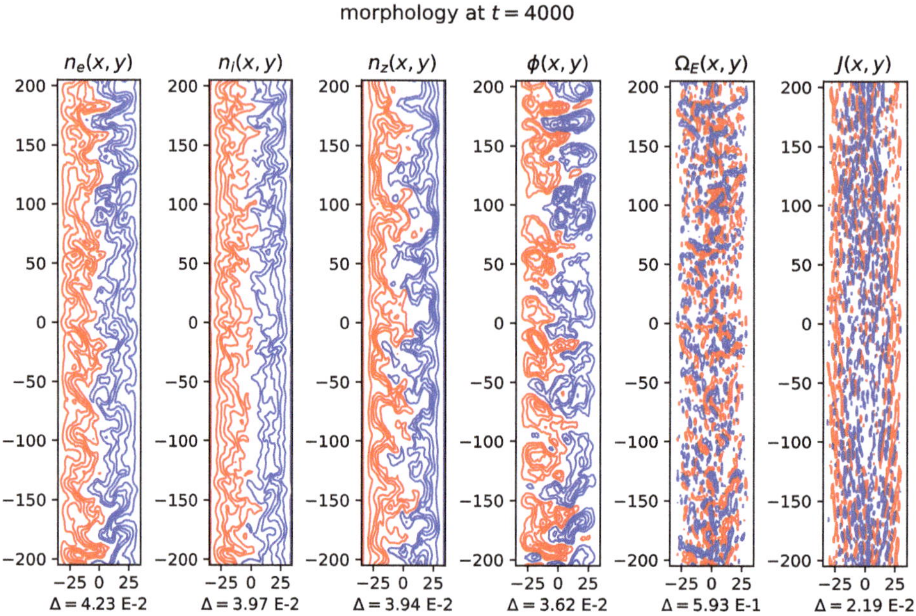

Figure 13.7. Spatial morphology at $t = 4000$ of the densities with their gradients, the ExB potential and vorticity, and the parallel current, for the hot plasma case with $\omega_n = 0.5$. Passing and trapped electrons are with the 'e' and 'z' subscripts, respectively. The ExB vorticity in units of $c_s/L_\perp$ is given by Ω_E, which for this model is no longer the same as $\delta_0^{-1}\varpi$ in normalised units. Positive contours are in red; negative, blue. The large scale structure in n_z is clear, and ϕ and Ω_E follow. The current remains at small scale.

This is an exploratory view of the effect of trapped electrons on the turbulence, with a case somewhat intermediate between ITG and pure trapped electron mode turbulence emphasised. The ion channels adjust to what the trapped electrons do, especially to break the adiabatic constraint on anything other than the passing electrons. This shows the indirect energetic character often on display in the self-consistent dynamics. The turbulence self adjusts within its constraints to produce a self consistent nonlinear dynamics, which is not obvious from any of its pieces taken in isolation.

13.6 On the kinetic details

In a kinetic computation the evolution of f is followed in the five-dimensional space represented by $\{xys\}$ in space and $\{zw\}$ in phase space. For electromagnetic cases one needs to cover spatial scales from $L_\perp$ to the smaller of d_e or ρ_s, while keeping the proper dimensionless ratios for the drift parameter $\rho_s/L_\perp$, the mass ratio m_e/m_i, the domain scale ratio L_y/L_x following $2\pi\rho_0/qL_\perp$, and the gradient scale ratio $R_0/L_\perp$. And as we have seen the need to capture the kinetic shear Alfvén Landau resonance is somewhat more severe than is generally realised.

We refer to computations in the literature both with the fluid approximations using adiabatic passing electrons and with kinetic models. These are work in progress mostly, because the resolution demands which we have been facing are not met. Which is understandable given the expense. One either vastly underestimates the needs for kinetic Alfvén responses at all scales from $L_\perp$ to d_e in velocity space, or abandons the long wavelengths needed to regulate decorrelation of the energy containing range of the turbulence, or, much worse in my opinion, uses artificial mass ratios for m_e/m_i, in many cases extreme enough to change the nature of the problem. Results can be expected to change in the future.

13.7 Summary

Particle trapping is a fundamentally kinetic problem. Even if we avoid specifically discussing individual particles and their dynamics, and concentrate upon the self consistent physical system of plasma and electromagnetic fields, one has the characteristics of the phase space advection following the structure of the Hamiltonian. These can be regarded as orbits whose properties arrange the dynamics even if individual gyrocentres do not stay on those orbits. The dependence of the amplitude of the magnetic field with position along the field lines leads to the existence of orbits which are closed in phase space, finitely extended in space. This part of velocity space does not experience field line connection because the orbits do not extend far enough for the periodicity constraints to become relevant. As a result, plasma elements in this region are free to rearrange phase shifts f to ϕ for maximal free energy release within the gradients, with coupling by curvature forcing. This is much as in an MHD ballooning system in which the interchange forcing overcomes the Alfvénic responses along field lines. So it is customary to treat the linear eigenmodes with a ballooning formulation.

However, a nonlinear system must always remain consistent with the constraints on each of the pieces. So if trapped electron turbulence is to be faithfully reproduced, the same standards as in pure drift Alfvén turbulence also apply here. The trapped species is incorporated into the whole, and it brings its own constraints. The entire model must be a superset of all these. So we have a three-species model with field line connection, boundary conditions, and aspect ratio considerations taken into account as before, with the envelope functions and their consistency constraints applied to the trapped species. When this is done, we have a viable gyrofluid model for the entire system.

The computational demands on kinetic models are sufficiently great that it can be judged that sufficiently well resolved computations to provide definitive results on the combination of trapped electron and drift Alfvén dynamics have not yet been presented for the kinetic case. The mass ratio is usually changed strongly towards unity, or the collisional diffusion scale in velocity space is increased, and/or the resonances involved in the kinetic damping mechanisms we have seen are not addressed. Published results report substantial progress, but the topic is still work in progress.

This is comparable to the situation from the previous Chapter simultaneously involving dynamics at the electron gyroradius scale and drift Alfvén and MHD dynamics up to the system scales. Resolution demands are very challenging for both cases even for gyrofluid models, let alone gyrokinetic ones. As stated there, the problems which involve communication across several different types of disparate dynamics is still wide open and will provide fertile ground for future generations of practitioners. Even the theory is not all done yet.

Further Reading

The toroidal drift of trapped particles in a dipole field summarised here is from the lectures by R Fitzpatrick, *Plasma Physics*, at `http://farside.ph.utexas.edu/teaching/plasma/lectures/Plasmahtml.html` especially the pages Van Allen Radiation Belts and Ring Current.

Trapped electron modes were done by B B Kadomtsev and O P Pogutse, in 'Plasma instability due to particle trapping in a toroidal geometry,' *Sov. Phys. JETP* **24** (1967) 1172, and in 'Plasma diffusion in toroidal systems,' *Sov. Phys. JETP* **31** (1970) 898, and by B Coppi and G Rewoldt, 'New trapped-electron instability,' *Phys. Rev. Lett.* **33** (1974) 1329. The relationship between the interchange drive of trapped electron turbulence and the toroidal precession drift was given in the first reference, which cites 1958 work in Russian by Kadomtsev for the proportionality of the precession drift to the radial derivative of the second adiabatic invariant.

The calculation using drift kinetic equations for general axisymmetric equilibria, yielding Eq. (13.63), was given by M N Rosenbluth and M L Sloan, 'Finite-β stabilization of the collisionless trapped particle instability,' *Phys. Fluids* **14** (1971) 1725. The small-ϵ version in Kadomtsev and Pogutse was generalised to finite pressure gradients by J W Connor, R J Hastie, and T J Martin, 'Effect of pressure gradients on the bounce-averaged particle drifts in a tokamak,' *Nucl. Fusion* **23** (1983) 1702. Further generalisation to noncircular flux surfaces was given by C M Roach, J W Connor, and S Janjua, 'Trapped particle precession in advanced tokamaks,' *Plasma Phys. Contr. Fusion* **37** (1995) 679.

Trapped electron equations in systems within a fluid drift model for turbulence were done by R E Waltz, 'Numerical simulation of electromagnetic turbulence in tokamaks" *Phys. Fluids* **28** (1985) 577, and M A Beer and G W Hammett, 'Bounce averaged trapped electron fluid equations for plasma turbulence,' *Phys. Plasmas* **3** (1996) 4018, and Jiquan Li and Y Kishimoto, 'Role of collisionless trapped electron mode in tokamak plasmas with electron internal transport barrier,' *Plasma Phys. Contr. Fusion* **44** (2002) A479, and X Garbet *et al*, 'Turbulent particle transport in magnetized plasmas,' *Phys. Rev. Lett.* **91** (2003) 035001.

The gyrokinetic calculation of the bounce motion was given by B H Fong and T S Hahm, 'Bounce-averaged kinetic equations and neoclassical polarization density,' *Phys. Plasmas* **6** (1999) 188. This is a direct analogue between the bounce motion for a trapped orbit and the gyromotion for a particle, and then between the drift of a gyrocentre and the toroidal drift of a bounce centre, as well as its radial drift for non-axisymmetric equilibria. For equilibrium relaxation phenomena it also includes the polarisation density resulting from the finite radial width of trapped orbits. The uneven poloidal distribution of the charge density of the trapped population was one of their main points.

Initial gyrokinetic simulations of trapped electron turbulence were done by T Dannert and F Jenko, 'Gyrokinetic simulation of collisionless trapped-electron mode turbulence,' *Phys. Plasmas* **12** (2005) 07239.

IOP Publishing

Turbulence and Instabilities in Magnetised Plasmas, Volume 2
Gyrokinetic theory and gyrofluid turbulence

Bruce Scott

Chapter 14

Turbulence and test particles

14.1 Introduction—trace species

In the context of magnetised plasma turbulence, a trace species is defined as one whose particles are advected by the turbulence but do not have a back reaction onto the ExB eddies and flows of the turbulence. In a two-dimensional setting this is equivalent to the model of test particles. Test particles are tracers which are individually advected by the turbulence. The similarity to a trace species is clear, and the model is basically the same in a simple 2D Eulerian fluid model, in which the only process is nonlinear advection. The only difference is subgrid dissipation. Test particles moving by themselves follow the flow lines of the turbulence but there is no application of anything which would otherwise affect the statistics of their motion. Their role is to measure statistical properties of the turbulence. One often assumes that the test particles have the same dynamics as the turbulence itself. At least, in our case, they should have the same dynamics as a passively advected scalar, which in 2D is what the trace species is. Apart from the self consistent dynamics of the electrons and main ions, we envisage a population of impurities. The test particles should have enough of a randomised start that they all evolve away from each other and therefore sample independently the degrees of freedom of the turbulence.

We have already seen, however, that the main feature of turbulence in a magnetised plasma is self-consistency. Especially the electrons do not behave as a proper test particle species or as a trace species. The reason to say this is the way in which the infinitesimally changed electron density distribution rapidly changes and directs the immediate and future evolution of the turbulent flows which do the advection of the electrons. Especially the adiabatic response leads to the strong back reaction effects onto the flow that we have been covering. As soon as the density fluctuation field changes, the potential field follows suit. The electrons in a sense direct and control this motion. The ions evolve with their nonlinear polarisation current so that they, too, do not follow a simple passive scalar advection. A trace species may indeed have solely a simple ExB advection process with nothing else in play.

doi:10.1088/978-0-7503-3855-4ch14

This is true at least in a 2D model. In a more complicated setting, the toroidal drifts as well as the parallel dynamics are also involved, which can give significant differences to the properties of the reaction of each species to the turbulence.

An investigation into trace species dynamics serves to illuminate the issues associated with either a test particle or trace species model which assumes passive advection. Differences to the electron and main ion responses can be measured with the mode structure diagnostics to provide a clear picture. The simple 2D model serves as a control case. The 3D model is the one which has the more realistic outcome. The results of a test particle or trace species equation running in time alongside the equations of the turbulence can be compared to those of the turbulence. This will ascertain whether the test particle or trace species model adequately represents the correlations and dynamics of the turbulence, or not.

The easiest way to do this is within the gyrofluid model, in which extra species couple to the rest solely through the polarisation and induction equations for the fields. The extra species, being a trace species, is assumed to have a density low enough that it can be neglected in the gyrocentre charge density and current so that there is no back reaction onto the turbulence, which remains controlled by the electrons and main ions. Then we can investigate the properties of the density fluctuations and correlations of each species with the electrostatic potential.

14.2 A two-dimensional trace species model

We covered the 2D dissipatively coupled turbulence model re-cast in gyrofluid language in Chapter 9. We can take this model and add a third species of trace ions. The equations for this in normalised units are

$$\frac{\partial n_e}{\partial t} + [\phi, n_0 + n_e] = D(\phi - n_e) \tag{14.1}$$

$$\frac{\partial n_i}{\partial t} + [\phi, n_0 + n_i] = 0 \tag{14.2}$$

$$\frac{\partial n_z}{\partial t} + [\phi, f_0 + n_z] = 0 \tag{14.3}$$

one for each of the electron (n_e), ion (n_i), and trace impurity (n_z) normalised densities. Each is normalised to its own species background density, which is n_{e0} for both the electrons and ions, and negligibly small for the impurities. The coefficient D is the dissipative coupling parameter, which we can set to run in a more adiabatic or more hydrodynamic regime. The background gradient drive represented by n_0 or f_0 gives an assumed shape of the background profile, which for impurities can be different from that of the main plasma. For 2D cases, the $k_y = 0$ component is suppressed in the time derivative, so fluctuations are followed but not transport, hence the background is just the gradient drive.

The field equation for polarisation and the main plasma background gradient drive are given by

$$\nabla_\perp^2 \phi = n_e - n_i, \qquad n_0 = -\omega_n x \tag{14.4}$$

The gradient drive for impurities is given by

$$\frac{\partial f_0}{\partial x} = -\omega_z x, \qquad \text{or} \qquad \frac{\partial f_0}{\partial x} = \exp[-(x/10)^2] \tag{14.5}$$

The units are normalised in terms of ρ_s and $c_s/L_\perp$. If $\omega_z = 0$ we have a localised background gradient drive, while if it is given a value we have the same shape for all of the drive terms. Each equation has the same subgrid dissipation operator, with strength $\nu_0 = 1$ as in all the other models we have considered.

The absence of n_z in the polarisation equation is what defines the impurities as a trace species. If a fluid model is reconstructed by subtracting the equations for $n_{e,i}$,

$$\frac{\partial \varpi}{\partial t} + [\phi, \varpi] = D(\phi - n_e) \qquad \text{where} \quad \varpi = n_e - n_i \tag{14.6}$$

the trace species is defined as one which has no forcing term in the vorticity equation (i.e., the D term involves only the electrons, and the polarisation itself involves only the main ions). So the electron and main ion plasma evolves according to its own rules, and the trace species is passively advected. This means that there is no back reaction from n_z back to ϕ, which presents a prescribed, if in an elaborate manner, time dependent turbulent flow field to the trace species.

For a standard case we initialise $n_e = n_i = \phi$ in an adiabatic initial state with one random-phase distribution, and n_z independently with another distribution with the same spectrum. The initial RMS amplitude is unity. The domain size is $L_y = 4L_x = 80\pi$, and the grid is 64×256 in x and y. The boundary conditions are the Dirichlet ones which are necessary to avoid large-scale down-gradient flows. The timestep is 0.01 in normalised units. Runs are taken to $t = 1000$ with statistics measured over $500 < t < 1000$. The coupling parameter is varied over a range of $10^{-3} < D < 1$, from hydrodynamic to adiabatic through intermediate cases.

Figure 14.1 shows the cross coherence and phase shift distributions between n_z, n_e and ϕ, for the $D = 0.001$ case. These show a very close relationship between the gyrocentre densities (the ions and electrons are also very close, broken only by the finiteness of ϖ). The corresponding measurements for n_e and ϕ, in the same figure, yield a turbulent state with hydrodynamic characteristics. The phase shifts are largest for the most strongly driven piece although they are not close to $\pi/2$, and the correlation is neither unity nor zero. Even at this small value, D is sufficient to affect the larger scales as it is equal to three factors of $k_\perp \rho_s$ if the latter is close to 0.1. It is understandable that the electrons and impurities should be highly correlated, since with the weak dissipative coupling the equations are nearly the same, even with the unequal gradient forcing.

The case with $D = 1$ is shown in figure 14.2. Here we find the electron relationship very strongly adiabatic, with correlation near unity and phase shifts strongly peaked near zero. But the relationship between n_z and n_e is much the same as in the $D = 10^{-3}$ case. So it is not a given that the nonlinear advection as the strongest piece

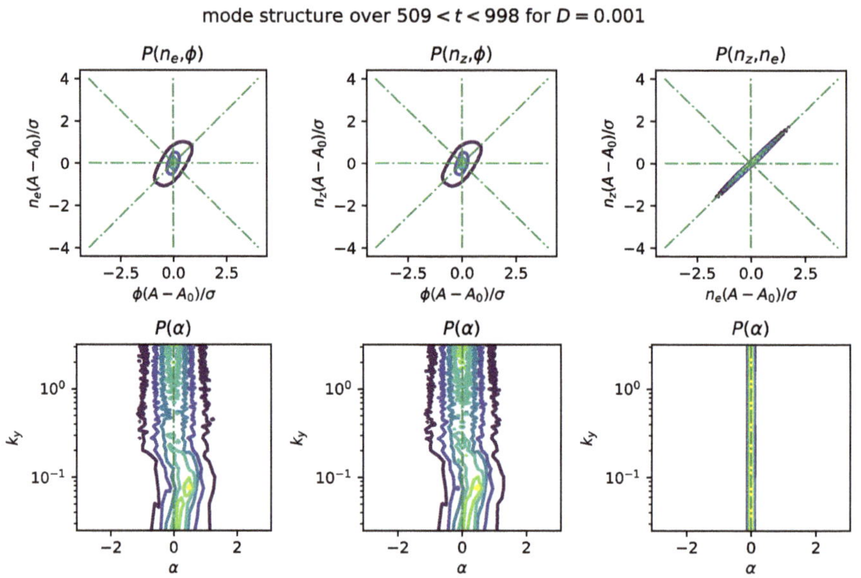

Figure 14.1. Cross coherence (top row) and phase shift distributions (bottom), between n_e (electron density) and ϕ (left), n_z (impurity density) and ϕ (center), and n_z and n_e (right), for the case $D = 0.001$ with mostly hydrodynamic characteristics. The cross coherence is relatively weak due to the nearly unopposed turbulent vorticity and advective ExB nonlinearities. Despite the loose correlation between n_e and ϕ, the densities follow each other extremely closely.

determines the passive advection statistics. In the $D = 1$ case the electrons are strongly adiabatic, indicating that the dissipative coupling process is very strong in their dynamics, and it is absent in the dynamics of the impurities. Nevertheless, we find both densities strongly correlated anyway. This is another form of dynamical alignment, and it suggests that the dynamics is not close to any linear relationship, and nor does it suggest that the turbulence is anything like a random walk process. In the latter we would not expect to see this sort of correlation between dependent variables.

In figure 14.3 we find more or less the same in the $D = 0.03$ case, which is more dissipative than either of the other two shown. The cross correlations are about 3×1 elliptical for either density to ϕ, but the densities are much more closely correlated between each other. The same is seen in the phase shifts, where we see a wider distribution relative to ϕ and the bend to slightly larger shifts for $k_y\rho_s \sim 0.3$ where the drive is strongest, but then a tight distribution near zero shift between the densities. This serves a control case for changes in the drive shape at this same D value.

The situation of completely different sourcing for the impurities and main plasma is of interest. We run again with $D = 0.03$, and the gradient drive for the electrons and ions is still done with $n_0 = -\omega_n x$ but now we drive the impurities with the narrow Gaussian localised in the middle of the domain. We now have outward transport away from the source layer in both directions, so the statistics are measured only

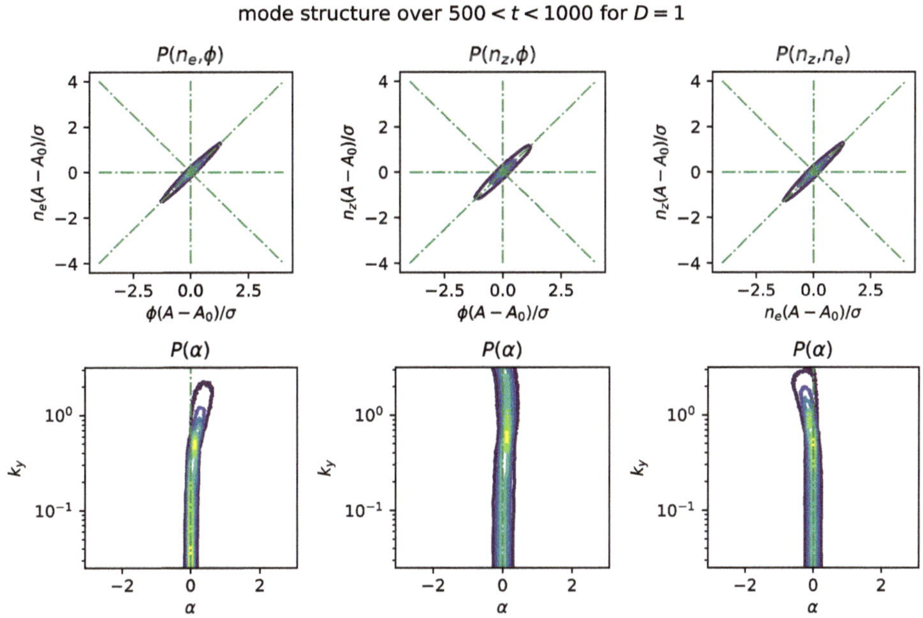

Figure 14.2. Cross coherence (top row) and phase shift distributions (bottom), between n_e (electron density) and ϕ (left), n_z (impurity density) and ϕ (center), and n_z and n_e (right), for the case $D = 1.0$ with mostly adiabatic characteristics. The cross coherence is very close due to the strong dissipative coupling, and the phase shifts are narrow and close to zero positive, compared to a more hydrodynamic state. Despite the physical coupling between n_e and ϕ, and its absence for n_z, the correlations and phase shifts between these two are largely independent of the strength of the coupling.

over the outer region, which is defined as $8 < x < 24$ in these units (x/ρ_s). This is down-gradient from the drive, or radial in the parlance of plasma confinement. In this case one has to wait somewhat longer for the correlations to emerge since at early times the linear responses are more prominent and these are now very different. Hence the measurement only after $t = 500$ for all cases. The spatial morphology of the four variables ϕ, n_e, n_z, and ϖ at $t = 1000$ is shown in figure 14.4. The densities n_e and n_z are plotted together with the background terms n_0 and f_0, respectively. The differences in gradient forcing are visible. The impurity fluctuations are well spread out over most of the domain. The subgrid dissipation coefficient allows them to survive long enough for this. The small-scale structure in n_e and n_z is visibly correlated. The potential and vorticity evolve in the same way as in the original 2D case, since that part of the dynamical system is unchanged.

The cross coherence and phase shift distributions between n_z, n_e and ϕ for this case are shown in figure 14.5. The fluctuations tend to be correlated on the right side ($x < 0$) and anti-correlated on the left side ($x > 0$), following in both cases the direction of the fluxes. Note that the electron flux is to the right everywhere by contrast, since that is the down-gradient side for the electrons everywhere. With this consideration we find somewhat sloppier statistics, but basically the same result. The phase shifts are smallest and the cross correlation is the tightest, between n_z and n_e,

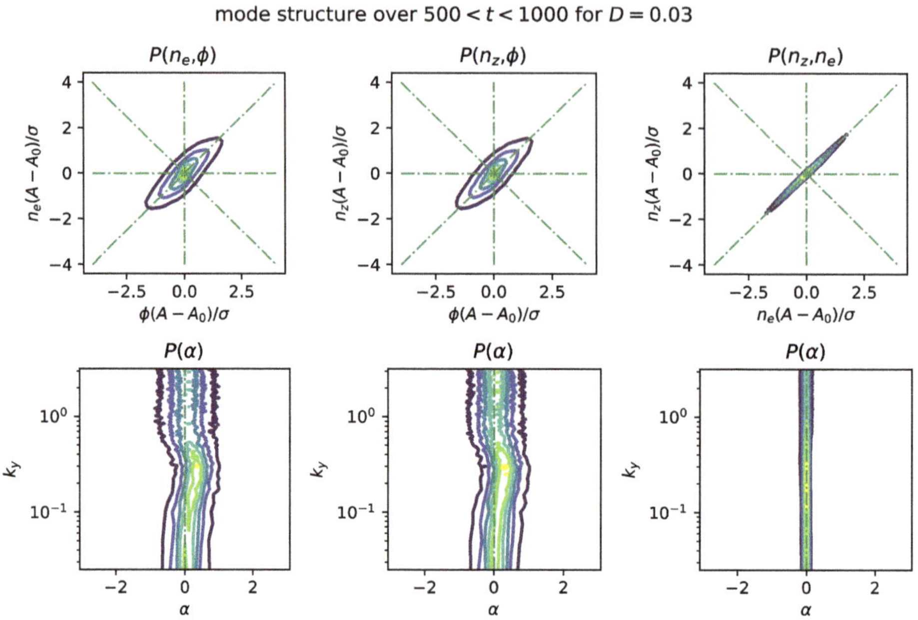

Figure 14.3. Cross coherence (top row) and phase shift distributions (bottom), between n_e (electron density) and ϕ (left), n_z (impurity density) and ϕ (center), and n_z and n_e (right), for the case $D = 0.03$ with strong nonadiabatic dynamics but still with enough adiabatic coupling to affect the statistics. In this regime as well, the densities follow each other extremely closely.

compared to either relative to ϕ. Again, this shows the nonlinear statistical nature of this dynamical alignment response.

These results from the simpler 2D model are less surprising than they could be, because the nonlinear advection is the strongest mechanism in the dynamical system for both ions and impurities. While the electrons are dissipatively coupled, the direction of causality is that the electrons determine how the potential should evolve, and then after that the potential advects each species in the same way. The electrons are the director for ϕ and not the other way around, while for the ions of either species the nonlinear advection represents the strongest (here, the only) term in the continuity equation. The situation is more complicated in the 3D toroidal case, in which the curvature terms also enter. We turn to that next.

14.3 A three-dimensional three-species gyrofluid model

We now repeat this investigation within the gyrofluid model with six moments per species. The procedures are as given in Chapter 11. The equations for the third species are the same as for the main ions, with the possibilities for differing τ_z and μ_z from the main ions ($\tau_i = \mu_i = 1$ usually), where for the electrons we have the assigned values $\tau_e = -1$ and $\mu_e = -m_e/m_D = 2.72 \times 10^{-4}$. Gyroaveraging is done for the impurities in the same way as for the ions, with the species gyroradii given by $\rho_z^2/\rho_s^2 = \mu_z\tau_z$ and $\rho_i^2/\rho_s^2 = \mu_i\tau_i$, respectively. The impurities do not appear in the polarisation or

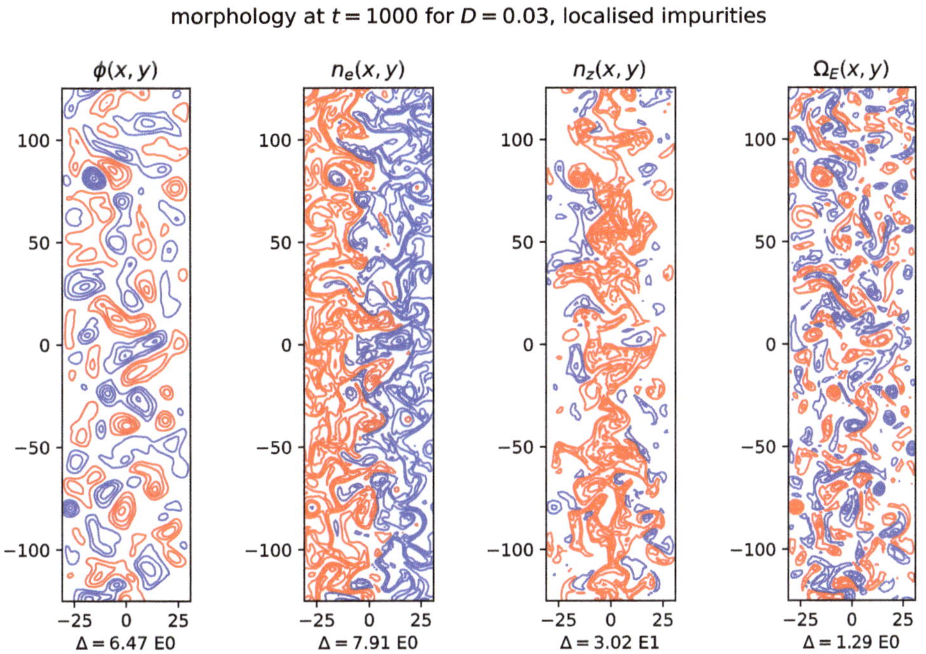

Figure 14.4. Spatial morphology of the turbulence, for the case $D = 0.03$ with the localised sourcing for impurities. This is the form at late time, $t = 1000$, after the spatial correlation between the electron (n_e) and trace impurity (n_z) densities has had time to emerge.

induction equations, which are unchanged from the single component singly charged case. In the background we assume $n_{e0} = n_{i0}$ and a value n_{z0} which is negligible. Masses are normalised to m_0 which is set to m_D for these cases. Temperatures are normalised to T_{e0} for the electrons, velocities are in terms of c_s with $c_s^2 = T_{e0}/m_0$. Other details are as given in Chapter 11. The question to be asked is whether the dynamical alignment we just saw in the 2D model persists here, or whether the impurities follow more closely the responses they would have in linear theory. Such an approach—given turbulence, linear responses—is referred to as the quasilinear model and since it has wide usage it is of interest to check it with available diagnostics.

14.3.1 Warm plasma case

The warm plasma case is the same as the one given in Chapter 11 and the reference cited there. The parameters following the scale ratios are $\hat{\epsilon} = 27200$ and $\omega_B = 0.0424$ for parallel dynamics and interchange forcing, respectively. The parameters controlling the electron responses are

$$\hat{\beta} = 1.75, \qquad \hat{\mu} = 7.41, \qquad C = 3.11 \tag{14.7}$$

This is from a base case reflective of strong, electromagnetic turbulence in the steep gradient regime (both $\hat{\beta}$ and $\hat{\mu}$ over unity). The timestep is 0.01 and the subgrid dissipation is done with $\nu_0 = 1$ and $s_f = 0.1$ as per Eq. (2.95).

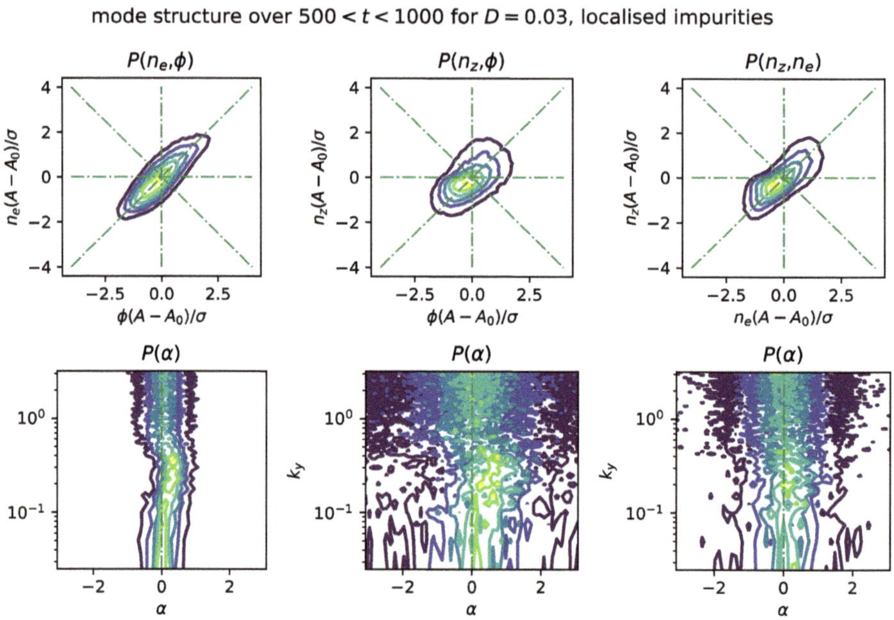

Figure 14.5. Cross coherence (top row) and phase shift distributions (bottom), between n_e (electron density) and ϕ (left), n_z (impurity density) and ϕ (center), and n_z and n_e (right), for the case $D = 0.03$ with the localised sourcing for impurities. The measurement is taken over $8 < x < 24$, isolating the region to the $x > 0$ side of the effective source and to later time. When this is done the correlation between the two densities becomes clear.

The grid is the same 64×256 in the perpendicular plane, with 32 in the parallel domain over one field line connection length. In 3D cases there is no special handling of the zonal ($k_y = k_\parallel = 0$) component, and the effect of transport can be ascertained by the small but visible changes in the overall zonal gradients. As in the localised 2D case, the gradient drive for electron and ion densities is $-\omega_n x$, while it is the localised Gaussian for the impurities. Here, though, $\omega_n = 0.5$ while the electron and ion temperatures are driven with $\omega_t = \omega_i = 1$, and the ion to electron temperature ratio is $\tau_i = 1$. The impurity to ion mass ratio is set to $\mu_z = 1.5$, reflecting tritium trace ions, and the impurity to electron temperature ratio is given three different values,

$$\tau_z = 0.5, \qquad \tau_z = 1, \qquad \tau_z = 2 \tag{14.8}$$

with 1 nominal, to control for the effects of forcing by the curvature terms. These latter have the factor τ_z in front of all the diamagnetic pieces (densities and temperatures for the state variables and velocities and heat fluxes for the flux variables). Varying τ_z will indicate the strength of these terms to affect the responses of the impurities, including their fluxes. The ion parallel dynamics is also affected, but for these steep gradient cases it is always subdominant.

The runs are initialised with the same 3D initial state as in Chapter 2, with the same field put in for all three species a_{00} moment variables, with the others started at zero. The Alfvén responses fill out the other variables, which are then forced upon by the gradient drive channels. Saturation is very rapid, as the turbulence is very robust

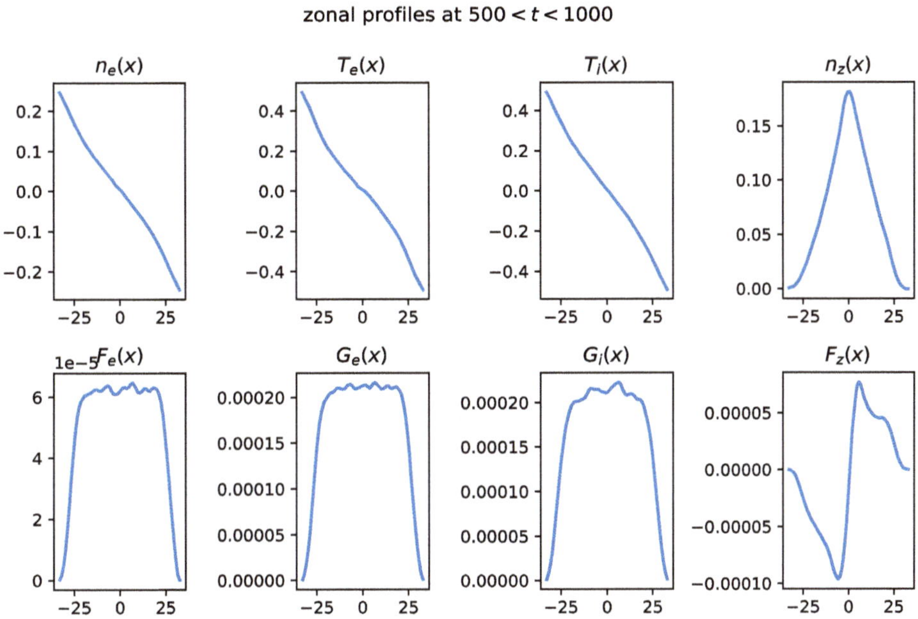

Figure 14.6. Profiles averaged over the angles and time between $500 < t < 1000$ for (top row) the electron density (n_e), the electron and ion temperatures ($T_{e,i}$), and the impurity density (n_z). The bottom row shows the respective turbulent ExB fluxes for the corresponding quantity in the top row. The symmetric impurity density profile with the antisymmetric impurity flux profile stand out.

compared to hot plasma cases. As in the 2D cases, the statistical measurements are over $500 < t < 1000$ for the mode structure. For the electrons and ions the mode structure is the same as before, since the impurities are a trace species ($a_z = 0$) so they do not affect the main plasma dynamics.

The time averaged profiles are shown in figure 14.6, for the electron and impurity densities along with the electron and ion temperatures, for which in each case $n = a_{00}$ and $T = (2^{1/2}a_{20} + 2a_{01})/3$ (isotropic average). These n's and T's are not to be confused with the background densities and temperatures defining the normalised parameters. Each appears in the top row with the corresponding turbulent ExB transport flux in the bottom row. The main gradients are slightly depleted following the robust turbulence. The strongest fluxes are the conductive heat fluxes G_{ei} for electrons and ions, owing to the ITG (ion temperature gradient driven) character which in turn follows the gradient drive combination The specific fluxes are given in terms of velocities normalised to c_s, so that the fluxes themselves are found using this factor and n_{z0} and $n_{z0}T_{z0}$ for the particle and heat flux components due to each species. Note that the background partial charge densities are given by a_z in terms of $n_{e0}e$ and the partial pressures are given by $a_z\tau_z$ in terms of $n_{e0}T_{e0}$. The specific particle flux for each species is given by

$$F_z = a_{00}[\phi_0, x] + a_{01}[\phi_1, x] \tag{14.9}$$

and the specific conductive heat flux for each species is given by

$$G_z = (2^{-1/2}a_{20} + a_{01})[\phi_0, x] + (a_{00} + 2a_{01})[\phi_1, x] \tag{14.10}$$

For electrons $\phi_0 = \phi$ and $\phi_1 = 0$ since they are not gyroaveraged. The impurity profile is somewhat broadened away from the drive profile (which has about half this width). The impurity flux is antisymmetric about $x = 0$, the effective source point. The averaged profiles show robust but not overly strong turbulence in the presence of persistent gradients (without too much relaxation), and an impurity flux away from the effective source region towards each radial boundary ($x = \pm 0.5$).

The spatial morphology of the four variables ϕ, n_e, n_z, and ϖ at $t = 1000$ is shown in figure 14.7, which we can compare to figure 14.4 for the simplified 2D model. Each quantity is combined with its background gradient profile, forming the entire quantity advected in each case. The vorticity is more sheet-like than in the 2D case which shows a few isolated coherent vortices. The differences in gradient forcing are visible here as well. However, the impurity and electron densities are still visibly correlated, especially on the right hand side where the gradients and the fluxes have the same sign.

The cross coherence and phase shift distributions between n_z, n_e and ϕ for this case are shown in figure 14.8, which we can compare to figure 14.5. The measurements are taken in the same way as in the 2D case, with a radial domain of

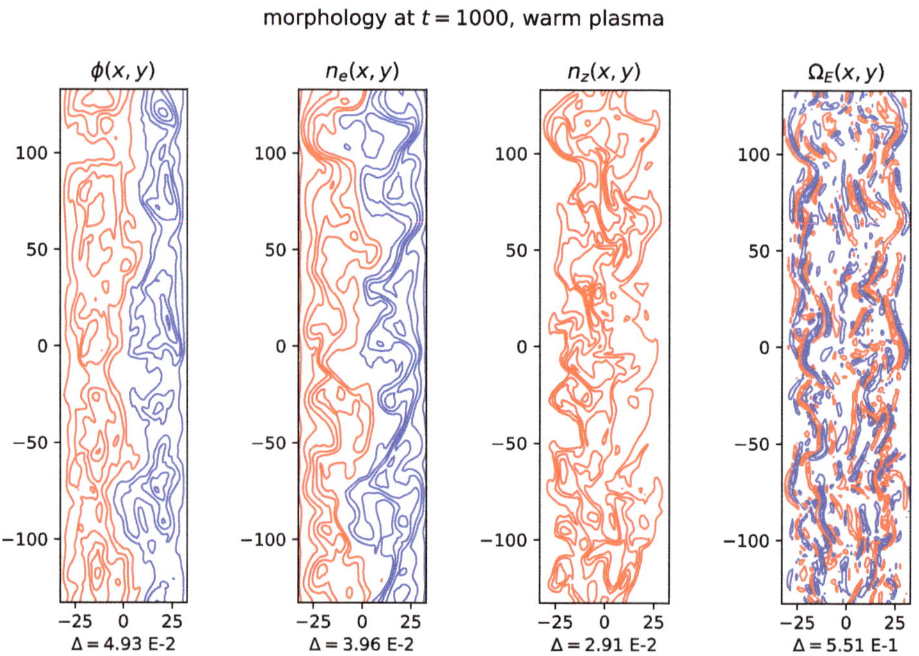

Figure 14.7. Spatial morphology of the turbulence, for the electromagnetic gyrofluid case. Compare with figure 14.4 for the 2D case. This is the form at late time, $t = 1000$, after the spatial correlation between the electron (n_e) and trace impurity (n_z) densities has had time to emerge. The vorticity sheet layers indicate strong hydrodynamic character, in which ExB advection is the dominant process. The differences in the densities are mostly owing to their background gradient profiles, which are incorporated into the structure shown.

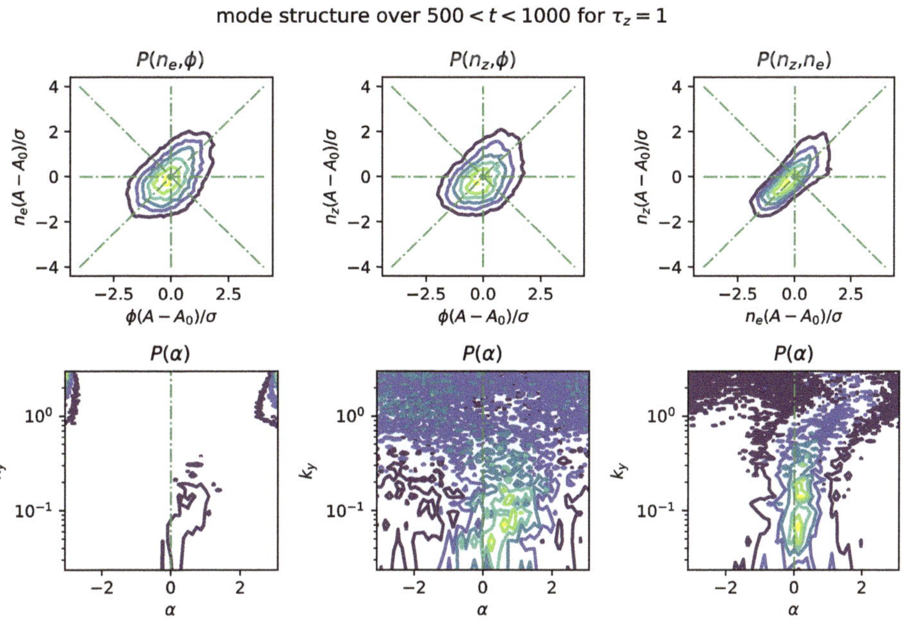

Figure 14.8. Cross coherence (top row) and phase shift distributions (bottom), between n_e (electron density) and ϕ (left), n_z (impurity density) and ϕ (center), and n_z and n_e (right), for the electromagnetic gyrofluid warm plasma case with $\tau_z = 1$. The coherence is weakened by the turbulent vorticity and advective ExB non-linearities, and the phase shifts are positive, more broadened than in the 2D case, but nevertheless close to zero until the finite Larmor radius (FLR) range ($k_y\rho_s \sim 1$) is reached. Despite the relation between n_e and ϕ, the densities follow each other closely. The sloppiness of the distribution in phases between n_z and ϕ does, however, stand out. In this and the next two figures, note this signal and also the slight shifts in the mean of the distribution of phases between n_z and n_e in the lower right frame in each case.

$0.125 < x < 0.375$ compared to the entire domain of $-1/2 < x < 1/2$. The first measurement shows moderately hydrodynamic character of electromagnetic drift Alfvén turbulence in the warm plasma regime. The turbulence is robust and opposes the electromagnetic responses effectively. The phase shifts are positive but not so strong as to show interchange character. The migration towards $+\pi$ is a manifestation of the ion polarisation, as seen in the cases shown in Chapter 3. The correlation between n_e and n_z, by contrast, shows a stronger correlation and smaller phase shifts, compared to those between either and ϕ. This indicates that the trace species follows the electrons effectively despite the large differences in their linear responses. The ExB advection is that dominant in the trace ion response to the fluctuating potential, to which it has no back reaction.

If we set τ_z different from unity, then the linear responses of the impurities are also different from those of the main ions. So we show the mode structure statistics for the $\tau_z = 1/2$ case in figure 14.9 and for the $\tau_z = 2$ case in figure 14.10. Similar responses are found here as well. The cross correlations have the same degree of sloppiness among all the cases, with the correlation between the densities somewhat tighter than that of either with respect to the ExB potential. The distribution of the phase shifts between n_z and ϕ is, however, a lot sloppier than especially that between

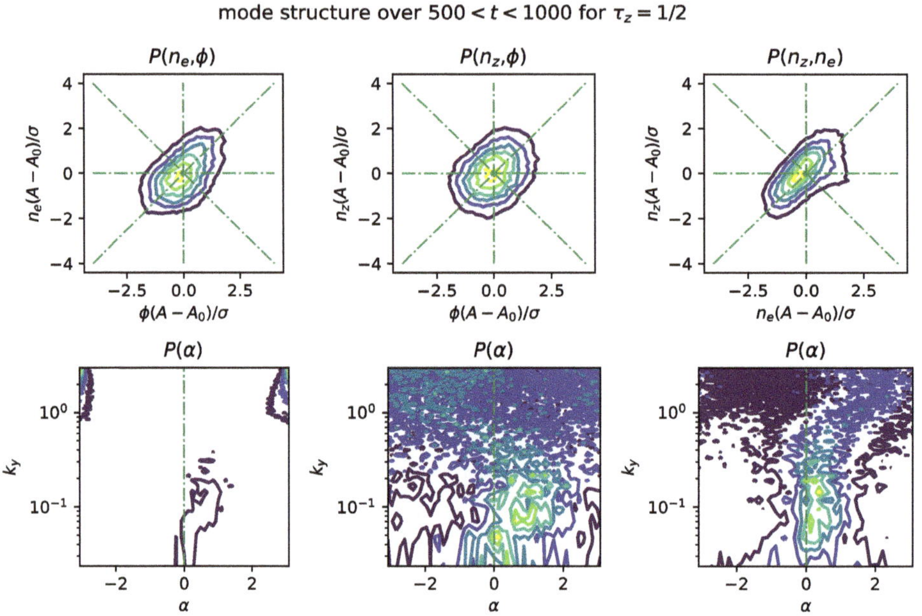

Figure 14.9. Cross coherence (top row) and phase shift distributions (bottom), between n_e (electron density) and ϕ (left), n_z (impurity density) and ϕ (center), and n_z and n_e (right), for the electromagnetic gyrofluid warm plasma case with lower $\tau_z = 1/2$. The cross correlations are similar to the $\tau_z = 1$ case, including between the densities. However, a slight tendency to shift the mean of the phase shift distribution between n_z and n_e towards the positive side is found (see text).

n_e and ϕ, but also than that between n_z and n_e. At first sight there is no strong positive or negative shift of the phases caused directly by changes in τ_z, which would follow if the linear responses were to have more effect. However, there is a finite shift which is seen if one looks closely. Especially between n_z and n_e, in the lower right frame of each figure, one can see among all three cases that there is a small tendency of increasing τ_z to drive the phase shift to be more negative, and a smaller one to leave it more positive. While the nonlinear responses for any trace ion are larger than the linear forcing or constraints, which is perhaps understandable for a steep gradient case, the linear response is not completely absent. Accordingly, we have to see if this result prevails when we look at hot plasma cases as well, which for larger $L_\perp/R$ and $K_\parallel L_\perp$ have correspondingly larger linear constraining.

14.3.2 Hot plasma case

For the hot plasma case we take a generic ITG regime case with scale ratio parameters $\hat{\epsilon} = 367$ and $\omega_B = 0.2$ with $\varepsilon_a = 0.2$, for parallel dynamics, interchange forcing, and magnetic pumping, respectively. Since this type of turbulence is not as nonlinearly robust as the more hydrodynamic, warm plasma cases with $\hat{\mu} > 1$, we take a smaller value of $\delta_0 = 0.008$ to allow more space in terms of ρ_s for the turbulence to develop. The magnetic shear value is $S = 0.5$. The parameters controlling the electron responses are

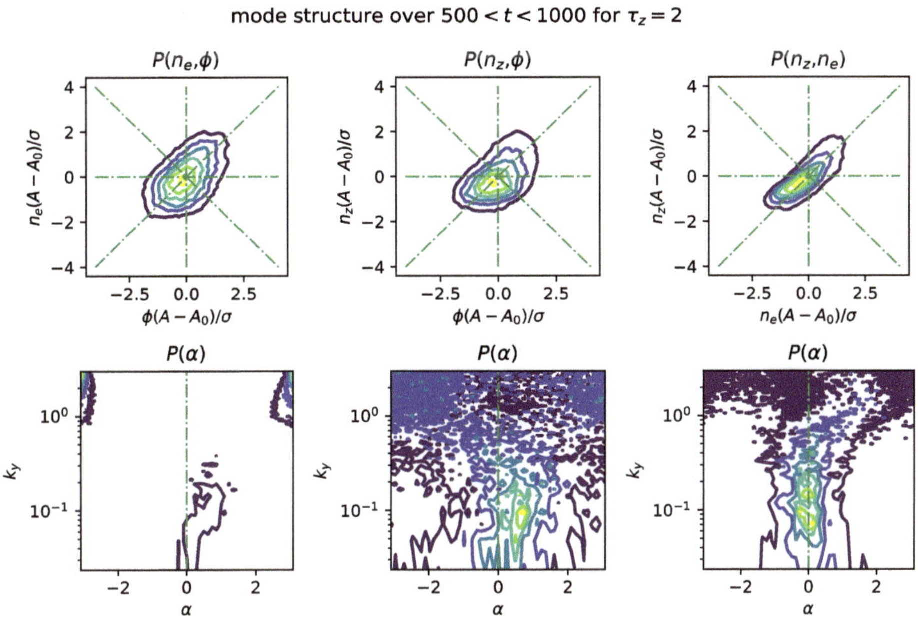

Figure 14.10. Cross coherence (top row) and phase shift distributions (bottom), between n_e (electron density) and ϕ (left), n_z (impurity density) and ϕ (center), and n_z and n_e (right), for the electromagnetic gyrofluid warm plasma case with higher $\tau_z = 2$. The cross correlations are similar to the $\tau_z = 1$ case, including between the densities. However, a slight tendency to shift the mean of the phase shift distribution between n_z and n_e towards the negative side is found (see text).

$$\hat{\beta} = 0.1, \qquad \hat{\mu} = 0.1, \qquad C = 0 \tag{14.11}$$

The electrons are moderately electromagnetic and collisionless. The timestep is 0.002 and the subgrid dissipation is done with $\nu_0 = 1$ and $s_f = 0.1$ as per Eq. (2.95).

The grid is 128 × 512 in the perpendicular plane, with 32 in the parallel domain over one field line connection length. The gradient drive is $-\omega_n x$ for the electron and ion densities, and $-\omega_z x$ for the impurities, with $\omega_n = 0.3$ and $\omega_z = 1.0$. The temperature gradient drive is $\omega_t = 1.0$ for all three species. The ion-to-electron temperature ratio is $\tau_i = 1$. The impurity mass and temperature ratio parameters are set as in the warm plasma cases just done.

The initialisation is also done the same way. Once again we test the linear responses by varying the impurity temperature parameter τ_z, with $\tau_i = 1$ for the main ions. For this case away from threshold effects, the turbulence is not stabilised by flows and saturates quickly. The statistical measurements are over $500 < t < 1000$ for the mode structure.

For the case with equal temperatures, we find the result given in figure 14.11. The appearance is different because the long wavelength portion of the spectrum is wider. The activity still peaks in the vicinity of $k_y\rho_s = 0.3$ even though the spectral minimum is just above 0.01. The electron relationship to ϕ is as expected, with phase shifts narrow and just positive. The impurities, on the other hand, are practically uncorrelated with

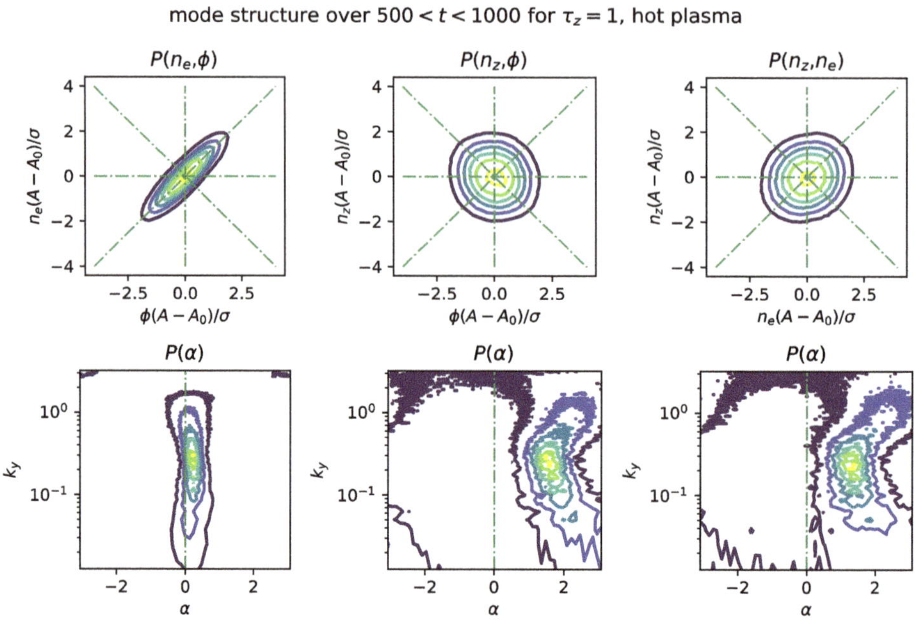

Figure 14.11. Cross coherence (top row) and phase shift distributions (bottom), between n_e (electron density) and ϕ (left), n_z (impurity density) and ϕ (center), and n_z and n_e (right), for the electromagnetic gyrofluid hot plasma case with $\tau_z = 1$.

either n_e or ϕ. The large impurity gradient gives rise to large phase shifts, near $\pi/2$ for this case, with a significant migration towards π as $k_y\rho_z$ approaches unity.

The presence of phase shifts between n_z and ϕ near $\pi/2$ is not universal though. For the case with $\tau_z = 0.5$, we find the result given in figure 14.12. The impurities are actually anticorrelated, with the cross coherence ellipses tilted towards the left rather than the right. The phase shifts are beyond $\pi/2$ for all wavelengths. The curvature term in the impurity density equation has the pressure components dropped by two due to this τ_z value, so that the ExB compression is stronger. This term forces its own relationship between n_z and ϕ. It is not the same consideration as in the ExB compressional correction to the ballooning mode, however (Chapter 13, Volume One), because there is no back reaction of ϕ to n_z or the temperature components.

More information is found from the case with $\tau_z = 2$, whose result appears in figure 14.13. The phase shift of the impurities is reduced in much the same way as in the warm plasma cases. The migration towards π for higher $k_y\rho_z$ is more prominent, since the unity point is closer to the rest of the spectrum. Many of these features point towards the importance of linear forcing. On the other hand, the very weak cross coherence between impurities and the electrons or the ExB eddies points the other way. Moreover, one does not find significant activity at long wavelength, which might have been expected from the lack of any constraining from back reaction. The activity in the impurity spectrum is where the ExB flux peaks, not where the state variables peak. There is slightly more long wavelength signal in the electron phase shifts than in the impurities.

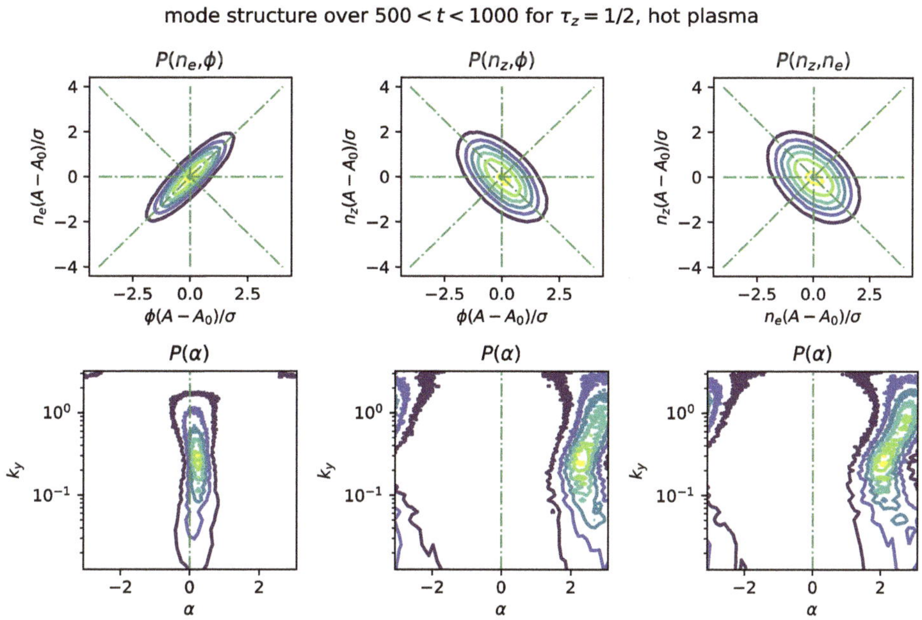

Figure 14.12. Cross coherence (top row) and phase shift distributions (bottom), between n_e (electron density) and ϕ (left), n_z (impurity density) and ϕ (center), and n_z and n_e (right), for the electromagnetic gyrofluid hot plasma case with $\tau_z = 0.5$.

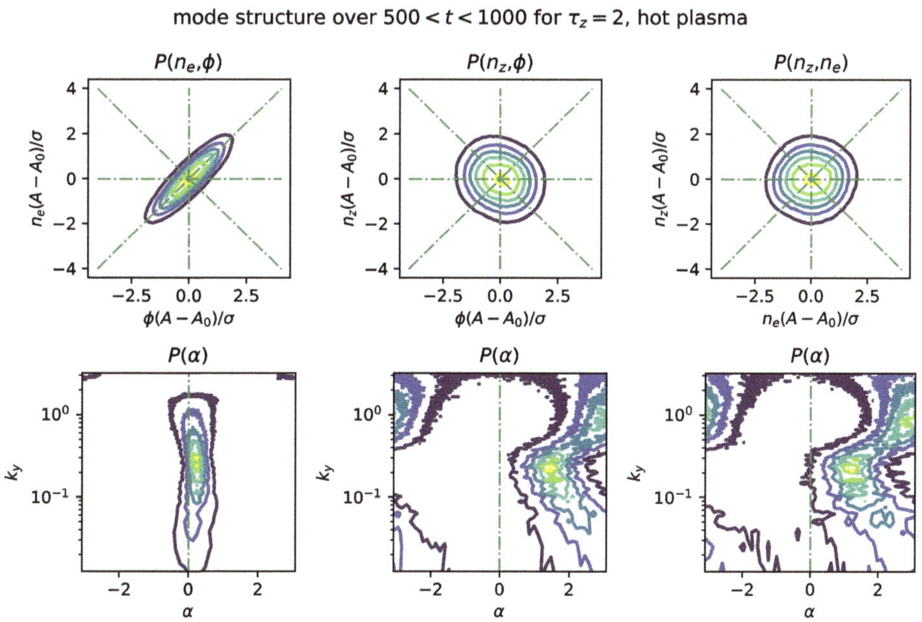

Figure 14.13. Cross coherence (top row) and phase shift distributions (bottom), between n_e (electron density) and ϕ (left), n_z (impurity density) and ϕ (center), and n_z and n_e (right), for the electromagnetic gyrofluid hot plasma case with $\tau_z = 2$.

14.3.3 Cases with pure advection

In the foregoing, the trace species has the same equation set as do the main ions, with the sole exception that the mass and temperature parameters can differ, hence also the FLR corrections. As a control case we can compare to the alternative situation in which the trace species is just test particles, namely, only the ExB terms are present, not the rest of the equations, which give the linear responses. So we replace the six equations for the trace species with a single ExB advection equation,

$$\frac{\partial n_z}{\partial t} + [\phi, f_0 + n_z] = 0 \tag{14.12}$$

with the passive scalar case given by setting the background gradient term f_0 to zero.

We run the warm plasma case again with this equation substituting for the gyrofluid treatment of the trace species. The same rules for the run itself are used, just with the different test particle treatment. There is only one case to check, though, since the species parameters τ_z and μ_z are absent for the pure ExB advection (test particle) model. The result is given in figure 14.14. The result is not dissimilar to the trace species cases in the same warm plasma case, but of course the dependence of the

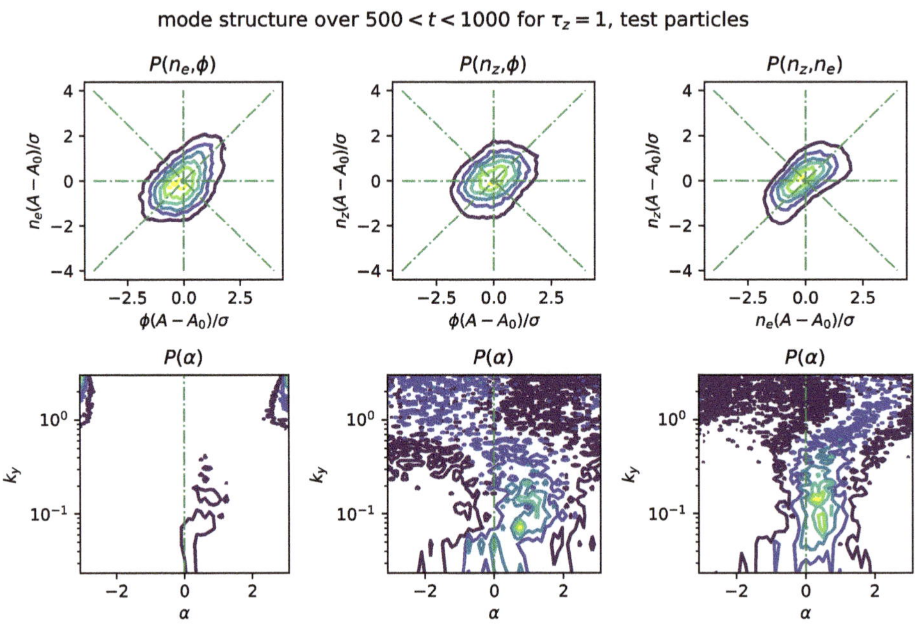

Figure 14.14. Cross coherence (top row) and phase shift distributions (bottom), between n_e (electron density) and ϕ (left), n_z (impurity density) and ϕ (center), and n_z and n_e (right), for the electromagnetic gyrofluid warm plasma case with the test particle model. There are no τ_z or μ_z parameters in this model. Compare this with the three cases in figures 14.8, 14.9, and 14.10. It is not dissimilar, but of course the slight dependence on τ_z in the phase shifts cannot be reproduced. The correlation between n_z and n_e is still significant, and their phase shifts are close to zero. Purely uncorrelated test particle diffusion would be expected to have strong, positive phase shifts and no cross coherence, but the signatures found in the trace species cases are different from that and also different from each other.

phase shifts on τ_z cannot be reproduced. It is significant that the cross correlation between n_z and ϕ is weaker but as in the other warm plasma cases it is stronger with n_e. The phase shifts relative to n_e are also small. This is not exactly what one would expect from a pure advection, or pure test particle case, in which no correlation to the other variables would be expected.

14.4 Summary

One finds a range of mode structure signatures in the cases with a trace species present, one which reacts to the turbulence, especially ExB advection but also diamagnetic fluxes, but which has no back reaction to the turbulence itself. The turbulence proceeds as if the trace species were not present. But the activity of the trace species is not a random forcing by turbulence. The forcing by linear mechanisms, especially the curvature forcing representing diamagnetic fluxes, is purely coherent. It is within the species except for the ExB compression in the curvature and the electric field forcing in the parallel momentum. By contrast, the ExB advection is usually expected to be a random forcing effect for test particles, and by extension, trace species. In these cases, however, the phase shifts between the densities was always relatively narrow, even when the cross correlation was found to be small. Wide phase shift statistical distribution (width at least $\pi/2$ even if not centred upon $\alpha = \pi/2$), as in the 2D interchange cases from Chapter 13 of Volume One, were not found in any of the cases. What was really unexpected is the contrast between the moderate cross correlation between the impurities and the other variables in the more hydrodynamic warm plasma case, and the weak to nonexistent cross correlation in the hot plasma case with relatively stronger linear forcing. This is something which bears further investigation. In any case, the combination of nonlinear, self-consistent responses, and even vestiges of the linear control parameters are all results which would not be likely seen if the reaction of a trace species were merely that of particles which follow flow streamlines, even given that these are time dependent.

My previous work on this using the isothermal version of the gyrofluid equations (two moments per species) did not find this feature, so the provisional conclusion is to cite the temperature gradient and diamagnetic heat flux physics as the principal cause. It will be interesting to see the results from a wider range of cases in future work.

Further Reading

Test particle theory is a single particle statistical treatment, in which the particles react to forcing (such as carry by a turbulent flow) through a kinetic transport equation. The original random walk is done this way, with a random forcing given by a Gaussian distribution. The main point is now to give increased emphasis to large but seldom steps in what is called a Levy flight. We do not treat these in detail herein because of the considerations involving self-consistency and small scale dissipation mentioned in the Introduction, and moved to consider a trace species following the same ExB dynamics as the electron distribution which forces it through the adiabatic response. Nevertheless ... interested readers can start with the following references, the latter ones of which provide a context for test particle statistics in turbulent flows.

The background references usually cited are P P Lévy, *Théorie de l'Addition des Variables Aléatoires* (1937, repr. 2nd ed. 1954); and B B Mandelbrot, *The Fractal Geometry of Nature* (Freeman, New York, 1982). Specific to Hamiltonian flows, a treatment in the context of chaos and islands of stability was given by G M Zaslavsky, D Stevens, and H Weitzner, in 'Self-similar transport in incomplete chaos,' *Phys. Rev.* E **48** (1993) 1683.

Test particle analysis in Hasegawa-Mima (*op cit*, Chapter 9) turbulence was done by G Manfredi and R O Dendy, in 'Transport properties of energetic particles in a turbulent electrostatic field,' *Phys. Plasmas* **4** (1997) 628.

Transitions from diffusive to ballistic test particle transport in this same model were found by S V Annibaldi and G Manfredi and R O Dendy and L O'C Drury, in 'Evidence for strange kinetics in Hasegawa-Mima turbulent transport,' *Plasma Phys. Contr. Fusion* **42** (2000) L13.

The statistical properties were further investigated by S V Annibaldi and G Manfredi and R O Dendy, in 'Non-Gaussian transport in strong plasma turbulence,' *Phys. Plasmas* **9** (2002) 791.

Similar analyses were done for Hasegawa-Wakatani (*op cit*, Chapter 9) turbulence by R Basu and T Jessen and V Naulin and J J Rasmussen, in 'Turbulent flux and the diffusion of passive tracers in electrostatic turbulence,' *Phys. Plasmas* **10** (2003) 2696.

Finite Larmor radii corrections to test particle analysis were studied using a decorrelation trajectory method by M Vlad, F Spineanu, S-I Itoh, M Yagi, and K Itoh, in 'Turbulent transport of ions with large Larmor radii,' *Plasma Phys. Contr. Fusion* **47** (2005) 1015.

Trace tritium transport was studied on a large tokamak by K-D Zastrow and J M Adams and Yu Baranov *et al*, in 'Trace Tritium Experiments on the JET Tokamak,' *Plasma Phys. Contr. Fusion* **46** (2004) B255.

The concept of dynamical alignment covered herein is from B Scott, 'Dynamical alignment in three species tokamak edge turbulence,' *Phys. Plasmas* **12** (2005) 082305.

Non-randomness of test particle motion in this type of turbulence in terms of Lagrangian coherent structures was studied by K Padberg, T Hauff, F Jenko, and O Junge, in 'Lagrangian structures and transport in turbulent magnetized plasmas,' *New J. Phys.* **9** (2007) 400.

IOP Publishing

Turbulence and Instabilities in Magnetised Plasmas, Volume 2
Gyrokinetic theory and gyrofluid turbulence
Bruce Scott

Chapter 15

Current driven MHD instabilities

15.1 Introduction

We review the well known magnetohydrodynamic (MHD) instabilities in the context of gyrofluid theory, not only to show how they are incorporated but also to highlight the role of the adiabatic response and thermal gradients for these modes. The most detail is given to the kink mode in tokamak-like plasmas with a cylindrical treatment, that is, neglecting curvature. We also cover tearing modes, which are related to magnetic reconnection. The ballooning modes were covered in fluid drift language in Volume One, so we give their setup within MHD herein. For the tearing and kink mode demonstrations, gyrofluid equations are used, showing how these MHD mode types are part of the gyrofluid physical content.

We briefly mention toroidal Alfvén eigenmodes and peeling modes at the end, but prefer to leave them to a future work on tokamak physics as they are mostly specific to tokamaks. These are both of the kink mode type. The first couples two global Alfvén waves through the toroidal curvature and can be driven by the pressure gradient of a third species. The second arises from a current layer at the radial edge of the plasma column. For more detail we refer to Further Reading.

15.2 Ideal MHD and the energy principle

Ideal MHD is called such because it is the natural generalisation of an ideal fluid to a current carrying plasma. The main properties, the lack of dissipation and the conservation of mass, momentum, and energy, remain.

The ideal MHD equations were covered in Chapter 6 of Volume One and appear as

$$\frac{\partial \rho}{\partial t} + \nabla \cdot \rho \mathbf{v} = 0 \qquad (15.1)$$

doi:10.1088/978-0-7503-3855-4ch15

$$\rho\frac{\partial \mathbf{v}}{\partial t} + \rho\mathbf{v}\cdot\nabla\mathbf{v} + \nabla p = \mathbf{J}\times\mathbf{B} \tag{15.2}$$

$$\frac{\partial p}{\partial t} + \mathbf{v}\cdot\nabla p + \Gamma p\nabla\cdot\mathbf{v} = 0 \tag{15.3}$$

$$\frac{\partial \mathbf{B}}{\partial t} = \nabla\times\mathbf{v}\times\mathbf{B} \tag{15.4}$$

the latter as Faraday's law, together with the field equation

$$\nabla\times\mathbf{B} = \mu_0\mathbf{J} \tag{15.5}$$

as Ampère's law. The ratio of specific heats, Γ, is usually left general in the theory. In the absence of a current or magnetic field, these revert to the equations for an ideal fluid, and $\Gamma = 5/3$ would signify an ideal gas.

The momentum theorem is found by combining the first two equations and invoking Ampère's law to eliminate $\mathbf{J}$,

$$\frac{\partial \mathbf{M}}{\partial t} + \nabla\cdot\mathbf{T} = 0 \tag{15.6}$$

where the momentum density and flux tensor are

$$\mathbf{M} = \rho\mathbf{v} \qquad \mathbf{T} = \rho\mathbf{v}\mathbf{v} + p\mathbf{g} + \frac{B^2}{2\mu_0}\mathbf{g} - \frac{1}{\mu_0}\mathbf{B}\mathbf{B} \tag{15.7}$$

where $\mathbf{g}$ is the metric tensor (the divergence of $a\mathbf{g}$ is the gradient of a, for any scalar a).

The fact that $\mathbf{T}$ is symmetric indicates conservation of vector angular momentum. With $\mathbf{x}$ the position vector relative to an origin, the momentum equation is crossed with $\mathbf{x}$ to obtain

$$\frac{\partial}{\partial t}\mathbf{x}\times\mathbf{M} + \nabla\cdot\mathbf{T}\times\mathbf{x} = \boldsymbol{\varepsilon}:\mathbf{T} = 0 \tag{15.8}$$

where $\boldsymbol{\varepsilon}$ is the Levi-Civita pseudotensor of rank three, which is completely antisymmetric in all its indices (cf. Chapter 4). The torque term in the momentum equation vanishes if $\mathbf{T}$ is symmetric,

$$(\varepsilon:\mathbf{T})_i = \varepsilon_{ijk}T^{jk} = 0 \quad \text{if} \quad T^{jk} = T^{kj} \tag{15.9}$$

due to the antisymmetry of $\boldsymbol{\varepsilon}$. The fact that $\mathbf{T}$ is symmetric for ideal MHD is clear from its definition.

The conserved energy is found by considering kinetic energy, thermal energy, and magnetic energy separately first. Multiply the mass conservation by $v^2/2$ and contract the velocity equation with $\mathbf{v}$, and add them, to find the kinetic energy equation

$$\frac{\partial}{\partial t}\rho\frac{v^2}{2} + \nabla\cdot\rho\frac{v^2}{2}\mathbf{v} = -\mathbf{v}\cdot\nabla p + \mathbf{v}\cdot\mathbf{J}\times\mathbf{B} \tag{15.10}$$

Multiply the mass conservation by T/m, use and add to the temperature equation, to find the thermal energy equation

$$\frac{\partial}{\partial t}\frac{p}{\Gamma - 1} + \nabla \cdot \frac{\Gamma}{\Gamma - 1} p\mathbf{v} = \mathbf{v} \cdot \nabla p \tag{15.11}$$

Contract the Faraday's law equation (for $\mathbf{B}$) with $\mathbf{B}/\mu_0$ to find the magnetic energy equation

$$\frac{\partial}{\partial t}\frac{B^2}{2\mu_0} + \nabla \cdot \frac{1}{\mu_0}\mathbf{B} \times \mathbf{v} \times \mathbf{B} = -\mathbf{v} \cdot \mathbf{J} \times \mathbf{B} \tag{15.12}$$

using Ampère's law to eliminate $\nabla \times \mathbf{B}$ on the right hand side. These are continuity equations for the three pieces, with exchange terms on the right hand side of each. They are seen to be equal and opposite in pairs, so when the equations are added the exchanges cancel, and we find

$$\frac{\partial}{\partial t}\left(\rho\frac{v^2}{2} + \frac{p}{\Gamma - 1} + \frac{B^2}{2\mu_0}\right) + \nabla \cdot \left(\rho\frac{v^2}{2}\mathbf{v} + \frac{\Gamma}{\Gamma - 1} p\mathbf{v} + \frac{B^2}{\mu_0}\mathbf{v}_\perp\right) = 0 \tag{15.13}$$

which is the conservation law for total energy. The perpendicular velocity is defined as

$$B^2\mathbf{v}_\perp = B^2\mathbf{v} - \mathbf{B}(\mathbf{B} \cdot \mathbf{v}) \tag{15.14}$$

in which B and $\mathbf{B}$ are fully nonlinear. Later, this will be linearised.

15.2.1 Energy of disturbances

The way the traditional form of disturbance analysis, either instability or oscillation, works in MHD or in fluid dynamics is to consider the displacement instead of the velocity. One has

$$\mathbf{v} = \frac{d\boldsymbol{\xi}}{dt} \tag{15.15}$$

where $\boldsymbol{\xi}$ is a spatial displacement field whose time derivative gives the velocity. No distinction is made in this expression between total and partial time derivative, because the system of equations is linearised with no flow in the equilibrium. The equilibrium is the MHD one,

$$\mathbf{J} \times \mathbf{B} = \nabla p \tag{15.16}$$

where here, $\mathbf{J}$, $\mathbf{B}$, and p are the undisturbed equilibrium quantities. These all have disturbances, which satisfy the linearised versions of the ideal MHD equations. In these equations the time derivative is cancelled on each side. For example, with

$$\frac{\partial}{\partial t}\delta p + \mathbf{v} \cdot \nabla p + \Gamma p \nabla \cdot \mathbf{v} = 0 \tag{15.17}$$

the nonlinear terms $\mathbf{v} \cdot \nabla(\delta p)$ and $(\delta p)\nabla \cdot \mathbf{v}$ are neglected. In this equation p is the equilibrium quantity and δp is the disturbance. The equation for $\boldsymbol{\xi}$ is used to recast it as

$$\delta p + \boldsymbol{\xi} \cdot \nabla p + \Gamma p \nabla \cdot \boldsymbol{\xi} = 0 \tag{15.18}$$

Similarly, the magnetic disturbance from Faraday's law is

$$\delta \mathbf{B} = \nabla \times \boldsymbol{\xi} \times \mathbf{B} \tag{15.19}$$

and the current disturbance is

$$\nabla \times \delta \mathbf{B} = \mu_0 \delta \mathbf{J} \tag{15.20}$$

from Ampère's law.

The MHD force equation, which is also the momentum equation, is then written as

$$\rho \frac{\partial^2 \boldsymbol{\xi}}{\partial t^2} = (\delta \mathbf{J}) \times \mathbf{B} + \mathbf{J} \times (\delta \mathbf{B}) - \nabla(\delta p) \tag{15.21}$$

in which we note the two time derivatives on the displacement rather than the one on the velocity. The combination on the right hand side can be regarded as a symmetric tensor operator on $\boldsymbol{\xi}$ itself. There is no problem with Fourier transforming time, since all the pieces on the right hand side are spatial operations and the equation is linear in $\boldsymbol{\xi}$. So we may assume a dependence given by

$$\boldsymbol{\xi} = \boldsymbol{\xi}_\omega \exp[-i\omega t] \tag{15.22}$$

and then the disturbance equation is

$$-\omega^2 \rho \boldsymbol{\xi}_\omega = (\delta \mathbf{J})_\omega \times \mathbf{B} + \mathbf{J} \times (\delta \mathbf{B})_\omega - \nabla(\delta p)_\omega \tag{15.23}$$

One normally solves this equation interior to a boundary surface (a flux surface of the equilibrium), on which $\boldsymbol{\xi}$ vanishes. It is then an eigenvalue equation. Once the spatial domain is finite differenced, it is an elaborate matrix equation for the components of $\boldsymbol{\xi}$.

There is a standard treatment called 'delta-W' according to which one constructs a set of energies out of the general equation. The kinetic energy is given by

$$K = \int d\mathcal{V} \frac{1}{2} \rho \, |\mathbf{v}|^2 = \int d\mathcal{V} \frac{1}{2} \rho \left| \frac{d\boldsymbol{\xi}}{dt} \right|^2 \tag{15.24}$$

The time derivative is

$$\frac{dK}{dt} = \int d\mathcal{V} \rho \mathbf{v} \cdot \frac{d\mathbf{v}}{dt} = \int d\mathcal{V} \rho \frac{d\boldsymbol{\xi}}{dt} \cdot \frac{d^2\boldsymbol{\xi}}{dt^2} \tag{15.25}$$

or

$$\frac{dK}{dt} = \int d\mathcal{V} \rho \frac{d\boldsymbol{\xi}}{dt} \cdot \mathbf{F}(\boldsymbol{\xi}) \tag{15.26}$$

where

$$\mathbf{F}(\boldsymbol{\xi}) = (\delta\mathbf{J}) \times \mathbf{B} + \mathbf{J} \times (\delta\mathbf{B}) - \nabla(\delta p) \tag{15.27}$$

with $\delta\mathbf{J}$, $\delta\mathbf{B}$, and δp given in terms of $\boldsymbol{\xi}$. It can be shown that $\mathbf{F}$ is an Hermitian operator, namely, that

$$\int d\mathcal{V}\, \rho\boldsymbol{\eta} \cdot \mathbf{F}(\boldsymbol{\xi}) = \int d\mathcal{V}\, \rho\boldsymbol{\xi} \cdot \mathbf{F}(\boldsymbol{\eta}) \tag{15.28}$$

It follows that

$$\frac{dK}{dt} = \frac{1}{2}\frac{d}{dt}\int d\mathcal{V}\, \rho\boldsymbol{\xi} \cdot \mathbf{F}(\boldsymbol{\xi}) \tag{15.29}$$

Since the left hand side gives the change in an energy, the right hand side is formulated in terms of another energy, namely

$$\delta W \equiv -\frac{1}{2}\int d\mathcal{V}\, \rho\boldsymbol{\xi} \cdot \mathbf{F}(\boldsymbol{\xi}) \tag{15.30}$$

so that

$$\frac{d}{dt}(K + \delta W) = 0 \tag{15.31}$$

which describes the conservation of a total energy. Then, K as the kinetic energy is balanced by a potential energy, which is what δW is.

For the complex displacement, the counterpart of this is the contraction of Eq. (15.23) with the complex conjugate of $\boldsymbol{\xi}$, to find

$$-\omega^2\rho\, |\boldsymbol{\xi}|^2 = \boldsymbol{\xi}^* \cdot \mathbf{F}(\boldsymbol{\xi}) \tag{15.32}$$

where we drop the subscript ω. Integrating over the spatial domain, we have

$$-\omega^2 \int d\mathcal{V}\, \rho\, |\boldsymbol{\xi}|^2 = \int d\mathcal{V}\, \boldsymbol{\xi}^* \cdot \mathbf{F}(\boldsymbol{\xi}) \tag{15.33}$$

Using the Hermitian property of $\mathbf{F}$, we then have

$$-\omega^2 \int d\mathcal{V}\, \rho\, |\boldsymbol{\xi}|^2 = \int d\mathcal{V}\, \frac{1}{2}(\boldsymbol{\xi} \cdot \mathbf{F}(\boldsymbol{\xi}^*) + \boldsymbol{\xi}^* \cdot \mathbf{F}(\boldsymbol{\xi})) = -\delta W(\boldsymbol{\xi}) \tag{15.34}$$

Therefore, the quantity on the left hand side is real, and so is δW. Since ρ is real and the squared magnitude of $\boldsymbol{\xi}$ is real, it follows that ω^2 is also real. The disturbance is either purely growing or purely oscillatory. The sign convention is that of an energy cost. If δW is positive, then ω^2 is also positive, and the displacement represents a stable oscillation with positive and negative frequencies which have zero imaginary part. If δW is negative, then there is one growing and one damped mode component, with equal and opposite real growth rates (imaginary frequency).

Leaving $\boldsymbol{\xi}$ in real space and time, we can define what the monograph on MHD instabilities by Mannheimer and Lashmore-Davies calls the Virial,

$$I = \int d\mathcal{V} \frac{1}{2}\rho \, |\boldsymbol{\xi}|^2 \tag{15.35}$$

which is a positive definite quantity. Differentiating this twice with time, we have

$$\frac{d^2 I}{dt^2} = \int d\mathcal{V}\left(\rho \left|\frac{d\boldsymbol{\xi}}{dt}\right|^2 + \rho\boldsymbol{\xi} \cdot \frac{d^2\boldsymbol{\xi}}{dt^2}\right) \tag{15.36}$$

In terms of the kinetic and potential energies of the disturbance, this is

$$\frac{d^2 I}{dt^2} = 2(K - \delta W) \tag{15.37}$$

If a disturbance is initialised at $\boldsymbol{\xi} = \boldsymbol{\eta}$, with $\boldsymbol{\xi}$ time dependent but $\boldsymbol{\eta}$ fixed, such that $K = 0$ but $\delta W(\boldsymbol{\eta}) = \delta W_0$ is finite, then the evolution starts with

$$\frac{d^2 I}{dt^2} = 4K = -4\delta W_0 \tag{15.38}$$

due to the conservation of the total disturbance energy $K + \delta W$. The original $I = I(\boldsymbol{\eta})$ then accelerates away from its original value. As K grows δW intensifies more negative accordingly. Energy can be released by a disturbance field in the particular equilibrium setup if one can be found for which δW is negative. If on the other hand it costs energy to grow the disturbance, it will not grow, but oscillate.

Term by term, we evaluate the expressions in $\mathbf{F}(\boldsymbol{\xi})$ using the forms for $\delta\mathbf{B}$ and δp in terms of $\boldsymbol{\xi}$, so that we have for the pressure term

$$\boldsymbol{\xi} \cdot \nabla(\delta p) = \nabla \cdot [\boldsymbol{\xi}(\delta p)] - (\delta p)\nabla \cdot \boldsymbol{\xi} \tag{15.39}$$

Under the integration, we always assume the divergence terms so constructed will vanish. So when they become more complicated than this, we can denote them schematically. Putting in for δp we have

$$\boldsymbol{\xi} \cdot \nabla(\delta p) = \nabla \cdot () + (\boldsymbol{\xi} \cdot \nabla p + \Gamma p \nabla \cdot \boldsymbol{\xi})\nabla \cdot \boldsymbol{\xi} \tag{15.40}$$

The term containing the disturbed current is

$$\boldsymbol{\xi} \cdot (\delta\mathbf{J}) \times \mathbf{B} = -(\delta\mathbf{J}) \cdot (\boldsymbol{\xi} \times \mathbf{B}) \tag{15.41}$$

using the vector identity

$$\mathbf{B} \cdot \nabla \times \mathbf{A} = \nabla \cdot (\mathbf{A} \times \mathbf{B}) + \mathbf{A} \cdot \nabla \times \mathbf{B} \tag{15.42}$$

and the Ampère's law for the disturbances, we obtain

$$\boldsymbol{\xi} \cdot (\delta\mathbf{J}) \times \mathbf{B} = \nabla \cdot () - \frac{1}{\mu_0} |(\delta\mathbf{B})|^2 \tag{15.43}$$

The term with the equilibrium current is

$$\boldsymbol{\xi} \cdot \mathbf{J} \times (\delta \mathbf{B}) = -\mathbf{J} \cdot \boldsymbol{\xi} \times (\delta \mathbf{B}) \tag{15.44}$$

So that we have

$$\delta W = \int d\mathcal{V} \left[\frac{1}{2\mu_0} |(\delta \mathbf{B})|^2 + \Gamma p(\nabla \cdot \boldsymbol{\xi})^2 + (\boldsymbol{\xi} \cdot \nabla p)(\nabla \cdot \boldsymbol{\xi}) + \mathbf{J} \cdot \boldsymbol{\xi} \times (\delta \mathbf{B}) \right] \tag{15.45}$$

The first two terms are positive definite and therefore always stabilising. The last two are the ones which can cause instability. Flow down a pressure gradient into an expansion is one, and a helical magnetic disturbance along the current is the other. Generally, these are the pressure driven and current driven modes, respectively, and all the rest is details.

This expression is traditionally recast. The magnetic disturbance and the equilibrium current are split into parallel and perpendicular components,

$$\delta \mathbf{B} = (\delta \mathbf{B})_{\parallel} + (\delta \mathbf{B})_{\perp} \tag{15.46}$$

with

$$B^2 (\delta \mathbf{B})_{\parallel} = \mathbf{B}\mathbf{B} \cdot (\delta \mathbf{B}), \qquad B^2 (\delta \mathbf{B})_{\perp} = \mathbf{B} \times (\delta \mathbf{B}) \times \mathbf{B} \tag{15.47}$$

and

$$\mathbf{J} = J_{\parallel} \mathbf{b} + \frac{1}{B^2} \mathbf{B} \times \nabla p \tag{15.48}$$

reckoned against the background magnetic field. Then, the ∇p factor is recast again as

$$\nabla p = \mathbf{J} \times \mathbf{B} = \frac{1}{\mu_0} \mathbf{B} \cdot \nabla \mathbf{B} - \frac{1}{2\mu_0} \nabla B^2 \tag{15.49}$$

The curvature vector is defined as

$$\boldsymbol{\kappa} = \mathbf{b} \cdot \nabla \mathbf{b} \tag{15.50}$$

Then, the δW expression is recast in these terms. Some authors prefer to keep $\nabla \cdot \boldsymbol{\xi}$ while others prefer to eliminate it in terms of other factors. Mannheimer and Lashmore-Davies (1989) write

$$\begin{aligned} \delta W = \frac{1}{2} \int d\mathcal{V} \Bigg[& \frac{1}{\mu_0} (\delta \mathbf{B})_{\perp}^2 + \frac{1}{\mu_0} \left| (\delta \mathbf{B})_{\parallel} - \frac{\mu_0}{B^2} \boldsymbol{\xi} \cdot \nabla p \right|^2 + \Gamma p (\nabla \cdot \boldsymbol{\xi})^2 \\ & - 2(\boldsymbol{\xi} \cdot \nabla p)(\boldsymbol{\xi} \cdot \boldsymbol{\kappa}) + J_{\parallel} \mathbf{b} \cdot \boldsymbol{\xi} \times (\delta \mathbf{B}) \Bigg] \end{aligned} \tag{15.51}$$

eliminating $\nabla \cdot \boldsymbol{\xi}$ everywhere except in pressure compression (third term), while White (2001) writes

$$\delta W = \frac{1}{2}\int d\mathcal{V}\left[\frac{1}{\mu_0}(\delta\mathbf{B})_\perp^2 + \frac{B^2}{\mu_0}\,|\nabla\cdot\boldsymbol{\xi}_\perp + 2\boldsymbol{\xi}_\perp\cdot\boldsymbol{\kappa}|^2 + \Gamma p(\nabla\cdot\boldsymbol{\xi})^2 \right.$$
$$\left. -\ 2(\boldsymbol{\xi}_\perp\cdot\nabla p)(\boldsymbol{\xi}_\perp\cdot\boldsymbol{\kappa}) - \sigma\mathbf{B}\cdot\boldsymbol{\xi}\times(\delta\mathbf{B})_\perp\right] \tag{15.52}$$

where $\sigma = J_\parallel/B$. In this version it is preferred to keep $\nabla \cdot \boldsymbol{\xi}$ everywhere except in the curvature (fourth term). The parallel displacement appears only in the pressure compression term. Finally, the kink term appears with **B** instead of **J** so as to clarify the contraction of a vector $\boldsymbol{\xi} \times \mathbf{B}$ and its curl, so that it is explicitly a vector twist of the disturbance.

In either case, the five terms represent field line bending, magnetic compression in a finite beta ($2\mu_0/B^2$) plasma with a pressure gradient, pressure compressibility (of the fluid, not the magnetic field), and pressure and current driven destabilisation, the latter strictly by the parallel current.

The displacement growth equation, Eq. (15.23), describes an eigenvalue problem, with real values for ω^2 as noted. Techniques which can find the largest or smallest eigenvalue of such systems can be applied. If the largest value of $\gamma^2 = -\omega^2$ is negative, then the system is stable against any disturbance. This is very useful in designing confined plasma devices which are MHD stable, a process which has a long and successful history. Those who want to go further with this can explore the texts mentioned in Further Reading. The reason we will stop here is that the advantages of the delta-W analysis, especially its mathematical rigour, become disadvantages as soon as the complications we are most interested in arise. These are the combination of the adiabatic response and the density and temperature gradients. Minimally, they give a real frequency to any instability, and if the limit of small rather than zero collisionality is taken, damping rates to any oscillation. Once this is present, any mode eigenvalue is complex, and such a result cannot be delivered by the delta-W theory. There have been attempts in the past to treat this with a complex resistivity, but then the phenomenology is all wrong and the result must be known before it can be anticipated. In most actual applications a similar series of assumptions to ours, i.e., low frequency or small-scale localisation, is taken. In the ballooning theory one usually takes incompressible motion and restricts to short wavelength modes. This becomes even more restrictive than the assumptions we are using, namely, reduced equations (reduced MHD, fluid drift dynamics, or gyrokinetics) and flux tube geometry. We could get the ballooning theory simply from linearised flux tube equations with one further assumption: that of dynamics on one field line. The delta-W theory uses a different derivation path to get the same equation once the same curvature vector is used.

As linear theory shifted its focus to two-fluid equations in the late 1970s the normal mode approach got more of the attention, i.e., solving an equation like Eq. (15.23) directly once one Fourier component in space is isolated. In this context, many MHD modes become translated into a different description for the same thing. We saw the ballooning and interchange modes in Chapter 13 of Volume One. We will turn to the kink modes below. In the meantime there is another class of modes, involving the tearing or reconnection of magnetic field lines. It is automatically a non-ideal instability since it requires that the flux conservation in ideal MHD be broken. We turn to these modes next.

15.3 Tearing modes and reconnection

This is a large topic which, like nonlinear MHD itself, has occupied entire texts (referenced below). We will introduce the gist of it and focus at first on a collisionless variant which requires electron inertia for its existence. There are also drift tearing modes which incorporate some of the properties of drift waves in their structure. Once the adiabatic response and the thermodynamic gradient are involved, one is well outside of the familiar MHD treatments. This combination will be addressed in the context of turbulence in a current layer with a pressure layer, afterwards.

The tearing mode is a basic manifestation of a current driven mode, but one which needs a mechanism to break flux conservation to function. In the reduced MHD system we have the vorticity and induction equations,

$$\frac{\partial \varpi}{\partial t} = \nabla_{\|} J_{\|}, \qquad \frac{\partial A_{\|}}{\partial t} = -\nabla_{\|}\phi - \eta_{\|} J_{\|} \tag{15.53}$$

where

$$\varpi = \frac{\rho}{B^2}\nabla_{\perp}^2\phi, \qquad \mu_0 J_{\|} = -\nabla_{\perp}^2 A_{\|} \tag{15.54}$$

in a sheared-slab geometry. The equilibrium is zero beta, with the equilibrium described by a sheet current along a main background guiding magnetic field, in a layer of finite width perpendicular to the guide field. Finite beta and finite resistivity corrections to the equilibrium

$$\mathbf{B} = B_0\nabla z + B^y(x)\nabla y \qquad \mathbf{J} = J_{\|}(x)\nabla z \qquad \frac{\partial B^y}{\partial x} = -\mu_0 J_{\|} \tag{15.55}$$

are neglected. The coordinate system $\{xyz\}$ satisfies $\nabla x \times \nabla y \cdot \nabla z = 1$. The conventional problem set the boundary conditions such that y is periodic but x extends to infinity such that $J_{\|} \to 0$ as $x \to \pm\infty$. An alternative has been done in which x is also periodic, so $J_{\|}$ and therefore B^y are sinusoidal. The latter is easier to solve, so we will work with it.

The fundamental principle from electromagnetism that one should have in mind is that like currents attract. Two parallel wires of current I placed a transverse distance a apart from each other experience a mutual $\mathbf{J} \times \mathbf{B}$ force. Simply calculate the magnetic field due to one wire (poloidal, if the current is toroidal), $B = \mu_0 I/2\pi a$, and then the force given by IB on one wire is toward the other. Tearing is thought of as current filamentation. In the limit of zero resistivity the wires cannot coalesce because the flux lines between them would have to disappear. The field lines are required to tear and reconnect to do that, hence the name 'tearing instability' which was given.

In this case we are considering a current layer of finite width given by geometrical factors times a perpendicular scale a, along a magnetic field of strength B_0, in the setup given in Eq. (15.55). The plasma beta is assumed small and the current ratio $\sigma = J_\parallel/B_0$ is also small enough so that the equilibrium corrections to B_0 are neglected. One takes a small resistivity ordering according to which the time scale given by resistive diffusion of the current layer is arbitrarily long. Due to Ampère's law we have

$$\frac{\partial A_\parallel}{\partial t} + \nabla_\parallel \phi = \frac{\eta_\parallel}{\mu_0} \nabla_\perp^2 A_\parallel \tag{15.56}$$

so that $\eta_\parallel/\mu_0$ is called the 'resistive diffusivity'. If $\eta_\parallel \to 0$ the remaining part of the equations represent shear Alfvén oscillations, as we did before, finding

$$\nabla \cdot \frac{1}{v_A^2} \frac{\partial^2}{\partial t^2} \nabla_\perp \phi = \nabla_\parallel \nabla_\perp^2 \nabla_\parallel \phi \tag{15.57}$$

in Eq. (15.3) from Chapter 15 of Volume One. So it is reasonable to set parameters around v_A as a velocity and a as a scale. We also have the shear length given by

$$L_s^{-1} = \left| \frac{1}{B_0} \frac{\partial B^y}{\partial x} \right| = \mu_0 \left| \frac{J_\parallel}{B_0} \right| \tag{15.58}$$

(Eq. 11.51 from Chapter 11 of Volume One) noting that it is defined positive. We require

$$L_s \gg a \tag{15.59}$$

to keep this description of the equilibrium reasonable.

The dimensionless parameter associated with these scales and the resistive diffusivity is called the 'Lundquist number', also called the magnetic Reynolds number. We set this equal to the ratio between the resistive diffusion time and the perpendicular Alfvén transit time,

$$S_L = \frac{v_A}{a} \frac{a^2}{\eta_\parallel/\mu_0} \gg 1 \tag{15.60}$$

with the inequality required to solve the problem the way we will do. Typical values in magnetic plasma confinement experiments are above 10^6, and even for a thin layer in a steep gradient region it is still 10^4 or above.

Due to this inequality, the problem allows solution by means of scale separation. The solution in most of the domain is found by neglecting the inertia term in the vorticity equation, demanding Alfvén force balance. In terms of the disturbances we have

$$\tilde{B}^x \frac{\partial J_\parallel}{\partial x} + B^y \frac{\partial \tilde{J}_\parallel}{\partial y} = 0 \qquad \text{with} \quad \tilde{B}^x = \frac{\partial \tilde{A}_\parallel}{\partial y}, \quad \mu_0 J_\parallel = -\frac{\partial B^y}{\partial x} \tag{15.61}$$

The simplest model for this is a periodic background magnetic field with wavenumber α, so that

$$B^y = B_0 \, S \sin \alpha x \tag{15.62}$$

where $\alpha = 2\pi/a$ in this model, and S gives the normalised shear a/L_s. It is assumed that S≪1 so that $\partial/\partial z$ may be neglected and we use a two-dimensional model. The disturbances are assumed to have a dependence given by

$$\tilde{A}_{\parallel} = \psi(x)\exp[\gamma t + ik_y y] \tag{15.63}$$

with ψ a complex amplitude whose x-dependence is to be solved for, along with γ. The wavenumber k_y is a parameter. This means that the equation away from $x = 0$, the singular surface where $k_{\parallel} = k_y x/L_s$ vanishes, is given by

$$\nabla_{\perp}^2\psi = -\alpha^2\psi \qquad \text{or} \qquad \psi'' = \left(k_y^2 - \alpha^2\right)\psi \tag{15.64}$$

where the prime $'$ denotes $\partial/\partial x$. The condition on this is that if $\alpha = 0$ then the solution decays away from $x = 0$ to effectively vanish before encountering the next singular surface, which is $\pi\alpha^{-1}$ away. In that case the solution is concave upwards, a decaying exponential. The derivative is finite, with ψ'/ψ oppositely signed to x. But if $\alpha^2 - k_y^2$ is positive, then the solution is concave downward. The derivative is finite, with ψ'/ψ signed with x. In the marginal case the wavelength of the solution is longer than $2\pi/\alpha$ so that ψ is minimum and finite at the singular surface and bowed upward and then downward to arrive at the same value at the next singular surface. In either case, the derivative at $x = 0$ is discontinuous and we must solve the approximate solution inside a narrow layer of half-width Δ (which we also solve for) on either side of the singular surface.

In the vicinity of the singular surface we approximate ψ by a constant value and then solve for the variation as a correction, and for ϕ (synonymous with $\tilde{\phi}$ since there is no equilibrium ϕ). The derivative discontinuity is given by

$$2\Delta'\psi = \left.\frac{\partial\psi}{\partial x}\right|_{x<0}^{x>0} \tag{15.65}$$

which is positive for our case if $\alpha^2 > k_y^2$. In terms of the parallel electric field we have

$$\tilde{E}_{\parallel} = -\left(\gamma\psi + ik_{\parallel}\phi\right) \quad \text{expect:} \quad E_{\parallel} \to 0 \quad \text{as} \quad |x| \gg \Delta \tag{15.66}$$

so that the match to the ideal MHD solution in the outer region is reasonable. The Ohm's law is

$$\eta_{\parallel}\tilde{J}_{\parallel} = \tilde{E}_{\parallel} \tag{15.67}$$

which is integrated through the inner layer to set

$$\frac{\eta_{\parallel}}{\mu_0}\frac{\Delta'}{\Delta}\psi = \int_{-\infty}^{\infty}\frac{dx}{\Delta}\left(\gamma\psi + ik_{\parallel}\phi\right) \tag{15.68}$$

We therefore expect that

$$\gamma \sim \frac{\eta_{\parallel}}{\mu_0}\frac{\Delta'}{\Delta} \tag{15.69}$$

which gives the growth rate ordering. We are assuming

$$\Delta' a \sim 1, \qquad \frac{\eta_\parallel/\mu_0}{a^2} \ll \frac{\eta_\parallel}{\mu_0}\frac{\Delta'}{\Delta} \sim \gamma \ll \frac{\eta_\parallel/\mu_0}{\Delta^2}, \qquad \Delta \ll a \tag{15.70}$$

by means of scale separation. So the amplitude $\psi \approx \psi(0)$ is approximately constant within the inner layer $x \sim \Delta$. All of these assumptions are taken *a priori* and are to be checked *a posteriori*, in any calculation for a specific parameter set.

The vorticity equation in the layer becomes

$$\gamma\frac{\mu_0\rho}{B^2}\phi'' = ik_\parallel\mu_0\tilde{J}_\parallel = ik_\parallel\frac{\mu_0}{\eta_\parallel}\tilde{E}_\parallel = -ik_\parallel\frac{\mu_0}{\eta_\parallel}(\gamma\psi + ik_\parallel\phi) \tag{15.71}$$

assuming $k_y^2\Delta^2 \ll 1$. Note that

$$k_\parallel = k_y\frac{x}{L_s} = k_\parallel' x \tag{15.72}$$

so that the coefficient on ϕ on the right hand side has two factors of x, giving the structure of the eigenmode. We can now normalise

$$y = -ik_\parallel'\Delta\phi/\gamma\psi(0), \qquad x \leftarrow x/\Delta, \qquad \text{so that} \quad \Delta^4 = \frac{(\eta_\parallel/\mu_0)\gamma}{(k_\parallel' v_A)^2} \tag{15.73}$$

which implicitly assumes that

$$\frac{(\eta_\parallel/\mu_0)}{\Delta^2}\gamma = (k_\parallel v_A)^2 \tag{15.74}$$

with $k_\parallel$ evaluated at the width of Δ. So with the ordering on γ we also have

$$\gamma \ll k_\parallel v_A \tag{15.75}$$

since with $\Delta \ll a$ the resistive layer diffusion is larger than γ. Using this ordering and normalisation we arrive at

$$y'' - x^2 y = -x \qquad \text{with} \quad y(0) = 0, \quad y(\infty) = x^{-1} \tag{15.76}$$

which determines the functional form of ϕ and therefore $E_\parallel$ and $J_\parallel$ within the layer. Although the homogeneous part of this equation (the left hand side) has known solutions (the parabolic cylinder functions, see Further Reading), a solution in closed form has not been found with this right hand side and these boundary conditions. On the other hand, there is an exact solution, given in terms of an integral. One can show by direct substitution that

$$y = \frac{x}{2}\int_0^{\pi/2} \exp\left(-\frac{x^2}{2}\cos\theta\right)\sin^{1/2}\theta \, d\theta \tag{15.77}$$

is an exact solution (given by Goldston and Rutherford, 1995, adapted from an equivalent form given by Rutherford and Furth, 1971). The asymptotic behaviour is $y'(0) = 1/2$ and $(1 - xy) \to 0$ for large x.

With this in hand, the rest can be computed. The eigenfunction in the inner layer is a combination of the relation $y(x)$, and also a correction to $\psi = \psi(0)$ which is used to check the assumptions

$$\psi'' = \frac{\mu_0}{\eta_\parallel}\tilde{E}_\parallel = -\frac{\mu_0}{\eta_\parallel}(\gamma\psi + ik_\parallel\phi) \tag{15.78}$$

This is integrated to a point x_L at which one still has $\Delta' x_L \ll 1$, since $\tilde{E}_\parallel$ reduces close to zero after only a few Δ. Further detail can be obtained from, e.g., the book by Goldston and Rutherford (1995).

The reason we will not go further than this is that as soon as density and temperature gradients, the adiabatic response, and the electron inertia are re-introduced all the results change. Not so much qualitatively, but in the details and numbers. This is why, nowadays, it is sufficient to do the resistive reduced MHD analysis to understand the basic mode. It depends on resistivity, or some other mechanism to break flux conservation, for finite growth, since $\gamma \to 0$ as $\eta_\parallel \to 0$ as was already clear after the set-up but before actual solution of the problem.

In the real world, flux conservation can be broken by electron inertia by itself. In the years after 1990 one finds significant interest in the collisionless reconnection problem as applied to these cases. From a gyrofluid starting point (Chapter 9), one can set the assumptions to neglect the parallel ion motion (slow sound waves), so that in a straight background magnetic field in the absence of a pressure gradient, one can neglect the ion gyrocentre variables (the importance of the gradient for this was also seen in the kinetic responses in Chapter 8). In a simple model neglecting temperature physics we are then down to two moment variables, the electron density $\tilde{n}_e$ and velocity $-\tilde{J}_\parallel/n_e e$, namely, the parallel current. Under these conditions one can equate vorticity and electron density, since the polarisation equation reduces to

$$n_e\frac{e^2}{T_i}(\Gamma_0 - 1)\phi = \tilde{n}_e e \equiv \varpi \tag{15.79}$$

Substituting ϖ in for $\tilde{n}_e$ everywhere it occurs, one finds

$$\frac{\partial\varpi}{\partial t} = \nabla_\parallel \tilde{J}_\parallel \tag{15.80}$$

$$\frac{\partial}{\partial t}\left(\tilde{A}_\parallel + \frac{m_e}{n_e e^2}\tilde{J}_\parallel\right) = \frac{T_e}{n_e e^2}\nabla_\parallel\varpi - \nabla_\parallel\phi \tag{15.81}$$

with $\tilde{A}_\parallel$ and $\tilde{J}_\parallel$ related through Ampère's law. This is the two-field model of Schep, Pegoraro, and Kuvshinov (1994), which was extensively used to study this problem during that decade. The FLR effects studied by Grasso (2000) are equivalent to the Padé approximant $b/(1 + b)$ to $1 - \Gamma_0$ (cf. Chapter 10). Recalling that $\tilde{n}_e$ and ϖ satisfy the same equation, they construct

$$\frac{\partial G_\pm}{\partial t} + [\Phi_\pm, G_\pm] = 0 \tag{15.82}$$

from the 2D version of these equations, where

$$G_{\pm} = \left(1 - d_e^2\nabla_{\perp}^2\right)\tilde{A}_{\|} \pm \frac{m_e V_e}{n_e e^2}\varpi, \qquad \Phi_{\pm} = \phi \pm V_e\tilde{A}_{\|} \tag{15.83}$$

with $d_e^2 = m_e/\mu_0 n_e e^2$, and the bracket is

$$[a, b] = \frac{1}{B_0}\left(\frac{\partial a}{\partial x}\frac{\partial b}{\partial y} - \frac{\partial b}{\partial x}\frac{\partial a}{\partial y}\right) \tag{15.84}$$

This depends on both finite ρ_s and d_e, since ϖ drops out of $G_{\pm}$ for $\rho_s \to 0$. As both $\rho_s \to 0$ and $d_e \to 0$, this reverts to flux conservation given by $d\tilde{A}_{\|}/dt = 0$. The dynamics proceeds by preserving the contour lines in $G_{\pm}$ and so for $\tilde{A}_{\|}$ they are broken and reconnected. Growth of the instability there depends on both ρ_s and d_e finite.

From the structure of the $\Phi_{\pm}$ this is seen as a manifestation of the perturbed Hamiltonian (proportional to $\phi - v_{\|}\tilde{A}_{\|}$) of the underlying gyrokinetic theory. From the electron gyrokinetic equation

$$\frac{\partial g}{\partial t} + [H, h] = 0, \qquad H = \phi - v_{\|}\tilde{A}_{\|}, \qquad h = g - eH(F^M/T_e) \tag{15.85}$$

for this model, taking the $v_{\|} \pm V_e$ moment and multiplying by $m_e/n_e e$ will recover Eq. (15.82) (try these!).

We will illustrate the development of this instability with the two-moment gyrofluid model described in Chapter 9. The curvature terms are absent, no magnetic pumping terms are present, and in the basic case the collisions are set to zero. One has solely the ExB and FLR advection terms and the parallel gradient terms through brackets with the potentials. Normalisation is in terms of a and c_s (use the setup of the previous Chapters with $L_{\perp}$ replaced by a). The drift parameter, and the normalised drift scale and ion gyroradius are

$$\delta_0 = \frac{\rho_s}{a}, \qquad \rho_s^2 = \delta_0^2, \qquad \rho_i^2 = \tau_i\rho_s^2 \tag{15.86}$$

where τ_i is the background ion/electron temperature ratio. The background magnetic field is $B = B_0\nabla z$ only, with both equilibrium and disturbance B^x and B^y through

$$B^x = [-A, x], \qquad B^y = [-A, y] \tag{15.87}$$

The bracket is

$$[a, b] = \delta_0\left(\frac{\partial a}{\partial x}\frac{\partial b}{\partial y} - \frac{\partial b}{\partial x}\frac{\partial a}{\partial y}\right) \tag{15.88}$$

Note that $A = \tilde{A}_{\|}/B_0\rho_s$ so that a factor of δ_0 appears in the magnetic field components.

The normalised equations are

$$\frac{\partial n}{\partial t} + [\phi_0, n] + [-A, u] = 0 \tag{15.89}$$

$$\frac{\partial A^*}{\partial t} + [\phi_0, A^*] + \tau_z[-A, n] = 0 \tag{15.90}$$

with

$$A^* = A + \mu_z u \tag{15.91}$$

for each species. The field equations are

$$P(\phi) = G_0(n_i) - n_e, \qquad -\rho_s^2\nabla_\perp^2 A = \beta_e(u_i - u_e) \tag{15.92}$$

The polarisation operator is

$$P = \tau_i^{-1}\left[1 - \left(1 - \rho_i^2\nabla_\perp^2\right)^{-1}\right] \tag{15.93}$$

and gyroaveraging operator and gyroaveraged potential are

$$G_0 = \left(1 - \frac{1}{2}\rho_z^2\nabla_\perp^2\right)^{-1}, \qquad \phi_0 = G_0(\phi) \tag{15.94}$$

with $\rho_z = \mu_i\tau_i$ for ions and zero for electrons. We also define

$$\varpi = n_e - n_i, \qquad J = u_i - u_e \tag{15.95}$$

The operators and the procedure for the induction equation for A are the same as in Chapter 11. The species constants a_z, μ_z and τ_z are -1, $-\mu_e$ and -1 for electrons and 1, 1, and τ_i for ions, respectively, Note that $\nabla_\|\phi_0$ and $[\phi_0, \mu_z u]$ have been combined in the equation for A^*.

The spatial domain is $L_x = L_y/2 = 1.0$, periodic in both directions. The initial state for the magnetic field due to the current layers is

$$J(x, k_y = 0) = 2\pi\, S\, \frac{\delta_0}{\beta_e}\cos(2\pi x - 0.37), \qquad A(x, k_y = 0) = \frac{\beta_e}{\delta_0^2}(2\pi)^{-2} J \tag{15.96}$$

where the 0.37 shift is applied to make both singular surfaces visible on the plots. This is then loaded into the dependent variables as $A_e^* = A - \mu_e J$ for the electrons and $A_i^* = A$ for the ions.

All y-dependence is excited by setting $n = n_{2D}(x, y)$ with n_{2D} the normalised 2D random-phase field given in Eq. (2.126) from Chapter 2. The leading constant is chosen so that the square root of the spatial-domain average of the square (RMS) amplitude of n is $A_0 = 10^{-8}$. The same n is initialised for each species, as in the turbulence cases. All activity proceeds according to the equations. We have both electron and ion moment variables, but we can measure the smallness of the ion ones, so long as no background gradients are set. Most of the initial activity decays away, except for the $k_y = 0$ component of A, whose decay is zero for zero resistivity, and the component of the initial state which forms the growing mode. The growing mode eventually contains almost only the electron moment variables, with all the ion effect in the polarisation. We do not assume any of this, just notice it as a main result. Recall this is the reason to excite all available degrees of freedom—we do not assume the form of any mode which emerges.

To get a short run to understand what happens, we set $S = 0.1$ and $\delta_0 = \rho_s/a = 0.1$, while setting $\beta_e = 10^{-3}$, with the mass parameters $\mu = 1$ for ions and 5.45×10^{-4} for electrons (i.e., proton ions). This is not realistic scale separation, but it allows a moderate grid $N_x \times N_y$ of 128×256, sufficient to qualitatively capture the linear stage and the early transition to nonlinearity, within a few minutes on a laptop. The timestep is 2×10^{-3} in terms of a/c_s. The largeness of S and δ_0 is not realistic, just a useful tool to optimise learning (one could then reduce δ_0 and increase N_x both by ten to get a more realistic case). The temperature disturbances are measured to be small, as are the ion variables. This case grows to nonlinear amplitude and then quickly runs out of resolution by $t = 46a/c_s$, or 23 000 time steps.

Much of the initial bath of fluctuations decays away, leaving the instability eigenmode to form and then grow. The growing mode exists almost only near the two singular surfaces where the main wave of $A(k_y = 0)$ has its extrema. This initial phase is over by about $t = 10$ and the structure of the instability becomes clear by $t = 15$. The form at $t = 20$ is shown in figure 15.1. These are cuts in x of the $k_y \neq 0$ component of the variables, dominated by $k_y = \pm 1$, through the position in y at

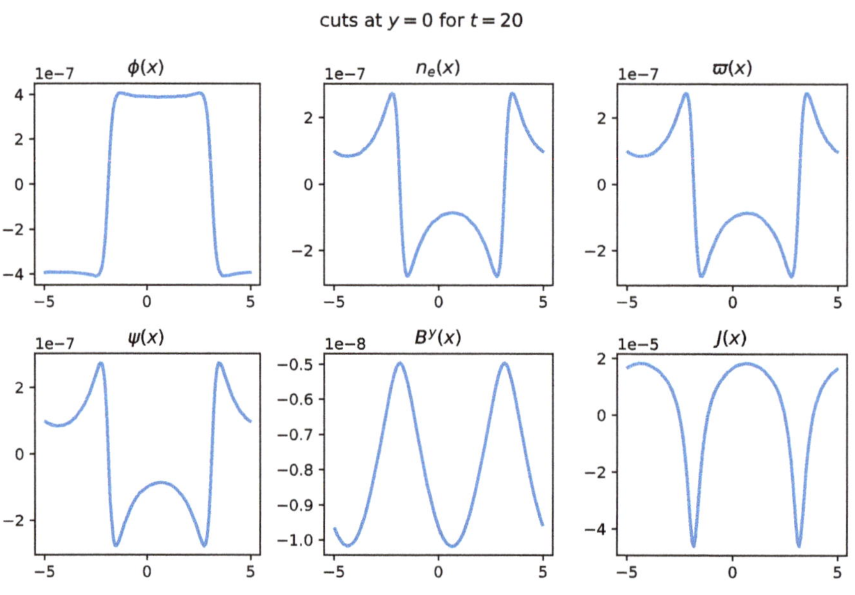

Figure 15.1. Cuts at $t = 20$ in the x-direction of the $k_y \neq 0$ component of the variables, dominated by $k_y = \pm 1$, through the position in y at which A is maximum. The form of A is concave down, with minima and slope disconinuity at the singular surfaces. The width of the inner layer is seen in the sharp transition from positive to negative values in any of the dependent variables. Note that n_e and ϖ are closely equal, reflecting the small value of n_i. The $1/x$ tails in ϕ expected from the theory do not have the necessary space to form in this case, with $\delta_0 = 0.1$ not small enough. They are visible if δ_0 is an order of magnitude smaller.

which the flux disturbance A is maximum. The concave down nature of A, with minima and slope discontinuity at the singular surfaces, is visible. The thinness of the inner layer is shown by ϖ and J which form spikes there. The equality of n_e and ϖ, not assumed by the model, is a check on the validity of the two-field model mentioned above. The structure of ϕ is interfered with by the proximity to the next singular surface.

The transition to nonlinearity is extremely rapid. The form at $t = 42$ is shown in figure 15.2, with the same format as in figure 15.1. The eigenfunction becomes asymmetric, with one side rising and the other falling, compared to the amplitude level. Once this change in the structure becomes visible, the nonlinear regime has been entered. Transition to nonlinearity can always be measured by changes in the structure of the eigenmode. In this case it corresponds to the opening of the magnetic island structure. The classic island structure is described by

$$A(x, y) = -S\,\delta_0^{-1}\,\frac{x^2}{2} + \psi\cos y \tag{15.97}$$

in analytical theory, where the first term is the expansion of the background flux around the singular surface (see Eq. (15.96); the constant value and the shift of position are deleted when doing the inner solution), and the second term is a slowly varying amplitude (constant in the analytical solution) times the wavelike dependence in y. Here we have its generalisation. There is a value of A, here equal to $-\psi$, whose contour forms a separatrix. The flux lines within it are closed, while outside

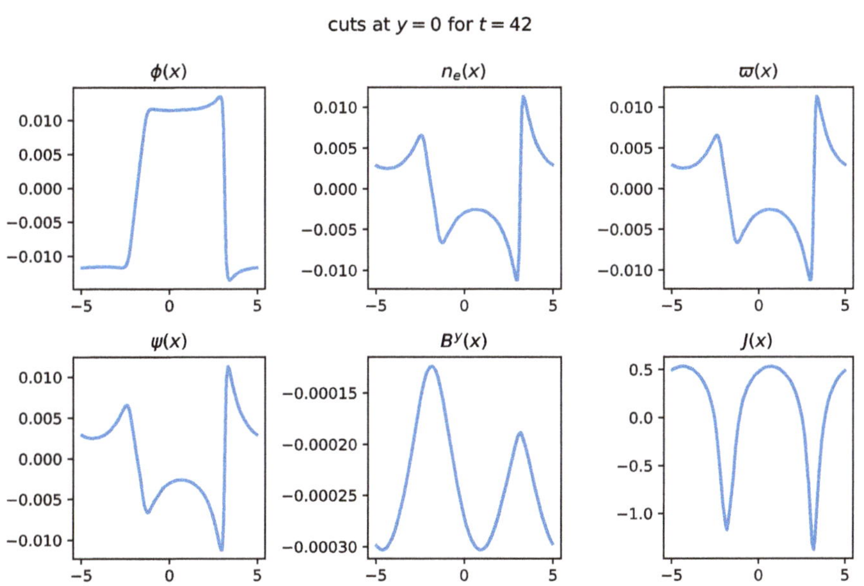

Figure 15.2. Cuts at $t = 42$ in the x-direction of the $k_y \neq 0$ component of the variables, dominated by $k_y = \pm 1$, through the position in y at which A is maximum. The form of the dependent variables is similar to the linear phase, but asymmetry corresponding to the difference in the O-point and X-point of the magnetic island is now visible (see the following figures).

they are deformed about their equilibrium positions but still extended periodically in y. As the island grows, a larger region becomes enclosed, which requires the tearing and reconnecting of the field lines. In Eq. (15.97), the island width is given by setting $A = -\psi$ and $y = 0$, then solving for x, so that

$$w^2 = 4S^{-1}\delta_0\psi \tag{15.98}$$

with w giving the half-width of the island. Note once more the factor of δ_0 from the normalisation.

This island structure is seen in the flux and current as the disturbances become nonlinear enough for visibility. The morphology of A and J is shown in figure 15.3, at $t = 42$, just before run termination. The O-points are where the closed contour lines reduce in enclosed area to zero. These are on the original singular surface lines in the middle of the islands. They are not precisely at $y = 0$ because the initialisation randomised the phase. The X-points are also on the singular surface lines but where the contour lines cross. The separatrix is the limit of the largest closed contour, while anything outside of it forms a contour which extends periodically in y. Note that $J = J(\psi)$ is approximately held in the nonlinear

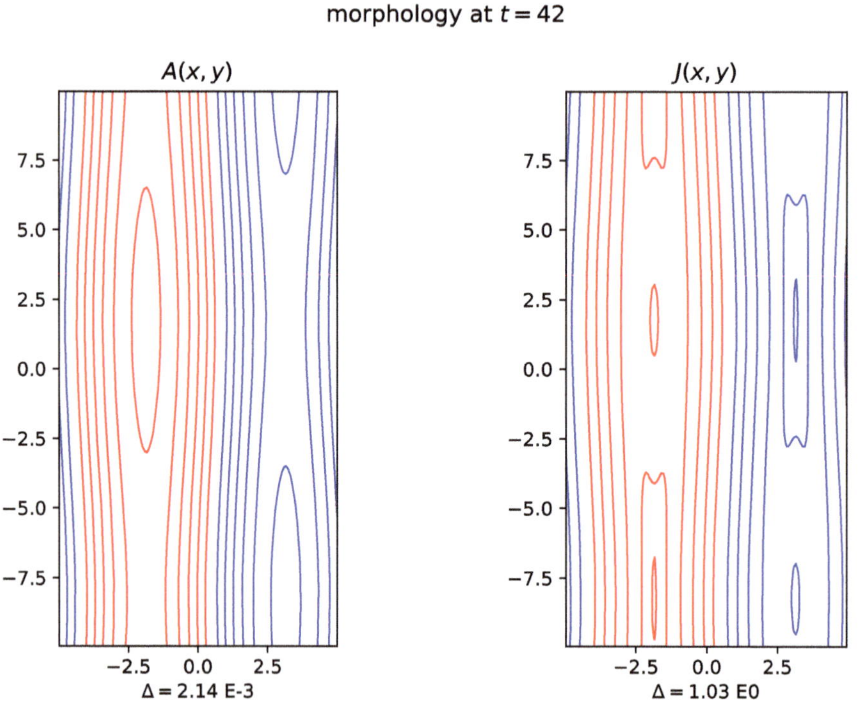

Figure 15.3. Spatial morphology of the flux and current in the xy-plane, at $t = 227.7$, just before the case runs out of resolution. The island formation is visible. The X-points are where the contours from either side of the singular surface cross. Field lines interior to the separatrix have been torn and reconnected. The current builds a spike at the X-points and is maximum on the separatrix.

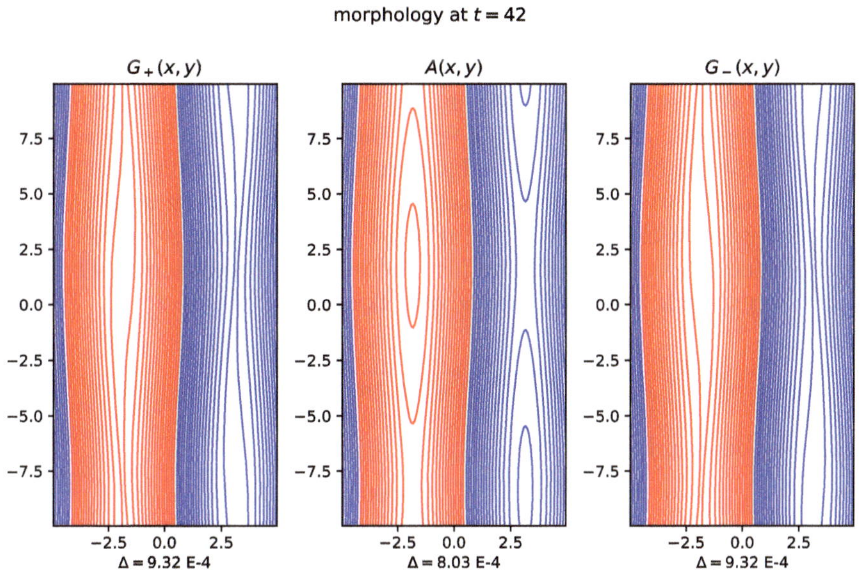

Figure 15.4. Spatial morphology at $t = 227.7$ of the flux and the $G_{\pm}$ quantities which are conserved in the limit of sufficient resolution. In practice the separatrix and X-point would require infinite resolution for this to be perfectly true for all flux lines of the $G_{\pm}$ quantities. However, one can see that ψ itself has undergone more reconnection than the $G_{\pm}$, which would tend to form deformed sheets near the island X-points rather than torn and reconnected lines.

stage, except at the X-point. This is a result of the asymptotic theory in which $\Delta \ll w \ll a$ is assumed (Rutherford, 1973).

In this normalisation the conservation law in Eq. (15.82) applies to the electron variables with $G_{\pm} = A^{*} \pm \mu_e^{-1/2} n$ and $\Phi_{\pm} = \phi \pm \mu_e^{-1/2} A$ giving the normalised version of Eq. (15.83). Flux conservation (of A) is therefore broken by finite μ_e and δ_0 (from the polarisation relating n to ϕ). The three fields $G_{\pm}$ and A are shown in figure 15.4, at $t = 42$, just before run termination. The clear separation of the field of A into closed and open regions is not mirrored in $G_{\pm}$, because the latter are conserved. Up to this point, this physics is faithfully reproduced. But already on the next frame the $G_{\pm}$ fields begin to reconnect, indicating the loss of effective resolution. The $x = 0$ separation line is deformed rather than reconnected into a separatrix for these quantities, until the sharp 'escape' of the line at the X-points of the island is no longer well resolved.

The reason that the run has to be terminated is that the current forms a spike at the X-points of the magnetic island structure—the sum of the $k_y L_y = 1$ and $k_y = 0$ components of $\tilde{A}_{\|}$. The spectrum quickly broadens to the limits of the grid. It becomes asymptotically difficult to resolve the current and the island structure at some finite point into the nonlinear regime. In a realistic case, this would then require a three-dimensional nonlinear model resolving both ρ_s and d_e. But even here, one can see the property of the collisionless reconnection process in the qualitative difference between A versus $G_{\pm}$.

15.4 Ballooning modes

We give a brief capture of the ideal ballooning mode using the delta-W formulation. In Chapter 13 of Volume One, this mode was found from the MHD vorticity equation keeping the curvature forcing term, the MHD Ohm's law, and a linearised pressure advection equation with the background gradient term. The strict MHD limit with all other curvature terms absent is equivalent to keeping the diamagnetic current's compression effect in the $\mathbf{J} \times \mathbf{B}$ force but neglecting it as well as $\nabla \cdot \mathbf{v}_E$ in the continuity equations. This can be justified as an approximation when $L_\perp \ll R_0$ since then the gradient term is much larger. Then one obtains

$$K_\parallel^2 \frac{\partial}{\partial s} g^{yy} \frac{\partial \phi}{\partial s} = \left(\frac{\gamma^2}{v_A^2} g^{yy} - \frac{2\mu_0 p_0}{B^2 R L_p} \mathcal{K}^y \right) \phi \tag{15.99}$$

as the eigenmode equation, where

$$g^{yy} = 1 + S^2 s^2, \qquad \mathcal{K}^y = \cos s + Ss \sin s \tag{15.100}$$

define the two salient pieces of geometric information: the curvature through $\mathcal{K}^y$ and the metric in the Laplacian through g^{yy} [these are Eqs (13.88/13.89) from Chapter 13 of Volume One].

For the delta-W formulation we follow the lectures by White. Given the expression he uses for δW (Eq. 15.52), the following approximations are made. Compressibility is neglected, so that the second and third terms are absent. The last term is neglected as the background current is not treated. The fourth term is the destabilisation through field line curvature. The first term is the energy cost of bending the field lines. These are the two terms in the simplified dispersion relation

$$\gamma^2 = \gamma_I^2 - k_\parallel^2 v_A^2 \tag{15.101}$$

respectively (this is Eq. 13.90 from Chapter 13 of Volume One). These two terms are further processed by considering the limit $k_\perp a \gg 1$ and expanding in powers of $(k_\perp a)^{-1}$. The resulting delta-W energy perturbation is

$$\delta W = \frac{1}{2} \int d\mathcal{V} \left[k_\perp^2 \, |\mathbf{b} \cdot \nabla \Phi|^2 - \frac{2}{B^2} (\mathbf{b} \times \mathbf{k}_\perp \cdot \nabla p)(\mathbf{b} \times \mathbf{k}_\perp \cdot \boldsymbol{\kappa}) \, |\Phi|^2 \right] \tag{15.102}$$

where the displacement and eigenfunction are defined by

$$\boldsymbol{\xi} = \frac{1}{B} (\mathbf{b} \times \mathbf{k}_\perp) \Phi \tag{15.103}$$

The two terms in δW represent the energy cost of field line bending and the potential energy gain through motion in a curved magnetic field and an ambient pressure gradient. Note that Φ is proportional to $(-1)^{1/2}$ times the electrostatic potential in an ExB displacement.

The same treatment yields for the kinetic energy

$$K = \int d\mathcal{V} \frac{\rho}{2} \left| \frac{d\boldsymbol{\xi}}{dt} \right|^2 = -\frac{1}{2} \omega^2 \int d\mathcal{V} \frac{\rho}{B^2} k_\perp^2 \, |\Phi|^2 \tag{15.104}$$

The eigenmode equation is obtained from Eq. (15.23) in the form of $\omega^2 K = \delta W$, taking the variation of both sides $\Phi \to \Phi + \delta\Phi$, yielding

$$\mathbf{B}\cdot\nabla\left(\frac{k_\perp^2}{B^2}\mathbf{B}\cdot\nabla\Phi\right) + \frac{\rho}{B^2}\omega^2 k_\perp^2\Phi + \frac{2}{B^2}(\mathbf{b}\times\mathbf{k}_\perp\cdot\nabla p)(\mathbf{b}\times\mathbf{k}_\perp\cdot\boldsymbol{\kappa})\Phi = 0 \quad (15.105)$$

which is his ballooning equation. Now, if we identify

$$k_\perp^2 \leftrightarrow k_y^2\, g^{yy}, \qquad \frac{2}{B^2}(\mathbf{b}\times\mathbf{k}_\perp\cdot\nabla p)(\mathbf{b}\times\mathbf{k}_\perp\cdot\boldsymbol{\kappa}) \leftrightarrow \frac{2\mu_0 p_0}{B^2 R L_p}\mathcal{K}^y \quad (15.106)$$

(note $\mathcal{K}^y$ results from the radial derivative of $\log R$), and see the definition of $v_A^2 = B^2/\mu_0\rho$ in the second term, we can identify White's ballooning equation with Eq. (15.99). We can also identify the ballooning delta-W expression with what we get from multiplying Eq. (15.99) by the eigenmode and integrating the outermost $\partial/\partial s$ term by parts,

$$\frac{\partial}{\partial s}K_\parallel^2 g^{yy}\phi\frac{\partial\phi}{\partial s} - K_\parallel^2 g^{yy}\left|\frac{\partial\phi}{\partial s}\right|^2 = \left(\frac{\gamma^2}{v_A^2}g^{yy} - \frac{2\mu_0 p_0}{B^2 R L_p}\mathcal{K}^y\right)|\phi|^2 \quad (15.107)$$

[this is Eq. (13.93) from Chapter 13 of Volume One]. The integration over this annihilates the first term, leaving

$$\frac{\gamma^2}{v_A^2}\int_{-\infty}^{\infty} ds\, g^{yy}\,|\phi|^2 = \frac{2\mu_0 p_0}{B^2 R L_p}\int_{-\infty}^{\infty} ds\,\mathcal{K}^y\,|\phi|^2 - K_\parallel^2\int_{-\infty}^{\infty} ds\, g^{yy}\left|\frac{\partial\phi}{\partial s}\right|^2 \quad (15.108)$$

[this is Eq. (13.94) from Chapter 13 of Volume One]. The term on the left hand side is positive definite. On the right hand side, the second term is the energy cost of field line bending, and the first term represents energy release if L_p and $\mathcal{K}^y$ have the same sign, as in the delta-W analysis. This is as far as we need to go, since we did the computations with this mode in Volume One.

We see why it is advantageous for these pressure modes to reformulate the delta-W equation to eliminate $\nabla\cdot\boldsymbol{\xi}$ everywhere it occurs except in the pressure compressibility term. The pressure modes live from quasistatic compressibility (the curvature terms with $\mathcal{K}$ in our case), but are distinct from fast-wave compressional modes which are neglected (the neglect of $\mathbf{A}_\perp$ and the low frequency approximations in our case).

15.5 Kink modes

The MHD mode type with which we want to spend the most effort in this Chapter is the kink mode. Not least because the ability to capture this mode with gyrofluid or gyrokinetic equations which do not contain fast compressional Alfvén dynamics is not widely known, despite formative work in the 1990s (cited below).

The basic setup is similar to the tearing mode, with ideal dynamics outside of a singular layer, where $k_\parallel$ vanishes, and the corrections inside of it. The principal difference is that the form of the outer solution is different. The displacement is finite on one size of the layer but forced by boundary conditions to vanish outside of it. One has a plasma column treatable as a cylinder, wherein curvature effects are neglected. The equilibrium is axially and azimuthally symmetric. The topology should be toroidal, so the axial direction is periodic, as well as the azimuthal (these correspond to toroidal and poloidal, in a torus). The equilibrium magnetic field is described by a radially dependent pitch parameter $q = q(r)$, which for cylindrical geometry can be defined by the expression

$$\mathbf{B} = B_0\nabla z + B_0\frac{r^2}{qR_0}\nabla\theta \tag{15.109}$$

where $2\pi R_0$ is the periodicity length of the axial direction.

Both tearing and kink modes are of the current driven type. In the tearing mode analysis, we saw that in the outer solution the gradient of the equilibrium parallel current J_0 is what appears. So there is the issue of the current versus the current gradient. The delta-W energy principle has J_0 only, through

$$\delta W_J = -\int d\mathcal{V}\,\sigma\mathbf{B}\cdot\boldsymbol{\xi}\times(\delta\mathbf{B})_\perp \tag{15.110}$$

where $\sigma = J_\| B$, from Eq. (15.52). Using Eq. (15.19) we set

$$\delta W_J = \int d\mathcal{V}\,\sigma(\boldsymbol{\xi}\times\mathbf{B})\cdot\nabla\times(\boldsymbol{\xi}\times\mathbf{B}) \tag{15.111}$$

Then, integrating the curl by parts produces the same expression because the $\nabla\sigma$ term has the form $\mathbf{a}\times\nabla\sigma\cdot\mathbf{a}$ which vanishes. So in the delta-W theory the current always appears by itself, not through its gradient. In reduced MHD theory, however, one takes the curl of the J-cross-B force in the form of

$$\nabla\cdot\frac{1}{B^2}\mathbf{B}\times(\mathbf{J}\times\mathbf{B}) = B\nabla_\|\frac{J_\|}{B} \tag{15.112}$$

since both **B** and **J** are divergence free. This vanishes in equilibrium for a straight cylinder, which is why it forms the outer solution for current driven modes. And here, we have the equilibrium current through its gradient.

The kink mode exists in principle anywhere $q(r)$ is a rational number. In practice, since $k_\| = (m - nq)/qR_0$ for a Fourier component with dependence $\exp(im\theta)$ and $\exp(-in\zeta)$, where $\zeta = z/R_0$, this parallel wavenumber becomes large very quickly as r departs the singular layer (rational surface $q = m/n$). So for kink modes in this setup, one looks for the lower mode numbers, ideally $n = 1$ and $m = 1$.

In a general analysis, one has in principle any type of unstable mode or oscillation for any choice of Fourier mode number pair (m, n), and much of the effort in the early literature was involved in sorting this out. In hindsight we can avoid much of it, and for illustrative purposes assume that interchange type dynamics or any pressure driven mode is not present, and focus solely on the current driven modes. We saw the tearing modes just above. These exist as well at all mode numbers, but are significant only where a fundamental rational surface with integer q, hence $n = 1$, is present, most significantly the (2, 1) mode. The (1, 1) mode is a special case for tearing, since the nature of the ideal solution disallows use of the constant-ψ approximation (the outer solution drops to zero as one crosses the singular surface from interior outward). Both the kink mode and tearing mode are in principle present here, and the situation in the real world is complicated by thermal and toroidal geometry complications. But to understand the instability we take the simplest case of the (1, 1) kink mode and briefly address the tearing mode situation afterward.

The outer solution is given as in the tearing mode, demanding that

$$\tilde{B}^r \frac{\partial J_\|}{\partial r} + B^\theta \frac{\partial \tilde{J}_\|}{\partial \theta} + B^\zeta \frac{\partial \tilde{J}_\|}{\partial \zeta} = 0 \tag{15.113}$$

For the Fourier mode dependences, with $\tilde{J}_\|$ and $\tilde{A}_\|$ defined as real,

$$\tilde{J}_\|(r, \theta, \zeta) = J(r) \exp[i(m\theta - n\zeta)] + (\text{c. c.}) \tag{15.114}$$

$$\tilde{A}_\|(r, \theta, \zeta) = \psi(r) \exp[i(m\theta - n\zeta)] + (\text{c. c.}) \tag{15.115}$$

this equation becomes

$$\frac{m}{r}\psi \frac{\partial J_\|}{\partial r} + B_0 k_\| J = 0 \tag{15.116}$$

We are using

$$B^r = \frac{m}{r}\frac{\partial \tilde{A}_\|}{\partial \theta}, \qquad k_\| = \frac{m - nq}{qR_0}, \qquad \nabla_\perp^2 = \frac{1}{r}\frac{\partial}{\partial r} r \frac{\partial}{\partial r} - \frac{m^2}{r^2} \tag{15.117}$$

The equilibrium and disturbed current are given through Ampère's law by

$$\mu_0 J_\| = \frac{B_0}{R_0}\frac{1}{r}\frac{\partial}{\partial r}\frac{r^2}{q}, \qquad \mu_0 J = -\nabla_\perp^2 \psi \tag{15.118}$$

Putting these into Eq. (15.116), we have

$$\frac{1}{r}\psi \frac{\partial}{\partial r}\frac{1}{r}\frac{\partial}{\partial r} r^2 \frac{m}{qR_0} - k_\| \nabla_\perp^2 \psi = 0 \tag{15.119}$$

In the first term the r^2 is treated separately, and it is noted that

$$\frac{\partial}{\partial r}\frac{m}{qR_0} = \frac{\partial k_\|}{\partial r} \tag{15.120}$$

These are worked through to find

$$k_\| \nabla_\perp^2 \psi = \left(\nabla_\perp^2 k_\| + \frac{2}{r}\frac{\partial k_\|}{\partial r}\right)\psi = 0 \tag{15.121}$$

In the outer region, ideal MHD applies and hence with $E_\| = 0$ we have

$$\gamma\psi + ik_\parallel\phi = 0 \qquad \text{hence} \quad \psi = \frac{-i}{\gamma}k_\parallel\phi \tag{15.122}$$

The equation for ψ is then rewritten as one for ϕ, obtaining a cancellation of the $\nabla_\perp^2 k_\parallel$ term, and arriving at

$$k_\parallel\nabla_\perp^2\phi = \frac{\partial k_\parallel}{\partial r} r\frac{\partial}{\partial r}\frac{\phi}{r} \tag{15.123}$$

For any other (m, n) this equation has separate solutions on either side of the singular surface, and matching the amplitudes leaves a discontinuity in the slope. The kink itself is usually stable due to the energy cost of the field line bending term, but for $(m, n) = (1, 1)$ only, the two sides of the equation can be made to vanish by setting

$$\phi = c_0 r \tag{15.124}$$

where c_0 is a complex phase factor.

Let us define the singular surface for the (1, 1) mode as

$$q(r) = 1 \quad \text{at} \quad r = r_1 \tag{15.125}$$

On the inside of the layer ($r < r_1$) the solution $\phi = c_0 r$ satisfies the requirement that $\phi \to 0$ as $r \to 0$. Note that, as in the ballooning mode, the ExB velocity of the disturbance is

$$\gamma\boldsymbol{\xi} = \frac{1}{B_0}\mathbf{b} \times (i\mathbf{k}_\perp\phi) = -\frac{1}{rB_0}i\phi\nabla r \qquad \text{so that} \quad \phi \propto ir\xi^r \tag{15.126}$$

since $\mathbf{k}_\perp = \nabla\theta$. The displacement is radial, and since ϕ/r is constant, it is homogeneous except for the phase factor $i(\theta - \zeta)$. The linearity of ϕ in r hence constancy of ξ^r denotes a rigid displacement of the plasma elements in the horizontal $r\theta$ plane with a helical $\cos(\theta - \zeta)$ structure. The direction of the displacement is given by the phase. Due to the helical dependence, the direction of the displacement rotates by one cycle over the entire periodicity length in the axial direction. On the outside of the layer ($r > r_1$), the solution $\phi = c_0 r$ does not satisfy the requirement that ϕ vanish for $r = a$, the conducting wall outer boundary. So in the ideal region for $r > r_1$, the only solution satisfying Eq. (15.123) is the trivial one, $\phi = 0$. This is the basic structure of the kink mode displacement.

15.5.1 Kink mode with gyrofluid equations

The kink mode forms a test case as to whether the gyrofluid equations, or some subset of them, can capture this class of global MHD dynamics. As there is a literature where this is done with reduced MHD (RMHD), and since RMHD is a subset of the four-field drift Alfvén model, which in turn is a subset of the gyrofluid equations, it follows that the kink mode should indeed be capturable by these equations just like the tearing mode (resistive or collisionless) is. Since 1991 there is work by Porcelli, Pegoraro and colleagues on the collisionless $m = n = 1$ instability using the two-field model mentioned above, which is the simplest form of gyrofluid equations for instabilities controlled by electrons. Indeed since 1995 there is the work by Naitou and colleagues on the kink instability with gyrokinetic computations. So we should be able to (a) capture the MHD mode and (b) see what happens in a hot plasma situation with temperature gradients in both electrons and ions. We can also find out if the ions and FLR corrections have any effect, which in the case of the tearing mode is relatively weak.

Since we treat temperature gradients, we use the six-moment gyrofluid equations, and for this purpose we set the collisions to zero. The collisionless skin depth d_e becomes the scale one needs to resolve, along with ρ_s. For MHD studies this procedure is unfamiliar, but as we have seen, it is native for turbulence studies. We need only pick a realistic situation and set the grid properly.

This being a linear study, we revert to having the equilibrium magnetic field in the background keeping $A_\parallel$ only for the disturbances. So ∇J_0 becomes another of the background gradient drive effects. In the MHD case we keep this gradient and set all the others to zero.

The geometry is cylindrical, with minor radius a_0 and periodicity length $2\pi R_0$. The magnetic field is given in Eq. (15.109). The coordinates follow

$$r\,\nabla r \times \nabla\theta \cdot \nabla z = +1 \tag{15.127}$$

The double periodicity of toroidal topology is expressed as

$$R_0\zeta = z, \qquad f(r, \theta, \zeta + 2\pi) = f(r, \theta, \zeta), \qquad f(r, \theta + 2\pi, \zeta) = f(r, \theta, \zeta) \tag{15.128}$$

The magnetic pumping and curvature terms are neglected. The equations are linearised by keeping the bracket terms (ExB advection and parallel derivatives) only where they act on the equilibrium quantities with gradients. The angle dependence is set using Fourier amplitudes. The single mode $m = n = 1$ is taken, so that the variables become complex amplitudes. The parallel dynamics is evaluated using

$$k_\parallel = \frac{m - nq}{qR_0} \tag{15.129}$$

which following q is a function of r.

Normalisation follows the scales a_0/c_s and a_0, the electron density and temperature n_e and T_e, and the magnetic field B_0. The scale ratio as a parameter, the normalised radius, and its domain are

$$\varepsilon_a = \frac{a_0}{R_0}, \qquad r_a = \frac{r}{a_0}, \qquad 0 < r_a < 1 \tag{15.130}$$

The boundary conditions are that the dependent variables vanish at both endpoints, so the grid is a power of two plus one with endpoints on the boundaries. The operators then are computed only in the interior, avoiding the coordinate singularity at the axis.

With the density and temperature gradients present as well as the current gradient, we have for each species

$$\frac{\partial n}{\partial t} + [\phi_0, \log n_0] + [\phi_1, \log T_0] + ik_\parallel u + [-A, u_0] = 0 \tag{15.131}$$

$$\begin{aligned}&\frac{\partial}{\partial t}(A + \mu_z u) + \mu_z[\phi_0, u_0] + ik_\parallel(\phi_0 + \tau_z(n + T_\parallel)) \\ &+ \tau_z[-A, (\log n_0 + \log T_0)] = 0\end{aligned} \tag{15.132}$$

$$\frac{1}{2}\frac{\partial T_\parallel}{\partial t} + \frac{1}{2}[\phi_0, \log T_0] + ik_\parallel(u + q_{\parallel\parallel}) + [-A, u_0] = 0 \tag{15.133}$$

$$\frac{\partial T_\perp}{\partial t} + [\phi_0, \log T_0] + [\phi_1, (\log n_0 + 2\log T_0)] + ik_\parallel q_{\perp\parallel} = 0 \tag{15.134}$$

$$\mu_z\frac{\partial q_{\parallel\parallel}}{\partial t} + \frac{3}{2}ik_\parallel(\tau_z T_\parallel) + \frac{3}{2}\tau_z[-A, \log T_0] = -\sqrt{3}\mu_z V_z\frac{\varepsilon_a}{q}q_{\parallel\parallel} \tag{15.135}$$

$$\mu_z\frac{\partial q_{\perp\parallel}}{\partial t} + ik_\parallel(\phi_1 + \tau_z T_\perp) + \tau_z[-A, \log T_0] = -\mu_z V_z\frac{\varepsilon_a}{q}q_{\perp\parallel} \tag{15.136}$$

where the species constants μ_z and τ_z are $-\mu_e$ and -1 for electrons and 1 and τ_i for ions, respectively, and V_z is the species thermal velocity given by $V_z^2 = \tau_z/\mu_z$, controlling the Landau damping. Note that there are no terms with a background heat flux gradient, since that is assumed to be zero. The field equations are

$$P(\phi) = \sum_{sp} a_z[G_0(n) + G_1(T_\perp)], \qquad -\nabla_\perp^2 A = \beta_e \sum_{sp} a_z u \tag{15.137}$$

where the species constant a_z is ±1 for ions and electrons, respectively. The gyroaveraging operators are

$$G_0 = \left(1 - \frac{1}{2}\rho_z^2\nabla_\perp^2\right)^{-1}, \qquad G_1 = \left(1 - \frac{1}{2}\rho_z^2\nabla_\perp^2\right)^{-1} G_0 - G_0 \tag{15.138}$$

and the polarisation operator is

$$P = \tau_i^{-1}\left[1 - \left(1 - \rho_i^2\nabla_\perp^2\right)^{-1}\right] \tag{15.139}$$

The gyroaveraged potentials are

$$\phi_0 = G_0(\phi), \qquad \phi_1 = G_1(\phi) \tag{15.140}$$

For electrons, $G_0 = 1$ and $G_1 = 0$ with the above forms used for the ions. The drift parameter, and the normalised gyroradii are

$$\delta_0 = \frac{\rho_s}{a_0}, \qquad \rho_z^2 = \mu_z\tau_z \tag{15.141}$$

The perpendicular Laplacian and the brackets with a quantity acting upon a profile are

$$\nabla_\perp^2 = \delta_0^2\left(\frac{\partial^2}{\partial r_a^2} + \frac{1}{r_a}\frac{\partial}{\partial r_a} - \frac{m^2}{r_a^2}\right), \qquad [a, L_0] = -\delta_0\frac{im}{r_a}a\frac{\partial L_0}{\partial r} \tag{15.142}$$

where L_0 is one of $\log n_0$, $\log T_0$, or u_0 for each species. The only place finite differences appear is in $\nabla_\perp^2$ which uses second order centred differences.

The basic MHD case sets $\log n_0$ and $\log T_0$ to zero and keeps only u_0 for the electrons. The profile $q(r)$ and hence the current are given by

$$q = q_0 + q_1 r_a^2, \qquad J_0 = -u_0 = \varepsilon_a\frac{\delta_0}{\beta_e}\frac{q_0}{q^2} \tag{15.143}$$

The geometric parameters are

$$\varepsilon_a = 0.3, \qquad q_0 = 0.7, \qquad q_1 = 3.3 \tag{15.144}$$

and the physical parameters are

$$\delta_0 = 0.005, \qquad \beta_e = 10^{-3}, \qquad \mu_e = 2.72\times 10^{-4}, \qquad \tau_i = 1 \tag{15.145}$$

The inverse skin depth ratio is $a/d_e = \delta^{-1}(\mu_e/\beta_e)^{1/2}$ which is about 400 for these parameters, so we run up to $N_r = 1025$ radial grid points to check the resolution. The timestep is 10^{-3} in terms of a_0/c_s.

The initial state is given by setting

$$n = 10^{-10}\exp[-100(r_a - 0.6)^2] \tag{15.146}$$

for each species, hence zero gyrocentre charge density. Note that the singular surface r_1 is not at 0.6 for this case, but close to 0.3. This value of β_e produces strong growth. By about $t = 100a_0/c_s$ the eigenmode solidifies and the growth rate is thereafter constant. The convergence test gives $\gamma = 0.101$, 0.102, 0.103, and 0.136 for $N_r = 1025$, 513, 257, and 129, all with errors below 10^{-6} and constant within about 10^{-8} over $100 < c_s t/a_0 < 200$. The value 0.102 is equivalent to $3.22\times 10^{-3} v_A/a_0$. That is slow in Alfvén units but quite fast in sound speed units.

The energetics is due solely to J_0 in this case. The drive through the parallel dynamics is in the equations for n and $T_\parallel$ for the electrons, with rate given by

$$\Gamma_r = \frac{1}{2E_T}\mathcal{V}^{-1}\int d\mathcal{V}\,(n + T_\parallel - \phi)(\nabla_\parallel J_0) \tag{15.147}$$

('r' for reconnection). The volume integration is given by the complex amplitude integrated over radius with volume element r_a. With this proviso the free energy and its pieces are given as in Chapter 9. The drive through the ExB advection of the current is

$$\Gamma_j = \frac{1}{2E_T}\mathcal{V}^{-1}\int d\mathcal{V}\,\mu_e u[\phi, J_0] \tag{15.148}$$

again solely for the electrons. We find the latter negligible and the former practically equal to γ, with the Landau damping channels down by 10^{-4} for electrons and 10^{-7} for ions.

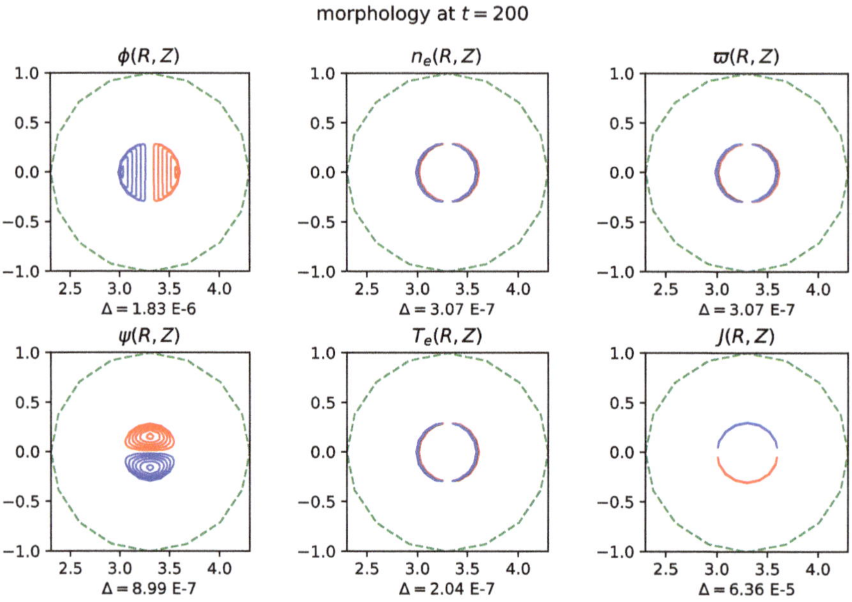

Figure 15.5. Spatial realisations in the $r\theta$-plane at $z = 0$ for ϕ and $A_{\|}$ (left, the MHD mode pieces), the vorticity and parallel current (right), and the electron density and temperature (center). This is for the case without thermal gradients, with $\beta_e = 10^{-3}$ and $N_r = 1025$ grid points. The classic MHD structure in ϕ and $A_{\|}$ is seen, and all the nonideal activity is localised to the singular surface. The initial state begins with zero phase for n_e and n_i (starting equal), so the growing mode has zero phase in ϕ and the quarter circle shift in $A_{\|}$. The large dashed circle is the domain boundary at $r_a = 1$.

The eigenmode structure is seen in the spatial realisations in the $r\theta$-plane at $z = 0$. These are given in figure 15.5, for ϕ, $A_{\|}$, ϖ, $J_{\|}$, as well as n_e and T_e, where $\varpi = n_e - n_i$ and $J_{\|} = u_i - u_e$, and for each species $T = (T_{\|} + 2T_{\perp})/3$. The classic MHD structure in ϕ and $A_{\|}$ is seen, and all the nonideal activity is localised to the singular surface. The initial state begins with zero phase for n_e and n_i (starting equal), so the growing mode has zero phase in ϕ and the quarter circle shift in $A_{\|}$. The large dashed circle is the domain boundary at $r_a = 1$.

The case with $\beta_e = 10^{-2}$ was also run. This is a fairly extreme case, which with $\tau_i = 1$ corresponds to a total β of four percent. Due to the slower growth found we have to run longer. The eigenmode solidifies by about $t = 240a_0/c_s$ and the run carries to $c_s t/a_0 = 500$. Convergence is more difficult, with $\gamma = 0.0205$, 0.0249, 0.0249, and 0.0404 for $N_r = 1025$, 513, 257, and 129. Interestingly, we would have judged this converged, had we not done the finest case. This is because the $N_r = 1025$ case is the only one which begins to resolve the electron skin depth layer. The growth rate drops slightly faster than $\beta_e^{-1/2}$, indicating a stabilising effect within MHD units of v_A rather than c_s. The mode structure and run integrity are the same as in the previous case.

We now incorporate the thermal gradients into the $\beta_e = 10^{-3}$ case. The basic structure of each profile is

$$\log T_0 = 1 + \cos \pi r_a \qquad \log n_0 = \frac{1}{2} \log T_0 \tag{15.149}$$

for each species. There are new energetic effects. These are the drive mechanisms we have covered for the turbulence studies, and follow the expressions given in Section 11.3 from Chapter 11. For the present purpose we collect all the ExB advective drives together into Γ_p (and this includes the Γ_j above), and all the magnetic flutter drives into Γ_v as we have done previously. The growth rate is almost the same as in the basic MHD case, with no indication of a tendency for $\nabla \log T_0$, especially, to stabilise. The convergence test gives $\gamma = 0.103$, 0.104, 0.106, and 0.138 for $N_r = 1025$, 513, 257, and 129, all with errors below 10^{-6} and constant within about 10^{-8} over $100 < c_s t/a_0 < 200$. The interesting part of the result is the energetics. We find significant thermal gradient drive, with $\Gamma_r = 0.0634$ then $\Gamma_p = 0.0385$ and $\Gamma_v = 0.0036$, with the Landau damping channels $\Gamma_l = 9.8 \times 10^{-4}$ for electrons and $\Gamma_L = 2.0 \times 10^{-4}$ for ions.

The eigenmode structure is seen in the spatial realisations in the $r\theta$-plane at $z = 0$. These are given in figure 15.6, for ϕ, $A_\parallel$, ϖ, $J_\parallel$, as well as n_e and T_e, with the same format as figure 15.5. The vorticity and current are still confined to the layer around the singular surface, but the density and temperature now exist everywhere ϕ does. The structure is slightly deformed except for the magnetic disturbance tracked by $A_\parallel$. It is observed to rotate in the counterclockwise direction (the electron drift direction) by a full cycle over about $240 a_0/c_s$ units in time. This is the indication that the eigenfrequency and growth rate are both nonzero, so that ω is complex. This is the indication alluded to above, that the eigenmode dynamics is not amenable to an

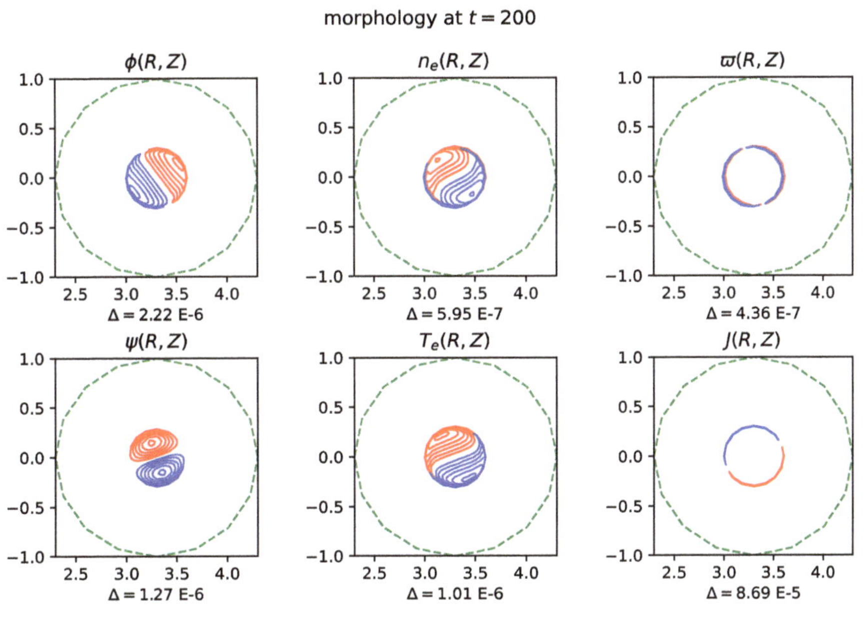

Figure 15.6. Spatial realisations in the $r\theta$-plane at $z = 0$ for ϕ and $A_\parallel$ (left, the MHD mode pieces), the vorticity and parallel current (right), and the electron density and temperature (center). This is for the case with thermal gradients, with $\beta_e = 10^{-3}$ and $N_r = 1025$ grid points. Now the displacement is only approximately rigid, with visible deformation away from $\phi \propto r$ due to the twisting. Only the magnetic disturbance ($A_\parallel$) keeps its original shape, despite the rotation of the eigenmode structure in the electron drift direction (counterclockwise, here).

energy principle. Nevertheless, the conserved free energy exists, and the calculations conserve it to high accuracy. The free energy principle is stronger and more general than the delta-W one from MHD.

15.6 Mode types not covered

Two types of MHD modes which I will not cover herein due to their tokamak specificity are briefly mentioned. The starting point of the study of global Alfvén modes is Eq. (15.57). One does similar things in a toroidal system, except that the curvature term is part of the dynamical mix. If we go back and re-derive Eq. (15.57) while keeping the curvature terms, still in a pure reduced MHD setting in the absence of thermal gradients, we find

$$\nabla \cdot \left(\frac{\omega^2}{v_A^2} - k_\parallel^2\right)\nabla_\perp \phi = \left(k_\parallel \nabla_\perp^2 k_\parallel\right)\phi + \mathcal{K}\left[\Gamma \mu_0 p_0 \mathcal{K}(\phi)\right] \tag{15.150}$$

in the limit that sound waves can be neglected. The curvature term represents finite beta compressibility of the ExB flow in a torus. It also couples poloidal modes within the basic Alfvén branch, due to the $\sin\theta$ and $\cos\theta$ dependences of the curvature terms. Coupling between (m, n) and $(m \pm 1, n)$ modes results, and one of the new mode types is the toroidal Alfvén eigenmode. The finite pressure gives rise to a large zoo of modes which draw one into work specific to one or more types of toroidal confinement concepts, which are beyond our scope. But they are of serious interest, due to the prospect that energetic fusion product particles can drive these modes and contribute to the premature loss of these particles before they deposit their energy onto the plasma.

The other mode type is called a peeling mode. The concept is that the plasma current can shed layers from its edge, due to a sort of localised kink mode that becomes unstable if a vacuum region between the edge of the plasma and a conducting wall is considered. Related to these are the resistive wall modes. They are not part of magnetised plasma turbulence *per se* but present a similar sort of operational parameter boundary as do ballooning and kink modes. For both these types, some references are cited in Further Reading.

15.7 Turbulence in a current channel

We come full circle when we consider the question of the Kelvin–Helmholtz instability in this context. We consider a straight background guide field in the z-direction. Without a current layer or a thermal gradient the situation is one of two-dimensional hydrodynamics, kept incompressible by the magnetic field. We need to show that we get this mode in its own context, in much the same way as the Euler vorticity equation was considered as a control case against various flavours of MHD in Chapter 3 of Volume One. Then we add a current layer, which gives magnetic shear to the background. This can stabilise the Kelvin–Helmholtz mode. It, however, allows tearing or reconnection, especially at finite collisionality. Then we can also consider a finite pressure enhancement layer coincident with the current layer. In this case, all three types of free energy source are present: a flow layer, a current layer, and a pressure layer. The situation is simple but the energetics is very interesting to sort out, and it gives any practitioner a setting in which to become familiar with how that is done.

15.7.1 Equations

The equations form a simplified four-field model in the gyrofluid language, as with the tearing example given in Eqs (15.89, 15.90), except now we restore the third dimension. This parallel direction is labelled z, not s, and the metric is straight Cartesian since there is no anticipation of how B^y will evolve. All gradient effects are self-consistent and set by the initial conditions; hence profiles are the slowly varying part of the $k_y = k_z = 0$ component. We have

$$\frac{\partial n}{\partial t} + [\phi_0, n] + \frac{\partial u}{\partial z} + [-A, u] = 0 \tag{15.151}$$

$$\frac{\partial A^*}{\partial t} + \mu_z[\phi_0, u] + \frac{\partial \Phi}{\partial z} + [-A, \Phi] = -C\, J \tag{15.152}$$

with

$$A^* = A + \mu_z u, \qquad \Phi = \phi_0 + \tau_z n \tag{15.153}$$

for each species. The rest is as in the tearing mode case, with Eq. (15.92) for the fields, (15.93) for the polarisation operator, (15.94) for the gyroaveraging including ϕ_0 in terms of ϕ, and (15.95) for the definitions of the vorticity variable and the current. The parameter definitions are also as in the tearing mode case. One thing of interest is to find out how three-dimensional the turbulence ever becomes, in a similar way as in the study of the role of the flute mode ($\partial/\partial z = 0$) in the case of drift wave turbulence in Chapter 11 of Volume One.

15.7.2 Initialisation

The layer profiles are set according to an initial state which contains

$$\varpi_0 = \delta_0 V' p_0, \qquad p_0 = \operatorname{sech}^2 10\, x, \qquad n_0 = -\omega_n\, p_0 \tag{15.154}$$

$$\psi_0 = 0.1 \frac{S}{\delta_0}(\log\cosh 5 - \log\cosh 10\, x) \tag{15.155}$$

with controlling parameters V', S, and ω_n for the vorticity, magnetic shear and current, and density gradients. These are set such that they are zero or small at the boundary.

The initial disturbances are a 2D random bath of fluctuations given by Eq. (2.126) from Chapter 2, multiplied by an envelope function in the parallel direction so that

$$n_{3D}(x, y, z) = n_{2D}(x, y)\exp[-(z/2.5)^2] \tag{15.156}$$

The parallel gradient ($\partial/\partial z$) in the absence of finite A guarantees excitation of ϕ and hence vorticity due to the adiabatic response, and therefore all degrees of freedom are eventually excited. A small amount of initial vorticity also arises from the FLR effects in polarisation for $\tau_i \neq 0$ cases.

These profiles and fluctuations are loaded into the variables at $t = 0$ according to

$$n_e = n_{3D} + n_0, \qquad n_i = n_{3D} + n_0 - \varpi_0, \qquad A^* = \psi_0 \tag{15.157}$$

for the densities and A^* for both species. The field equations are then solved for ϕ and A at $t = 0$ according to the gyrofluid procedure in Chapter 11. The grid is $64 \times 256 \times 64$, and the spatial domain is 1.0, 4.0, and 2π, in x, y, and z, respectively. The timestep is $\tau = 0.002$. The parameters are

$$\begin{array}{lll} \hat{\beta} = 5, & \hat{\mu} = 2.72, & C = 1.39, \\ X = 10, & \hat{\varepsilon} = 10^4, & \delta = 0.016 \end{array} \tag{15.158}$$

as in the cases for drift Alfvén turbulence given in Chapter 2. The collisionality is set lower to avoid substantial resistive decay of the current layer.

15.7.3 Energetics

The energetic analysis is done differently from the cases with externally imposed gradient drive terms. The dynamics is entirely self consistent, so the gradient drive is given by the conservative energy transfer from the $k_y = 0$ component to the $k_y \neq 0$ ones. The initial profiles contain most of the energy unless they are destroyed by the nonlinear dynamics, which can be fairly violent, especially in the Kelvin–Helmholtz case. We did the analysis for self-consistent transfer between flows and turbulence for the fluid cases in Chapter 16 of Volume One. In the gyrofluid case, with FLR corrections to ϕ, we cannot directly combine the n_e and n_i contributions into ϖ but must compute them first and then add them. So the Reynolds stress appears in a different form. We express for each variable a 'zonal' and 'fluctuation' component according to

$$\phi_Z = \langle \phi \rangle, \qquad \tilde{\phi} = \phi - \phi_Z \tag{15.159}$$

using subscript capital Z rather than 0 since these are dynamic components. In this case the zonal component is a simple average over the yz-plane.

It is easiest to compute the energy loss in the zonal component and note that its negative is then the drive term for the fluctuations. We have for each density,

$$\frac{\partial n_Z}{\partial t} + \left\langle [\tilde{\phi}_0, \tilde{n}] \right\rangle + \left\langle [-\tilde{A}_\parallel, \tilde{u}_\parallel] \right\rangle = 0 \tag{15.160}$$

The change to the zonal component of the thermal free energy is

$$\sum_{sp} \left[a_z \tau_z n_Z \frac{\partial n_Z}{\partial t} \right] = -\sum_{sp} \left[n_Z \left(\left\langle [\tilde{\phi}_0, \tilde{n}] \right\rangle + \left\langle [-\tilde{A}_\parallel, \tilde{u}_\parallel] \right\rangle \right) \right] \tag{15.161}$$

where we recall that $a_z \tau_z$ is the partial pressure in the background due to each species. The change to the zonal ExB energy is

$$\sum_{sp}\left[a_z\langle\phi_0\rangle\frac{\partial n_Z}{\partial t}\right] = -\sum_{sp}\left[\langle\phi_0\rangle\left(\langle[\tilde{\phi}_0, \tilde{n}]\rangle + \left\langle[-\tilde{A}_\parallel, \tilde{u}_\parallel]\right\rangle\right)\right] \tag{15.162}$$

In each equation the gradient drive terms are significant only if the zonal pieces in n and ϕ_0 are large. This will be the case only if the density layer n_0 (for each species) and/or the flow layer given by ϖ_0 (the difference among the n_0) are initialised. The drive terms for the fluctuations are the terms on the right hand sides of these equations written with the positive sign. In the ExB energy the species sum of the term due to the zonal ExB flux yields the Reynolds stress, and the species sum of the zonal magnetic flutter flux yields the Maxwell stress. The reader can show this as an exercise using

$$\varpi = -\sum_{sp} a_z n, \qquad J = \sum_{sp} a_z u \tag{15.163}$$

in the zero-FLR case with

$$\nabla_\perp^2\phi = \varpi, \qquad -\nabla_\perp^2 A = \hat{\beta}J \tag{15.164}$$

Sum the energetics over species to obtain the form with

$$\frac{\partial\varpi_Z}{\partial t} + \langle[\tilde{\phi}, \varpi]\rangle + \left\langle[\tilde{A}_\parallel, \tilde{J}_\parallel]\right\rangle = 0 \tag{15.165}$$

and then do the resulting $\nabla_\perp^2$ in the second argument of each bracket by parts (twice) to have it outside an expression with the Reynolds and Maxwell stress terms written explicitly. The forms given in Chapter 16 of Volume One are then recovered.

The free energy drive by reconnection of the current is found from the two velocity equations. We have

$$\frac{\partial A_Z^*}{\partial t} + \mu_z\left\langle[\tilde{\phi}_0, \tilde{u}_\parallel]\right\rangle + \left\langle[-\tilde{A}_\parallel, (\tilde{\phi}_0 + \tau_z\tilde{n})]\right\rangle = -C\, J_Z \tag{15.166}$$

The change to the zonal magnetic energy is

$$\begin{aligned}\sum_{sp}\left[a_z u_Z\frac{\partial A_Z^*}{\partial t}\right] = -\sum_{sp}\Big[a_z\, u_Z\Big(\mu_z\left\langle[\tilde{\phi}_0, \tilde{u}_\parallel]\right\rangle \\ + \left\langle[-\tilde{A}_\parallel, (\tilde{\phi}_0 + \tau_z\tilde{n})]\right\rangle\Big)\Big] - C\, J_Z^2\end{aligned} \tag{15.167}$$

where we recall that $a_z\mu_z$ is the mass density in the background due to each species. The resistive decay is given by the last term. The drive terms for the fluctuations are the ones under the sum, written with the positve sign. When these are dominant, we have current driven dynamics, as opposed to flow driven or pressure driven. The three current driven terms (following the species sum of $a_z u_Z$ as J_Z) are the inertial (with $\mu_z\tilde{u}_\parallel$), the Alfvénic (with $\tilde{\phi}_0$), and the adiabatic (with $\tau_z\tilde{n}$) drive channels. The Alfvénic one is the usual MHD reconnection current drive, while the adiabatic one is the two-fluid one often called semi-collisional, and the inertial one is due to ExB advection of the inertial current gradient (usually smaller). We find out what is taking place by evaluating all these drive terms since all three drive types (flow, pressure, and current) are present.

15.7.4 Dynamics

The first case set up is the Kelvin–Helmholtz case used as a control. The parameters are $V' = 1$ while $S = \omega_n = 0$. The breakdown of the flow shear (vorticity) layer is very rapid, mostly occurring before $t = 70$. Running to $t = 200$ is sufficient for this analysis. Coarsely taken (the sample interval is $\Delta t = 10$) time traces of the zonal free energy components and the energetic channels described above are shown in figure 15.7. The zonal energy is nearly all in the ExB channel E_e, with the small density one E_n present due to the ion gyrocentre density which represents the vorticity (in fluid drift language, this represents the diamagnetic component of the ion flow). It drops by about a factor of two over this run length. The energetics shows the Reynolds and Maxwell stresses (labelled R_E and R_M), and also the vestige ion gyrocentre density profile drive (N_n) due to the vorticity. The others, magnetic flutter density gradient drive (M_v), and the current drive terms due to the inertia (J_j), Alfvénic coupling (J_A), and adiabatic coupling (J_a), are essentially zero since there is no current layer. The morphology at the visible beginning of the vorticity layer breakup into a vortex street, the classic result, is shown in figure 15.8. Note the tilt of the disturbances to upper left and lower right, which is the manifestation of the

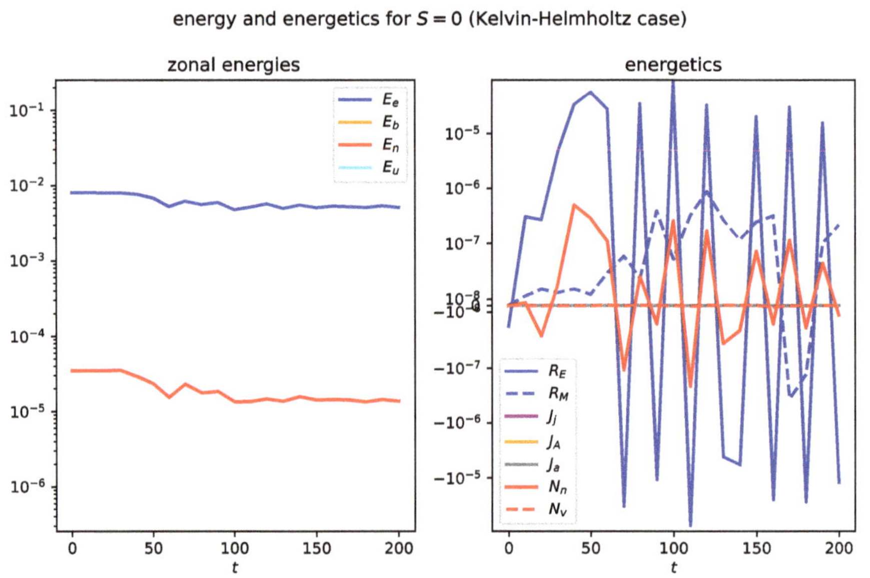

Figure 15.7. Time traces of the zonal free energy components (left) and their energetics (right), for the Kelvin–Helmholtz case $V' = 1$ with $S = \omega_n = 0$, used as the control case for turbulence in a channel. Only the flow layer is present, so the flow is the only source of free energy available to be converted into fluctuations. The zonal ExB energy (E_e) is seen to deplete, and the main drive for the fluctuations is, by default, the flow (R_E). Depletion occurs until about $t = 70$, after which the flow and the turbulence exchange energetically.

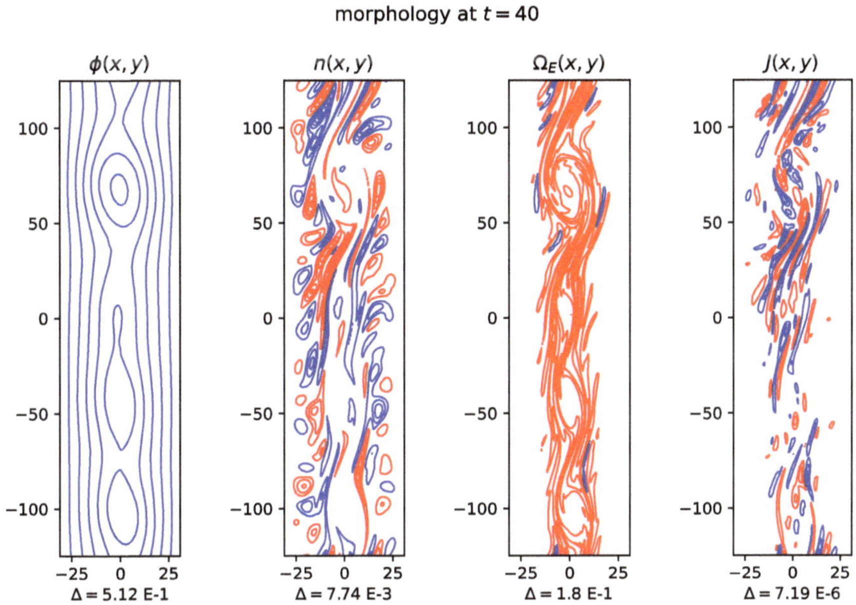

Figure 15.8. Spatial morphology of the flow potential, the density, the vorticity, and the current, for the Kelvin–Helmholtz case ($V' = 1$) at $t = 40$. The emerging vorticity disturbances are visible, and are tilted back toward the left so that the correlation between flow shear and Reynolds stress is negative, indicating energetic transfer out of the flow and into the eddies.

negative correlation between zonal vorticity and zonal Reynolds stress which indicates energy transfer from the flow to the eddies (cf. Chapter 16 of Volume One).

Before considering nonzero ω_n we increase the strength of the current layer, from $S = 0.1$ to $S = 1$, taking 0.3 as the nominal value for later study. The vorticity layer is left at $V' = 1$ to find the effect of the increasing magnetic shear. The result for $S = 0.3$ is shown in figure 15.9, with the same format as the one before. The energetics is less robust than before although dynamics is still present. The flow instability is essentially eliminated, but now the flow and current combination effectively taps the current layer. The Alfvénic coupling is the main form of the drive, as would be expected from an MHD picture. However, the energetics to and from the flow is significant although it is in close balance, with the drive due to the Maxwell stress positive and the Reynolds stress negative. This is the Alfvénic dynamical signature, since with $\phi \approx \pm v_A A$ in physical units, this balance is the result. It accounts for the complicated dynamics seen in the morphology, shown in figure 15.10. Both flow eddies in ϕ and a structure in A appearing as a set of magnetic islands are visible.

For other values of S the dynamics shows a change from the flow to the current as a source of energy. For $S = 0.1$ the ExB energy is still the largest and the zonal component still drops significantly, but after about $t = 70$ the Maxwell stress comes up as a drive channel and then the Reynolds stress shifts negative to balance it. The Alfvén coupling drive is present but subdominant for $t < 70$ and unsteadily

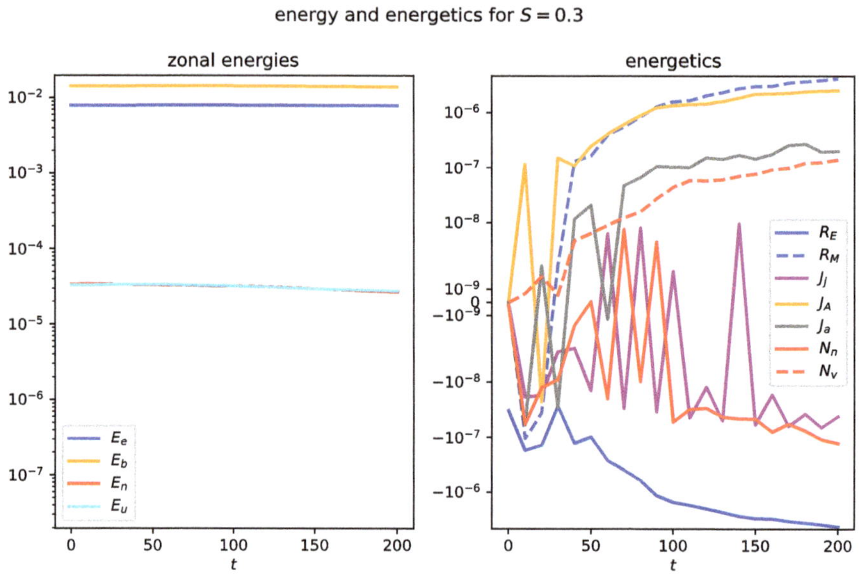

Figure 15.9. Time traces of the zonal free energy components (left) and their energetics (right), for the current channel case $V' = 1$ and $S = 0.3$ with $\omega_n = 0$. Now both the flow layer and the current layer are present. The Reynolds (R_E) and Maxwell (R_M) stress channels are strong but balance closely, indicating Alfvénic dynamic character. The Alfvénic drive (J_A) due to the current layer is strong at all times. This is the energetics of reconnection in the presence of a flow.

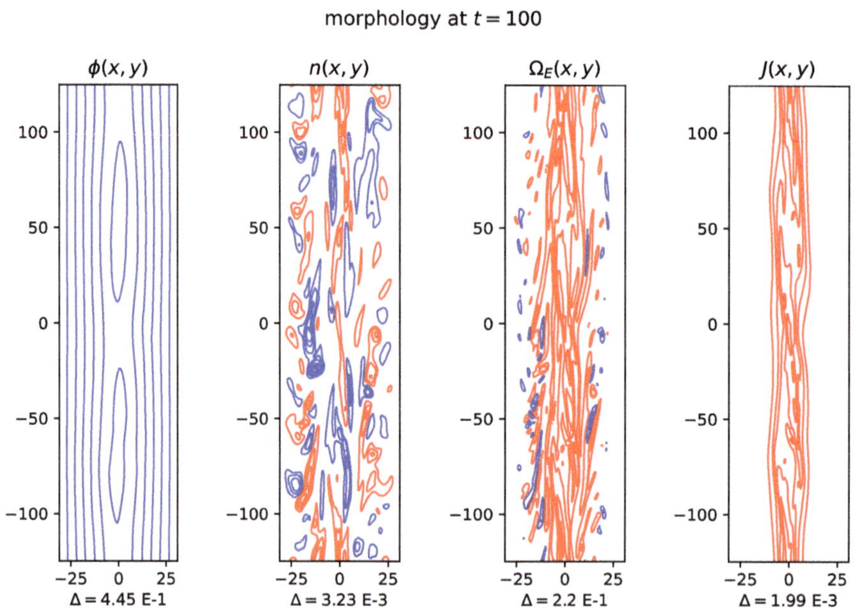

Figure 15.10. Spatial morphology of the flow potential, the density, the vorticity, and the current, for the reconnecting case ($V' = 1$ and $S = 0.3$) at $t = 100$. The double island structure in the current is visible, and the vorticity is now in sheets rather than eddies. This is now a reconnection case, not a flow dominant case, though the energetic exchange with the flow is significant.

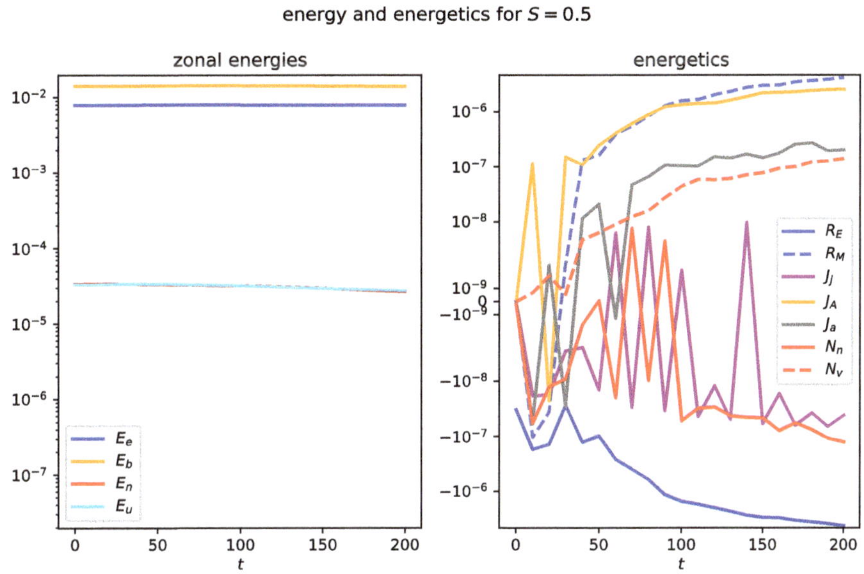

Figure 15.11. Time traces of the zonal free energy components (left) and their energetics (right), for the current channel case $V' = 1$ and $S = 0.5$ with $\omega_n = 0$. Now the current layer is stronger, and its partial depletion is more visible. The Reynolds (R_E) and Maxwell (R_M) stress channels remain in balance, even as they switch transfer direction at $t = 130$. The Alfvénic drive (J_A) due to the current layer is the strongest at all times. This dynamics becomes progressively stronger as S increases.

averaging to zero thereafter. For $S = 0.5$ the Alfvén coupling drive is strongest followed by the Maxwell stress, but then after $t = 130$ the Maxwell and Reynolds stress channels switch signs, still in close balance but now with R_E positive. This is shown in figure 15.11. The $S = 1$ case gives a similar result, just stronger and faster. As the current layer breaks up, progressively more small scale structure is produced as S increases.

Now we set a finite $\omega_n = 1$, so that all three free energy sources are present. The time traces of the zonal free energy and energetics are shown in figure 15.12. Not only is the density layer the principal free energy source (E_n), it also produces the strongest energetics (N_n), although the energy source pattern from the $\omega_n = 0$ case is still present underneath. The morphology for $t = 60$ is shown in figure 15.13. This time point is before the island structure becomes more predominant at later times, but its beginnings are visible here. Nevertheless, this is near the temporal peak in the pressure drive, and the Alfvénic balance between the Reynolds and Maxwell stresses has been broken up. Hence, when the drive by a pressure gradient becomes larger than the drive by flows or currents, the adiabatic response and elements of drift wave mode structure become significant, and we quickly depart the MHD description of the dynamics.

The wavenumber spectra for the Kelvin–Helmholtz case is shown in figure 15.14. The labels are the same as in the time traces. The energy components shown are for the disturbances, with $k_y \neq 0$. The energetics are the spectra of the drive pieces we have been examining, for example, rearrange

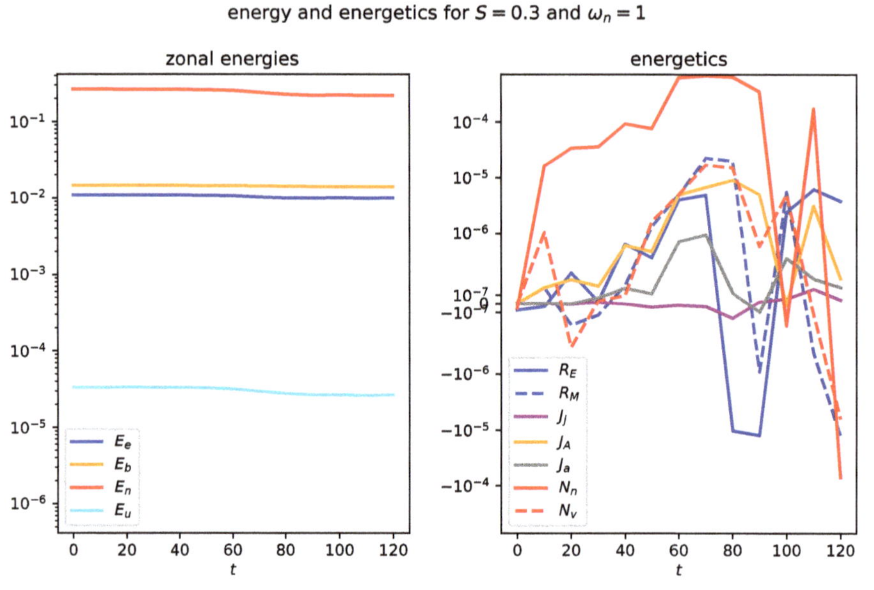

Figure 15.12. Time traces of the zonal free energy components (left) and their energetics (right), for the case $V' = 1$ and $S = 0.5$ with $\omega_n = 1$. Now all three layers are present, the flow, the current, and the pressure. The pressure (E_n) is the strongest zonal free energy component, and its ExB advection drive (N_n) is the strongest at all times. The magnetic flutter (Nv) and Maxwell stress (R_M) drive channels are concurrent with the Alfvén (J_A) drive, and even the adiabatic (J_a) drive due to the current layer through coupling to the pressure is visible. The Reynolds and Maxwell stresses no longer balance, indicating the effectiveness of the adiabatic response in breaking up the pure Alfvén dynamics.

$$\langle\phi_0\rangle\left\langle[\tilde{\phi}_0, \tilde{n}]\right\rangle = \frac{\partial}{\partial x}(\cdots) - \langle\tilde{n}u_E^x\rangle\frac{\partial}{\partial x}\langle\phi_0\rangle \tag{15.168}$$

and then Fourier decompose the quadratic piece giving the flux (recall that $u_E^x = [\phi_0, x]$). The x-derivative of a zonal quantity has only the $k_y = 0$ piece anyway, so all $k_y > 0$ dependence comes from the flux piece. All the energetic channels are done this way and then given in symmetric-log form, with both signs visible. The Kelvin–Helmholtz control case shows what we expect: most of the activity in the first two k_y, with dominant transfer out of the zonal vorticity layer for these modes, and then an inverse transfer in the middle of the range. Long wavelengths are driven, and short wavelengths are damped, following the analysis given in Chapter 16 of Volume One.

The wavenumber spectra for the current channel case is shown in figure 15.15. The dominant drive is the magnetic nonlinearity, with the Alfvén process (J_A) driving and the Reynolds and Maxwell stress processes, with R_M positive, nearly equal and opposite in the driving range (first two modes). The energetic spectrum is wider, and we find a strong positive drive directly from the flow for most of the short wavelengths in this time range. The Alfvén balance $R_E + R_M \approx 0$ no longer holds in this range. The magnetic and flow energy components are the largest and are well

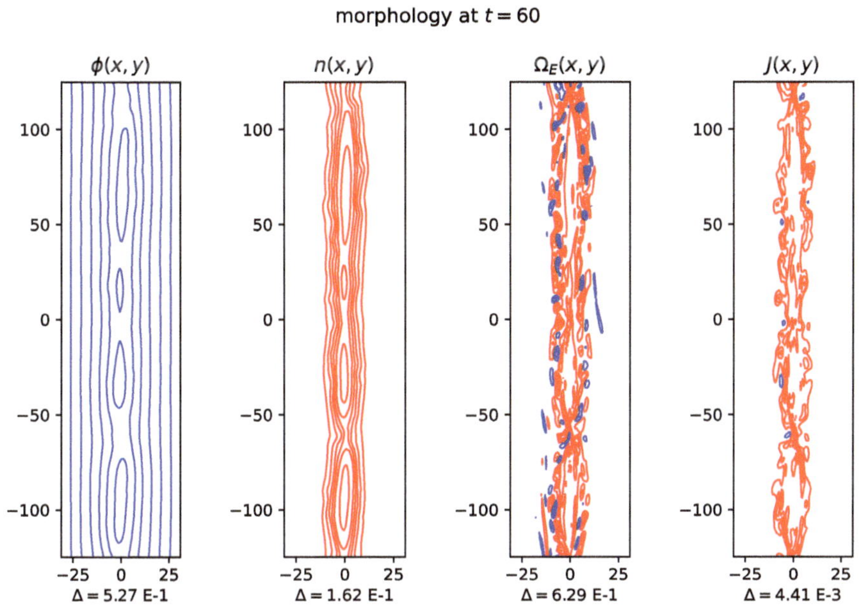

Figure 15.13. Spatial morphology of the flow potential, the density, the vorticity, and the current, for the pressure layer case ($V' = 1$ and $S = 0.3$ with $\omega_n = 1$) at $t = 60$. The small scale vorticity and current disturbances are more prominent with this degree of adiabatic response in the dynamics. The island structure is still visible but it no longer represents the dominant process.

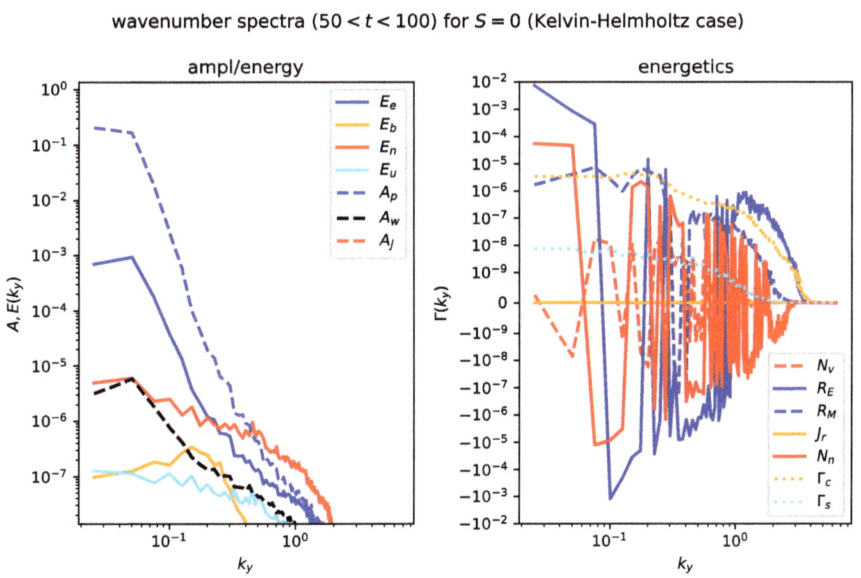

Figure 15.14. Wavenumber spectra of the free energy and energetic channels for the Kelvin–Helmholtz case for $50 < t < 100$. The k_y is normalised in terms of ρ_s. The longest two wavelengths are dominant, with the flow energy (E_e) the largest and the amplitude of ϕ (A_p) very large. The Reynolds stress process (R_E) has most of the energetics and turns negative in the middle of the spectral range.

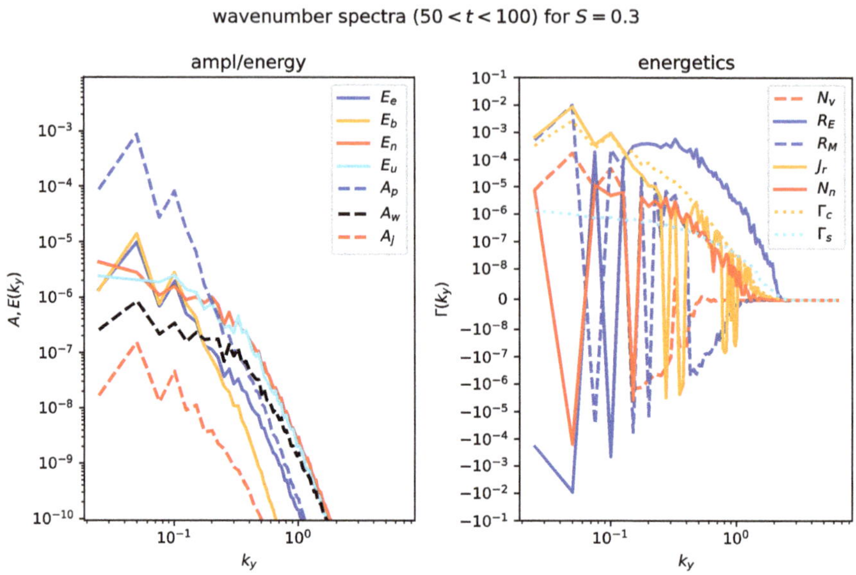

Figure 15.15. Wavenumber spectra of the free energy and energetic channels for the current channel case for $50 < t < 100$. The first few wavelengths are dominant, with the flow (E_e) and magnetic (E_b) energy largest and nearly equipartitioned. The Alfvénic nature of the energetics is shown with the Alfvén drive (J_A) and Maxwell stress (R_M) processes largest and positive, with the Reynolds stress (R_E) process negative. The near balance $R_E + R_M \approx 0$ is indicative of the Alfvén wave structure, especially in the first two modes. This is a reconnection process assisted by the flow. Note that the flow is driving the shorter wavelengths over this time range.

equipartitioned for the first several modes. The amplitude of ϕ given by A_p is still very large.

The wavenumber spectra for the pressure/current channel case are shown in figure 15.16. The dominant drive is now the pressure, which is largest in the third mode for this time range, and it affects much more of the spectrum than did the drive in the other cases. The pressure has the largest free energy and the amplitude of ϕ is now close to that of n_e. The flow and magnetic energy are still nearly equipartitioned, and the vorticity is now robust and well spread. These are all drift wave turbulence signatures, even though the morphology shows localisation to the channel and the persistence of island-like magnetic structures in the current.

The foregoing is a revealing study of the energetics of magnetised plasma turbulence in a basic scenario in which the geometry is reduced to a simple channel case, but one or more of the classes of drive are present: flow, current, and pressure. With the flow and current together, the dynamics is driven by the current but the flow layer serves as a conduit for energetic exchanges with the rest of the system. With all three present the pressure drive can easily be the largest but the structure is a complicated combination of all three mode types: Kelvin–Helmholtz, tearing and magnetic island formation, and drift waves. We could not have found these results without the energetic analysis as it is not clear from the evolution of the morphology or even the spectra.

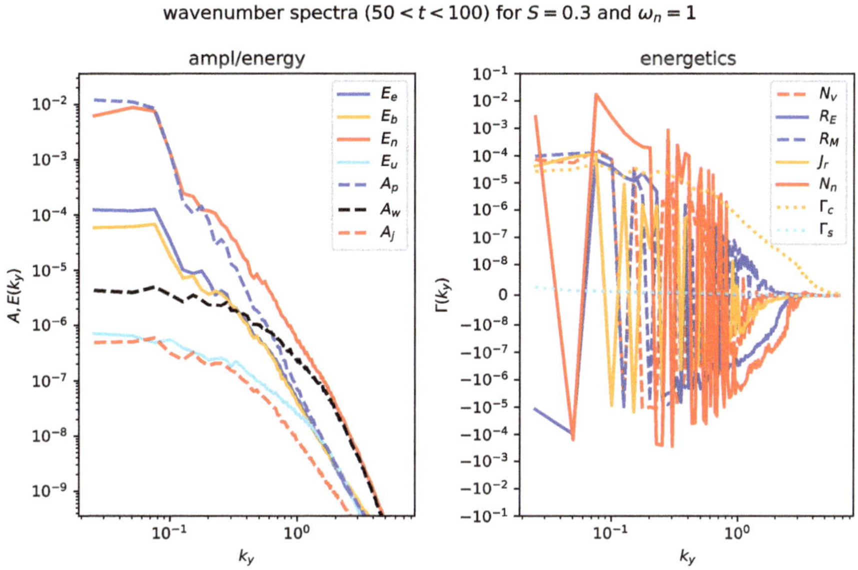

Figure 15.16. Wavenumber spectra of the free energy and energetic channels for the pressure/current channel case for $50 < t < 100$. A wider spectral range controls the energetics than before, with the thermal free energy (E_n) the largest, and the flow (E_e) and magnetic (E_b) energy still nearly equipartitioned. The vorticity is much flatter and is stronger than in the other cases. The amplitude of ϕ is nearly the same as the electron density in the disturbances. The energetics shows the pressure drive to be the strongest process, and the Alfvén balance is broken up. Although the morphology still looks like reconnection, these are drift wave turbulence energetic signatures.

15.8 Summary

Most of the basic MHD instabilities are rather well known and this coverage is cursory. Its main purpose is not so much to review the situation around MHD instabilities, as others have done that far better than I could, but to show their capture using gyrofluid equations in any situation where the strong guide field limit holds. So in principle, any electromagnetic computation applied to a general parameter situation (of course, always within this regime), MHD processes as well as drift wave and drift Alfvén turbulence are all potentially active. One must be ready for any of them, and so it is better practice not to tailor computations around any specific mode type or process. This is why we use general initial conditions (e.g., don't initialise a magnetic island, rather, let it emerge naturally), taking care only to excite enough of the system's degrees of freedom, ideally, all of them. Also, this is why we do not concentrate grid points near a resonant or singular surface. In a nonlinear situation, the location of these resonances may shift anyway, and the extent to and whether that occurs is a result of the problem, not an input.

Traditional ideal MHD instabilities are analysed using the delta-W method. While very powerful, especially when complicated geometry precludes strong insight, these methods are restricted to Sturm-Liouville problems. Since drift dynamics in a gradient, let alone turbulence, is not amenable to these, we had to take the path of establishing the delta-W theory, exploring a few of its basics which are well known, and then capturing the same instabilities using our quite different approach: gyrokinetics → delta-f → gyrofluid → recovery of drift dynamics as a superset of reduced MHD. RMHD can do these instabilities, and therefore so can the gyrokinetic and gyrofluid approach. But in a global electromagnetic computation of something like a tokamak, one has to be ready for anything, not just a particular class of linear instability.

Computation of turbulence and the $m = n = 1$ type of MHD instabilities in combination in a nonlinear setting has been tried, using RMHD type models. The difficulties of computational cost mostly get in the way, and in several cases the computations become non-viable just as they are about to get interesting. Treating the linear stages of the kink instability requires the domain to extend all the way to the magnetic axis, $r_a = 0$ in a cylinder. So nonlinear computations have to do the same, and this is a serious difficulty especially in a torus where different quantities shift their global maxima away from the axis in different directions (e.g., out/in versus up/down). Traditional turbulence computations, on the other hand, get away with placing an inner boundary, usually at $r_a = 0.1$ or 0.2, since the dynamics in a global setting is strongest in the middle third or outer half of the domain. One cannot do this for current driven modes. The reason is that the necessary relationship between the profiles of the current, the pitch parameter, and the poloidal flux is correct only if the domain includes the axis. One can wriggle out of this for tearing modes but not for anything of the kink mode type.

Note that in the above the toroidal versions of the current driven instabilities are not addressed. They introduce an entirely different world to the problem and in recent years there has been a lot of activity. On the other hand, many if not most of the effects treated are very specific to the type of device they are done in. As such, they are outside the scope of this work. But it is pertinent to mention such things as toroidal Alfvén eigenmodes and their relevance to fusion plasmas with highly energetic charged particles which are fusion products, and then point to some of their treatments in the literature.

One major shortcoming of any of the large scale global MHD calculations, even when they are two-fluid, is that such a dearth of the available authors have faced the need to capture dynamics, especially nonlinear vorticity, at the drift scale and current dynamics at the skin depth. These ingredients are interesting precisely because of the difficulty they present in terms of scale separation. This makes it necessary, even vital, to set the relevant parameters accordingly, so that the range of phenomena at the system size (global MHD), the drift scale (drift wave vorticity), and the region in between (mesoscale MHD and long wavelength drift Alfvén turbulence), has a reasonable interaction range for each of the pieces.

This means that the dimensionless parameters such as the drift scale ratio or the mass ratio in combination with the parallel/perpendicular scale ratio, must retain their actual values. We did this above for the kink mode but not for the collisionless reconnection problem, precisely showing the difficulty. The growth rates become very slow for realistic cases and one must run for a long time. This is not so helpful in an introduction to processes like this one but becomes vital to actual situations in the real world, and one must pay the price to do this properly. Hence, definitive results require patience and many of them are still something for the future. There is still much to do, especially properly.

Further Reading

Ideal MHD instabilities provide a topic with a vast literature. I was helped by the following texts: J P Freidberg, *Ideal Magnetohydrodynamics* (Plenum Press, 1987). W M Mannheimer and C N Lashmore-Davies, *MHD and Microinstabilities in Confined Plasma* (CRC Press, 1989) R J Goldston and P H Rutherford, *Introduction to Plasma Physics* (CRC Press, 1995) R B White, *The Theory of Toroidally Confined Plasmas* (World Scientific, 2001) and a lecture by White from Princeton Plasma Physics Laboratory, from 2012, which gives his derivation of the delta-W principle and the ballooning equation.

The astrophysical context of tearing and reconnection is given by E N Parker, in *Cosmical magnetic fields. Their origin and their activity* (Oxford: Clarendon Press, 1979). In the fusion research world we start with the linear theory including the constant-ψ approximation for the inner solution and the derivative discontinuity Δ' from the outer solution, from H P Furth, J Killeen, and M N Rosenbluth, 'Finite-resistivity instabilities of a sheet pinch,' *Phys. Fluids* **6** (1963) 459. The exact analytical solution to the equation $y'' - x(xy - 1) = 0$ is from P H Rutherford and H P Furth, Princeton Plasma Physics Laboratory Report No. MATT-872, 1971, slightly recast in Chapter 20 of the book by Goldston cited above. The cylindrical version, where the relationship between the equilibrium current, q profile, and the flux profile defines the outer solution, was given in H P Furth, P H Rutherford and H Selburg, 'Tearing mode in the cylindrical tokamak,' *Phys. Fluids* **16** (1973) 1054.

The parabolic cylinder functions referred to in the analysis of the tearing mode equation (Eq. (15.76)) are detailed in Chapter 19 of M Abramowitz and I A Stegun, *Handbook of Mathematical Functions* (9th pr., Dover, 1970).

The nonlinear solution using the $\nabla_{\|}J_{\|} = 0$ approximation inside the island with polarisation as a correction giving the algebraic growth was given in P H Rutherford, 'Nonlinear growth of the tearing mode,' *Phys. Fluids* **16** (1973) 1903. Nonlinear two-fluid effects including the background thermal gradient were described by R B White, D A Monticello, M N Rosenbluth, and B V Waddell, 'Saturation of the tearing mode,' *Phys. Fluids* **20** (1977) 800, and by B Scott and A B Hassam and J F Drake, 'Nonlinear evolution of drift-tearing modes,' *Phys. Fluids* **28** (1985) 275, and by M Kotschenreuther, 'Nonlinear dynamics of magnetic islands with curvature and pressure,' *Phys. Fluids* **28** (1985) 294. Stabilisation and nonlinear stability issues specific to the case with a temperature gradient were given by B Scott, J F Drake and A B Hassam, *Phys. Rev. Lett.* **54** (1985) 1027; B Scott and A B Hassam, 'Analytical theory of nonlinear drift-tearing mode stability,' *Phys. Fluids* **30** (1987) 90. Reduced MHD 3D nonlinear tearing was described by B V Waddell, B Carreras, H R Hicks, and J A Holmes, 'Nonlinear interaction of tearing modes in highly resistive tokamaks,' *Phys. Fluids* **22** (1979) 896.

The idea of collisionless tearing and reconnection is that the electron inertia causes decoupling of the plasma motion from the magnetic flux, here interpreted as the breaking of purely MHD flux conservation by the finite skin depth, provided by the electron inertia. See H P Furth, in Plasma Physics and Controlled Nuclear Fusion Research, 1962 (IAEA, Vienna, 1962), Vol. I, p. 169; B Coppi, Phys. Lett. 11, 226 (1964); Phys. Fluids 8, 2273 (1965); R Pellat, G Laval, and M Vuillemin, in Plasma Physics and Controlled Nuclear Fusion Research, 1965 (IAEA, Vienna, 1966), vol. II, p. 259; M A Cross and G Van Hoven, *Phys. Fluids* **19** (1976) 1591. Collisionless tearing in the conventional constant-ψ sense B Coppi and J W-K Mark and L Sugiyama and G Bertin, 'Reconnecting modes in collisonless plasmas,' *Phys. Rev. Lett.* **42** (1978) 1058 The relationship to the kink modes and $m = 1$ modes is addressed below.

The two-field model setting vorticity equal to electron density is given in T J Schep and F Pegoraro and B N Kuvshinov, 'Generalized two-fluid theory of nonlinear magnetic structures,' *Phys. Plasmas* **1** (1994) 2843. Conserved quantities correcting the flux by finite skin depth factors are given in F Pegoraro, B N Kuvshinov, J Rem, and T J Schep, 'Electron inertia and small-scale magnetic structures in a nonuniform collisionless plasma,' *Adv. Space Res.* **19** (1997) 1823. The linear and nonlinear properties of the two-field model similar to what is shown here are described in M Ottaviani and F Porcelli, 'Nonlinear collisionless magnetic reconnection,' *Phys. Rev. Lett.* **71** (1993) 3802, and in M Ottaviani and F Porcelli, 'Fast nonlinear magnetic reconnection,' *Phys. Plasmas* **2** (1995) 4104. The Hamiltonian properties are described in D Grasso, F Pegoraro, F Porcelli, and F Califano, 'Hamiltonian magnetic reconnection,' *Plasma Phys. Contr. Fusion* **41** (1999) 1497. The FLR polarisation correction in the two-field model was given by D Grasso, F Califano, F Pegoraro, and F Porcelli, 'Ion Larmor radius effects in collisionless reconnection,' *Plasma Phys. Rep.* **26** (2000) 512. My version of this problem within the four-field fluid drift model is in B Scott and F Porcelli, 'Two dimensional fast reconnection in a fluid drift model,' *Phys. Plasmas* **11** (2004) 5468.

The pressure driven ballooning instability was described by V D Shafranov and E I Yurchenko, 'Influence of ballooning effects on plasma stability in closed systems,' *Nucl. Fusion* **8** (1968) 329. The description of how the flute mode type functions in finite magnetic shear was given by B Coppi, 'Topology of ballooning modes,' *Phys. Rev. Lett.* **39** (1977) 939, and D Dobrott, D B Nelson, J M Greene, A H Glasser, M S Chance, and E A Frieman, 'Theory of ballooning modes in tokamaks with finite shear,' *Phys. Rev. Lett.* **39** (1977) 943. The image mode construction that allows the extended domain along the magnetic field was given by J W Connor, R J Hastie, and J B Taylor, in 'Shear, periodicity, and plasma ballooning modes,' *Phys. Rev. Lett.* **40** (1978) 396.

Resistive ballooning computations within the reduced MHD framework were given by B A Carreras and L Garcia, 'Theory of resistive pressure-gradient-driven turbulence,' *Phys. Fluids* **30** (1987) 1388. The two fluid additions were done by H Sugama, M Wakatani and A Hasegawa, 'Study of resistive drift and resistive interchange modes in a cylindrical plasma with magnetic shear,' *Phys. Fluids* **31** (1988) 1601.

In a tokamak the $m = 1$ kink and tearing modes are closely related. The pure ideal mode is neutrally stable in the simplest case (zero layer width), as explained in the text by Freidburg cited above. The kink mode description is due to V D Shafranov, in *Zh. Tekh. Fiz.* **40** (1970) 421 [V D Shafranov, *Sov. Phys.-Tech. Phys.* **15** (1970) 175]. Modes fitting this description were observed in tokamak plasmas by S von Goeler, W Stodiek, and N Sauthoff, 'Studies of internal disruptions and $m = 1$ oscillations in tokamak discharges with soft-x-ray techniques,' *Phys. Rev. Lett.* **33** (1974) 1201.

Inside the layer a finite width is provided by resistivity or electron inertia, as in the tearing mode. The physical picture of an island filling up the region interior to the original $q = 1$ surface is given by B B Kadomtsev, *Fiz. Plazmy* **1** (1975) 710 [*Sov. J. Plasma Phys.* **1** (1975) 389]. For the role of collisionless reconnection see G Ara, B Basu, B Coppi, G Laval, M N Rosenbluth and B V Waddell, 'Magnetic reconnection and $m = 1$ oscillations in current carrying plasmas,' *Ann. Phys.* **112** (1978) 443, and B Basu and B Coppi, 'Theory of $m = 1$ modes in collisionless plasmas,' *Phys. Fluids* **24** (1981) 465. The additional role of the finite ρ_s was added to this by F Porcelli, in 'Collisionless $m = 1$ tearing mode,' *Phys. Rev. Lett.* **66** (1991) 425.

Nonlinear resistive two-fluid saturation effects due to flows were considered by D Biskamp, in 'Nonlinear theory of the $m = 1$ mode in hot tokamak plasmas,' *Phys. Rev. Lett.* **46** (1981) 1522.

The gyrokinetic treatment of the kink mode using gyrokinetic equations mentioned in the text was done by H Naitou *et al*, in 'Gyrokinetic simulation of internal kink modes,' *Phys. Plasmas* **2** (1995) 4257. Initial nonlinear development, including acceleration of the growth, is in T Matsumoto *et al*, 'Nonlinear acceleration of the electron inertia-dominated magnetohydrodynamic modes due to electron parallel compressibility,' *Phys. Plasmas* **12** (2005) 092505.

The simultaneous computation of the $m = 1$ instability and resistive ballooning mode turbulence was attempted by Y Nishimura, J D Callen, and C C Hegna, 'Onset of high-n ballooning modes during tokamak sawtooth crashes,' *Phys. Plasmas* **6** (1999) 4685. The spectrum was full just as the nonlinear stage produced turbulence. But this is the type of computation whose flourishing phase will be found in the future.

On dynamos, one can start with two papers concerning magnetic helicity conservation during plasma current relaxation due to reconnection: H Ji, S C Prager, and J S Sarff, 'Conservation of magnetic helicity during plasma relaxation,' *Phys. Rev. Lett.* **74** (1995) 2945, and E T Vishniac and J-Y Cho, 'Magnetic helicity conservation and astrophysical dynamos,' *Astrophys. J.* **550** (2001) 752.

Global two fluid MHD attempts including turbulence are being done despite the resolution and conservation issues. Nevertheless, the ideas are often strong, as in A Thyagaraja, M Valovič and P J Knight, 'Global two-fluid turbulence simulations of L-H transitions and edge localized mode dynamics in the COMPASS-D tokamak,' *Phys. Plasmas* **17** (2010) 042507.

Toroidal kink modes were studied by O P Pogutse and E I Yurchenko, in 'Energy principle and kink instability in a toroidal plasma with strong magnetic field,' Nucl. Fusion 18 (1978) 1629. Ballooning mode (high-$k_\perp$) counterparts were described by these same authors, in JETP Lett. 31 (1980) 449.

Toroidal Alfvén Eigenmodes (TAE) arise from the toroidal coupling between shear Alfvén waves resonant at neighboring rational surfaces at the same toroidal mode number (i.e., between m, n and $m \pm 1$, n) They were described by C Z Cheng, L Chen, and M S Chance, in 'High-n ideal and resistive shear Alfvén waves in tokamaks,' Ann. Phys. 161 (1985) 21; and C Z Cheng and M S Chance, in 'Lown shear Alfvn spectra in axisymmetric toroidal plasmas,' Phys. Fluids 29 (1986) 3695. Finite-β (pressure ratio) modifications including a transition to shorter wavelength were described by G Y Fu and C Z Cheng, Phys. Fluids B 2 (1990) 985.

These modes are of interest in fusion plasma research due to the possibility that they can be driven by energetic particles such as externally applied beam ions (for heating) or fusion products (such as α-particles). See G Y Fu and J W Van Dam, 'Excitation of the toroidicity-induced shear Alfvén eigenmode by fusion alpha particles in an ignited tokamak,' Phys. Fluids B 1 (1989) 1949; and F Zonca, 'Alfvenic instabilities driven by energetic alpha particles,' Plasma Phys. Contr. Fusion 35 (1993) B307.

Toroidal Alfvén eigenmodes and their excitation by energetic particles have been reviewed by L Chen and F Zonca, in 'Physics of Alfvén waves and energetic particles in burning plasmas,' Rev. Mod. Phys. 88 (2016) 015008.

For observations in plasma confinement experiments, see K-L Wong, in 'A review of Alfvén eigenmode observations in toroidal plasmas,' Plasma Phys. Contr. Fusion 41 (1999) R1; and W W Heidbrink, in 'Basic physics of Alfvén instabilities driven by energetic particles in toroidally confined plasmas,' Phys. Plasmas 15 (2008) 055501.

Peeling modes arise from the introduction of a vacuum region between the plasma column, at whose radial edge ($r = a$) the plasma vanishes, and an outer wall ($r = b > a$) at which the fields are assumed to vanish. They were described by J W Connor, R J Hastie, H R Wilson, and R L Miller, in 'Magnetohydrodynamic stability of tokamak edge plasmas,' Phys. Plasmas 5 (1998) 2687. This reference describes peeling modes, which exist in cylindrical geometry without curvature terms, and edge ballooning modes, with 'edge' in this context defined by the existence of the vacuum region, and most specifically their coupling in a finite-β toroidal plasma. Their application is to the issue of the edge plasma performance (achievable density and temperature in the outer core of fusion tokamaks). See P B Snyder, H R Wilson, et al, ELMs and constraints on the H-mode pedestal: peeling-ballooning stability calculation and comparison with experiment,' Nucl. Fusion 44 (2004) 320.

IOP Publishing

Turbulence and Instabilities in Magnetised Plasmas, Volume 2
Gyrokinetic theory and gyrofluid turbulence
Bruce Scott

Chapter 16

Gyrokinetic gauge transform for large scales

16.1 Background and introduction

The background of the theory as a gauge transform is the work originally done for a drift kinetic system (i.e., motion of particles and evolution of the distribution function with prescribed fields) by Littlejohn in 1983 cited in Chapter 5. In this approach, the Lie transform is not used, and within the ordering scheme a set of differentials is added to the fundamental one-form to obtain another one in a different representation, in which (1) the magnetic moment coordinate is defined and shown to be conserved, (2) the resulting motion of the drift centre is decoupled from the fast gyromotion, and (3) a closed one-form is obtained in which there is no dependence upon the gyrophase angle. The gyromotion consists of the conserved magnetic moment coordinate, whose precise meaning depends on the level of the ordering and the depth of the expansion, and the fast gyration of the gyrophase angle appears isolated from the rest of the dynamics.

In the gyrokinetic Lie transform approach this is all done within the first step in which the coordinates are set up but before the fields are then added in a perturbation expansion. In the gauge transform approach, the more familiar drift kinetic treatment is expanded to cover the entire transformation, incorporating the essential step to make it a gyrokinetic representation, namely, turning the polarisation drift into a polarisation density by re-forming the ExB energy into the Hamiltonian. The Littlejohn approach actually includes this step but does not take it to its conclusion. That is what we will do herein. Once again, doing the operations at the level of the fundamental one-form guarantees exact phase space symmetry and energetic consistency of the resulting equations.

The ordering done in this approach is different. The original idea of gyrokinetic ordering is that the dynamics was supposed to be at larger scale than the ion gyroradius, even as finite Larmor radius (FLR) corrections are included, while the amplitude of the fields could be left arbitrary, subject only to the main constraint that any dynamics be slow compared to the gyrofrequency (i.e., the ExB vorcitity must satisfy $\Omega_E \ll \Omega_i$). The approach we used in Chapter 5 allows arbitrary scale (specifically, $k_\perp \rho_i$) but requires small amplitude in terms of $e\phi/T_e \ll 1$ in the dynamics. A magnetised plasma equilibrium with thermal gradients contains flows at the level of the basic ion force balance, $ne\nabla_\perp \phi \sim \nabla_\perp T$, generalised to allow for the density and temperature of any species. In this case $e\nabla\phi$ is the same size as ∇T, so one must allow $e\phi/T$ formally at unit order. The gyrokinetic ordering is still satisfied, however, since the large amplitude exists at $k_\perp L_\perp \sim 1$, where $L_\perp$ is the temperature scale length. The larger scale implied for the unit order potential combines with ρ_i to make each application of the derivatives smaller. A force balance ϕ potential yields an ExB flow of order not c_s but $\mathbf{v}_D$, the diamagnetic velocity. The energy in this flow is comparable to the ExB energy of the turbulence. Then, the vorticity implied by this flow is actually smaller than that of the turbulence, and smaller than Ω_i by two factors of $\rho_i/L_\perp$.

Either way, small ϕ or small $k_\perp$, the vorticity ordering is maintained. The interest in doing this for our purposes, including both large and small scales in a computation, is in the investigation into how these two ordering approaches overlap. Do they recover the same forms in the appropriate limit? That is, does the low amplitude ordering become isomorphic to the large scale ordering if the flow amplitude is subsonic? If the answer is yes, then the small-scale ordering is actually valid for large scales as well, and can be used there if the spectrum 'cooperates' in the sense that the spectrum of the amplitude results in satisfaction of the vorticity ordering at all scales. This consideration provides the point of doing the calculations contained in this Chapter.

16.2 Gyrokinetic theory as a gauge transform

Gyrokinetic theory is not an orbit average over equations, but a set of operations on a Lagrangian which involves a change of representation from particle to gyrocentre variables. What those gyrocentre variables actually are is the result of choices made during the gauge transformation. A gauge transformation is a combination of operations involving a coordinate change and the addition of several total differentials to the original Lagrangian to produce another one which reflects the same dynamics in different language as the original one. The low-frequency approximations enter through a chosen ordering scheme, in which the only really essential element is the smallness of the ExB vorticity or parallel transit frequencies compared to gyrofrequencies. Since it is the slowest, the ion gyrofrequency is usually considered. Since they are the fastest, the electron and/or shear Alfvén transit frequencies are considered. The ExB vorticity is considered since it underlies any turbulent dynamics involving ExB motion. These operations imply

$$\omega \sim \Omega_E \sim k_\parallel V_a \ll \Omega_i \tag{16.1}$$

all of which is part of the low frequency description (here, V_a is the scale-dependent kinetic shear Alfvén velocity, encompassing both $\mathbf{v}_A$ and V_e, and we also assume $k_\perp v_A \ll \omega$ and $\beta_e \ll 1$). These approximations are well satisfied in magnetised plasmas, usually by at least two orders of magnitude, even in steep gradient regimes.

The procedure we will use closely follows Littlejohn's variational method as outlined in Chapter 5. The main difference is that we take the dynamical role of the field potentials on equal footing to the gyrocentre motion and treat the field/particle system as a whole. We do not split the potential into a background flow piece and small fluctuations either, but treat it as a unit assuming the gyrokinetic ordering is eventually satisfied across the spectrum. By not separating the equilibrium ExB flow out, we expect the electromagnetic version of the theory to recover not only equilibrium flows but also the MHD (Grad–Shafranov) equilibrium self-consistently.

The ExB velocity is handled using canonical form in the Lagrangian. The gauge operations move the field potential into the Hamiltonian as part of the transform. With the time dependent field potential kept out of the geometric part of the Lagrangian, canonical form is maintained, and we have the same basic structure as we did for the Lie transform version of the theory. The procedures from Chapters 4 and 5 are maintained throughout.

The usual assumptions of low-frequency theory are quasineutrality (by which $\varepsilon_0 E^2/2$ is neglected while allowing ϕ to evolve), and shear-Alfvén magnetic responses (by which only $A_\|$ is kept in the magnetic potential). The same parameter restrictions as before also apply here. In the dynamics only the parallel electric field involves inductive responses, and dynamical magnetic compressibility is therefore disallowed. The magnetic compressibility implied by the existence of diamagnetic flows and heat fluxes and polarisation currents is kept in the theory, so that any low-frequency compressibility effects are retained.

We will show the details of how fluid drift equations and the MHD Grad–Shafranov equilibrium for axisymmetric systems are recovered. The time derivatives of the two field equations recover the polarisation current balance and the parallel electron force balance, which are the two equations also serving as anchors for the fluid drift system. The equilibrium state of these equations then contains the Grad–Shafranov equation as a subset. The result of the conventional gyrokinetic approach is also recovered when for subsonic ExB flow the long-wavelength expression of the conventional Lagrangian is found to yield the one we will find herein. The ease of the correspondence follows from preserving canonical form throughout.

16.3 Gauge transform to get the Lagrangian

We consider generally a particle Lagrangian L_p which gets transformed to a gyrocentre one. The structure is

$$L_p\, dt = \mathbf{p} \cdot \mathbf{dz} - H\, dt \tag{16.2}$$

cast as a fundamental one-form, where the components of $\mathbf{p}$ are canonical momenta, the components of $\mathbf{z}$ are the phase space coordinates, $\mathbf{p} \cdot \mathbf{dz}$ is the symplectic part, and H is the Hamiltonian (the time component). In general this is six-dimensional (6D) dynamics, but in gyrokinetics the gyromotion involving perpendicular velocity space components is separated away so that the actual dynamics covers the 4D space of $\{\mathbf{R}, p_z\}$ consisting of gyrocentre positions and the parallel canonical momentum. Collisions bring in the fifth coordinate, usually the magnetic moment μ conserved by the drift motion. The sixth coordinate is the gyro-phase angle ϑ which only appears in the gyromotion since H and $\mathbf{p}$ and the rest of L_p is gyro-phase independent.

Starting with the textbook treatment, we restrict to non-relativistic situations with the time not being varied. We have the free-particle and interaction Lagrangians as

$$L_p = \frac{m}{2}\dot{\mathbf{x}} \cdot \dot{\mathbf{x}} + e\mathbf{A} \cdot \dot{\mathbf{x}} - e\phi \tag{16.3}$$

for the particles, and the free-field piece $\mathcal{L}_f$ which we treat later. The Legendre transform is applied as

$$\mathbf{p} \equiv \partial L_p/\partial \dot{\mathbf{x}} \qquad H \equiv \mathbf{p} \cdot \dot{\mathbf{x}} - L_p, \qquad L_p = \mathbf{p} \cdot \dot{\mathbf{x}} - H \tag{16.4}$$

We then turn L_p into a fundamental one-form by considering the differential action

$$L_p\, dt = \mathbf{p} \cdot \mathbf{dx} - H\, dt \quad \text{where} \quad H = m\frac{U^2}{2} + e\phi, \quad m\mathbf{U} = \mathbf{p} - e\mathbf{A} \tag{16.5}$$

where U, a vector, is not a coordinate but a function of dependent variables. Since we will use $\mathbf{A}$ as an anchor for low-frequency drift motion we change coordinates back to $\{\mathbf{x}, \mathbf{v}\}$ so that

$$L_p\, dt = (e\mathbf{A} + m\mathbf{v}) \cdot \mathbf{dx} - H\, dt \qquad \text{where} \qquad H = m\frac{v^2}{2} + e\phi \tag{16.6}$$

This gets us to the structure which is the usual starting point for this theory as outlined in Chapters 4 and 5.

To get a low-frequency low-beta kinetic Lagrangian we assume ϕ is a dynamical field but that $\mathbf{A}$ evolves through small shear-Alfvén disturbances parallel to $\mathbf{B}$. We re-cast $\mathbf{A}$ generally in terms of an equilibrium piece $\mathbf{A}$ assumed to be static, and add to it the dynamical piece $A_{\|}\mathbf{b}$ assuming $A_{\|}$ to be the other dynamical field. Now, $\mathbf{A}$, $\mathbf{b} = \mathbf{B}/B$, and $B = |\mathbf{B}|$ with $\mathbf{B} = \nabla \times \mathbf{A}$ are assumed to be static quantities describing the background magnetic field, while the particle coordinates and ϕ and $A_{\|}$ constitute the set of dependent variables which represent the dynamical system. This sets up

$$L_p\, dt = \left(e\mathbf{A} + A_{\|}\mathbf{b} + m\mathbf{v}\right) \cdot \mathbf{dx} - H\, dt \tag{16.7}$$

where $\mathbf{A}$ now represents the equilibrium only. It is the anchor for the drifts.

We assume the gyromotion is a fast frequency to be eliminated, retaining dynamics on the time scale of the ExB vorticity and the parallel transit frequencies. This leads to $1/e$ as the formal small parameter for expansion (which tracks the ratio Ω_E/Ω_i between the ExB vorticity and ion gyrofrequency).

We expand the velocity in terms of parallel motion, gyromotion, and drift motion

$$\mathbf{v} = v_{\|}\mathbf{b} + \mathbf{w} + \mathbf{u}_0 \tag{16.8}$$

Before further changes we consider $A_{\|}$ and then change representation from the parallel velocity $v_{\|}$ to the parallel canonical momentum p_z defined by

$$mv_\parallel + eA_\parallel = p_z \tag{16.9}$$

The one-form is now given by

$$L_p\, dt = \left(e\mathbf{A} + p_z\mathbf{b} + m\mathbf{w} + m\mathbf{u}_0\right) \cdot \mathbf{dx} - H\, dt \tag{16.10}$$

$$H = m\frac{U^2}{2} + \frac{m}{2}\,|\mathbf{w} + \mathbf{u}_0|^2 + e\phi \qquad mU = p_z - eA_\parallel \tag{16.11}$$

where U, now a scalar, is not a coordinate but a function of dependent variables. Hence $\mathbf{A}$ represents the background, and the perturbation $A_\parallel$ is moved out of $\mathbf{p} \cdot \mathbf{dz}$ and into H, and $v_\parallel$ is replaced by p_z as a coordinate, re-establishing canonical form after $A_\parallel$ is introduced.

The next step is to set up the expansion. The small parameter is formally any factor of $1/e$ which tracks drifts. Flows (gradients of ϕ) will enter through $\mathbf{u}_0$. The conventional drift kinetic representation treats $\mathbf{w}$ but leaves $\mathbf{u}_0$ in the symplectic part, along with the geometry. The time derivative of this piece in the Euler–Lagrange equations represents the polarisation drift. In a gyrokinetic representation we treat $\mathbf{w} + \mathbf{u}_0$ together, which moves ϕ out of the symplectic part and into H, maintaining canonical form. Polarisation no longer enters as a drift, but as a density in the field equation for ϕ which we will see later. This step is what makes the treatment gyrokinetic rather than drift kinetic. It depends on this change of representation, not the finite-gyroradius (FLR) effects. A model with zero-FLR and polarisation density is still a gyrokinetic one. The distinction becomes obvious only when the self-consistent field equations are at hand, which we necessarily defer to that time.

We introduce an arbitrary spatial coordinate change in which

$$\mathbf{x} = \mathbf{R} + \mathbf{a} \tag{16.12}$$

where we can choose $\mathbf{a}$ according to how we want to arrange the representation. We expect its magnitude to satisfy $a \ll L_B$ where L_B is the scale of variation of the magnetic field. We then Taylor-expand $\mathbf{A}$ and ϕ in powers of $\mathbf{a}$ and arrange them by order. The one-form in Eq. (16.10) splits according to orders as

$$L_0\, dt = e\mathbf{A} \cdot \mathbf{dR} - e\phi\, dt \tag{16.13}$$

$$\begin{aligned} L_1\, dt = {} & e\mathbf{a} \cdot \nabla\mathbf{A} \cdot \mathbf{dR} + e\mathbf{A} \cdot \mathbf{da} + \left(p_z\mathbf{b} + m\mathbf{w} + m\mathbf{u}_0\right) \cdot \mathbf{dR} \\ & - \left(e\mathbf{a} \cdot \nabla\phi + m\frac{U^2}{2} + \frac{m}{2}\,|\,\mathbf{w} + \mathbf{u}_0\,|^2\right) dt \end{aligned} \tag{16.14}$$

$$L_2\, dt = e\mathbf{a} \cdot \nabla\mathbf{A} \cdot \mathbf{da} + \left(p_z\mathbf{b} + m\mathbf{w} + m\mathbf{u}_0\right) \cdot \mathbf{da} - \frac{1}{2}e(\mathbf{a} \cdot \nabla)^2\phi\, dt \tag{16.15}$$

where the dependence of $\mathbf{A}$ and ϕ and ∇ is now upon $\mathbf{R}$ after the Taylor expansion. Due to the factors of $1/e$ the spatial variation of $\mathbf{b}$ enters one order lower than that of $\mathbf{A}$ so we do not expand it.

The lowest-order one-form is $L_0\,dt$. Varying $\mathbf{R}$ by contravariant components δR^i we find

$$e\big(A_{j,i} - A_{i,j}\big)\dot{R}^j - e\phi_{,i} = 0 \tag{16.16}$$

where the sum over repeated indices, one up and one down, is taken as understood (Einstein convention). The subscript after a comma denotes differentiation, e.g.,

$$A_{i,j} = \frac{\partial A_i}{\partial x^j} \tag{16.17}$$

the derivative of covariant component i with respect to the jth coordinate. Solving for $\dot{R}^j$, we find

$$\dot{R}_0^i = \frac{1}{B^2}\big(A_{i,j} - A_{j,i}\big)\phi_{,j} + \frac{1}{B^2}B^iB_j\dot{R}_0^j, \qquad B^k = \varepsilon^{kij}\big(A_{j,i} - A_{i,j}\big) \tag{16.18}$$

where indices $\{ijk\}$ range over the spatial dimensions, and ε^{ijk} is the Levi-Civita pseudotensor of rank three with units $g^{-1/2}$, the spatial Jacobian (Chapter 4). This solution describes the lowest-order ExB drift. We note that at this level the parallel component $B_j\dot{R}_0^j$ is indeterminate. Since the parallel dynamics enters at the next order, however, we leave this component at zero and specify

$$\mathbf{u}_0 = \mathbf{u}_E = \frac{1}{B^2}\nabla\phi\cdot\mathbf{F}, \qquad \mathbf{F} = \nabla\mathbf{A} - (\nabla\mathbf{A})^T \tag{16.19}$$

with superscript T denoting the transpose. This operation is how the drift tensor $\mathbf{F}$ enters the problem. Specifying $\mathbf{u}_0$ as $\mathbf{u}_E$ is the main choice in this step.

The next-order one-form is $L_1\,dt$. We first subtract the total differential

$$d(\mathbf{a}\cdot\mathbf{A}) = \mathbf{da}\cdot\mathbf{A} + \mathbf{dR}\cdot\nabla\mathbf{A}\cdot\mathbf{a} \tag{16.20}$$

where we note that $\mathbf{A}$ depends on $\mathbf{R}$ but is static. The subtraction yields

$$L_1\,dt - d(\mathbf{a}\cdot\mathbf{A}) = e\mathbf{a}\cdot\mathbf{F}\cdot\mathbf{dR} + \big(p_z\mathbf{b} + m\mathbf{w} + m\mathbf{u}_0\big)\cdot\mathbf{dR} + \cdots \tag{16.21}$$

where we have written only the terms in the symplectic part. We now choose $\mathbf{a}$. Customary operations in the theory set $\mathbf{a}$ to cancel $\mathbf{w}$ out of the symplectic part, leaving $\mathbf{u}_0$ there as a background term. In our case we choose $\mathbf{a}$ to cancel both $\mathbf{u}_0$ and $\mathbf{w}$ out of the symplectic part. This sets

$$e\mathbf{a}\cdot\mathbf{F} + m(\mathbf{w} + \mathbf{u}_0) = 0 \tag{16.22}$$

and the solution is

$$\mathbf{a} = \frac{m}{eB^2}(\mathbf{w} + \mathbf{u}_0)\cdot\mathbf{F} \tag{16.23}$$

where we arbitrarily choose $\mathbf{b}\cdot\mathbf{a} = 0$. This $\mathbf{a}$ can be called a gyro-drift radius, and if $\mathbf{u}_0 \to 0$ it reverts to the standard gyroradius. In both cases it is the directed version, not the magnitude. We also find

$$\mathbf{a} \cdot \nabla\phi = -m\mathbf{u}_0 \cdot (\mathbf{w} + \mathbf{u}_0) \tag{16.24}$$

This piece helps a cancellation in the Hamiltonian,

$$\mathbf{a} \cdot \nabla\phi + m\mathbf{w} \cdot \mathbf{u}_0 + m\frac{u_0^2}{2} = -m\frac{u_0^2}{2} \tag{16.25}$$

Discarding the total differential term and substituting $\mathbf{u}_0 \to \mathbf{u}_E$, the first-order one-form correction is

$$L_1\, dt = p_z \mathbf{b} \cdot \mathbf{dR} - \left(e\phi + m\frac{U^2}{2} + m\frac{w^2}{2} - m\frac{u_E^2}{2}\right)dt \tag{16.26}$$

Note the sign change for the last term, which is a result of the aforementioned cancellation. This operation is how $mu_E^2/2$ appears as a field energy term. It is actually at this point that we can sensibly drop the $\varepsilon_0 E^2/2$ term from the field Lagrangian as small compared with $mu_E^2/2$, which is where the assumption of quasineutrality is applied.

In fact the result for $L_0 + L_1$ is almost good enough to build the model, since we have obtained the quadratic field term in ϕ necessary to build its field equation. However, we haven't specified $\mathbf{w}$ yet, and at minimum we need w^2 in H. Moreover, to get FLR effects and to have a conserved magnetic moment we have to proceed to the next order and consider the details of gyromotion as in the conventional treatment. We set up an auxiliary basis $\mathbf{e}_1$ and $\mathbf{e}_2$ for the plane perpendicular to $\mathbf{b}$, with restrictions and coordinate sense

$$\mathbf{e}_1 \cdot \mathbf{e}_2 = 0, \qquad \mathbf{b} \cdot \mathbf{e}_{1,2} = 0, \qquad \mathbf{b} \cdot \mathbf{e}_1 \times \mathbf{e}_2 = 1 \tag{16.27}$$

giving the signs (with $\mathbf{b}$ toward the observer the sense of the motion is clockwise for ions). We introduce the gyrophase angle ϑ hence covering the plane with w and ϑ and relating $\mathbf{da}$ to $\mathbf{w}$. Due to the large gyrofrequency, Ω, the fast part of $\mathbf{da}$ is solely due to $d(\mathbf{w} \cdot \mathbf{F})$ with contributions due to $\mathbf{u}_0$ down an order. Contributions due to $\nabla\mathbf{e}_{1,2}$ give gyrophase invariance, and here we neglect all the others due to the spatial variation of $\mathbf{B}$. The motion ($\Omega\, dt$) is described locally as a geometric circle with angle variation $d\vartheta$ where we identify $\mathbf{w}$ as the directed gyration velocity and ϑ the gyrophase angle

$$\mathbf{w} = -w(\mathbf{e}_1 \sin\vartheta + \mathbf{e}_2 \cos\vartheta) \tag{16.28}$$

The sense of the gyromotion is signed with the charge e, as it appears in Ω. We will identify the magnetic moment coordinate with the conserved quantity multiplying $d\vartheta$ at the end. We will have $(\mathbf{w} + \mathbf{u}_0) \cdot \mathbf{da}$ with $e\mathbf{a} \cdot \mathbf{F} + m(\mathbf{w} + \mathbf{u}_0) = 0$. Large Ω says keep only the $\mathbf{dw}$ part of $\mathbf{da}$ so that $\mathbf{da} \to \Omega^{-1}\mathbf{b} \times \mathbf{dw}$ with the $\mathbf{u}_0 \cdot \mathbf{da}$ piece averaging to zero. This leaves $\mathbf{w} \cdot \mathbf{b} \times \mathbf{dw}$ which is then worked through as $\mathbf{dw} \cdot \mathbf{w} \times \mathbf{b}$ according to

$$\mathbf{dw} = -w(\mathbf{e}_1 \cos\vartheta - \mathbf{e}_2 \sin\vartheta)d\vartheta - w\mathbf{dR} \cdot (\nabla\mathbf{e}_1 \sin\vartheta + \nabla\mathbf{e}_2 \cos\vartheta) \tag{16.29}$$

$$\mathbf{w} \times \mathbf{b} = w(\mathbf{e}_2 \sin \vartheta - \mathbf{e}_1 \cos \vartheta) \tag{16.30}$$

finally, we use $\nabla \mathbf{e}_1 \cdot \mathbf{e}_2 + \nabla \mathbf{e}_2 \cdot \mathbf{e}_1 = 0$ to express it as

$$\mathbf{w} \cdot \mathbf{da} = \mathbf{dw} \cdot \mathbf{w} \times \mathbf{b} = \frac{w^2}{\Omega}(d\vartheta - \mathbf{dR} \cdot \nabla \mathbf{e}_1 \cdot \mathbf{e}_2) \tag{16.31}$$

This is also the minimal description of gyromotion preserving gyrogauge invariance under spatially dependent rotations of the $\mathbf{e}_{1,\,2}$ unit vectors.

At second order the Lagrangian correction is

$$L_2\, dt = \left(e\mathbf{a} \cdot \nabla \mathbf{A} + p_z \mathbf{b} + m\mathbf{w} + m\mathbf{u}_0\right) \cdot \mathbf{da} - \frac{1}{2}e(\mathbf{a} \cdot \nabla)^2 \phi\, dt \tag{16.32}$$

The first step is to use $\mathbf{b} \cdot \mathbf{da} = 0$ to strip the p_z term and then subtract $dS_2 = (e/2)d(\mathbf{a} \cdot \nabla \mathbf{A} \cdot \mathbf{a})$ to symmetrise the form with $\nabla \mathbf{A}$ obtaining

$$L_2\, dt + dS_2 = \left(\frac{1}{2}e\mathbf{a} \cdot \mathbf{F} + m\mathbf{w} + m\mathbf{u}_0\right) \cdot \mathbf{da} - \frac{1}{2}e(\mathbf{a} \cdot \nabla)^2 \phi\, dt \tag{16.33}$$

The definition of $\mathbf{a}$ through $\mathbf{a} \cdot \mathbf{F}$ in Eq. (16.22) combines the terms such that

$$L_2\, dt + dS_2 = \frac{1}{2}m(\mathbf{w} + \mathbf{u}_0) \cdot \mathbf{da} - \frac{1}{2}e(\mathbf{a} \cdot \nabla)^2 \phi\, dt \tag{16.34}$$

Using the gyro-drift motion approximation on the $\mathbf{da}$, we obtain

$$L_2\, dt + dS_2 = \frac{1}{2}\frac{mw^2}{\Omega}(d\vartheta - \mathbf{dR} \cdot \nabla \mathbf{e}_1 \cdot \mathbf{e}_2) - \frac{1}{2}e(\mathbf{a} \cdot \nabla)^2 \phi\, dt \tag{16.35}$$

Averaging the gradient components in $(\mathbf{a} \cdot \nabla)^2$ over ϑ, we obtain

$$L_2\, dt + dS_2 = M(d\vartheta - \mathbf{dR} \cdot \nabla \mathbf{e}_1 \cdot \mathbf{e}_2) - e\frac{a^2}{4}\nabla_\perp^2 \phi\, dt \tag{16.36}$$

where we have identified

$$M = \frac{m}{e}\frac{mw^2}{2B} \tag{16.37}$$

as the conserved quantity multiplying $d\vartheta$. Since ϑ-dependence has been eliminated everywhere else in L_p, we now note that M is a constant of the motion and suitable for use as a coordinate. As in Chapter 5, we write it as M for generality. It is usually written as $M = (m/e)\mu$ but here it is important to note that μ is usually the lowest order magnetic moment, while in actual fact M is conserved exactly to all orders, since it multiplies $d\vartheta$ and the rest of L is arranged order by order to eliminate gyrophase angle dependence.

The quantity given by

$$\mathbf{W} = \nabla \mathbf{e}_1 \cdot \mathbf{e}_2 \tag{16.38}$$

can be called the twist of the gyrophase plane. In a straight magnetic field it vanishes. It is formally small since the spatial scale is L_B, but it is formally important since the $d\vartheta$ piece by itself is not gyrogauge invariant. If we rotate the unit vectors describing the gyrophase plane by a spatially dependent angle α,

$$\vartheta' = \vartheta + \alpha(\mathbf{R}) \tag{16.39}$$

with the prime denoting the new coordinate system after the rotation of the unit vectors in the gyrophase plane. then $d\vartheta$ does not equal $d\vartheta'$ since there is a contribution from α. However, the combination $d\vartheta - \mathbf{W}\cdot\mathbf{dR}$ is invariant. To show this, use the dependence of the unit vectors $\mathbf{e}_{1,\,2}$ on ϑ, use $d\alpha = \mathbf{dR}\cdot\nabla\alpha$, and then combine the set to show that

$$d\vartheta - \mathbf{W}\cdot\mathbf{dR} = d\vartheta' - \mathbf{W}'\cdot\mathbf{dR} \tag{16.40}$$

In practice the gyromotion drops out of the kinetic equation anyway, since $\partial/\partial\vartheta$ and dM/dt vanish. The $M\mathbf{W}$ piece is a small $O(a/L_B)^2$ correction to the (a/L_B) drifts and does not introduce any new effects of charge separation. In practice no numerical simulation at present day keeps it. For profile scales $L_\perp \ll L_B$ the higher-order terms from the contribution of u_E^2 to a^2 are larger and these receive the attention.

We have now accounted for w^2 and a^2 appearing only through their magnitudes so we combine the result $L_{0,\,1,\,2}\,dt$ as

$$L_p\,dt = \left(e\mathbf{A} + p_z\mathbf{b} - M\mathbf{W}\right)\cdot\mathbf{dR} + M\,d\vartheta - H\,dt \tag{16.41}$$

with Hamiltonian

$$H = m\frac{U^2}{2} + M\Omega + \left(1 + \frac{a^2}{4}\nabla_\perp^2\right)e\phi - m\frac{u_E^2}{2} \tag{16.42}$$

where

$$mU = p_z - eA_\parallel, \qquad u_E^2 = \frac{1}{B^2}\,|\nabla_\perp\phi|^2, \qquad a^2 = \frac{2M\Omega + mu_E^2}{m\Omega^2} \tag{16.43}$$

where the extra differentials $dS_{1,\,2}$ are dropped (formally changing the representation). This is the result for the gyrocentre Lagrangian up to second order in the expansion. In a field theory it is embedded into the field Lagrangian and then the Euler–Lagrange equations, the evolution equation for the gyrokinetic distribution function, and the field equations are found in the same way as given in Chapters 4 and 5. The procedure is the same for any version of the theory.

16.4 Correspondence among the Lagrangians

Of interest is to compare the result just obtained with the conventional Lagrangian using the Lie transform method. Besides the ability to get the same result with different methods, it is also of interest to see how well the match is in the large-scale regime, which is the one this Lagrangian represents. The usual ordering assumes $e\phi/T_e$ small, but we want to know how well the usual result can capture the force balance result in the profiles, $n_i e\nabla\phi + \nabla p \approx 0$ for a singly charged ion species, where ∇ carries the profile scale $L_\perp$. Although $e\phi/T_e$ is formally unity here, the associated ExB velocity is still subsonic so long as $\delta_0 = \rho_s/L_\perp$, the drift parameter, remains small. Of course the vorticity ordering, $\Omega_E \ll \Omega_i$, is then satisfied to order δ_0^2. So it seems the theory should remain valid here, even if it assumed small amplitude at the outset, because the scale was formally ordered with ρ_s, not $L_\perp$.

16.4.1 Low Mach number limit of the Lagrangian

Since the profile scale is larger than the drift scale, unit amplitude of the ϕ profile implies a velocity subsonic by one factor of δ_0. So we can take the $mu_E^2 \ll T$ ordering on the gauge transform Hamiltonian (the only place ϕ occurs) in Eq. (16.42), to obtain

$$H = m\frac{U^2}{2} + M\Omega + \left(1 + \frac{M\Omega}{2m\Omega^2}\nabla_\perp^2\right)e\phi - m\frac{u_E^2}{2} \tag{16.44}$$

Note that the only change is to drop the u_E^2 term in what is already the FLR correction to ϕ. We note at the outset that

$$1 + \frac{M\Omega}{2m\Omega^2}\nabla_\perp^2 \rightarrow 1 - \frac{\rho_L^2}{4}k_\perp^2 \tag{16.45}$$

in wavenumber space, where in the absence of the inertial Mach correction to it we re-label the gyroradius a as ρ_L. This corresponds to the expansion of $J_0(\phi)$ to the first correction, as given in Eq. (5.122). Hence we have the *a posteriori* limitation that the Lagrangian in this case is already limited to $k_\perp^2\rho_L^2 \ll 1$, so that it is a large-scale description of the theory. The form of the Lagrangian itself in Eq. (16.41) remains, as does the field term in Eq. (16.51) for the electromagnetic case.

16.4.2 Long wavelength limit of the conventional Lagrangian

We have already done the expansions and their effect on the gyroaveraging operator in Chapter 5, specifically Eqs (5.122 to 5.124). Summarising these, we have

$$H = m\frac{U^2}{2} + M\Omega + \left(1 + \frac{\rho_L^2}{4}\nabla_\perp^2\right)e\phi - m\frac{u_E^2}{2} \tag{16.46}$$

$$mU = p_z - eA_\parallel, \qquad u_E^2 = \frac{1}{B^2}\,|\nabla_\perp\phi|^2 \tag{16.47}$$

as the long wavelength limit of the conventional gyrokinetic Lagrangian, still keeping the nominal FLR correction on ϕ for the comparison. The main point is, these two Lagrangians obviously coincide, so that we have the same model in the appropriate limit.

This cements the point that a derivation like this is often better than its original formulation, but not always so. In the gauge transform case we started with unit order $e\phi/T_e$, did the derivation, but didn't get the limitation on scale, $k_\perp\rho_L \ll 1$, until afterward. In the Lie transform case we started with small $e\phi/T_e$ but left the scale in maximal ordering though nominally with $k_\perp\rho_L \sim 1$. Afterward, we now find that the entire model remains valid in the large scale limit, but only so long as the resulting ExB flow remains subsonic there. However, since in a magnetised plasma, which already assumes $\delta_0 \ll 1$, we always have the large-scale flow in near force balance and well into the subsonic regime. Finally, we can state that the Lie transform Lagrangian and field theory is always valid so long as the plasma is well magnetised, even for the equilibrium and its flow regime.

16.5 The field Lagrangian

The field theory embeds this into a phase space

$$L = \sum_{sp} \int d\Lambda f\, L_p + \int d\mathcal{V}\, \mathcal{L}_f \tag{16.48}$$

where for shear-Alfvén conditions the field Lagrangian density is

$$\mathcal{L}_f = \varepsilon_0 \frac{E^2}{2} - \frac{B^2}{2\mu_0} \qquad \to \qquad \mathcal{L}_f = -\frac{1}{2\mu_0}\left|\nabla_\perp A_\parallel\right|^2 \tag{16.49}$$

following the quasineutrality approximation. In the Lagrangian this is consistent with considering $A_\parallel$ as a perturbation to the equilibrium, with

$$\mathbf{B} = \nabla \times \mathbf{A}, \qquad \mathbf{B}_T = \mathbf{B} + \nabla \times (A_\parallel \mathbf{b}) \tag{16.50}$$

as the equilibrium and total magnetic field, respectively.

With these approximations we re-write the system Lagrangian as Eq. (16.48) above, now specifically defining the field Lagrangian as

$$\mathcal{L}_f = -\frac{1}{2\mu_0}\left|\nabla_\perp A_\parallel\right|^2 \tag{16.51}$$

and the gyrocentre Lagrangian as

$$L_p = \left(e\mathbf{A} + p_z \mathbf{b}\right) \cdot \dot{\mathbf{R}} + \frac{m}{e}\mu(\dot{\vartheta} - \mathbf{W} \cdot \dot{\mathbf{R}}) - H \tag{16.52}$$

$$H = m\frac{U^2}{2} + \mu B + \left(1 + \frac{a^2}{4}\nabla_\perp^2\right)e\phi - m\frac{u_E^2}{2} \tag{16.53}$$

$$mU = p_z - eA_\parallel, \qquad u_E^2 = \frac{1}{B^2}\left|\nabla_\perp \phi\right|^2, \qquad a^2 = \frac{2\mu B + m u_E^2}{m\Omega^2} \tag{16.54}$$

where we re-write $M\Omega$ in terms of the more familiar (and unsigned) μB, with $\mu = (m/e)M$ the gauge-transformed magnetic moment. This is a complete description of the dynamical system. In following, we recap the derivation of the Euler–Lagrange equations for the gyrocentres, the gyrokinetic equation for their distribution function, and then the Euler–Lagrangian equations for the field variables giving their self-consistent equations in the model.

16.5.1 The Euler–Lagrange equations for gyrocentres

Gyrocentre motion itself arises from L_p only. We note that derivatives arise from variations with respect to the phase space coordinates holding each other fixed. Hence, we note again that the spatial gradient operator is taken with respect to gyrocentre positions $\mathbf{R}$ holding p_z fixed. We define geometric quantities

$$\mathbf{A}^* = \mathbf{A} + \frac{1}{e}p_z\mathbf{b} - \frac{m}{e^2}\mu\mathbf{W}, \qquad \mathbf{B}^* = \nabla \times \mathbf{A}^*, \qquad \mathbf{F} = B\epsilon \cdot \mathbf{b} \tag{16.55}$$

such that $B^{-1}\mathbf{F} \cdot \mathbf{G} = \mathbf{G} \times \mathbf{b}$ for any vector $\mathbf{G}$ (this is the origin of the drifts). Note that $\mathbf{B}$, $\mathbf{b}$, and B are all kept separate as they can be independently approximated. The derivatives of H are

$$\nabla H = e\nabla\phi_E + \mu\nabla B_E - eU\nabla A_\parallel \tag{16.56}$$

$$\frac{\partial H}{\partial p_z} = mU \qquad \frac{\partial H}{\partial \mu} = B_E \qquad \frac{\partial H}{\partial \vartheta} = 0 \tag{16.57}$$

where

$$e\phi_E = e\phi - m\frac{u_E^2}{2}\left(1 - \frac{\Omega_E}{2\Omega}\right) \quad B_E = B\left(1 + \frac{\Omega_E}{2\Omega}\right) \quad \Omega_E = \frac{1}{B}\nabla_\perp^2\phi \tag{16.58}$$

Varying $\mathbf{R}$ and p_z together in $L_p\, dt$ and setting the coefficients of the variation components to zero yields the drift motion

$$B_\parallel^*\dot{\mathbf{R}} = \nabla H \cdot \frac{1}{e}\frac{\mathbf{F}}{B} + \frac{\partial H}{\partial p_z}\mathbf{B}^*, \qquad B_\parallel^*\dot{p}_z = -\mathbf{B}^* \cdot \nabla H \tag{16.59}$$

and varying ϑ and μ yields the gyromotion

$$\dot{\mu} = 0 \qquad \dot{\vartheta} = \frac{m}{e}\frac{\partial H}{\partial \mu} + \mathbf{W} \cdot \dot{\mathbf{R}} \tag{16.60}$$

In the drift motion the general form of the phase space volume element is

$$B_\parallel^* = e\frac{\partial \mathbf{A}^*}{\partial z} \cdot \mathbf{B}^* \tag{16.61}$$

where in our case the z-coordinate is p_z. It is because p_z enters $\mathbf{A}^*$ only through $\mathbf{b}$ that the usual formula $B_\parallel^* = \mathbf{b} \cdot \mathbf{B}^*$ holds. In fact this is not general. For example, using an energy representation $mv^2/2$ for the z-coordinate yields $B_\parallel^* = U^{-1}\mathbf{b} \cdot \mathbf{B}^*$ where U

is the parallel velocity function of $mv^2/2$ and μ and spatial dependence through B and also $A_\|$. But the form in Eq. (16.61) is general.

16.5.2 The gyrokinetic equation

The operations to get to this are familiar. We first observe that both

$$\nabla \cdot \mathbf{B}^* = 0 \qquad \nabla \cdot \frac{1}{e}\frac{\mathbf{F}}{B} + \frac{\partial \mathbf{B}^*}{\partial p_z} = 0 \tag{16.62}$$

hold, since

$$\frac{\mathbf{F}}{B} = \epsilon \cdot \mathbf{b} \qquad \frac{\partial \mathbf{B}^*}{\partial p_z} = \frac{1}{e}\nabla \times \mathbf{b} \tag{16.63}$$

It is necessary to formulate $\mathbf{F}$ in this fashion because most models approximate $\mathbf{b}$ while keeping the full $\mathbf{B}$. Using this, we find that the requirement of phase space incompressibility

$$\nabla \cdot B_\|^* \dot{\mathbf{R}} + \frac{\partial}{\partial p_z} B_\|^* \dot{p}_z = 0 \tag{16.64}$$

is satisfied. The distribution function f is just the density of gyrocentres in phase space, so we have its continuity equation

$$\frac{\partial f}{\partial t} + \frac{1}{B_\|^*}\nabla \cdot \left(fB_\|^* \dot{\mathbf{R}}\right) + \frac{1}{B_\|^*}\frac{\partial}{\partial p_z}\left(fB_\|^* \dot{p}_z\right) = 0 \tag{16.65}$$

in divergence form, or equivalently

$$\frac{\partial f}{\partial t} + \dot{\mathbf{R}} \cdot \nabla f + \dot{p}_z \frac{\partial f}{\partial p_z} = 0 \tag{16.66}$$

in advection form using the phase-space incompressibility. There is no μ term since $\dot{\mu} = 0$ and no ϑ term since $\partial f/\partial\vartheta = 0$. This is our application of Liouville's theorem. Using the results for the gyrocentre motion we have

$$B_\|^* \frac{\partial f}{\partial t} + \nabla H \cdot \frac{1}{e}\frac{\mathbf{F}}{B} \cdot \nabla f + \mathbf{B}^* \cdot \left(\frac{\partial H}{\partial p_z}\nabla f - \frac{\partial f}{\partial p_z}\nabla H\right) = 0 \tag{16.67}$$

This is our gyrokinetic equation. One thing to note is that the only appearance of $\mu \mathbf{W}$ is in its contribution to $\mathbf{B}^*$ and $B_\|^*$ and it is small there and introduces no new effects. This is why it is usually dropped in computations.

16.5.3 The field equations

Given the system Lagrangian in Eq. (16.48) we find the equations for ϕ and $A_\|$ by varying them in $L_p\, dt$ and setting the coefficients of the variations to zero.

The induction equation for $A_\parallel$ is

$$-\nabla_\perp^2 A_\parallel = \mu_0 J_\parallel \qquad J_\parallel = \sum_{sp} \int d\mathcal{W}\, eU\, f \tag{16.68}$$

where $J_\parallel$ is the gyrocentre current. Eq. (16.68) is the gyrokinetic Ampère's law.

The polarisation equation for ϕ is

$$\nabla \cdot \left(N_E \nabla_\perp \phi + \nabla_\perp^2 P_E \right) + \rho_G = 0 \tag{16.69}$$

where

$$\rho_G = \sum_{sp} ne \quad N_E = \sum_{sp} \frac{nm}{B^2}\left(1 - \frac{\Omega_E}{2\Omega}\right) \quad P_E = \sum_{sp} \frac{m}{2eB^2}\left(p_\perp + nm\frac{u_E^2}{2}\right) \tag{16.70}$$

are the gyrocentre charge density, the polarisability, and the generalised FLR correction to the gyrocentre charge density, with moment quantities given by

$$n = \int d\mathcal{W} f \qquad p_\perp = \int d\mathcal{W}\, \mu B\, f \tag{16.71}$$

These represent the gyrocentre density and perpendicular pressure. Eq. (16.69) is the gyrokinetic Poisson equation.

Together, Eqs (16.67), (16.68), (16.69) describe the complete self-consistent dynamical system which arises from the description due to the Lagrangian in Eq. (16.48).

16.6 MHD and MHD equilibrium

In general magnetohydrodynamics (MHD) the relevant low frequency approximation is the combination $\beta \ll 1$ and $\omega \ll k_\perp v_A$ in the ordering, just as in the more general case keeping the adiabatic response. Low frequency MHD is reduced MHD (labelled RMHD). Its basis is an equation for the ExB vorticity which includes the field line bending term, and an Ohm's law which is an equation for the magnetic flux, equivalent to the statement that the parallel electric field vanishes, in the ideal case, or is balanced by resistivity. The ExB vorticity equation is the same in RMHD as it is in the fluid drift model, and the Ohm's law is a subset of the one in the fluid drift model which retains the adiabatic response.

The small parameter for expansion is the effective aspect ratio, $K_\parallel L_\perp$, between the perpendicular and parallel scales. In a tokamak this parameter is the aspect ratio of the field line pitch, given by a/qR where a and R are the toroidal minor and major radii, and q is the field line pitch parameter given by the ratio of the contravariant angle components of the magnetic vector field. The ordering expects r/qR to be in the range of 0.1 at the largest, which for a reasonable profile $q(r)$ occurs in the outer third, where $r/a = 1$ at the edge.

The small parameter in gyrokinetic theory is ρ_L/L_B for the background magnetic field, which is ρ_L/R in a tokamak. When the fields are present in self-consistent dynamics, one requires $\rho_L/L_\perp$ since that makes $c_s/L_\perp \ll \Omega_i$, namely the effective vorticity ordering. These statements look different but they have the same purpose, namely, to keep the low frequency Alfvén ordering by which the parallel shear Alfvén response is retained but the perpendicular Alfvén response is fast enough that force balance is maintained perpendicular to the background magnetic field and the ExB vorticity dynamics is non-inductive. Hence, $A_\parallel$ is retained but $\mathbf{A}_\perp$ is dropped in the theory.

The ordering scheme underlying gyrokinetic theory is the same as the one underlying RMHD even if the statement of the small parameter is different. It therefore follows that RMHD should be captured by gyrokinetic theory, and in a total-f setting so should the equilibrium. We should be able to derive the RMHD vorticity equation and Ohm's law and then also the Grad–Shafranov equation for equilibrium, all starting from the equations of gyrokinetic field theory. We have addressed some of this in the delta-f limit. It remains to do so generally, with a gyrokinetic Lagrangian which is valid for equilibrium scales.

The gist of this is that the two main RMHD equations are found from the time derivatives of the two field equations, yielding the vorticity equation and the Ohm's law, respectively. These are essentially gyrofluid moment equations, since the terms are evaluated using velocity-space moments of the terms in the gyrokinetic equation. This is facilitated by using the divergence form of the gyrokinetic equation (Eq. 16.65) to evaluate the expressions of $\partial f/\partial t$ which result. We will be dealing with moments over unity and over p_z, in the vorticity equation and Ohm's law, respectively.

The zero-FLR limit of Eq. (16.69) is given by

$$\varpi + \rho_{\rm G} = 0 \qquad \varpi = \nabla \cdot \frac{\rho}{B^2}\nabla_\perp\phi \qquad \rho_{\rm G} = \sum_{sp}\int d\mathcal{W} fe \tag{16.72}$$

Its time derivative is

$$\frac{\partial\varpi}{\partial t} + \sum_{sp}\int d\mathcal{W}\, e\frac{\partial f}{\partial t} = 0 \tag{16.73}$$

Substituting for the time derivative of f, we have

$$\frac{\partial\varpi}{\partial t} = \nabla\cdot\mathbf{J}_{\rm G} \qquad \mathbf{J}_{\rm G} = \sum_{sp}\int d\mathcal{W} fe\dot{\mathbf{R}} \tag{16.74}$$

where $\mathbf{J}_{\rm G}$ is the gyrocentre current including all the drift motion, and we note that $d\mathcal{W}/B_\parallel^*$ commutes past $\nabla\cdot$ and the integration annihilates $\partial/\partial p_z$ on any quantity. In ϖ we have either version, with ρ reflecting the dependent variable f or the Boussinesq approximation in which is it prescribed (usually constant). The main terms in $\dot{\mathbf{R}}$ are the ExB drift and the parallel motion, with the grad-B drift in ∇H and the curvature drift due to $\nabla \times \mathbf{b}$ from $\mathbf{B}^*$ in an inhomogeneous magnetic field. These are also the terms in the RMHD vorticity equation. In a situation with only one ion species, we have

$$\frac{\partial\varpi}{\partial t} + \nabla\cdot(\varpi\mathbf{v}_E - nm_i\mathbf{U}_E) = \nabla\cdot\left(\frac{J_\parallel}{B}\mathbf{B}_T\right) + \nabla\cdot\left(p_\perp\mathbf{u}_{\nabla B} + P_\parallel\mathbf{u}_C\right) \tag{16.75}$$

where $\mathbf{U}_E$, $\mathbf{u}_{\nabla B}$, and $\mathbf{u}_C$ are the ion inertial drift and the grad-B and curvature drifts and $P_\parallel$ and $p_\perp$ are the total-f parallel and perpendicular energy moments of f, and $\mathbf{B}_T$ is the total magnetic field, all defined in Chapter 6 (following and after Eq. 6.118). With $\mathbf{U}_E$ ordered small, and $P_\parallel$ and $p_\perp$ ordered as the isotropic pressure p, these are also the terms in the RMHD vorticity equation.

The time derivative of Eq. (16.68) is given by

$$-\nabla_\perp^2\frac{\partial A_\parallel}{\partial t} = \mu_0\sum_{sp}\int d\mathcal{W}\, eU\frac{\partial f}{\partial t} + ef\frac{\partial U}{\partial t} \tag{16.76}$$

We express

$$\frac{\partial U}{\partial t} = -\frac{e}{m}\frac{\partial A_\parallel}{\partial t} \tag{16.77}$$

and pull this term to the left hand side, so that

$$(d_e^{-2} - \nabla_\perp^2)\frac{\partial A_\parallel}{\partial t} = \mu_0\sum_{sp}\int d\mathcal{W}\, eU\frac{\partial f}{\partial t} \tag{16.78}$$

where d_e is the skin depth given by

$$d_e^2 = \sum_{sp}\int d\mathcal{W}\,\frac{\mu_0 fe^2}{m} \tag{16.79}$$

in which the dominant contribution is from the electrons. Substituting for the time derivative of f, we have

$$\left(d_e^{-2} - \nabla_\perp^2\right)\frac{\partial A_\parallel}{\partial t} = -\mu_0\sum_{sp}\int d\mathcal{W}\, eU\frac{1}{B_\parallel^*}\left(\nabla\cdot fB_\parallel^*\dot{\mathbf{R}}\right) - \mu_0\sum_{sp}\int d\mathcal{W}\,\frac{e}{m}(f\dot{p}_z) \tag{16.80}$$

where in the last term we use $m(\partial U/\partial p_z) = 1$. There are several terms in this equation, as in even the fluid drift model, but the ones which occur in RMHD are the first one and the contribution by ϕ to the last one. In addition, the contribution in $\dot{\mathbf{R}}$ due to U yields the parallel pressure term and a correction from the action of the gradient upon $A_\parallel$ in the integration by parts. These three terms are

$$d_e^2\frac{\partial A_\parallel}{\partial t} = -\mu_0\sum_{sp}\int d\mathcal{W}\,\frac{e}{m}mU^2 f - \mu_0\sum_{sp}\int d\mathcal{W}\,\frac{e^2}{m}f\nabla_\parallel\phi + \cdots \tag{16.81}$$

Noting that the signed e is negative for electrons, and that due to the $1/m$ factor they provide the main contribution, we can divide through by d_e^2 and obtain

$$\frac{\partial A_\parallel}{\partial t} + \nabla_\parallel \phi = \frac{1}{n_e e} \nabla_\parallel p_\parallel + \cdots \tag{16.82}$$

where the parallel pressure moment is

$$p_\parallel = \int d\mathcal{W}\, mU^2 f \tag{16.83}$$

namely $P_\parallel$ in the neglect of the parallel Mach number correction $nmu_\parallel^2$, which, however, is very small for electrons, and the entire ion contribution appears with the $1/m$ factor and therefore is negligible. The left side of this Ohm's law is the same as the one in RMHD, and without dissipation indicates the ideal limit. The ellipsis represents all the other terms, which can include the resistivity. The $-\nabla_\perp^2 A_\parallel$ factor on the left hand side, which was dropped here, is the electron inertia. The pressure gradient term is the same one as in the fluid drift equations providing the adiabatic response.

So we see that RMHD emerges from the time derivatives of the field equations in this model, already enveloped as a subset of the fluid drift equations which are represented by the other main terms in these two equations and then also the continuity equations for densities and temperatures as we have already seen in the delta-f model.

The equation of MHD equilibrium is the equilibrated state of the vorticity equation, which is Eq. (16.75) with ϕ and ϖ set to zero. The two main terms are the parallel current (field line bending) term and the diamagnetic current (all the magnetic drifts together). The MHD equilibrium sets the flows, hence ϕ, to zero and considers the balance between parallel current and the magnetic drifts. The dominant terms are

$$B\nabla_\parallel \frac{J_\parallel}{B} + \nabla \cdot \frac{1}{2B} \mathbf{b} \times \nabla (P_\parallel + p_\perp) = 0 \tag{16.84}$$

In a tokamak magnetic field under the conventional ordering we set

$$\mathbf{b} \to R\nabla\varphi \qquad B \to \frac{I}{R} \tag{16.85}$$

where φ is the geometric toroidal angle, and I is constant. The magnetic field is done with

$$\mathbf{B} = I\nabla\varphi + \nabla\psi \times \nabla\varphi \tag{16.86}$$

in two dimensions in an axisymmetric equilibrium. This form preserves $\nabla \cdot \mathbf{B} = 0$ and relates $A_\parallel R$ to ψ through

$$\nabla \times (A_\parallel \mathbf{b}) \to \nabla(A_\parallel R) \times \nabla\varphi \to \nabla\psi \times \nabla\varphi \tag{16.87}$$

whose square $|\nabla\psi|^2/R^2$ appears in the field Lagrangian. Ampère's law is re-derived from the Lagrangian as

$$-\Delta^* \psi = \mu_0 J_\parallel R \qquad \text{where} \qquad \Delta^* = R^2 \nabla \cdot \frac{1}{R^2} \nabla_\perp \tag{16.88}$$

Hence, $J_\|/B$ becomes $I^{-1}(J_\| R)$. Neglecting the Mach correction to $P_\|$ and the anisotropy $p_\| - p_\perp$, we have

$$B\nabla_\| \left(-\frac{1}{\mu_0 I}\Delta^* \psi \right) = \nabla \cdot \frac{R^2}{I}\nabla\varphi \times \nabla p \tag{16.89}$$

The final statement of equilibrium is the one usually quoted first, namely $p = p(\psi)$. With this taken into account, we have

$$\mathbf{J} \times \mathbf{B} = \nabla p \implies p = p(\psi) \qquad \text{hence} \quad \nabla p = p' \nabla \psi \tag{16.90}$$

where the prime denotes the derivative with respect to ψ. Then we have

$$B\nabla_\| \left(-\frac{1}{\mu_0 I}\Delta^* \psi \right) = \nabla \cdot \frac{p' R^2}{I}\nabla\varphi \times \nabla \psi \tag{16.91}$$

With the representation for $\mathbf{B}$, and assuming axisymmetry, we recast this as

$$B\nabla_\| \left(\frac{1}{\mu_0 I}\Delta^* \psi + \frac{p' R^2}{I} \right) = 0 \tag{16.92}$$

Now multiply through by the constants $\mu_0 I$ to find

$$B\nabla_\| \left(\Delta^* \psi + \mu_0 p' R^2 \right) = 0 \tag{16.93}$$

This is the statement of the Grad–Shafranov equation along the magnetic field. The equation is satisfied up to addition to the quantity inside parentheses by an arbitrary flux function. We can therefore state without loss of generality

$$II' + \Delta^* \psi + \mu_0 p' R^2 = 0 \tag{16.94}$$

This is the Grad–Shafranov equation for an axisymmetric equilibrium. It is also the statement giving the correction to $I=$ constant. Conventional tokamak ordering assumes that this correction is small, and therefore this equation can be used to check that *a posteriori*. One usually finds that I varies by no more than one to three percent across the poloidal cross section, since not only a/qR but also β is small. This is the standard justification for these approximations, and also for RMHD in general for a tokamak-like magnetic field (mostly toroidal, even if not necessarily axisymmetric). Ultimately, both RMHD and gyrokinetic theory are dependent upon related assumptions (i.e., never having to face the dynamical $\mathbf{A}_\perp$), and both require $a/qR \ll 1$ and $\beta \ll 1$ in a toroidal case.

These operations summarise the capture of reduced MHD and the Grad–Shafranov equilibrium by gyrokinetic theory. The time derivative of the gyrokinetic Poisson equation yields the reduced MHD vorticity equation, and the time derivative of the gyrokinetic Ampere's law yields the reduced MHD Ohm's law. The equilibrium of the reduced MHD vorticity equation in the absence of strong flows yields the parallel gradient of the Grad–Shafranov equation, or the RMHD equilibrium.

16.7 Summary

This outline has shown how the gyrokinetic representation of low frequency plasma dynamics can be derived in a manner more transparent than the ones which use Lie transforms and differential form calculus. A straightforward re-casting of Littlejohn's 1983 gauge transform method for drift kinetics is sufficient to derive the gyrokinetic Lagrangian allowing for shear-Alfvén electromagnetic electron responses and for finite-amplitude ExB flows. The re-casting is a matter of keeping to canonical form so that ϕ lands in the Hamiltonian, and then the field theory version gives a self consistent dynamical description. It is the change of representation which makes it a gyrokinetic rather than a drift kinetic theory. Instead of a polarisation drift in the Euler–Lagrange equations, we have polarisation as a density in a gyrokinetic Poisson equation for the self-consistent field potential.

If the flow amplitude is small, such that the ExB contribution to gyroaveraging is negligible, and the plasma beta is low enough that electromagnetic induction in electron responses is negligible, then the model reduces to the long wavelength limit of the conventional small-fluctuation one that is most familiar. As long as the large-scale ExB flow is subsonic, the conventional model then expands to cover the entire available spectrum. This is the usual case in a magnetised plasma.

The equations derived from the gyrokinetic Lagrangian are not only the kinetic equation of motion (one per species) but also the self-consistent electric and shear-Alfvén magnetic field potentials, to which each species contributes through gyrocentre charge and current density. The symmetry principles of a Lagrangian field theory guarantee energetic consistency in the model. The equations derived from the gyrokinetic Lagrangian can then serve as the basis for deriving a total-f gyrofluid model within the same representation which then has the same level of consistency. These methods then form the basis for well-constructed numerical models in computations of turbulence, MHD, energetic particle dynamics, and equilibrium and transport in magnetised plasmas.

Further Reading

The large scale gyrokinetic Lagrangian given herein was originally done by Lie transforms, in N Miyato and B Scott and D Strintzi and S Tokuda, 'A Modification of the Guiding-Centre Fundamental 1-Form with Strong ExB Flow,' *J. Phys. Soc. Japan* **78** (2009) 104501. The route back to MHD from the gyrokinetic equations was also given in this paper.

The demonstration that the two ordering approaches converge for subsonic flows at large scale was given in B Scott, 'Gyrokinetic Theory and Dynamics of the Tokamak Edge,' *Contrib. Plasma Phys.* **56** (2016) 534.

IOP Publishing

Turbulence and Instabilities in Magnetised Plasmas, Volume 2

Gyrokinetic theory and gyrofluid turbulence

Bruce Scott

Chapter 17

Lie–Poisson bracket for gyrokinetics

17.1 Introduction

Not every reader will have been satisfied with the brief comparison of the gyrokinetic evolution equation to a Poisson bracket formulation given in Chapter 6. There, it was used to motivate phase space conservation and the symmetry behind the conservation laws of the system superset of the evolution equation and the field equations.

Herein we will explore the bracket structure further, first as an outgrowth of the theory as we have it, and then more formally in terms of a bracket algebra. We are used to the distribution function f as a phase space density, as the fundamental variable. We also have the idea of a functional, such as the Noether energy given for an electrostatic model as the sum over species and integral over phase space of fH, where H is the Hamiltonian for a single gyrocentre. We will build on this to expand the idea to any possible functional of f, which includes the field equations. Naturally, we will use the information already obtained from the Lagrangian field theory in terms of the field equations and the Noether energy. But in doing this we find a deeper structure which the theory satisfies, which may be more interesting to the purist.

The Noether energy is taken over as the Hamiltonian functional of f, as the most basic of these functionals. The functional derivatives with respect to the fields already vanish, and the one with respect to f with the fields fixed is just H. We can back into the idea of a functional bracket using some straightforward operations on the gyrokinetic evolution equation after having set that up in a bracket structure with general properties. Once the Lie–Poisson functional bracket is set up in this fashion, members of that set are shown to satisfy the properties of an algebra. The evolution equation is then recovered once again using the functional methods. Then we do the same for the simplest form of these equations, namely the two-dimensional (2D) Euler equation for incompressible fluid dynamics (the vorticity equation), and the 2D gyrofluid interchange model. The field equations serve as helpers so they already must be known. The basic formalism is not new, but the application of it to the gyrokinetic field theory is unusual. In the words of a colleague, it is helpful to know the theory has this structure. And so it is relevant to include this material herein.

17.2 Poisson bracket formulation

We have already seen the basic structure of the Poisson bracket in Section 6.4.1. The simplest definition is the 2D spatial bracket given by

$$[a, b] = \frac{\partial(a, b)}{\partial(x, y)} = \frac{\partial a}{\partial x}\frac{\partial b}{\partial y} - \frac{\partial a}{\partial y}\frac{\partial b}{\partial x} \tag{17.1}$$

for two scalars with dependence $a = a(x, y)$, which is also the same as the Jacobian of a, b with respect to x, y. The coordinates x, y form a conjugate pair. In general the structure is

$$[a, b] = \sum_{ij} \frac{\partial a}{\partial x^i} T^{ij} \frac{\partial b}{\partial x^j} \tag{17.2}$$

where T^{ij} are the elements of a coupling matrix. The sum is over all coordinate indices, and hereafter the sum over repeated indices, one up and one down, is taken as understood (Einstein convention). We can re-write this as

$$[a, b] = a_{,i}\, T^{ij}\, b_{,j} \tag{17.3}$$

where the subscript after a comma denotes partial differentiation with respect to the coordinate of that index. Usually, the T^{ij} are real numbers, though even in some of our cases they have included geometry and so $T^{ij} = T^{ij}(x^k)$. But they are not functions of the dependent variables a, b in common cases. The key property of the matrix is antisymmetry

$$T^{ij} = -T^{ji} \qquad \text{hence} \quad [a, b] = -[b, a] \tag{17.4}$$

and the divergence property (cf. Eqs. 17.11 and 17.12, below). The bracket also satisfies the Jacobi identity,

$$[a, [b, c]] + \text{cyc} \equiv [a, [b, c]] + [b, [c, a]] + [c, [a, b]] = 0 \tag{17.5}$$

As Goldstein points out (see Further Reading), there is no simple form to prove this, but what can be done first is to collect all occurrences of the second derivative of one of the arguments, say c, and show that set of terms vanish individually. Then, by extension, so do all the others. For the xy-bracket in Eq. (17.1), we have

$$[a, [b, c]] + \text{cyc} = a_{,y} b_{,y}\, c_{,xx} - b_{,y} a_{,y}\, c_{,xx} + \cdots \tag{17.6}$$

from the first two terms in Eq. (17.5), the only ones which contain $c_{,xx}$, the second derivative with respect to x. For the other second derivatives on c we have

$$\begin{aligned}[a, [b, c]] + \text{cyc} = {} & (a_{,x} b_{,x} - b_{,x} a_{,x}) c_{,yy} \\ & - \left(a_{,x} b_{,y} - b_{,x} a_{,y}\right) c_{,xy} - \left(a_{,y} b_{,x} - b_{,y} a_{,x}\right) c_{,yx} + \cdots \end{aligned} \tag{17.7}$$

in which both sets of the mixed derivatives are needed to show cancellation.

For the general bracket in Eq. (17.2) with coefficients T^{ij} constant, we have

$$a_{,i}T^{ij}(b_{,k}T^{kl}c_{,l})_{,j} + \text{cyc} = a_{,i}T^{ij}b_{,k}T^{kl}c_{,lj} + b_{,i}T^{ij}c_{,kj}T^{kl}a_{,l} + \cdots \quad (17.8)$$

for all the terms involving any second derivatives of c. Re-label indices $i \leftrightarrow l$ on the first set and $i \leftrightarrow k$ and then $j \leftrightarrow i$ on the second set to find

$$a_{,l}\, b_{,k}\, c_{,ij}(T^{lj}\, T^{ki} + T^{ki}\, T^{jl}) = 0 \quad (17.9)$$

by the anti-symmetry of T^{lj} and the symmetry of $c_{,ij}$. This is the proof given by Goldstein.

For the case with non-constant coefficients T^{ij} it is more complicated to prove the Jacobi identity. From Eq. (17.2) we integrate the first of the two derivatives by parts to find

$$[a, b] = \left(aT^{ij}b_{,j}\right)_{,i} - ab_{,j}T^{ij}{}_{,i} - aT^{ij}b_{,ij} \quad (17.10)$$

The last term vanishes by antisymmetry of T^{ij}, so that to obtain the divergence property,

$$[a, b] = \left(aT^{ij}b_{,j}\right)_{,i} \quad (17.11)$$

we require the divergence condition,

$$T^{ij}{}_{,i} = 0 \quad (17.12)$$

For a space with non-constant metric determinant g, these become

$$\frac{1}{\sqrt{g}}\left(\sqrt{g}\,T^{ij}\right)_{,i} = 0 \qquad [a, b] = \frac{1}{\sqrt{g}}\left(\sqrt{g}\,aT^{ij}b_{,j}\right)_{,i} \quad (17.13)$$

The Jacobi identity in Eq. (17.5) then depends on these through

$$[a, [b, c]] = a_{,i}T^{ij}(b_{,l}T^{lm}c_{,m})_{,j} = a_{,i}b_{,l}c_{,m}T^{ij}T^{lm}{}_{,j} \quad (17.14)$$

in which only the terms with differentiation on the T's are written. Once we have the *abc* isolated with single derivatives in this way, we can just as easily permute the indices *ilm* in this expression as the scalars *abc*, so we have

$$[a, [b, c]] + \text{cyc} = a_{,i}b_{,l}c_{,m}T^{ij}T^{lm}{}_{,j} + \text{cyc} \quad (17.15)$$

where since index j is contracted we permute the other three. Now, with *abc* arbitrary, we simply have to show that

$$T^{ij}T^{lm}{}_{,j} + \text{cyc} = 0 \quad (17.16)$$

to prove the Jacobi identity for any given Poisson bracket. We may write this as the elements of a vector

$$P_k = \varepsilon_{kilm}\, T^{ij}T^{lm}{}_{,j} = 0 \quad (17.17)$$

If this vector vanishes, then the Jacobi identity is satisfied.

17.3 Phase space Jacobian and the four-bracket form

We did the reduction of the gyrocentre Lagrangian into the equations of motion for gyrocentres before. We recap it here in slightly different language than in Chapters 5 and 16.

The gyrocentre Lagrangian is in general

$$L_p = e\mathbf{A}^* \cdot \dot{\mathbf{R}} + M\dot{\vartheta} - H \tag{17.18}$$

The gyromotion separates out since M is conserved and there is no ϑ dependence in the rest of the system dynamics. This leaves the four active coordinates which can be labelled Z^a with with indices a one of $\{ijkz\}$ with z the fourth coordinate. That is, in this section and the next, the $\{abcd\}$ are phase space indices and not scalar variables. We rewrite L_p in terms of the resulting phase space vectors

$$L_p = eA_a^*\dot{Z}^a + M\dot{\vartheta} - H \tag{17.19}$$

In general the component A_z^* is zero but the A_a^* can have derivatives in any coordinate except ϑ. Varying the differential action $L_p\, dt$ we have

$$\delta L_p\, dt = \delta Z^a(eA^*_{b,\,a}dZ^b) + eA_b^*d(\delta Z^b) - \delta Z^a\, H_{,a}\, dt + \delta M(\cdots) + \delta\vartheta(\cdots) \tag{17.20}$$

concentrating only on the four active coordinates spanned by $\{abcd\}$. Subtracting the relevant differential we obtain

$$\begin{aligned}\delta L_p\, dt - d(eA_b^*\, \delta Z^b) = \delta Z^a\Big[e\Big(A^*_{b,\,a} - A^*_{a,\,b}\Big)dZ^b - H_{,a}\, dt\Big] \\ + \delta M(\cdots) + \delta\vartheta(\cdots)\end{aligned} \tag{17.21}$$

This yields as Euler–Lagrange equations for the four coordinates $\{\mathbf{R},\, z\}$

$$e\Big(A^*_{p,\,a} - A^*_{a,\,p}\Big)dZ^p = H_{,a}\, dt \tag{17.22}$$

and re-labelling the existing summed index b as p for convenience with what happens next.

We anticipate solving this by operating from the left with the 4D Levi-Civita pseudotensor without units, denoted ϵ^{abcd}, which takes values ± 1 or 0 according to positive/negative permutation or repetition of indices $[abcd]$ (cf. Chapter 4). Also using $A^*_{c,\,d}$ we have

$$eA^*_{c,\,d}\,\epsilon^{abcd}\Big(A^*_{p,\,a} - A^*_{a,\,p}\Big)dZ^p = \epsilon^{abcd}H_{,a}A^*_{c,\,d}\, dt \tag{17.23}$$

In each step we will use the fact that there is no A_z^* so that index z must be among the derivatives.

First considering index b is z so that $\{acd\}$ are the spatial $\{ijk\}$ indices (the sign remains since it is two exchanges between $ijkz$ and $izjk$), we have

$$eA^*_{j,\,k}\,\epsilon^{ijk}\Big(A^*_{p,\,i} - A^*_{i,\,p}\Big)dZ^p = \epsilon^{ijk}H_{,i}A^*_{j,\,k}\, dt \tag{17.24}$$

where the notation $\varepsilon^{ijk} = [ijk]$ is the 3D Levi-Civita pseudotensor without units. We note that

$$\varepsilon^{ijk} = \sqrt{g}\,\varepsilon^{ijk} \qquad \varepsilon^{ijk}A^*_{j,k} = -(B^*)^i \qquad A^*_{l,i} - A^*_{i,l} = \varepsilon_{ilm}(B^*)^m \tag{17.25}$$

where $\{lm\}$ are also spatial indices and $\sqrt{g}$ is the spatial coordinate volume element (g is the determinant of the covariant metric tensor components), given by

$$\frac{1}{\sqrt{g}} = \nabla x^1 \times \nabla x^2 \cdot \nabla x^3 \tag{17.26}$$

We expand in terms of index p being spatial or z, so that since A^*_z is zero, we have

$$eA^*_{j,k}\varepsilon^{ijk}\Big[\big(A^*_{l,i} - A^*_{i,l}\big)dZ^l - A^*_{i,z}dz\Big] = \varepsilon^{ijk}H_{,i}A^*_{j,k}\,dt \tag{17.27}$$

The spatial terms cancel mutually since

$$\varepsilon^{ijk}A^*_{j,k} = -\sqrt{g}(B^*)^i \qquad A^*_{l,i} - A^*_{i,l} = \varepsilon_{ilm}(B^*)^m \tag{17.28}$$

so that the combination is proportional to $(B^*)^i\varepsilon^{ilm}(B^*)^m = 0$. This leaves

$$\sqrt{g}(B^*)^i eA^*_{i,z}\dot{Z}^z = \varepsilon^{ijk}H_{,i}A^*_{j,k} \tag{17.29}$$

and here we define

$$B^*_{\|} = eA^*_{i,z}(B^*)^i \tag{17.30}$$

so that

$$\dot{Z}^z = \varepsilon^{izjk}H_{,i}A^*_{j,k} \tag{17.31}$$

now establishing the units of ε^{abcd} to be $(\sqrt{g}B^*_{\|})^{-1}$. Here, Eq. (17.31) is the counterpart to the second expression in Eq. (6.10) from Chapter 6, and Eq. (17.30) to Eq. (6.12), respectively.

Next we choose index b to be spatial so that one of the others is z, but c cannot be z, which leaves only a and d as choices. We expand according to whether index a is spatial or z and the same for index p, so that

$$\begin{aligned} &eA^*_{j,k}\varepsilon^{zijk}\big(A^*_{l,z}\big)dZ^l + eA^*_{k,z}\varepsilon^{ijkz}\Big[\big(A^*_{l,i} - A^*_{i,l}\big)dZ^l - A^*_{i,z}dZ^z\Big] \\ &\quad = \varepsilon^{ijkz}H_{,i}A^*_{k,z}\,dt \end{aligned} \tag{17.32}$$

noting that any occurrence of A^*_z drops out. Noting that $\{zijk\}$ is 3 exchanges away from $\{ijkz\}$ we have

$$\begin{aligned} &-eA^*_{j,k}\varepsilon^{ijk}\big(A^*_{l,z}\big)dZ^l + eA^*_{k,z}\varepsilon^{ijk}\Big[\big(A^*_{l,i} - A^*_{i,l}\big)dZ^l - A^*_{i,z}dZ^z\Big] \\ &\quad = \varepsilon^{ijkz}H_{,i}A^*_{k,z}\,dt \end{aligned} \tag{17.33}$$

The $A^*_{i,z}$ term drops due to antisymmetry of ε^{ijk} leaving

$$e(B^*)^i\big(A^*_{l,z}\big)dZ^l + eA^*_{k,z}\varepsilon^{ijk}[\varepsilon_{ilm}(B^*)^m dZ^l] = \varepsilon^{ijkz}H_{,i}A^*_{k,z}\,dt \tag{17.34}$$

Finally, contracting the 3D Levi-Civita pseudotensors, we find

$$eA^*_{i,\,z}(B^*)^i dZ^j = \frac{1}{\sqrt{g}}\varepsilon^{ijkz} H_{,i} A^*_{k,\,z}\, dt \tag{17.35}$$

which is the the counterpart to the first expression in Eq. (6.10) from Chapter 6. Collectively, we have proven

$$eA^*_{c,\,d}\varepsilon^{abcd}\left(A^*_{p,\,a} - A^*_{a,\,p}\right) = \sqrt{g}\,B^*_\parallel \delta^b{}_p \tag{17.36}$$

and hence

$$eA^*_{c,\,d}\varepsilon^{abcd}\left(A^*_{p,\,a} - A^*_{a,\,p}\right) = \delta^b{}_p \tag{17.37}$$

with $B^*_\parallel$ defined in Eq. (17.30). The solution to the Euler–Lagrange equations is

$$\dot{Z}^b = \varepsilon^{abcd} H_{,a} A^*_{c,\,d} \tag{17.38}$$

which combines both expressions in Eq. (6.10) from Chapter 6. The gyrokinetic evolution equation is then

$$\frac{\partial f}{\partial t} + \varepsilon^{abcd} H_{,a} f_{,b} A^*_{c,\,d} = 0 \qquad \text{or} \qquad \frac{\partial f}{\partial t} + [H, f] = 0 \tag{17.39}$$

for any choice of coordinates under canonical form. We can call this the four-bracket form of the evolution equation. It has the same structure as a Poisson bracket, as we can identify

$$T^{ab} = \varepsilon^{abcd} A^*_{c,\,d} \tag{17.40}$$

The divergence condition is satisfied since multiplication of T^{ab} by the factor $\sqrt{g}\,B^*_\parallel$ cancels the same factor in ε^{abcd}, leaving the coordinate dependence in A^*_c as the only remaining one. The resulting contraction with $A^*_{c,\,db}$ vanishes due to the antisymmetry.

17.4 Setting up a Lie–Poisson functional bracket

The foregoing is all for scalars, and in terms of vector/tensor spaces these are all considered as scalar fields. The gyrokinetic equation itself (Eq. 17.39) can represent the equivalence of two scalars, or of the condition that one scalar vanishes. A functional is different from this. It takes functions as arguments and returns real numbers. The simplest way to create one out of a scalar in the phase space covered by coordinates $\mathbf{z}$ is to multiply by a delta function, and then integrate to find the value of the scalar at a given point, which is a real number,

$$f_0 = f(\mathbf{z}_0) = \int d\Lambda\, f(\mathbf{z})\, \delta(\mathbf{z} - \mathbf{z}_0) \tag{17.41}$$

This is a special case of a test function, $\alpha(\mathbf{z})$ used to create the functional

$$\mathcal{F}[f] = \int d\Lambda\, f(\mathbf{z})\, \alpha(\mathbf{z}) \tag{17.42}$$

The functional derivative of this with respect to f is then just the test function,

$$\frac{\delta \mathcal{F}}{\delta f} = \alpha(\mathbf{z}) \tag{17.43}$$

Doing this with Eq. (17.39), we create what is called the weak form of the original equation,

$$\int d\Lambda\, \alpha(\mathbf{z})\left(\frac{\partial f}{\partial t} + [H, f]\right) = 0 \tag{17.44}$$

or, substituting the definition of $\mathcal{F}[f]$ into the time derivative,

$$\frac{\partial \mathcal{F}}{\partial t} + \int d\Lambda\, \alpha(\mathbf{z})[H, f] = 0 \tag{17.45}$$

The Poisson bracket is also evaluated at $\mathbf{z}$ within the integration. The functional appears under the time derivative because the test function is fixed. Using the Leibnitz property of the bracket we have

$$\frac{\partial \mathcal{F}}{\partial t} = -\int d\Lambda\, f[\alpha,\, H] \tag{17.46}$$

It happens that we already have another functional to think of, namely, the Noether energy given by

$$\mathcal{E} = \sum_{sp} \int d\Lambda\, fH - \int d\mathcal{V}\, \mathcal{L}_f \tag{17.47}$$

where the sum is over species, noting that f and H are species quantities while the field Lagrangian density $\mathcal{L}_f$ is a spatial quantity dependent on a set of field variables $\phi_a(\mathbf{x})$. If we consider the variation of this with respect to f, we have the general variation with respect to the particular f of each given species plus the variation due to each ϕ_a. But for $\mathcal{E}$ the variations due to the ϕ_a,

$$\frac{\delta \mathcal{E}}{\delta \phi_a} = 0 \qquad \text{for each} \quad \phi_a \tag{17.48}$$

are already zero by definition. This is one of the benefits of having set up the system Lagrangian in canonical form. The Noether energy is simply the negative of the time component of this Lagrangian, and since that is the only place the fields appear, then their Euler–Lagrange equations are just this condition that the functional derivative of this time component with respect to each field variable vanishes. This leaves

$$\frac{\delta \mathcal{E}}{\delta f_s} = H_s \tag{17.49}$$

where the subscript s denotes the given species (i.e., one particular f_s among all the fs). So in Eq. (17.45) we can now rewrite both arguments of the bracket in the second term as functional derivatives,

$$\frac{\partial \mathcal{F}}{\partial t} = -\int d\Lambda f\left[\frac{\delta \mathcal{F}}{\delta f}, \frac{\delta \mathcal{E}}{\delta f}\right] \tag{17.50}$$

in which each of $\mathcal{F}$, f, and H are for the given species. The right hand side is another functional, given by the dependent variable f multiplied by a Poisson bracket between two functional derivatives. By convention we write this with the dependent variable functional first and the energy functional second, necessitating the minus sign.

This is a particular example of a Lie–Poisson functional bracket defined by

$$\{\mathcal{F}, \mathcal{G}\} = \int d\tau \, \frac{\delta \mathcal{F}}{\delta f_i} O^{ij} \frac{\delta \mathcal{G}}{\delta f_j} \tag{17.51}$$

where the integration is over a domain given by a spatial volume or a phase space. The indices ij are to the set of dependent variables (in this case, species), and the O^{ij} are operators acting on the first argument to the left and the second argument to the right. The operators can include derivatives. The required properties of the functional bracket are antisymmetry,

$$\{\mathcal{F}, \mathcal{G}\} = -\{\mathcal{G}, \mathcal{F}\} \tag{17.52}$$

the Leibnitz rule,

$$\{\mathcal{F}, \mathcal{G}\mathcal{H}\} = \mathcal{H}\{\mathcal{F}, \mathcal{G}\} + \mathcal{G}\{\mathcal{F}, \mathcal{H}\} \tag{17.53}$$

and the Jacobi identity,

$$\{\mathcal{F}, \{\mathcal{G}, \mathcal{H}\}\} + \text{cyc} = 0 \tag{17.54}$$

The first is trivial, given the presence of the inner Poisson bracket in Eq. (17.50). The second follows from the fact that a functional is just a number and commutes past derivatives and integrations. We have

$$\{\mathcal{F}, \mathcal{G}\mathcal{H}\} = \int d\tau \, \frac{\delta \mathcal{F}}{\delta f_i} O^{ij}\left(\mathcal{H}\frac{\delta \mathcal{G}}{\delta f_j} + \mathcal{G}\frac{\delta \mathcal{H}}{\delta f_j}\right) \tag{17.55}$$

and it remains only to pull $\mathcal{H}$ out of the first term and $\mathcal{G}$ out of the second. This follows from the Leibnitz property of the functional derivatives themselves. The third, like the Poisson bracket, depends on the properties of the operators O^{ij}.

The functional bracket in our case is the integral over the Poisson bracket between two functional derivatives,

$$\{\mathcal{F}, \mathcal{G}\} = -\sum_s \int d\Lambda f_s\left[\frac{\delta \mathcal{F}}{\delta f_s}, \frac{\delta \mathcal{G}}{\delta f_s}\right] \tag{17.56}$$

where the appearances of f in the operator (first factor) and in each of the functional derivatives are all of the same species (subscript s). The Poisson bracket on the inside is given by Eq. (17.39). Eq. (17.56) defines what can be called the gyrokinetic

Lie–Poisson functional bracket. The operator O is diagonal with respect to f but contains the anti-symmetric combination of derivatives with respect to the phase space coordinates. To prove the Jacobi identity, we have to take the functional derivative of this structure. Through the Leibnitz rule it is given by the three terms due to variation of f, $\mathcal{F}$, and $\mathcal{G}$. The variations of the functional derivatives contribute sets of terms like the second derivatives in the Poisson bracket, and they mutually cancel in the same fashion. So only the term due to the variation of f in the operator contributes.

In setting up the Jacobi identity, we have

$$\{\mathcal{F}, \{\mathcal{G}, \mathcal{H}\}\} + \text{cyc} = -\sum_s \int d\Lambda f\left[\frac{\delta\mathcal{F}}{\delta f_s}, \left[\frac{\delta\mathcal{G}}{\delta f_s}, \frac{\delta\mathcal{H}}{\delta f_s}\right]\right] + \text{cyc} \tag{17.57}$$

That is, proof of the Jacobi identity for the inner Poisson bracket is sufficient for proof of the Jacobi identity for the functional bracket. We have already done much of the work, in the run-up to Eq. (17.17) and to Eq. (17.39). The four-bracket in the latter is defined by the T in Eq. (17.40), so proof of the Jacobi identity rests on the vanishing of the vector

$$P_p = \varepsilon_{pacd}\, T^{ab} T^{cd}{}_{,b} \tag{17.58}$$

Since this T satisfies the divergence condition, the derivative with respect to index b becomes a divergence of the whole expression

$$P_p = \left(\varepsilon_{pacd}\, T^{ab} T^{cd}\right)_{,b} \tag{17.59}$$

Putting in

$$\varepsilon_{pacd} T^{cd} = \varepsilon_{pacd} \varepsilon^{cdij} A^*_{i,\,j} \tag{17.60}$$

we anti-symmetrise it for what comes later,

$$\varepsilon_{pacd} T^{cd} = \frac{1}{2} \varepsilon_{pacd} \varepsilon^{cdij} (A^*_{i,\,j} - A^*_{j,\,i}) \tag{17.61}$$

The cd contraction on the two Levi-Civita pseudotensors yields

$$\varepsilon_{pacd} T^{cd} = \frac{1}{2} (A^*_{p,\,a} - A^*_{a,\,p}) \tag{17.62}$$

Putting in the definition of T^{ab}, we have

$$P_p = \frac{1}{2}\left(\varepsilon^{abcd} A^*_{c,\,d} (A^*_{p,\,a} - A^*_{a,\,p})\right)_{,b} \tag{17.63}$$

Using the result in Eq. (17.37), we have

$$P_p = \frac{1}{2}\left(\frac{1}{e}\delta^b{}_p\right)_{,b} = 0 \tag{17.64}$$

The Jacobi identity

$$\{\mathcal{F}, \{\mathcal{G}, \mathcal{H}\}\} + \{\mathcal{G}, \{\mathcal{H}, \mathcal{F}\}\} + \{\mathcal{H}, \{\mathcal{F}, \mathcal{G}\}\} = 0 \tag{17.65}$$

is therefore satisfied for our case.

With this setup we return to Eq. (17.50), and rewrite it as

$$\frac{\partial \mathcal{F}}{\partial t} = \{\mathcal{F}, \mathcal{E}\} \tag{17.66}$$

so that with the definition in Eq. (17.56), we have

$$\frac{\partial \mathcal{F}}{\partial t} = -\int d\Lambda\, f[\alpha, H] \tag{17.67}$$

with only the same species as the specified f in the functional $\mathcal{F}$ enters. Using the Leibnitz rule on the Poisson bracket

$$\frac{\partial \mathcal{F}}{\partial t} = \int d\Lambda\, \alpha[f, H] \tag{17.68}$$

recovering the weak form of the gyrokinetic evolution equation.

Then, specifying

$$\alpha(\mathbf{z}) = \delta(\mathbf{z} - \mathbf{z}_0) \tag{17.69}$$

recovers the evolution equation with both sides evaluated at the point $\mathbf{z}_0$. Since the choice of $\mathbf{z}_0$ is arbitrary, we have the evolution equation valid at all points $\mathbf{z}_0$ covering the space. Then, we can replace $\mathbf{z}_0 \to \mathbf{z}$. This is a mathematical nicety, but it avoids issues of f itself not being a functional. This yields

$$\frac{\partial f}{\partial t} + [H, f] = 0 \tag{17.70}$$

with the Poisson bracket defined in Eq. (17.40) so that the gyrokinetic evolution Eq. (17.39) is recovered.

17.5 Lie–Poisson brackets for two-dimensional models

Herein we recapture the same result for two much simpler cases, the simplest 2D models, one given by a pure Euler vorticity equation and then one given by the gyrofluid 2D interchange model.

The Euler vorticity equation and its field equation are given by

$$\frac{\partial \varpi}{\partial t} + [\phi, \varpi] = 0 \qquad \nabla_\perp^2 \phi = \varpi \tag{17.71}$$

where here we have the Poisson bracket returning to the form of Eq. (17.1). We have the energy functional

$$\mathcal{E} = \frac{1}{2} \int dx\, dy\ |\nabla_\perp \phi|^2 \tag{17.72}$$

It is equivalent to

$$\mathcal{E} = \frac{1}{2} \int dx\, dy\, (-\phi\varpi) \tag{17.73}$$

The functional derivative of this uses the field equation, considering ϕ as dependent upon ϖ as we did above for the fields in terms of f. It is given by

$$\frac{\delta \mathcal{E}}{\delta \varpi} = -\phi \tag{17.74}$$

As before, we define

$$\mathcal{F}[\varpi] = \int dx\, dy\, \varpi(\mathbf{x})\, \alpha(\mathbf{x}) \tag{17.75}$$

so that

$$\frac{\delta \mathcal{F}}{\delta \varpi} = \alpha(\mathbf{x}) \tag{17.76}$$

as in Eq. (17.43). This leads to the Lie–Poisson functional bracket given by

$$\{\mathcal{F}, \mathcal{G}\} = \int dx\, dy\, \varpi \left[\frac{\delta \mathcal{F}}{\delta \varpi}, \frac{\delta \mathcal{G}}{\delta \varpi}\right] \tag{17.77}$$

Stating

$$\frac{\partial \mathcal{F}}{\partial t} = \{\mathcal{F}, \mathcal{E}\} \tag{17.78}$$

we have with this definition

$$\frac{\partial \mathcal{F}}{\partial t} = \int dx\, dy\, \varpi[\alpha(\mathbf{x}), -\phi] \tag{17.79}$$

Using the Leibnitz rule so that

$$\frac{\partial \mathcal{F}}{\partial t} = \int dx\, dy\, \alpha(\mathbf{x})[\varpi, \phi] \tag{17.80}$$

Then, setting $\alpha(\mathbf{x}) = \delta(\mathbf{x} - \mathbf{x}_0)$ recovers Eq. (17.71) at $\mathbf{x} = \mathbf{x}_0$. Alternatively, use of Eq. (17.71) to solve for $[\varpi, \phi]$ in Eq. (17.80) recovers the time derivative of Eq. (17.75).

The gyrofluid 2D interchange model has a bracket with a similar structure to the Euler vorticity one, except now the curvature terms are set by an extra piece in the bracket due to the geometry. We define the free energy as

$$\mathcal{E} = \frac{1}{2} \int dx\, dy\, (n_e^2 + \tau_i n_i^2 + |\nabla_\perp \phi|^2) \tag{17.81}$$

and the field equation as

$$\nabla_\perp^2 \phi = \varpi = n_e - n_i \tag{17.82}$$

The free energy is therefore equivalent to

$$\mathcal{E} = \frac{1}{2}\int dx\, dy\, [(n_e - \phi)n_e + (\tau_i n_i + \phi)] \tag{17.83}$$

The functional derivatives are

$$\frac{\delta\mathcal{E}}{\delta n_i} = \tau_i n_i + \phi \qquad \frac{\delta\mathcal{E}}{\delta n_e} = n_e - \phi \tag{17.84}$$

We define

$$\{\mathcal{F}, \mathcal{G}\} = -\sum_s \frac{1}{a_s}\int dx\, dy\, (n_s + \omega_B x)\left[\frac{\delta\mathcal{F}}{\delta n_s}, \frac{\delta\mathcal{G}}{\delta n_s}\right] \tag{17.85}$$

where $a_e = -1$ and $a_i = 1$ are the normalised charge for the two species. Then using

$$\frac{\delta\mathcal{F}}{\delta n_s} = \alpha(\mathbf{x}) \tag{17.86}$$

for each species, we find that

$$\frac{\partial\mathcal{F}}{\partial t} = \{\mathcal{F}, \mathcal{E}\} \tag{17.87}$$

corresponds to

$$\frac{\partial n_e}{\partial t} + [\phi, n_e] = \omega_B[x, (\phi - n_e)] \tag{17.88}$$

$$\frac{\partial n_i}{\partial t} + [\phi, n_i] = \omega_B[x, (\phi + \tau_i n_i)] \tag{17.89}$$

having used $[n, n] = 0$. This is the same as the simplest gyrofluid 2D interchange model with ω_B the curvature parameter and τ_i the ion temperature parameter.

Setting the brackets up in this fashion serves as an anchor for this type of model within the more general Lie–Poisson functional bracket theory. The signs are of interest. Note that in Eq. (17.77) we have a plus sign, while in the gyrokinetic case in Eq. (17.56) we have a minus sign, while in the 2D interchange model in Eq. (17.85) we have a minus sign and the $1/a_s$ factor for the normalised charge. These are all the same thing in different language, since the energy functional derivative for vorticity has the minus sign, and in the gyrokinetic case the $1/e$ factor is present in the operator, since $A^*_{k,z} = (1/e)b_k$. In a gyrokinetic case with a curved magnetic field, the geometry terms come from the finite spatial derivatives of $\mathbf{A}^*$ through $\mathbf{b}$ as well as the factor of B in the Hamiltonian. The models are basically all members of the same set.

17.6 Casimir invariants

There is a class of functionals of the dependent variables whose property is to vanish in a Lie–Poisson functional bracket with any other functional of the dependent variables. In our case the most basic examples are in the class of functionals mentioned in Chapters 5, 6 and 7, the functionals which are integrals of any function $G(f)$ integrated over phase space,

$$\mathcal{G} = \int d\Lambda \, G(f) \qquad \frac{\delta \mathcal{G}}{\delta f} = G' \equiv \frac{\partial G}{\partial f} \tag{17.90}$$

The bracket of $\mathcal{G}$ with any other functional $\mathcal{F}$ is given by

$$\{\mathcal{G}, \mathcal{F}\} = -\int d\Lambda f \left[G', \frac{\delta \mathcal{F}}{\delta f} \right] \tag{17.91}$$

Using the Leibnitz property and the divergence property of the inner Poisson bracket, we have

$$\{\mathcal{G}, \mathcal{F}\} = \int d\Lambda \, G' \left[f, \frac{\delta \mathcal{F}}{\delta f} \right] \tag{17.92}$$

Inversion of the chain rule on derivatives of G yields

$$\{\mathcal{G}, \mathcal{F}\} = \int d\Lambda \left[G, \frac{\delta \mathcal{F}}{\delta f} \right] = 0 \tag{17.93}$$

which vanishes by the divergence property. Note that this doesn't depend on what the functional $\mathcal{F}$ is, but is a property of the definition of the bracket. Examples of such Casimirs include

$$N = \int d\Lambda f \qquad L_2(f) = \int d\Lambda f^2 \qquad S = -\int d\Lambda f \log f \tag{17.94}$$

the density, the L_2 norm of f, and the entropy, respectively.

The usefulness of these Casimirs is in the construction of minimisation principles. The one we used in Chapter 7 to define the free energy is, in this context,

$$\mathcal{F} = \mathcal{E} - TS \tag{17.95}$$

the free energy given by the energy (here, the Noether energy functional) and the entropy, with a system temperature T serving as a Langrange multiplier. The resulting second variation of $\mathcal{F}$ becomes the free energy of a delta-f gyrokinetic dynamical system.

Further Reading

Poisson brackets are covered in *Classical Mechanics*, by H Goldstein (Academic Press, 1960). The Jacobi identity is proved on pages 399–400 of this reference.

The Lie–Poisson bracket written as a contraction between the dependent variable and a Poisson bracket is attributed to S Lie, *Theorie der Transformationsgruppen*, 2nd Ed. (B G Teubner, Leipzig, 1890), repr. Chelsea, New York (1970). See the review by J-L Thiffeault and P J Morrison, in 'Classification and Casimir invariants of Lie–Poisson brackets,' *Physica D* **136** (2000) 205. The 2D Euler vorticity equation is done in Section 2.2.2 and the 2D incompressible MHD equations are done in Section 2.2.3, both of this reference.

The Lie–Poisson bracket approach to fluid and plasma dynamics is reviewed by P J Morrison, in 'Hamiltonian description of the ideal fluid,' *Rev. Mod. Phys.* **70** (1998) 467. I was particularly helped by an earlier lecture, P J Morrison, 'Poisson brackets for fluids and plasmas,' in *Mathematical Methods in Hydrodynamics and Integrability in Dynamical Systems* (ed. M Tabor and Y M Treve), *AIP Conf. Series* **88** (1982) 13. Another capsule review with emphasis on Casimir invariants is P J Morrison and S Eliezer, 'Spontaneous symmetry breaking and neutral stability in the noncanonical Hamiltonian formalism,' *Phys. Rev. A* **33** (1986) 4205. The minimisation principle as built from Lie–Poisson brackets and Casimirs is given on page 4208.

Index

www.ingramcontent.com/pod-product-compliance
Ingram Content Group UK Ltd.
Pitfield, Milton Keynes, MK11 3LW, UK
UKHW051910100926
8556IPUK00005B/830

9 780750 338530